"What we from our point of view call colonization, missions to the heathen, spread of civilization, etc., has another face—the face of a bird of prey seeking with cruel intentness for distant quarry—a face worthy of a race of pirates and highwaymen. All the eagles and other predatory creatures that adorn our coats of arms seem to me apt psychological representatives of our true nature."

—Carl G. Jung, *Memories, Dreams, Reflections*, 1961

extraction

empire

Authors & Contributors: A Tribe Called Red, Allan Adam, Howard Adams, Yassin 'Narcy' Alsalman, Christopher Alton, Pedro Aparicio, Margaret Atwood, Aaron Barcant, Réal V. Benoit, Justice Thomas Berger, Hernán Bianchi Benguria, Paula Butler, David Chancellor, Lianne Marie Leda Charlie, Jean Chrétien, Tiffany Kaewen Dang, Alain Deneault, Eriel Tchekwie Deranger, Diaguitas Huascoaltinos, Mary Eberts, Genevieve Ennis Hume, Georges Erasmus, Pierre Falcon, Evan Farley, Alex Golub, David Hargreaves, Daniel Hemmendinger, Gord Hill, James Hopkinson, Hume Atelier, Michael Ignatieff, Thomas King, Naomi Klein, Erica Violet Lee, Kari Polanyi Levitt, Nina-Marie Lister, Ryan McMahon, Chris Meyer, Ossie Michelin, Kent Monkman, Doug Morrison, James Murray, Joan K. Murray, Charmaine Nelson, Eli Nelson, George Osodi, Maryanne Pearce, Barry Pottle, Moura Quayle, Louis Riel, RVTR, Olga Semenovych, Michelle St. John, Maurice Strong, Ashley C. Thompson, Anna Lowenhaupt Tsing, John Van Nostrand, Mel Watkins, Sally M. Weaver, Patrick Wolfe, Rita Wong, The Wyrd Sisters, Kate Yoon, Suzanne Zeller
Editor: Pierre Bélanger
Editorial Team & Lead Researchers: Ghazal Jafari, Tiffany Kaewen Dang, Hernán Bianchi Benguria, Christopher Alton, Zannah Matson, Sam Gillis
Design & Mappings: unless otherwise noted, all graphic visualization ©2018 OPSYS / Landscape Infrastructure Lab

This book was set in Helvetica World and Adobe Garamond Pro by OPSYS Media and was printed and bound in Altona, Manitoba (Canada).

Library of Congress Cataloging-in-Publication Data is available.

ISBN 978-0-262-53382-9

This book is dedicated to all the contributors, supporters, and visitors of the EXTRACTION exhibition project at the Venice Architecture Biennale held in 2016 (Italy) and those who fight and struggle every day against ongoing colonial oppression and territorial dispossession perpetuated by the settler-state of Canada. Systems of domination upheld by imperial rules of law such as the British *Magna Carta* and the Christian *Doctrines of Discovery* that entrench ideologies of "terra nullius" and "terra incognita" must be rejected by each generation to retroactively honor the past, present, and future of Indigenous lands and peoples. Royalties from the sales of this book will be donated to the Athabasca Chipewyan First Nation and related land-based programs of Denendeh. *Mahsi Cho. Existence is Resistance.*

EXTRACTION EMPIRE

Undermining the Systems, States, and Scales of Canada's Global Resource Empire

2017 – 1217

Over 95% of the total 10 million km^2 of land in Canada are Crown Lands.

Territories of extraction cover more than 80% of the planet, extending across lands on which the majority of the world's 7.5 billion people live.

More than half of the mining operations in the world are Canadian-owned, operated, equipped, or engineered.

2/3 of the world's mining companies are headquartered in Canada, thanks to low royalty rates set by the Crown and favorable investment environments of the Toronto Stock Exchange.

Canada operated for more than 200 years as the largest, and longest lasting, transnational corporation in the world, before Confederation in 1867.

Canada is one of the few, remaining constitutional monarchies in the world. Citizenship depends on sworn allegiance to the Queen and recognition of the Monarch's supremacy.

Canada's Head of State is the same as England's, Queen Elizabeth II, the longest lasting Sovereign in the history of the British Monarchy.

With over 6.6 billion acres—1/6 of the earth's surface—Queen Elizabeth II is the single, largest landlord in the world.

Indigenous peoples —including First Nations, Métis, and Inuit— represent nearly 5% of the total population in Canada.

Yet, the total amount of land occupied and ceded—Indian Reservations and Aboriginal lands—totals less than 0.35% of the country's entire landmass.

The power of this *Extractive State* lies in the violent separation of surface rights above ground and mineral rights below ground.

As the supreme law of the land, the British *Magna Carta* of the 13th century entrenched the oppressive idea of absolute, centralized, monarchical power, eight centuries ago.

The *Magna Carta* delineated the contours of colonization between those who are within the boundaries of the kingdom and the modern settler-state, from those lying outside of it.

There are dangerous tensions and inequalities between the paper worlds of extraction and the oral histories of the land and life. The map is simply no longer the territory.

Beyond 2017, the 150th year of colonial Confederation in Canada marks 800 years of imperial domination, racial violence, colonial oppression, and territorial dispossession.

This is a story of empire—recounted here in the following 800 pages—across eight centuries of domination that needs to change, and finally come to an end in this generation.

In the future, as in the not-so-distant past, the real power is land.

maxime valet terra

"Not only do imperial colonial powers redefine territories, they also breed new empires, replaying their cycles of dissemination and domination over and over again."

—Suzanne Zeller, "The Colonial World as Geological Metaphor: Strata(gems) of Empire in Victorian Canada," 2001

What Is Empire?

If everything comes from the ground, then extraction not only redefines our understanding of urbanization, it exposes the territorial risks of the 21st century. From gold to gravel, copper to coltan, iron to uranium, geological resources represent the invisible mineral media—below the visible surface of the earth—that supports technological aspects of so-called modern life. In subway tunnels or on suburban streets, in electronic manufacturing or information media, on stock exchanges or in commodity markets, the geological materiality of contemporary urbanism may seem inescapable, but the marks of technological imperialism seem even more indelible. Either in the assembly of consumer goods like smartphones or the construction of concrete highways, contemporary life is mediated through mineral extraction: it is the urban, political, and cultural ore of urbanization. As Maurice Charland wrote in the late 1980s on technological nationalism, "space binding technologies extend power as they foster empire."

Enter geography. Where do these materials come from? Who do they belong to? Under whose jurisdiction? How are they moved and removed? Where do they go? Who processes them? What energies are required? What do they leave behind?

Often perceived as remote, and deftly marginalized as hinterlands, the sites and systems of resource mining not only expose the scales and states of industrial extraction, but they also redraw the boundaries of urban economies and extents of patterns of consumption, well beyond the gray zones and footprints of cities. From property rights above ground to mineral rights below ground, every single dimension of urban life is not only mediated by resource extraction, it implicates and imbricates vast territorial infrastructures and massive credit systems of financial and environmental consequence. Those effects are not only place-specific or global in scope, they are telescopic and transnational: across boundaries of states and nations, traversing political spaces, affecting inhabited lands, and transgressing cultural spaces.

If the Eurocentrism of the Anthropocene has become a new Bronze Age buzzword, then the threading of geology with geography is where technology finally meets territory.

Canada is undeniably at the heart of this massive, international, and geopolitical resource infrastructure. In spite of its humanitarian and peacekeeping image as great middle power of the 20th century, it is the most active mining nation in the world today, with more than half of the globe's mining companies headquartered in Canada and listed on the

Toronto Stock Exchange. Over half of the world's mines are operated, serviced, financed, or engineered by Canadians thanks to a sophisticated fiscal system threaded through the financial hinterlands of the Caribbean. Yet, ironically, Canada is no longer "bound up with the interplay of world empires" as George Grant asserted back in 1969, Canada has become an empire in its own right, quietly asserting its reach around the planet. This raises issues of profound social, logistical, environmental, and political significance with global repercussions of empire building today.

What fuels extraction? Why does it continue to dominate? How did this empire emerge? What did it grow from? How far does it extend? Who does it impact? What, if any, alternatives exist?

These are the pressing questions and public debates that face Canada in the next urban century, as it assumes the unspoken role as global resource giant and planetary staple supplier. While extraction is often associated with technological means of resource exploitation, it is Canada—as colonial settler-state—that was, and still is, the exclusive and dominant tool of the British-cum-Canadian Crown. Located in between the lines of the law of the single, largest constitutional monarchy in the world, its power lies in a simple, subtle, yet remarkably systemically violent force: the imposed and enforced separation of surface rights from subsurface rights. Although this regal fracture is imagined as legal, the territorial repercussions of this simplified divide, when seen from the side, not only edifies the disembodiment of the ground from what is below ground, it enmeshes the bureaucratic partitioning of water from land—the indivisible systems of river valleys, muskegs, aquifers, estuaries, and atmospheres. If territory of the State is its supreme technology, then the underground mineral material not only carries a dismembering political message, it is the colonial media itself and all the imperial infrastructure it helps to build and support.

Whereas the object of territory is usually seen as a domain of a nation, it is the reflection of the State's extractive impulse—expressed primarily through the monarchistic structures of the Crown. The duplicitous natures and architectures of its sustained influence for centuries now, in Canada and around the world, are cloaked in the white space of an extensive bureaucratic system of centralized state regime, racial supremacy, police enforcement, planned settlement, metropolitan concentration, Indigenous assimilation, and corporate transnationalism. As Jane M. Jacobs wrote in *Edge of Empire* during the mid-1990s, these extrastate infrastructures are extensive, "imperial expansions established specific spatial arrangements in which imaginative geographies of desire hardened into material spatialities of political connection, economic dependency, architectural imposition, infrastructures of segregation, and landscape transformation."

Life and language, locked in imperial media and machination, to the benefit of the settler-state majority. Dislocated, for unsettled, repressed minorities. The State of Canada is therefore more than just the 150-year declaration of Confederation dating back to 1867, more than the 300-year old legacy of the transnational corporation of the Hudson's Bay Company, it is the stark imagination of a system of control whose roots lie deep in the colonial headwaters of an imperial center, formed centuries earlier. At its core, Canada is the supreme outcome of an age of territorial discovery and land dispossession initiated across the non-European world under papal bulls and religious doctrines from 15th-century chancery of the Christian State—the Vatican. The Christian Church bureaucratically sanctioned theft and enslavement of Africans, as well as the violence and extermination of Indigenous peoples, across lines of the non-Christian world.

On the surface of the State, this colonial ground is a complex instrument of imperial communication and racial domination whose voice speaks through the engineered entrenchment of legal principles; white spaces whose echoes are virtually unrecognizable under the binary premise and naïve promise of imperial bilingualism and contemporary multiculturalism. The face of colonial explorers, surveyors, and governors may have changed, but the shadows of its predominantly pro-British, Victorian power structure remain the same, with very few exceptions, darker and darker over time.

Internally and externally, the bureaucratic apparatus and institutional architects of this Extractive State are now the accountants and lawyers of government, the financial analysts, the policy dispensaries, and the geomatic experts of federalism, that include the CEOs and CSR officers of the extractive industry.

Across and in between borders, through the nationalization of nature and naturalization of resources, the extractive ethos has crafted a foreign and domestic policy that mythologizes the North and subjugates the South in a neoliberal world of binaries. This transnational stronghold has locked-in systemic patterns of urbanization where consumption lies at the end of the extraction spectrum, held together by extensive property systems, financial mechanisms, class and labor regimes, complex legal structures, and wilderness imaginaries. And if the North is shackled to the South by the complex pipelines of powerful extractive myths, then Indigenous emancipation is imprisoned by the neocolonial entanglements of contemporary constitutional empire building. Not only are these super-colonial powers doomed to perpetuate histories and cycles of domination over and over again, Canada is knowingly and unknowingly committing the greatest atrocity of all: the cultural genocide of Indigenous populations in the territorial prison of its resource hinterland, while resisting forms of assimilation and acculturation in the dense metropolitan centers originally built to displace the non-White, and to exclude the non-Christian in any way possible.

To delineate the contours of empire, and the edges of its ongoing world-making, is therefore to reveal its colonial origins and inherent oppressive ideologies that are historically-rooted and based in Victorian, colonial supremacy.

Within these contexts of settler-state colonialism and spheres of extractive influence, this book retroactively engages and projectively stimulates action vis-à-vis the frictions, oppositions, contradictions, and possibilities that exist in between land and law, land and language, land and hand.

Moving towards the next generation of the 22nd century, the state of extraction is a project that requires a different kind of imaging and imagination, action and retroaction. New forms of representation, as well as forms of reclamation, are needed to undermine colonial strategies of land dispossession and usurp uneven forms of development that extinguish different sovereignties. If it is to do so justly and intelligently, Canada will not only have to confront the technological challenges of scale—upscaling and downscaling—but the political complexities and environmental realities of transnational operations, while fully facing the economic myths of open pit mining culture and mass, industrial economies of scale. On the eve of its colonial commemoration of Confederation in 2017, Canada will therefore have to recognize and reexamine its own imperial role, beyond itself and within itself, for the foreseeable future and the living legacy of next generations. Through a process of transfiguration, Canada will have to reeducate itself by listening, documenting, and translating stories of atrocities it has committed over the past four to five centuries and act swiftly and justly on those histories of the past and of the present through reflective and reflexive cognition, as well as through new legal, curricular, and pedagogical inscriptions. Profiling both the historic and contemporary culture of extraction from a lateral political-ecological lens, this book therefore aims to cast light on these new horizons and develop a deeper discourse on the ecologies of extraction, nationally and transnationally.

Between the rows of financial information and in between the lines of the leases and agreements, lies a historic provocation. It was the nearly three hundred-year old Hudson's Bay Company, incorporation and territorialization together in utero, that developed into contemporary Canada amidst colonial representation at the height of the British Empire: now it is the corporate extension of its Victorian ghost, materialized in the Midas touch of territorial incorporation known as Canada's Confederation in 1867. Marking Canada's sesquicentennial, the declaration made by Senator and Constitutional Expert Eugene Forsey in his 1967 speech *Our Present Discontents* at Acadia University, resonates louder than ever before: "The British North America Act was designed by British overlords; from which it follows, of course, that we must now scrap it and give ourselves a homemade one."

Through the voices of leading thinkers and activists across fields of political influence and action, this book confronts Canada's colonial histories of extraction—seen through strategies of assimilation, integration, and extermination—by excavating historic, unpublished, and rare materials from a variety of archives. Reexamining the sources, evolutions, and transfers of imperial resource roles and colonial logics—from colonial outpost to global storehouse, from empire to empire—the book redefines and reformulates Canada's history as "Extraction Empire" that emerges out of eight hundred years of history since the birth of the British Magna Carta of 1215—the law of the land whose roots underlie the constitutional foundation of Canada. As histories of imperialism continue to grow and dispossession likely progresses, the varying scales of domination that the Extractive State imposes and inflicts will not only require transparency of its actions and disclosure of its underlying bureaucracy, it will require a gradual unbuilding of the Crown's control through territorial transcriptions on the surface and superficies of the State that rightfully belong to Indigenous peoples. Eventually, citizens of this country will have to learn to see themselves as guests and visitors on these lands.

Deterritorialization of extractive states will require, a priori, the decolonization of the State of Canada through the displacement of the Crown itself and the displacement of the power of its settler-colonial state. If it intends to avoid new histories of colonial domination and extractive oppression from repeating themselves, the legacy of the next generation will have to be rebuilt and regrown—one skin at a time, one mind at a time, from continent to continent, from coast to coast, from one nation to the next. On the shorelines of conflict that characterize the project of deterritorialization and decolonization of the extractive state lies new possibilities and new alliances, but more importantly, reclaimed freedoms and reasserted sovereignties that will emerge from the unbuilding, abandonment, and fall of extractive empires. ×

Oceanus occidentalis
Terra del Rey de portugall
Has antilhas del Rey de castella:
Este he o marco dantre castella y portugall
Linha equinocialis:
Tropicus capricorni
Pollus antarticus:
Mare oceanus:

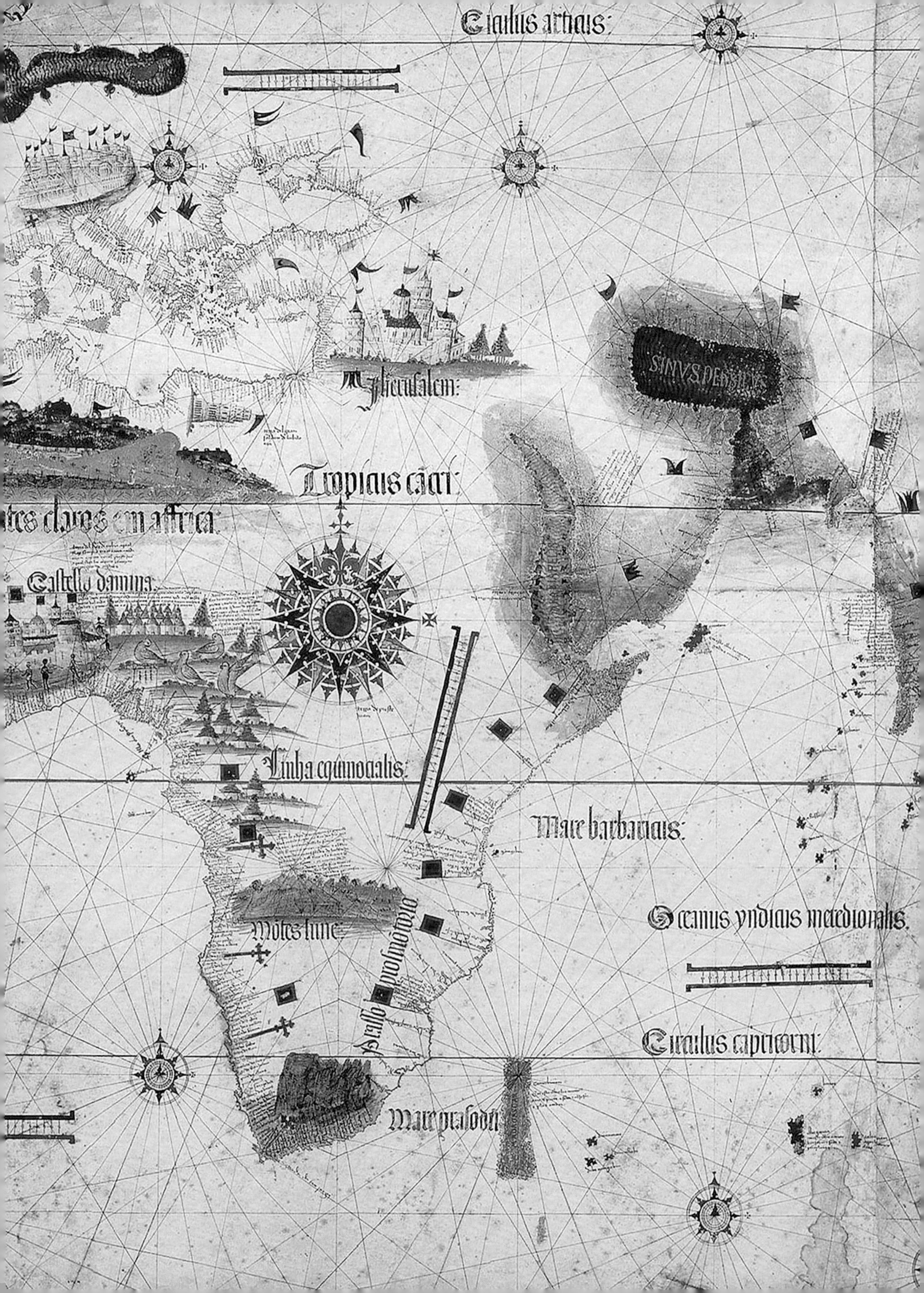

Circulus articus:
Jherusalem:
SINUS PERSICUS
Tropicus cacer:
tes claros em affrica
Castello damina:
Linha equinocialis:
Mare barbaricus:
Oceanus yndicus meridionalis.
Mõtes lune
Circulus capricorni:
Mare prasodu

What Is Extraction? While extraction is often associated with mining, machines, and other mechanical means of resource exploitation, it is the State of Canada that is, a priori, the preeminent extractive technology. The Crown, its original source of power. Whereas territory is usually seen as a tool of the state, it is the State itself—primarily through the royal imagination of the monarchy and enforced management of lands under its jurisdiction—that mobilizes territory as a bureaucratic system of political control, legal empowerment, cultural domination, Indigenous dispossession, settlement imposition, resource acquisition, environmental engineering, and international disposition. Across a country of 10 million km^2, over 95% of lands in Canada fall under the power of the Crown. Seen geographically and understood politically, the process of extraction is a profound and deep-rooted ideology of self-entitlement, exercised through forceful financial penetration, both overtly and covertly, most often felt spatially and temporally downstream, years, decades, and generations later. Its effects are often irreversible and widespread across the intertwined living world of Indigenous peoples, plants, soils, waters—ecologically, corporeally, and psychologically. To track and trace the depths of this Extractive State is to reveal its colonial origins and inherent oppressive ideologies of Victorian supremacy. Through this process of grounding power, this book aims to raise awareness, stimulate action, and engage retroaction in the face of the frictions, oppositions, and contradictions between land and law, land and language, land and hand. To do this, requires a series of preliminary assumptions of the extractive nature of the modern nation state: *I. State:* Guided by a pervasive frontier mentality, the State itself is a territorial technology that requires multiple levels of legal and political encirclement, as well as complete cultural containment to exist, function, and control. *II. Infrastructure:* The State is a process that fixes and mobilizes power through accumulation, extraction, and consumptive demand. In this way, the State functions like a continuously operating, open pit mine. *III. Metropolis:* As colonial métropoles purposely innovated by the British Empire, concentrations of people witnessed by the processes of populating cities dangerously conflates the premise and complexity of processes of urbanization by fracturing relations between people and territory through displacement and dispossession of life from land. *IV. Incorporation:* Through the federation of nationhood, the State requires a geographic surface through which a political territory can be mapped and mobilized. Here, the map is not only the territory, it is the technology. *V. Crown Control:* Through the power of the Crown, the State preys on the legal division, dissection, partition, and separation of the land beneath our feet; it must separate surface rights above (lots) from subsurface rights below (minerals) to produce and perpetuate property. *VI. Subjugation:* The Extractive State is a colonial power that requires civic submission, racial supremacy, and psychological indoctrination through a cycle of multi-generational dependence on the system of the Welfare State to operate. *VII. Habeas Corpus:* In the politico-scientific bureaucratization of the environment, the Crown undermines and extinguishes the body politic of Indigenous peoples through a suite of segregationist, assimilationist, and exterminationist measures. *VIII. Bodies:* By fracturing the bond between land and water, humans and animals, life and land, the State wages a war on the body, and therefore on the mind, as territory and space of colonial violence. *IX. Banks:* Through the virtual creation of credit, the industrial division of time, and the digital consolidation of capital, the commodification of resources is operationalized by isolating and distancing value in an uneven financial diagram of market systems. *X. Paper Worlds:* Through acts, stats, and maps, the power of the Crown is continuously upheld by the lines of bureaucracy and rules of governance, in between agencies and institutions. *XI. Arms:* Resource extraction corporations are simply extensions of government policy and Crown law by creating a third space that distances the state from between the private realm of the individual and public realm of the government and sub-governments. *XII. Extra-States:* The expansion of the State creates new territories, extra-states, sub-states, in between and within the borders of existing nation states. Abroad, it functions like a virtual platform: foreign and fiscal policies enable resource extraction without territory, working in between and beyond borders of nations and states, through displacement and dispossession. Conditions of borderlessness and processes of offshoring actually require the presence of state borders. *XIII. Deception:* The claimed decolonization of the British State (and earlier, that of the French State) through the declaration of the Commonwealth of Nations in the mid-20th century has erroneously, but strategically confounded political independence as corollary to post-colonialism by tightly coupling both concepts as one and the same… Thus, the process of the transformation of extraction, if it is to occur within this generation, therefore requires a priori the retroactive deterritorialization of the State. Since the project of deterritorialization—and that of the underlying decolonization it requires—is a multidimensional and entangled problématique, there is first a representation and remapping of the underlying colonial ideologies of land dispossession embedded in the contemporary practices of extraction that are critically needed in order to better understand the spatial and physical oppression of the Extractive State, to reimagine it, and ultimately, to usurp it. If the revolutionary process of decolonization has historically been violent, as Frantz Fanon suggests in *The Wretched of the Earth* (*Les Damnés de la Terre*) in the early 1960s, then political discourses must urgently reclaim, recuperate, and reinstitute access to land and water as central to human life broadly, and Indigenous peoples specifically. This is an essential and basic right upon which contemporary living is dependent, and for more than 80% of land on which over a third of the world's population relies on—what Shuswap Chief George Manuel referred to in the 1970s as "The Fourth World"—for ecological regeneration and cultural replenishment. But, it requires a gradual weakening and abandonment of the overlapping extractive states to ensure full and unconditional reciprocities. In an oppressive system, the cultural engagement of reciprocity, as Malcolm X so appropriately put it in the mid-1960s, may be between "the ballot or the bullet." So, within the deep relations and manipulative inconsistencies that exist between the map and the territory, between law and land, between the image and memory of territory, between the mind and the mine of a nation's economy; lies stories, histories, prehistories, and neohistories of the states, structures, signs, systems, and scales of how Canada has become a dominant, global resource empire. As a process of transfiguration then, this historical evolution away from past predatory practices and from abstractions of paper worlds, signals a transgenerational revolution in the reclaimed sovereignties and freedoms of yet unborn generations to come. This is the Trojan horse of state deterritorialization. If, the "main battle of imperialism" that is entrenched in the systems of extraction and empire building "is over land," as political theorist Edward Said proposed in the early 1990s and iterated by Indigenous peoples time and time again, then the process of undermining those systems of domination, oppression, and violation—either by rejection, resistance, subversion, or transfiguration—must necessarily and absolutely begin from the ground up.

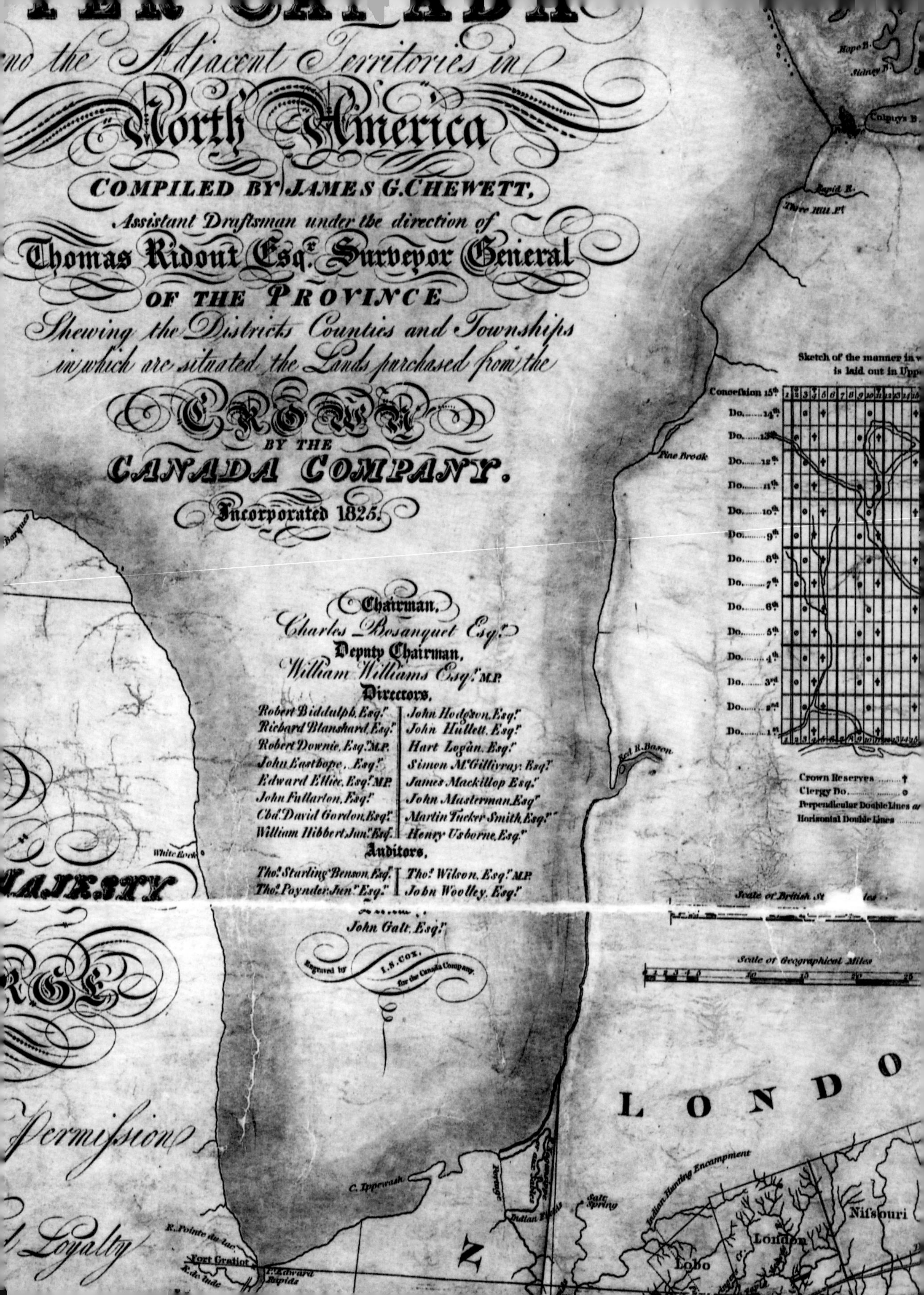

and the Adjacent Territories in
North America
COMPILED BY JAMES G. CHEWETT,
Assistant Draftsman under the direction of
Thomas Ridout Esq.r Surveyor General
OF THE PROVINCE
Shewing the Districts Counties and Townships
in which are situated the Lands purchased from the
CROWN
BY THE
CANADA COMPANY.
Incorporated 1825.
Chairman,
Charles Bosanquet Esq.r
Deputy Chairman,
William Williams Esq.r M.P.
Directors,
Robert Biddulph, Esq.r
Richard Blanshard, Esq.r
Robert Downie, Esq.r M.P.
John Eastbope, Esq.r
Edward Ellice, Esq.r M.P.
John Fullarton, Esq.r
Cha.s David Gordon, Esq.r
William Hibbert Jun.r Esq.r
John Hodgson, Esq.r
John Hullett, Esq.r
Hart Logan, Esq.r
Simon M.c Gillivray, Esq.r
James Mackillop Esq.r
John Masterman, Esq.r
Martin Tucker Smith, Esq.r
Henry Usborne, Esq.r
Auditors,
Tho.s Starling Benson, Esq.r
Tho.s Poynder Jun.r Esq.r
Tho.s Wilson, Esq.r M.P.
John Woolley, Esq.r
John Galt, Esq.r
Engraved by I. S. Cox, for the Canada Company.
Sketch of the manner in w
is laid out in Upp
Concession 15th
Do. 14th
Do. 13th
Do. 12th
Do. 11th
Do. 10th
Do. 9th
Do. 8th
Do. 7th
Do. 6th
Do. 5th
Do. 4th
Do. 3rd
Do. 2nd
Do. 1st
Crown Reserves
Clergy Do.
Perpendicular Double Lines
Horisontal Double Lines
Scale of British St
Scale of Geographical Miles
Hope B.
Sidney B.
Colpuys B.
Rapid R.
Three Hill Pt.
Pine Brook
Red R. Bason
White Rock
MAJESTY
Permission
Loyalty
C. Ippewash
Salt Spring
Indian Hunting Encampment
Indian Fields
R. Pointe du Lac
Fort Gradiot
Ft. Edward
Rapids
LONDO
London
Lobo
Nissouri

Zero
Euphrasy
Alta
Java
Artemesia
Ospry
Merlin
Melancthon
Proton
Luther
Amaranth
Mono
Adjala
Tecumseth
Sunnidale
Flos
Vespra
Tiny
Medonte
Orillia
Matchedash
Rama
Mara
Thorah
Eldon
Innisfil
Georgina
Brock
Mariposa
Gwillimbury North
Gwillimbury East
Gwillimbury West
Scott
Reach
Uxbridge
Whitchurch
King
Whitby
Markham
Pickering
York
Albion
Caledon
Chinguacousy
Toronto
Trafalgar
Nelson
Flamboro East & West
Beverly
Dumfries
Waterloo
Wilmot
Woolwich
Garrafraxa
Nichol
Eramosa
Blandford
Blenheim
Burford
Oakland
Ancaster
Glanford
Binbrook
Niagara
Queenston
Lewiston
Stamford
Clergy Reserve for the Six Nations Lands
Crown Reserve for the Six Nations Lands
Clergy Reserve for the County of Lincoln
Dundas Street
S I M C O E
H O M E
W E N T W O R T H
L A K E
Upper Landing
Holland R.
Newmarket
Black R.
Talbot R.
Matchedash B.
Welch Pool
Low Ground
Second Fall 10 Ft
Upper Fall
Third Fall 13 Ft
Fourth Fall 10 Ft
Fifth Fall 10 Ft
Sixth Fall 15 Ft
Seventh Fall
Severn
Griffith I.
Cloud I.
Campbell Cliff
Owen's Sound
Hope I.
Christian I.
Appearance of an extinguished Volcano
Salt Spring
Village

Unsettling the Mining Frontier

Mel Watkins

"The idea that mines of gold and silver are the sources of national wealth was at that time [during the seventeenth century] singularly prevalent in Europe; a fatal delusion, which has done more to impoverish the nations which adopted it, and has cost more lives in America, than the united influence of war and bad laws."[1]

—Alexis de Tocqueville,
Democracy in America, 1838

Metals matter. Once upon a time, there was a Bronze Age (Early, Middle, and Late), a Copper Age, an Iron Age. Only then comes History without the overt materialism. There have been golden ages of this and that, but no Gold Age.

Mining matters—a phrase that deserves a better fate than to be a Canadian government slogan promoting mining. Running out of wood for charcoal, Britain turned to coal. Water in mine shafts had to be pumped out. The steam engine emerged to meet that need. Coal was used as fuel for the steam engine, a case of positive feedback, feeding economic growth. The steam engine lent itself to many uses like railways and steamships. So it is that mining became a critical trigger of the Industrial Revolution, and thereby of the carbon emissions that we now know set off the escalating crisis of climate change. But, as anthropologist Sidney Mintz mentions, "in understanding the relationship between commodity and person, we unearth anew the history of ourselves."[2] It is a stunning example of what the great economist Joseph Schumpeter called "creative destruction."[3]

○ ○ ○

In recent times, there has entered the discourse of political economy the term "resource curse," meaning that a surfeit of resources may not be a blessing overall. To some, if not many, this may defy logic. Surely to have resources is better than not to have them. And is Canadian economic history itself not written around the great staple trades from fish and fur forward that culminates in our present prosperity relative to most of the world?

Indeed, such is the richness of our resources that in each era of global history our elites have been able to find the perfect way, from their perspective, to lock ourselves into the global economy the better to reward extraction and promote staple exports, albeit while lining their own pockets.

Canadians may want not to dwell on our export of asbestos and of the uranium that the Manhattan Project used to develop the atomic bomb first dropped on Japan in 1945; there may even have been Canadian uranium in the bomb dropped on Hiroshima. We may prefer to close our eyes to the scholarly writings on how the prairies were consciously cleared of Aboriginal peoples to make way for wheat as another of our great staple exports. That was then, this is now.

But—perhaps "but" should be the first word in the Great Canadian Narrative—what are we to say now with the mounting evidence, albeit mostly ignored in the mainstream media, about the human rights violations and environmental degradation committed by Canadian mining companies that have evolved into global players with more than a little help from our elected governments?

ↄ ɕ ɕ

Meanwhile—a word beloved by historians for it contains everything to be known—let's consider how we got here. The title for this essay is that of a neglected 1936 book by Harold Innis, the renowned economic historian who famously wrote about Canada's resource exports a.k.a. staple trades, particularly fur and cod. ("Staple" in this context means a natural resource developed to be exported typically in an unprocessed form.) Innis was a founder of the staple approach to Canadian economic history, that morphed into the "staple theory of economic growth,"[4] of how resource exports generated economic growth in Canada, and influenced the nature of the economic development that resulted.

Innis' book on mining was published as part of a series on *Canadian Frontiers of Settlement* edited by another renowned Canadian economic historian, W.A. Mackintosh, who was the co-founder with Innis in the 1920s and 30s of the staple approach, and is well known for his writing on wheat as a staple. Thus these two scholars left their stamp on the study of Canada down to the present day.

> "Nothing is more typical of colonial development than the restless, unceasing search for staples which would permit the pioneer community to come into close contact with the commercial world and leave behind the disabilities of a pioneer existence."[5]
>
> —W.A. Mackintosh, "Economic Factors in Canadian History," 1923

> "Concentration on the production of staples for export to more highly industrialized areas in Europe and later in the United States had broad implications for the Canadian economic, political and social structure. Each staple in its turn left its stamp, and the shift to new staples invariably produced periods of crises in which adjustments in the old structure were painfully made and a new pattern created in relation to a new staple."[6]
>
> —Harold Innis, *Empire and Communications*, 1950

Innis' book restricted 'mining' to hard rock mining, that is, hard metals, e.g. gold, silver, iron, copper, zinc, nickel, tin, lead, plus placer mining in the case of gold. The focus was on the Yukon and the Klondike (gold), British Columbia's Kootenay region (gold, silver, copper, lead) and Northern Ontario (silver, nickel, copper, gold). Innis excluded coal (soft rock), in which Canada was not well endowed and was a net importer rather than exporter, and asbestos which is a fibrous ore. Since Innis wrote back in the 1920s, 30s, and 40s, uranium, iron ore, diamonds, potash, and bitumen (strip-mined from oil sands a.k.a. tar sands) have emerged as staple exports. There is also the interesting case of importing bauxite to refine into aluminum with Canada's abundance of hydro-electric power and then export it.

What is today called Canada, as a creation of the last half-millennium, is part of the so-called "New World" (the Americas and Australasia by settlers), frontier societies in the dual sense of frontiers of settlement—pushing aside Aboriginal peoples—and frontiers of staple exploitation. These two frontiers do not always coincide.

The specific staple comes to shape settlement, in some cases even inhibiting it. Such was true of both the cod fisheries and the fur trade. In the fisheries during the initial stage it was in the interest of European fishermen to block settlement, which was in any event inhibited where a winter fishery was not possible. The fur trade was forced to retreat in the face of settlement, so fur traders discouraged it. In both cases, in the long run, migration could not be stopped and settlement triumphed.

Still, at the beginning of Canada, as a fragment thrown off by Europe, trade preceded settlement rather than settlers searching for things to trade. Canada was literally a "trading nation"—to use today's overworked phrase—before it was a settler colony; and it is significant that Canada still prefers to export unprocessed staples rather than build a diversified economy capable of generating more employment.

At the beginning of contact with the New World, everywhere there was the search for precious metals. The pie in the sky was already on the ground, or in the rivers, or underground, or on display by local power, there for the taking.

When one links settlement and mining in the discussion of the New World, what comes to mind? Looting—by the 'settlers' from the Old World—is highly probable. (The word "loot" enters the English language, according to William Dalrymple, from, revealingly, Hindustani slang for plunder, and did so at the time that the East India Company was plundering south Asia; a new book on resource exploitation in present-day Africa by Tom Burgis of The Financial Times is, so it happens, titled *The Looting Machine*.) Marx called this seizing of the colony's land and its resources, its privatization in the name of imperial capital, the subjugation, even extermination, of Indigenous people, "primitive accumulation." The discovery of silver at Potosí, now in Bolivia, "the biggest richest strike in history,"[7] created an instant city, the most populous in the continent, though mostly consisting of forced laborers, and one of the largest in the world, bigger than Paris, Rome, Madrid, London. The shock waves spread back across the Atlantic and gave a powerful assist to the growth of European capitalism—though, oddly enough, not for Spain, which became itself a victim of a resource curse with silver flowing in and mostly just flowing out again—and all that has followed therefrom. It was like a cyclone long before Innis used the telling phrase "cyclonic," conjuring up a devastating storm wreaking havoc, to describe the gold rushes of the nineteenth century.

> "Two planetary scientists believe that large objects passed the moon repeatedly over tens of millions of years and tipped its orbit, some hitting Earth and leaving precious metals behind."[8]
>
> —*The New York Times*, November 27, 2015

As Innis knew when he let his mind wander creatively—hither and thither, one thing leading to another—the storm stirred up by the gold rushes partook of today's 'perfect storm':

> "The cyclonic effects of the gold rushes in the Pacific region were evident in the expansion of shipping and trade on the Atlantic and the Pacific and in the development of Great Britain as a metropolitan centre of the world. It is significant that [the economist Alfred] Marshall suggested that after 1873 the economic history of one country could not be written. At a later date the gold-rushes had profound effects on continental development. Transcontinental railways were built to San Francisco and in turn from Montreal to Vancouver to link up the economic areas based on the discovery of gold with the eastern seaboard of North America ... The Klondike gold-rush had its effect in hastening construction of two additional transcontinental railways which became the basis of the Canadian National Railways. The vast resources of a continent were opened up with transcontinental lines."[9]

We are tempted to think of the North American frontier of settlement as methodically marching westward, but the gold rushes were like giant leapfrogs. Frederick Jackson Turner, the American historian who famously invented the "frontier thesis" to explain American history, saw the settlement frontier as first and foremost, with the mining frontier coming second.

Locally, in the Yukon, the rush started slow and then picked up speed. In the beginning there was a sense of community among the miners, of looking after each other. With the great rush, it was each man for himself "so humanity, and nearly every other consideration which enters into everyday life, were utterly wanting."[10] Gold as a staple in Canada was, from its outset, the ultimate manifestation of the greed that permeates capitalism.

ↄ c ᴖ

> "Prospecting is an ugly business: it makes a man start thinking like a thief."[11]
>
> —Eleanor Catton, *The Luminaries*, 2013

The word "rush" is revealing: miners rushed in, exiting a prior rush, and shortly thereafter rushed out, leaving not that much by way of permanent settlement.

The globalizing crisis-prone economy that gold helped create was itself linked for purposes of stabilization and control by the international gold standard. Gold was, magically, both problem and solution, both cause and effect. Innis wrote of gold's "position in the monetary system by which it was able to command most effectively the resources of modern civilization."[12] In his magisterial book on the nineteenth century, the German historian Jürgen Osterhammel writes of "The Gold Standard as Moral Order"; it "universalized the values of classical liberalism"—like property rights and free trade. For some time, gold co-existed with silver as a monetary standard, but the great gold discoveries in the United States, Australia, South Africa, and Canada gave gold the advantage. With the British empire dominant, the gold standard was in practice a sterling standard.

Fittingly, the heyday of the gold standard in the late nineteenth and early twentieth centuries has come to be called "the golden age of capitalism." The passing of the gold standard has, however, not made gold obsolete; demand being presently driven by the growing middle classes of industrializing India and China, which buy three-quarters of newly minted gold—70% of which is for jewelry. And there have always been those who hold gold as a hedge against inflation, though this is discouraged by the present deflationary times.

Mining in British Columbia was less remote—closer to 'civilization'—than the Yukon. There was some infrastructure there and more that could be built, like railways and hydro generating plants that could have their own further effects. There was coal for smelting. In short, there was a base for settlement and industrialization.

In northern Ontario, with railways being built, blasting exposed rich lodes of ore. Call it luck, a neglected factor in the writing of Canadian economic history. The building of the first transcontinental, the Canadian Pacific, unearthed the rich nickel-copper ore in Sudbury. The building of a railway north into the agricultural land of the clay belt—thought to be a new Saskatchewan in Ontario—yielded ore bodies in the Cobalt area. Such serendipity—in the context of proactive resource development policies by the Ontario government—promised more diversified development.

In spite of this, mining country was unable to escape the fate of the ghost town, of that rush for resources that comes and goes. "Canada is full of ghost towns" according to the great scholar Northrop Frye, "visible ruins unparalleled in Europe." British Columbia has the most ghost towns of any state or province in North America; you can visit them in coffee table picture books.

> "[As death nears] I am … surrounding myself, as I did when I was a boy, with metals and minerals, little emblems of eternity."[13]
>
> —Oliver Sacks, "My Periodic Table," 2015

Ponder the story told by Tom Walkom, columnist for the *Toronto Star*, under the brilliant headline "You can't go home again in this raw and casual country":

> "South Porcupine—I went home this weekend and it wasn't there.
>
> There was nothing.
>
> The house where I grew up is gone. Burned and bulldozed.
>
> The cedar hedge my father and I planted in 1962 is gone…
>
> The yard has returned to bush…
>
> All that remains to show that anything was there is a 'No Trespassing' sign.
>
> My public school is gone…
>
> They are all gone. Finished. As if they were never there. Razed, spit out and chewed up. Or covered in cyanide and mud.
>
> Part of me looks at this and says it's all for the best … because the mines have stayed open, because the owners rejected sentimentality and instead dug and razed wherever they could find a vein of gold.
>
> But in my darker moods, I see the back road between South Porcupine and Timmins as bleak metaphor.
>
> We are a raw and casual people, we Canadians—a people with little respect for the things we have built, be they houses or primitive curling rinks or social institutions.
>
> That which we have we are apt to crush and mix with cyanide. We sell the good stuff and keep the dross.
>
> In the name of vague progress, we set out to destroy what we have so painstakingly created.
>
> We forget that once everything has been bulldozed, there will be nothing left."[14]

Today, in remote areas, in the era of the airplane, rather than creating a settlement, a company town, the company can build an enclave, with a fly-in fly-out work force—technology permitting of the return of the nomad.

A positive long-run outcome for a resource community based on a staple is the good luck of a sequence of staples. Consider Kenora in northwest Ontario. It began as a point of portage in the fur trade, morphed into a settlement to mine gold, then became the site of a pulp and paper mill, and now is poised to participate in the potentially huge ore discoveries in northwestern Ontario.

Back to Innis: "The later development of metallic mining in a young country [New World] is primarily isolated from agriculture. The discovery of mines assumes accessibility to vast stretches of mineral-bearing formations which preclude agricultural development"—and hence limited settlement since even food has to be imported.

Yet Innis saw the minerals, the smelters, the railways, the hydro-electricity—and foresaw, more industrialization, more development than actually happened. Peter George, economic historian and later President of McMaster University in his survey of Ontario's mining history, writes: "Innis believed that the mining industry would reduce Ontario's dependence on the export of staple commodities, by contributing to a highly integrated, advanced industrial economy."[15] This runs counter to a view frequently encountered in the post-Innis political economy literature; that Innis was pessimistic about the prospects for growth in a staples economy. Ironically, we now know that Innis was wrong in his optimism with respect to mining.

○ ○ ○

Innis' book was published over eighty years ago, a nanosecond in the history of the planet, but a huge chunk of lived time. Much has changed, but the mining mode of extraction and depletion—of corporate hit and run—has only become more entrenched. We can only offer some vignettes and reflections.

Consider realities that Innis—all too typical of his time—neglected. Settlement means white settlement—meaning non-Aboriginal. Aboriginal peoples, who were already there, are not to be seen, their land rights ignored. They are excluded from benefits but not from costs. It seems odd that an historian of the fur trade who understood their fundamental contribution now left them out of the narrative. Matthew Evenden (in an excellent essay "The Northern Vision of Harold Innis," in the book edited by William J. Buxton) notes Innis' indifference to Aboriginal peoples—and to women (see below)—as part of what Buxton summarizes as Innis' "close relationship with members of the elite responsible for … development."[16] That elite, clearly, was white, male, educated, privileged European descendants.

A major development affecting mining is that Aboriginal peoples have put themselves back on the stage. Development projects must pay heed to their rights and concerns. Deals can be made, in the context of Aboriginal rights of governance, beneficial to Aboriginal peoples, as is demonstrated by Suzanne Mills and Brendan Sweeney in general and with particular reference to nickel mining at Voisey's Bay in Labrador. The companies have had to abandon their colonial attitude, though the question of the ability of Aboriginal peoples to veto a resource development remains legally and politically uncertain.[17]

Aboriginal peoples inside Canada may say "No" because they have a relationship to the land, to nature, that transcends it being real estate, a commodity. Therein lies the promise, for them and for us, of a profoundly different view of extraction and of responsibility.

A revealing story from the Northwest Territories in the seventies (to which the author can personally attest). In Yellowknife, arsenic was leaking into the water from gold mining and its tailing ponds, poisoning fish and risking cancer for those who ate them. The government issues an advisory not to eat fish which is a major part of the Dene diet. The Dene chief in a nearby community can't understand how this could happen. The workers in the mine would not poison the water in which they fish themselves for recreation. He appears not to understand that workers do not run the mine, but rather the owner who lives in faraway Texas. The chief in his wisdom sees the world as it might have been, imagines a degree of democracy that surely, the white man, with his swagger and imperious nature, must have.

Consider too the remarkable story of the Dene widows—whose husbands died from radioactivity—told by Peter C. van Wyck in a highly original and refreshing essay "Innis and I on the Highway of the Atom" in Buxton's book. A uranium mine is operating on Great Bear Lake in the Northwest Territories during the Second World War. Van Wyck writes of "unstable staples," of "leakages" (not linkages; see below) "in the form of cancers, stories, addictions, and depressions" that radioactivity wrought upon the Dene in their innocence, but who nevertheless went to Japan after the war to apologize for the contribution of their people to the bombing of Hiroshima and Nagasaki, something that the owners and managers of the mine, the Canadian government, and the scientists who worked on the Manhattan Project did not see fit to do.[18]

There is also no mention in Innis' book of environmental matters, of damages done, of failure to redress, of mines closed and abandoned, of tailing ponds left, of catastrophes waiting to happen. Tailing ponds themselves are inherent to the business. They are what are dug up other than the ore desired, or poisons used to free the ore.

A most revealing case of runaway mining capitalism, a truly cautionary tale, is that of the massive release of contaminants into the waterways of Colorado and beyond in the summer of 2015 when the concrete plug on an abandoned mine gave way. The mustard yellow rivers were a thing of beauty on TV news. Long ago, in the 1890s, the Gold King Mine prospered. In the early 1920s, almost a century ago, it was closed, abandoned by the then owners with no cleanup. The federal Environmental Protection Administration (EPA), while monitoring the situation, accidentally caused the release. Nearby state governments threatened to sue the EPA for damages while the present owner, who had only recently bought the site for a song, said his deed relieved him from liability and that he was holding on, hoping that remnants of gold and silver would be worth mining at some point. It is estimated there are in the order of 500,000 abandoned mines in the United States.[19]

Note in the Colorado case the owner's hope that "remnants of gold and silver would be worth mining." In fact, as the best ore bodies are mined, the corporations have developed technologies that enable them to mine ore, particularly gold, where there are literally only slivers of metal in vast quantities of rock and dirt. It's like trying to find the head of a pin in a haystack, meanwhile poisoning the water and despoiling the countryside.

Only weeks after the Colorado spill, on November 5, 2015, Brazil had, according to its environment minister "the worst mining accident in the country's history,"[20] with 12 people dead and 11 missing.[21] A dam burst at an iron ore mine and vast quantities of sludge, said to include arsenic and other toxic metals, poured out and quickly spread, as far as 450km. The mining company—jointly owned by two of the world's largest mining companies, Vale of Brazil (which now owns Inco) and Anglo-Australia's BHP—claims the sludge is not toxic. Mining has been going on in the area of the mine since colonial days when gold was extracted and sent to Portugal, yet the local community has to this day no other activity than mining. There is a terrible constancy—of maltreating both nature and people—in this story.

> "It's our biggest issue in the county. Mine tailings. Run off … We're constantly finding new contamination like in this area here. A lot of these mines have been closed for decades. Companies are bankrupt … State doesn't have the resources."[22]
>
> —*True Detective*, HBO, July 12, 2015

Yet, lest we are misled, the power of Innis' writing in Canadian economic history resulted from his fundamental interest in rivers, mountains, plains, forests, the mineral-rich Canadian Shield. Think not of the environment but of something deeper, ecology, the planet itself, things mysterious, beyond our control. Innis was the prose counterpart of the Group of Seven's landscapes. Nations were created out of land and water—and rocks. In Innis' masterful conclusion to *The Fur Trade in Canada* he writes: "The present Dominion emerged not in spite of geography but because of it. The significance of the fur trade consisted in its determination of the geographic framework. Later economic developments in Canada were profoundly influenced by this background."[23] Geography pervades Innis' writing—and today his legacy is more firmly established at universities in geography departments than in economics departments or history departments.

Sociology is welcoming, like geography. Eric Pineault's exquisitely titled "The Panacea: Panax Quinquefolius and the Mirage of the Extractive Economy" is a profound meditation on Innis on staples from the perspective of today's discourse and concerns—as an extended riff on, would you believe, ginseng as a neglected staple in New France.[24]

An extractive economy, relying on export of primary resources, is for Pineault a 'primarized' economy, and he sees Quebec and Canada as presently being re-primarized—sinking into what some of us have called a "staples trap" (see below). Exploitation and deepening depletion results in "collapse of the stock (whether ginseng, beavers, cod, spruce, or soon, oil)."[25] This is a compelling list of how most all of Canada' great staple trades can be properly described as extractive: indeed, a staples economy is by and large an extractive economy, and the consequences of that destructive relationship with nature must be understood. Pineault's concluding sentence: "If we do not free ourselves from the extractivist mentality that conditions our relationship to our development, our re-primarized economy is doomed to fall… like ginseng."[26] Pineault makes explicit what is implicit in Innis, of what needs to be borne in mind as we study mining.

○○○

Where Innis was guilty—and representative of the patriarchal times—was in his utter exclusion of women in his writings. They were always there but it has taken feminist scholars to find them. In the mining town of Flin Flon—copper-zinc, on the Manitoba-Saskatchewan border but mostly in Manitoba—the dichotomy of men at work outside the home and women in the home was even more evident than in Canada generally.[27] It was not a pretty picture: for example, "for some women depression and periodic nervous breakdowns become a chronic response to the irresolvable contradictions of their lives."[28]

Emma Jackson—in an overview of the literature since Luxton and with particular reference to Fort McMurray, the strip mining capital of Canada—finds that the gender bias in mining, and in oil and gas, persists—albeit in nuanced ways, qualitative as well as quantitative, with specificity as to time and place.[29] Ironically, Innisian-style, women are both dissatisfied with their marginality, while reinforcing it by making family and community life more tolerable. The effect of globalization on corporate practice has had a homogenizing effect in the developed world, arguably lessening overt gender discrimination.

Staple dependence entrenches gendered patterns in the labor market and the culture, but mining towns are not islands unto themselves and everywhere women have become more empowered and men more empathetic.

But extractive industries still have their stamp. Fort McMurray, Jackson finds, has a frontier, masculine, gendered, sexist culture.[30]

"The earth means something different now. It never heals, upturned constantly."[31]

—Anne Michaels, *Miner's Pond*, 1991

○ ○ ○

We have seen Innis' concern with the spread effects from mining. Innis' staple approach has subsequently been made more precise by the application of the taxonomy of 'linkages,' developed by the renowned economist Albert O. Hirschman, by which economic growth may spread from the export sector to other sectors of the domestic economy. The most obvious linkages are forward, backward, and consumption.[32]

Forward linkage is the further processing of the staple prior to export. Historically, it came to be called in Ontario the "manufacturing condition." This has been a weak, though not non-existent, linkage in mining. To quote the economist Trevor Fast: "It is perhaps the worst kept secret that the vast majority of Canadian exports take the form of either raw or nearly raw resources."[33]

Since further processing of the resource would seem the most obvious of linkages, it may surprise that this should be so a half millennium after exports of resources began. It hints at decisions made in the metropolis importing the staple that appropriates the benefits from processing to the disadvantage of the hinterland. It also hints at a preference in the hinterland to pay the bills for imports by pushing the raw resource and not bothering with processing. With such a mentality, it does not surprise that mine owners and operators would be disinclined to push for the broader industrialization that Innis hoped for.

A fascinating case in point is the refining of the nickel ore of Sudbury. In spite of years of agitation in Ontario, Inco, an American company with close ties to monopoly power, resolutely refused to move the refinery from New Jersey to Ontario. H.V. Nelles, in his thorough examination, refers to "the dark and intricate network of self-interest" that included J.P. Morgan and U.S. Steel.[34] Nothing happened until it became public knowledge that nickel refined from Sudbury ore, that could be used to build tanks, had been sold to Germany in 1916, when on two occasions German U-boats had docked in New Jersey and been loaded with refined nickel. The US had not yet entered the First World War against Germany but Canada had. The full consequences of foreign ownership of a Canadian resource were finally intolerable. Refining was moved to Port Colborne, Ontario.

Backward linkage from mining means the potential for domestic manufacture of mining machinery. It is another worst kept secret that Canada is generally deficient in machinery manufacture, though there has been some late innovation in mining machinery. Geographer Iain Wallace cites "the growth of Finning Ltd. from a supplier of heavy equipment to BC resource industries to being one of the biggest such firms globally."[35] An early example from the Kootenay is the manufacture of explosives.

Consumption linkage is the spending of income earned in mining. In general, this linkage depends on the equality or inequality of its distribution. Historically, the striking contrast is that between slave-using plantation staples like cotton or sugar with very unequal distribution of income, and a family farm staple like wheat with a more equal distribution of income and hence a larger market for mass manufactured goods with potential for domestic production. In the absence of slavery, wages of labor were relatively high in the New World compared with much of the Old, though not without valiant union struggles.

Innis notes that "success of the flour-milling industry and the pork-packing industry at Calgary and Edmonton were, in large part, a result of Kootenay demands."[36] Typically, he detected long-run consequences: "Demands of mining contributed definitely to the trend of agriculture in Alberta, which emphasized livestock rather than grain, and did much to develop the outlook [right wing Social Credit] which differentiates that province from the province of Saskatchewan which has been dominated by wheat [left-wing CCF]."[37]

To pursue the political for one more moment: miners have tended to be politically progressive, reflecting the high level of unionization and hinterland resentment of metropolitan dominance. Famously in Sudbury there was the intense, sometimes violent, struggle between the Communist-led Mine-Mill and the social democratic/CCF/NDP-led steelworkers.

ↄ ᴒ c

Add financial linkage, or more accurately, the intimate relationship between mining and the stock exchanges.

Mining is a gambling life. The prospector lives by throwing the dice. The miner risks being buried alive as part of his job. Peddlers of penny stocks pick the pockets of those who crave instant wealth.

Mining companies require equity capital, and the Toronto Stock Exchange, now Toronto-Montreal, is the greatest source for mining capital in the world. Likewise, the extractive sector is the greatest recipient of financing from the Canadian crown corporation Export Development Canada.

Mining, being particularly speculative by its nature, has facilitated exchanges that resemble casinos. Still, Toronto with its banks and stock exchange has become, with a powerful assist from mining, a major financial center creating many jobs.

In his comprehensive history of Canadian business, Michael Bliss offers a telling anecdote: "A brief gold rush in Eastern Ontario in 1866 created the boom town of Eldorado and an important Canadian tradition of promoting worthless mining stock; total production from Eldorado's mines was eighty-five ounces … By 1898 Toronto's financial district housed more specialists in mining shares than there were members of the Toronto Stock Exchange."

With the political power base of mining and finance, regulation of mining itself is notable for its absence. Mining companies from outside Canada that do not own any mining claim on Canadian soil find Canada an inviting place to be registered.[38]

○ ○ ○

The existence of economic rents, which so frighten the big resource companies which insist any 'surplus' above costs are simply profits, has been known since the time of David Ricardo in the early nineteenth century. They consist of a return to 'land' after the wages of 'labor' have been paid and 'capital' has received a normal profit. They are there for the taking by government imposing royalties. Hirschman called this fiscal linkage. The companies do their damnedest to label the rents profits and minimize royalties. The third worst kept secret is that Canadian governments are not serious about collecting royalties or taxing the resource corporations.

In 1973, Manitoba's NDP government asked Eric Kierans—economist, former president of the Montreal Stock Exchange, Liberal politician—to investigate the matter of rents and resources in Manitoba nickel mining. Kierans took his assignment seriously, too much so for the mining companies and, it turned out, for the government of Manitoba. The rents were there for the calculating, which Kierans did, and they were substantial. Unless appropriated by the government through taxes and royalty rents were for the taking by the companies, which meant that when the companies were foreign-owned the rents and the revenues left Manitoba. Kierans saw the resources as belonging to the public and advocated the public ownership of the companies. This was all too much for a modest social democratic government.

Which does not mean that royalties don't make sense. They increase government revenues. If there's a flaw in the case it's that governments get dependent on them and are tempted to push extraction, though in Canada there seems little scope for any more pushing than has already taken place.

It's a different matter, however, in resource-rich poor countries, as in Latin America. Their resources have long-been looted by companies from the developed world, very much including Canadian companies in recent years; the historic record is the core of the powerful writings of the late Eduardo Galeano, notably his *Open Veins of Latin America*. In Latin America, it has led historically to underdevelopment rather than development.

As resources have become more important in the present stage of globalization, notably with the industrialization of China and its enormous appetite for resources, some Latin American countries, typically the more politically progressive like Ecuador and Bolivia—some would include Venezuela under Hugo Chavez—see resource exports as the engine of growth which has long eluded them. This is a left nationalist variant of the Canadian model where the rents are captured by the government and used to rid the country of its terrible poverty. For that to be possible, resource development has to be pushed.

This passes under the name of the "new extractivism."[39] The downside is that while governments will say they are pursuing the infinitely flexible "sustainable development," they may be tempted to pay too little heed to environmental matters—which, as we've noted, is true also of the Canadian model.

Are there ways out of this conundrum? The country that has the resources can say "No" to mining. That may seem improbable but both Costa Rica and El Salvador in Central America have done just that: Costa Rica in the name of protecting ecotourism which creates many more jobs than mining, and El Salvador to stop the pollution of water essential for agriculture in a densely populated country.

The existing paradigm of politics and of ideas may pretend that there is a fine-tuning solution to this. In fact, as Naomi Klein insists in her book, *This Changes Everything*, this progressive conventional wisdom must be transcended by abandoning what is at its heart, namely, the domination of nature by humans, the belief that 'nature' is there for the taking, that the Earth is there to be torn up, without even any compulsion to repair it. Rather, we need to see ourselves as part of nature, part of the planet.[40] (Such matters were eloquently addressed by Pope Francis in the summer of 2015.)

Spatial empires, like mining, need to take account of time, of the billions of years required to create ores which are then depleted ASAP, confident that future technology can solve any problems that result; if they don't we'll take spaceships to somewhere else, or maybe just the 1%. This starts to sound like fantasy: absurd, insane, not even good science fiction, a terrifying way to treat our home planet.

○ ○ ○

Back to Hirschman for a moment. Think not about linkages from staples but about modes of production, in this case of staples, or micro-modes of production.[41] "Mode of production"—a term we owe to Marx—means a way of producing, in particular, the relationship between capital and labor, and technological change. The sociologist Wallace Clement shows how Innis' "dirt economics" (to use Innis' own term) is rich in detail about both capital–labor relations and technology. One can trace the transition, as in gold mining, from independent commodity production (placer mining) to capitalist mining proper, first entrepreneurial and then corporate. Clement can be read as complementing the staples approach, at least when the latter is cast in the reductive, albeit policy-rich form, of a staple *theory* of *economic* growth.[42]

A striking statistic today about what Clement is talking about: in a world dominated by large corporations, it is estimated that there are 16 million small scale miners and artisans who account for 10 percent of gold production and 90 percent of the jobs. Large-scale mechanized gold mines are "vastly more efficient … safer for workers … don't provide many jobs."[43] The direct benefits of corporate mining, through wages of local labor, are limited. The companies should be required to pay taxes and royalties locally on a large scale which are then spent for the public benefit; better still they should be locally owned. Or labor using smaller-scale technology needs to be created and applied. All of which is, of course, easier to prescribe than make happen. What of present artisan mining? Finnegan quotes one of the artisan miners in Peru: "I hope the gold price falls and the mines close and we all move to towns where we don't have to live like animals."[44]

There has, quite recently, been a major development: Canadian mining, in a first for Canadian staples, has gone abroad. There has been investment abroad in the past in infrastructure building (railways and hydro) and banking. The novelty now is that comparative advantage in trade has evolved to become comparative advantage in direct investment abroad. Long labeled a dependent Canadian economy in terms of foreign ownership, direct foreign investment by Canada now exceeds direct foreign investment in Canada, and mining is an important part of that significant turnabout. There's an empire of mining and Canada is not just part of it, it's the biggest player, with three-quarters of the world's mining companies operating out of Canada.[45]

> "Mines were the major pioneers in opening up the world to imperialism."[46]
>
> —Eric Hobsbawm, *The Age of Empire: 1875–1914*, 1987

Fast explains very well why Canadian mining companies have gone abroad in such an impressive way. It's the corporate drive for resource (staples) exploitation in the present era of neoliberal a.k.a. neoconservative, corporate globalization.[47] (The very word globalization was born during that time.) And, when Canadian companies go abroad they take Canadian-style limited regulation with them; it's starting to sound like imperialism.

Canadian owners benefit. Jobs are created in head offices and on Bay Street (Canada's Wall Street), in project management services and in geophysical expertise that can operate abroad.[48] Fast shows that the concentration of ownership in Canada increases the return to capital and hence inequality in wealth and income distribution, a major point in the time of Piketty.[49]

Settlement and the Mining Frontier now means impact on existing settlements abroad, the foreign versions of our Aboriginal peoples who are seen as blocking development, a nuisance, a hindrance. Frontier has morphed into empire, notably in Latin America and also in eastern and southern Europe. "Mining companies operate [abroad] with a staggering sense of entitlement. They actually identify 'resource nationalism' as the greatest threat for limiting 'their rights and profits.'"[50]

Canada has now become an important violator of human rights. "Mining corporations, often Canadian, are said to have contributed in one way or another to conflicts that have placed millions of people in jeopardy and have led to deaths, systematic rape, the forced recruitment of child soldiers, and legions of refuges."[51]

The Canadian state reshapes, warps, foreign policy—including human rights—to assist Canadian companies abroad. John Baird, Minister of Foreign Affairs in the recent Conservative government, left to become an instant advisor to Canadian-based multinational Barrick Gold, the biggest gold miner in the world. Former University of Toronto president David Naylor is named to Barrick's board of directors replacing former Prime Minister Brian Mulroney—but resigns a year later to return to the University of Toronto on a full-time basis. The extraordinary power and appeal of the corporation is evident in Barrick's ability to command the services of these people, who then risk lending legitimacy to mining—and to the extractive industry in general—with its dubious reputation, above all with respect to human rights.

Mining capitalists, long in the vanguard of imperialism (think Cecil Rhodes) have something of a reputation in the US, in Africa, in general, as the rogues of capitalism, with a slight sense of morality—except as philanthropists to cleanse themselves for eternity. The miners, the workers, are literally between a rock and a hard place, their best hope being in unions, which capital resisted historically in the Global North and resists today in the Global South.

Governments, at the center and the margins of the global economy, have been of slight avail in coping with corporations. Corporations insist on foreign investor rights that permit them to sue governments for monetary compensation that attempt to curtail corporate rights. As for human rights at the margin, their protection depends on the United Nations, on the proliferation within civil society of NGOs like Mining Watch, Amnesty, Above Ground, and the role of independent writers and scholars.

Finally, a gem of a linkage, rarely if ever noted, from which all the English-reading world benefited:

> "On this day [July 25] in 1897, American writer Jack London set out to join the Klondike Gold Rush in the Yukon, a remote and unforgiving region in northwest Canada and Alaska. His cabin was situated at a busy crossroad and was a popular stop for other miners and adventurers, who regaled London with stories. The Yukon wasn't kind to London in terms of health or wealth; he never found gold and lost four teeth to scurvy. But he wrote down what the prospectors, miners and adventurers told him, and when he returned to California, he set about writing a series of stories. They became *The Sun of Wolf* (1900). In a few short years London was the most successful writer in America."[52]

In sum, the specific nature of Canadian capitalism has been well described by the scholar Daniel Drache as "advanced resource capitalism."[53] Canada's trade, as we know, has a strong resource export bias. A list of Canada's top industrial firms is likewise biased toward the resource-based.

In what might be thought of as the conventional mature capitalism, there is an alliance of industrial capital and finance capital. In Canada, the alliance is between a staples fraction within industrial capital and finance; we have seen the symbiotic relationship in mining. Manufacturing per se is not powerful—not even primary manufacturing.

Canada's persistence over the long-run as a staple exporter is indicative of a staple trap, a major cause of which is the undue influence of staple exporters. In Canada, we are acutely conscious of the bitumen trap; add mining and you get an unambiguous extraction trap.

There is, of course, a demand side to this story. Minerals and metals are an essential input of industrialization and have grown with it and followed its path, from the US to China. With technological change and new products there are new resource requirements: think cell phones and rare metals, with the latter believed to exist in the major ore discovery made in northwestern Ontario.

And then there's war—which requires weapons, which require metals; all of which has consequences. The classic textbook in Canadian economic history by Tom Easterbrook and Hugh Aitken tells us that "the First World War was directly responsible for very considerable expansion in Canadian non-ferrous metal refining"[54] and the Second World War had a similar though quantitatively much greater effect. The Cold War was a large and sustained boost. The United States, now the world's dominant power, faced shortages of a number of minerals of strategic military importance. Canada was well endowed with many of them and well situated to supply them and more than willing to do so.[55] Thus we came fully into the embrace of the American military-industrial complex, with all its consequences for morality.

○ ○ ○

This then is the story—some stories from the story—of Canadian mining since the assault by the Old World on the New. There is a story behind the story in the writings of four Canadian prophets of the twentieth century: Harold Innis, George Grant, Marshall McLuhan, and Ursula Franklin. Their writings share a deep understanding of, and prescience about, technology and power. Heed them as voices wary of the globalization characteristic of mining.

Innis has already had our attention, but here is one more quote that shows the sweeping power of his pen: "Under the stimulus of treasure from the new world the price system ate its way more rapidly into the economy of Europe and into economic thought"[56]—thereby reducing our very ability to understand the destructive source of it all. His *The Cod Fisheries* bears the sub-title *The History of an International Economy*. Today, it's *Pick-Your-Staple: The History of a Global Economy* as history itself increasingly escapes national limits and goes global. Or, in recognition of the later Innis as a founder of the Toronto School of Communications, the title of his book *Empire and Communications*, being his phrase for what became, with America and the digital revolution, so-called globalization in corporate speak.

The Canadian philosopher and professor of religious studies, George Grant, anticipated the phenomenon of 'globalization' with what he called the "universal homogenous state," which he warned us against. His *Lament for a Nation* in 1965 on the dying of Canada, that so offended the conventional mind, could now be the title for how Canada's government buys into globalization at the sacrifice of Canadian sovereignty and decency, with a Prime Minister (Stephen Harper) for almost a decade—a neo-conservative, not a Tory and Red Tory like Grant—who was in practice a climate-change denier in the era of extreme climate change. Grant foresaw what Michael Hardt and Antonio Negri simply call (rather than Imperialisms) in their book of that title.

Marshall McLuhan, with his wild, manic, free association (linkages galore) wins the prize for remarkable foresight: the Internet, social media, the digital revolution—a revolution in communication; perhaps the biggest since the printing press. It has made all of us McLuhan's children.

McLuhan's understanding ended up mostly being poured into the same institutional world, a fate he willfully ignored. The multinational, now global, corporation rules, its rights carved in the stone of trade agreements that pass for globalization, with the same old military-industrial complex permeating outer and inner space.

We need to retrieve McLuhan as portending a new global consciousness. Call it deep globalization, like deep ecology and Bill McKibben's deep economics. The old paradigm has now placed Earth itself in jeopardy.

> "Everything lurched: a man with a cane was crossing the street, a dull groan suddenly surged through the asphalt … and then the earth opened up beneath his feet: it swallowed the man, and with him a car and a dog, all the oxygen around and even the screams of passers-by … Tunnels bored by five centuries of voracious silver lust, and from time to time some poor soul accidentally discovered just what a half-assed job they'd done of covering them over."[57]
>
> —Yuri Herrera, *Signs Preceding the End of the World*, 2009

We live in catastrophic times. McLuhan knew that to be a possible fate. Consider this neglected quote that anticipates Klein: "Renaissance man came to North America with gunpowder and printing, and he determined to conquer or transform, or subdue nature … In North America, only, we conquered nature … We broke it, smashed it to pieces … Now we're trying to put it back together again."[58]

Which brings us to the caring voice of the late Ursula Franklin. A woman. A pacifist. An emeritus professor of engineering with a hands-on knowledge of technology. Her thinking, her meditations on technology, are permeated by the deep conviction that technology must, at every stage from creation to application, be—in E.F. Schumacher's phrase so risible to the clever—"as if people matter."[59]

There are benefits from technology, but for whom? There are costs, but borne by whom? Obvious questions, but rarely asked in the corridors of power. People are presently not governed but managed, by managers whose most important skill is messaging. The matter of scale, of size, is determined by the narrow standards of corporate efficiency and profitability, not by justice and human rights and compassion. Nature, in spite of its independent power, is likewise 'managed,' meaning mismanaged, rather than respected.

We must think outside the technocratic paradigm of companies and the governments that serve them; in the words of Abraham Rotstein, scholar and punster *extraordinaire*, "Buddy, can you spare a paradigm." Read Franklin and imagine a humane form of mining appropriate qualitatively to the Global South and quantitatively to the planet.

We have had enough by way of cautionary tales. We must—in a marvelous phrase of the American writer Lisa Belkin—make of such tales a paradigm of healing. We must both mend our ways and mend the planet.

The New World, where extraction took on a life of its own, is no longer new. Once looted by others we now loot it ourselves and loot the world beyond us. We must learn to respect nature and to respect the human rights of ourselves and of others. They are linked symbiotically, as Wallace Clement notes: "The 'cyclonics' of mining are still with us, and the ones blown about the most are the wage laborers employed—or unemployed—by the industry."[60] To heed one is to help the other. Heed both and there is healing.

Notes

This essay is written in memory of the late Abraham Rotstein. The author is indebted to a number of people for assistance in the writing of this chapter: Kelly Crichton, Matt Watkins, Denis Smith, Emma Jackson, Karen Keenan, Tom Walkom, Wally Clement, John Watson, Barry Corbett—and in particular Christopher Alton and Pierre Bélanger.

1 Alexis de Tocqueville, "Origin of the Anglo-Americans, and its importance in relation to their future condition," chap. 2 in *Democracy in America*, trans. Henry Reeve (New York, NY: George Dearborn & Co., 1838), I:13.

2 Sidney W. Mintz, *Sweetness and Power: The Place of Sugar in Modern History* (New York, NY: Penguin Books, 1985): 214

3 See Joseph A. Schumpeter, "The Process of Creative Destruction," chap. 7 in *Capitalism, Socialism & Democracy* (London, UK: Routledge, [1943] 2003), 81–86.

4 See Melville H. Watkins, "A Staple Theory of Economic Growth," *The Canadian Journal of Economics and Political Science* 29, no. 2 (May 1963): 141–58; reprinted in *Staples and Beyond: Selected Writings of Mel Watkins*, ed. H.M. Grant and David A. Wolfe (Montreal, QC: McGill-Queen's University Press, 2006). See also David McNally, "Staple Theory as Commodity Fetishism: Marx, Innis and Canadian Political Economy," *Studies in Political Economy* 6 (Autumn 1981): 35–63.

5 W.A. Mackintosh, "Economic Factors in Canadian History," [1923] in *Approaches to Canadian Economic History*, Carleton Library Series, ed. W.T. Easterbrook and M.H. Watkins (Montreal, QC: McGill-Queen's University Press, [1967] 2003), 4.

6 Harold A. Innis, introduction to *Empire and Communications* (Toronto, ON: University of Toronto Press, [1950] 1972), 5–6.

7 Charles C. Mann, *1493: Uncovering the New World Columbus Created*, 1st ed. (New York, NY: Alfred A. Knopf, 2011), 26.

8 "Scientists Link Moon's Tilt and Earth's Gold," *The New York Times*, November 27, 2015, https://www.nytimes.com/2015/11/28/science/scientists-link-moons-tilt-and-earths-gold.html.

9 Harold A. Innis, "Liquidity Preference and the Specialization of Production in north America and the Pacific," chap. 6 in *Staples, Markets, and Cultural Change: Selected Essays*, ed. Daniel Drache (Montreal, QC: McGill-Queen's University Press, 1995), 115.

10 *Sessional Papers of Canada, 1900*, no. 33w, as quoted in Harold A. Innis, *Settlement and the Mining Frontier*, vol. 9, part 2 of *Canadian Frontiers of Settlement*, ed. W.A. Mackintosh and W.L.G. Joerg (Toronto, ON: Macmillan, 1936), 211.

11 Eleanor Catton, "Saturn in Libra," in *The Luminaries* (Toronto, ON: McClelland & Stewart, 2013), n.p.

12 Innis, *Settlement and the Mining Frontier*, 270.

13 Oliver Sacks, "My Periodic Table," *The New York Times*, July 24, 2015, https://www.nytimes.com/2015/07/26/opinion/my-periodic-table.html.

14 Thomas Walkom, "You can't go home again in this raw and casual country," *The Toronto Star*, October 4, 1994, A17.

15 Innis, *Settlement and the Mining Frontier*, 403, as referenced in Peter George, "Ontario's Mining Industry, 1870-1940," in *Progress Without Planning: The Economic History of Ontario from Confederation to the Second World War*, ed. Ian M. Drummond (Toronto, ON: University of Toronto Press,1987), 72.

16 William J. Buxton, ed., introduction to *Harold Innis and the North: Appraisals and Contestations* (Montreal, QC: McGill-Queen's University Press, 2013), 33.

17 See Suzanne Mills and Brendan Sweeney, "Employment Relations in the Neostaples Resource Economy: Impact Benefit Agreements and Aboriginal Governance in Canada's Nickel Mining Industry," *Studies in Political Economy* 91, no. 1 (Spring 2013): 7–34.

18 Peter C. van Wyck, "Innis and I on the Highway of the Atom," chap. 15 in Buxton, ed., *Harold Innis and the North*, 328; see also Ibid., 326–53.

19 *Hearing on Abandoned Mines in the United States and Opportunities for Good Samaritan Cleanups, Before the Subcommittee on Water Resources and Environment*, October 21, 2015 (opening statement of Chairman Bob Gibbs, R-OH).

20 Bruce Douglas, "Anger rises as Brazilian mine disaster threatens river and sea with toxic mud," *The Guardian*, November 22, 2015, https://www.theguardian.com/business/2015/nov/22/anger-rises-as-brazilian-mine-disaster-threatens-river-and-sea-with-toxic-mud.

21 Dom Phillips, "Brazil's mining tragedy: was it a preventable disaster?" *The Guardian*, November 25, 2015, https://www.theguardian.com/sustainable-business/2015/nov/25/brazils-mining-tragedy-dam-preventable-disaster-samarco-vale-bhp-billiton.

22 Nic Pizzolatto and Scott Lasser, "Down Will Come," *True Detective*, season 2, episode 4, directed by Jeremy Podeswa, aired July 12, 2015 (USA: HBO), TV.

23 Harold A. Innis, *The Fur Trade in Canada: An Introduction to Canadian Economic History*, rev. ed. (Toronto, ON: University of Toronto Press, [1930] 1956), 393.

24 Éric Pineault, "The Panacea: Panax Quinquefolius and the Mirage of the Extractive Economy," in *The Staple Theory @ 50: Reflections on the Lasting Significance of Mel Watkins' "A Staple Theory of Economic Growth,"* ed. Jim Stanford (Ottawa, ON: Canadian Centre for Policy Alternatives, March 2014), 84–90.

25 Ibid., 90.

26 Ibid.

27 See Meg Luxton, *More Than a Labour of Love: Three Generations of Women's Work in the Home* (Toronto, ON: Women's Educational Press, 1980).

28 Ibid., 198.

29 See Emma Jackson, "Gendering Canada's Staple Economy: Investigating Women's Experiences of Fort McMurray and Alberta" (BA Honours Thesis, Department of Geography and Environment, Mount Allison University, 2015).

30 Ibid.

31 Anne Michaels, "Miner's Pond," [1991] in *Poems: The Weight of Oranges / Miner's Pond*, 1st ed. (Toronto, ON: McClelland & Stewart, 1997), 53.

32 Watkins, "Staple Theory of Economic Growth."

33 Travis Fast, "Stapled to the Front Door: Neoliberal Extractivism in Canada," *Studies in Political Economy* 94, no. 1 (Autumn 2014): 37, 59n15. "Near raw" includes mildly processed natural resources such as pulp, lumber, upgraded bitumen etc., basic metals, and basic chemicals. It should be noted that the category "basic metals" includes, for example, steel and aluminum production, which are not considered staples.

34 H.V. Nelles, *The Politics of Development: Forests, Mines and Hydro-Electric Power in Ontario, 1849-1941*, 2nd ed. (Montreal, QC: McGill-Queen's University Press, [1974] 2005), 87.

35 Iain Wallace, "Restructuring in the Canadian Mining and Mineral-Processing Industries," chap. 7 in *Canada and the Global Economy: The Geography of Structural and Technological Change*, ed. John N.H. Britton (Montreal, QC: McGill-Queen's University Press, 1996), 135.

36 Innis, *Settlement and the Mining Frontier*, 309.

37 Ibid., 317.

38 See Alain Deneault and William Sacher, *Imperial Canada Inc.: Legal Haven of Choice for the World's Mining Industries*, trans. Robin Philpot and Fred A. Reed, (Vancouver, BC: Talonbooks, 2012).

39 See Henry Veltmeyer and James Petras, *The New Extractivism: A Post-Neoliberal Development Model or Imperialism of the Twenty-First Century?* (London, UK: Zed Books, 2014).

40 See Naomi Klein, *This Changes Everything: Capitalism vs. The Climate* (Toronto, ON: Alfred A. Knopf Canada, 2014).

41 Albert O. Hirschman, "A Generalized Linkage Approach to Development, with Special Reference to Staples," *Economic Development and Cultural Change* 25 (1977): 67–98.

42 See Wallace Clement, *Class, Power and Property: Essays on Canadian Society* (Toronto, ON: Methuen, 1983).

43 William Finnegan, "Tears of the Sun: The gold rush at the top of the world," Letter from Peru, *The New Yorker*, April 20, 2015, http://www.newyorker.com/magazine/2015/04/20/tears-of-the-sun.

44 Ibid.

45 Deneault and Sacher, *Imperial Canada Inc.*, 1.

46 Eric J. Hobsbawm, *The Age of Empire, 1875–1914* (New York, NY: Vintage, [1987] 1989), 63, as quoted in Kornel Chang, *Pacific Connections: The Making of the U.S.-Canadian Borderlands* (Berkeley, CA: University of California Press, 2012), 97.

47 See Fast, "Stapled to the Front Door."

48 Wallace, "Restructuring in the Canadian Mining," 135.

49 See Fast, "Stapled to the Front Door."

50 Judith Marshall, "Contesting Big Mining from Canada to Mozambique," in *State of Power 2015: An annual anthology on global power and resistance*, ed. Nick Buxton and Madeleine Bélanger Dumontier (Amsterdam: The Transnational Institute, 2015), 73, https://www.tni.org/files/download/tni_state-of-power-2015.pdf.

51 Deneault and Sacher, *Imperial Canada Inc.*, 1.

52 "The Writer's Almanac," Minnesota Public Radio, July 25, 2015, podcast, 5 min, http://writersalmanac.org/episodes/20150725/. Note that Donald Trump's paternal grandfather started the family down the path to infamy and fortune when he joined the rush to the Klondike and opened a tent restaurant serving horsemeat from animals that had died on the long and arduous trek there. See Michael D'Antonio, *Never Enough: Donald Trump and the Pursuit of Success* (New York, NY: St. Martin's Press, 2015).

53 Daniel Drache, "Staple-ization: A Theory of Canadian Capitalist Development," in *Imperialism, Nationalism, and Canada: Essays from the Marxist Institute of Toronto*, ed. Craig Heron (Toronto, ON: New Hogtown Press, 1977), 16.

54 W.T. Easterbrook and Hugh G.J. Aitken, *Canadian Economic History* (Toronto, ON: University of Toronto Press, [1956] 1988), 519.

55 See Hugh G.J. Aitken, *American Capital and Canadian Resources* (Cambridge, MA: Harvard University Press, 1961).

56 Harold A. Innis, "The Penetrative Powers of the Price System," *The Canadian Journal of Economics and Political Science* 4, no. 3 (August 1938): 299.

57 Yuri Herrera, "The Earth," chap. 1 in *Signs Preceding the End of the World*, trans. Lisa Dillman (London, UK: And Other Stories, [2009] 2015), par. 1–3.

58 Marshall McLuhan, in *Marshall McLuhan: The Man and His Message*, ed. George Sanderson and Frank MacDonald (Golden, CO: Fulcrum, 1989), 16.

59 Ursula Franklin, *Real World of Technology*, rev. ed. (Ottawa, ON: House of Anansi Press, [1990] 2004), 84.

60 Clement, *Class, Power and Property*, 129.

Mineral Map of the Dominion of Canada, Department of Mines, 1924

Cree Lake
1530'
Geikie
R.
Pink
Land
R.
Methy Por.
Methy L.
Co.
Mudjatick
R.
Foster L.
1600'
Foster R.
Haultain
R.
Churchill
L.
Pelican L.
Snake L.
H.B.Co.
Churchill
Stanley
Peter Pond L.
1330'
H.B.Co.
Lake
Ile à la Crosse
Nemeiben
L.
Lac la
Lac la
Plonge
Montreal
R.
Primrose L.
Doré
L.
SASKATCHE
Montreal L.
Beaver
River
H.B.Co.
Green
L.
Big River
Ft. Pitt
Prince Albert

Paper Worlds

A Conversation with Alex Golub

Alex Golub is a political anthropologist and associate professor at the University of Hawai'i at Mānoa, focusing on the relationship between grassroots people and the mining and hydrocarbon industries in Papua New Guinea. He wrote *Leviathans at the Gold Mine* (2014).

X: Your work reveals the difference—a type of white space—between what's being said in policy and what's being done on the ground. And that problematic relationship—this white space between state, corporation, and people, that you collapse—is unique for us.

AG: Well, these are deep waters. We're all working in our own fields with different backgrounds, and projects like this create an opportunity because these big questions about land and writing are deep and perennial.

With this project on extraction, you're dealing with such big themes that you could go back to the Magna Carta, which of course was overturned months after it was signed. So this dynamic of the difference between what exists on paper and what the actual politics are, is enduring. The original document was just one piece of medieval politics that happened to become the justification for the Commonwealth tradition.

The big thing for me to point out is, in generic Western Anglo-Protestant culture, we tend to make the assumption that there is a distinction between saying and doing. We tend to separate those off. It's a very old and deep Western tradition—the soul is free from the body, word is less important than action. But one of the things that anthropologists know is that this is wrong. These text artifacts we're dealing with are always caught up, they're always deployed in action, and we can never just look at their *contents*; we always have to look at their use in *context*. We don't want to talk about what the Magna Carta says in the abstract. We want to understand how it was used by John Softsword as a sop to his lords in the summer of 1215.

One of my central points in writing the book on Porgera (*Leviathans at the Gold Mine: Creating Indigenous and Corporate Actors in Papua New Guinea*), is that the Porgera case is more than just a case of misrecognition, of there being a white space between the documents and the practice. My point is that, that world of documents is, in itself, incoherent. You could not actually put together a single systematic organized account of land rights from the documents, because each of them was created by a historical conjuncture in which they were required for a certain purpose. So documents never describe things in themselves, but are always interpreted, entextualized and reentextualized in different situations.

The first thing to do is move past the idea that there is an official story somewhere, and then there are facts on the ground. We need a more nuanced understanding of how concrete people in concrete situations read and use documents. How does that affect action at different scales? Who is reading these documents? Where? What do they think the documents say? When we say, the "official story," whose story do we mean by that? Documents help mediate those two different social processes.

Anthropologists want to understand how that mediation happens through documents, and how the story circulates. We always want to directly observe social processes in a particular time and place.

James Scott's *Seeing Like a State* takes the other view, in which there's this thing called the state, and it's not anybody in particular, it's not made of people, but this abstract living entity which observes the world. That's what I am trying to get away from.

"'...In that Empire, the Art of Cartography attained such Perfection that the map of a single Province occupied the entirety of a City, and the map of the Empire, the entirety of a Province. In time, those Unconscionable Maps no longer satisfied, and the Cartographers Guilds struck a Map of the Empire whose size was that of the Empire, and which coincided point for point with it. The following Generations, who were not so fond of the Study of Cartography as their Forebears had been, saw that that vast Map was Useless, and not without some Pitilessness was it, that they delivered it up to the Inclemencies of Sun and Winters. In the Deserts of the West, still today, there are Tattered Ruins of that Map, inhabited by Animals and Beggars; in all the Land there is no other Relic of the Disciplines of Geography.' (Suarez Miranda, Viajes de varones prudentes, Libro IV, Cap. XLV, Lérida, 1658)"

—Jorge Luis Borges, "On Exactitude in Science," Collected Fictions, 1998

X: So you describe the relationship between individuals on the ground and the state in terms of the Leviathan that is created. You say that the state doesn't function because Papua New Guineans themselves can't "put a face to the state." Can you explain that a little more?

AG: Well, again, everyone is on the ground. The state is on the ground, people are on the ground, gravity keeps on pulling us toward the center of the Earth. So I always wonder, whenever people say "state," what is the functioning of the state? Who is the state? Where are the documents circulating?

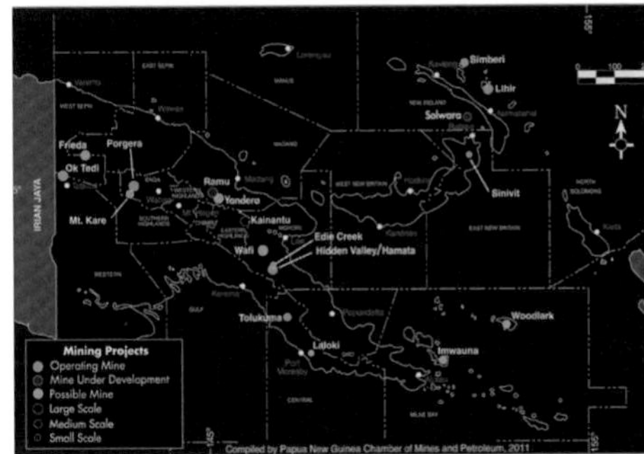

Mining Lease Districts, Papua New Guinea (Papua New Guinea Chamber of Mining and Petroleum, 2011)

In the case of that example, I think one of the problems is that in America and British settler colonies, we are socialized to the idea that institutions exist, that it's normal for humans to hold an office, and that they should execute their operating procedure in that office as part of a bureaucracy. We don't ask our parents to embezzle. It's also in the material culture that surrounds us. As the saying goes, all the world's a stage. You wake up every morning, and, given the limited options of props around you, you do the job that you do.

On the other hand, in Papua New Guinea, they're much more free to imagine the world, and to try out different opportunities. A lot of people portray Papua New Guinea as being timeless, traditional, or static, whereas Western institutions are dynamic, progressive, and changing. In reality, we're the ones that get up, eat the same thing, go to the same office, do the same thing, over and over again. You go to the store, you give the cashier your groceries, they ring it up, you pay for it, then you leave, because that's how it goes. Whereas it's much more common in Papua New Guinea for people to show up to the store and say, "I've got an idea, why don't you give me these groceries for free?" People are constantly wondering how they can do things differently. They're very inventive and dynamic. Why does it have to be this way and not another way?

It's very difficult for them to see somebody who they grew up with, and see them as part of an impersonal institution, or merely as part of a bureaucracy. It's hard for everyone. It takes a lot of practice and training. They see that person as a person, and they want to poke around, innovate, and try new things. That makes it really hard to run a bureaucracy.

X: In terms of being able to "put a face to the state," was there a moment when that relationship between individuals and the state became obvious to you?

PORGERA

GOLD

Association Mark
(Porgera SML Landowners Association, 2017)

AG: The sociologist Pierre Bordieu talks about synoptic accounts—accounts which, in systematic forms, present all of social life. As an ethnographer, you always want to do a synoptic account. It takes a lot of sensitivity and self-confidence to be willing to write messy accounts, and recognize that the world is not perfectly orderly.

But in my case, I was constantly encountering these looping effects, where people would say it was orderly, and say "there's a system in place, and it works like this," and then there were documents that would say that, but then that didn't seem to be really how it worked, but then people *did* assume it was orderly, and they were in fact using the official account. There were these levels of reflexivity where I felt like I could go on forever in a house of mirrors, trying to separate the "official" account from the "real" account.

So I was forced to re-conceptualize what was going on theoretically, because the empirical material demanded it. I think that's one of the reasons people like this book—it's trying to move these questions forward in order to explain the facts as I encountered them, not develop new theory for its own sake. The situation called for it.

X: In the standard telling of the Porgera story, do you think there has been a tendency towards reductive and simple political positions?

AG: Yes, I think that a lot of the "folk theory" that politicians have in framing debates is very simplistic. In what's happening in Porgera, as we grow to a global audience and groups such as Mining Watch Canada get involved and connect with the UN Forum on Indigenous Issues, people attempt to frame what's happening in the valleys in a certain way.

As a result, it's sort of inevitable that what's really going on gets lost. I think that's just the nature of politics. If you read the blog of the Porgera Landowner Association (PLA), they denounce how the mine is destroying their land. They use the idioms that they need to mobilize support based on stories that first world audiences are sympathetic to. It's been very effective. People see pictures of Porgerans, standing next to First Nations peoples, holding anti-Barrick signs. But what is actually happening in Porgera is a bit more complicated than the story you get on that website. This is another reason you always want to put the text in context. You want to put the denotational content of these texts, in this case a blog entry, into the social interactions in which it circulates.

In fact, the PLA doesn't want mining to stop, they want it to continue. Porgerans say they've "traded their mountain for development." That's literally what one of the original negotiators told me. And they're not happy about the pollution, by any means. But the real issue is how much money they're getting from it, and how they're being treated. Respect, autonomy, fair compensation and prosperity—that's what they want. For instance, they want security forces to stop killing and raping people. And they want a remedy for the wrongs that have been committed against them. Who could disagree with that?

So I think it's important to cover the natural history of these discourses as they move around, to get a sense of the complexities of this situation. Because in general, the further away that you get from Porgera, the less and less detail you're going to get. The story that is told at a distance will be more and more motivated by political concerns in the location where the story is entextualized, and it will be turned into good guys versus bad guys. There's nothing wrong with that—that's just standard politics. But it's not a fully adequate scholarly account of what's going on.

One Hundred Kina Bill, Papua New Guinea (banknotenews.com)

X: In the letter from the PLA to the CEO of Barrick, the PLA voices concerns over the sale of the mines to the Chinese company, and its central concern is one of maintaining the MOA commitments that were signed in the first place. That speaks back to the fact that what the landowners want is due democratic recognition for the mining.

AG: Yes. And human rights discourse is how they are explaining this to people. And it's absolutely appropriate that the PLA use this discourse in the UN Forum on Indigenous Issues to pursue its interests. At the same time, it's just not empirically satisfying to me when people say that these people are colonized Indigenous peoples, just like every other kind of colonized Indigenous peoples. Or else there are two kinds of Indigenous peoples, the helpless victims and the brave resisters. But most of life is about middlemanship and mediation, and stories are not quite as cut-and-dry. We can't really understand the reality of what Indigenous peoples want until we are ready to move past our stereotypes and listen to them on their own terms.

For instance, in the comparison between Canada and Porgera, one of the differences between Papua New Guinea and Canada is that in Papua New Guinea, there is no large, centralized bureaucracy. The government has historically lacked military power to impose itself on grassroots people. There's something that looks like the government, they have offices and civil servants, but the kind of efficacy that comes

"A vision of the world is a division of the world, based on a fundamental principle of division which distributes all the things of the world to the complementary classes. To bring order is to bring division, to divide the universe into opposing entities…The limit produces difference and the different things 'by an arbitrary institution'… This magical act presupposed and produces collective belief, that is ignorance of its own arbitrariness… the group constitutes itself as such by instating what unites and what separates it. The cultural act par excellence is the one that traces the line that produces a separate, delimited space…"

—Pierre Bourdieu, The Logic of Practice, 1990

from bureaucratic organization is lacking in Papua New Guinea in a way that isn't in Canada. In Papua New Guinea, customary land cannot be alienated, and most of it has not been. First Nations peoples in Canada have had a very different experience, to put it mildly. I think that explains the position of a lot of First Nations peoples in Canada. On the other hand, the Papua New Guinea Defense Force has been deployed twice in states of emergency—in 2009 and 2014. But they're not good at exerting a kind of prolonged control over bodies of Indigenous peoples in such a way that mining can proceed without a hitch. As far as I can tell, they're just good at running wild. So these cases really are different. We need to be attuned to these differences and the way they shape mining politics.

X: As you write, when Barrick ended up buying the mine (in 2006) things started to go downhill. Would you be interested in talking about what that change represented?

AG: When I first got to Porgera, it seemed to be a very successful place, because people told me it was successful. The Porgerans felt like they had brought the mine in and they were now holding it down, and sucking the blood out of it. This was really before the World Wide Web was widespread, and when a lot of people, including myself, just didn't know about the mining industry. When the operators dumped waste rock into the water, it just seemed like standard practice. The connections that activists made in the early 2000s hadn't been made yet. So Porgera seemed special to me when I started researching it because people felt it was successful, a lot of money went into the valleys, and it looked good. We didn't feel like people had gotten what they wanted, but we felt like there was a good chance that they would get it in the future.

After the Barrick negotiation, which I documented and resulted in a failed resettlement plan that did not move people out, things began getting worse. Barrick's arrival was a part of that, because Barrick was much less transparent; it's basically a financial company with mines attached. Barrick as a company was part of the problem, but it wasn't the main or only reason that things seemed to be going downhill. There were social factors. There was more and more in-migration, children were coming of age who didn't get any of the benefits that their parents had; they were basically forced to live with the mine that they didn't agree to, and were not getting as many of the benefits. The government was not giving its citizens the services they were entitled to. Especially, the people who were responsible for the hard work that had previously kept a small-world connection—like the mine manager who could chat with the district administrator or phone up the head of the Porgera Development Authority—sort of busted apart as some of them got old and some of them just couldn't sacrifice their lives any longer to keep the valley together. So the personal networks which integrated and connected institutions, disintegrated.

Protest & Petition by Porgera Landowners to PNG Government regarding Barrick's violations of Memorandum of Agreement at the Porgera Mine, 2015 (Porgera Alliance)

The stuff that had been going wrong for some time—like the activities of the security guards, their shooting and raping people, and the open pit—got worse; or at least more visible. I hadn't heard about that stuff before, but due to the Web, and due to the work of Catherine Coumans (of Mining Watch Canada), we could see that these problems were real and very, very serious.

Then the gold price increased after 9/11, which means that the company was willing to do more and more to keep the mine going—so instead of planning for closure, now they were planning for expansion. So I think that from 2001 to the present, a lot of the problems intensified and didn't get better, and that's sort of where we are now.

X: This goes back to your point about mediation and the documentations of what you conceptualize as a "paper world." Your observation on identification on these documents—for instance, on contracts, the signatures of the corporate elite, and then the thumbprints of the PLA—is somewhat indicative of an ongoing and existing colonial relationship.

The paper world mistakenly flattens the relationship, and seems to render both parties on the same playing field. A lot of your work is about exposing the white space on which the relationship operates. Your anthropology brings the lives forward, at both the levels of the landowners and the corporate elite.

Where does that come full figure for you?

AG: I think what anthropology has learned from Melanesian philosophy is that boundaries connect people, rather than separate them. The documents portray two distinct groups of peopled called "the mine" and "the Ipili," and make the two groups discrete, so that they can be related to each other. You couldn't have the kind of relationship connecting the two of them that enables the mine to exist if those groups were not already separate. So the documents, and the social processes that produce them, differentiate those groups and create a simple story of two distinct groups. In reality, there weren't two groups previously, and there won't be two groups in the future, but rather a series of social connections between a variety of stakeholders; stakeholders who are connected by their shared story of being disjointed.

Riots at Porgera Mine, Papua New Guinea, after Barrick's security contractor reportedly killed local, non-licensed miners, 2013 (Jethro Tulin, Akali Tange Association)

That binary distinction then gets mapped on to this binary distinction in Canada, because there seems to be some kind of elective affinity between the two of them. To me, this is fascinating.

X: Based on the lives that you've studied, is there an alternative to mining negotiations that you propose?

AG: I think one of the reasons that the Porgera agreements initially worked is that, first, it was democratic in the sense that the Ipili people more or less had their say—and there's a lot of *less* in there as well as *more*—in the negotiations. They really were partners in negotiation, at least as I can tell form my interviews and the historical records. But secondly, the very genre of interaction they used to negotiate was similar to other genres that they had used in the past and felt comfortable with.

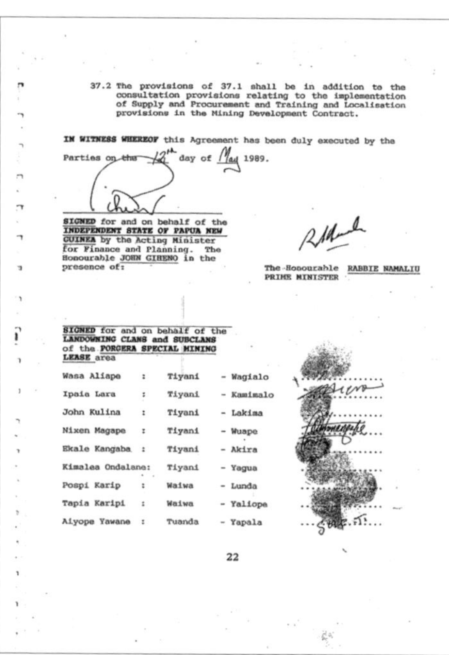

37.2 The provisions of 37.1 shall be in addition to the consultation provisions relating to the implementation of Supply and Procurement and Training and Localisation provisions in the Mining Development Contract.

IN WITNESS WHEREOF this Agreement has been duly executed by the Parties on the 12th day of May 1989.

SIGNED for and on behalf of the **INDEPENDENT STATE OF PAPUA NEW GUINEA** by the Acting Minister for Finance and Planning. The Honourable **JOHN GIHENO** in the presence of:

The Honourable **RABBIE NAMALIU**
PRIME MINISTER

SIGNED for and on behalf of the **LANDOWNING CLANS and SUBCLANS** of the **PORGERA SPECIAL MINING LEASE** area

Wasa Aliape	:	Tiyani	- Wagialo
Ipaia Lara	:	Tiyani	- Kamimalo
John Kulina	:	Tiyani	- Lakima
Nixen Magape	:	Tiyani	- Wuape
Ekale Kangaba	:	Tiyani	- Akira
Kimalea Ondalane	:	Tiyani	- Yagua
Pospi Karip	:	Waiwa	- Lunda
Tapia Karipi	:	Waiwa	- Yaliope
Aiyope Yawane	:	Tuanda	- Yapala

22

Memorandum of Agreement for the development of Porgera gold mine between the Independent State of Papua New Guinea and Porgera Landowners, May 1989 (Columbia Center on Sustainable Investment - Extractive Industries)

You should never assume that just because there's a thumbprint on the signature, that the people who couldn't write were poor negotiators. Although of course, if you look at what happened on the long term, things have not worked out well. But at the time, these people were savvy operators. Like I said, it was a world without institutionalization, so they were great negotiators. They take a lot more responsibility for their life than we do, because they have to make everything.

So, I think it's important that not just the content, but also the structure of the negotiations, come out according to their demands. One of the interesting trends in Indigenous communities in Canada is the concept of refusing the state and its terms, refusing the multicultural settlement, and demanding that Indigenous genres structure the negotiation between the individual and the state. Here I think of Audra Simpson's work on this topic.

X: Are these the politics of refusal and the social expressions of blockades at work, from present and into future?

AG: What I see happening—this is my perspective as a settler (or perhaps Hegel)—is that refusal of recognition on one level is a demand for recognition at a higher level. When Indigenous or grassroots people say 'no' to state regimes of recognition, compensation, and reconciliation, they are insisting that we don't substitute policy for politics. They want to have a say in what decisions are made, and how they are being made. They don't want to be told "the law says you are entitled to amount X if you fill out form Y." We live in a world where democracy is increasingly crippled by bigness. Decisions made in one office affects the lives of thousands or millions. Better customer service will not make the problem go away. This is not a blockade or blockage in the sense of pure negation. It's a positive demand that people be allowed to participate in the making of their world. This is an invitation that is hard to refuse. ◆

This conversation took place between Alex Golub, Christopher Alton, and Pierre Bélanger between 2015 and 2017.

Barrick in Papua New Guinea

The Story of Porgera

The Porgera Joint Venture (PJV) gold mine is located in the Papua New Guinea Highlands at the terminus of the Highlands Highway, which cuts 700km through the jungle to connect the Porgera Valley to the port city of Lae. Gold was discovered in the Porgera Valley as early as 1938 by two Australian prospectors, who returned in 1948 to begin small-scale alluvial mining in the valley. Until the 1980s the economy of Porgera was dominated mostly by subsistence farming.

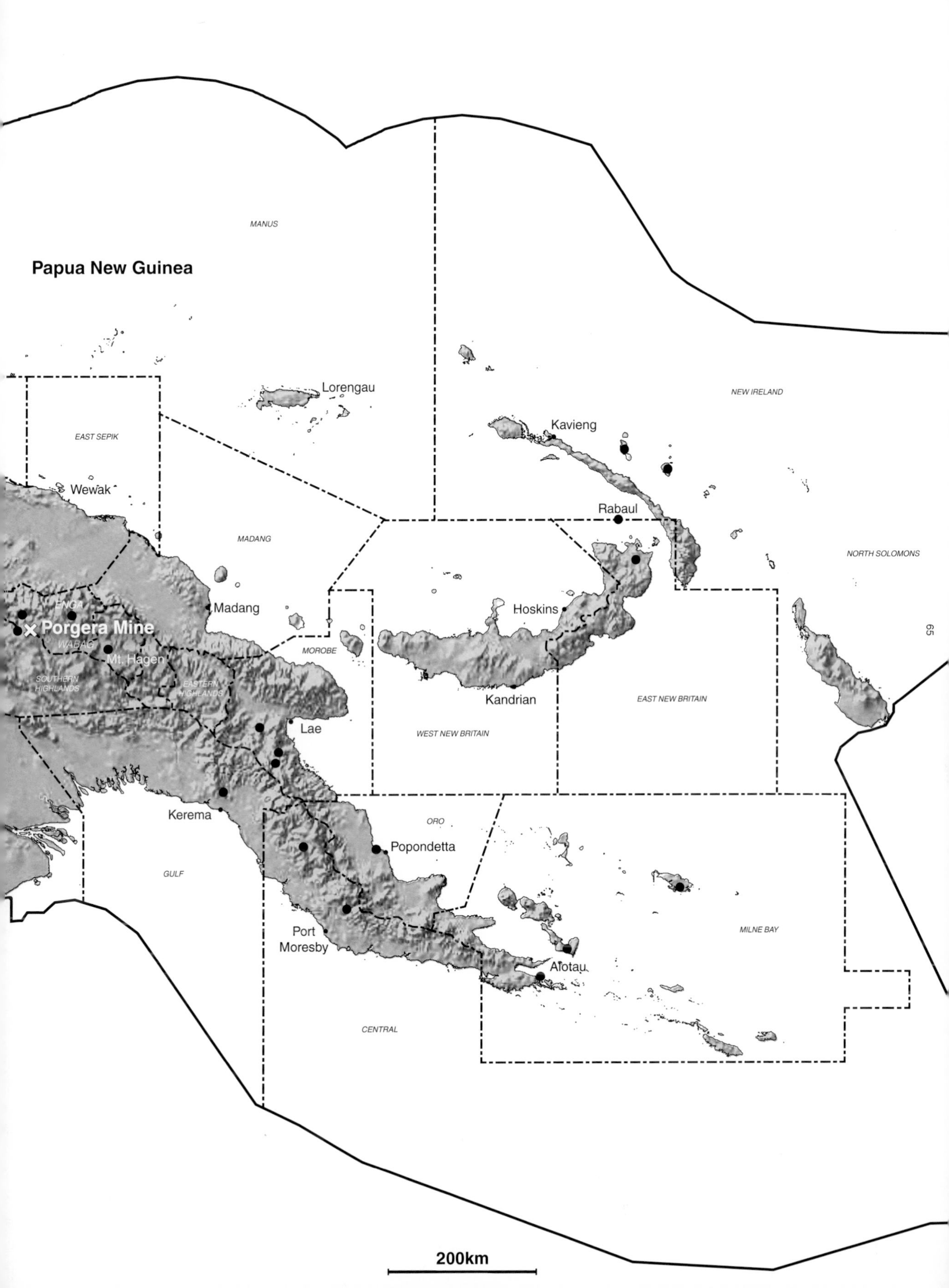
Papua New Guinea
MANUS
Lorengau
NEW IRELAND
Kavieng
EAST SEPIK
Wewak
Rabaul
MADANG
NORTH SOLOMONS
ENGA
Madang
Hoskins
Porgera Mine
WABAG
MOROBE
Mt. Hagen
SOUTHERN HIGHLANDS
EASTERN HIGHLANDS
Kandrian
EAST NEW BRITAIN
Lae
WEST NEW BRITAIN
Kerema
ORO
Popondetta
GULF
Port Moresby
MILNE BAY
Alotau
CENTRAL
200km

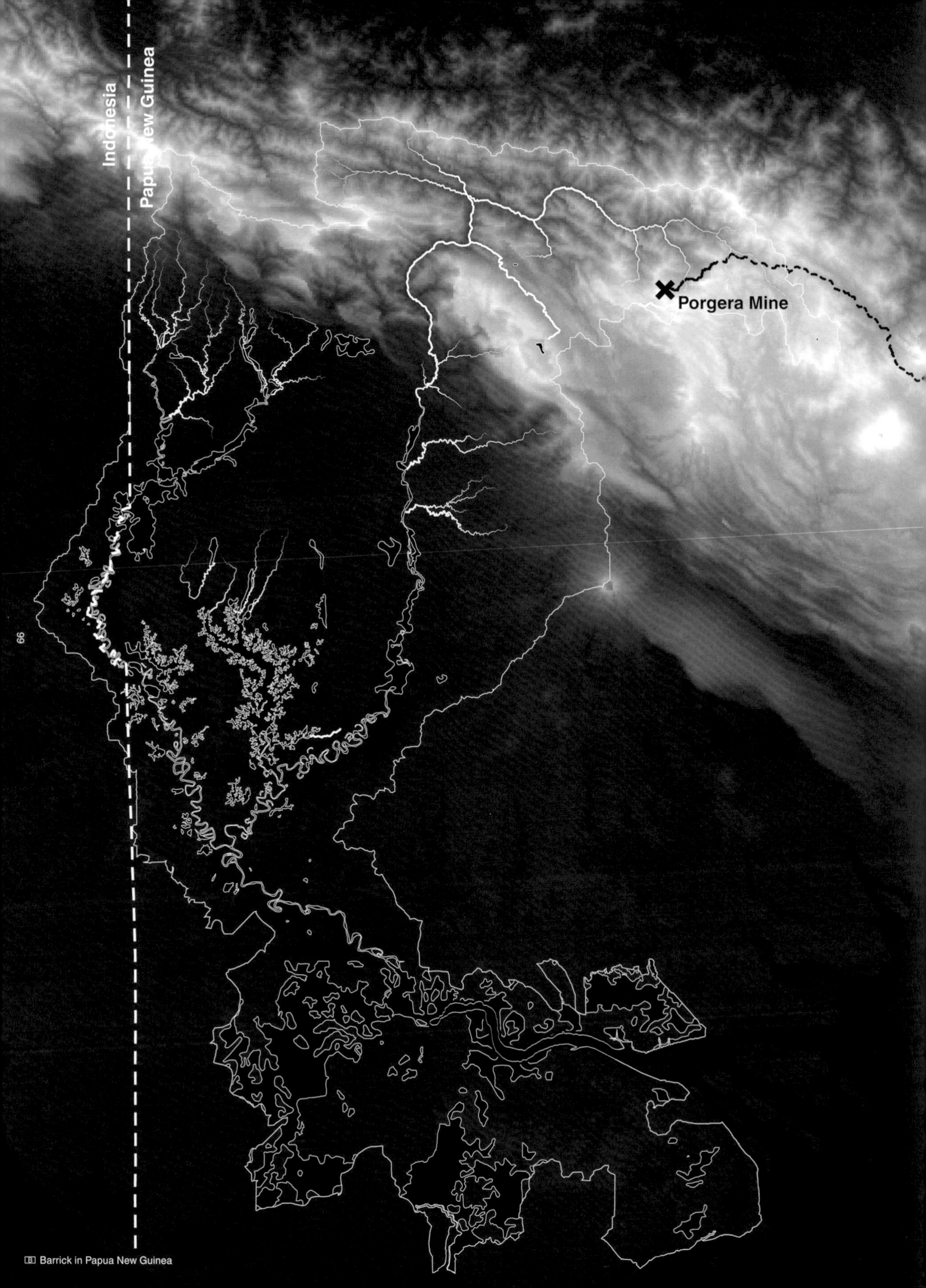

Barrick in Papua New Guinea

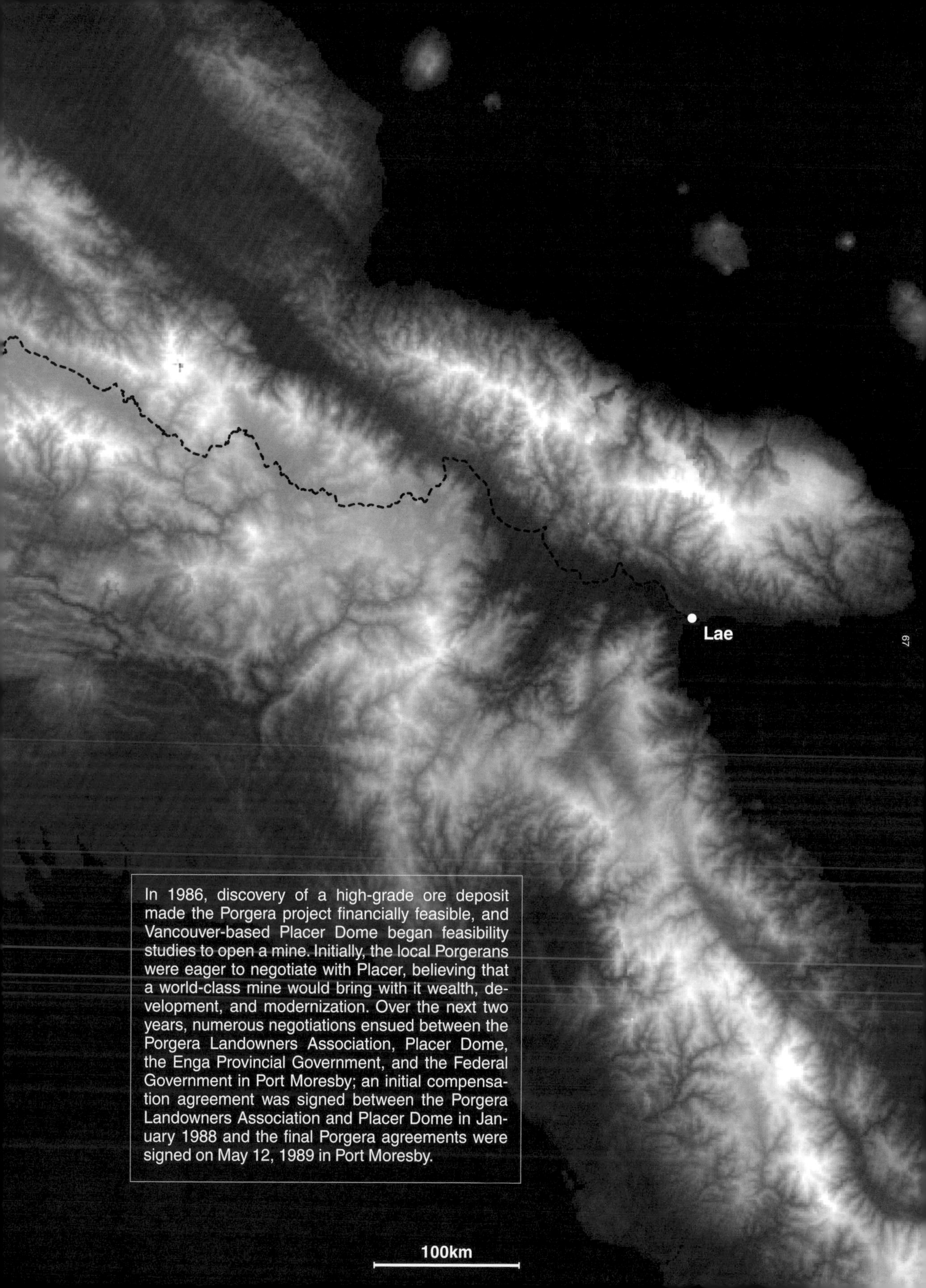

Lae

In 1986, discovery of a high-grade ore deposit made the Porgera project financially feasible, and Vancouver-based Placer Dome began feasibility studies to open a mine. Initially, the local Porgerans were eager to negotiate with Placer, believing that a world-class mine would bring with it wealth, development, and modernization. Over the next two years, numerous negotiations ensued between the Porgera Landowners Association, Placer Dome, the Enga Provincial Government, and the Federal Government in Port Moresby; an initial compensation agreement was signed between the Porgera Landowners Association and Placer Dome in January 1988 and the final Porgera agreements were signed on May 12, 1989 in Port Moresby.

100km

According to Jerry K. Jacka, the author of *Alchemy in the Rain Forest*, the PJV began producing gold in 1990, and within a decade, the initial success of the Porgera agreements was already a thing of the past. Between 1990 and 2000, the population of Porgera more than doubled from just over 9,000 to almost 23,000. Placer's decision to dump acid tailings into the local river system not only brought an end to local alluvial mining practices, but also contaminated the headwaters of the Strickland River, the second largest river in the country. Placer expanded the mine in 1994, increasing its footprint to move 100,000 tonnes of earth each day. Though the mine began as an underground operation, through the 1990s it eventually became predominantly an open pit mine. In 2006, Barrick Gold Corporation acquired a 75% share of PJV from Placer-Dome, along with an additional 20% interest (then owned by Emperor Mines) the following year.

Porgera Mine Site

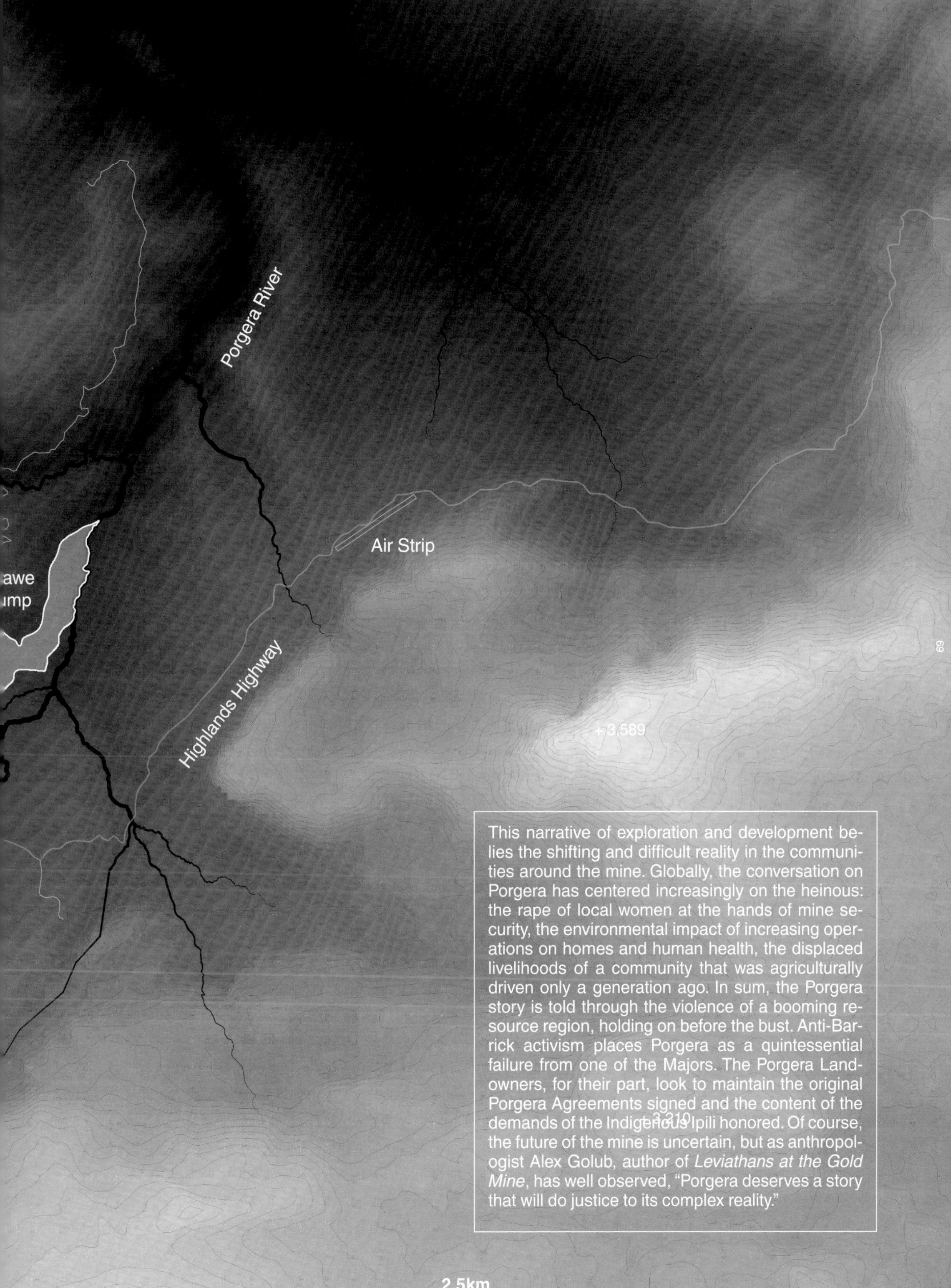

This narrative of exploration and development belies the shifting and difficult reality in the communities around the mine. Globally, the conversation on Porgera has centered increasingly on the heinous: the rape of local women at the hands of mine security, the environmental impact of increasing operations on homes and human health, the displaced livelihoods of a community that was agriculturally driven only a generation ago. In sum, the Porgera story is told through the violence of a booming resource region, holding on before the bust. Anti-Barrick activism places Porgera as a quintessential failure from one of the Majors. The Porgera Landowners, for their part, look to maintain the original Porgera Agreements signed and the content of the demands of the Indigenous Ipili honored. Of course, the future of the mine is uncertain, but as anthropologist Alex Golub, author of *Leviathans at the Gold Mine*, has well observed, "Porgera deserves a story that will do justice to its complex reality."

At the Financial Heart of the World Mining Industry

In which the majority of the world's extractive corporations flock to Toronto to enjoy investment tax incentives

Alain Deneault

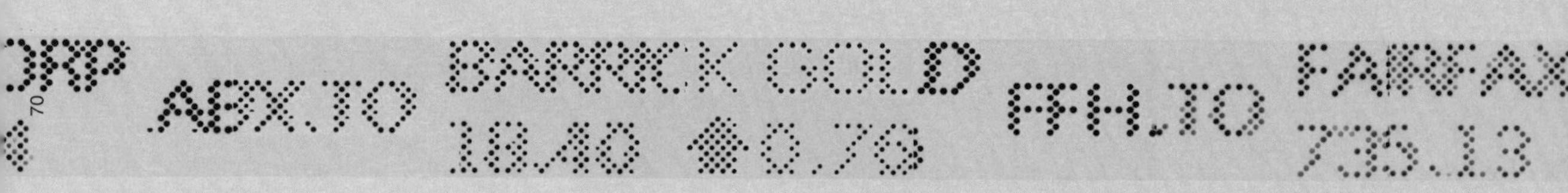

> "All cities can be described as colonial: at the local level, the powers that form them organize their hinterland and live off the surplus the non-urban realm provides. At the global level, existing cities organize the surplus both of their own society as well as that of others overseas; the local relationship of town-to-country becomes the metropolis–colony connection on a word scale."[1]
>
> —Anthony D. King, *Urbanism, Colonialism, and the World-Economy*, 1990

René Nollevaux, Belgian manager of a copper mine in the Democratic Republic of the Congo, clearly believed he was stating the obvious when he remarked that "in the mining industry, venture capital usually comes from Canada." The comment was made to filmmaker Thierry Michel, who was shooting a documentary about Belgian neocolonial relations in the African Great Lakes area.[2] Canada, as we know, is the favorite jurisdiction of mining companies.[3] According to a 2012 publication of the Toronto Stock Exchange, 75 percent of the world's mining companies choose to register their activities in Canada, and 60 percent of those who issue stock market shares register in Toronto, which is far ahead of its direct rival, the London Stock Exchange. In 2011, 90 percent of the shares issued in the mining sector throughout the world were administered by the Toronto Stock Exchange (TSX), and through shares, Toronto financed the mining sector to the tune of $220 billion between 2007 and 2011.[4]

The romance of Canada's universal goodness, partly arising from Canadian peacekeeping operations, begins to unravel as soon as we are confronted with the sorry environmental, social, political, safety, and tax record of the extractive industry hosted by Canada. Throughout the world, parliamentary commissions, courts, UN experts, independent observers, specialists in the economy of the South, and experienced journalists have testified to the injustices—and sometimes the crimes—committed or massively supported by Canadian mining companies active in countries in the South. When we say 'crimes,' we are not using anthropological nuance; we mean that people are reporting cases of corruption, tax fraud, institutionalized plundering, massive pollution, injury to public health, violent expropriation, murdered demonstrators, complicity in the assassination of people opposed to mining projects, strategic lawsuits against public participation (SLAPPs) and criminalization of political opposition, arms trafficking, and collusion with warlords and other belligerents involved in conflicts centered on control of mining deposits. If you take a look at the abundant literature supporting allegations against the mining industry, you will have enough information to grasp this reality.[5]

With the claim that extractive industries "make a major contribution to Canadian prosperity,"[6] the Federal Government bends, twists, or stretches every one of the principles that it promotes in international forums. Ably assisted by the Ontario government, it has made Toronto into the nerve center of the mining haven that Canada represents at the global level. Mining corporations defined as "Canadian" often do not have any substantial activity in Canada: they come from other countries and operate mines in other countries. They are incorporated in Canada in order to benefit from the regulatory, legal, and tax advantages provided by Canada.[7] Hence, an abnormally high number of mining managers from Australia, Israel, Sweden, Belgium, the United States, among others, converge on Ontario to create prospecting or mining companies. These companies work mining claims obtained in Ecuador, Chile, Zambia, Haiti, Burkina Faso, Indonesia, Romania, or elsewhere. Close to half the mining projects registered on the Toronto Stock Exchange (TSX) are located outside of Canada and many companies registered in Canada do not have a single claim on Canadian soil.[8]

The TSX is first and foremost an institution fostering the frenzied speculation that the industry loves. On this exchange it is notoriously easy for a company to list presumed deposits and magnify their value. The great majority of the 1,600 mining companies registered in Toronto are "juniors," which means that they are exclusively involved in finding new deposits.[9] Lacking the financial, technical, human, and political resources to run mines themselves, these small companies make their profits from stock market speculation. Their goal is to discover deposits and sell them to the majors who alone are in a position to oper-

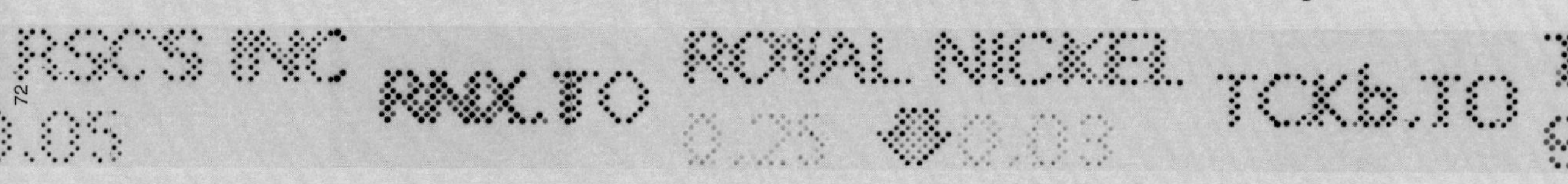

ate mining facilities; a sale will cause a junior's share price to go up. The juniors sometimes invest more in marketing campaigns aimed at potential investors than in the actual search for mineral deposits.

In giving juniors more leeway than anywhere else to cultivate ambiguity around the true potential of their mining deposits, the Toronto Stock Exchange favors the interests of mining companies. The TSX advocates disclosure of both a mine's *reserves* and its *resources. Reserves* are supposed to be a detailed and precise estimate of a deposit's exploitable potential, while *resources*—this is the thorny concept—are a crude estimate of everything the deposit may contain. The information on *resources*, essentially based on possibly altered excerpts from geological studies, provides investors with a glowing picture of a potential for exploitation that is higher than the real potential, based on the idea that a mineral's price may rise or extraction techniques may become more sophisticated. Disclosure of *resources* encourages stock market speculation and makes mining stocks go up.

In this context, there is nothing surprising about the many frauds and scandals that have characterized the Toronto Stock Exchange throughout its history.[10] Now-forgotten stories include the 1960s Windfall affair, in which the "discovery" of copper, silver, and zinc caused an uproar at the TSX, but turned out to be nothing but unfounded rumor.[11] The Bre-X swindle in the 1990s is another example. The company had sprinkled gold over its rock samples to give the impression it had acquired a high-quality deposit.[12] Toronto stood revealed as a temple of the casino economy, this according to the statements of financial advisers themselves. The reality is that no one can say what is in the ground until a mine is actually dug, and outcomes depend on such a wide range of factors that buying shares in a junior is like gambling at the casino. The actual geology of the claim, how easy it is to access, variations in world prices, technological advances, and the local political climate may all have an impact. Mining experts estimate that the odds of success are somewhere between 1 in 500 and 1 in 1,000.[13] Mining exploration is a risky business, by essence speculative, as French taxpayers learned to their cost in 2007, when a publicly owned company, Areva, bought a Toronto-based junior, UraMin, only to find that the company's reserves were far more difficult to exploit than had been foreseen. Areva also discovered that UraMin's key deposit had been overestimated by 42 percent, and all this was taking place at a time when the price of uranium was dropping. How much did this affair cost French taxpayers? All we can say is that having acquired UraMin for 1.8 billion euro, Areva in 2011 announced a 1.46 billion euro write-down on the UraMin assets.[14]

According to Wilfrid Laurier University economists William J. McNally and Brian F. Smith, both the Toronto Stock Exchange and the Ontario Securities Commission (OSC, that is supposed to regulate it) are negligent when it comes to illegal insider trading.[15] Unlike practice in the United States, such trading is rarely investigated. As a matter of fact, investors themselves worry about Toronto's weak enforcement of regulations. In 2004, the governor of the Bank of Canada, David Dodge, said that his colleagues in New York, Boston, and London spoke of Toronto as a financial Far West because of its lax regulatory framework. In 2006, a Harvard law professor, Howell Jackson, observed that in terms of effectiveness, Canadian measures did not bear comparison with the United States. The following year, even Finance Minister Jim Flaherty expressed similar reservations about the powerlessness of Canadian regulatory agencies.[16] Also in 2007, the president of the Ontario Teachers' Pension Plan, Claude Lamoureux, and Barbara Stymiest, former head of the Toronto Stock Exchange and chief operating officer of the Royal Bank of Canada, deplored the fact that securities laws are simply not enforced in Canada.[17] On the rare occasions when guilty parties are arrested, penalties are sometimes less than the proceeds of the crime. According to McNally and Smith, "almost half of the repurchasing firms fail to report their trades to the OSC."[18] The Ontario regulator is certainly not trying to frighten anyone: it recently established a new way of punishing wrongdoers—while making sure that cases remain confidential.[19] The OSC's particular brand of compassion may derive from the fact that it has been headed by people who are particularly indulgent to the financial sector, including (from 2005 to 2010) David Wilson, an investor who was previously vice-president of Scotiabank.[20]

Connecting to Caribbean Tax Havens

The shares of a company registered in the Toronto Stock Exchange may be bought and sold in Ontario while being held by people with accounts in the Bahamas, Cayman Islands, Barbados, or the Turks and Caicos Islands. The TSX, and the Ontario and Canadian governments, work to convince companies in the extractive sector to incorporate or establish subsidiaries in Canada or register in Canadian stock exchanges, but they do nothing to induce these companies to accumulate their profits either in the country where they sell their shares or in the countries where they carry out their prospecting or run their mines. Companies in the extractive sector appreciate Ontario laws because they can easily raise capital through stock market speculation; this way of raising money has nothing to do with the actual profits they record. As a consequence, they may pay almost no taxes in Canada.

Canadian mining companies often use subsidiaries in Caribbean tax havens to negotiate the acquisition of mining claims. The tax havens' secrecy laws make it impossible to find out how widespread this practice is, but a few documents provide a kind of sampling. When the terrible war in the African Great Lakes region officially came to an end in 2003, Democratic Republic of the Congo MP Christophe Lutundula was appointed by the Congolese Parliament to investigate the value of the mining contracts signed during the conflict. Lutundula found that many contracts involving mining deposits, which had been the basis for a war that killed millions of people, had been signed by belligerent parties with mining company subsidiaries incorporated in tax havens.[21] According to the Commission chaired by Lutundula and several independent organizations, Kinross registered its business partnership in the British Virgin Islands,[22] the Lundin Group's subsidiaries were located in Bermuda,[23] and Emaxon, a Montreal company belonging to Israeli diamond magnate Dan Gertler, was run from Panama.[24] Agreements between the Congolese state and Canadian mining companies such as Anvil Mining (whose tax rate was zero percent) were so advantageous in terms of taxes that DR Congo itself was now, in fact, a tax haven. The term *contrat léonin* was used to describe these agreements: one-sided contracts in which the most powerful party took the lion's share. Resources were sold to offshore entities during a war in which the fighters were trying to get money and arms. Other examples of mining contracts with offshore companies occasionally come to light, such as the first industrial contract negotiated by the president of Mali, Moussa Traoré: the agreement was signed with a Canadian exploration company, AGEM, whose Barbados subsidiary was officially responsible for closing the deal.[25]

This phenomenon sometimes leads to obscene situations. Congolese decision-makers, for example, signed an agreement with a company called Vin Mart Canada: nobody knows to whom this company belongs, nor even where, finally, it is incorporated.[26]

Under this system, it is very easy for companies to corrupt their partners in the South. Raymond W. Baker, a Harvard-trained consultant and businessman active in Africa, estimated at the time of the conflict in the Great Lakes area that spending related to political corruption deprived poor countries of $500 billion a year. Baker also found that the World Bank and the IMF were doing nothing to block corruption or money laundering, even though it was obviously the money they were providing to Southern countries that was being diverted to tax havens. When corruption is associated with the corrupted party, the corrupter is easily forgotten. So far, nothing has been done to punish the Western actors with whom the bribes originate. "For the World Bank and also the IMF, discussion of illicit

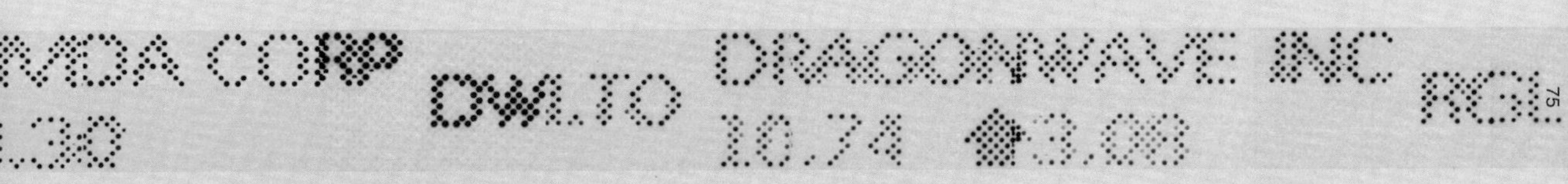

cross-border transfers is addressed almost exclusively to developing and transitional countries and occasionally offshore financial centers but virtually never to Western banks and corporations."[27] James Henry of the Tax Justice Network also notes the flow of money from poor countries to tax havens:

> "We estimate that the amount of funds held offshore by individuals is about $11.5 trillion—with a resulting annual loss of tax revenue on the income from these assets of about $250 billion. This is five times what the World Bank estimated in 2002 was needed to address the UN Millennium Development Goal of halving world poverty by 2015. This much money could also pay to transform the world's energy infrastructure to tackle climate change. [The World Bank has reported] the cross-border flow of the global proceeds from criminal activities, corruption, and tax evasion at $1–$1.6 trillion per year, half from developing and transitional economies."[28]

Over the past ten years, a very large number of serious documents have been published deploring the harm caused by tax havens in the South. In 2009, Oxfam International estimated that developing countries faced a shortfall of $124 billion because of offshore activities.

"This money could pay for health and education services, for protection against the deepening impact of the economic crisis such as safety nets to help those who have lost jobs and for projects to protect poor people already affected by climate change. $16 billion a year would be sufficient to give every child a school place and $50 billion a year is needed to help poor countries protect their people from climate change"[29]

Canadian Laxism

In 2011, the OECD blamed Canada for reneging on its commitment to fight corruption. Canada has done nothing within its own territory to punish any company for political corruption in other countries, even though it attracts the majority of the world's mining companies, which, as the OECD points out, are notorious for participating in influence peddling abroad.[30] An anonymous representative from "one of Canada's most well-known and respected companies globally" told the OECD that Canadian companies abroad face "pressures … to engage in corrupt practices."[31] As a true legal and regulatory haven, Canada protects mining companies incorporated under its laws from the consequences of their wrongdoing abroad.

UN experts have asked Canadian political authorities to investigate the role of Canadian companies in DR Congo, after having duly cited these companies on a list of businesses considered to be in violation of OECD's "Guidelines for Multinational Enterprises."[32] As the UN observers state without any equivocation, "the OECD Guidelines offer a mechanism for bringing violations of them by business enterprises to the attention of home Governments, that is, Governments of the countries where the enterprises are registered. Governments with jurisdiction over these enterprises are complicit themselves when they do not take remedial measures."[33] The only response by Stephen Harper's Conservative government has been to issue a directive in 2009, entitled *Building the Canadian Advantage*, providing the mining industry with a Corporate Social Responsibility counselor who is explicitly deprived of any real power: "The Counsellor will not review the activities of a Canadian company on his or her own initiative, make binding recommendations or policy or legislative recommendations, create new performance standards, or formally mediate between parties."[34]

Rules on financial disclosure in Canada explicitly force companies to make public any information about moral crises, political instability, and ecological crises that they provoke in the South—but only if these phenomena might have an effect on the price of their shares in the stock market.[35]

Oxford Pro Bono Publico is surprised at how difficult it is for citizens to launch civil lawsuits against Canadian corporations in cases of alleged abuse occurring outside Canadian borders.[36] The situation is different in the United States where the Alien Tort Claims Act authorizes lawsuits for serious cases. We hope that the rules of the game may be changed by a precedent set in 2013, when an Ontario Superior Court judge accepted a claim filed by Guatemalans against a Canadian company (Hudson's Bay Company).[37]

Colonizing the South through Mining Policies

Over the past twenty or thirty years, the great ore-consuming areas of the world—China, Europe, and North America—have joined in a race to control metals, including base metals such as aluminum, copper, zinc, and iron, rarer metals such as lithium, tantalum, and molybdenum, and precious metals such as gold, silver, and platinum. While OECD countries' high consumption patterns were stable, China's consumption of refined metals increased seventeen-fold between 1990 and 2011.[38] In other words, the current mining boom is driven by Chinese growth. These metals are used in products we use every day, from electronic gadgets to cars, and to produce energy (such as nuclear energy) and weapons.

Southern states are subjected by the World Bank, the IMF, and the World Trade Organization (WTO) to a competitive logic that forces them to open their borders. The juniors have taken advantage of these open-border policies to establish the "free mining" system, which provides them with guaranteed unlimited access to underground resources. The free mining principle is largely based on a colonial model rooted in Canadian history. Once a profitable deposit has been discovered by a junior and ceded, for a high price, to a major, the major generally sets a large-scale mining project in motion.[39]

An exponentially growing demand combined with the deteriorating quality of deposits lead to a system that produces hundreds of millions of tonnes of waste every year and uses massive quantities of toxic processing agents (such as the cyanide used to extract gold). The use of open-pit mining techniques, involving craters several kilometers in diameter and hundreds of meters deep, is one of the stigmata most representative of the system's predilection for intervention on a gigantic scale. In terms of the environment and public health, devastated lunar landscapes and accumulated waste have consequences that will be felt for dozens or, more likely, hundreds of years to come. The damage we can anticipate, and the accounts we receive from areas close to the mining sites, are alarming: ecosystems destroyed; species eliminated; soil, air, and water massively polluted; as well as increased incidence of cancers, respiratory ailments, and spontaneous miscarriages.[40]

In relation to this long list of serious issues, Canada presents itself to the international mining sector as a regulatory haven enabling multinationals to act with impunity in Southern countries. Canada's diplomats and co-operation agencies spare no effort in pressuring these countries' governments to carry out the acts of dispossession required by industrial mining activity: expropriating—violently if need be—people living on mining claims often acquired by corporations through dubious means; developing tailor-made mining codes; and providing corporations with facilities and infrastructure such as access to energy, water, and transportation networks.[41]

The Canadian International Development Agency (CIDA), merged into the Department of Foreign Affairs, Trade, and Development (today Global Affairs Canada), funded the reform of Peru and Colombia's mining codes, as well as a major dam in Mali that provides energy for mines in the western part of the country.[42] Between 2011 and 2013, CIDA also undertook to fund projects identified as "development" that went no further than providing compensation for communities abused by large-scale mining projects. As an example of such abuse, in the 1990s, Canada's High Commissioner in Tanzania pressed Tanzanian officials to expropriate from people living on land that had been ceded to the Sutton corporation. The result was the massive eviction of the population of Bulyanhulu, a process during which dozens of artisanal miners are alleged to have been buried alive.[43] Documents

disclosed in 2013 by American whistleblower Edward Snowden, also indicate that Canada has spied on the Brazilian Ministry of Mines and Energy,[44] confirming once again the tight connections between the Canadian government and the mining industry.

Cashing in on the Stock Market

In 1994, the mining lobby and the Canadian government undertook to attract even more capital to the Toronto Stock Exchange, sending every possible signal to seduce investors. The stock market is "an important source of capital for the mineral industry in general, but is the *only* source of capital for the 'junior' sector."[45] Pension funds, insurance companies, and banks to which Canadians entrust their assets, as well as Canadian individuals (wealthy and less wealthy), were encouraged to invest massively in shares issued by mining companies in Toronto. Investors of all kinds were drawn into creating the mining industry's nest egg in Toronto, regardless of where this industry was acting in the world. The industry learned that it was easy to get venture capital in Toronto. In 2011, some 185 corporations active in Africa, 286 in Latin America, 315 in Europe, and 1,275 in the United States went through Toronto specifically to raise money for their often controversial projects.[46] The Canadian government actively supports investments in the mining sector, particularly by drawing on its employees' pension funds or the budgets of its public development agencies.

The mining lobby is clearly the driving force in the game. In 1994, the Canadian government was so profoundly subjugated that it allowed the industry to speak on its behalf in its own press releases. This was the year the mining industry redefined the concept of government regulation in terms that were endorsed by political authorities. In a spirit of 'good governance,' the mining lobby's leading voices in Canada, the Mining Association of Canada, and the Prospectors and Developers Association of Canada, organized a series of discussions with various social partners as part of what they called the Whitehorse Initiative.[47] This kind of 'discussion,' legitimized by government participation, is a specious form of political deliberation, providing a basis for political decisions that are supposed to reflect a so-called 'partnership' between highly unequal partners.[48] In this context, the state is seen as just one partner among many. The mining industry's goal was an obligatory 'consensus': it wanted the principles, measures, and guidelines it was promoting to be endorsed by everyone in sight, including representatives of "federal, provincial and territorial governments; business, including the banking community; aboriginal groups; environmentalists; and labor."[49]

The Canadian government made itself one with the process, adopting the mining companies' rhetoric to the point where it completely lost any semblance of autonomy. The industry became the speaker making statements on the government's behalf, with the federal Department of Natural Resources explaining that "the mining industry concluded that it

ARDIER B 0.00 HOU.TO HRZ BT CRD OL BL 3.13 ▲0.05 SMF

needed support, assistance and advice within a non-adversarial framework to help it develop a new strategic vision and to create solutions for the 21st century."[50] Industry, in other words, is the source of all action and decisions, as is explicitly stated by political authorities themselves. Political discourse bows down before the unrelenting logic of the globalized market: the "challenges" faced by industry are "outside Canada's control." The department also states that, "we cannot escape the reality of the nature of global competition"—leaving it unclear to whom "we" might refer. "Numerous mineral-rich countries have liberalized their economic and political systems to attract investment,"[51] adds the Department (or the mining lobby that causes it to speak). Domestically, these exchanges led to an agreement between the Canadian government and Aboriginal peoples on mining issues. Overall, as a British and Uruguayan research group observed, "these deliberations seem to have moved little beyond the industry's opening gambit,"[52] which was to "see its opportunities expanded, instead of reduced" in Canadian territory (as John Carrington, vice-president of Noranda Minerals, candidly observed).[53]

A year later, in 1995, a representative of Natural Resources Canada speaking to a House of Commons Standing Committee went one better by expressing himself exactly as if he were a mining industry lobbyist. Keith Brewer placidly remarked: "Canada has lost some ground in attracting mining investment as a result of an increasing burden of mining taxation."[54] Government had fully adopted industry demands: "improving the investment climate for investors" and "streamlining and harmonizing regulatory and tax regimes."[55] And yet, during the same House of Commons session, a participant had drawn attention to two exorbitant tax deduction programs, the "Canadian exploration expense" (CEE) in which mine exploration and some development costs may be written off at 100 percent, and the "Canadian development expense" (CDE) in which mine development and some operating costs are deducted at 30 percent.[56] But the government was not going to listen to anyone but the mining industry.

Another Federal Government program involves what are known as *flow-through shares*. This is a tax program specifically encouraging investors to put money in mining shares or, to put it another way, a program that indirectly subsidizes stock market speculation on the securities of junior companies. Flow-through shares are a way for mining companies to provide shareholders with tax deductions for expenses that the companies themselves cannot claim because they have already deducted over 100 percent of their income.[57] (This testifies, of course, to the incredible generosity of Canadian tax laws for mining

companies.) The goal is to induce investors both big and small to support mining securities. There is no way this program can be justified according to the arguments used when it was introduced in 1954 as a way of stimulating investment in mining. Today, the program allows mining, gas, and oil companies to issue shares that are entirely tax-deductible in Canada. In this country, investing in a mining company is tax-free just as a charity donation is tax-free.

In 2000, the Federal Government, judging that investment in the mining sector was slowing down, increased the rewards provided by the program through what were now known as *super-flow-through shares*. This involved an additional tax deduction of 15 percent for investors holding shares in a mining exploration company incorporated in Canada. Most Canadian provinces then also gave investors an additional tax credit, Quebec being an exception in this respect.[58] This improved Canadian program did not even require that the exploration and mining projects be carried out in Canada. For those investors who earn a living by playing the market, this is a rare and beautiful opportunity for tax avoidance. Thanks to double flow-through share programs, spending on mining exploration went from $300 million in the late 1990s to $1.72 billion in 2006, and the discovery of new deposits rose from 15 in 1999, to 268 in 2005.[59] While these tax incentives may not always lead to actual mining projects, there is absolutely no doubt that they are keenly enjoyed by speculators. This way of organizing capital leads to severe social and environmental damage of the kind inflicted by Barrick Gold in Papua New Guinea—damage that led both the government of Norway and Quebec's credit union federation, the Mouvement Desjardins, to withdraw their investments from this company.[60]

In any case, Canadian companies dedicated to exploiting mining wealth outside the country are fiscal shapeshifters. All they have to do to avoid taxes in Canada is to incorporate themselves as income trusts. Theoretically, beneficiaries of the trust will be landed with the tax bill—but as long as these beneficiaries open accounts and incorporate entities in tax havens, no tax is paid by anyone. This is a quirk of the Canadian model adopted in 2011: income trusts are taxed like any other entity, except when they do not have any assets in Canada.[61]

Canadian government rhetoric can no longer be distinguished from that of the Caribbean tax havens that it did so much to develop. The closeness between the two makes perfect sense: Canada, as a jurisdiction providing mining companies with legal protection

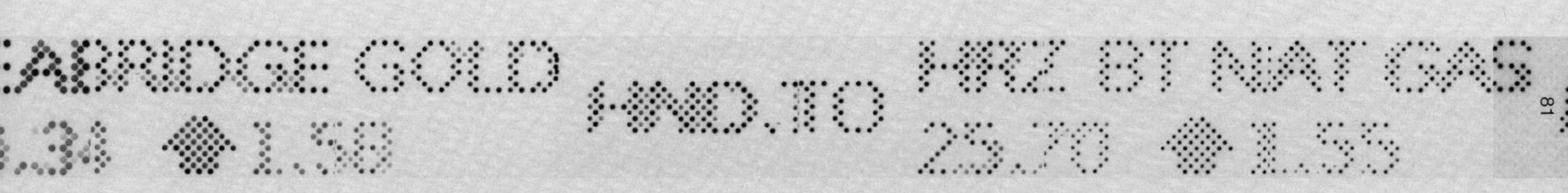

and open financial channels to the tax havens of the Caribbean and elsewhere, has more similarities than dissimilarities today with offshore states. The international offshore order makes it possible, for example, for national or foreign arms merchants to establish a mining corporation in Canada. Through a subsidiary in a tax haven, this corporation will then be able to sign a 'lion's share' contract with a Southern state, manage its tax evasion strategies, and handle the cost of political corruption and arms deals. It can transport commodities or minerals on ships using flags of convenience, and eventually process raw materials in free zones. It will also be in a position, using offshore accounting practices beyond the reach of any law, to trade in dangerous or strategic materials such as uranium. Populations injured or wronged in other countries will find very few channels open to them in Canada to oppose what the corporation is doing in their country. In Toronto, the corporation's stock may well be rising throughout.

□□□

Notes

1 Anthony D. King, *Urbanism, Colonialism, and the World-Economy: Cultural and Spatial Foundations of the World Urban System* (London, UK: Routledge, [1990] 2015), 15; referencing Raymond Williams, *The Country and the City* (New York, NY: Oxford University Press, 1973); see also R.J. Johnston, *City and Society: An Outline for Urban Geography* (Harmondsworth, UK: Penguin Books, 1980), 67–76.

2 *Katanga Business*, directed by Thierry Michel (Belgium: Les Films De La Passerelle, Les Films D'Ici and RTBF, 2009), author's translation.

3 Canada's relation to the international mining industry is discussed at length in Alain Deneault and William Sacher, *Imperial Canada Inc.: Legal Haven of Choice for the World's Mining Industries*, trans. Robin Philpot and Fred A. Reed, (Vancouver, BC: Talonbooks, 2012).

4 Toronto Stock Exchange, *A Capital Opportunity: Mining*, 2012. This document is no longer available online and the TSX has also withdrawn more recent versions.

5 See Deneault and Sacher, *Imperial Canada Inc.*; Alain Deneault with Delphine Abadie and William Sacher, *Noir Canada: Pillage, corruption et criminalité en Afrique* (Montreal, QC: Écosociété, 2008); Alain Deneault and William Sacher, *Paradis sous terre: Comment le Canada est devenu la plaque tournante de l'industrie minière mondiale* (Paris: Rue de l'échiquier, 2012); Alain Deneault and William Sacher, "L'industrie minière reine du Canada: La Bourse de Toronto séduit les sociétés de prospection et d'extraction," *Le Monde Displomatique*, September 2013, http://www.monde-diplomatique.fr/2013/09/DENEAULT/49598.

6 "Canada's Enhanced Corporate Social Responsibility Strategy to Strengthen Canada's Extractive Sector Abroad," Global Affairs Canada, last modified September 16, 2016, http://www.international.gc.ca/trade-agreements-accords-commerciaux/topics-domaines/other-autre/csr-strat-rse.aspx?lang=eng.

7 Natural Resources Canada, *Exploration and Mining in Canada: An Investor's Brief* (Ottawa, ON: Government of Canada, 2016), http://www.nrcan.gc.ca/sites/www.nrcan.gc.ca/files/mineralsmetals/pdf/mms-smm/poli-poli/pdf/Investment_Brief_e.pdf.

8 Al Hudec, "The Canadian Advantage - International Mining Finance," (Farris presentation, Lex Mundi Annual Conference, Vancouver, BC, September 30, 2010), http://www.farris.com/images/uploads/ALH_-_The_Canadian_Advantage_-_International_Mining_Finance.pdf.

9 Ibid. See also Stephanie Loiacono, "Strike Gold with Junior Mining," Investopedia, June 11, 2007, http://www.investopedia.com/articles/stocks/07/junior_mining.asp.

10 See Deneault and Sacher, *Imperial Canada Inc.*, 87–125; and Deneault and Sacher, *Paradis sous terre*, 33–61.

11 See Christopher Armstrong, "Windfall," chap. 6 in *Moose Pastures and Mergers: The Ontario Securities Commission and the Regulation of Share Markets in Canada, 1940–1980* (Toronto, ON: University of Toronto Press, 2001), 155–200.

12 Deneault and Sacher, *Imperial Canada Inc.*, 118, 122. See also Anna Lowenhaupt Tsing, "Inside the Economy of Appearances," in this volume and *Public Culture* 12, no. 1 (Winter 2000): 115–44.

13 Stephen McIntosh, "Rio Tinto Exploration," speech, Citigroup Exploration Conference, Sydney, June 27, 2016, 3, http://www.riotinto.com/documents/160627_Presentation_Citigroup_exploration_conference_Stephen_McIntosh_script.pdf.

14 Peter Koven, "Uraemia assets a nearly $2-billion drag on Areva," *Financial Post*, http://business.financialpost.com/commodities/mining/uramin-assets-a-nearly-2-billion-drag-on-areva.

15 William J. McNally and Brian F. Smith, "Do Insiders Play by the Rules?" *Canadian Public Policy / Analyse De Politiques* 29, no. 2 (June 2003): 125.

16 Bruce Livesey, *Thieves of Bay Street: How Banks, Brokerages and the Wealthy Steal Billions from Canadians* (Toronto, ON: Random House Canada, 2012), 226.

17 "Ont. Teachers' Pension Plan boss rails against handling of white-collar crime," *CBC News*, September 11, 2007, http://www.cbc.ca/news/business/ont-teachers-pension-plan-boss-rails-against-handling-of-white-collar-crime-1.652971; and "Les crimes financiers sont ignorés au Canada: Claude Lamoureux estime que ce type de délit est traité avec légèreté," *Le Devoir*, September 12, 2007, http://www.ledevoir.com/economie/actualites-economiques/156478/les-crimes-financiers-sont-ignores-au-canada.

18 McNally and Smith, "Do Insiders Play by the Rules?" 137.

19 François Desjardins, "La CVMO devra être prudente dans l'application de sa nouvelle politique, dit Fair Canada," *Le Devoir*, October 25, 2011, http://www.ledevoir.com/economie/actualites-economiques/334397/regler-sans-avouer-la-cvmo-devra-etre-prudente-dans-l-application-de-sa-nouvelle-politique-dit-fair-canada.

20 Livesey, *Thieves of Bay Street*, 233.

21 See RAID-UK, *Rights and Accountability in Development's Unofficial Translation of Lutundula Report*, December 2005, http://www.raid-uk.org/sites/default/files/lutundula-excerpts.pdf; regarding English-translated excerpts of République démocratique du Congo, *Assemblee Nationale Commission Spéciale chargée de l'examen de la validité des conventions signées pendant les guerres de 1996-1997 et de 1998-2003* [Rapport Lutundula], June 2005, http://www.droitcongolais.info/files/rapport-lutundula_269e4v5f.pdf.

22 "Kinross Gold and Katanga Mining: Part of the Pillage of the Democratic Republic of Congo?" Mining Watch Canada, April 8, 2006, https://miningwatch.ca/blog/2006/4/8/kinross-gold-and-katanga-mining-part-pillage-democratic-republic-congo.

23 "Rapport Lutundula," section identified within the document as "page 150," 94–95.

24 Amnesty International, *Democratic Republic of Congo—arming the East*, AI index AFR 62/006/2005, July 4, 2005, 41, https://www.amnesty.org/en/documents/AFR62/006/2005/en/.

25 Gilles Labarthe and François-Xavier Verschave, *L'Or africain : Pillages, trafics & commerce international* (Marseilles: Agone, 2007), 56. See also DanWatch and Concord Danmark, *Golden Profits on Ghana's Expense - An Example of Incoherence in EU Policy*, May 2010, https://ida.dk/sites/default/files/PCD%20case%20study_may%202010.pdf.

26 "Rapport Lutundula," section identified within the document as "page 121–22," 76.

27 Raymond W. Baker, *Capitalism's Achilles Heel: Dirty Money and How to Renew the Free-market System* (Hoboken, NJ: Wiley, 2005), 252.

28 Kamran Mofid, "Neo-liberalism & Austerity = Privatisation of Profits and Benefits & Socialisation of Costs and Consequences," Globalisation for the Common Good Initiative, August 8, 2012, subheading "A Background reading," par. 1, http://www.gcgi.info/index.php/blog/277-neo-liberalisma-austerity-privatisation-of-profits-and-benefitsa-social-

isation-of-costs-and-consequences, based on data from the Tax Justice Network, www.taxjustice.net/cms/front_content.php?idcatart=2&lang=1. See also the graphic presentation produced by *National Geographic*, May 2010, reproduced by *Democratic Underground*, www.democraticunderground.com/discuss/duboard.php?az=view_all&address=439x413398; and UN Office of Drug and Crime and The World Bank, *Stolen Asset Recovery (StAR) Initiative: Challenges, Opportunities, and Action Plan*, June 2007, https://siteresources.worldbank.org/NEWS/Resources/Star-rep-full.pdf.

29 Sebastien Fourmy, Policy and Advocacy Director from Oxfam France, as quoted in Oxfam International, "Tax haven crackdown could deliver $120bn a year to fight poverty," March 13, 2009, par. 8, https://www.oxfam.org/en/pressroom/pressreleases/2009-03-13/tax-haven-crackdown-could-deliver-120bn-year-fight-poverty.

30 See OECD, *OECD Working Group on Bribery 2011 Annual Report*, 2012, http://www.oecd.org/daf/anti-bribery/AntiBriberyAnnRep2011.pdf. See also OECD, *Phase 3 Report on Implementing the OECD Anti-Bribery Convention in Canada*, March 2011, http://www.oecd.org/canada/Canadaphase3reportEN.pdf; and Julian Sher, "OECD slams Canada's lack of prosecution of bribery offences," *The Globe and Mail*, March 28, 2011, https://beta.theglobeandmail.com/report-on-business/economy/oecd-slams-canadas-lack-of-prosecution-of-bribery-offences/article580736/?ref=http://www.theglobeandmail.com&.

31 OECD, *Phase 3 Report*, 9.

32 See UN Security Council, "Companies on which the Panel recommends the placing of financial restrictions," annex I in *Final report of the Panel of Experts on the Illegal Exploitation of Natural Resources and Other Forms of Wealth of the Democratic Republic of the Congo*, no. S/2001/357, April 12, 2001, 35–37, http://www.pcr.uu.se/digitalAssets/96/a_96819-f_congo_20021031.pdf; and "Business enterprises considered by the Panel to be in violation of the OECD Guidelines for Multinational Enterprises," annex III in *Illegal Exploitation of Natural Resources*, 41–44. See also "OECD Guidelines for Multinational Enterprises," OECD, 2017, http://mneguidelines.oecd.org/guidelines/.

33 UN Security Council, *Illegal Exploitation of Natural Resources*, 31.

34 "Building the Canadian Advantage: A Corporate Social Responsibility Strategy for the Canadian Extractive Sector Abroad," Global Affairs Canada, March 2009, last modified April 22, 2016, under subheading "Office of the Extractive Sector CSR Counsellor," par. 6, http://www.international.gc.ca/trade-agreements-accords-commerciaux/topics-domaines/other-autre/csr-strat-rse-2009.aspx?lang=eng.

35 See the Federal Government's National Instrument 51-102, as referenced in Claire Woodside, *Lifting the Veil: Exploring the Transparency of Canadian Companies*, November 2009, 14–15, https://resourcegovernance.org/sites/default/files/Lifting_the_veil-Nov2009.pdf.

36 See Oxford Pro Bono Publico, *Obstacles to Justice and Redress for Victims of Corporate Human Rights Abuse*, November 3, 2008, http://www2.law.ox.ac.uk/opbp/Oxford-Pro-Bono-Publico-submission-to-Ruggie-3-Nov-2008.pdf.

37 Bertrand Marotte, "Guatemalan mine claims against HudBay can be tried in Canada, judge says," *The Globe and Mail*, July 23, 2013, https://beta.theglobeandmail.com/report-on-business/industry-news/energy-and-resources/guatemalan-mine-claims-against-hudbay-can-be-tried-in-canada-judge-says/article13360800/?ref=http://www.theglobeandmail.com&.

38 The World Bank, "Commodity Annex," in *Global Economic Prospects*, January 2012, 6, http://siteresources.worldbank.org/INTPROSPECTS/Resources/334934-1322593305595/8287139-1326374900917/GEP2012A_Commodity_Appendix.pdf.

39 See, for example, Eva Liedholm Johnson, "Mineral Rights: Legal Systems Governing Exploration and Exploitation," (Doctoral diss. in Real Estate Planning, KTH Architecture and the Built Environment, 2010), https://www.kth.se/polopoly_fs/1.131782!/Menu/general/column-content/attachment/FULLTEXT01(2).pdf; and Dawn Hoogeveen, "Sub-surface Property, Free-entry Mineral Staking and Settler Colonialism in Canada," *Antipode* 47, no. 1 (January 2015): 121–38.

40 For a brief summary, see Nina Shen Rastogi, "Production of Gold Has Many Negative Environmental Effects," *The Washington Post*, September 21, 2010, http://www.washingtonpost.com/wp-dyn/content/article/2010/09/20/AR2010092004730.html; and "About Coal Mining Impacts," Greenpeace, July 1, 2016, http://www.greenpeace.org/international/en/campaigns/climate-change/coal/Coal-mining-impacts/.

41 For example, see Wanyee Kinuthia, "'Accumulation by Dispossession' by the Global Extractive Industry: The Case of Canada," (MA diss., School of International Development and Global Studies, Faculty of Social Sciences, University of Ottawa, 2013), http://www.collectionscanada.gc.ca/obj/thesescanada/vol2/OOU/TC-OOU-30170.pdf.

42 See José de Echave C., *Guests at the Big Table? Growth of the Extractive Sector, Indigenous/Peasant Participation in Multi-Partite Processes, and the Canadian Presence in Peru* (Ottawa, ON: The North-South Institute and CooperAcción, [2010] November 2011), http://www.nsi-ins.ca/wp-content/uploads/2012/10/2011-Guests-at-the-Big-Table-Growth-of-the-Extractive-Sector.pdf; and CIDA, *Bamako-Ségou Very High Tension Connector Line Project, Mali*, ca. 1999, http://publications.gc.ca/collections/collection_2017/amc-gac/CD4-94-2001-eng.pdf.

43 See Denault and Sacher, *Imperial Canada Inc.*, 70–77; and Deneault and Sacher, chap. 6 in *Paradis sous terre*.

44 Colin Freeze and Stephanie Nolen, "Charges That Canada Spied on Brazil Unveil CSEC's Inner Workings," *The Globe and Mail*, October 7, 2013, http://www.theglobeandmail.com/news/world/brazil-spying-report-spotlights-canadas-electronic-eavesdroppers/article14720003/.

45 Finance and Taxation Issue Group, *Whitehorse Mining Initiative*, October 1994, 1, http://epe.lac-bac.gc.ca/100/200/301/nrcan-rncan/mms-smm/whitehorse_finance-e/fintxiss.pdf, emphasis in original.

46 Toronto Stock Exchange, "A Capital Opportunity: Mining."

47 For more information on the Whitehorse Mining Initiative, see Mary Louise McAllister and Cynthia Jacqueline Alexander, *A Stake in the Future: Redefining the Canadian Mineral Indutry* (Vancouver, BC: University of British Columbia Press, 1997).

48 See Alain Deneault, *Gouvernance: Le management totalitaire*, Lettres libres (Montreal, QC: Lux, 2013).

49 Finance and Taxation Issue Group, *Whitehorse Mining Initiative*, (Ottawa, ON: Mining Association of Canada, 1994), iii. See also Leadership Council, *The Whitehorse Mining Initiative: Leadership Council Accord, Final Report*, October 1994, http://commdev.org/userfiles/files/721_file_WMI_Accord_en.pdf.

50 "Whitehorse Mining Initiative," Natural Resources Canada, last modified April 20, 2017, under subheading "Introduction," par. 4, http://www.nrcan.gc.ca/mining-materials/policy/government-canada/8698.

51 Ibid., par. 3.

52 Forest Peoples Program, Philippine Indigenous Peoples Links, and World Rainforest Movement, *Undermining the Forests. The Need to Control Transnational Mining Companies: A Canadian Case Study*, January 2000, 18, http://miningwatch.ca/sites/default/files/undermining_the_forests.pdf.

53 *Northern Miner*, September 26, 1994, as quoted in *Undermining the Forests*, 18, 88n136.

54 House of Commons of Canada, 35th Parliament, 1st session, Standing Committee on Environment and Sustainable Development, meeting no. 158, November 28, 1995, under subheading ".0950," par. 3, www.parl.gc.ca/content/hoc/archives/committee/351/sust/evidence/158_95-11-28/sust158_blk-e.html.

55 Leadership Council, *The Whitehorse Mining Initiative*, 1–2.

56 House of Commons of Canada, 35th Parliament, 1st session, under subheading ".0920," par. 7.

57 See "Flow-through shares (FTSs)," Government of Canada, last modified April 28, 2008, www.cra-arc.gc.ca/tx/bsnss/tpcs/fts-paa/menu-eng.html; and House of Commons of Canada, 35th Parliament, 1st session, under subheading ".0920," par. 7–10.

58 Under this system, an investor who buys $50,000 worth of shares in a Canadian mining company can deduct the full amount from his or her annual taxable income, as well as an extra 15 percent of this amount at the federal level (15% x $50,000 = $7,500) and another percentage at the provincial level (for example, in Saskatchewan, 10%: 10% x $50,000 = $5,000). In all, the investor can deduct $62,500.

59 David Ndubuzor, Katelyn Johnson, and Jan Pavel, *Using Flow-Through Shares to Stimulate Innovation Companies in Canada: A Research Project Presented to the Greater Saskatoon Chamber of Commerce*, December 2009, 13, http://www.saskatoonchamber.com/ckfinder/userfiles/files/Research%20Papers/2009/Using%20Flow%20Through%20Shares%20to%20Encourage%20Innovation.pdf.

60 See Guy Taillefer, "Barrick Gold - Liste Noire," *Le Devoir*, February 5, 2009, http://www.ledevoir.com/international/actualites-internationales/231640/barrick-gold-liste-noire; and Jean-François Barbe, "Desjardins bénéficie de la popularité des fonds de fonds et de l'ISR," *Finance Et Investissement*, September 5, 2013, http://www.finance-investissement.com/nouvelles/produits/desjardins-beneficie-de-la-popularite-des-fonds-de-fonds-et-de-l-isr/a/53181.

61 See Dennis Hoesgen and Eric Hoesgen, "Energy Income Trusts: A Comeback in the Making," *Oil and Gas Investments Bulletin*, May 20, 2011, http://oilandgas-investments.com/2011/investing/energy-income-trusts/.

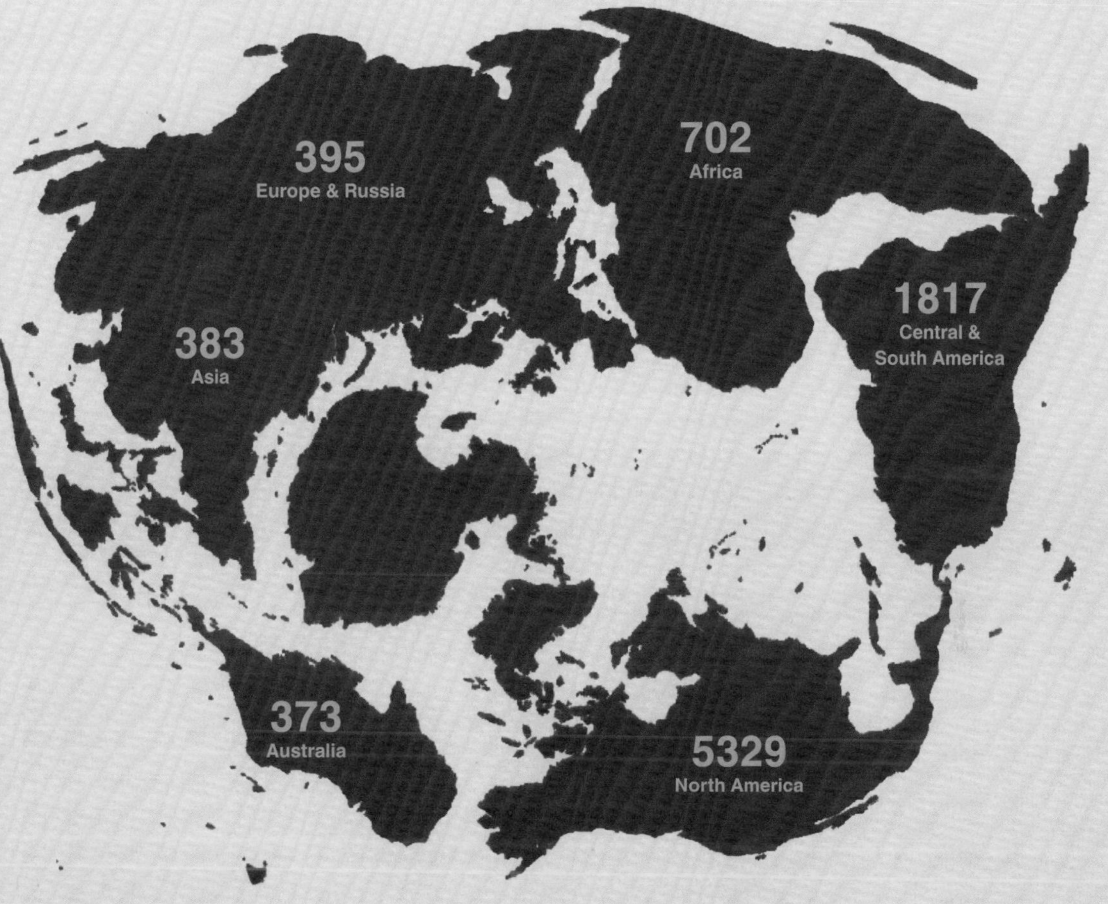

Canadian transnational corporations operate over 9,000 mining projects in every continent and corner of the world

Caribbean Canadian Insurance Complex

A Conversation with Kari Levitt & Aaron Barcant

Kari Polanyi Levitt is a development economist and professor emeritus at McGill University, researching the impact of foreign direct investment in host countries and dependent development and industrialization in the Caribbean. Kari also dedicated her career on the literary legacy of her father, economic historian Karl Polanyi.

Aaron Barcant is a Trinidadian researcher at Concordia University and assistant to Kari Levitt. He worked as a researcher with Alain Deneault on the Caribbean offshore.

"Canada's contemporary problems cannot be understood without the dimension of historical perspective, any more than can the 'structural' problems of Latin America, or those of the Caribbean plantation economies."

—Kari Levitt, Silent Surrender: The Multinational Corporation in Canada, 1970

Part I: Kari Levitt

X: How would you describe the economic/financial relationship between Canada and the Caribbean?

KL: You have to be careful not to assume things. There is little Canadian mining in the Caribbean. There is bauxite mining in Jamaica, but only one of the companies was Canadian (Alcan, aluminum mining and manufacturing corporation), the other three were American. And now the facilities of Alcan in Jamaica are owned by Rio Tinto. Alcan was also active in Guyana. There are also petroleum industries in Trinidad, but none of the Canadian companies are involved in extraction. Canadians are involved in exploration.

There is a new potential mining project being developed in Trinidad, around Pitch Lake, where there are deposits similar to tar sands. Pitch Lake is a natural asphalt lake that used to be a key source of asphalt, which was used for paving roads in the UK.

With regard to the Canadian banks, they are very present in the English-speaking Caribbean. The three most important ones include RBC, Scotiabank, and CIBC. The original banks were British, such as Barclays and the Colonial Bank. The Canadian banks originally went to the Caribbean in support of the sugar industry. There was an active trade between Canada and the Caribbean with fish and timber moving south, and sugar and rum moving north.

Today we have a different picture. It may seem that Canada has very extensive investments in the Caribbean. However, what looks like Foreign Direct Investment in the Caribbean is almost all in the finance sector and almost all involve shell companies involved in tax havens or similar financial transactions. There are huge amounts of investment but very little of this has a counterpart in the real economy.

X: Why do you think the Caribbean has become so prominent in the offshore economy?

KL: Perhaps because it is accessible, English-speaking, secure. But it is hard to tell.

X: In your work, you describe the contemporary process of globalization as new mercantilism; do you think there are similarities between the historical situation (especially the relationship with Canadian banks) and what is happening now in the Caribbean?

KL: In my book, *Silent Surrender*, there is a chapter that discusses the transition from old mercantilism to new. In that book I focused mostly on Canada and comparing the situation in Canada with that of the US, considering Canada's colonial heritage. And then I started working in the Caribbean and I could see similarities there as well. The old mercantilism was a system developed between the seventeenth and

Royal Mutual Funds Inc.
Royal Insurance Holdings
RBC General Insurance Company
RBC Insurance Company of Canada
RBC Life Insurance Company
RBC Direct Investing
RBC Phillips, Hager & North Investment Counsel Inc.
RBC Global Asset Management Inc.
RBC Investor Services Trust
RBC Dominion Securities Limited
RBC Capital Trust
Royal Bank Mortgage Corporation
RBC Covered Bond Guarantor Limited Partnership
Royal Trust Corporation of Canada

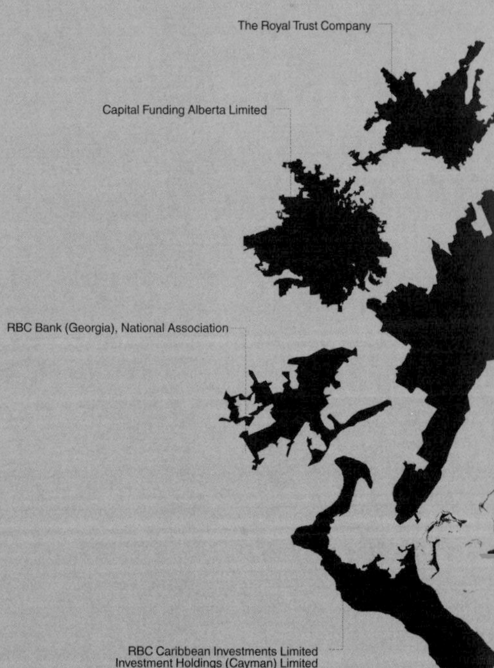

RBC Holdings (Bahamas) Limited

RBC Financial (Caribbean) Lim

eighteenth centuries around the slave trade and sugar plantations, beginning with the Portuguese, then Dutch, then English.

This system was made profitable by the prosperity of the Caribbean plantation. Jamaica and Haiti were the two important islands at the time. My study of old mercantilism involved studying the charter companies and understanding the mercantile systems, and how communication, transportation, and other processes were all connected within the system. Everything that came out of England was connected with that system.

Commemorative 50-cent postage stamp, (Universal Postal Union / Bahamas Postal Administration, 2008)

The merchant–planter relationship was at the core of this system, with the merchant positioned in the metropole and the planter in the hinterland, where the production took place. Therefore, the risk always stayed at the production end, with the planter.

In describing the new mercantilist system, I noted similar approaches in the structures and systems of contemporary transnational corporations, where they have the controlling center and they have the peripheries, and where risk is managed across production sites. In this system, the profit occurs at the point of the sale; the new mercantilist system can be described as neocolonial.

X: So Canadian banks, being themselves corporations, do you think they behaved in the same way in the Caribbean? Do you feel that the Caribbean had played a role in Canada's own economic development?

KL: Canada has had a historic relationship with the Caribbean, considering their common colonial ties. Canada felt that the Caribbean was their little corner of the world, where Canada matters; where Canadian money is accepted. Even now, when you go to the US just a little past the border, Canadian money is no longer accepted, but when you go to these little islands, Canadian money is as good as American.

For the Caribbean, they consider Canadian banks to be safer. And there were many cases when Canadian banks "bailed out" local banks when they were in a bad situation.

But in terms of the Caribbean contribution to Canada, we have to look more at the diaspora, the migration patterns. Human relationships were very important in the Canada-Caribbean relationship. The Caribbean people were just more comfortable with Canada. You can say that Canada has had a special relationship with the English-speaking Caribbean.

The Caribbean is of declining importance to Canada in terms of trade. When I was writing in the 80s, the trade was already declining and now it is even more so. CIDA used to be important in the Caribbean but now it has become much less important.

X: In your writing, you said that the original mercantilist system was a joint project between state and capital. What do you think is the role of state under new mercantilism as you described it?

KL: Staying with the old/new mercantilism idea, in the old mercantilism, state and capital had a very close connection. In the time of the nascent nation-state, this system played a huge role in state development. There was a close connection between the state's military power and its trade. In the present, it is not so close. In fact, you could say it is the opposite; the large TNCs operate outside of the state, and have diverse sources of profit. The producer-merchant relationship is now more like a monopsony, with many sellers and few buyers, so the price is set by the buyers.

USA Holdco Corporation
Capital Markets, LLC

BlueBay Asset Management (Services) Ltd.
RBC Europe Limited

RBC Finance S.A.R.L./B.V.

RBC Holdings (Channel Islands) Limited

RBC Capital Markets Arbitrage S.A.
RBC Investor Services Bank S.A.
RBC Holdings (Luxembourg) S.A R.L.
RBC Luxembourg (Suisse) Holdings S.A R.L

Royal Bank of Canada (Suisse) S.A.

RBC Holdings (Channel Islands) Limited

Royal Bank of Canada Insurance Company Ltd.
RBC (Barbados) Funding Ltd.
RBC (Barbados) Trading Bank Corporation
RBC Holdings (Barbados) Ltd.

Royal Bank of Canada (RBC), Caribbean Banking Network

Scotiabank, Caribbean Banking Network

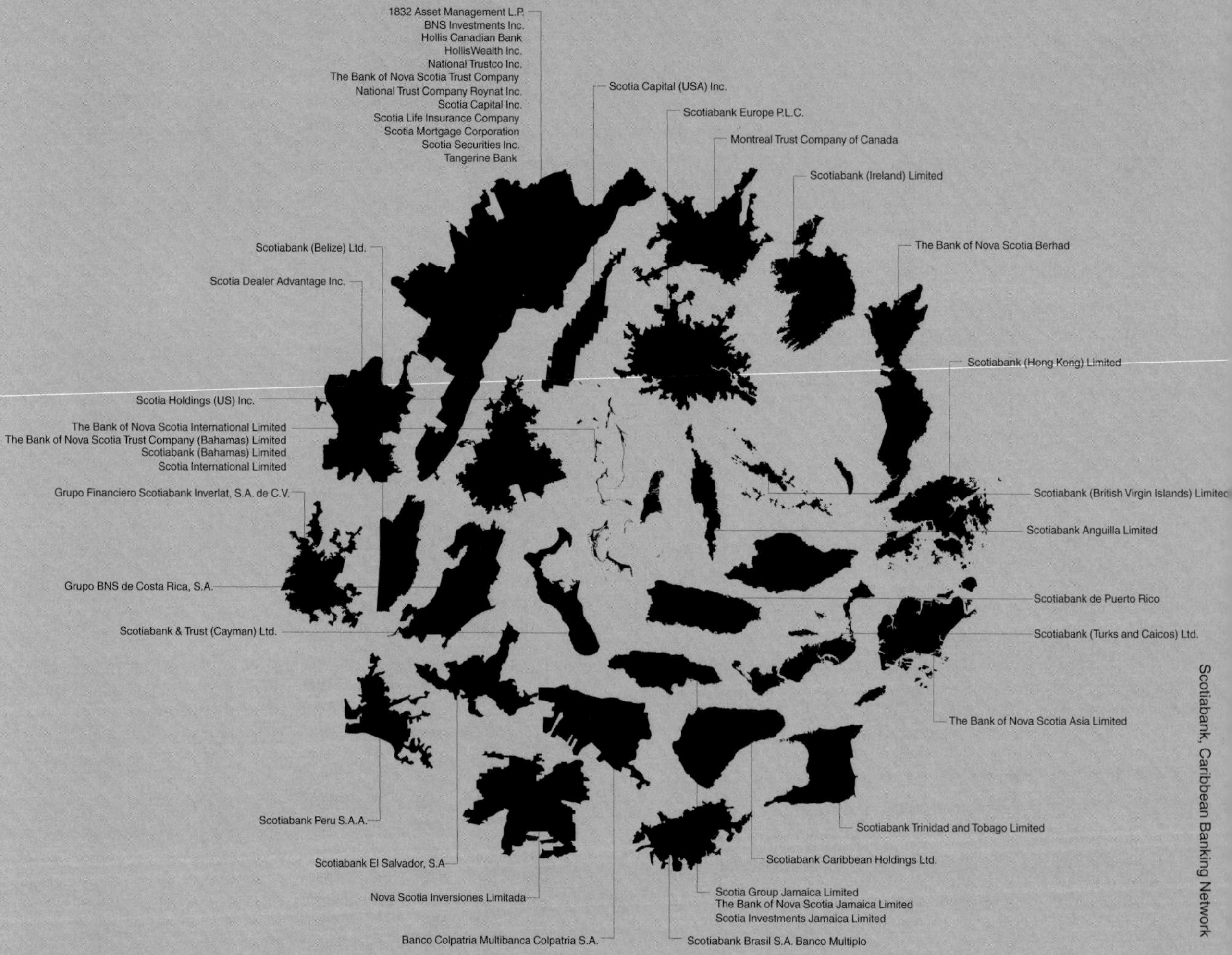

1832 Asset Management L.P.
BNS Investments Inc.
Hollis Canadian Bank
HollisWealth Inc.
National Trustco Inc.
The Bank of Nova Scotia Trust Company
National Trust Company Roynat Inc.
Scotia Capital Inc.
Scotia Life Insurance Company
Scotia Mortgage Corporation
Scotia Securities Inc.
Tangerine Bank

Scotia Capital (USA) Inc.
Scotiabank Europe P.L.C.
Montreal Trust Company of Canada
Scotiabank (Ireland) Limited
The Bank of Nova Scotia Berhad
Scotiabank (Belize) Ltd.
Scotia Dealer Advantage Inc.
Scotiabank (Hong Kong) Limited
Scotia Holdings (US) Inc.

The Bank of Nova Scotia International Limited
The Bank of Nova Scotia Trust Company (Bahamas) Limited
Scotiabank (Bahamas) Limited
Scotia International Limited

Grupo Financiero Scotiabank Inverlat, S.A. de C.V.
Scotiabank (British Virgin Islands) Limited
Scotiabank Anguilla Limited
Grupo BNS de Costa Rica, S.A.
Scotiabank de Puerto Rico
Scotiabank & Trust (Cayman) Ltd.
Scotiabank (Turks and Caicos) Ltd.
The Bank of Nova Scotia Asia Limited
Scotiabank Peru S.A.A.
Scotiabank Trinidad and Tobago Limited
Scotiabank Caribbean Holdings Ltd.
Scotiabank El Salvador, S.A
Nova Scotia Inversiones Limitada

Scotia Group Jamaica Limited
The Bank of Nova Scotia Jamaica Limited
Scotia Investments Jamaica Limited

Banco Colpatria Multibanca Colpatria S.A.
Scotiabank Brasil S.A. Banco Multiplo

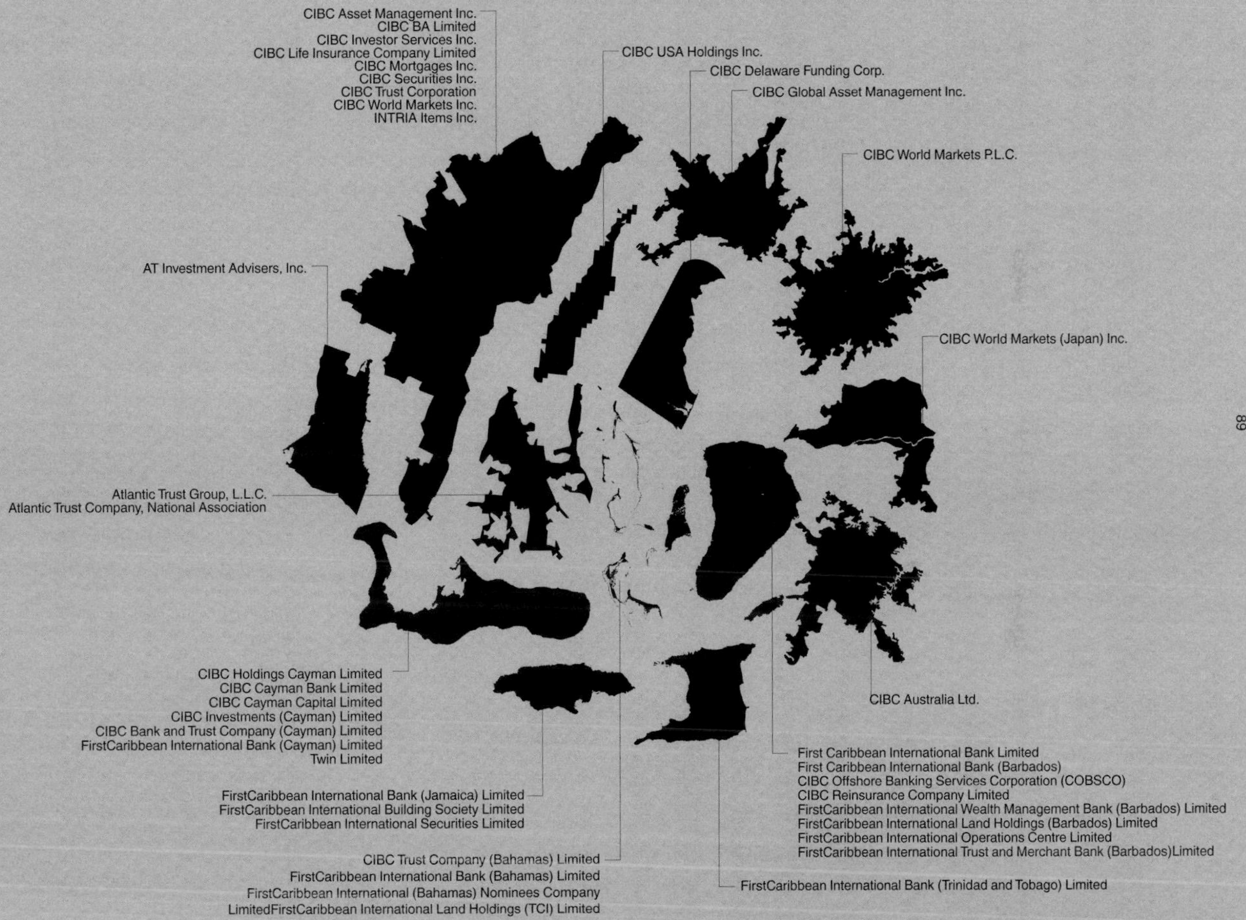
CIBC Asset Management Inc.
CIBC BA Limited
CIBC Investor Services Inc.
CIBC Life Insurance Company Limited
CIBC Mortgages Inc.
CIBC Securities Inc.
CIBC Trust Corporation
CIBC World Markets Inc.
INTRIA Items Inc.
CIBC USA Holdings Inc.
CIBC Delaware Funding Corp.
CIBC Global Asset Management Inc.
CIBC World Markets P.L.C.
AT Investment Advisers, Inc.
CIBC World Markets (Japan) Inc.
Atlantic Trust Group, L.L.C.
Atlantic Trust Company, National Association
CIBC Holdings Cayman Limited
CIBC Cayman Bank Limited
CIBC Cayman Capital Limited
CIBC Investments (Cayman) Limited
CIBC Bank and Trust Company (Cayman) Limited
FirstCaribbean International Bank (Cayman) Limited
Twin Limited
CIBC Australia Ltd.
First Caribbean International Bank Limited
First Caribbean International Bank (Barbados)
CIBC Offshore Banking Services Corporation (COBSCO)
CIBC Reinsurance Company Limited
FirstCaribbean International Wealth Management Bank (Barbados) Limited
FirstCaribbean International Land Holdings (Barbados) Limited
FirstCaribbean International Operations Centre Limited
FirstCaribbean International Trust and Merchant Bank (Barbados)Limited
FirstCaribbean International Bank (Jamaica) Limited
FirstCaribbean International Building Society Limited
FirstCaribbean International Securities Limited
CIBC Trust Company (Bahamas) Limited
FirstCaribbean International Bank (Bahamas) Limited
FirstCaribbean International (Bahamas) Nominees Company
LimitedFirstCaribbean International Land Holdings (TCI) Limited
FirstCaribbean International Bank (Trinidad and Tobago) Limited

Canadian Imperial Bank of Commerce (CIBC), Caribbean Banking Network

Canada has a bilateral relationship with individual countries in the Caribbean, which does enable the type of business practices involved in the offshore economy.

I was in one of the islands where they had 150 head offices of companies, but none of them were producing anything. But even so, this economy serves a purpose, and they provide some benefits locally—for instance, by providing renter income.

X: Do you think the Caribbean countries are taking advantage of the economic system or are they being re-colonized?

KL: It's both.

The large Canadian presence is not sustainable for the local economy, and makes it financially unstable. For example, the Caribbean was very much affected by the global financial crisis of 2008, more so than other places, due to its dependence on these financial services. The impact of the 2008 financial crisis included, for example, the collapse of one of the largest insurance companies, the Colonial Life Insurance Company (CLICO). Almost everywhere in the Caribbean, IMF support was required, and of course as we know those come with specific restrictions and conditionalities that are not favorable to the local economy.

Part II: Aaron Barcant

AB: To add to Kari's comments, the involvement of Canadian banks in the Caribbean was a historical opportunity. It basically took place at a period of time when there was the retreat of the former colonial powers and the advance of new imperialism (coming from the US). Canada took advantage of this opportunity to position itself as something new. For Canada, this was a unique opportunity because it did not have much world power and it was not seen as one at that time, so it was able to position itself to build a special relationship with the Caribbean. This contributed to the development of offshore. And with offshore, it is all about public perception, so this kind of approach continues now.

Offshore centers are very specialized. In his first book (*Paul Martin & Companies*, 2005, in French), Alain Deneault calls them "legal havens," because each plays a different role. For example, some cater more to the mining companies, and some like Panama are a legal haven for ships. It is important to understand this distinction and not to generalize.

The Tax Justice Network is a good resource for documentation of offshore activities.

X: Does the RBC moving its headquarters to Trinidad and Tobago have anything to do with the new mining explorations?

AB: It definitely has a lot to do with the presence of the extractive industries there. Trinidad is considered to be the most stable island because of its petroleum industry. Trinidad also has a very unique and special relationship with Canada, in connection with the East Indian population. There were missionaries from Canada who provided social support and education in Trinidad, primarily to the East Indian population, so there was a high regard for Canada. This ties in to the political landscape of Trinidad because it is still very much split along race lines (black and East Indian). Speaking in very general terms, historically the East Indian population has been more closely connected with Canada (including with migration), while the black population has been more connected with the US. The Harper Government, that was recently in power, established an affiliation with the Trinidadian political party that is said to have an East Indian support base; the leaders of this party actually went to one of these missionary schools and had publicly expressed

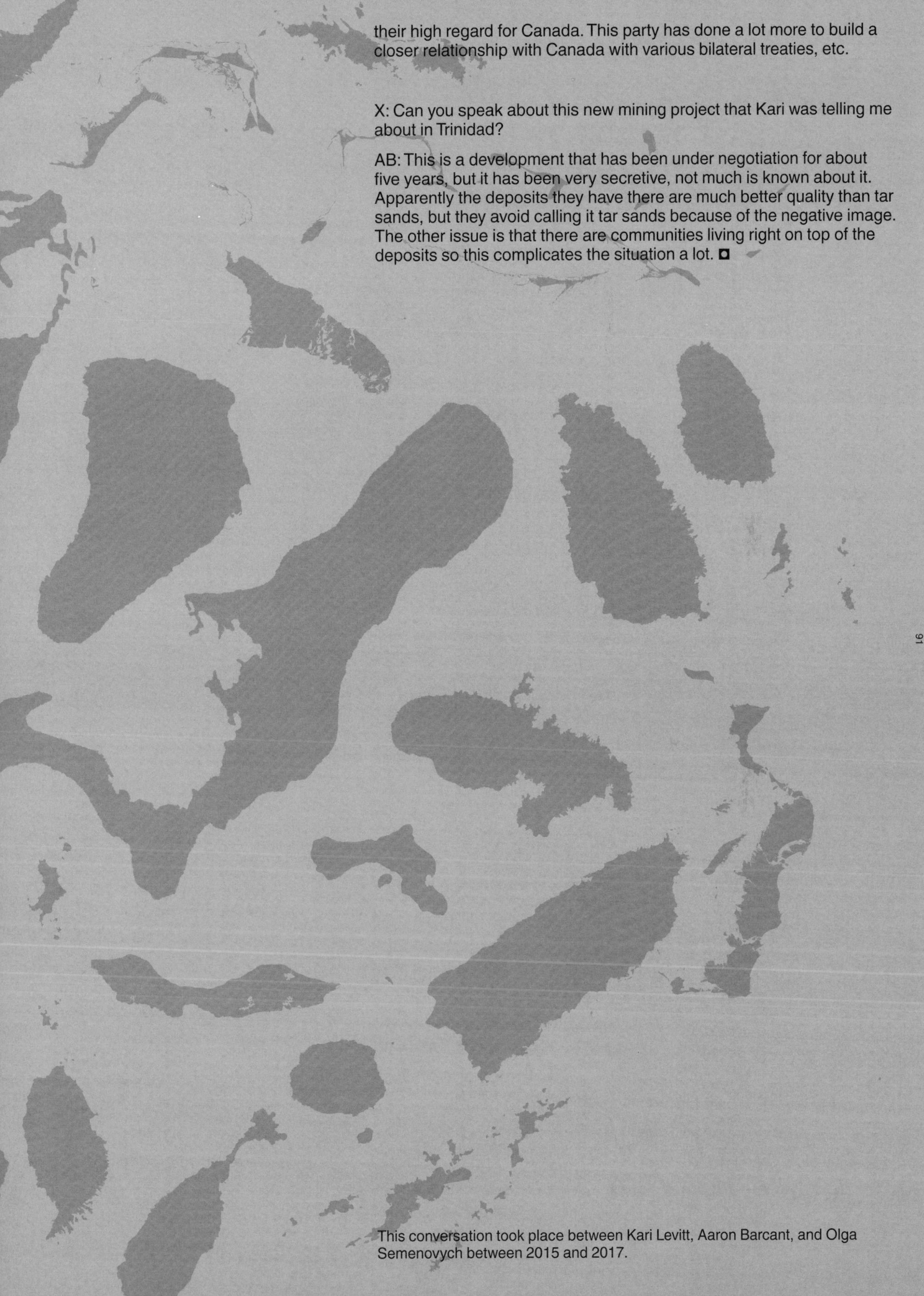

their high regard for Canada. This party has done a lot more to build a closer relationship with Canada with various bilateral treaties, etc.

X: Can you speak about this new mining project that Kari was telling me about in Trinidad?

AB: This is a development that has been under negotiation for about five years, but it has been very secretive, not much is known about it. Apparently the deposits they have there are much better quality than tar sands, but they avoid calling it tar sands because of the negative image. The other issue is that there are communities living right on top of the deposits so this complicates the situation a lot. ◘

This conversation took place between Kari Levitt, Aaron Barcant, and Olga Semenovych between 2015 and 2017.

Colonial Extractions

Race and Canadian Mining in Contemporary Africa

Paula Butler

"Most descriptions of resource frontiers take for granted the existence of resources; they label and count the resources and tell us who owns what. The landscape itself appears inert: ready to be dismembered and packaged for export. In contrast, the challenge I've set myself is to make the landscape a lively actor. Landscapes are simultaneously natural and social, and they shift and turn in the interplay of human and nonhuman practices. Frontier landscapes are particularly active: hills flood away, streams are stuck in mud, vines swarm over fresh stumps, ants and humans are on the move. On the frontier, nature goes wild."[1]

—Anna Lowenhaupt Tsing,
Friction: An Ethnography of Global Connection, 2005

Since the formal end of the colonial era in Africa, many analyses have been produced to explain why mining has not benefited African populations and what policies need to be changed to address that reality. These analyses are expressed eloquently and powerfully in three classical texts: Kwame Nkrumah's *Neo-Colonialism: The Last Stage of Imperialism* (1965); Walter Rodney's *How Europe Underdeveloped Africa* (1974); and Greg Lanning's *Africa Undermined* (1979). The accounts these authors present of the African political economy, with mining as central feature, are eerily familiar, describing circumstances that are not substantially different today.

Rodney sets out in graphic detail how African societies' endogenous processes of development were brought to a halt by the European colonial and early post-colonial presence. He describes how European entities—mining companies, shipping companies, cocao[2] and coffee firms, and the governments and banks of the colonizer states—engaged in harsh modes of primitive accumulation, extracting extreme degrees of "surplus value" from African peoples through forced or grossly underpaid labor, including what amounted to slave labor half a century after slavery was officially abolished. This extraction of value from African lands and people was achieved through racist coercion and brutality. Most notoriously, the peoples of the Congo, an enormously wealthy territory, under the Belgian King Leopold II, suffered horrendous brutality that included torture, whippings, maimings, rapes, and indiscriminate massacres.[3] This surplus value generated enormous profits for companies and thus high rates of return, enriching some European investors at enormous cost to Africans. Rodney offers many examples of the extreme, race-based disparities in wages paid to African and white/European workers. Nigerian coal miners at the British-owned Enugu colliery worked six days for the same wage that a British miner earned in an hour,[4] and in Zambia, European lorry drivers employed in the copper mines earned ten times what a Zambian was paid for the same job.[5] Rodney asserts that there was no objective reason to grossly underpay highly productive African workers; rather, this was a case of overt racist exploitation backed up by the authoritarian clout of the colonial state—its police and its arbitrary laws—the goal of which was to ensure maximum profits for European private companies. Rodney identifies the colonial state as functioning primarily to "guarantee optimum conditions under which private companies could exploit Africans."[6] In this regard, a key role of colonial states was to seize land from Indigenous populations and sell it at nominal prices to European individuals or private companies. These various forms of coercion, exploitation, and seizure generated very high profits for European investors in Africa—returns on investments started at 15 per cent and increased from there.[7] Rodney states: "The returns on colonial investment were consistently higher than those on investments in the metropoles, so the financiers stood to benefit from sponsoring colonial enterprise."[8]

The picture Rodney paints of a continent being callously drained of its resource wealth and human productivity is grim; indeed, he coins the term "underdevelopment" to counter the story of "development," which he regards as a misnomer or lie. A decade earlier, Nkrumah had coined the term "neo-colonialism" to describe the network of Euro-American mining and financial corporations that continued to dominate the resource extraction sector throughout the African continent even while formal or "flag" independence was being achieved by most former colonies. He commented caustically: "The whole of the economy is geared to the interests of the foreign capital that dominates it."[9] What is particularly striking about Nkrumah's book is the clarity with which he identifies the conditions required to harness resource extraction to beneficial development outcomes for African populations—and the extent to which these prescriptions remain relevant but unpracticed almost half a century later:

> "The African countries are faced with the need to turn subsistence economies into organisms that will generate viable and improved conditions of living for their populations. However, many African governments … are granting concessions for the working of mineral, agricultural, and forestry resources whose purpose is the drawing off of output to sustain and enlarge the industries and economies of the imperialist countries. Not one of the investing syndicates has any intention of founding in any one of these countries an integrated industrial complex that would give impetus to genuine economic growth."[10]

Nkrumah does not object in principle to foreign investment in African resource extraction; he *does* object to its occurring in ways that deepen poverty and exacerbate the North-South income gap. He asserts the agency of formally independent African states, noting that "however little real power the government of a neocolonial state may possess, it must have, from the fact of its nominal independence, a certain area of [*sic*] manoeuvre."[11] Nkrumah's book is written with a view to spurring African leaders to develop a united front for using the continent's resources to benefit the greatest number of people.

On this point, Greg Lanning points to the formidable constraints on African leaders that have thus far largely prevented them from heeding Nkrumah's call. For various reasons, African governments have been compelled to support the presence of foreign investors and foreign interests. African politicians know from the experience of leaders like Julius Nyerere and Robert Mugabe that actions to prioritize national interests risk driving away Western aid, debt relief, and investment.[12] Lanning examines each of the typical claims made by Western countries to justify the terms of foreign-dominated African mining; then, in measured, almost understated terms, he exposes these claims as false. Industrial mining tends to create *un*employment rather than employment; it does little to increase the technological competencies of local people, given that foreign professionals continue to arrive to do such work; it does not create linkages with the local economy; and it generates minimal tax revenues for African host governments. The grim conclusion he reached in 1979, after a detailed study, was that the power of Western capitalist interests—both corporate and state—to determine the terms of mining in post-colonial African states is such that there is little likelihood that African states will be able to redirect mineral wealth to the benefit of their populations. Rather, Lanning sees African industrial mining as having been developed since the late nineteenth century to serve the financial interests of foreign investors. There is little sign of any substantial change in this historical pattern. In the decades since Nkrumah, Rodney, and Lanning, many works of scholarship, policy, and activism[13] [14] [15] [16] [17] [18] have reiterated that resource extraction has failed to benefit African nations and peoples. The reasons why include the high level of exportation of mineral resources from the continent, mainly in unprocessed form; the hurdles erected against establishing "value-added" industries on the continent; the Balkanization of the continent such that a united strategy on foreign investment is lacking; the highly favorable terms granted to foreign investors; a lack of transparency on the part of mining companies regarding their earning and payments to governments; evidence of cheating on the part of some foreign companies to minimize payments to host states; the failure to integrate the mining sector with broad-based national development strategies; the failure to develop downstream and upstream economic linkages with the mining industry; the inefficiency of a "mineral rent based on taxes" for "optimizing government revenues and sustaining economic growth";[19] minimal employment creation and, arguably, employment loss; and a host of complaints related to social and environmental harms. As a consequence of all this, the continent's wealth is being stolen—affluent people in other parts of the world are benefiting from the extraction of minerals from territories where African peoples live in poverty and social malaise.[20] That, at least, is the persistent sense. Respected economists Mkandawire and Soludo observed with dismay that a liberalized, free-market-driven mining sector was being actively (even coercively) introduced and advanced by the leading institutions in the West such as the World Bank:

> "Ghana is back to gold, cocoa, and timber, and Zambia is reviving its copper mining. Most of the new investments in Africa is going into mining, not so much because of policy changes as because of the global demand for minerals ... Unless policies are deliberately introduced to relax the focus on natural-resource exports and to conscientiously use mineral wealth to develop human resources and more diversified, labour-absorbing activities, African countries will be embarking on a course of structural inflexibility and vulnerability to terms-of-trade instability."[21]

This point was reiterated twelve years later in the UNECA 2011 report, "Minerals and Africa's Development": "The greater trade liberalization process has accentuated the structural vulnerabilities of mineral-producing developing countries ... UNCTAD (2010) noted that 'By the end of the 1990s the production structure of the [Sub-Saharan African] sub-region had become reminiscent of the colonial period, consisting overwhelmingly of agriculture and mining.'"[22]

This identification of what is not right with foreign-dominated mining on the African continent has been accompanied by recommendations for solving problems. These prescriptions are by now well known, are not extreme, and are based often on how other jurisdictions have made mining work for national development. The most recent comprehensive articulation of a constructive strategy for African mining was produced by the UN Economic Commission on Africa (UNECA) and the African Union (AU) with a view to operationalizing the "African Mining Vision" developed by the AU in 2009. This 2011 report, titled *Minerals and Africa's Development: The International Study Group's Report on Africa's Mineral Regimes*, set out a vision that reprises that of Nkrumah's forty-six years earlier:

> "Africa has to shift focus from simply mineral extraction to much broader developmental imperatives in which mineral policy integrates with development policy. This is the central thinking in this report—that the continent's vast mineral resources can play a transformative role in Africa's development only if it builds appropriate social and economic development linkages that meet national and regional developmental objectives."

The report suggests host country strategies for "capturing" and "sharing" mineral revenues. And it offers comprehensive policy recommendations, including these: that linkages be fostered with the wider economy, including the development of value-added processing; that artisanal mining be supported; that approaches to consultation with local communities be improved; and that human rights (i.e., social and environmental well-being) be protected.

One question this raises is whether there should be foreign-owned mining in African countries at all, given the existing barriers to harnessing resource wealth to broad development objectives. For instance, in Canada, the Ontario Chiefs of the Nishinawbe Aski Nation called for a moratorium on mining until the provincial mining legislation could be amended. Some African communities have reportedly called for the departure of Canadian mining companies. Paul Jourdan, the former CEO of Mintek, a South African mining research organization, has insisted that until mining can be developed as part of an integrated economic plan with backward and forward linkages in the national economy, it would be preferable to leave the minerals in the ground.[23]

"Not mining" is the least favored scenario for countries like Canada that are heavily invested in resources extraction. Those countries seek strategies that will cast mining in a better light by linking it to sustainable development. They challenge the notion that the mineral resource industry is necessarily a "curse." Attempts to mitigate the resource curse have led to various policy prescriptions, although, according to Arellano-Yanguas, certain prescriptions have recently predominated. This new orthodoxy features the following: "(a) decentralization of government; (b) a greater role for direct citizen participation in deciding on how to spend mining revenue; and (c) cooperation between state agencies and commercial organisations (public-private partnerships)."[24] Having conducted a detailed study of Peru, Arellano-Yanguas critiques that set of prescriptions and concludes that regulation by the state is key: "Peru requires capable and stable public institutions in order to overcome the resource curse. Mining operations need robust regulation and enforcement by the state."[25] This view corroborates the analysis presented by the author[26] regarding the potentially nationalistic, development-oriented use of state power—particularly in "global South" states—to counter free trade globalization. In his study of the Congo and of the literature on the resource curse, Matti[27] similarly notes that countries such as Botswana, Chile, and Malaysia provide evidence that rich mineral endowments *can* be managed in ways that benefit the nation as a whole; like Arellano-Yanguas, he concludes there is a "political component" to the resource curse. The thesis I advance in the book[28] is that this "political component" can be named more precisely as international neo-colonialist-neoliberal domination and control of the sector—a process in which Canada, as a country with a globally active corporate mining sector, is directly implicated.

Does Canadian Mining Presence Develop or Underdevelop African Countries?

If the mining sectors in African countries have historically delivered strong returns for foreign investors and only meager returns—and considerable social and environmental harm—to African states and populations, what is the nature of the Canadian presence in contemporary African mining? In the late 1990s, political economist Bonnie Campbell began pointing out the lack of coherence between Canada's claims to support African development and poverty reduction or "basic needs," on the one hand, and its simultaneous support of structural adjustment programs featuring foreign-investment-favoring liberalization of African mining sectors in conjunction with various forms of public assistance to the private Canadian mining sector, on the other.[29] [30] This raises a basic question: Could Canada act in solidarity with African country citizens to support and advance the kinds of policies that have been clearly and repeatedly identified as necessary in order to harness mining to development? Or does Canada act principally in its own self-interest, with little regard—and even contempt—for the right of Africans to life and well-being?

This is an unsettling question. To begin to answer it, I turn to some of the interview data I collected from Canadian mining professionals. Contemporary Canadian mining and exploration companies have much to gain from ventures in African countries, and the people I interviewed spoke frankly and often enthusiastically of these benefits. Asked why they had gone to African countries, interviewees stated:

> "It's where the rich deposits are.
>
> Very high grade for the world. You can't find those deposits anywhere else in the market. It's really got some of the last places on the planet that haven't been explored … there's no data base … the opportunities are just huge to go in and find something.
>
> A company our size would have a very difficult time getting the projects we have now in a country that was First World. In Africa, where most people

sort of fear to tread, it's a lot easier for us, which allows us to acquire things and develop things that we normally wouldn't be able to do.

We were able to get superior projects."

Many interviewees referred to the favorable financial aspects of mining in African jurisdictions. Access to "low cost" or "cheap" labor was the most common "attraction" named[31]:

"You've got a large workforce that's cheap labour, and most of them are fairly educated.

You have a lower-cost labour pool … probably the same operation in North America would not have been profitable.

You know we don't pay them much, we pay them a couple of hundred bucks US a month for the low skill levels."

Such references to low-wage labor are supported by other sources on wages in African countries' mining sectors. An ILO study observed that in countries with a 'labor surplus'—which describes most African countries—increased labor productivity leading to greater mining industry profits can occur without wage increases.[32] The high-profile Bomani Commission[33] report in Tanzania drew attention to a case in which an expatriate employee was paid 6 million Tanzanian shillings (Tsh) while a Tanzanian worker was paid only 800,000 Tsh for the same work. The Tanzanian Mines and Construction Workers Union cited wages of Tanzanian mine employees as ranging from US$200 to $4,000 per month, compared to expatriate monthly salaries of US$6,000 to $20,000.[34] Even more strikingly, Curtis and Lissu, while noting that the average monthly wage for Tanzanian mine workers, US$128 to $240, was a good salary in the Tanzanian context, conclude with the following sardonic observation: "However, by contrast, Barrick's chief executive, Greg Wilkins, received US$9.4m in 2006, including basic salary, bonus, and stock options. It would take an average Tanzanian miner over 500 years to make this amount of money."[35]

In Botswana, statistics gathered on the diamond mining sector showed that 210 expatriate employees earned a total annual income in 2010 of US$14,700,000 (i.e., $70,000 per person on average), while the 2,790 local employees earned collectively just under US$8,000,000[36] (i.e., $2,867 per person on average). Such figures are strikingly reminiscent of Walter Rodney's identification of a common tenfold difference in race-based wages; indeed, at the present time, the disparity is far greater. In the case just cited, the expatriates were earning on average twenty-three times more than Botswana employees.

From a classical economist's perspective, low wages give countries a competitive advantage that they would not otherwise have. However, feminist and critical race scholars have analyzed gendered and raced wage inequities as techniques of subordination and exploitation. As Mohanty[37] and Galabuzi[38] suggest, such inequities are more readily explained as the effects of social power relations—sexism and racism—that categorize and rank the value of the same bodies differently than others, than as a function of different capabilities or needs.[39] Moreover, capitalism has thrived in part through its ability to appropriate the value of human labor by casting some bodies as less entitled to the value of their labor than others. While David Harvey[40] analyzed capitalism's propensity to survive its inherent contradictions through geographical expansion (what he calls the "spatial fix"), I suggest the term "racial fix" to describe capitalism's expansion into the policies of racially inferiorized bodies ('cheap labor'), some of which have been made more vulnerable, and thus more readily exploitable, through being territorially displaced and dispossessed of land. In this way, discriminatory wages are normalized by an everyday assumption of white supremacist societies: that "Third World" others do not need the same lifestyle (housing, health, public infrastructure, post-secondary education, vacations, etc.) as First World selves. These stereotypes of the "African Other" who can live on less, who needs less, and who is implicitly worth less, then reproduce and re-secure the structural dimension of white supremacy.

However, African mine workers are vigorously contesting their 'lower cost' or 'worth less' status. In recent years, employees at many Canadian-owned mines in African countries have gone on strike or engaged in work stoppages or public demonstrations to demand higher wages and/or a reduced wage gap between local and expatriate employees. As one Canadian mining professional among those I interviewed stated:

> "You can't be paying people a dollar a day and then flying in your executives three at a time in three different chartered planes, three different people arriving at three different times and the plane is costing US$1,000 an hour, you know people are looking. How come they can't afford to give me this? So when they quickly start to say, 'pay me what you pay your workers in other countries,' it causes problems for companies."

Interviewees referred to the benefits of mining in African countries as having a great deal to do with the investment incentives provided by the host governments. The generous incentives for foreign investment in mining that were introduced in the context of World Bank-led economic liberalization policy advice to African countries have been extensively documented.[41] Interviewees' comments simply confirmed what has already been identified in the literature:

> "There are no import restrictions.
>
> We don't pay any TV tax for ten years.
>
> We get a special rate for the railway transportation, we also have a special rate for the electrical power."

Many similar comments were made. Such advantages were then associated with very favorable outcomes that kept company shareholders and institutional investors happy. Among those I interviewed, only one reported having lost investors' money. All the others reported financial success:

> "Our money goes a lot farther there.
>
> The operations were lower-cost there and the chances for profit were much greater in terms of the lower operating costs.
>
> As far as return on capital, return on investment, it's outrageously good. I mean we paid back our initial investment in eight months.
>
> You've got this great thing going on in Africa and the reason it's good is because it's in Africa and you don't pay as much money, and things like that, because of the nature of the country."

Such statements are consistent with the attractive profits and high rates of return reported by Canadian banks' and investment houses' precious metal funds, as it is discussed in more detail by the author.[42]

When I inquired about the possibility of African secondary or "value-added" industries becoming established, interviewees acknowledged there was little sign of this happening in any significant manner and that the mining industry supplies and services business was dominated by Western countries. One interviewee referred to mining contractors and heavy equipment companies coming from "Europe and Australia." In fact, the Canadian state and its mining industry actively promote exports of Canadian mining services, equipment, and technology. One interviewee painted a vivid picture of the dearth of spin-off economic opportunities for local African ore-sampling laboratories:

> "The labs in the area where we are, are more or less, they're struggling for business. They were doing just fine before when they were dealing with a lot of small companies like ours, junior exploration companies, and as soon as the big companies were established—[name of three major companies], these guys are large enough that they have their own labs. So suddenly the labs—I mean I used to go there and they couldn't promise you anything for three weeks. Now you come in with 14 samples and it's like Christmas. And you're the only customer around. So it's not really—these big companies are becoming self-sufficient—which is having the opposite effect on secondary industry. But once you get going, there are things you can do locally, but a lot of it would be pretty elementary I would think—you know, get some basic hand tools locally made, which we do, but it's not

a big-ticket item. We have to get all these trays made for our logs, for the diamond drill core logs, so we're going to generate some work for some carpenters in [district] or some carpenters in [district] because we need hundreds and hundreds of these things, so we'll generate some work there, but again it's not high-tech stuff."

It has been well documented that in the mining industry, there is more money to be made in processing than in extraction. In Canada, for instance the $32 billion contribution from mining to the Canadian GDP in 2009 was composed of $7 billion from production (getting the resource out of the ground) and $25 billion from mineral manufacturing.[43] The AU's "African Mining Vision"[44] emphasizes the need for "downstream value-addition" (i.e., the establishment of mineral processing industries, or "beneficiation"); "upstream value-addition" (i.e., the manufacture and provision of goods and services required by the mining sector, such as are currently being offered by Canada's secondary mining industry); and "sideways value-addition" (i.e., training, human resource development, and knowledge acquisition). The 2011 UNECA report on African mining asserts that the mining sector must "build appropriate social and economic development linkages that meet national and regional development objectives."[45]

In light of the value of such processes in relation to broader development goals, it is significant that there has been so little prioritization given to these directions by donor governments and multilateral agencies. Many ordinary African people have expressed their frustration at this. Rayford Mbulu, president of the Mineworkers Union of Zambia, complained publicly in 2010 about the lack of a "trickle down effect" on the economy and the people; he raised the question of "how far the mines and their operators are adding value for the national economy despite them exploiting our natural resources."[46] In a recent manifesto, the African National Congress Youth Wing's demands for greater national control over the mining sector and its wealth called for:

- Local beneficiation and industrialization of a minimum of 60% of the minerals extracted from beneath South Africa's soil. The beneficiation should happen in the communities where the mining happens.

- Provision of education, skills and expertise to South African youth in order to capacitate them to play a meaningful role in the entire mining value-chain.[47]

Although the ANC Youth Wing has been cast as "militant," such "demands" are in fact hardly extreme. They are, rather, exactly what most "developed country" jurisdictions have done in order to benefit from resource endowments. Demands for the construction of local processing facilities and thus employment and skills acquisition opportunities were exactly what was negotiated by the Government of Newfoundland and Labrador and the Innu with regard to the development of nickel at Voisey's Bay. When diamonds were discovered in the Northwest Territories, it was deemed important to establish a diamond cutting and polishing industry in Yellowknife where one has never existed. Another such industry was established in Sudbury after diamonds began to be mined at the Victor mine in Northern Ontario.[48] Within a few years of the discovery of lucrative new metals and gemstones, Canadian jurisdictions had successfully demanded and obtained value-added industries. Yet for African countries, with the possible exceptions of Botswana and South Africa, despite the clear value of beneficiation and other value-added processes, and the clear demands for such secondary industries from within Africa, influential donor agencies and states have placed little priority on such objectives.[49]

⊠⊠⊠

Notes

This essay was originally published as "'Something from Nothing': Generating Wealth in the Racialized Mining Economy," chap. 4 in *Colonial Extractions: Race and Canadian Mining in Contemporary Africa* (Toronto, ON: University of Toronto Press, 2015), reproduced with permission from The University of Toronto Press.

1 Anna Lowenhaupt Tsing, *Friction: An Ethnography of Global Connection* (Princeton, NJ: Princeton University Press, 2005), 29.

2 *Oxford English Dictionary*, (online version), s.v. "cacao," http://www.oed.com/.

3 Canadian officer, William Stairs, was associated with that violence, as noted in Paula Butler, *Colonial Extractions: Race and Canadian Mining in Contemporary Africa* (Toronto, ON: University of Toronto Press, 2015), 56.

4 Walter Rodney, *How Europe Underdeveloped Africa* (London, UK: Bogle-L'Ouverture Publications, 1973), 163–64.

5 Ibid., 165.

6 Ibid., 179.

7 Ibid., 162–89.

8 Ibid.,177.

9 Kwame Nkrumah, *Neo-Colonialism: The Last Stage of Imperialism* (London, UK: Thomas Nelson, 1965), 12.

10 Ibid., 234.

11 Ibid., xiv.

12 Greg Lanning with Marti Mueller, *Africa Undermined: A History of the Mining Companies and the Underdevelopment of Africa* (Hammondsworth, UK: Penguin, 1979), 498–89.

13 Faysal Yachir, *Mining in Africa Today: Strategies and Prospects* (New York, NY: Zed Books, 1988).

14 Benedict Mushingwe, "Tanzanian Mining Industry: An Investigation on Its Potential to Attract Metallurgical and Engineering Industries" (Master's thesis, Queen's University at Kingston, ON, 1995).

15 Thandika Mkandawire and Charles C. Soludo, *Our Continent Our Future: African Perspectives on Structural Adjustment* (Dakar: CODESRIA, 1999).

16 Thomas Akabzaa, "African Mining Codes: A Race to the Bottom," *African Agenda* 7, no. 3 (2004): 8–10.

17 Thomas Akabzaa, "Mining Legislation and Net Returns from Mining in Ghana," in *Regulating Mining in Africa: For Whose Benefit?* ed. Bonnie Campbell (Uppsala: Nordiska Afrikainstitutet, 2004), 25–29.

18 Tundu Antiphas Lissu, "'Conducive environment' for development? Globalization, national economy, and the politics of plunder in Tanzania's mining industry," (2004), accessed November 10, 2014, http://www.tanzaniagateway.org/docs/conductive_envirnments_for_development.pdf.

19 Mushingwe, "Tanzanian Mining Industry," 41.

20 See UNECA (UN Economic Commission for Africa and African Union), *Minerals and Africa's Development: The International Study Group Report on Africa's Mineral Regimes* (Addis Ababa: 2011), 92: "But the widespread sense that Africa has not obtained commensurate compensation from exploitation of its mineral resources is impossible to ignore."

21 Mkandawire and Soludo, *Our Continent Our Future*, 35.

22 UNECA, *Minerals and Africa's Development*, 116.

23 Paul Jourdan, "Creating Resource Industry Linkages in Mining (2010), accessed May 14, 2004, http://www.trademarksa.org/node/1603.

24 Javier Arellano-Yanguas, "A Thoroughly Modern Resource Curse? The New Natural Resource Policy Agenda and the Mining Revival in Peru" (working paper no. 300, Institute of Development Studies, March 2008), accessed April 19, 2013, www2.ids.ac.uk/futurestate/pdfs/wp300.pdf, 3.

25 Ibid., 12.

26 See Butler, *Colonial Extractions*, 39–42, on contrasting theorizations of the nature of the state and state power in the global neoliberal economy. The author argues that strongly nationalistic states are required in the global South in order to maximize the benefits of resource industries for local populations and citizens.

27 Stephanie Matti, "Resources and Rent Seeking in the Democratic Republic of Congo," *Third World Quarterly* 31, no. 3 (2010): 401–13.

28 Butler, *Colonial Extractions.*

29 See Bonnie Campbell, "Canadian Mining Interests and Human Rights in Africa in the Context of Globalization" (Montreal, QC: International Centre for Human Rights and Democratic Development, 1999).

30 See also Bonnie Campbell, ed., *Regulating Mining in Africa: For Whose Benefit?* (Uppsala: Nordiska Afrikainstitutet, 2004).

31 Canadian mining companies' public relations materials routinely assert that wages paid to workers in African countries are higher than most other types of waged employment available in the country. This may well be the case, but they remain much lower than wages/salaries for the same work done in Canada.

32 International Labour Organization (ILO), *African Briefing: Global Wage Report 2010/11*, accessed January 22, 2014, http://www.ilo.org/wcmsp5/groups/public/---dgreports/---dcomm/---publ/documents/publication/wcms_150025.pdf.

33 This was a presidential commission set up by Tanzania President Kikwete in 2007 to investigate accusations of "theft" of natural resources by predominantly foreign mining companies. It was chaired by a highly respected Tanzanian legal expert and former Supreme Court Justice, Judge Mark Bomani.

34 Mark Curtis and Tundu Lissu, *A Golden Opportunity: How Tanzania is Failing to Benefit from Gold Mining* (Dar es Salaam: Christian Council of Tanzania, National Council of Muslims in Tanzania, and Tanzania Episcopal Conference, 2008), 38–39.

35 Ibid., 39.

36 Mike Morris, Raphael Kaplinsky, and David Kaplan, "Commodities and Linkages: Industrialisation in Sub Saharan Africa," (occasional paper no. 13, Making the Most of Commodities Program (MMCP), October 2011), 21.

37 Chandra Talpade Mohanty, *Feminism without Borders: Decolonizing Theory, Practicing Solidarity* (Durham, NC: Duke University Press, 2003).

38 Grace-Edward Galabuzi, *Canada's Economic Apartheid: The Social Exclusion of Racialized Groups in the New Century* (Toronto, ON: Canadian Scholars' Press, 2006).

39 Cynthia Enloe suggests that the more precise term should be "labour made cheap"; see Cynthia Enloe, *The Curious Feminist: Searching for Women in a New Age of Empire* (Berkeley, CA: University of California Press, 2004), 2; see also Galabuzi, Canada's Economic Apartheid, 40–54.

40 David Harvey, *Spaces of Capital: Towards a Critical Geography*, (New York, NY: Routledge, 2001).

41 See Bonnie Campbell, "Canadian Mining in Africa: 'Do As You Please' Approach Comes at High Cost," *Canadian*

Dimension, May 24, 2011, accessed September 23, 2014, http://canadiandimension.com/articles/3982; Cambpell, *Regulating Mining in Africa*; Thomas Akabzaa, "Mining Legislation in Ghana." Documenting the same phenomenon, Morris, Kablinsky, and Kaplan in "Commodities and Linkages" note with regard to Tanzania that "the low-tax regime introduced to foster the gold industry meant that the contribution of the gold sector to total government revenue was only $46.5m in 2004–5 (the most recent year for which data is available), contributing only 1.4 percent of total government revenue" (44).

42 Butler, *Colonial Extractions*.

43 Mining Association of Canada, "Facts and Figures 2010," http://mining.ca/sites/default/files/documents/FactsandFigures2010.pdf, 8.

44 African Union, "African Mining Vision," February 2009, accessed November 3, 2014, http://www.africaminingvision.org/amv_resources/AMV/Africa_Mining_Vision_English.pdf.

45 UNECA, *Minerals and Africa's Development*, xiv.

46 Kapembwa Sinkamba, "Zambia Lacks a Comprehensive Trickle Down—Mine Workers Union," *SteelGuru*, July 28, 2010, accessed January 22, 2014, http://www.steelguru.com/metals_news/Zambia_lacks_a_comprehensive_trickle_down_-_Mine_Workers_Union/157208.html.

47 Geoffrey Candy, "ANC Youth League memorandum to chamber of mines," October 27, 2011, accessed January 22, 2014, https://www.moneyweb.co.za/archive/the-speech-the-economic-freedom-marchers-should-ha/.

48 Andy Hoffman, "Skilled immigrants staff Sudbury gem plant," *Globe and Mail Update*, October 14, 2009, accessed January 22, 2014, http://www.theglobeandmail.com/report-on-business/industry-news/energy-and-resources/skilled-immigrants-staff-sudbury-gem-plant/article562007. Ironically, while this cutting and polishing industry was created as an employment-generating venture in light of Sudbury's 10 per cent unemployment rate at that time, all initial employees were experienced Vietnamese diamond cutters brought in on special work permits. The same multinational company was also employing Vietnamese workers at its diamond polishing plant in Yellowknife. The hourly wage for these workers is not reported, although logically, it would have to be low enough to offset the cost of transporting workers from Vietnam. The company is reportedly working with a Sudbury college to develop a diamond polishing and cutting training program for local residents.

49 Both South Africa and Botswana have attempted to develop stronger upstream and downstream industries in connection with mining, and both countries continue to struggle to achieve this. See Morris, Kablinsky, and Kaplan, "Commodities and Linkages"; Christian M. Rogerson, "Mining Enterprise, Regulatory Frameworks, and Local Economic Development in South Africa," *African Journal of Business Management* 5, no. 35 (December 2011): 13,377–82.

Kingdoms and States situated on and between the Rivers Senegal and Gambia. Africa, detail c.1807 (Arrowsmith Map of Africa)

Sensing Like a State

A Conversation with David Hargreaves

David Hargreaves is Vice President of Surveillance and Intelligence at MacDonald, Dettwiler, and Associates Ltd. (MDA), a British Columbia based company offering information services and systems solutions to customers in maritime defense and security, land surveillance and intelligence, space and remote sensing, aviation, energy, and mining.

X: How does your department, in surveillance and intelligence, work within MDA (MacDonald, Dettwiler, and Associates Ltd.)?

DH: There are two big parts to MDA: communications, and surveillance and intelligence. I am in the surveillance and intelligence part. My scope includes our radar satellites and radar satellite missions. It includes our airborne program, so airborne radar and unmanned aircraft programs, ships and naval systems, and then a variety of other programs that are permanently defense-related.

X: How do you position these technologies in resource extraction and the resource industry?

DH: So you're talking about sectors like mining and oil and gas. Those sectors use different parts of our technology; for the most part they are using the outputs. To give you an example, we supply space radar data into the oil and gas sector for two primary areas: one is oil on water, which is both monitoring and detecting oil spills and detecting oil seeps used more in the exploration phase. In the mining area we use it for spatial mapping and geological mapping, but principally for monitoring the stability of mines and their safety. Arctic monitoring is a big application. Global maritime monitoring is a big application.

"The cadastral survey was but one technique in the growing armory of the utilitarian modern state. Where the premodern state was content with a level of intelligence sufficient to allow it to keep order, extract taxes, and raise armies, the modem state increasingly aspired to 'take in charge' the physical and human resources of the nation and make them more productive. These more positive ends of statecraft required a much greater knowledge of the society."

—James C. Scott, Seeing like a State: How Certain Schemes to Improve the Human Condition Have Failed, 1998

X: I'm very intrigued on this eye from above that is seeing the major geologic transformations in the planet right now. How do you think the RADARSAT technology relates, prevents, or embraces that notion of surveying a new geologic era made by the process of urbanization?

DH: I don't have the answer for that question. With geologic changes you would probably be better off talking to the scientific community that use data from RADARSAT, and use the data for those long-term research projects. Our business is more around changes that are happening on a far more rapid basis. The huge advantage of monitoring in general, and radar is a segment of that, is the possibility to see the global—the entire earth—on a frequent basis, only a few days apart. That is our big value to add from a space monitoring ability. Geologic changes are on a longer timeframe. So probably there are scientific and academic people who are looking at that type of changes, but in our case, that is not the case from a business perspective.

RADARSAT Constellation Mission (©Canadian Space Agency / Communications Research Centre Canada)

X: So from a business perspective it would be more about observing the day-to-day expansion or formation of new exploitable deposits. Is it a tool that determines the first moves before entering to build an oil platform or a coal mine, for example?

DH: Yes. They could certainly use it for mapping and looking at an area on a preliminary planning cycle. They also use it for monitoring the mine. They can use radar technology to monitor very small shifts in the stability of its structure, so they can see for example the size of the mine changing very slowly in very small amounts, centimeters, millimeters, over time. You can then identify points where there may be the potential for collapse.

The big business is mainly around applications where you need monitoring. Our technology is used to monitor mines, oil fields, and open seas for ship traffic.

24-hour coverage (©Canadian Space Agency / MDA)

X: The RADARSAT in the Canadian 100-dollar bill shows an interesting historic timeline of exploration. On that, I wonder how does it feel to be on that level of pioneering and exploration?

DH: We certainly are very proud of what we've done with radar, and obviously to be on the bill is a big thing. I think RADARSAT, which is what is on the bill, and then RADARSAT-2 and the RADARSAT Constellation, are breakthrough capabilities and technologies, so from that perspective it is pretty interesting. I don't think we have ever stopped and thought about how we fit in this very large picture you are talking about.

X: What are the fundamental changes in the transitions from RADARSAT to RADARSAT-2, and now RADARSAT Constellation? Specifically with the RADARSAT Constellation, what are we expecting to see, for the resource industry, in the years to come?

DH: RADARSAT-1 was a pioneering thing when it was first conceived. The idea developed mostly around monitoring the Arctic. Canada was most interested in having some kind of capability or ability to monitor what was going on in the Arctic. When it was built and launched, MDA as well as others in Canada spent at least five, maybe even ten years, developing what could be done with RADARSAT data. Obviously arctic ice monitoring was a core application, but then we started to work on what else you could do with that.

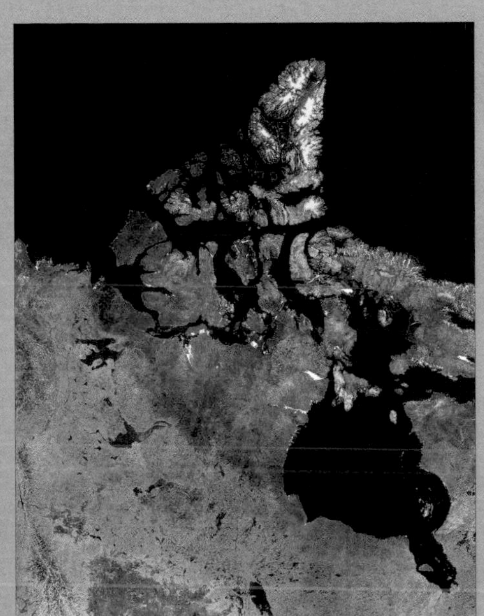
RADARSAT, Arctic & Sub-Arctic Regions (©Canadian Space Agency / MDA)

X: Was a lot of information on the big oil reserves of the Arctic uncovered then?

DH: No, at that time the application was mostly around ice and ice hole monitoring, for navigation-related purposes. For example, one of the big users is the Canadian Ice Service, which produces ice charts. RADARSAT-1 had very little to do with the oil and gas sector at all, other than maybe navigation.

X: What types of new markets will MDA open for the resource industry with the RADARSAT Constellation?

DH: The RADARSAT Constellation Mission contemplates three more satellites and it has some additional capabilities as well, but the biggest purpose of the RADARSAT Constellation Mission is to increase the frequency of monitoring. Because of orbit mechanics and how satellite orbits work, they can only see a certain part of the Earth at certain times. So if you have more satellites up there, you can see parts of the earth more frequently than you could with only one satellite. With the RADARSAT Constellation Mission, we're able to monitor the entire Canadian Arctic every day.

The RADARSAT Constellation Mission (RCM) is able to provide more timely and higher frequency information to customers. That drives up the value of that information to the customers. For example, the agricultural sector is very interested in some of the capabilities that the RCM will have. I don't think there is any sort of step function in the market; it is a growth and evolutionary market enlargement that is largely based on increased value proposition, from improved information content and the frequency revisit of the cover area.

A lot of the technologies we create enable the mining and oil and gas sectors to operate in a more efficient, more environmentally safe, more generally safe manner. That is big part of the value proposition that we offer.

X: Your recent partnership with OneWeb opens questions on how information extraction and distribution will be redefined when the Internet reaches the next billion users?

DH: A huge focus for us is certainly data processing, data information, data extraction; I guess the current buzzword is big data. The images that we generate are absolutely massive, huge quantities of data. Radar is a very interesting data source because it is not a visible image or an optical image; you have to do some heavy duty mathematical processing to extract stuff from it. We are spending a lot on that. We also have a lot of effort around things like data fusion, so we take the radar data and combine with other sources of data to get overall better information products, information contents for our customers.

100-dollar note, *Journey Series*, Bank of Canada Specimen, front/back (©2006 Bank of Canada via cdnpapermoney.com)

We are involved in a variety of satellite communications, driven by the need for more and more global bandwidth. We're involved in both traditional communication programs and programs like OneWeb, where we are involved from a communication side rather than a radar side. We are building satellites for a company called Skybox, which got sold to Google. We are building not radar but optical satellites for them. At this point I don't know, and I don't think anybody at MDA knows exactly what Google plans to use them for. That's an internal secret.

We are more about the infrastructure.

X: Is it fair to assume then that MDA is a company that provides the infrastructure for data mining to be possible?

DH: Right, exactly, and then in certain specific verticals we would do our own data mining. But absolutely, we are much more around providing infrastructure solutions for other organizations to do data mining. Google is a good example, we are building their satellite infrastructure, and then Google will do whatever they are going to do with it, including data mining. And that is true of other companies. There are some exceptions to that, where we do service-type things, but we are more of an infrastructure kind of company. That is what we are good at. We understand how to build complicated systems. We are not experts on internet marketing.

X: I came upon the fact that MDA provided the cartographic bases for the Plan Colombia. From your position is that surveying or is that surveillance? How do those things come together?

DH: I don't know that exact Colombia example, but I know we do have projects with Colombia. But you are right, there is a continuum. There is pure resource monitoring, for instance monitoring mines. Then there are also security applications that tend to be for organizations that are not defense organizations. There is a continuum of all of that we supply and support. Typically those kinds of organizations have their own capability so they are buying data from us, like RADARSAT data, and we don't really get into understanding, nor do they tell us, what exactly they are doing with the data, what the outcomes are, it's not our job to understand that.

La Chasse-Galerie, 1892 (Musée National des Beaux-Arts du Québec)

We have to live within the regulatory environment, and within the export environment that is closely tied to Canada's allies, including the United States. But within the scope of that, we supply data to everybody from purely commercial companies like mining companies, to security type organizations, all the way to defense organizations.

X: So if we put RADARSAT again in that historical lineage of the 100-dollar bill, is exploration moving from a previous surveying industry to a more surveilling kind of industry?

DH: Generally speaking, I would say yes. There is definitely a move to use these kinds of technologies for things that were previously done with people on the ground. There is no question that 10 years ago we supplied virtually no data into the mining sector or the oil and gas sector, and now both of those sectors are using a fair amount of data from us. Oil and gas and mining are the fastest growing sector for us, but they are still not the largest sectors.

X: So why do you think that turn is occurring?

DH: It's driven by business factors, which are primarily economic. It all comes down to the value proposition for their business. If we make the business more efficient, or more profitable, or safer, or more environmentally friendly, that factors into their business in some valuable way. It depends on the application: for instance, the Deepwater Horizon incident (Gulf of Mexico, 2010) was huge, so monitoring oil spills has become a huge deal for companies.

X: Is the increase of RADARSAT demand related to an expansion of the mining sector to geographies where there is political and social turbulence for instance?

DH: Some of that, for sure. The reality is that we don't do a lot of work in countries that are really problematic. We have limitations on those. I guess it depends on the degrees, I mean we work in Colombia. We work in other South American countries, all over Asia of course. But we don't work in any country that has sanctions or any security reasons or risks. But I don't think that this drives the market.

X: I'll close with a more speculative question, since you are seeing things move on a global scale… How do you see the political map of the world in fifty years from now, specifically when we bring in factors like climate change or an increase in migrations?

DH: I don't know, that is an interesting question. This is my opinion, not MDA'S opinion—I don't think MDA has an opinion on this—but what I do see is what everybody else does. Technology is moving at a seemingly faster pace, so the amount of people that have access to technology and are able to use technology is increasing incredibly rapidly. My observation would be that traditional national boundaries—not just the physical boundaries, but also boundaries regarding information and policy—are going to be broken down and undermined by the access to information and technology. To a large extent that is already happening. Even the financial sector is a great example: money moves around the world so much faster than it used to. I think that threats to the world are changing rather rapidly, but at least in Canada, there is a public denial of some of the changes that are taking place. If you went to your average person in Canada, and ask them their views on the threats in the world, I think they would kind of dismiss the threats. But I think the emerging threats are going to affect everybody on a worldwide scale. ◈

This conversation took place between David Hargreaves and Pedro Aparicio in 2015 and 2016.

Into Thin Air

A Conversation with [redacted]

This interviewee is an undisclosed government official working on foreign investment projects in Chile, including Barrick Gold's Pascua-Lama binational gold mine, located in the border with Argentina.

X: What's so different about Pascua-Lama compared to other mining projects?

A: The mining project is the first binational mine in the world, with an isolated location in the mountains, at high altitude. It operates within special regulations between borders. To enable the project, a special treaty was signed before 2000 through significant, diplomatic negotiations.

X: Is there a unique, geopolitical complexity?

A: With the binational mining agreement between Chile and Argentina that came into effect in 2000, Barrick Gold—the world's largest gold company, headquartered in Toronto—initiated a binational mining project for gold, silver, and copper, that spans across the border between Chile and Argentina. It was called "Pascua" on the Chilean side and "Lama" on the Argentinian side. Pascua-Lama is less than 10km away from Barrick's Veladero mine in Argentina, which began operations in 2005. However, Pascua-Lama has not begun operations, despite the go-ahead for construction back in 2009. These sites are a major strategic project for Barrick since Pascua-Lama gold reserves are an estimated 15.4 million ounces—more than triple those of Veladero. Veladero already accounts for a fifth of the San Juan Province economy share; and in Chile, Pascua-Lama is seen as an opportunity for new jobs.

"The global corporation, adrift from its national political moorings and roaming an increasingly borderless world market, is a myth."

—William Keller & Louis Pauly, The Myth of the Global Corporation, 1999

X: Was there a recent event that changed the course of operations?

A: On September 21st, 2015, Veladero was ordered temporary closure by an Argentinian local Justice, due to a recent cyanide spill. There are major concerns in the downstream agricultural communities regarding the potential contamination of aquifers with cyanide, and destruction of glaciers as a result of mining operations. Although Barrick claims that they have strict water management plans and processing methods, in addition to a participatory approach towards local communities; Chilean courts have sanctioned Pascua-Lama to the point that, with decreasing metal prices, it has halted the project indefinitely.

Protest against Pascua Lama in front of Chile's presidential palace, Santiago, 2006 (AP Images)

X: What are the implications of environmental infractions?

A: If Barrick will be fined for every infraction, it will be around 16 million USD, before the reversal of the trial process. The new revised fine could be up to 200 million USD. It's going to be that way, and it could potentially kill the project. But, keep in mind, this is an 8 billion USD project, so 200 million USD is relatively little. So, Barrick could pay the fine, repair all damage that they're ordered to. But, even if they take measures requested by the environmental authority, there is no guarantee that there will not be another lawsuit. I think the problem here—and not only for Barrick, but for many investments in Chile—is that there are unresolved issues within the regulatory frameworks. An important one is a missing policy for the Indigenous and Tribal Peoples (ILO) Convention Nº169.

X: Is regulation endangering the project?

A: No. In mining, this is one aspect among many others. Environmental permitting is extremely complicated in Chile. And, there is no guarantee if the project will be launched. Appeals are possible, but then there are sectoral, technical permits, which are complicated too.

For example, Canadian companies, need a valid permit for exploring underground aquifers for water. They have to pass through state-owned land, where the National Resources Department will often prohibit the right-of-way. There is no coordination, nor staff, nor structure between the national and regional governments which makes permitting very difficult.

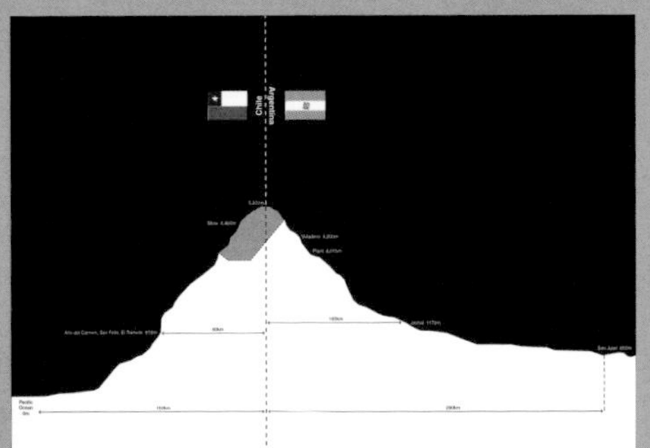

High altitude location of Pascua Lama mining project along border of Chile and Argentina

Needless to say, there is a need for a Major Project Management Office for these large domestic and international projects, like Pascua-Lama, Codelco's Andina, Colbún, and HidroAysén, among others.

X: Is it easier for a mining company like Barrick to extract resources across nation-state borders?

A: No, on the contrary, it's very complex. Imagine they have trouble with the Argentinian workers, they have to respond to the Argentinian side according to their own laws. The same applies for the Chilean side. While they extract the ore on the Chilean side, where most of the ore body is, it will be processed in the Argentinian side. That also carries a lot of complexity, because for every single operation, you have to go by the regulatory framework of each respective country.

Barrick does have a transnational accord for the movement of the ore and people. But still, they have to respond in many aspects—especially the environmental ones as well as social and employment—according to each country's policies. Bottom-line, the treaty has existed for many years—I think the agreement was signed between 1992 and 1999.

Although they have this binational agreement, which includes special taxes, it's more like two separate operations in reality. So whatever they do on the Chilean side, it is subject to Chilean policies. Whatever they do on the Argentinian side, is subject to Argentinian regulations. So Barrick still has to deal with two different governments, two different projects, two different sets of regulations and policies.

X: What are the economic implications of the mountain environments of Veladero and Pascua-Lama? Is this a big factor?

Former presidents of Chile, Eduardo Frei R., and Argentina, Carlos S. Menem, shaking hands after signing the binational mining treaty in 1997 (El Librepensador)

A: On the Argentinian side, Veladero is on the very same mountain range latitude. As long as Barrick is capable of developing technology that allows the exploration of these remote places, it doesn't matter how inaccessible they are. It's still worth it because the project is economically viable. But what can happen, is that at one point, man won't be able to have technologies that'll allow these projects to be profitable; especially if they become too complex to operate. Frontiers—economic, technological, or operational—are always relative. I don't know about Barrick; that question might be answered differently by René Muga, the current Director of Barrick Chile.

X: So, these mines are not really remote?

A: Look, no project is isolated. Everything is related. Mining companies not only have to worry about their extraction operations, but also have to consider what impacts they have altogether, something called 'externalities.' Think of it, it's not the same to have one mining project in the Huasco Valley, than having three or four operations, where three or four companies don't even talk to each other. In this case, there are too many projects, that are very badly managed.

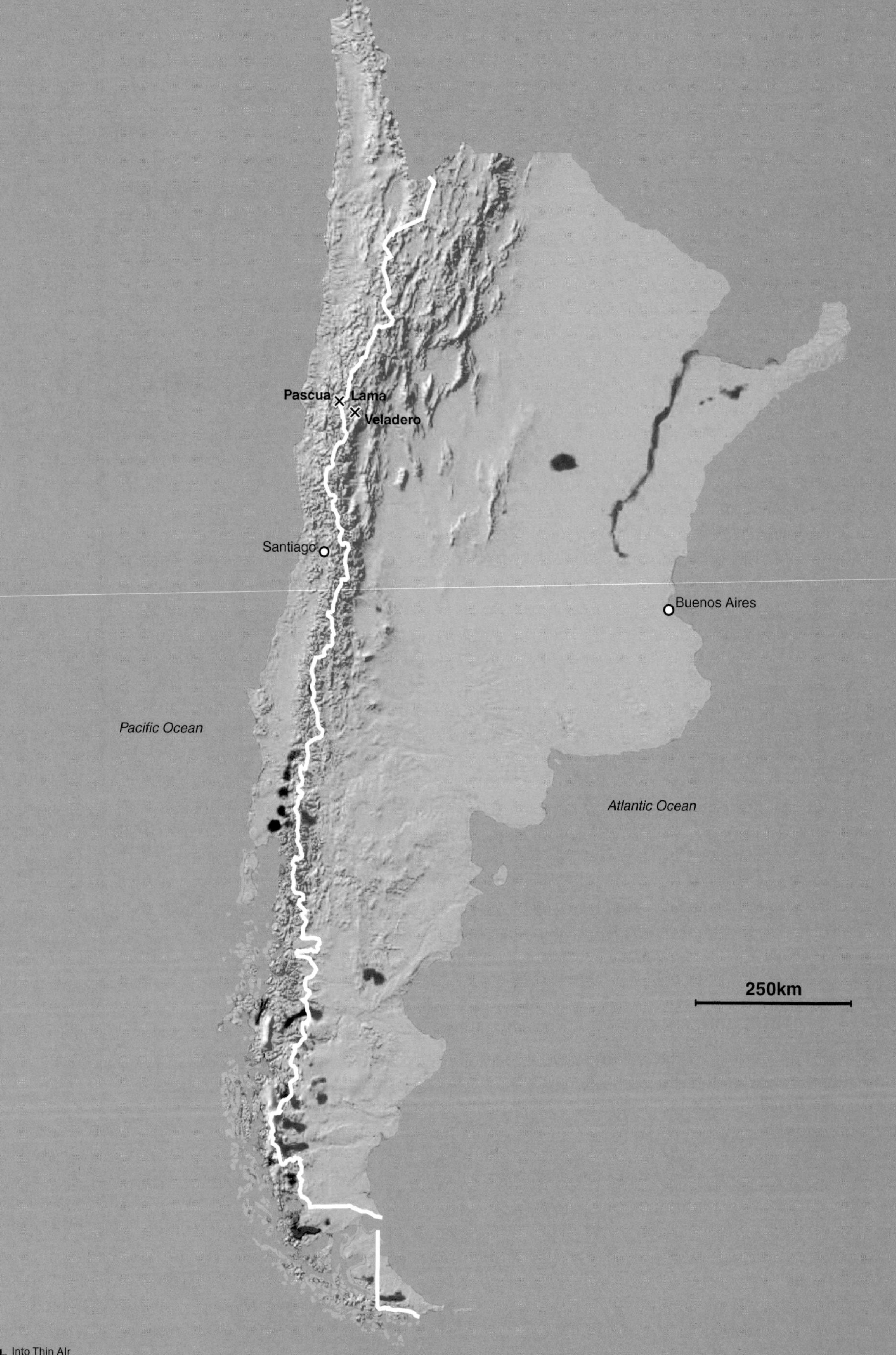
Pascua
Lama
Veladero
Santiago
Buenos Aires
Pacific Ocean
Atlantic Ocean
250km

X: Are there any advantages at all to massive projects of this scale?

A: The advantage of these large projects is that they are allowed to perform a single mining operation across multiple borders, that is all. That gives Barrick the possibility to move people and material between the borders, within limits of the agreements. A little further away, Barrick's Veladero mine on the Argentinian side has its own issues.

Barrick's mining machinery cutting through mountain glacier at Pascua Lama, ca. 2012 (Anonymous source via center-hre.org)

X: Are the glaciers affected?

A: Let's place Pascua-Lama into context and give you an example that is taking place in Chile now. One of the factors that is making resource extraction more and more expensive in Chile is the lack of water resources due to the desert conditions of this mountainous environment. It's impossible that any upcoming project in Chile is to be done without desalinizing and pumping the water up to the mountains. That is extremely expensive, especially with the prices of energy in Chile. So water is becoming a limiting factor now, for mining extraction in Chile, at least in the desert and in higher altitudes.

"You don't need to know the industry you're going into. If you apply yourself, you can always find the experts."

—Donald Trumbell, Peter Munk: The Making of a Modern Tycoon, 1996 (Rule 7 of his 34 Golden Rules)

X: What is next for Barrick and Pascua-Lama?

A: I think Barrick is very silent now. They've opted not to talk much about it, perhaps because they are not sure of what they're going to do with the project. They must be waiting to see what happens with copper prices, and what will happen with the revision of the environmental fines that are pending.

X: Is there any other particular project in the region that you find significant?

A: Well, more recently, Goldcorp is working with Teck. For the first time, two Canadian companies with two relatively nearby projects, are working together. They are putting together two mining operations: the Relincho Project with El Morro. It will be operating together as the Corredor (now Nueva Unión) Project in the Atacama Region. Instead of building two separate tailing disposal dams, two separate desalinization plants, and two separate seaports, they decided to make a joint venture and build the infrastructure together.

Alto Del Carmen Church mural, portraying Barrick eating and destroying the Huasco Valley (www.fotolog.com/sinperdon/)

X: Pascua-Lama has been widely contested by communities on both sides of the border between Chile and Argentina for its impacts, namely to water resources. Is there any way to rebuild trust from the border communities?

A: There have been a series of events that have undermined and divided community trust across the region of the Huasco Valley. You have the agriculturally-based Huascoaltinos Peoples that are part of the Diaguitas Indigenous peoples across the valley region. The valley begins in Chile where the big 'ag' industry of vineyards confronts the mining operations. An 8-year long lasting drought has compounded this problem since all these industrial operations require a lot of water.

So there you have the scope of the problem coalescing in the Huasco Valley: Indigenous and non-Indigenous peoples, mining communities, mining opponents, big agribusinesses. And then, you have mining: water-consumptive and tailings heavy. No one wants to share a single drop of water, nor risk water pollution that would jeopardize their farm products and exports. It's going to be very difficult to reestablish trust.

X: So, does Convention Nº169 address any of these issues?

A: Regulatory policy doesn't prevent companies from conducting early consultation. In Chile, ten years ago, no one even knew what early consultation was. Today, it's essential. For example, a junior exploration company comes in with a pickaxe, starts to talk to people, and then tells them about the project. Then, they sell the project to a large mining company, at a much better price.

Companies in general, including Barrick and others, have not really taken hold of that consultation process, especially with Indigenous communities. It's very difficult because sometimes they reach an agreement, but government authorities are hesitant. So when the junior company steps back, they say "look, we have done early consultation, we've been speaking with the Atacameños for six months, and we have reached these agreements, so please, when you conduct the Indigenous consultation, please take this into account."

Protest outside Atacama Regional Court after Pascua-Lama was sanctioned (©Antutalla)

X: What is the future of large mining projects like Pascua-Lama?

A: Barrick has invested considerable money in the project already and we can't forget that the company has to respond to the shareholders. With all they've invested in spite of a very complicated conjuncture, it is very difficult now to pull out, so possible partnerships may be on the horizon. Partnerships help share and dilute financial risks, especially when the price of gold is not adequate to assume more risk. It's common to see Canadian companies seek co-investment with other mining companies, from another countries. That is not rare in itself, especially with a very large project. In fact, it has happened in Chile with other mega-projects.

Graffiti in Alto Del Carmen reading, "Pa$cua Lama: Bread for Today, Hunger for Tomorrow" (Daniela Estrada/IPS)

In Chile, as in anywhere else, it's important to take into account the environment in which investments are being made. These investments must consider the rich history and vibrant presence of the communities and their territorial rights that live in these regions. Human rights are not just another factor in mining. Since the Huasco Valley is like a microcosm of the industrial world, water and human rights *are*, and *will continue to be*, the defining context and contest of extraction. ■

This conversation took place between a government official supervising the Pascua-Lama project (whose anonymity has been maintained on their request) and Hernán Bianchi Benguria in 2015. As of mid-2017, Barrick Gold entered into agreement with Shandong Gold for a shared, 50/50 joint venture, to operate the Veladero mine in Argentina, for $960 million.

Barrick Now

Peter Munk
ca. 2004–08

Greg referred to this mind boggling-number which I don't know if you all picked up. The mind-boggling number that our Pascua-Lama mine, one of the foremost mines probably in the whole universe, we're talking about a two and a half billion dollar investment—in a country who's total investment in the history of mining has not achieved that—requires, or required five hundred separate approvals. Five Hundred.

Just think of the number of people that have to produce those documents, the number of people that have to check those documents, the number of people that those people who check it have to report to. And think how non-productive that activity is. And a large part of that, a large part of that, is caused by the pressure applied on politicians by groups who Greg earlier referred to, who exercise pressure, and believe that by stopping development they are serving a lofty and higher cause.

The NGOs and the people who are demonstrating against us, be that at Pascua-Lama, and spread information and publish reports that are full of falsehoods do not have to play the game by the same standards that we do. We play the game by being accountable. We play the game by being transparent. And most importantly, we play the game by providing a means to alleviate poverty and to alleviate that cycle of poverty in which thousands, tens of thousands of people have been caught.

The good lord did not put gold deposits in the middle of Manhattan or Paris. The good lord picked for some unique and obscure reason to put gold into areas like the middle of the Tanzanian jungle, on top of the Andes mountains in remote communities where the options to escape poverty are nil. And Barrick comes, and other mining companies come—and I'm talking about the responsible mining community—and provides tens of thousands of jobs, and that means a hundred thousand opportunities for people who otherwise would not have those. The demonstrators, they should really ask themselves, what alternatives do they offer people to escape this cycle of poverty?

You know today, Barrick is one of the few remaining Canadian champions in the industrial scene. We've heard and we've read about the demise of many Canadian champions. We lost just in the last 18 months, the standard bearers of the Canadian mining industry, that was the industry that built this country. The Falconbridges, the Incos, the Norandas, they're all gone. Barrick is here to carry the flag. Barrick operates in more countries that any other Canadian company.

And I must say to you—and you know that as well as I do—that we've done that in contrast to many others of our Canadian international companies, with neither protection from the government nor subsidies from taxpayers. We've done it on your brains and Greg's brains and our efforts on our own.

Thank you for being here today.

Peter Munk is founder and chairman emeritus of Barrick Gold Corporation.

transcribed from *Mirages d'un Eldorado*, Film (Montréal, Productions Multi-Monde, 2008)

Toronto, Canada

Re: Letter Presented at the Barrick Gold 2010 Annual General Meeting

We have come from the Huasco Valley in Chile, representing the Diaguita Huascoaltinos Indigenous and Agricultural Community. We are the direct heirs of the Native People of Huasco Alto, and we have inhabited this land since time immemorial. Our Community consists of 250 families of Indigenous peasants, farmers and herders; we are the only Diaguita community that remained organized after the Spanish colony in Huasco Valley and we also have title to our lands.

Huasco Valley is the last unpolluted valley of northern Chile. Our lands guard important natural and cultural resources, and they hold the major fresh water reservoirs of this valley. That is why in 2006, we decided to make our Community territory a Natural and Cultural Reserve. This is incompatible with Barrick's Pascua Lama and future Pachuy megaproject.

Barrick Gold, without respect for our traditions, our plans, and our right to self-determination, wants to force us to accept the mega mining in our Reserve. In 1998, Barrick Gold seized about 124,000 acres of ancestral lands that belong to our Community. Then, Barrick installed a locked gate that prevents the passage of herders through our own land. This gate is illegal as this road is public, but Barrick continues to refuse public access.

The Pascua Lama project was approved by the State of Chile in 2001, without permission from our community. So we sued the State of Chile in the Inter-American Commission on Human Rights, and this demand was admitted for processing in February of this year.

Although the project officially began this year, Barrick exploration has led to the degradation of the glaciers near the Pascua Lama project. In 2005, the General Directorate of Water of Chile issued a report that blames the company for the loss of 50-75% of glaciers in the area. Recently, on November 11, 2009, the Chilean Government fined Barrick Gold for, among other things, continuing to damage the glaciers, drawing water from unauthorized sites and breaking occupational health and air quality commitments.

Courtesy of Comunidad Agricola Diaguita Huascoaltinos

Now, Barrick has illegally extended its work to other sectors of our domain title. In those areas we can already see the destruction of wetlands and forests, and the extraction of water from unauthorized sites, among other damages. These actions have led us to bring two lawsuits against the company in the courts of Chile which are now being processed.

Also, in seeking to better its image, Barrick Gold, in conjunction with the National Indigenous Development Corporation of Chile, has promoted the creation of Diaguita Communities with no territorial base. With financial support from the company, Barrick has manipulated and corrupted our culture. They have denied that we, Huascoaltinos, are an Indigenous people, they have raised false community leaders, and they have brought professionals to teach the Huascoaltinos about our own culture. What right do you have to come to teach us about our own traditions? What right do you have to manipulate our traditions, inventing costumes, dances, forms of weaving and pottery that are not our own? With this, the company has divided and confused the identity of our people, and has caused us great damage.

We have always been aware that in the land of Huascoaltinos there is great mineral wealth, but our real wealth is its landscapes, the pure water rising in the Andes, with its unique animals and plants. It is our responsibility to protect this precious legacy, as a mark of respect to our ancestors, as a gift to our children and grandchildren, and also as a contribution to the care of Mother Earth and the heritage of all mankind. Therefore, as Huascoaltinos, we are going to defend the Valley. We will not allow Barrick to destroy our land and our culture. We will not allow you to appropriate the legacy left by our ancestors. Today, we come here to order the closure of Pascua Lama. Shareholders, if you continue to mine in our lands, you will remain complicit in the pollution and destruction of our culture and you will be enriched in return for the death of our people. We are here to tell you again that we do not need your money to develop and we are not seeking compensation, because there is not fair compensation for the death of our Mother. We just want you to leave our lands and allow us to live in peace.

Idolia del Carmen Bordones Jorquera
Jaime Nibaldo Ardiles Ardiles
María Inés Bordones Jorquera
Daniela Guzmán González (Interpreter & Advisor)
CADHA - Comunidad Agrícola Diaguita Huascoaltinos

Barrick in South America

The Story of Pascua-Lama

Deep in the Andean Mountain Range at the southern periphery of the Atacama desert, straddling the border between South American nations Chile and Argentina, lies the first binational gold mining project in the world. Named "Pascua" on its Chilean side and "Lama" on the Argentinian side, Pascua-Lama is owned by Barrick Gold Corporation. Situated at a high altitude of up to 5,200m ASL, 75% of the ore body is in Chile, while the remaining ore and processing facilities sit entirely in Argentina. A proposed 4km tunnel with a conveyor will transport the ore across the international border, from Chile into Argentina for processing.

Although discovery of the vast ore body dates back to 1977, political tensions between Chile and Argentina prevented the development of a mine project until the signing of the Friendship Treaty of 1984. Since the 1990s, a series of mining protocols were developed between the neighboring countries, along with a presidential declaration ordering the development of a mining treaty, which was finalized and signed by both Presidents in 1997, officially coming into effect in 2000. Meanwhile, in 1994, through the acquisition of Lac Minerals Ltd., Barrick had begun feasibility and baseline studies for the mine. Through this measure, Pascua-Lama became a politically and legally viable project at an international level, obtaining its respective environmental operating permits from both countries in 2006. At the time, Pascua-Lama was projected as a $1.5 billion operation.

However, the project has since become complicated by the interests and concerns of downstream agricultural settlements—on both sides of the Andes—over fears of massive use and contamination of aquifers and widespread destruction of glaciers. Barrick claims to have both considerations under strict management and control. Meanwhile, less than 10km to the southeast, Barrick's other Andean gold project, Veladero, began production in 2005. By the time Barrick gave Pascua-Lama its go-ahead for construction in 2009, with full support from Chilean, Argentinian, and Canadian heads of state, the projected budget had doubled to $3 billion.

Due to ongoing accusations from local communities and social organizations, along with numerous official inspections, Pascua-Lama has been persistently sanctioned by Chilean courts, making it not viable in current mineral market conditions. Furthermore, in 2015, an Argentinian judge ordered a brief closure of Veladero due to a cyanide spill during September 2015, a move that further fueled Barrick's notoriety in the region. Given today's market and operating costs and constraints—along with ballooning projected total costs of $10 billion over the past three decades —Barrick has requested an official temporary closure protocol.

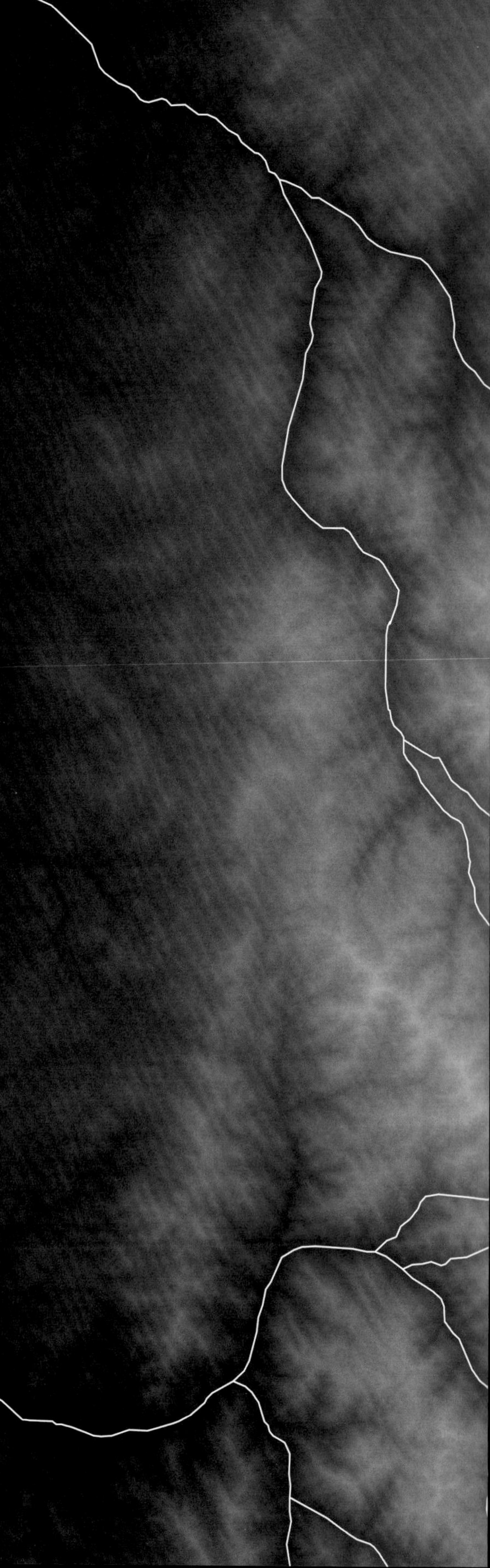

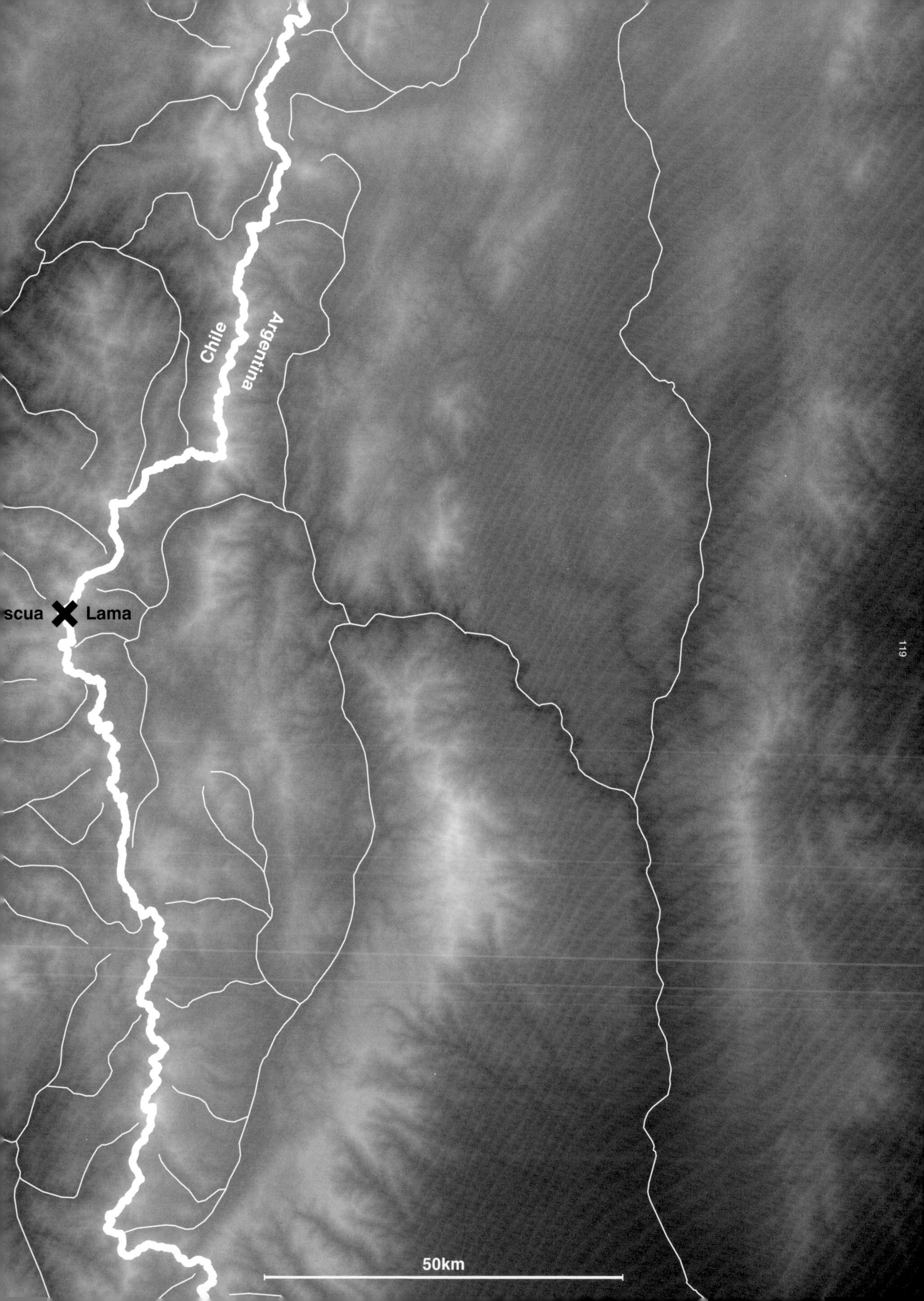
Chile
Argentina
scua
Lama
50km

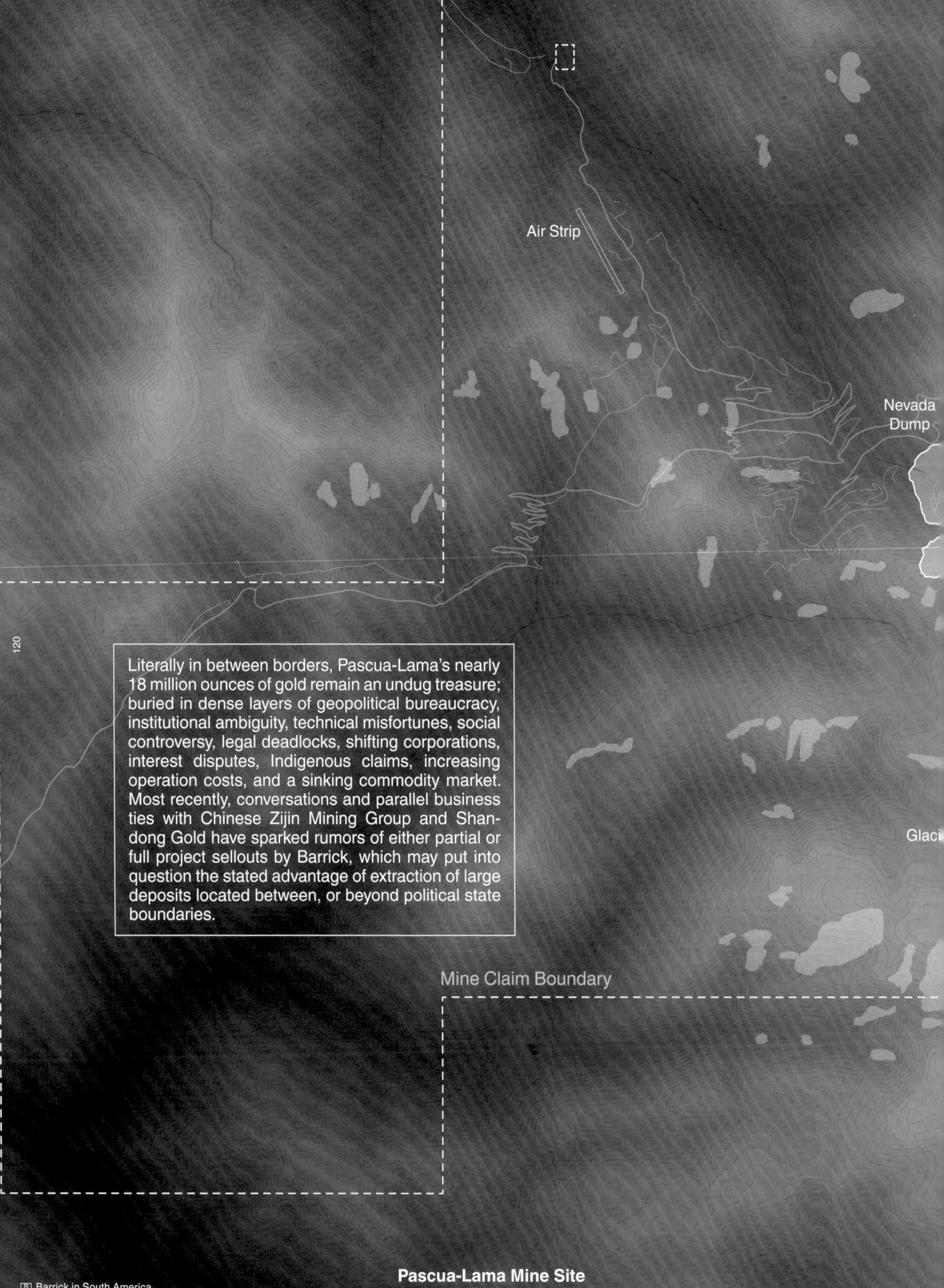

Air Strip
Nevada Dump
Literally in between borders, Pascua-Lama's nearly 18 million ounces of gold remain an undug treasure; buried in dense layers of geopolitical bureaucracy, institutional ambiguity, technical misfortunes, social controversy, legal deadlocks, shifting corporations, interest disputes, Indigenous claims, increasing operation costs, and a sinking commodity market. Most recently, conversations and parallel business ties with Chinese Zijin Mining Group and Shandong Gold have sparked rumors of either partial or full project sellouts by Barrick, which may put into question the stated advantage of extraction of large deposits located between, or beyond political state boundaries.
Glaci
Mine Claim Boundary

Pascua-Lama Mine Site

Chile
Argentina
Glaciers
Pascua
Lama
Morro
Dump
Conveyor
Open Pits
Proposed
Tailings
Proposed
Processing Plant
Veladero
5km

Mother Superior

Pierre Lassonde
Senate Chamber, Ottawa, Canada
26th day of May, 2017

Prime Minister, Senators, Distinguished Guests,

Thank you very much for your invitation so that I may talk to you about Arts, culture, and generosity. We weren't talking necessarily about patrons and philanthropy, but I wanted to talk to you about philanthropy today.

I grew up in a house where art was omnipresent thanks to my mother, who was a collector. I learned about philosophy, rhetoric, etc. during my schooling. I was not 9 years of age in 1967, I was slightly older than that. We played Bach, Beatles, as well as jazz in our household. I lived this as a child but I brought it all the way to my adulthood. I also want to transfer this to my children and grandchildren. When I think of all this, I believe that the key to all passion resides in this one simple question: what motivates us as people, and as a society? Yesterday, the former premier of Manitoba, Gary Doer, talked about our values as Canadians. Thirty to Forty years ago, our values were, simply, we're not American. But that is in the past. We're now much more than that. When I moved back to philanthropism, there was an excellent moment of my life that really made me think at the very beginning of my philanthropy.

My first wife, who unfortunately passed away, before she passed away, she had promised Mother Superior, who was trying to build a new college for girls, that she would contribute to her project. Unfortunately, she passed away a bit too early, and it fell to me to fulfill that promise. So, 6, 7, 8, 9 months after her death, I got a call from above that I should really show up at the school and talk numbers. So, at the appointed day, I did show up, and they led me into the big hall. I don't know if I can picture this for you, but it felt a bit like a tribunal, because there was a big, long table, and all the sisters were on the one side of the table, all with their full garment– the whole cornet and the whole habit–and me, they had put a little student chair, right across, with short legs. And I sat there, and Mother Superior then said to me, "Mr. Lassonde, we're really sorry about your loss, and we are praying with you." And she went on for 2 or 3 minutes about how the community was praying for me, and then she fell silent, and she looked at me, so now it was my turn. And I had really thought about this a lot. So, I told them, I thanked them, it was really helpful to me, their prayer. And finally, I got to the number, and I said, well, look: I'll be very happy to contribute half a million dollars to your new project.

Silence. Complete Silence.

And then Mother Superior turned left, and she was talking with "Mother Économe" (psst psst psst). And then she turned right, and talked to Mother Something-Whatever, (psst psst psst). And then she came back to me and said, "Mr. Lassonde, this is a very generous offer, but it's not God's number." And without missing a beat, she goes on, "we urge you to pray to the Holy Spirit, and you come back in a month, and hopefully you will have been éclairé by that time. It's amazing how the Holy Spirit is "in the times." I mean, it has a cell phone, or something... But Mother Superior cost me a million dollars.

It's a story I like to tell, because at the time, I must admit, I wasn't quite ready to give that much money, but then I realized that when you are invested in a project, you have to be wholly invested. It's kind of funny, ever since then, whenever our family has a project that we're looking at, that we're trying to wrap our head around putting money behind, we always go back to: well, what is our "God's number"? It became a reference. Our family philanthropy is directed mainly to three different fronts, first of all, education; second of all, the Arts; and, third of all, the community in which we live. When we talk of education, it's important to think about the institutions that contribute to our education. Educating our youth is one of our most important missions as a country. If you examine the GDP index, the highest correlation with GDP is the level of education amongst our youth.

I came from the natural resource business, and I'm often told, what are our greatest resources? Is it our oil, our mines, our forest, our lake? No, it's not. If you ask me, it's our millions of young people that are educated. Those are our natural resources. We do have, however, one thing in favour of Canada's natural resource industry. As a country, we will always be in the business of the natural resources, because we have one fact that is absolutely incredible: we're the second largest land mass country in the world. And because of that, we will always have an advantage over everybody else, but one country. But then, we have one of the most educated countries in the world, which makes us have an advantage over everybody else. It is also essential, when we talk about institutions... Universities are not businesses. They need help, not only from the government, but from philanthropists. Philanthropism is very important to support our institutions, not only artistic institutions, but also those of the educational variety, or any institution that is not a business one. About a year ago, I spoke to the Quebec Chamber of Commerce. I gave a presentation that I had called "Being Surrounded by the Good." Most of the time you want to do good. It just comes with it.

I talked about how important it was to invest in Arts and cultures to create creative cities. Not only this, but when you create creative towns and cities, you create an ecosystem where the Arts become an integral part of life. Where you can think of cultural tourism—these are all self-sufficient businesses, and that's what's the beauty of it all. So, of course, in Quebec City, I had just finished a new pavilion at the Fine Arts Museum. It's an incredible feat of architecture. The vision I had was very simple. If you create a building that looks like a Fabergé Egg from afar, you see this building and you want to look at it more closely. And, you know that the Fabergé Eggs were created to contain treasure. So what is your reaction? You want to see what treasures are inside, no doubt. The new building in Quebec City, if you examine the number of visitors, the number of visitors doubled over one year, and autonomous revenue has gone up by over 200% because now even businesspeople want to carry out events in the building. Not only that, but the general population wants to go to the museum, wants to appropriate the museum. There are 2 million visitors who come by boat every year. If we can obtain 5% of cultural tourism, that's already 100,000 or 200,000 people. Out of 1 million people, that is enormous. So, it's the kind of thing that we strongly support. The other project that I had shared at the time with the members of the Quebec Chamber of Commerce was what Zita Cobb did about Fogo Island. Fogo Island was a dying community off the coast of Newfoundland. And when I say dying, it was really dying. They were down to just over 1,000 people, from a community of 15,000 people. And the government really wanted to wipe it off the face of the Earth. They didn't want to have a postal station, so they took the postal station away, they didn't want to have a hospital so they took that away. They did everything to get rid of them, but the people there didn't want to leave. And this one lady came in and said, you know what, if we get together—we've been here for over 200 years. We should know how to do things around here. If you've lived here

for 200 years, what do you know that the rest of the world doesn't know? To make a long story short, with the help of the Arts and the community, she created the Fogo Island Inn, which is today one of the most prized destinations of tourism in the world. It's got a five-star hotel. If you haven't been there, I would highly encourage you to go. They have four artists' studios for an artists-in-residence program, and the artists have to live in the community in Fogo Island, and they have these studios that they go to during the day, and the only requirement is that at the end of their duty, they have to have an exhibition for the local residents of their work, or a lecture of their work, if they're poets or authors. This is a community now where the Arts are an integral part of the community. She has created an incredible community. The fact that they're Irish and they can tell stories also helps.

Charles Landry, a renowned urban development consultant who came up with the idea of creative cities, said, and I quote, "ordinary people can make the extraordinary happen if given the chance and the means to go beyond the boundaries of everyday life." It's an incredibly powerful statement, because the Arts is what makes that possible. If you're walking down the street, whether it's here in Ottawa or in Toronto, and you see a piece of art that moves you, you can go beyond the ordinary. And that's really what we're trying to do in Canada. Why are we different? Why are we as Canadians different? Maybe in part because we make Arts and culture a central part of who we are.

When we look at the ecosystem—I've talked a bit about the cities, but—whether it's the Art Gallery of Ontario in Toronto, or the Royal Ontario Museum, or the MBAM in Montreal or the Quebec National Museum in Quebec City, all these institutions create ecosystems around them that are vitally important for artists and also for communities. And what you see a lot of times is that the artists move into a community which then gets gentrified, because everybody wants to live there! Then it gets too expensive, and they have to move somewhere else, but that fact alone is what makes cities absolutely incredible in terms of diversity and in terms of people who want to live there—where you can walk at night and feel good. How many places in the world can you walk at midnight and feel that you're not going to get run over, or killed? In Canada, most of the cities, if not all, are very much like that. You have neighbourhoods where you feel totally safe. Why? Because, in part, that ecosystem that I talked about is so important.

The idea of beauty that I evoked in my talk in Quebec City is also talked about in a book by Pierre Thibault and Francois Cardinal. Their book is called Et si la beauté rendait heureux. Very much in the same theme that I just talked about. They talk about well-designed space that makes life more pleasant, and changes our relationship to the space, to others, and even to time. They made the point that if schools were designed as neighbourhoods, as places where you feel great, instead of prison, which most kids feel like they are. You could take kids from a disadvantaged neighbourhood, and imagine the joy that they would have in being in those schools. And the fact is that he did it, and it has a huge impact on the kids' learning. To achieve this, we need to be creative and bring the Arts in our lives. But we also need, I think, to do more.

I don't like to say this, but the only certainty in life, apart from taxes, is that it will end. And if we do not live our lives in a way that contributes to the collective good, and play a part in its progression and creative power, we will not benefit from its richness and we will fear its end.

There's one project that, as a philanthropist, I have at heart. When we talk about the greater good of a country, it's a project that for me personally encompasses the Arts, the history, the culture, philanthropy, and just about everything else—and it is the National Portrait Gallery. The idea's not new. It's been kicking around for over 20 years. But it hasn't happened yet. And

there's a great space for it right across the street, which has been vacant for about 20 years—guess what! And if you look at the last 150 years of Canada's history, Library and Archives Canada has collected, maintained, preserved over 20,000 drawings and paintings, including the four Indian kings dated from 1710. An incredible rare collection of oil-on-canvas portraits. It also has 4 million photographs, including the entire estate of Yousuf Karsh, the incredible photographer that we know from the 1960s. He left his entire collection to Archives Canada. These have never been shown! They are national treasures and they languish in Library and Archives Canada. And I'm not even talking about what's in the AGO in Toronto, or the MBAM in Montreal, and I'm not even talking about the private collections of Hudson's Bay, who's been around for 400 years, these incredible portraits and pictures, and the same with the CIBC, of the gold rush in the Klondike, and the BMO private collection, and the donors. These treasures deserve to be exhibited, for Canadians to look into the eyes of those who have shaped Canada's history, and those who are paving the path for tomorrow. From Indigenous people to early settlers, from inventors to athletes, from activists to artists, from scientists to capitalists, from feminist trailblazers to political movers and shakers, a collection of portraits offer a panoramic view on Canada's past and present, and shine a spotlight on the incredible diversity that makes up our identity as a country. The old adage "a picture is worth a thousand words" is still true today, even more so in the world of Instagram, Snapchat, and selfies. Why? Because they are a testament to our collective desire to put faces to our shared experiences. Portrait museums in London, with over 2 million visitors a year, Washington, Canberra, have proved to be highly popular. Today, they are even more avant-gardiste because they present through interactive technology a great way to engage visitors, historians, students, on the themes of history, society, ancestry, personality, and achievement. I honestly believe that the creation of a Canadian National Portrait Gallery would be a fitting realization for Canada's 150th anniversary.

All my life, I've been fortunate to come across inspiring people and institutions. I had that opportunity in business. I also was fortunate in the Arts, in this same respect. Not only was I the head of the Quebec City Art Gallery, I've also been head of the Canada Council for almost two years now. And this year, we are celebrating our 60th anniversary. Clearly I'm a bit older than that, but the Council is in much better shape than I am. Just to come back to philanthropy: even though the Council was a recommendation of the Massey Report, because Vincent Massey was the one who of course chaired, it was a Royal Commission on National Development in the Arts, Letters, and Sciences. It was better known as the Massey Report for Vincent Massey.

The Massey Report was the most comprehensive diagnosis of cultural life ever undertaken in Canada at that point. And it described a really bleak cultural landscape. Professional theatre was moribund, musical life was largely confined to the church basement, and professional artistic ventures were virtually nonexistent outside of the larger cities. To develop cultural and intellectual life in Canada, the Massey Report recommended that the Federal Government establish a Canada Council for the Encouragement of the Arts, Letters, Humanities, and Social Sciences.

But it took two philanthropists to make it happen. Those were Walter Killam and Sir James Dunn, who together gave 50 million dollars in 1957. Just to give you an idea of what 50 million dollars meant, 50 million dollars in 1957, if you took it in gold bullion—gold was 20 dollars. Today, gold is 1,200 dollars. So, that would be 3 billion dollars in today's money. This is incredible philanthropy, when you think about it, in 1957. And this is the real origin of the Canada Council for the Arts.

In presenting the Canada Council to the Parliament, the Prime Minister of the day, Louis St-Laurent, said, "our main objective in recommending the establishment of the Canada Council is to provide some assistance to universities, to the Arts, humanities, and social sciences, as well as to students in those fields, without attempting in any way to control their activities or to tamper with their freedom." That was incredibly farsighted. Government should, I feel, support the cultural development of the nation, but not attempt to control it. As a result, the Council is, in St-Laurent's words, "as free from state control as it is prudent for anyone entrusted with public funds to be." So, the Canada Council operates at arm's length from the government, through an 11-member board, and we report to the Minister of Heritage.

When the Council was founded in 1957, the total amount of money we were given in that year was 1.5 million dollars, and it was shared between the humanities and social sciences. Of course, by 1964, these funds were clearly insufficient. What's incredible is–in the Massey Commission Report, they stated that in 1951, according to the report, English Canada had produced a total of 14 works of fiction in an entire year. Isn't that incredible? Today, the Canada Council has a budget of $257 million, and we distribute to over 2,300 organizations and institutions across Canada, to over 17,000 writers, and thousands of artists. So, we have come a long way. Now, with the doubling of the budget of the Council over the next five years to $360 million (at least, that's what we're told), the Council now has more resources than ever to follow through on our commitments to the Arts in Canada.

It must not be forgotten that the Council's investment has an impact well beyond the projects that we fund. For example, a book publishing grant could lead to the production of a film. The best example of that is Life of Pi. We helped Mr. Martel. And guess what? An incredible movie came out of that. I can give you hundreds and hundreds of examples that the little seed that the Canada Council provided, whether it was Leonard Cohen, back 30 years ago, or Yann Martel today, has helped shaped Canada's culture.

The Canada Council is not the only organization that is playing in that field. We have a complementary role to other Arts support and government organizations, but we do play a very critical role in that ecosystem that produces the creative city, that ecosystem where cultural tourism becomes viable. The response by artists and stakeholders to the Canada Council's plan for the coming years has been unanimously received by the community. I can tell you that the board of directors that I chair will examine the deployment of what I see as an inspiring future. Inspiring, owing to its effort to ensure that the Arts become a central part to all our lives, and central to the discussion of our future. In its Act, when it was created–the Council–it says we want to create the Council for the creation of Arts and the enjoyment of the Arts by all Canadians. And I can tell you that when I chair the board, my question to the staff all the time is, how do we get to 36 million Canadians? I want to make sure that every Canadian has a chance to enjoy the Arts as much as I do, or you do. I have previously said that art is a reflection of our society and culture, and the various cultures that make up our society.

The Arts as supported by the Canada Council is a key driver of our cultural industries, and of our cultural heritage, because that is where many of the talents and innovations are developed without which the cultural industries would not survive. When the Arts are an integral part of community life and professional life, and even one's personal life, you can see the effects through the creativity, the spirit of inclusion, and the values of community. So, that brings me back to the original question: what moves us as individuals, and as Canadians? I think the answer for all of us is the same: a better Canada.

▸◂

Pierre Lassonde is current Chairman of the Board of the Canada Council for the Arts, Co-Founder & Chairman of Franco-Nevada Corporation, and former Chair of the World Gold Council.

A Fucking Hole in the Ground

Réal V. Benoit (1971)

Often, people would ask me why I never talked about the mines. The mines, that were for me, well, Almost ten years of my life. To tell you the truth, I really don't know why. But today, I'm really gonna try by starting to say that It's a fucking hole in the ground… I really have to say it like that, I can't do otherwise, and if some people don't like it, well, I'm very sorry But, I can't talk to you about the mines, and what it's like (or looks like, or feels like) Without telling you that it's a fucking hole in the ground… It rains on you, it rains on your head all day long, You can't see where you're going, Air, the air is no good to breathe, And, that's where there is some air (to breathe). And, you never know if you're going back up, With the others at night, Ah, no, it's like I said, it's a Fucking Hole in the Ground… Noise, there's a lot of noise, More than you want, Without talking about the smoke, Not only does it burn your eyes, It burns the inside of your body, It's dirty, it's moist, it's dangerous, I think there's nothing worse on Earth, Forced work (labour) is even worse, These fucking holes in the ground… Those…., those who've never had a chance to see them, Well, they're really lucky. Those… those who want to make it a career, Well, I don't know about them, I have no idea what they're thinking, Because there's on average at least 2 guys of out 2 looking to do something else, They want to get out to leave a little. These fucking holes in the ground.

Souvent on, on m'a demandé pourquoi je parlais jamais des mines. Les mines qui furent pourtant pour moi, ben, presque dix ans de ma vie. Pour vous dire vrai, je sais vraiment pas. Mais aujourd'hui, je vais le faire en commençant par vous dire que… C'est un ostie trou de calvaire. Il faut, hé, il faut vraiment que le dise comme ça. Non, je peux pas faire autrement. Et pis si il y en a qui aime pas ça, ben, je regrette énormément. Mais je peux pas vous parler des mines, pis de ça que ça de l'air, Sans commencer par vous dire que… C'est un ostie trou de calvaire. Il te mouille, il te mouille sur la tête à grand journée. Tu vois pas où tu va, L'air, l'air est pas bon à respirer Pis, ça c'est les places où c'est qu'il en a. Tu sais jamais, jamais si tu vas remonter Avec les autres le soir. Oh non c'est ben comme j'ai nommé: Un ostie trou de calvaire. Du bruit, du bruit il y en a plus que dans… Sans parler de la boucane qui elle

Barrick in Nevada
The Story of Goldstrike and Cortez

The Carlin Trend is a geologic formation in northeastern Nevada, home to one of the richest gold mining regions in the world. The Goldstrike Property, owned by Barrick Gold Corporation, is one of the largest and longest-producing projects in the region. Goldstrike includes the Betze-Post open pit and the Meikle and Rodeo underground mines, along with processing facilities.

In 1985, Franco-Nevada, the Toronto-based pioneering gold royalty and streaming company, placed a $2 million investment in Goldstrike, then a small exploration project jointly owned by PanCana Minerals and Western State Mining. This was the first royalty investment made by Franco-Nevada, and the first gold royalty investment of its kind. In exchange for capital costs of exploration and construction, Franco-Nevada would receive royalty payments from Goldstrike when production started, ranging between 2–6% of profits. Capitalizing on this royalty investment model, Franco-Nevada maintains status as one of the richest gold companies in the world, despite not owning any operating mines or exploration projects. After production began at the mine in 1987, the property was purchased by Barrick, who continues royalty payments to Franco-Nevada.

Parallel to the Carlin Gold Trend on which Goldstrike sits, lies the Eureka Trend, a mere 50km away on the other side of Interstate 80, where the Barrick-owned Cortez gold mine operates. Like Goldstrike, Cortez employs both open pit and underground mining, containing processing facilities. Cortez has two open pits, Pipeline and Cortez Hills, and an underground project, Cortez Hills Underground.

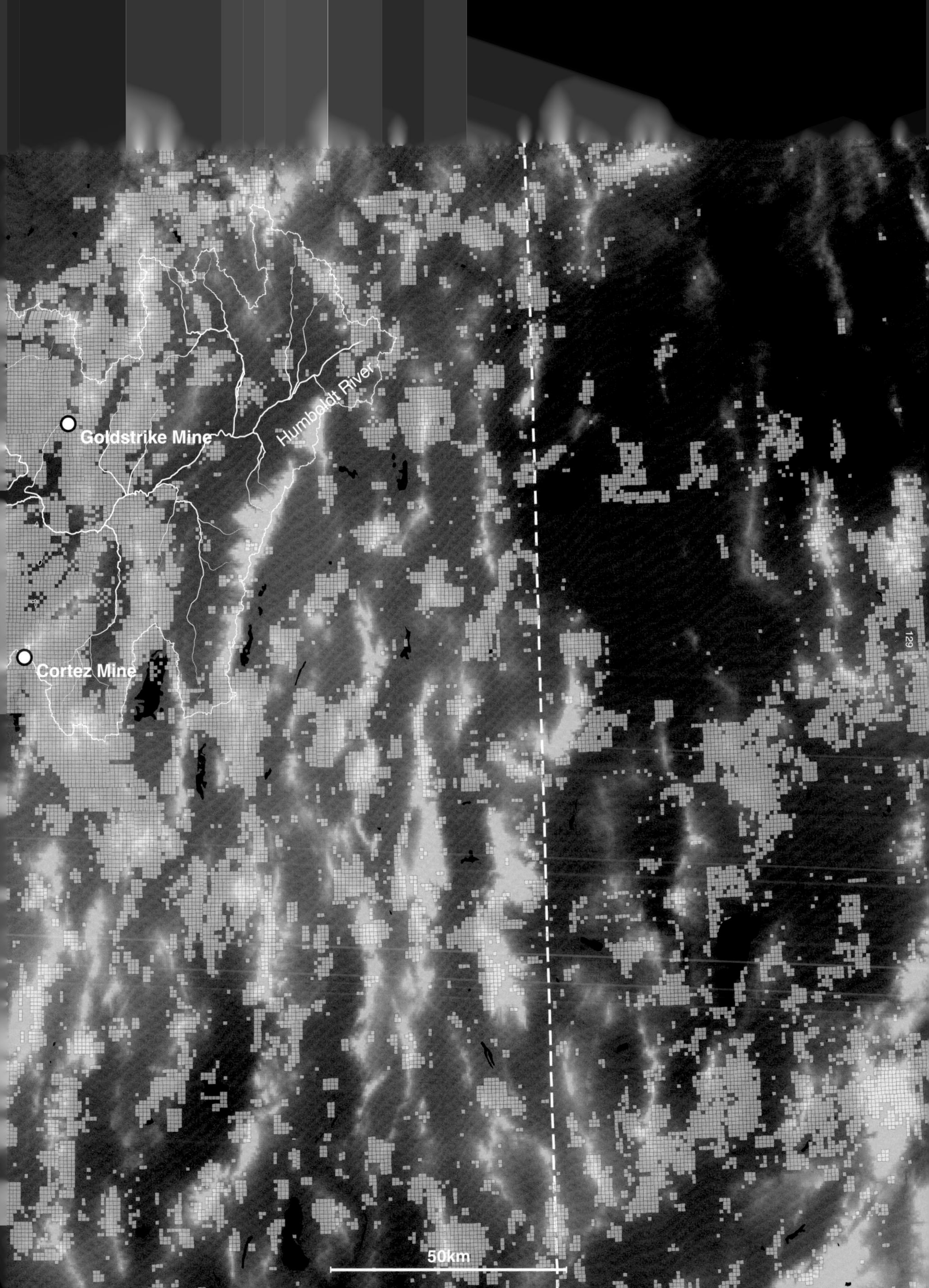
Goldstrike Mine
Humboldt River
Cortez Mine
50km

Goldstrike Mine Site

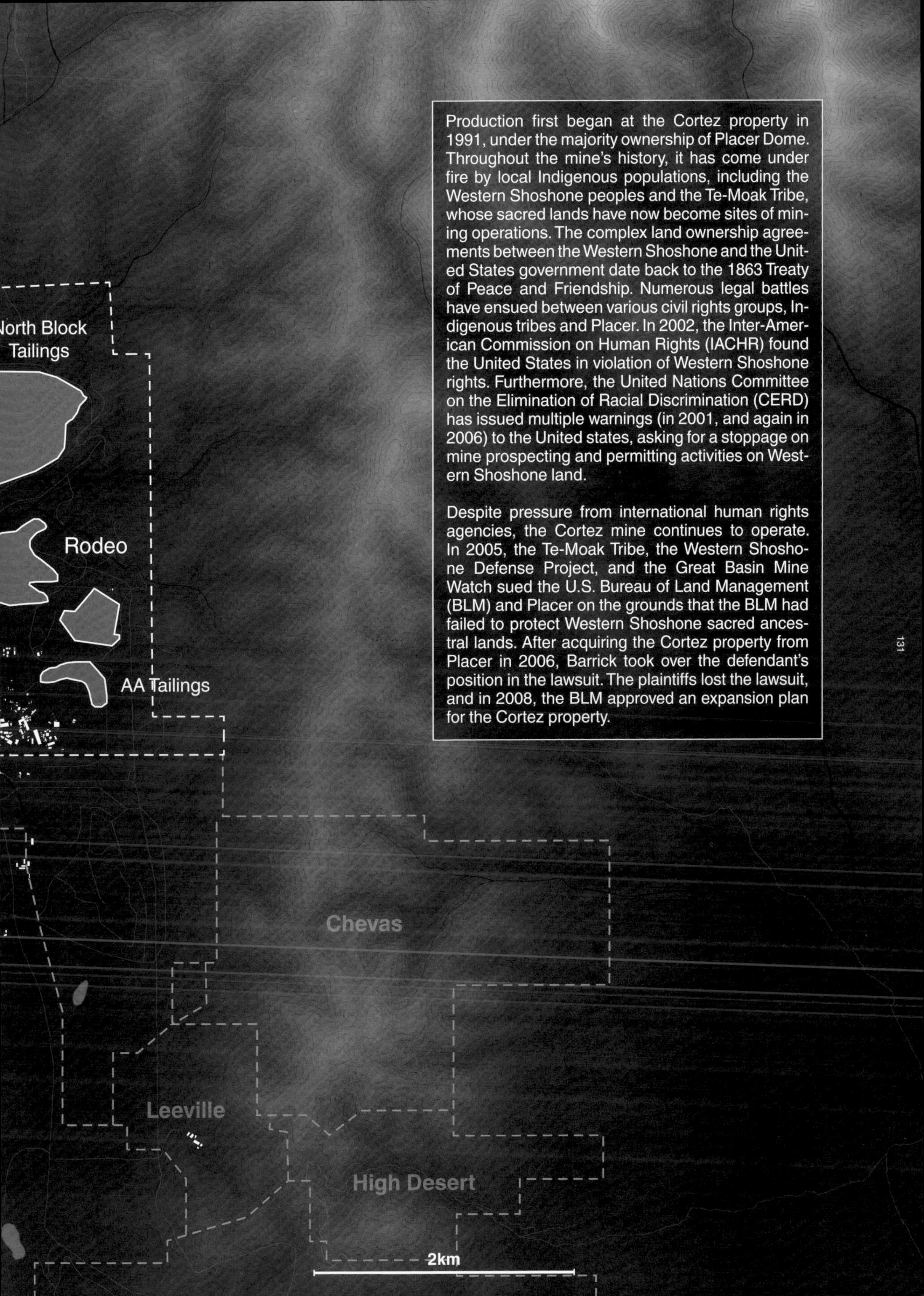

Production first began at the Cortez property in 1991, under the majority ownership of Placer Dome. Throughout the mine's history, it has come under fire by local Indigenous populations, including the Western Shoshone peoples and the Te-Moak Tribe, whose sacred lands have now become sites of mining operations. The complex land ownership agreements between the Western Shoshone and the United States government date back to the 1863 Treaty of Peace and Friendship. Numerous legal battles have ensued between various civil rights groups, Indigenous tribes and Placer. In 2002, the Inter-American Commission on Human Rights (IACHR) found the United States in violation of Western Shoshone rights. Furthermore, the United Nations Committee on the Elimination of Racial Discrimination (CERD) has issued multiple warnings (in 2001, and again in 2006) to the United states, asking for a stoppage on mine prospecting and permitting activities on Western Shoshone land.

Despite pressure from international human rights agencies, the Cortez mine continues to operate. In 2005, the Te-Moak Tribe, the Western Shoshone Defense Project, and the Great Basin Mine Watch sued the U.S. Bureau of Land Management (BLM) and Placer on the grounds that the BLM had failed to protect Western Shoshone sacred ancestral lands. After acquiring the Cortez property from Placer in 2006, Barrick took over the defendant's position in the lawsuit. The plaintiffs lost the lawsuit, and in 2008, the BLM approved an expansion plan for the Cortez property.

The New Africa

A Conversation with David Chancellor

David Chancellor is a documentary photographer based in South Africa, whose trajectory has increasingly focused on the commodification of wildlife. His work includes the photography series *With Butterflies and Warriors* (2014), *Intruders* (2011), and *Hunters* (2010).

X: Instead of operating at a global scale, from a map, or high above the ground, like other mine site photographers, your work examines spatial and environmental interactions at human scales. Can you describe your process or your motivations?

DC: My work looks very much at the roots of the commodification of wildlife, especially in game reserves. Geographically, that probably spans from Africa and then across to Asia. The same commodification of the environment happens with mineral reserves, in a similar way.

"From an Indigenous perspective, the genre of landscape painting [and of landscape photography] is one of the conceptual and visceral tools of colonization."

—Jolene Rickard, Decolonizing the Arts of Dispossession, 2016

X: Can you describe your series "Intruders"?

DC: The Tanzanian Mine series, "Intruders," came directly as a commission from a Canadian newspaper/magazine, so it was not something that I particularly wanted to look at. I think it is interesting that people conceive the various directions that the work goes in and then possibly attach their particular ideas and beliefs to that direction.

A commission might show the demand to commodify the exploitation of wildlife, through the poaching of ivory and rhino horn. This demand comes from Asia, in that Asian corporations extract mineral reserves from Africa, then possibly bring back Asian workers who work in Africa to manufacture roads, which allows the transport system to work faster and then those mineral reserves to leave the country faster. So it's very much a downhill spiral that I'm on, and I try not to think of it too much as it becomes very depressing.

X: Is there a significance to establishing a relationship with the people that you photograph?

DC: Gaining trust for what I do, as a photographer, is really important. You must have some buy-in and some relationship with the people you're photographing. Otherwise you are just duplicating the subject matter that you're looking at.

That duplication is what I am trying to look at in my photography. If I were to go into those environments and do the same, it would be very easy for someone to turn around and say, "what are you doing here? Maybe you're taking our image but you're also exploiting us to a degree." So I feel it's important to buy in and to relate. And that manifests itself in very simple ways. Wanting to sleep on the ground with people. Wanting to be with people in the environment that they're in. That's really simple and very easy to do and it can be as simple as sitting on the floor when you're talking to somebody rather than standing there over the top of them. Not wearing sunglasses. Eye contact.

X: How did these personal relationships grant you access to the borders of the North Mara mine?

DC: That's a good question. In North Mara, it was very difficult. I was with a journalist, and I very rarely work with journalists. I find that if you aren't careful, there will be an agenda, and you'll have to travel down a particular route. Even if there isn't an agenda, there's another person

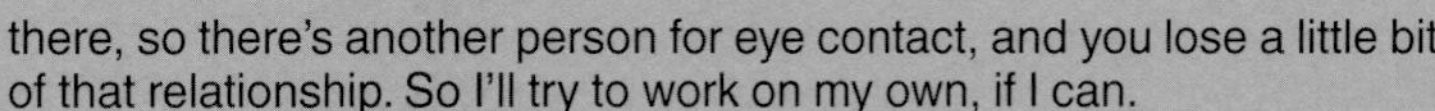
there, so there's another person for eye contact, and you lose a little bit of that relationship. So I'll try to work on my own, if I can.

Intruders, 2011 (David Chancellor)

X: When you're working with a journalist from a foreign newspaper, as you did in this case, there are probably moments of volatility because of your different ways of working. How did you see things differently?

DC: I spent a lot more time in the surrounding villages than he did. He was keen on looking at the clashes, the intruders who had been shot, and the mine workers who had been pelted by stones; but I spent time in the villages and looked at the pile of rocks *from* the villages rather than the other way around.

I noticed sympathy for a group of individuals who had an open-cast mine put in their backyard, then told "no, you can't touch it because it's actually owned by a Canadian company." I also looked at the hospitals where these guys were treated. This is a big company, so there's a lot of money for the workers. We spent two or three nights in the camp with the workers, and they were having nice dinners. You go into the hospitals and there was no electricity, no water, and people in beds who have been shot and injured. I documented those bloodstained beds, which are very graphic.

This particular mine has been rebranded and renamed several times. The continued purchase by new companies meant that the villages suffered, and different constraints were put on the village. Some companies would decide to fence the mines, some wouldn't. Some would let the workers come and go through the rubble, and some wouldn't. The only consistency is that these people live and stay there.

X: Your portrayal of the process of commodification makes it seem like a new form of colonization.

DC: Yes. I think that's exactly right. The ability to commodify resources in a rural, third world environment is shocking. For example, it's the ability to produce a herd of an animal where it's not naturally occurring. I want to document that, and question how it got there.

In Africa, people want to go and see the Big Five animal species in an environment where they think it should exist naturally. Let's use South Africa for example: an American would come over, and they would like to drink wine on a Tuesday, go to Table Mountain on Wednesday, and see the Big Five on a Thursday. So a lot of effort goes into producing an artificial environment that foreign visitors are not aware is artificial, to make them feel that they are now in Africa.

You've got to use all of those technologies that come from the First World in order to manage what should naturally occur in the Third World. So you need an ecologist. You need a breeder. You need someone to balance predator/prey ratios. And you're doing this in a fenced environment that can only survive if you manage it, because outside of it you've got man going about his daily business.

What I'm doing here is showing the equivalent of Mickey Mouse sitting backstage in Disneyland, leaning on his head, giving a cigarette to a four-year-old child.

X: In Texas, with your series "Hunters," you took some photographs of Big Five hunters and their rooms filled with animals, which illustrates the First World commodification of the Third World environment.

DC: Did you look at "Diamonds" as well?

X: Did that follow directly after your "Intruders" series?

DC: No, but it was possibly a result of it. Here you've got a very high-end jeweler who travels from the US by Learjet to come and choose which particular diamonds he wants from the diamond mines themselves. The most interesting picture out of that series for me is the workers cleaning and polishing the diamonds.

That's addressing a similar type of thing to what you're talking about. It's cause and effect. I did those hunters in Texas because I really like what you're hinting at, that there was no way to attribute what these guys were doing by photographing them next to a single animal. Particularly when you see a guy that has 142 animals, it starts to come full circle.

Diamonds, 2011 (David Chancellor)

X: It's ironic that in the context of the global conservation movement, the demand for other resources or the Big Five is growing.

DC: I think it's pointing to the fact that we are fooling ourselves in a lot of cases in believing that the conservation efforts are actually working in many parts of the world. Conservation is a very easy and acceptable cause to attach yourself to; particularly from a large corporate perspective or celebrity perspective. It's soft and cuddly, like the African animals I'm sure we've all had as cuddly toys at some point.

Therefore, we seem to want to continually believe that we are doing good by our actions. A lot of the work that I'm trying to do is to show that we really need to question that. We can't have knee-jerk reactions just because we feel it's politically correct to say that we shouldn't be shooting lions. We can't say that when we aren't considering the people that are standing shoulder to shoulder with those lions in a rural environment. The Cecil the Lion story was fascinating for me, because of the knee-jerk reactions of "I didn't realize people were still hunting lions, we need to stop this." There is no consideration of the income that is generated and could potentially go to rural communities to allow them to live with the lions. And why are we protecting those lions? We are protecting those lions so people can pay to go and see them.

X: Another aspect that we're really interested in your work, is the process itself. You've been talking about stopping, looking, and understanding the work while working with communities. One of your quotes that we came across was: "The instrument isn't the camera, it's the photographer."

DC: I've wanted to get it to the state where how you do it is instinctive. I use one particular piece of equipment, one particular type of film, and everything I shoot is generally hand metered. I work with a fixed lens, there are no zoom lenses, so I have to walk, and if I walk I have to have a dialogue with the person that I'm with. Nothing is shot from a long way away. And once it's all working, I never have to think about it. So, therefore, it does become a journey to understand what I don't understand.

X: Your view from inside, up close through your instrument without a zoom lens, has a fixed position that has to move with your body. Your work seems to be a critique of the whole idea of geography itself. You capture geographies beyond the state perspective, beyond the national box.

DC: I think that's extremely astute. For something like National Geographic, it's interesting that they use the word National in front of Geographic. I get that side of what you're saying, but what also rubs me the wrong way is the celebration or glorification of wildlife that we know, healthy in open spaces. People say, "yes, that's fine, photograph it, but make sure we don't see any fences." If I'm not going to see any fences, it's because there aren't any fences.

Hunters, 2010 (David Chancellor)

X: That is really revealing. I would assume you have never published in National Geographic, purposefully for this reason?

DC: No, I haven't. But, I spoke at their January seminar in Washington, D. C. [Laughs] ▣

This conversation took place between David Chancellor, Evan Farley, James Murray, and Pierre Bélanger in 2015. Since this interview, Chancellor published a series of photographs of the world's largest rhino farm in South Africa in a 2016 issue of *National Geographic*.

Barrick in Tanzania

The Story of North Mara

Situated in the northeast Tanzanian district of Tarime in the Mara River watershed, North Mara Gold Mine lies less than 100km upstream from Lake Victoria, the largest freshwater lake on the African continent, straddling three countries: Uganda, Kenya, and Tanzania. North Mara is a high-grade open pit and underground mine owned and operated by leading gold producer, Acacia Mining, a Tanzanian-based subsidiary of Toronto-based Barrick Gold Corporation.

Opened in 2002 by Vancouver-based Placer Dome, the mine was acquired in 2006 by African Barrick Gold (Barrick subsidiary and predecessor to Acacia Mining). Despite a strategic decision to rebrand in 2014, Acacia's majority shareholder (64%) is Barrick Gold Corporation. Reporting $94 million USD in after-tax profit in 2014, Acacia Mining continues to operate three gold mines in Tanzania, with active explorations in Kenya, Burkina Faso, and Mali.

The North Mara Gold Mine consists of a processing facility, two open pits, and an underground project located on two distinct sites—the Nyabigena open pit and Gokona underground project to the north, and the Nyabirama open pit and processing facilities to the south. According to the National Bureau of Statistics of Tanzania, a 5km haul road that connects the operations cuts through at least seven surrounding villages, inhabited by over 60,000 people. The region's population traditionally subsist on a combination of subsistence agriculture and small-scale artisanal mining; both activities are threatened by the mine's requirement for exclusive land and mineral rights and the subsequent indiscriminate acquisition, followed by the securitization of both.

Characterized by regular and violent clashes between local communities and mine security, the North Mara Gold Mine has consistently attracted international attention surrounding serious allegations of excessive use of force resulting in debilitating injury and death, gross human rights abuse that include sexual assault, as well as misconduct on the part of African Barrick Gold, and now Acacia, in investigating and redressing these injustices. Large groups of villagers frequently trespass across the relatively porous mine perimeter to engage in illegal mining by scavenging low-grade ore from waste tailings.

According to an October 2015 report produced by the UK-based Corporate Responsibility Coalition (CORE), from 2008 to 2014, 16 deaths and 11 serious injuries in 14 separate incidents at or near the mine sit and 14 charges of sexual assault involving mine security personnel were confirmed. Since that evaluation period concluded, 20 new cases of death or injury have been reported.

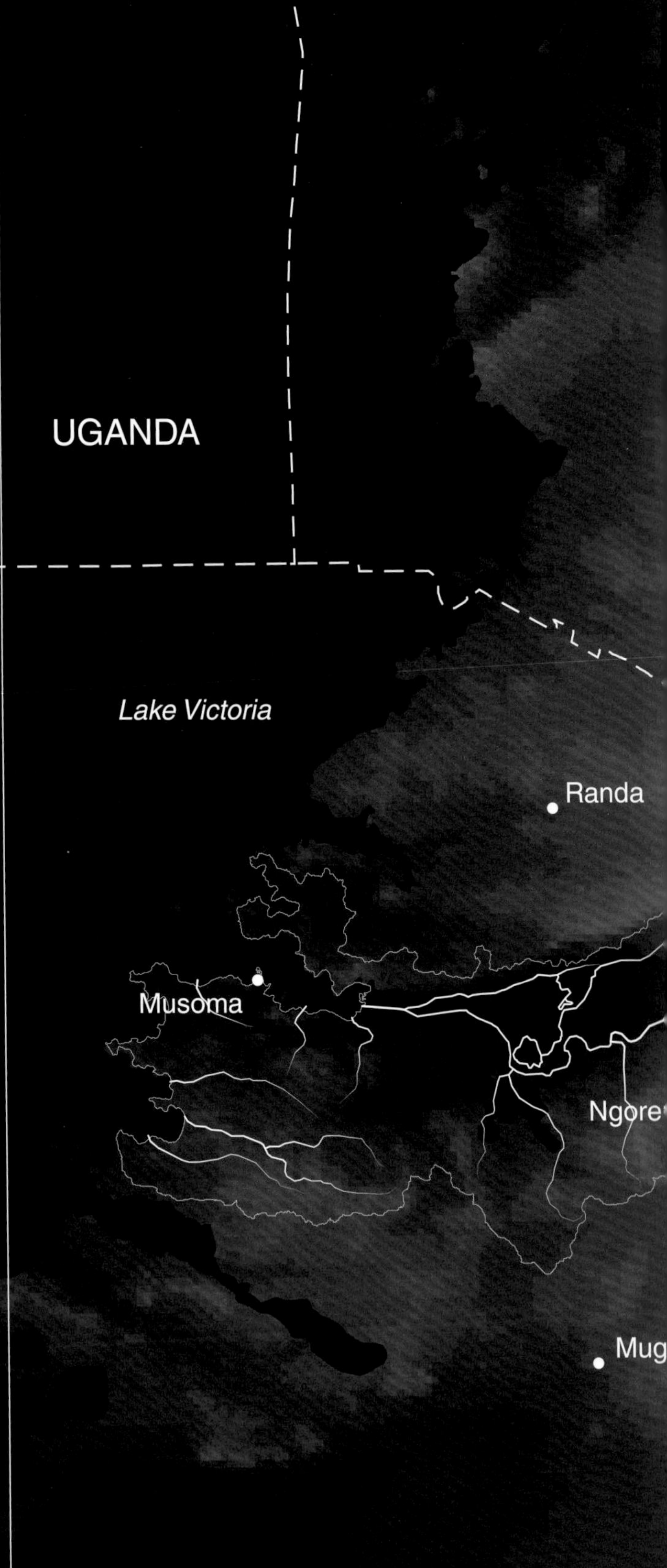

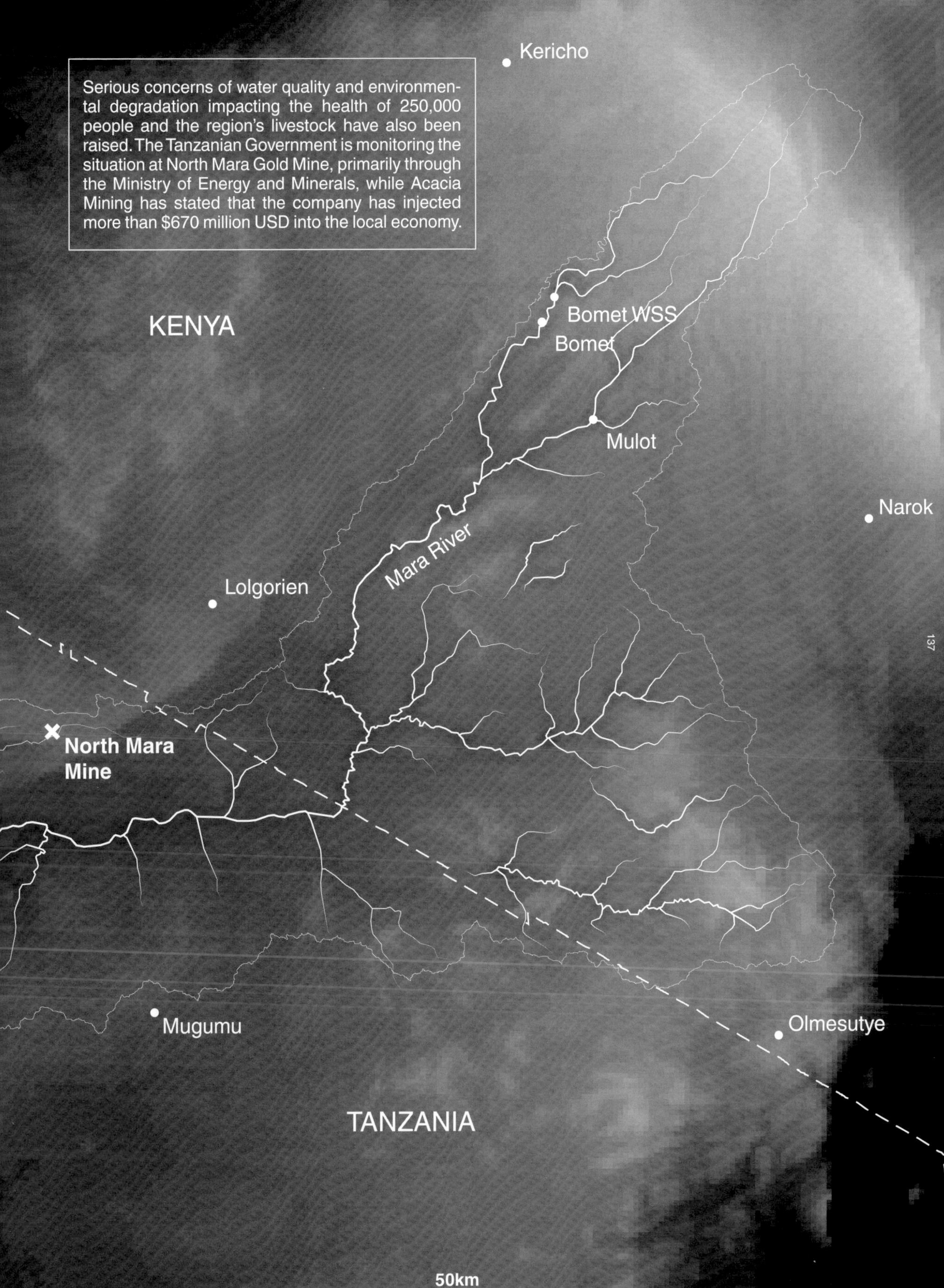
Serious concerns of water quality and environmental degradation impacting the health of 250,000 people and the region's livestock have also been raised. The Tanzanian Government is monitoring the situation at North Mara Gold Mine, primarily through the Ministry of Energy and Minerals, while Acacia Mining has stated that the company has injected more than $670 million USD into the local economy.
Kericho
KENYA
Bomet WSS
Bomet
Mulot
Narok
Mara River
Lolgorien
North Mara Mine
Mugumu
Olmesutye
TANZANIA
50km

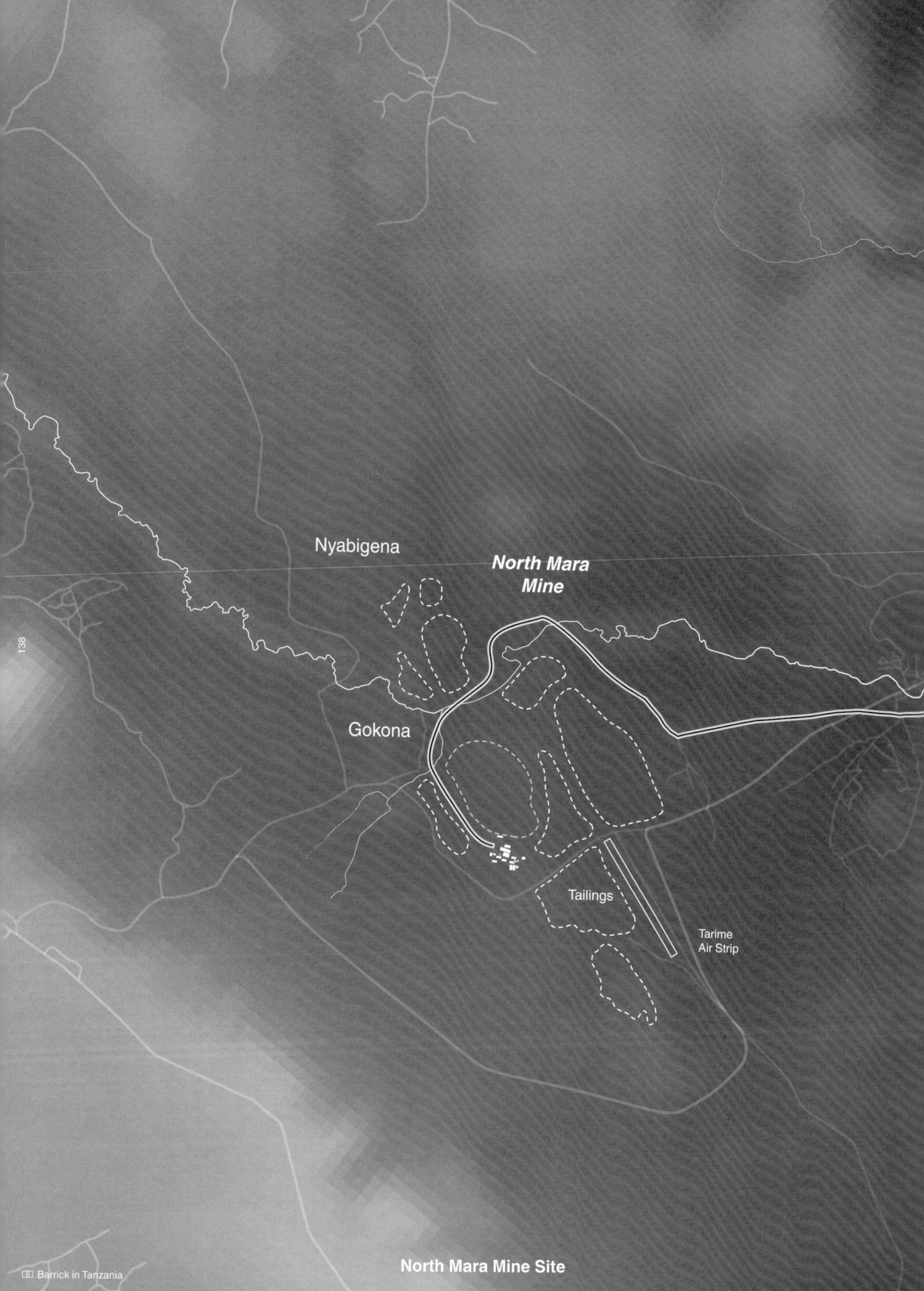

North Mara Mine Site

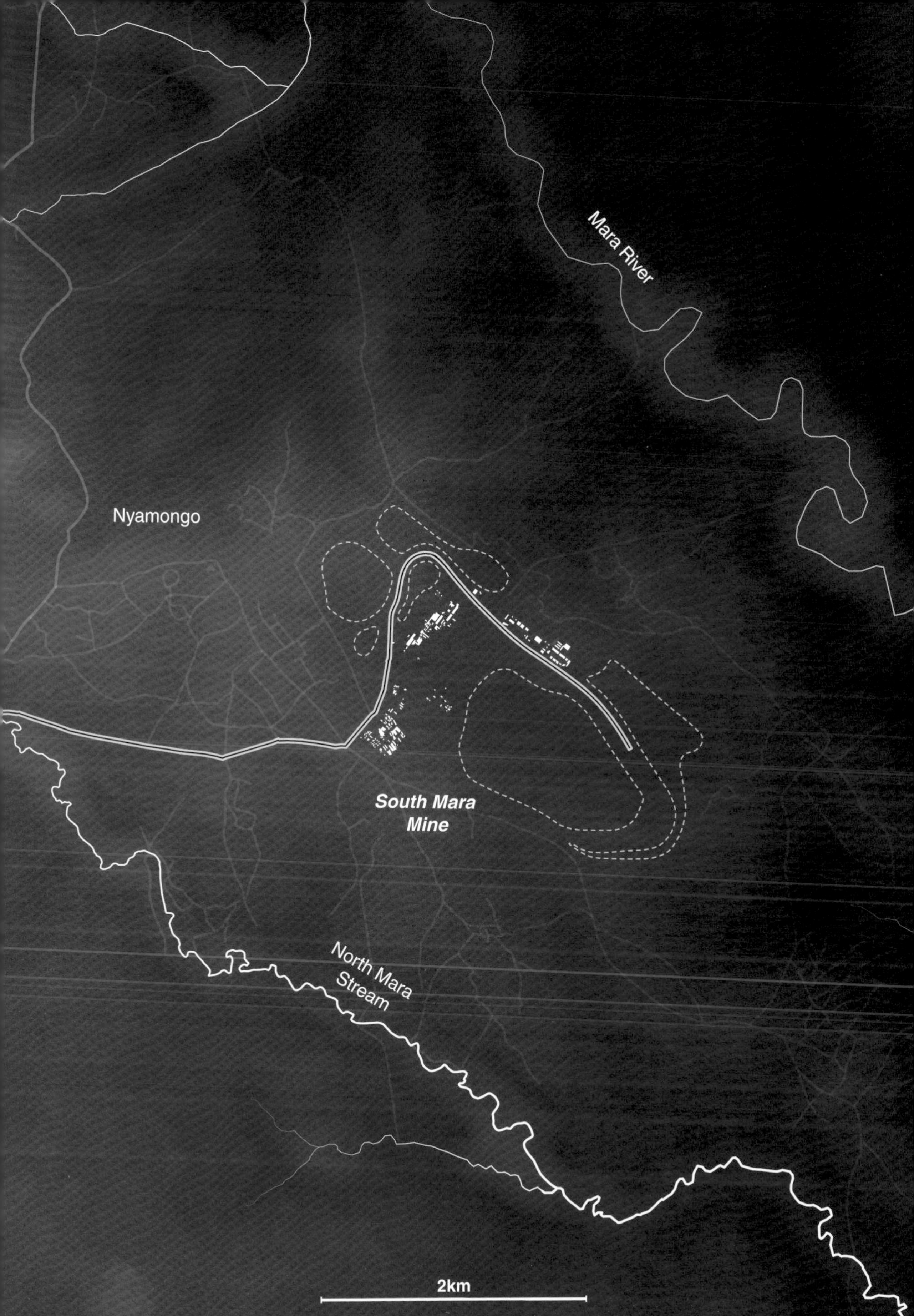
Mara River
Nyamongo
South Mara
Mine
North Mara
Stream
2km

State of Insecurity

A Conversation with James Hopkinson

James Hopkinson is a complex environment risk management advisor at Assaye Risk with experience in Africa, the Middle East, and South Asia. Following a career in the British military and the Ministry of Defense's Operations Directorate, he joined Blue Hackle, an international risk management provider, later becoming its Chief Operating Officer.

X: Can you describe your background and experience in the risk management sector?

JH: I spent 24 years in the Infantry in the British Army. I worked for a private security company called Blue Hackle, which ties back very directly to a Scottish Regiment. It provided euphemistically what I would term guns, guards, and gates in difficult places. Assaye Risk is broader, we don't term it a private security company. We term it a non-technical risk management company. So while security is a specialization at its core, what it seeks to do is take account and to integrate across the other elements of non-technical risk or surface risk as it manifests itself in emerging markets and high-risk environments. So that might be anything from dealing with regulatory, social, political, or indeed economic elements. Those elements all come to overlap, so Assaye Risk is really about taking a wider notion when it comes to security.

"The colonial world is a world cut in two. The dividing line, the frontiers are shown by barracks and police stations. In the colonies it is the policeman and the soldier who are the official, instituted go-betweens, the spokesmen of the settler and his rule of oppression."

—Frantz Fanon, The Wretched of the Earth, 1961

X: Turning to North Mara Mine in Tanzania, your company has oversight of risk management for the mine site?

JH: Yes, we support Acacia. We started the journey with an operation review of what they were doing at North Mara. We started working with them probably, it would've been September 2013. Subsequently, we've produced a security strategy for them, for their operations in Africa, and we have done quite a lot of operational design up at North Mara for them to make sure that it's better integrated and operates more effectively than it did previously.

Unlicensed miners searching gold mine tailings on periphery of North Mara Mine, 2011 (Trevor Snapp / Bloomberg/Getty)

X: Coming in 2013, you were well aware of the ongoing controversies surrounding the mine site regarding human rights abuses.

JH: Which is why we were brought in to have a look at it.

X: But since then, the allegations haven't really gone away. There are still frequent accusations and allegations of excessive use of force and a failure in the grievance and compliance system. Do you think those are warranted? From your perspective, what is the condition on the ground?

JH: There was a substantial backlog of those grievances at the site, which are in the process of being cleared up now. If you look at any of the episodes that have taken place more recently, you will find that none of that is down to Acacia's internal private security force—whether it's the contracted-out G4S element or whether it's the internal Acacia piece that's managed by ourselves.* Where there have been fatalities or injuries that take place on the site, those have tended to be through accidents, or directed by the Tanzanian Police Force. A good key performance indicator is that if we look at expenditure of ammunition on the site by Acacia's Security Forces, we are using 98% less ammunition than was used before we got there. It's much less about direct confrontation now.

*G4S is a British-based, global integrated security company.

X: Are you familiar with the border conditions of the mine site?

JH: That area is marked, it's demarcated with posts, large rocks, and so on. One of the challenges is pretty much anything you put there to demark it gets taken down, so if you were to put a fence there, that fence would disappear overnight. Then within that, you have a number of active mine sites. Gokona and Rama are clearly separated by a fairly substantial haul road, which goes over a public highway as well. So it's incredibly difficult to prevent trespassing on the mine lease area.

Border patrol & security guard on periphery of North Mara Gold Mine, Tanzania, 2012 (David Chancellor, Untitled # IV, *Intruders* Series / Kiosk)

What you're able then to do is build up a matrix or zonal security plan based on where you're prepared to admit penetration and where you're not prepared to admit penetration. Then what you do is, effectively, you put your troops to task set against your zonal security plan. Now one of the places where we try to prevent penetration is where they're actively mining.

Quite a lot of these intruding individuals, given the security plan that's in place, are what we would term 'young fighting males.' What we seek to do is keep them away from that. What we've been very successful at doing—not by directly confronting them but by making it much more difficult for them to penetrate those areas—is to lessen the down-time in production.

X: So since you can't stop them, as you said, and since there are corridors of acceptable intrusion, is one of the strategies to allow intruders to come to point as long as it doesn't interfere with those direct operations of the high-end ore?

JH: Yes, it's about managing people away from the operational areas. Having come up with a zonal plan, you then seek to manage them away with a lot less focus on direct confrontation. Previously the model was much more about direct confrontation, and that then results in greater fatalities and injuries. It's a vicious circle. What you're trying to do is manage them away, but you've got a lot of drivers there, whether it's land speculation, whether it's illegal milling of gold, there's a number of pieces that drive this, and I'm afraid, I can't see it going away. It's going to be a continuing requirement, and they're bright, so there is no difference between dealing with the Taliban or ISIS. Tactics evolve, and every time you put something in place, they evolve their tactics. So you're constantly trying to keep ahead of them or catch up, and close off that particular loophole that they've decided to work through.

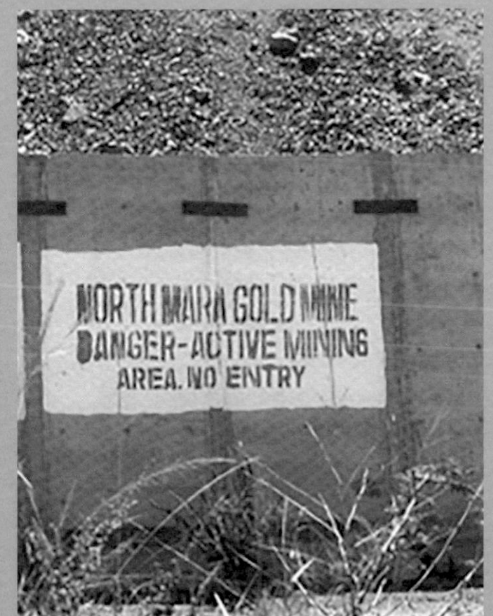

Border wall, North Mara Mine, 2014 (VICE News)

X: Given that the operational security is driving most of the issues, who is hiring the G4S and the Tanzanian Police? Do they directly sign a contract with Acacia, or is that managed through you?

JH: No, that's done directly by them. So what you've got with the Tanzanian Police Force is an MOU, or provision of policing.

X: And it's done directly with Acacia because Acacia has the rights to the leased area?

JH: Correct. So what we're effectively doing is supporting them with management delivery support. We've put in the management part in order to manage the overarching security plan that we've put in place. The security plan has a number of component parts: you've got the Police, then you've got the T-bar wall, then you've got elements of that T-bar wall that are electrified and connected to towers that have thermal imaging cameras. So when the fence is triggered, we'll zoom the camera in, and the cameras have exceptionally good coverage. That feed then goes into an integrated operations room, where a response is managed.

Because of limits on capacity, response is often a combination of Acacia security personnel and Tanzanian Police Force, depending on the numbers of trespassers and illegal miners seeking to gain access.

Each of those component parts are getting better, and a great deal of work has gone into enforcing compliance of voluntary principles on security and human rights. The reality is that much more practical scenario-based training goes on, so that those boys on the ground with a weapon are confident in the weapon that they're holding, and they're also confident in the rules of engagement that they can apply with that weapon. Given that Acacia doesn't hold any lethal use of force measures, they primarily use CS gas and beanbag rounds.

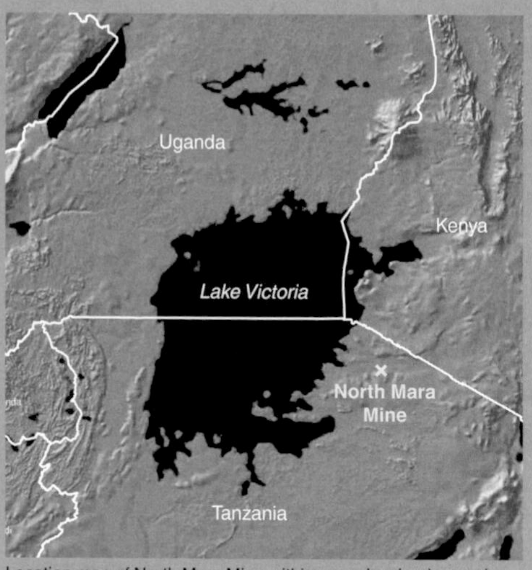

Location map of North Mara Mine within cross-border river region between Tanzania and Kenya

X: So if the Tanzanian Police Force is the only force capable of lethal force, are you able to support their training in these standard western protocols?

JH: When it comes to standard principles on security and human rights, you have to be quite careful about how that is done and how that is achieved. So all we can seek to do is to coordinate with the Tanzanian Police Force. One of the initiatives that has been taken forward is to introduce a liaison officer into the Acacia Integrated Security Control. And the premise behind that is that the response will be better, the joint work with the police will be better, and the ability to alter or influence the police's behavior should be better.

X: I don't understand why G4S is not integrated into one umbrella that is the holistic, comprehensive asset-management, risk-management plan?

JH: Acacia was unable or unwilling, at the management level, to put in place what was required in order to change. It was really an operational effectiveness issue. Previously, it was all Acacia. But their inability to deliver on what they were being asked to do, led to a conversation where we said, "we will provide the management support to deliver it for you." So what you have got then is a coordination of the Acacia elements. So G4S provides effectively, static man guarding, and then you have a mix of Acacia and Assaye Risk, and Assaye Risk will only do the management support. I don't do guns, guards, and gates. But Acacia will have a mobile response force (the MRF) trained to deal with intrusions of the mine site, either on its own or in conjunction with the Tanzanian Police when it requires additional support.

German East Africa Colony, 1906 (Deutsches Historisches Museum)

X: The metrics that we've already discussed seem to be very successful, but they're also very narrow when allegations continue for the excess use of force abuses, and I wonder, is there some sort of adaptive plan?

JH: I'm not seeing that same clarion call and actually when you look back at the fatalities that have occurred this year, none of those fatalities have been down to the actions of the Acacia security efforts. They've either been accidents or they've been direct action by the Tanzanian Police. That for us is a piece we need to be in position to affect, and we've been able to do that. Changing and affecting the behavior of the Tanzanian Police, that is very much a work in progress, and one needs to be particularly careful about how one approaches this issue, given the Voluntary Principles of Security and Human Rights. What you can do to support the police, and what you cannot do. It then ties back to training programs provided according to the capacity-building protocols in the US and UK, in areas such as public order.

There are programs, for example, of rule of law in Tanzania, that will take time. What we're trying to do is change the behaviors, for example, by bringing a liaison officer inside the operations room, so that he will see misbehavior on the screen and will be able to deal with it there; and then by use of direct communications with the police network. As we've found over the last fourteen years, training Indigenous forces and changing their behaviors to make them effective and to operate against best practice—western best practice, UN conventions on human rights, voluntary principles on security and human rights—is a long process that takes an awful lot of time and effort.

X: Do you think there are any success stories? I'm not so sure if you can point to a success of a western intervention to change that culture.

Sir Arthur Wellesley, 1st Duke of Wellington, 1815 (Apsley House, The Wellington Museum, London)

JH: [laughter] I would have said, yes we have. Has it been as effective or successful as we would wish? Well, clearly not; because we've seen what ISIS can do—a relatively small number of people, not particularly well-armed, or weren't at that time, able to take Mosul and other parts of Northern Iraq as though there was no one there at all. And I think that comes back to mentoring. And it's a mix. I'm not going to tell you… it's not an easy environment to manage. What's interesting is, with the measures that have been put in place, you're seeing a lot less women and children on the site. Because it's not accessible, it's difficult for them.

X: On the mine site or in the region?

JH: On the mine site, it's very difficult for them. They've managed onto low-grade ore piles where it's not an issue. The mass intrusion of able-bodied young men conversely almost creates a slightly larger escalation in terms of the level of violence.

Only men are present. And the way it's carved up, they've been channeled through particular intrusion routes where they can be managed. Each of the villages breaks down, so the Kikuyu are a warlike tribe; they are not pastoralists, they are not classic farmers. They either have cattle or they fight, so quite a lot of the backbone of the Army in Tanzania tends to come from the Kuria. And it's the Kuria that are the tribes in North Mara, which is an opposition stronghold. So what you're seeing is not only violence against Acacia and the Tanzanian Police, but you're seeing violence amongst the villages, which is blood feuds. So there's quite a lot of trespasses and illegal miner violence, which is then left on the mine site for Acacia to investigate.

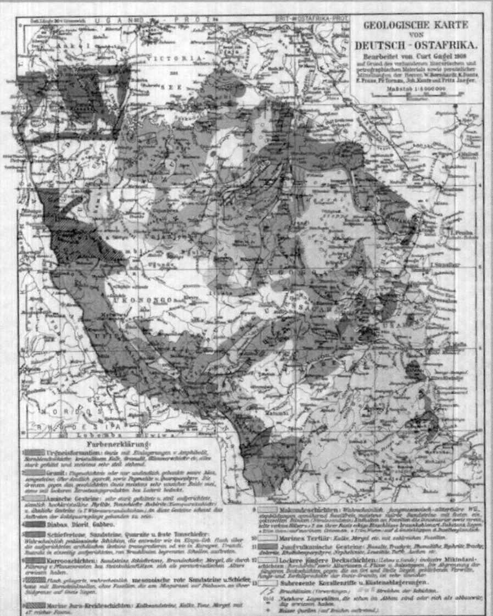

Geological Map of German East Africa, 1908 (Universität Greifswald)

X: Can you explain Assaye Risk and how it was named?

JH: Yes. If you look back to the Battle of Assaye, there were three British Regiments involved in that battle—two were Scottish. And I subsequently joined one of those regiments as a Highlander. That's where it comes from. And that particular battle was the battle of which the Duke of Wellington was most proud in later life, so he was asked: "what was the battle that you were most proud of winning?" And he said: "it was the Battle of Assaye."

X: My take was that while he was very proud of it, it was also incredibly traumatic for the Duke of Wellington because it was so bloody and he lost so many men.

JH: Yes, but it ended up basically toppling Tipu, and putting to bed with a fairly substantial force including quite a lot of French mercenaries. It was a notable victory for the East India Company very early on.

X: Someone in my class interviewed Michael Ignatieff, the previous Liberal Party Leader of Canada, and his concern was if mining can be considered a foreign policy; in that perspective, North Mara Mine and other controversial sites like Porgera in Papua New Guinea, can become national security risks. Do you agree with that?

JH: No. Something like Barrick Gold is a multinational company and I would say that's more a reflection on the Canadians and their political viewpoint, that they believe they're responsible for the actions of anything that's listed in their country. I think the problem with that approach is, what are they doing to assist? If you look at what Canada does and what is has done, for example, in Tanzania, it's not a great deal. You can get yourself into a real pickle because it's a multinational company, and how are you going to alter behaviors? Now none of these companies want to find themselves in risky situations, primarily because they have to respond to their shareholders. It is a risk/reward kind of relationship. You operate in emerging markets, and they are difficult, complex environments in which to operate. To try and apply the same norms as you would do operating in Canada, is plainly ludicrous to my mind.

Members of the Mwita family from Nyamongo, next to Barrick's North Mara Gold Mine, scavenging through waste rock from the mine's tailings, 2011 (Trevor Snapp / Bloomberg / Getty)

X: Well it seems to be an attempt to transfer sovereignty from the national structure to the multinational corporation.

JH: Yeah, and it doesn't transfer well, you know, it's patent hogwash. Don't quote that [laughs]. ❢

This conversation took place between James Hopkinson and Ashley C. Thompson in 2015.

Rusinga
Florence
Mfwanganu
Homa
Swusi 2230
Ugaja
Kossowa
Mau
Sanse
Sotik
Naiwuscha S.
Naiwuscha Stat.
Longonot
Suswa
Nairobi
Mohuru-Sp.
Schirati
Mori B.
Mara B.
Mori
Isuria Ebene
Lorogoti-
Burrung-Ebene
Sossian
Guasso
Dogilani
Kedong
Kapotei
Kapotei Ebene
Magadi
Legisartat-B.
Ndassekera
Schombole
Luonji 2027
Ikoma
Speke-Golf
Lamadi
Mwansa
Kiruwassile
Ssonjo
Ssale
Natron-S.
Selei 2930
Matambatu
Longido 2627
Kitumbeine 2600
Oldonjo l'Engai (tät. Vulkan)
Ngorongoro-Krater
Nera
Moamara
Simiju
Nduddamo
Igarja
Ikuru
Uduhe
Ssegerei
Mangwinu
Mantinne
Manjonga
Kussule
Igunda
Essimingori 2100
Mondul 2700
Meru-B. 4630
Aruscha
Hohenlohe-S.
Kissale
Ssogon
Sulatu
Ussongo
Njawa
Mkalama
Umbugwe
Balangda
Sambu
Massai
Neiborruri
Tambarale
Ussure
Hochplateau
Ndigira
Steppe
Ndala
Iwumba
Mangiira
Irangi
Kondoa Irangi
Kwa Fumbi
Baliba
Rubugwa
Tura
Djiwe la Mponda
Ssuna
Mtoro
Kilimba
Nguru
Gombe
Ikonge
Kilimatinde
Kirurumo
Mwumba

Un-Natural Resources

A Conversation with Moura Quayle

Moura Quayle is a landscape architect, educator, and institutional leader in British Columbia; currently serving as Chair of the Board for the Canadian International Resources and Development Institute (CIRDI) and Genome Canada. She is Director of the Liu Institute for Global Issues at the University of British Columbia.

X: Your career is very unique, and your achievements, unprecedented. I'd love to speak about certain moments in your career as a landscape architect, where there are very clear moves and developments that allow you to speak about the influence of landscape architecture as a territorial practice.

For instance, in 1989, you wrote *Portrait of a Profession* with colleagues Neil Guppy and Luc Roberge, where you sought to provide a systematic representation or portrait of the profession of landscape architecture "as practiced by a wide variety of individuals," in order to create a tangible base (where none has existed beyond anecdotal sources) to begin asking "analytic questions considering the causes and consequences of current professional practices" and to provide "prescriptions for improvements within the profession."[1]

With this in mind, our first question is, in essence, how did you move from the role of a professor in landscape architecture at the University of Guelph to your involvement in resource development, now as head of the Liu Institute for Global Issues?

MQ: I think the reason for my rather constant reinvention, if you will, really goes back to my design education. I've been patiently and not so patiently working on a book which is called *Design Leadership*, published in 2017 by Columbia University Press. It discusses how designers are prepared to shape leadership in a very different way because of how we think, our mindset, our toolkit. It describes my leadership experiences and how they were really shaped by my four years at the University of Guelph and my subsequent graduate work at Berkeley, but maybe even more so, through 15 years of teaching architects and landscape architects.

For me, the move from working as a landscape architect and teaching landscape architects really came when the landscape architecture program here at the University of British Columbia was in, what was then called the Faculty of Agricultural Sciences. It was essentially dying when I became Dean in 1997. I spent eight years in that faculty doing everything from going back to *tabula rasa* with our undergraduate programs and introducing a problem-based learning approach, to teaching science, establishing a number of new research centers; and finally, we changed the name.

Even though a lot of the research says not to do structural change, I thought in this case we needed structural change, because it shakes things up and it gets people actually thinking in a slightly different way. By the time I left to go to government we had become the Faculty of Land and Food Systems. Interestingly enough, one of the last things that I did as Dean was that I moved the landscape architects out of the faculty, which resulted in the creation of the School of Architecture and Landscape Architecture.

So right at the perfect time; and a lot of this is serendipitous, I was just handing my note to the Provost saying it's time for me to move on. I'm a transformer not a sustainer.

Sustaining, I can do it, but it's not as interesting to me. It was then that I got the call from Gordon Campbell asking if I would be interested in being the Deputy Minister of Advanced Education. And again, I was

"Landscape architecture reflects (and refracts) a larger culture in which most 'nationals' wish to distinguish themselves from the migrant and the indigene—wish themselves, that is, to displace both. Indigeneity is scarcely mentioned in the field's seminal texts nor discussed in its conference halls and online forums … As landscape architects strive to 'effect real world change' in a century defined by planetary disruption and mass migration, can there be any investigation more relevant than this?"

—Rod Barnett, "Designing Indian Country," 2016

1 See Moura Quayle, L. Neil Guppy, and Luc Raymond Roberge, *Portrait of a Profession: Landscape Architecture in 1988* (Ottawa, ON: Canadian Society of Landscape Architects, 1989).

very surprised, but the more I thought about it, the more I saw this as a fantastic opportunity for learning, a change of cultures. I was Deputy of Advanced Education for three years and then I worked with Gordon on something that he created called the Pacific Coast Collaborative. We created the Collaborative because so many of our challenges don't know political boundaries: ocean health, plant invasive species don't stop at the 49th parallel; marine debris does not stop at the 49th parallel. The Collaborative tackled things such as making sure that you can plug in your hybrid all the way from California to Alaska. They are still working on many of the challenges and it has been interesting to watch it morph. It is now also working on climate issues after the separate climate group that was in place at the time we started the Collaborative dissolved.

Having finished my government appointment and coming back to UBC I thought to myself: "what am I going to do? Will I go back and tackle architects and landscape architects again?" This was the time when Premier Campbell had his climate epiphany while in Beijing, which translated into a whole bunch of actions at the Ministry. So I started thinking, "who can actually change the world fast enough?" And the answer I arrived at is that it's actually business. It's ok for government to set policy frameworks, academia can help, but it's actually business that needs to tune in and can actually make change quickly. This realization also coincided with a meeting with one of my former Dean colleagues, Dan Muzyka, who wanted to bring design into the Sauder Business School. I spent the next five years creating the d.studio at Sauder. Meanwhile, I was even able to bring in my early professional experience as an owner of a start-up, a landscape architecture firm in Victoria, which I ran from 1975 until 1981.

I moved over to the Liu Institute in the summer of 2014 and have spent the last two and a half years there. We now have our program, the Master of Public Policy and Global Affairs (MPPGA) up and running. There, I've been pushing for changes especially around how we position professional education. We now have three such projects going on with our MPPGA students: one in India on microfinance; one in Indonesia on the problem of them burning the landscape and creating haze; and another one on Indigenous peoples and drinking water in Canada and in Peru, with Global Affairs Canada.

Sidewalk resistance during the 150th Year of Confederation, 2017 (Le Collectif No Borders)

Essentially, I'm running a studio. Everything I do, I'm trying to embed, without them knowing it, a set of design skills. Right now I'm experimenting with an idea of a Policy Studio, because I think if we keep doing policy the way we've been doing it, it's probably not going to change a lot. But if we could think differently about policy design, analysis, or implementation, then I think we have the capacity to make change. Here I see the role for the Liu Institute as a convener to bring together all the folks that need to be part of the policy process in a relatively neutral environment; as neutral as we can get.

X: There are a couple of things that come to mind here. First, an observation I had as you were mentioning how policy on the one end has been practiced, let's say in a particular vacuum, and one could argue an administrative and potentially bureaucratic vacuum, and then another hand where the status quo of business has been singularly focused on economic returns. You seem to insert yourself or slip in-between the divide that separates them. It seems that the grounding of both research on the one end, and policy, can sometimes seem removed or abstract from land, while business very often knows a lot about the ground; but essentially since it's looking to get things out of the ground very quickly. So, I don't think it's by coincidence that your background is based on land, living systems, and the growing knowledge of political systems that you've been gaining.

Building on this, my question lies in relationship to this third studio that you're doing on Indigenous affairs and water systems in Canada and Peru. At what point, or how through your background, do the political Indigenous complexities and advocacies play a part in your work? I'm not sure if the University of Guelph had any kind of course on political Indigenous societies… but ironically, any discussion of 'Native' or 'Indigenous' in landscape architecture is about plants, not people or culture, or law. So where does that understanding and acknowledgement come in? How have you learned it or lived it?

MQ: Some of my sensitivity came from during my years as a Deputy Minister, because we were responsible, along with the Ministry of Education, for trying to improve the learning outcomes and the education potential, getting First Nations children and young adults into a system, whether it's theirs, ours, didn't matter—giving them learning opportunities. Being careful about how we prepare ourselves to work appropriately with First Nations, for me is even more important as we continue to engage with climate change challenges. There are many examples of proposals coming from the University sector that are centered around the interest of the researcher, who then managed to get the First Nations to agree that it could be an interesting topic.

As opposed to doing what a good design researcher would do, which would be to say that it's not about me, it's actually about you. And asking really good questions. Right now I'm teaching, along with one of my PhD students, a new course in the Faculty of Arts called "Introduction to Strategic Design." This course is essentially trying to do just that, to work with these second-year students in a studio format to help them ask better questions; to define problems better.

X: You seem to understand your own self-development, but also your limitations. On the other hand, it also seems to potentially point towards a significant shortcoming in the field to not really politicize the understanding of land or territory. This, as I've grown to understand more and more, is the shortcoming of landscape architecture and related design disciplines: focusing their work in cities or metropolitan regions, and not really understanding matters of scale, especially the political scale, and one that also pre-dates colonial forms of planning.

MQ: I think it's a big gap. It's a huge gap. You're right.

Sidewalk protest against the presence of Gold Corp mining operations on Coast Salish Territory, 2011 (Murray Bush / Flux Photo / Vancouver Media Coop)

X: Could you elaborate a little more on why you think this is a gap and deficit of the design disciplines at large? Are the design disciplines themselves tools of the colonial settler-state that designers themselves are poorly knowledgeable of?

MQ: Like any discipline, but perhaps less than most, we as designers can fall into narrow and parochial thinking. I'm not sure I understand totally your reference to the "colonial settler-state"—however, I do think we don't prepare our design students for a full understanding of the cultures that have formed our societies and therefore our landscapes. Even metropolitan regions have a history of land ownership and stewardship that must be taken into consideration. Bridging the gap can perhaps happen through awareness—in continuing education of the design professions and in a review and 'updating' of our curriculum to build in the political and cultural understanding of the landscapes we work in.

X: You've also explained how you're working through, both pedagogically and academically, this relationship between business and academia, which I think is extremely important. I also agree with you about the lack of understanding about what is applied research. Does this also inform your current work as part of a number of different organizations, but amongst others I think an important and prominent one is the *Canadian International Resource Development Institute* (CIRDI), where you as the sole landscape architect, are the only person capable of understanding—or let's say redrawing—a problem or reformulating a question and comprehend that the process opens up new constituencies and agencies?

Maybe business doesn't like that because it does open up, let's say, a number of different sometimes unknown factors; and CIRDI has its own series of contentious and difficulties including its image since, let's face it, resource extraction is a very contentious and contested territorial practice.

MQ: Absolutely. CIRDI's Strategic Plan now talks about natural resources. The government isn't even using the extractive language anymore, but clearly any kind of mining activity has implications around integrated resource management and governance. So, to your question, I have been only peripherally using any kind of landscape architecture, land-based thinking, other than strongly supporting the integrated resource management language and calling for our projects to not be siloed and not be mining focused, but be landscape-focused. I'm hoping that through the kinds of projects that CIRDI works on, we can encourage more young academics to build their competencies in a content sense and in a breadth of geographies.

X: Could you elaborate a little more on your thought and the big vision? What is missing? What change is required?

In relation to the definition of integrated resource management given in the CIRDI Strategic Plan: "Integrated resource management builds knowledge on the interaction between resources within a defined region. Effective and sustainable integrated resource management requires a strong knowledge base, practical experience, and collaboration with stakeholders, especially local populations." Do you see this definition as the need for understanding the politics of land, especially when it comes to territorial projects and Indigenous populations?

MQ: I don't think I have much more to say on this. Sorry!

Sticker art & resistance on the 150th Year of Confederation, 2017 (Eric Ritskes)

X: There is something that is quite intrinsic to the core and perhaps it's somewhat so pervasive as part of your background that now it becomes a matter of doing, but the fact that you come back to an understanding of integrated resource management and international affairs which could also be seen as being an understanding of the politics of land that comes through with your background in landscape.

However, whereas the policy of CIRDI and also of Global Affairs has been to promote resource-based development, which has somewhat de-politicized natural resources, we see here domestically, compared to internationally, the entanglements, complexities, difficulties, and literally obstacles for the policy administrators to really understand and engage, to not only recognize but also to address matters of reclamation. These two perspectives are somewhat at odds.

Pulling back from a global international perspective, how does one reconcile that we are exporting a resource-based development policy based on the template of the one that we have been developing for 100-150 years here in Canada (and of course inherited from the past) and yet we are incurring not only difficulties but mounting opposition to this same way of doing business and same way of dispensing policy? It is part image and part communications, but it is also, to come back to what you mentioned early on, it is part structure. If you were to intervene, where do you see some of the issues or frictions that can be disentangled and addressed?

MQ: I can't help but think that the design mind that is open to inquiry and listening to all the different positions that exist on a lot of these what people call "wicked problems"—don't really like that framing—but these challenges that we're facing, which are really around trade-offs balancing. I think they could be advanced if we could get in the room with a process. We need to design a process that has a beginning and an end and a timeline, and puts all the right people in the room. By right people I mean people who are decision-makers and the receivers, the users. I hark it back to the good old days of the roundtables in British Columbia that really went a long way to resolving, not perfectly, but to getting at this integrated resource management puzzle and bringing together all the different voices to find the common ground and agreeing on common principles.

I wish CIRDI could do domestic projects, although I think we may not have capacity. But separating out what we're learning from doing these projects in Peru and Mongolia and in West Africa, it actually reflects back to our own resource governance which, you rightly pointed out, is wanting. We're supposedly taking Canadian expertise from a resource governance stand point and working on projects in these emerging economies. But sometimes I think we have it backwards. I'm kind of hoping that the kind of learning that happens with the Peruvians on their approach is too; so now they have watersheds and watershed management and so learning the other way would make sense. Because right now I don't know what's happening with a lot of the challenges that we've got in the mining sector for example, other than the advantage that CIRDI has is we have this interesting Advisory Council that the Mining Association of Canada is involved in. They certainly understand that there's huge challenges that are not going to go away. The fact that they are at least at the table learning and seeking advice sometimes, I think is positive. I'm not sure I've answered your question.

X: Could you be a little more specific about what kinds of domestic projects? And can you elaborate on specific examples or experiences that you have had so far in terms of lessons learnt, do's and don'ts?

MQ: Again—I really don't feel that I can be more specific. I am deep into getting the Policy School going—and I just don't have the energy to reflect on this right now. Just can't do it!

CIRDI ICIRD

CANADIAN INTERNATIONAL RESOURCES AND DEVELOPMENT INSTITUTE

INSTITUT CANADIEN INTERNATIONAL DES RESSOURCES ET DU DÉVELOPPEMENT

Mark of the Canadian International Resources & Development Institute (CIRDI)

X: To a certain extent, your understanding that "maybe we have it backwards," that maybe in fact we (metropolitan, city-based dwellers of the settler state) should be learning also in a more reciprocal way from both ground conditions here and also the acknowledgement first and foremost when you're working in Canada on any form of territory which 90% of it is Crown lands on both seated and unseated territories, that in fact you're working with a number of different political nations. If one embraces this, it becomes a sentinel for working in different parts of the world.

Since we're looking to build a level of advocacy, we've been speaking with Justice Thomas Berger who has been very influential when stating: "The problem with Canadians is the failure to recognize Indigenous populations as political societies." Period. And for us, this has certainly been important.

Historically and culturally, the colonial forms of administration and policy of the country are important to understand… especially when looking at the evolution of the Canadian International Development Agency, which has now become Global Affairs Canada. Given this, how do landscape architects, urban planners, and urban designers potentially address matters of integrated resource management, but also understand the very deep political, territorial, historical, ground that they're treading on?

It's a question of respect, recognition, but also there are significant levels of research associated with it that directly relate to what you're speaking about, that we need to design a process and understand where that process has failed in the past in order to transform it.

Landscape architecture in itself demonstrates a lived experience and commitment to matters on land, to people that live on those lands, and also at the same time that we have very real, urban, economic, and environmental challenges that need to be addressed, and it's not a question of being an opponent or proponent on either side, as opposed to understanding that there is a kind of third alternative which is outside of a somewhat colonial, administrative, or bureaucratic box.

"Discipline refers to the ways in which various actors (soldiers, geographers, etc.) used cartography and processes like mapping to exert control over geography, nature, and human inhabitants (often indigenous peoples) who were regularly seen as expendable or an obstacle to some Eurocentric notion of progress."

—Charmaine A. Nelson, "Interrogating the Colonial Cartographical Imagination," 2017

The question is: how one might include in the pedagogy or the curriculum of decision-makers—either in business, university settings, or at the level of even academia or the student level—or rather, how one begins to understand, learn, and live an understanding of the political societies and the political territory that we work in? That goes to the core of the very notion of what we consider to be "natural resources," since Canada's image is built on the exploitation of natural resources and Indigenous territories. Maybe the problem is that we—settler-based designers—think that they are natural when in fact there are deep historical narratives associated with these resources and territorial ones. I recognize a lot of people don't want to go there, but I also think that these issues are not going away.

MQ: No. It has been very interesting building this Master's Program of Public Policy and Global Affairs and we have an Energy and Resources stream. What I'm hoping is that we will start to be able to have dual degrees. Going to your question of how do we build capacity and understanding, wouldn't it be fabulous to have a dual degree between a Masters of Landscape Architecture and MPPGA, and an MBA and an MLA, and an MBA and MPPGA? I can see that as a real response… we need to be equipping our students with the ability of being innovators. I must say I get very frustrated when people say that design is that creative fluffy kind of stuff. Design is extremely rigorous. Designers have to balance the critical and the creative, which is one of our added values… it is the gift that I think we need to give to as many people as we can.

Mark of the student-led counter organization to CIRDI, 'Stop the Institute CIRDI' "holding the Universities of British Columbia and Simon Fraser University accountable to their human rights commitments" (Eviatar Bach)

X: Can you reflect on the student-organized opposition to CIRDI from the Mining Justice Alliance?

MQ: This has been a particularly frustrating aspect of my CIRDI work—in the sense that there was and still is a lot incorrect information about CIRDI. In the early days, communication in many forms was lacking. We now have an excellent web-site which is transparent about governance, our projects and how we work. I am very much in favor of students and citizens asking good questions about the roles of organizations in our society. That is important. However, I am not impressed when there is no attempt on their part of listen, to engage and to be part of a learning dialogue (from both sides). It makes it very difficult for organizations like CIRDI to be learning organizations themselves when critics refuse to seek accurate information and actively listen. ○

This conversation took place between Moura Quayle and Pierre Bélanger in 2016 and 2017.

Alberta to Texas

Prime Minister Justin Trudeau
9th day of March, 2017
Houston, Texas

Bonsoir, mes chers amis. Good evening. Thank you, Dan, for that kind introduction and thank you all for being here. To the many policymakers and industry leaders here today, thank you. And of course to the organizers of CERA Week, thank you so much for this award. It truly is an honour for me personally, for Canada's superb Natural Resources Minister Jim Carr who's been here all week. And for the Government of Canada and the people of Canada who elected us. I'd like to begin with a little family history, if I may. You may have heard that my father, Pierre Elliott Trudeau, was also Prime Minister of Canada back in the 70's and early 80's. In 1980, I was in my prime Star Wars years, at nine or so, so politics was not the first thing on my mind. But that year, my dad's government introduced a policy called the NEP, the National Energy Program. It lasted until 1985 and it didn't work. It was a failure. It didn't mean to be, but it ended up being the wrong policy at the wrong time. Now I'm not being disloyal when I say this. It's just a historical fact. The NEP introduced a level of state control over energy that hurt growth and jobs. It became hugely controversial. Consequently, when I began my run for the leadership in 2012, 30 years later, the Trudeau name was mud in Alberta, still. A Trudeau, it was said, could not get elected dog-catcher in Alberta. But here's the thing. I knew that there is no path to prosperity in Canada that does not include a thriving, vibrant energy sector, both traditional and renewable. I knew that our resource industries provide thousands of well-paying, middle-class jobs not just in Alberta but across Canada and around the world including, of course, here in Texas. The Enbridge Spectrum merger, which creates one of the largest energy infrastructure companies on the continent, is only the most recent example of an extraordinarily productive partnership. Last year alone, Texas trade with Canada was worth $35.1 billion. Texas exports to Canada worth nearly $20 billion. Imports from Canada, $15 billion. That translates into thousands of good Texan jobs, my friends. By our reckoning, 460,000 jobs. And the ties between our two countries are economic but also strategic. This is the most successful economic relationship in the world, supporting millions of middle-class jobs on both sides of the Canada-U.S. border. Canada buys more from the United States than it does from any other country. We are the number one customer of two-thirds of U.S. states and in the top three for 48 different states. Nothing is more essential to the U.S. economy than access to a secure, reliable source of energy, and Canada is that source. We have the third largest oil reserves in the world, and provide more than 40 per cent of America's imported crude. And this extends well beyond oil. We supply you with more electricity and uranium than any other country, too. My point is just this: all Canadians, cold weather dwellers that we are, get the importance of energy. So to emphasize that, immediately upon launching my leadership campaign, I went to Calgary, Alberta. I think quite a few people including quite a few of my own party thought I was crazy. But guess what? A little over four years later, with a number of strong Trudeau Liberals elected across Alberta, we're on our way to getting three new pipeline projects under way, which will help connect Canada's oil patch with energy markets around the world. The first, Kinder Morgan's Trans-Mountain line, will run from Alberta across the Rockies to the Pacific. The second, Trans-Canada's Keystone XL pipeline, recently approved by President Trump, will ship Canadian crude to refineries here, in Texas. And Enbridge's Line 3 replacement will also come south. These ambitious projects will go a long way towards ensuring North American energy security for years to come. I make no bones about it. We're very proud of this. It's progress. It's important. As I said on the

very first trip to the oil patch back in 2012, no country would find 173 billion barrels of oil in the ground and just leave them there. The resource will be developed. Our job is to ensure that this is done responsibly, safely, and sustainably. Which brings me to the second piece, equally critical. While developing our resources for the economic benefit of Canadians, we must also look to the future. Now what I just said was... there will come a day, far off, but inevitable at some point, when traditional energy sources will no longer be needed. In preparing for that day, we have two critical responsibilities. One is to sustain the planet between now and then so we can pass it on to our children better than we found it. And the second is to get ahead of the curve on innovation. And we in Canada are doing just that. Canadian companies are leaders in developing technologies such as carbon capture and storage, next-generation biofuels, advanced batteries for electric cars and cleaner oil sands extraction processes, among other advances. This creates good jobs and it also helps the planet. Here's the crux of it. In Canada, as I said, we know all about preparing for winter and for the long cold nights. So when we go camping—and we do love to camp, that particular stereotype is totally accurate... when we go camping we light our campfires before the sun goes down. And that doesn't mean we're anti-daylight. So it's exactly the same with energy. Innovating and pursuing renewables isn't somehow in competition with those traditional resources. It's common sense. It's wise preparation for the future. Our children and their children deserve no less. All of which brings me full circle. We would not be on this path, not even close, had we as a government not insisted that environmental protection and resource development go hand-in-hand. Our immediate predecessors tried a different route for 10 years—to ignore the environment. It didn't work any more than the NEP of the 1980s worked. They couldn't move forward on big energy projects. Our predecessors failed because in the 21st century Canadians will not accept that we have to choose between a healthy planet and a strong economy. People want both, and they can have both. It takes compromise. It takes hard work, but it is possible. The proof that we have to combine and we can combine environment and economy is that little more than a year into our first term of government, we have made progress both on pipelines and on a national carbon reduction plan which finally puts a price on carbon pollution. And this was worked out in cooperation with our provinces. It's a first step and more work needs to be done, but it is a clear, new path for our country after years of false starts. And let me be very clear. We could not have moved forward on pipelines had we not acted on climate. And we could have not acted on climate had we not also focused on jobs, that is on the needs of the Canadian people, especially those of the middle class and those working hard to join it. We are showing not only that it can be done but that it must be done, that this is the way forward that all of our citizens expect. And that's why, my friends, I am so very pleased and honoured to receive this award tonight. We are showing that environmental leadership and economic growth are inseparable, that they must go together. And today, because of this award, people around the world will take notice and hopefully join us on the path to a brighter future. Merci beaucoup, tout le monde.

▷◁

Justin Trudeau is the current and 23rd Prime Minister of Canada since 2015.

transcribed from *CERAWeek Global Energy and Environment Leadership Award Dinner*, Houston, Texas (Thursday, March 9, 2017)

At One Part per Billion

A Conversation with Douglas Morrison

Doug Morrison is President and CEO of Holistic Mining Practices at the Centre for Excellence in Mining Innovation (CEMI) non-profit in Sudbury, Ontario. Prior to joining CEMI in 2011, Doug had a long career in mining and engineering consulting industries.

X: Is resource mining really essential?

DM: Yes, it is. Material resources are important to the global-industrial complex: if, historically, we had stopped at coal, people would have nice warm houses and that would be the end of it. We actually moved on from coal to iron, and together, we make steel. So the whole industrial complex is ruling because of those things.

Materials change the ruling empire. Materials are always being replaced by something else. The Roman Empire replaced the Bronze Age because we discovered copper, then we discovered tin, then we went to the Iron Age where we had iron swords instead of bronze swords. But in those times, they didn't get to industrial-scale production of iron and steel. So it's important not only to get new materials going, but also to start playing with them on an industrial scale. Discovering the metals and how to use them are matters of technological innovation, but the scaling up is actually a business innovation.

"Technological nationalism's promise is suspect because the commodified culture it would constitute would have no stability, and would be but another instance of the culture of technological society."

—Maurice Charland, "Technological Nationalism," 1986

X: When you talk about a massive scale of metal production, are you referring to the present day?

DM: Yes. People say that all of Western society—the United States and Western Europe—has ended up in a post-industrial world. But we haven't. What we actually live in, is a world where the industrial components have gone elsewhere. There's a lot of steel production in China and Korea, where they can do it cheaper. Industrial-scale steel production is still happening.

In cities that you and I live in, people think, "well, we don't need this mineral stuff for anything." But you can't live without minerals in every post-industrial economy. Even in the cell phones we're using right now, the metals included are produced on an industrial scale. Most people in cities don't think very much about where their food comes from, even less so where their metals come from.

Queen Elizabeth II, Copper Cliff Mine, Sudbury, 1959 (*Inco Triangle* 20, no. 4, July 1959 / Sudbury Museums)

X: That addresses the question from a historical perspective. There's also an operative/logistical question you raise in relationship to the state of the mining economy today, with the grades of different ores, the depths we're reaching in mines, and the air circulation that is fundamental to mines.

DM: All those things that have to go into the industrial process are also on an industrial scale. So the movement of air through these mines, most people think it's just to supply the people who go down with breathing air. It's not—it's about cooling the whole system down. The total volume of air we consume is massive—far, far greater than we need for the people to breathe.

Certainly the power we use to operate the mines is also on an industrial scale. So a mine is not a mine—it's also a power plant. In order to operate mines in remote places, you have to carry these artifacts of industrialization to remote places.

In terms of depth, mining operations run from 8,000 feet to 2.5 kilometers underground, and the rock temperature there is approximately 60 degrees centigrade. Nobody can work in that, much less computers and equipment.

You have to think of depth and remoteness of mines as the same in terms of influence on capital. Both require heavier capital investments to get operations started. So when you have mines in remote stations that already have the infrastructure like power lines, if the grade of the material is declining, you have to get more efficient at digging.

X: There's a resonating ratio—for instance, in the case of gold, you mentioned one gram per ton is one part per billion?

DM: Yes, one gram per market ton. That's still what's achievable…if you get two or three grams, you're safer, but you can still survive on one gram. It's an astonishing number if you think about it.

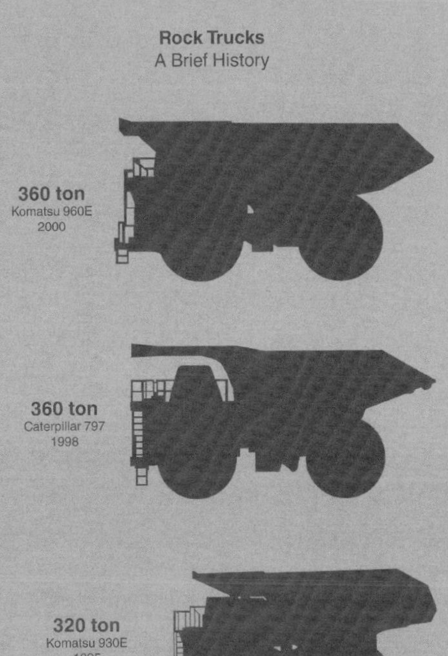

X: As a Scotsman, the legacy of the Scottish people is different from the English imperialist. Are there legacies of displacement in the highlands by the English Empire during its growth?

DM: That's exactly right. In Canada, resource extraction and the displacement of Indigenous populations is similar to what the English did for centuries with tribal lands in Northern England and Scotland.

The Highland tribal regions were seen as savage, just as the Romans saw them as savage too. It's a universal theme of development, it's just a way we've evolved. For the Scots, very soon after that colonization process had taken place, they began to participate in the industrial complex. What happened in Scotland was that the Highland people came to realize they're actually really effective at doing that kind of work. Once they accepted to participate in the industrial complex of the English Empire, they actually became very successful.

In Canada, companies are similarly focusing on how to co-opt some of those Indigenous populations to participate in mining operations or to eventually displace and relocate them.

X: Has Canada then become a global resource empire in the same way?

DM: I would say it has, but it's not because of steel, it's because of influence. This is the other part about conquering tribal societies: it's not about size so much as influence. Canada has influence over many industrial practices. For example, the guidelines on the margins of tailings ponds were written by Canadian mining consultants. Canadians are genuinely trying to be stewards of the environment, as well as stewards of the resource. A lot of the technological development was also from Canada, because they were trying to improve mining conditions.

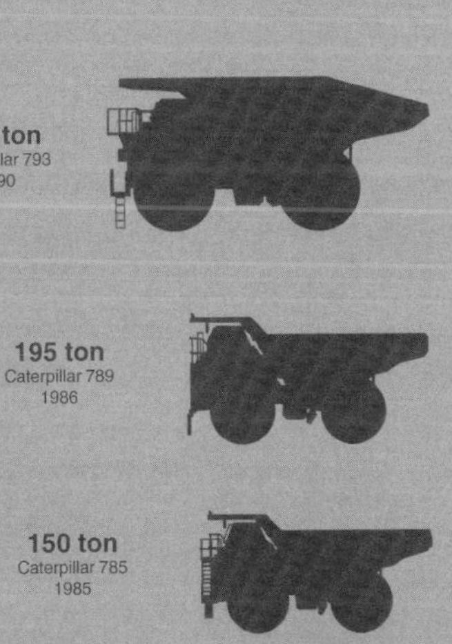

If Canada had a stronger imperial sense, like South Africa did, then you would have a black underclass that does all the work. In that case, there's no incentive to improve the working class of the workers, because they're seen as alien. But Canadian miners were Canadian. Canada has a more egalitarian social structure than most other countries.

X: Contrary to popular opinion, you mentioned that mining has apparently become the second-safest job after teaching…

DM: … statistically speaking, yes. The only segment of the economy in Algeria that's safer than the mines is the education system. This might be true in Canada as well.

X: How do mining operations that take place in other countries that don't share from the same egalitarian structure build in conditions in their mining leases?

DM: Canada is imposing its view of safety on mines around the world. There are definitely very unsafe mines around the world, but that's because they're not subject to rules by global financiers, and instead they're financed locally. If you get a financing firm, the big guys, you have to conform to higher safety standards. So that's a Canadian imperial export.

There is something called the *Equator Principles*—principles laid out by the major finance houses for capital in the world. Certainly the largest three, and many other mining companies, have to conform to these principles. These guidelines were produced by Canadian consultants and engineers, and became the international guidelines on tailings and mining facilities. So international standards are comparable all over the world.

Nickel Tailings #34, Sudbury, Ontario (©1996 Edward Burtynsky)

Variations occur either with small corporations, or corporations that are wealthy enough to own their own mine without financing from major banks. So the places where you find the worst conditions are the places where major corporations are not involved. You have blood diamonds in places like Sierra Leone, or small hybrid gold mines that are owned by local families or perhaps a syndicate.

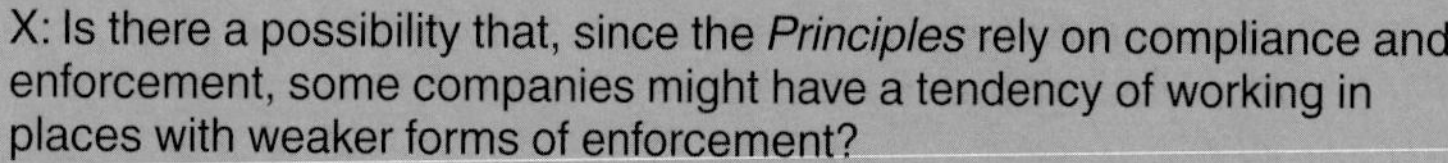

X: Is there a possibility that, since the *Principles* rely on compliance and enforcement, some companies might have a tendency of working in places with weaker forms of enforcement?

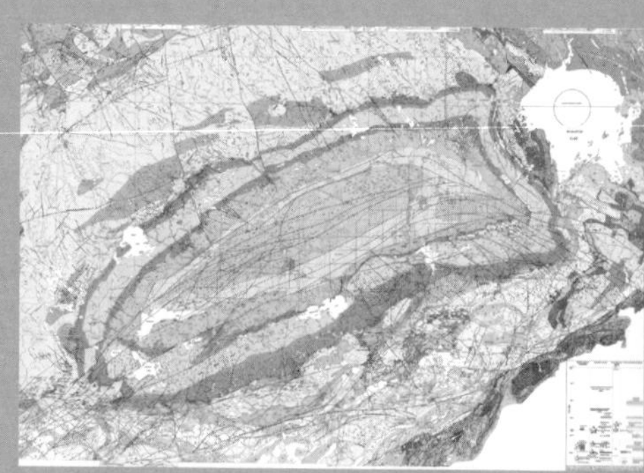

Sudbury's Geological Basin (Natural Resources Canada)

DM: Yes, absolutely. You have a failure with compliance when you have a certain level of corruption, and officials and regulators can be bribed to look the other way. That tends not to be true in Canada, the US, and Europe, but it tends to be much more true in South America and Africa. There is also, in Canada and European countries, cases in which companies' reputations have been damaged internationally by charges of corruption and bribery. So in addition to the Equator Principles, there are more regulations controlling business ethics.

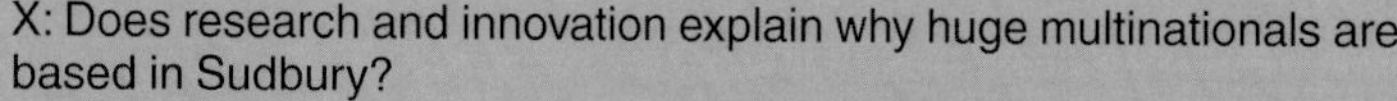

X: Does research and innovation explain why huge multinationals are based in Sudbury?

DM: Yes. Let's step back for a second and just say that Canada might be a resource empire, but it is only one of several, right? The size and population of Canada is comparable to those of Australia. Then you have the United States, which I don't know the numbers for, but it's also a major center for mining with a much larger population. In South America, Brazil and Chile are major mining empires as well, in the steel market.

Steel is two different commodities: iron ore and coal. The one people tend to forget is coal, and that's metallurgical coal, not the coal we burn in power stations. Metallurgical coal is very low in volatiles, and it is almost pure carbon. And then there are a whole bunch of other small metals that are included in steel to give it the particular properties you need.

Advertisement (THOR Global Enterprises)

The biggest mining companies in the world are, in order: BHP Billiton, Vale, and Rio Tinto. They are all dominant in the iron ore and coking coal businesses. The second biggest commodity, after steel, is copper. Copper is used for piping, electricity and power lines. Gradually, copper piping for water is being replaced by plastic, but copper for electricity is the single biggest use of copper. So the amount of copper used can be a measure of a society's state of social and economic development.

⊞ At One Part per Billion ☞

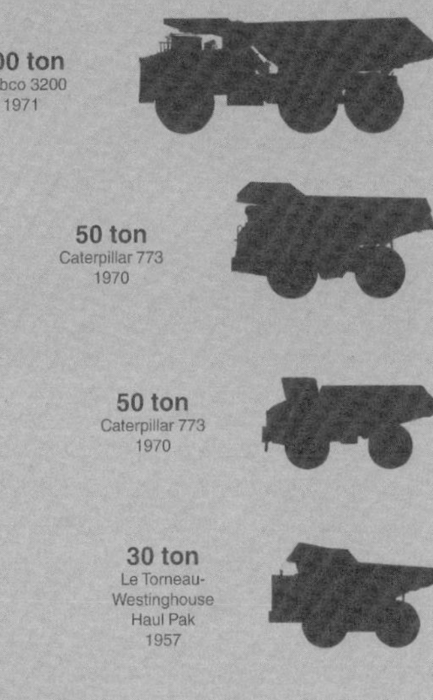

X: So there's a misunderstanding, it's not money that makes the world go around, it's copper?

DM: It is copper, yes, electrification too. And then all the things you want to build—buildings, cars, spaceships, satellites—have steel in them.

X: Is the speed of urbanization today occurring at a pace where we do not produce enough metal resources from recycling? Or, is mining raw, virgin resources and minerals really a necessity?

DM: Yes, absolutely. Re-use is a wonderful, cost-effective idea, but I would guess that the total percentage of reused metals in the economy is certainly less than 10%. My guess is that it is probably even closer to the range of 2%.

There is a huge number of people that are moving up from what we could call an underdeveloped economy to a developed economy. The Chinese government has raised the standards of living for hundreds of millions of people. That then means that you have a drastic increase in the consumption of metals, steel, copper, and all the alloys that go along with that.

X: Why is the Sudbury basin so significant?

DM: Look, the major companies in the Sudbury Basin—Vale, Rio Tinto, Glencore—are major international corporations, and they will move their castle to anywhere they can produce metals at a profit. Many of the deposits in the Sudbury Basin have been mined for the last hundred years and are getting progressively older and deeper. Older means technically more difficult and both of those things means more expensive. And so eventually you get to the point where operations say, "the rate of return on our investment in this asset is not sufficient for us to continue."

So, looking at it purely from the selfish point of view, Sudbury is a major producer of metals—and it's our primary business. Then yes, those of us living and working in Sudbury need to find new and higher-grade deposits in order to continue to attract the investment that needs to make those mines work.

Very large corporations have an internal financial hurdle that means they have to have a certain level of profitability working for them to make it worthwhile. When that number drops below the trigger number, they will sell that asset onto smaller companies who have lower internal financial hurdles and then that smaller company can make a profit where the larger corporation cannot make a profit. So at some point in the future, the mines in Sudbury may well change hands again. We went through this acquisition phase when China had a really over-heated economy, and now that it's cooled, the profitability of our mines are declining. That's why we need to find more ore bodies that are higher grade and can meet the needs of those large companies. If we fail to do that then Sudbury as a mining center may well decline.

X: So, is it about more profit, or more ore, or both?

DM: To go back to the basics, most metal mines that produce the small metals, like nickel and copper, are producing grades of less than 2%, less than 1% for copper. So over 95% of the material we produce from a metal mine is waste rock that has no value. Many nickel and copper mines produce very low grades.

What ultimately matters is the total volume of metal that you produce. That is a tradeoff between the total volume that you mine and the grade at which you mine it. The largest copper mines in the world today use

"I wanted to use the 170-tonne trucks because those were the largest in the industry at the time, and I knew that the unit cost per tonne mile of moving a tonne with those was going to be lower than say an 80-tonne truck or a 50-tonne truck, even though the conditions were very soft. The biggest challenge I had initially was convincing people there that we could do this successfully. Great Canadian Oil Sands at the time had tried the big 150-tonne trucks, and they didn't have much success with them. They were switching their fleet back down to 85-tonne truck, mechanical drive."

—Jim Carter, Former CEO, Syncrude, 2011

183 m W.A.C. Bennett Dam

20 m Hagersville CGC Mine ON
40 m Kearney Mine ON
60 m Cassidy Lake Mine NB
110 m Lake Tio Mine AU
240 m Madoc-Conley-Henderson Mine ON
250 m Edako Mine BC
300 m Beavery Brooke Mine NL
Bellekeno Mine YT
335 m Ojibway Mine ON
350 m Jeffrey "Johns-Manville" Mine QC
450 m Kansanshi Mine ZM
480 m Cigar Lake Mine SK
500 m Goldstrike Mine NV
Goderich Mine ON
533 m CN Tower
560 m Kalgoorlie Super Pit AU
640 m Ekati Mine NT
680 m Deep Geological Repository ON
Antamina Mine PE
700 m Candelaria Mining Complex CH
730 m Cantung Mine NT
800 m Akie Project BC
805 m Highland Valley Mine BC
950 m Werner Lake Project ON
975 m ZCA No. 4 "St Joe" Mine NY
1,000 m Brunswick No. 12 Mine NB
1,180 m McCreedy West Mine ON
1,200 m Cassidy Shaft at Kalgoorlie AU

How Low Can We Go?

The deepest mines in the world owned and operated by Canadian resource corporations

1,500 m — Lac de Iles Mine ON
1,508 m — 777 Mine MB
1,600 m — Belle Plaine Mine SK
2,000 m — Sudbury Neutrino Observatory ON
2,225 m — Macassa Mine ON
2,500 m — Creighton Mine ON
2,927 m — Kidd Creek Mine ON

Key to Regions
ON - Ontario
NB - New Brunswick
BC - British Columbia
NL - Newfoundland
YT - Yukon Territory
QC - Quebec
SK - Saskatchewan
NT - Northwest Territories
NS - Nova Scotia
MB - Manitoba
AU - Australia
FN - Finland
NY - New York
CH - Chile
PE - Peru
NV - Nevada
ZM - Zambia

a method called "block caving," which is a very inexpensive method. It means you can exploit very low-grade copper deposits, at a huge volume of material. The mines in Chile and the U.S. and elsewhere are producing ore at a level of 100,000 tons per day, with a low ore grade of around 1%. At a typical mine in Canada, a nickel mine will produce 3,000 to 5,000 tons per day. So it's very much smaller than the typical block cave mines.

The New Texas
A Comparison of Subsurface Mineral Deposits between Texas & Saskatchewan

X: In terms of finding more ore, are companies like Vale or Rio Tinto investing in research and development to find these ore bodies?

DM: Not really. Investment in those areas, as in across the country, is minimal. Until, let's say, a couple of years ago, many of the larger companies did invest in mining research. But we need to make a distinction here between mining research and process research. Mines produce ore, which goes into the processing part of the plant. A metallurgical operation will process the ore to roughly five times its original concentration. So you can see that for every dollar you spend in mining research, the same dollar will give you five times the value in the processing side. So gold mining companies have focused their attention on the research and innovation you need in the processing side, and they've invested relatively little in the mining side.

In the last four years, service sector companies have increasingly invested in research and innovation. In particular, they have invested in the innovation end of the spectrum, rather than the research. This part of something called the TRL—the technological readiness levels—going from 1 to 9 which provides a rating for fundamental research, or conceptual research, or innovation across a spectrum.

X: How would you characterize research and innovation in Canada today?

DM: In Canada, people still tend to use the words research and innovation as if they are interchangeable. For me, distinguishing the two is important: in simple terms, research is the creation of new knowledge, and innovation is the creation of new businesses as a result of that knowledge. Many politicians in Canada are using the two words as if they are synonymous, but they are not.

As is probably the case in the United States, there has been a decline for funding at Canadian universities, and there is a focus on the innovation process meaning that it places us back to staple resource extraction, with very little innovation or contribution. ⊞

This conversation took place between Douglas Morrison, Daniel Hemmendinger, Chris Meyer, and Pierre Bélanger in 2015.

⊞ At One Part per Billion ↑

Underground potash borer, Rocanville, Saskatchewan (Troy Fleece / The Canadian Press)

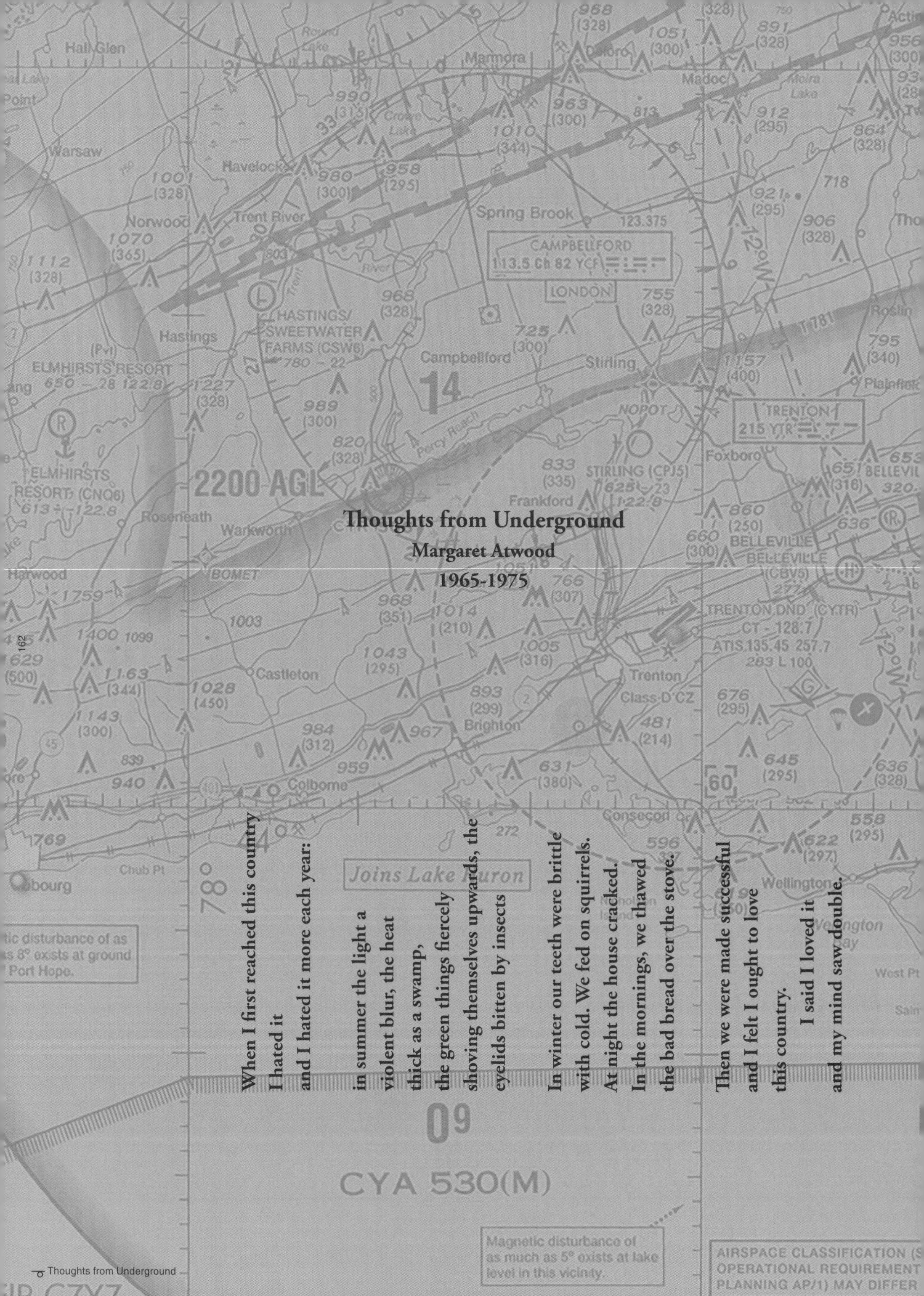

Thoughts from Underground

Margaret Atwood

1965-1975

When I first reached this country
I hated it
and I hated it more each year:

in summer the light a
violent blur, the heat
thick as a swamp,
the green things fiercely
shoving themselves upwards, the
eyelids bitten by insects

In winter our teeth were brittle
with cold. We fed on squirrels.
At night the house cracked.
In the mornings, we thawed
the bad bread over the stove.

Then we were made successful
and I felt I ought to love
this country.
    I said I loved it
and my mind saw double.

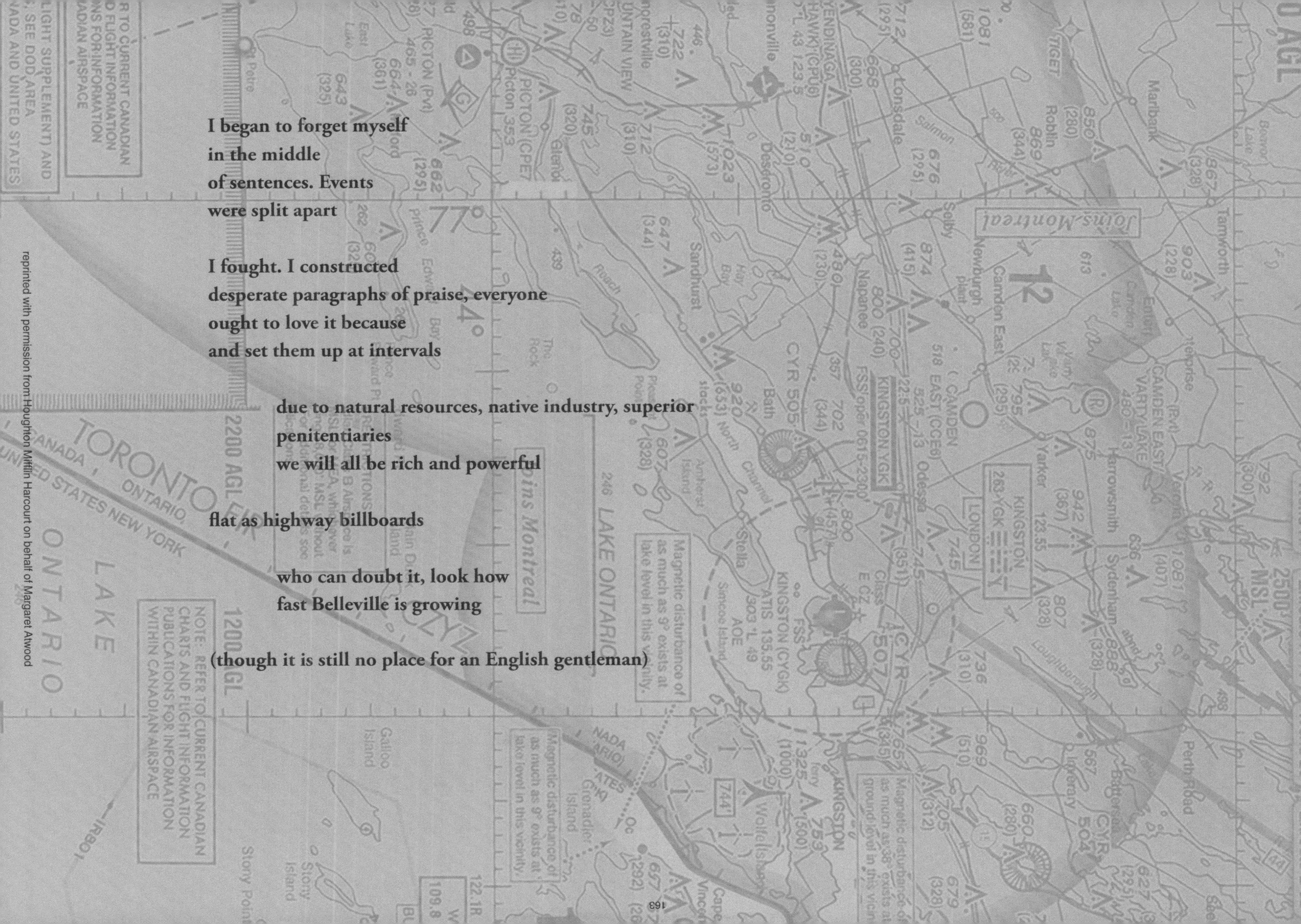

I began to forget myself
in the middle
of sentences. Events
were split apart

I fought. I constructed
desperate paragraphs of praise, everyone
ought to love it because
and set them up at intervals

due to natural resources, native industry, superior
penitentiaries
we will all be rich and powerful

flat as highway billboards

who can doubt it, look how
fast Belleville is growing

(though it is still no place for an English gentleman)

344 HARPER'S WEEKLY. [MAY 3, 1879.

CANADA'S REAL HOME RULE.

BEACONSFIELD (to Lorne). "Good boy, not to act on your own responsibility, but to come and ask your mamma first. But we are rather too busy just now with civilizing colonial savages."

Harper's Weekly, May 3, 1879

Canada as Emerging Energy Super Power

Notes for an Address by
The Right Honourable Stephen Harper, Prime Minister of Canada
14th day of July, 2006
London House, UK

Good evening, ladies and gentlemen. Thank you very much for your warm welcome. And special thanks to Mr. Dahdaleh for your generous introduction. I'd like to begin by acknowledging a few people here tonight. The acting Canadian High Commissioner in London, Guy Saint-Jacques. I understand he threw a great Canada Day party in Trafalgar Square this year. Her Majesty's High Commissioner to Canada, David Reddaway. And your Minister of State for Energy, Malcolm Wicks. Ladies and Gentlemen, this is actually my first speech to a business audience outside Canada since becoming Prime Minister. And it is only fitting that it's to your distinguished organization. Because the Canada-UK Chamber has been promoting commerce between our nations for almost 90 years. And because the business relationship between our countries dates back to the very founding of Canada. In fact, for two centuries prior to our confederation in 1867, much of Canada was effectively owned, operated, and governed under the red ensign of a London-based corporation, the mighty Hudson's Bay Company. Our co-sponsor tonight, the Canada Club, owes its founding in 1810 to the fur traders of the North-West Company, the main rival and eventual partner of the HBC. Still, business is but one aspect of our combined history. That history is built by layer upon layer of common experiences, shared values and ancient family ties. In my own case, the Harper family traces its known forefathers back to the northern England and southern Scotland of the 1600s.

Bonded by History

But a far greater orator than I—or any Harper of the past 400 years—once described Canada-U.K. relations this way: The ties which join [Canada] to the mother country are more flexible than elastic, stronger than steel and tenser than any material known to science. Canada bridges the gap between the old world and the new, and reunites the world with a new bond of comradeship. The speaker, as you might have guessed, was the incomparable Winston Churchill. The occasion was a speech in Ottawa in 1929, part of a cross-country tour of what he called "the Great Dominion." He gave 16 speeches in 9 cities. Every one of them was delivered to sold-out rooms and repeated standing ovations. On that same tour, Mr. Churchill reminded Canadians of what they owed to Britain. At the heart of our relationship, he said: "is the golden circle of the Crown which links us all together with the majestic past that takes us back to the Tudors, the Plantagenets, the Magna Carta, habeas corpus, petition of rights, and English common law; all those massive stepping stones which the people of the British race shaped and forged to the joy, and peace, and glory of mankind." How right he was. Britain gave Canada all that, and much more. Including: Parliamentary democracy; A commitment to basic freedoms; The industrial revolution; and the entrepreneurial spirit and free market economy. Not to mention Shakespeare, Dickens, Kipling, Lewis, and Chesterton. Of course, we haven't accepted all of our inheritance from Britain. The take-up rates on rugby and association football are certainly not as high as ice hockey. And Canadians remain utterly baffled by cricket. But seriously and truthfully, much of what Canada is today we can trace to our origins as a colony of the British Empire. Now I know it's unfashionable to refer to colonialism in anything other than negative terms. And certainly, no part of the world is unscarred by the excesses of empires.

But in the Canadian context, the actions of the British Empire were largely benign and occasionally brilliant. The magnanimous provisions of the Quebec Act of 1774 ensured the survival of the French language and culture in Canada–to the everlasting benefit of our country. And the treaties negotiated with the Aboriginal inhabitants of our country, while far from perfect, were some of the fairest and most generous of the period. This genius for governance shown by the mother country at the time no doubt explains in part why Canada's path to independence was so long, patient and peaceful. And it explains why your Queen is still our Queen, and why our "bond of comradeship" remains as sturdy today as it was in Mr. Churchill's time.

Eternal Allies

That bond, ladies and gentlemen, was forged in bad times as well as good. Sometimes in the flames of war. When Britain has bled, Canada has bled. A generation of our young men share eternity with British Tommies in the fields of France. Another generation of Britons and Canadians fought side by side against Nazi fascism. Yet another helped our American cousins prevail over the menace of Soviet communism. And ever since that brief, illusory moment when we thought we were witness to "the end of history," we have been allied in a new global conflict. This is a conflict without borders. A conflict fought abroad and at home. A conflict in which the aggressor stands for nothing yet seeks to impose its will. Through the destruction of terrorism. Through the slaughter of the innocent. And through the perversion of a faith. So once more we face, as Churchill put it, "gangs of bandits who seek to darken the light of the world." And once more we must appeal to our values, marshal our resources and steadfastly apply our will to defeat them. This war on terror will not be easy. Nor will it be short. But it must be won. And Canada's new national government is absolutely determined, once again, to stand shoulder to shoulder with our British allies, to stay the course and to win the fight.

Canada's New Defence and Foreign Policy

Ladies and gentlemen, during last winter's election campaign, I made it crystal-clear where my party stood on national defence, foreign policy, and the fight against terror. We promised to rebuild Canada's long-neglected armed forces. To reassert Canadian sovereignty over our Arctic territories. And to reclaim the modest leadership role we once held on the world stage. And this is exactly what we have been doing since Canadians gave us their trust on January 23rd. One of my first actions as Prime Minister was to visit our soldiers in southern Afghanistan–who are standing shoulder to shoulder with British forces in the Kandahar and Helmand provinces. Together, they're taking the fight to the Taliban and helping the Afghan government assert control over these areas. And they are helping the Afghan people rebuild their war-ravaged country. Canada, like Britain, has committed to this mission for at least two more years. And committed to doing our duty for global peace and security over the long term. Which is why my government increased defence spending by two and a half billion pounds (over $5 billion) in our first budget. We are expanding the Canadian armed forces by recruiting and training 23,000 new regular and reserve troops. And we are providing them with the tools they need to carry out their missions. Last month, we launched a major new military procurement program. We will be purchasing new transport ships, a new fleet of military trucks, medium–to heavy-lift helicopters, and large tactical and strategic-lift aircraft. All these acquisitions will make the Canadian forces bigger, stronger and better able to respond quickly to threats at home and abroad.

Domestic Security

As you know all too well, the terrorism threat is not just external. Canadians were shocked and saddened by the London subway attacks that coincided with last year's G8 meetings. We also take this threat very seriously. And we're acting on it. First, by increasing the financial and human resources needed to enhance domestic security. Our inaugural budget dedicated an additional three quarters of a billion pounds (close to $1.5 billion) to improving emergency preparedness and the security of our borders and transportation systems. We're plugging the holes, filling the gaps, and working hard to stay one step ahead of the agents of hate and terror. Second, we're working closely with our international allies to penetrate global terror's networks. We're sharing information and co-ordinating investigations. Last month's arrests of 17 persons in the Toronto area, for example, marked the culmination of a two-year investigation involving intelligence-sharing with authorities in the United Kingdom and the United States. Several subsequent arrests in Britain were triggered by the investigation and arrests in Canada. We're also taking a leadership role in the international effort to choke off terrorist financing. Last week Toronto was named the permanent headquarters of the secretariat of the Egmont Group, representing financial intelligence agencies from 101 countries. But our best weapon in the fight against terrorism is another gift of our British heritage—our open and democratic society, and, more specifically, our embrace of cultural diversity. It is true, of course, that the apostles of terror use the symbols of culture or faith to justify crimes of violence. They hate open, diverse, democratic societies like ours because they want the exact opposite, societies that are closed, homogeneous, and dogmatic. But they and their vision will be rejected, rejected by men and women of generosity and goodwill in all communities. And most importantly rejected by men and women in the very communities they claim to represent. We have already seen this in Canada since the recent arrests. Because Canadians—no matter what their religious, ethnic or cultural heritage—recognize that ours is a land of opportunity. Where everyone with the will to succeed can build a good future for themselves and their families. Where what matters most is where you're going, not where you came from. Ladies and gentlemen, our government will do all we can to make our society secure and ensure that terrorism finds no comfort in Canada. And we'll do it by preserving and strengthening the values we inherited from you, Freedom, democracy, human rights, the rule of law. The values that built Canada, the values that unite all Canadians, and the values that will keep both our countries strong and secure.

Canada/UK Trade

At the very foundation of a strong, secure country is, of course, a strong, stable economy. And thanks to our balanced budget, our tax cuts, debt reduction, the high demand for our natural resources, our competitive economy, our clean, safe cities, and our highly skilled workforce, Canada is an extraordinary country in which to do business. Canada's new national government, as demonstrated in our first budget, is committed to balanced budgets, low interest rates, debt reduction, lower taxes, a stronger economic union, and an open, competitive economy. We're building on a solid foundation. The fundamentals of our economy are strong. The cost of doing business in Canada is now among the lowest in the industrialized world. Our natural resources are in high demand. Our cities are clean and safe—and our environmental and criminal justice policies aim to make them safer and cleaner. Our people are hard-working, highly skilled and global in their outlook. In short, Canada is a great place to do business. But I hardly need to tell you that. British investment in Canada has doubled since 1999, from 7 billion pounds to 14 billion pounds ($15 billion to $30 billion).

A New Energy Superpower

One of the primary targets for British investors has been our booming energy sector. They have recognized Canada's emergence as a global energy powerhouse–the emerging "energy superpower" our government intends to build. It's no exaggeration. We are currently the fifth largest energy producer in the world. We rank 3rd and 7th in global gas and oil production respectively. We generate more hydro-electric power than any other country on earth. And we are the world's largest supplier of uranium. But that's just the beginning. Our government is making new investments in renewable energy sources such as biofuels. And an ocean of oil-soaked sand lies under the muskeg of northern Alberta–my home province. The oil sands are the second largest oil deposit in the world, bigger than Iraq, Iran, or Russia; exceeded only by Saudi Arabia. Digging the bitumen out of the ground, squeezing out the oil and converting it into synthetic crude is a monumental challenge. It requires vast amounts of capital, Brobdingnagian technology, and an army of skilled workers. In short, it is an enterprise of epic proportions, akin to the building of the pyramids or China's Great Wall. Only bigger. By 2015, Canadian oil production is forecast to reach almost 4 million barrels a day. Two thirds of it will come from the oil sands. Even now, Canada is the only non-OPEC country with growing oil deliverability. And let's be clear. We are a stable, reliable producer in a volatile, unpredictable world. We believe in the free exchange of energy products based on competitive market principles, not self-serving monopolistic political strategies. That's why policymakers in Washington–not to mention investors in Houston and New York–now talk about Canada and continental energy security in the same breath. That's why Canada surpassed the Saudis four years ago as the largest supplier of petroleum products to the United States. And that's why industry analysts are recommending Canada as "possessing the most attractive combination of circumstances for energy investment of any place in the world." British companies are already significant players in the Canadian energy sector. BP has been there for 50 years. It's already one of our leading producers of natural gas and it has a major stake in Canada's next huge gas development–the Mackenzie River Delta in the Northwest Territories. BG Group has also accumulated a large exploration stake in the Mackenzie River Valley. There are trillions of cubic feet of gas in the region, and we are hopeful that the huge pipeline needed to deliver it to southern markets will finally go ahead. British firms invested nearly three billion pounds (over $6 billion) in our energy and metals sectors last year. And I think we'll see even more British investment as word of Canada's stature as the West's most important energy storehouse gets out.

Canada/UK Trade Today

Of course, the energy sector is not the only source of British investment. There are already about 650 UK-based companies and subsidiaries operating in Canada. You employ more than 70,000 people in 20 different industries. British exports to Canada was close to five billion pounds ($10 billion) last year. And even if you're not doing business in Canada, chances are you're vacationing there. We welcomed over nearly a million visitors from the UK last year. And we look forward to seeing you all at the 2010 Winter Olympics in Vancouver and Whistler, British Columbia. Canada is not as big a player in Britain as we'd like to be, but we're getting there. I'm glad to see many of you sporting Blackberrys. They're made in Canada, you know. The Ontario manufacturer, Research in Motion, recently got security approvals from Whitehall and has started selling its marvelous devices to several ministries. Chances are the video card in your computer monitor was made by ATI Technologies of Toronto. One of our home-grown heroes is Quebec-based Bombardier, the aircraft and railcar manufacturer. Believe it or not, Bombardier is the largest full-time employer in Northern Ireland.

All this means we're not just hewers of wood and drawers of water anymore. Although we're still pretty good at those things too. The great granite plate known as the Canadian Shield is a vast storehouse of precious metals. We have long-been a major producer of nickel, gold, copper, potash, coal, and cement. But it may be news to you that Canada is now the world's third largest producer of diamonds. A decade ago Canadian diamonds were only a gleam in a prospector's eye. Today there are three producing mines and two more in development. And the Royal Bank predicts diamonds will bring over 30 billion pounds (almost $70 billion) to the Canadian economy over the next 25 years. The Shield also yields a third of the world's uranium supply. There aren't many hotter commodities–so to speak–in the resources market these days. The price is higher than it's been in three decades. Around the world, nearly 200 new reactors are proposed, planned, or under construction. And, as you know, Britain is one of the countries considering expansion of its nuclear generating capacity. We'll hope you'll remember that Canada is not just a source of uranium. We also manufacture state-of-the-art Candu reactor technology, and we're world leaders in the safe management of fuel waste. Which is one more reason to think of Canada as an energy superpower–and a strong candidate for British investment. I know Britain's trade orientation has successfully focused on the European Union in recent years. But the success of British enterprise, for centuries, has been its ability to spot opportunities and nimbly move to exploit them. That is something else we learned from you. So you'll forgive me if I remind you of it now, because the world is beating a path to our door. And we want Britain to be as much a part of our future as she has been of our past.

Conclusion

Ladies and gentlemen, let me conclude by saying that I have no doubt that the "bonds of comradeship" Mr. Churchill talked about in the early 20th century will remain just as strong throughout the 21st. The "little island" and the "Great Dominion" are eternally bonded by language, culture, economics, and values. That's why our business relationships are so strong and successful and why they will only be growing stronger in the future. It's why our troops are again serving side by side–this time in Afghanistan–defending freedom and building democracy. Why our intelligence services are working hand in glove to keep our homelands safe and secure. And why I am honoured to have had this opportunity to speak to your organizations today. Thank you. God bless Canada and God save the Queen.

Stephen Harper served as 22nd Prime Minister of Canada between 2006 and 2015.

Who Is Maurice Strong?

A Conversation with Maurice Strong

Maurice Strong was responsible for the creation of the UN Environment Programme, the organization of the 1972 Conference on the Human Environment in Stockholm, and the 1992 Rio de Janeiro Earth Summit. As an international businessman in the oil and mineral industry, he served as President of the Power Corporation of Canada, and CEO of Petro-Canada and Ontario Hydro. He published *Where on Earth Are We Going?* (2000).

X: Can you describe one of your experiences at Harvard with Robert McNamara that you remember as being strongly influential?

MS: I soon got to know Robert and I changed my view about him significantly to the point where I came to believe he was the best person to head the World Bank.

X: McNamara was Secretary of Defense during the disastrous Vietnam War, from 1961 to 1968. What made you change your mind or perspective about him?

MS: Well, I'm not sure it was a thing that happened at a particular moment. But, for example, when I was the Executive Director of the United Nations Environment Programme in the early 1970s, he came to the new UN Headquarters in Nairobi and delivered a speech in 1973 on poverty and development.[1] His speech really transformed my understanding and my appreciation of him. That was a real turning point in my own view of his role and potential as an international leader and global thinker.

Robert Strange McNamara, 5th President of the World Bank Group, 1968–1981 (Associated Press)

X: Did the Vietnam War influence your views leading to the 1972 United Nations (UN) Conference on the Human Environment in Stockholm?

MS: I don't really think so. By that time, McNamara had himself decided that his role in going to war under Kennedy was a mistake... the Vietnam War was not the best strategy against Communism, and mostly futile. So, he turned his back on that conflict during the Johnson Administration.

X: Did the Vietnam War influence your views on the environment in the late 1960s prior to the UN Conference on the Human Environment that you organized in Stockholm?

MS: No, I don't think there's much relation between those two events.

X: Is there any particular set of events that made it absolutely urgent for you to organize the Stockholm Conference in 1972, which clearly is a turning point in environmental history around the world?

MS: I agree with your last statement, but on the other hand, it wasn't really related to Robert McNamara views.

X: So the organization of the Stockholm Conference was more informed by your own perception of events, independent of the Vietnam War and Robert McNamara. Given that you were travelling extensively and seeing many changes throughout the world, can you point to any other international events that you saw, or were particularly urgent to address?

"I am convinced that the prophets of doom ought to be taken seriously. In other words, Doomsday is a possibility."

—Maurice Strong, BBC Interview, 1972

1 Robert S. McNamara, "The Nairobi Speech," (address to the Board of Governors by the President of the World Bank Group, Nairobi, Kenya, September 24, 1973).

MS: You've done your homework, but your questions are complicated to answer.

X: I'll try to make them simpler. Are there events that influenced the urgency for putting together the UN Conference?

MS: Well, yes, all over the world. Or, most over the world, there were concerns about the urgency with which the international community needed to come together to cooperate to address environmental issues.

Maurice Strong at the 1972 Stockholm Conference (UN Photo / Yutaka Nagata)

X: Did your meeting with Sverker Åström—the Swedish State Secretary of Foreign Affairs and Permanent Representative to the United Nations—at Harvard University, influence your thoughts on the organization of a conference on the environment at the United Nations?

MS: Well, I don't think so. I mean, I did meet him at Harvard. I did get to know Åström. He was a great influence on what I was doing, and a great help. But I don't recall that being coming out of our common interests at Harvard.

Maurice Strong cycling in Stockholm at the center of delegates of the 1972 UN Conference on the Human Environment (©Pressens Bild AB)

X: Were you familiar with the work of Jay Forrester and 'systems dynamics' at the Massachusetts Institute of Technology (MIT), while you were at Harvard?

MS: Yes, very much so.

X: Was Forrester's work influential?

MS: Very influential, at least on me. And, I think more generally, as well.

X: I spoke with Jay Forrester about a year and a half ago. He is very focused and knowledgeable. He was very influential with a number of people including his students who then went on to publish *The Limits to Growth* with the Club of Rome. Was that book an important influence?

MS: It was one of the best. There were several books or writings about that event, but the most authoritative, and the best, were his works.[2]

X: Did Forrester influence the book that you later commissioned after the Stockholm Conference, *Only One Earth*, about the future of the planet?

MS: Well, yes, I think all those things influenced me. These were the broader issues out of which my own concerns emerged.

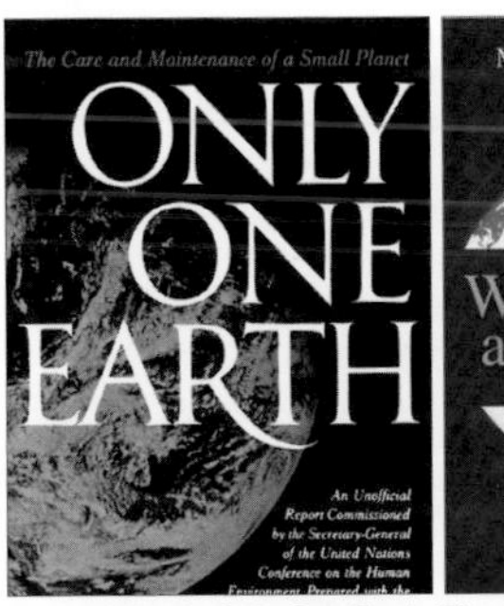

Barbara Ward, *Only One Earth: The Care and Maintenance of a Small Planet*, 1972; and Maurice Strong, *Where on Earth Are We Going?* 2001

X: Very clearly, in terms of your international career and business commitment towards publishing around the world, what impact do you see as being the most important coming out of the Stockholm Conference?

MS: Well, I think it put the issue on the international map. There were few people that did not hear or understand of the Stockholm Conference before that. So, its widespread recognition really put the whole issue related to the environment on the international agenda.

X: Is one of the strengths of development, as an international strategy, critically important to the future of the environment?

2 Jay Forrester, engineer and scientist, died on November 16, 2016, at the age of 98, one year after this conversation.

MS: It is, but more so under earlier Liberal Governments. In Canada, we currently have a very ideologically-negative Conservative Government, with an ideologically-negative approach to the environment.[3] So, Stockholm went beyond that. The Stockholm Conference had an enduring effect that was, to some degree, overlooked by the current Government in Canada and by other governments. Overall, the Stockholm Conference was a major event in the history of the environmental movement.

3 In reference to the Conservative Administration of Stephen Harper, then Prime Minister of Canada, from 2006 to 2015.

X: Could current leaders—both corporate and political—learn a lot more from understanding the historical importance of the Stockholm Conference?

MS: Yes, I think that is generally true. But, even corporate leaders weren't so interested at that time, but it was the Stockholm Conference that elicited interest on the environment and, in many cases, brought together support from the corporate community.

X: That's extremely interesting. Given the fact that there's not so much mention in current literature about the role of corporate leaders on the environment, are there any particularly interesting executives or influential corporate leaders that influenced you, prior to the Stockholm conference?

MS: I can't think at this moment of any individual. Corporate leaders in the 1970s were only beginning to get interested on the issue of the environment at that time so I can't remember any particular one that took the lead on the issue.

Maurice Strong as Head of Canada's Government External Aid Office, precursor to CIDA, the Canadian International Development Agency (Canadian Broadcasting Corporation, 1966–1970)

X: You served as an executive in a number of influential positions domestically and internationally as President of Power Corp. in the mid 1960s, CEO of Petro-Canada in the late 70s, Chairman of the International Energy Development Corporation in the early 1980s, President of the Canadian International Development Agency (CIDA) in the late 1980s, CEO of Ontario Hydro for its restructuring in the 1990s, and Secretary-General at the UN for the 1992 Rio Earth Summit. Can you discuss some of your influences and role models?

MS: I think my influence, well, I can't really say how influential I was with others, but those who influenced me were people like Jay Forrester.

X: Did you ever meet the Canadian-American economist John Kenneth Galbraith?

MS: I met him and I was influenced by him. But, I didn't have a direct or ongoing relationship with him. I was certainly influenced by his ideas and his writings.

Maurice Strong with Al Gore, 2009 (Maxim News Network)

X: Was Galbraith's 1967 book, *The New Industrial State*, about industry and environment, influential at the time?

MS: Yes, it was.

X: Although Galbraith seems outdated, does his work seem to be important again, to understand the role of industry in environment?

MS: Well, yes indeed. Industry is the principal means by which we impact on the environment and therefore industry has an essential role in dealing with the problems which it helps to create.

X: Was Ayn Rand's *Atlas Shrugged* from 1957, influential to you at all?

MS: Not directly, I mean, it was one of the things that I read and understood. But, I didn't have any direct relationship with Ayn Rand. Her works were interesting, I didn't agree with them all. They were certainly all important to know about, and it was in that sense that I was influenced by her.

John Kenneth Galbraith, *The New Industrial State*, 1967; and Donella H. Meadows et al., *The Limits to Growth: A Report for the Club of Rome's Project on the Predicament of Mankind*, 1972

X: Is there enough teaching about leadership—corporate-based or community-based—in Canadian universities today, to prepare young people for the challenges of industry and environment especially in activities that cross borders?

MS: Well, it's much better than it was, yes. And the Stockholm Conference played an important part in increasing the interest, at least here in Canada, in the universities and the corporations.

X: I've read that John Ralston Saul was a close colleague and friend of yours. He dedicated his book *Voltaire's Bastards* to you. Can you describe what impact his views of nationhood had on you?

MS: Well, I'm not quite sure how to answer that. He had a great influence on me but I think it was the other way around. I was interested in those issues (of nation, state, development, and environment) long before he was. I always admired him and appreciated the way he elaborated those issues and helped to promote them.

X: What book or experience early on, either with the Hudson's Bay Company, or learning Inuit, would you say was particularly influential to you?

MS: Well, I don't really know. I read every book I think I could find that had some relevance at that time, but I can't think of a particular one that was at the center of that.

Maurice Strong, 2003 (Calgary Herald File)

X: Regarding the changing international agenda and foreign development policy of Canada in the world that has shifted from aid to extraction, has Canada become a global resource empire?

MS: Well, Canada is the source of many strategic resources, but it decided to be an empire; the resources are just there, *they are strategic*. This is important in terms of the development of the Canadian economy and international interests in the Canadian economy.

X: To shape the future?

MS: It would be best for Canada to understand itself as a resource-based economy.

This conversation took place between Maurice Strong and Pierre Bélanger on July 16 and August 31, 2015; then transcribed posthumously following Mr. Strong's death in late 2015, on November 27, at the age of 86.

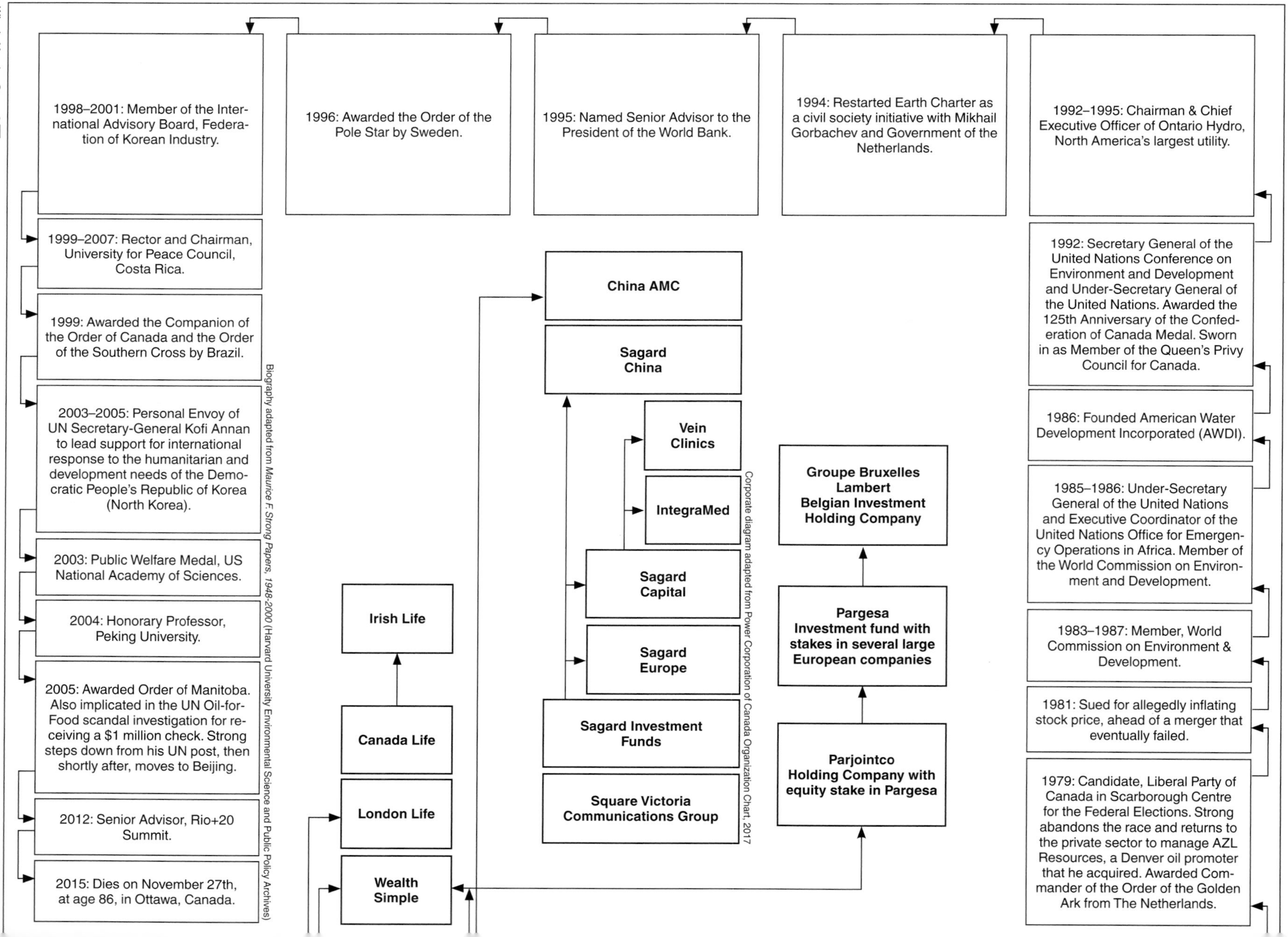

1979: Candidate, Liberal Party of Canada in Scarborough Centre for the Federal Elections. Strong abandons the race and returns to the private sector to manage AZL Resources, a Denver oil promoter that he acquired. Awarded Commander of the Order of the Golden Ark from The Netherlands.

1981: Sued for allegedly inflating stock price, ahead of a merger that eventually failed.

1983–1987: Member, World Commission on Environment & Development.

1985–1986: Under-Secretary General of the United Nations and Executive Coordinator of the United Nations Office for Emergency Operations in Africa. Member of the World Commission on Environment and Development.

1986: Founded American Water Development Incorporated (AWDI).

1992: Secretary General of the United Nations Conference on Environment and Development and Under-Secretary General of the United Nations. Awarded the 125th Anniversary of the Confederation of Canada Medal. Sworn in as Member of the Queen's Privy Council for Canada.

1992–1995: Chairman & Chief Executive Officer of Ontario Hydro, North America's largest utility.

1994: Restarted Earth Charter as a civil society initiative with Mikhail Gorbachev and Government of the Netherlands.

1995: Named Senior Advisor to the President of the World Bank.

1996: Awarded the Order of the Pole Star by Sweden.

1998–2001: Member of the International Advisory Board, Federation of Korean Industry.

1999–2007: Rector and Chairman, University for Peace Council, Costa Rica.

1999: Awarded the Companion of the Order of Canada and the Order of the Southern Cross by Brazil.

2003–2005: Personal Envoy of UN Secretary-General Kofi Annan to lead support for international response to the humanitarian and development needs of the Democratic People's Republic of Korea (North Korea).

2003: Public Welfare Medal, US National Academy of Sciences.

2004: Honorary Professor, Peking University.

2005: Awarded Order of Manitoba. Also implicated in the UN Oil-for-Food scandal investigation for receiving a $1 million check. Strong steps down from his UN post, then shortly after, moves to Beijing.

2012: Senior Advisor, Rio+20 Summit.

2015: Dies on November 27th, at age 86, in Ottawa, Canada.

Biography adapted from *Maurice F. Strong Papers, 1948-2000* (Harvard University Environmental Science and Public Policy Archives)

Parjointco
Holding Company with equity stake in Pargesa

Pargesa
Investment fund with stakes in several large European companies

Groupe Bruxelles Lambert
Belgian Investment Holding Company

Wealth Simple

London Life

Canada Life

Irish Life

Square Victoria Communications Group

Sagard Investment Funds

Sagard Europe

Sagard Capital

IntegraMed

Vein Clinics

Sagard China

China AMC

Corporate diagram adapted from Power Corporation of Canada Organization Chart, 2017

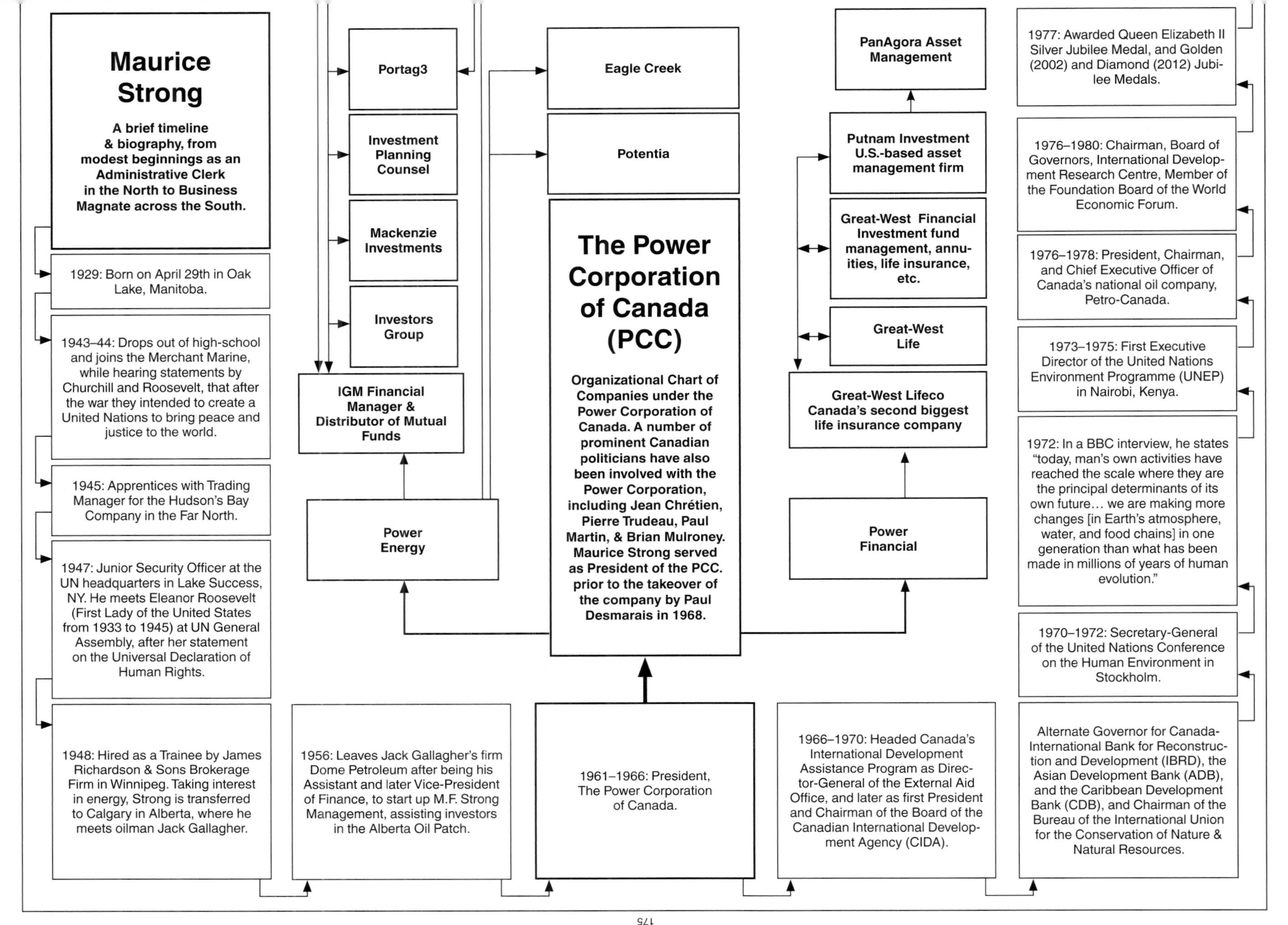
Maurice Strong
A brief timeline & biography, from modest beginnings as an Administrative Clerk in the North to Business Magnate across the South.
1929: Born on April 29th in Oak Lake, Manitoba.
1943–44: Drops out of high-school and joins the Merchant Marine, while hearing statements by Churchill and Roosevelt, that after the war they intended to create a United Nations to bring peace and justice to the world.
1945: Apprentices with Trading Manager for the Hudson's Bay Company in the Far North.
1947: Junior Security Officer at the UN headquarters in Lake Success, NY. He meets Eleanor Roosevelt (First Lady of the United States from 1933 to 1945) at UN General Assembly, after her statement on the Universal Declaration of Human Rights.
1948: Hired as a Trainee by James Richardson & Sons Brokerage Firm in Winnipeg. Taking interest in energy, Strong is transferred to Calgary in Alberta, where he meets oilman Jack Gallagher.
1956: Leaves Jack Gallagher's firm Dome Petroleum after being his Assistant and later Vice-President of Finance, to start up M.F. Strong Management, assisting investors in the Alberta Oil Patch.
1961–1966: President, The Power Corporation of Canada.
1966–1970: Headed Canada's International Development Assistance Program as Director-General of the External Aid Office, and later as first President and Chairman of the Board of the Canadian International Development Agency (CIDA).
Alternate Governor for Canada-International Bank for Reconstruction and Development (IBRD), the Asian Development Bank (ADB), and the Caribbean Development Bank (CDB), and Chairman of the Bureau of the International Union for the Conservation of Nature & Natural Resources.
1970–1972: Secretary-General of the United Nations Conference on the Human Environment in Stockholm.
1972: In a BBC interview, he states "today, man's own activities have reached the scale where they are the principal determinants of its own future… we are making more changes [in Earth's atmosphere, water, and food chains] in one generation than what has been made in millions of years of human evolution."
1973–1975: First Executive Director of the United Nations Environment Programme (UNEP) in Nairobi, Kenya.
1976–1978: President, Chairman, and Chief Executive Officer of Canada's national oil company, Petro-Canada.
1976–1980: Chairman, Board of Governors, International Development Research Centre, Member of the Foundation Board of the World Economic Forum.
1977: Awarded Queen Elizabeth II Silver Jubilee Medal, and Golden (2002) and Diamond (2012) Jubilee Medals.
The Power Corporation of Canada (PCC)
Organizational Chart of Companies under the Power Corporation of Canada. A number of prominent Canadian politicians have also been involved with the Power Corporation, including Jean Chrétien, Pierre Trudeau, Paul Martin, & Brian Mulroney. Maurice Strong served as President of the PCC. prior to the takeover of the company by Paul Desmarais in 1968.
Power Energy
IGM Financial Manager & Distributor of Mutual Funds
Portag3
Investment Planning Counsel
Mackenzie Investments
Investors Group
Eagle Creek
Potentia
Power Financial
Great-West Lifeco Canada's second biggest life insurance company
Great-West Life
Great-West Financial Investment fund management, annuities, life insurance, etc.
Putnam Investment U.S.-based asset management firm
PanAgora Asset Management

Desmarais Estate in Sagard, QC

Ian Austen, "The Name Is 'Power' and It Fits," *The New York Times*, January 26, 2007

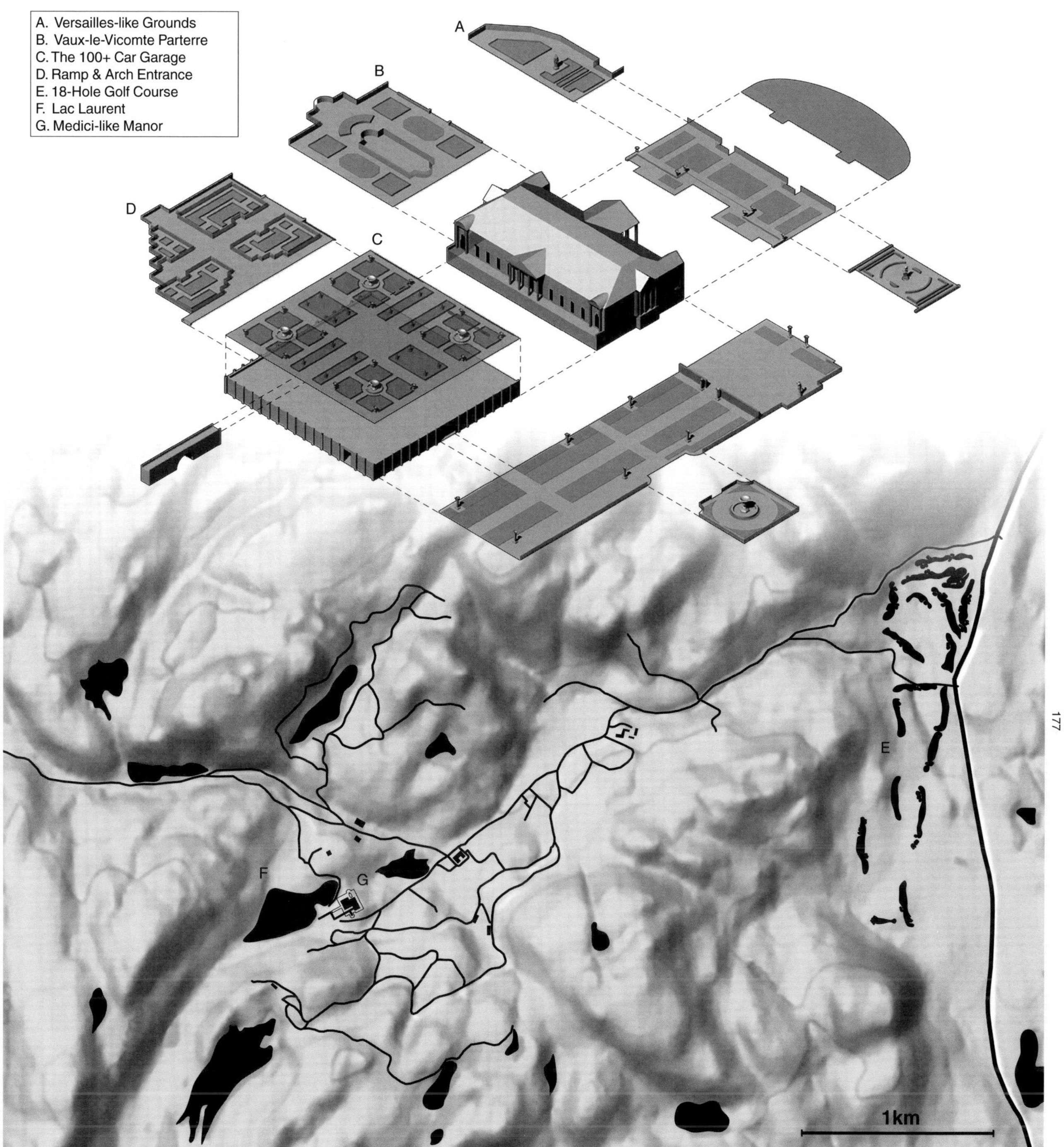

The Architectural Imperialism & Gardenesque Renaissance of the Desmarais Estate

Nestled in the low-lying hills of the Laurentians of Charlevoix in Quebec, the 21,000-acre forest-estate, commonly known as 'Domaine Laforest' or 'Domaine Sagard,' is home to an elite family of power brokers since the turn of the millennium. At the center of this family is the management of the Power Corporation of Canada (PCC). Headquartered in Montreal, the Power Corporation is a multinational holding corporation whose portfolio includes energy, resources, communications, and insurance assets with companies located from China to Belgium. Acquired by Paul Desmarais, Sr. (1927–2013) from its years under business magnate Maurice Strong, the Power Corporation has recently seen control move into the hands of sons Paul Desmarais Jr. and André Desmarais, following the death of the patriarch in 2013. As a clear and overt demonstration of empire building, the sprawling 21,000-acre estate aptly, yet abhorrently, combines vestiges of imperial garden architecture, from Palladio's sixteenth-century Italian Renaissance Villa La Malcontenta, to the seventeenth-century French parterres of Vaux-Le-Vicomte. Extending eastward, is the modern golf course architecture of an 18-hole golf course designed by Thomas McBroom, lining the shores of a river (Petit Saguenay) and a stream (Laurent) in a practically perfect, textbook display of Ruskinian picturesque. ◿

Royal Bank of Canada Plaza, Toronto

Duncan McDowall, *Quick to the Frontier: Canada's Royal Bank*, 1993

South Tower

40 RBC (executive suite)
39 RBC Dominion Securities
38 Norton Rose Fulbright Canada LLP
37 Norton Rose Fulbright Canada LLP
36 Norton Rose Fulbright Canada LLP
35 Norton Rose Fulbright Canada LLP
34 Norton Rose Fulbright Canada LLP
33 BDO Canada LLP
32 Wessex Capital Partners Ltd
32 Mitsubishi Power Systems Canada
32 Hugessen Consulting
31 New Gold Inc
31 Odgers Berndston
31 Ventus Energy
30 RBC
29 Stratagis Capital Advisors
29 Lloyds Canada Inc.
29 Aurora Msc
29 Alvarez & Marsal
28 Cormark Securities Inc.
27 Sprott Resource Corporation
26 Sprott Money Ltd
25 RBC Dominion Secuties
24 Transcanada: Canadian Mainline System
23 North American Palladium
22 Cormerica Bank
21 Detour Gold Corporation
20 OMERS Private Equity
19 RBC GTA Regional Office
19 RBC Real Estate Markets
19 RBC Group Advantage
19 RBC Career Sales Force
18 JP Morgan Chase Bank
17 Bank of Tokyo - Mitsubishi UFJ
16 Vale Canada Ltd
15 RBC Capital Trust
14 OMERS Ventures
13 Oxford Properties Management
12 RBC
11 RBC
10 RBC
09 RBC
08 RBC
07 RBC
06 RBC
05 RBC
04 RBC
03 RBC
02 RBC

North Tower

Stewart Title Guaranty Company 26
Axis Corporation 25
Interac Association 25
Interac Association 24
OMERS Capital Markets 23
Yamana Gold Inc 22
Borealis Infrastructure Management 21
AIP Venture Capital 20
Linden & Associates 20
China Investment Corporation 20
19
Vacant 18
Vacant 17
State Bank of India (Canada) 16
Cutfeld Freeman & Company 16
Berkshire Hathaway Specialty Insurance 15
Vacant 14
Vale Canada Ltd 13
Oxford Property Group (OMERS) 12
RBC Human Resources 11
Oxford Property Group (OMERS) 10
Oxford Property Group (OMERS) 09
Oxford Property Group (OMERS) 08
Oxford Property Group (OMERS) 07
Oxford Property Group (OMERS) 06
Oxford Property Group (OMERS) 05
RBC Dominion Securities Inc 04
RBC Capital Partners 04
RBC 03
RBC 02

Atrium

Trading Floor
Banking Hall

"Suspended Virtual Volume"
by Jesús Rafael Soto (1977)

50m

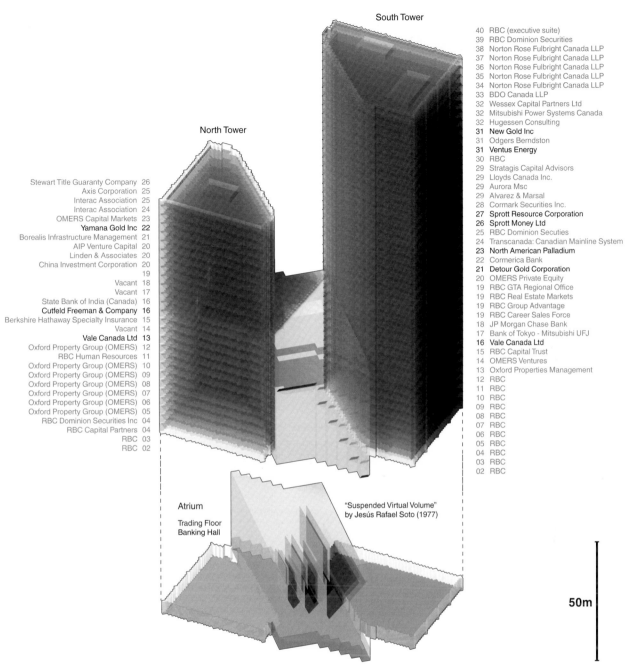

The Geological Real Estate of the Royal Bank Plaza

Originally built in 1976, the Royal Bank Plaza building is one of downtown Toronto's most iconic banking towers in the world. Designed by the Webb Zerafa Menkes Housden Partnership (WZMH) whose namesake was established with the design for the *CN Tower* in 1973, WZMH portfolio has grown to include *Scotia Plaza* (Toronto, 1989) and the new home for *Communications Security Establishment Canada* in Ottawa, dubbed "Canada's top-secret billion-dollar spy palace" by the Canadian Broadcasting Corporation. WZMH has gained a unique reputation for complex, large scale delivery of corporate, institutional, and government projects. Housing the international headquarters of the Royal Bank of Canada (one of Canada's big five banks), the tower's real estate is home to a suite of tenants that are some of the largest resource investment companies and legal consultancies in the world: New Gold, TransCanada, Sprott, and Norton Rose Fulbright. Formed by two adjoining, triangular office towers (26 and 41 levels), the building is entirely clad in gold reflective glass, linked by a 'banking hall' on the 12th floor, a 40-metre high, glass-enclosed atrium above a trading floor of more than 400 traders. According to Brookfield Properties, the building features, "14,000 windows of the towers are coated with a layer of 24-karat gold (2,500 ounces), worth about $70 per window. The total amount of gold used in the windows is worth over 1 million dollars." Although it houses the most gold leaf content of any other building facade in the world, lending spectacular reflections in sunlight, the process used in the manufacturing of the glass is a metaphor for the mineral extraction industry. The precious metal embedded in the glass facade of the tower is entirely unrecoverable. ΔΔ

Parliament Building Centre Block, Ottawa, ON

Saffron Blaze, via http://www.mackenzie.co

1. Exterior Walls - Nepean Sandstone (Nepean, ON) **2.** Decorative Trim, Exterior - Ohio - Sandstone (Wakeman, OH) **3.** Archway, Peace Tower - Wallace Sandstone (Wallace, NS) **4.** Decorative Trim, Courtyards - Wallace Sandstone (Wallace, NS) **5.** Decorative Trim, Penthouses - Wallace Sandstone (Wallace, NS) **6.** Cornerstones - Portage du Fort Marble (Portage du Fort, QC) **7.** Main Steps - Stanstead Granite (Beebe, QC) **8.** Gargoyles, Peace Tower (ext.) - Stanstead Granite (Beebe, QC) **9.** Decorative Trim, Library - Potsdam Sandstone (Malone, NY) **10.** Flying Buttresses, Library - Gloucester Limestone (Gloucester, ON) **11.** Walls, Lower Corridors - Tyndall Limestone (Garson, MB) **12.** Walls, Main Entrance - Tyndall Limestone (Garson, MB) **13.** Floor, House of Commons - Tyndall Limestone (Garson, MB) **14.** Floor, Senate Chamber - Tyndall Limestone (Garson, MB) **15.** Floor, Confederation Hall - Tyndall Limestone (Garson, MB) **16.** Floor, Hall of Honour - Tyndall Limestone (Garson, MB) **17.** Railway Committee Room - Bath Stone (Bath, UK) **18.** Floor, Entrance Hall - Missisquoi Bolder Grey Marble (Philipsburg, QC) **19.** Floor, Corridors - Missisquoi Bolder Grey Marble (Philipsburg, QC) **20.** Floor, Stairs - Missisquoi Bolder Grey Marble (Philipsburg, QC) **21.** Walls, Upper Corridors - Missisquoi Verte Gris Marble (Philipsburg, QC) **22.** Base, Main Corridors - Missisquoi Black Marble (Philipsburg, QC) **23.** Walls, 6th flr Corridor - Hoidge Light Pink Cloud Marble (Bancroft, ON) **24.** Border, 6th floor Corridor - Hoidge Dark Pink Cloud Marble (Bancroft, ON) **25.** Columns, Main Dining Room - Grand Antique Marble (VT) **26.** Balustrade, Commons - Rose Tavernelle Marble (Italy) **27.** Border, Main Corridors - Vert Antique Marble (Roxbury, VT) **28.** Columns, Senate - Red Verona Marble (Italy) **29.** Columns, Memorial Chamber - St-Anne Marble (France) **30.** Base, Memorial Chamber - Belgian Black Marble (Belgium) **31.** Walls, Memorial Chamber- Chateau Gaillard Stone (France) **32.** Walls, Memorial Chamber - Hoptonwood Limestone (Darbyshire, UK) **33.** Vaults, Senate - Peerless Indiana Limestone (Bedford, IN) **34.** Vaults, Commons - Peerless Indiana Limestone (Bedford, IN) **35.** Vaults, Foyer - Peerless Indiana Limestone (Bedford, IN) **36.** Walls, Lavatories - Pink Tennessee Marble (Knoxville area, TN) **37.** Floors, Lavatories - Pink Tennessee Marble (Knoxville area, TN) **38.** Various Ornaments - Cedar Tennessee Marble (Knoxville area, TN) **39.** Gargoyles, Peace Tower (int.) - Stanstead Granite (Beebe, QC) **40.** Circular Border, Main Entrance - Tinos no. 2 Serpentine (Tinos, Greece) **41.** Floor, Memorial Chamber - Battlefield Stone (France and Belgium) **42.** Mantles, Various Locations - Black and Gold Marble (Belgium) **43.** Countertops, Serving Room - Bianco Italian Marble (Italy) **44.** Columns, Library Entrance - Red Granite (St. George, NB) **45.** Columns, Confederation Hall - Windsor Green Syenite (Windsor, VT) **46.** Columns, Hall of Honour - Windsor Green Syenite (Windsor, VT) **47.** Flag - Nylon (Toronto, ON) **48.** Roof, Centre Block - Copper (MA) **49.** Carillon Bells, Peace Tower - Cast Bronze (Croydon, UK) **50.** Clock, Peace Tower - (OH)

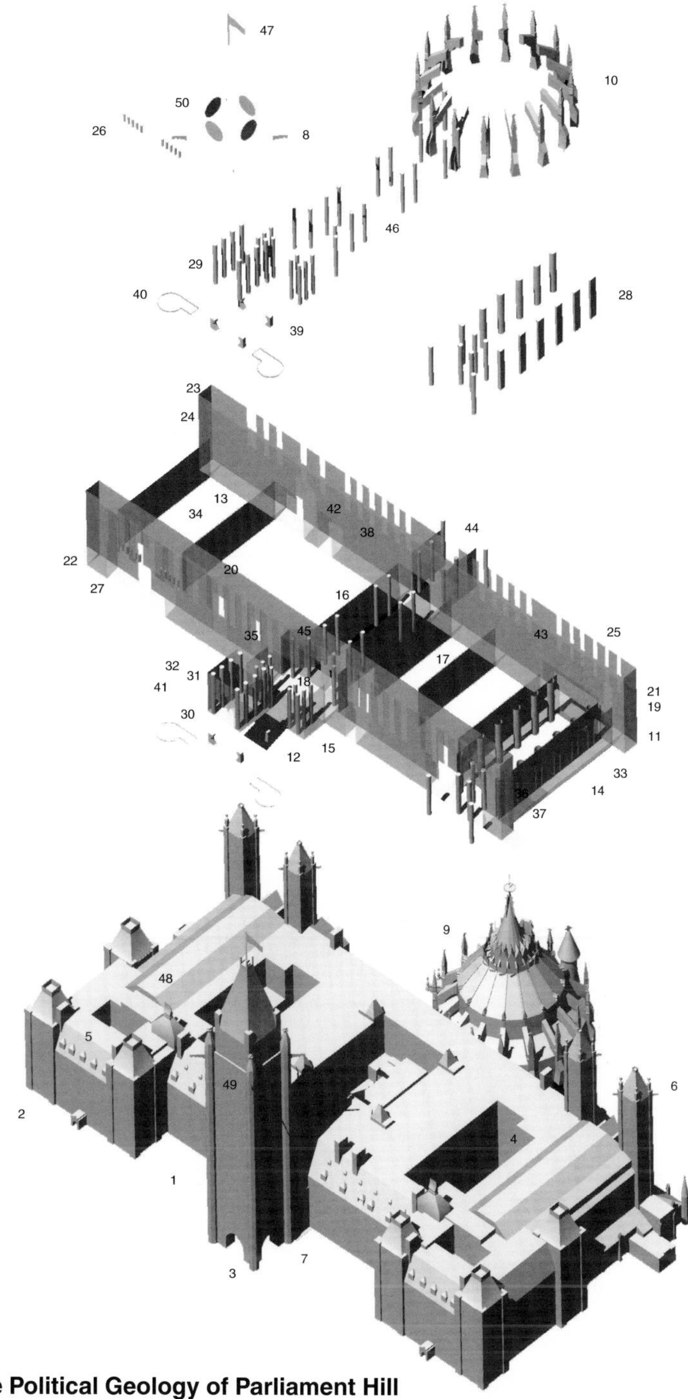

The Political Geology of Parliament Hill

Canada's Parliament Hill is host to the single largest architectural project in Canada today: a 20-year rebuilding project worth over 3 billion dollars that will be completed in the mid-2020s. According to the Geological Survey of Canada, there are stories written in the stones of the Parliament Building of the 'Dominion' ever since Queen Victoria designated Ottawa as the capital of the country in 1858. While the gray, neo-gothic architecture reportedly represents the forms of parliamentary democracy (potentially like that of Britain), it more accurately appropriates, reflects, and revives a Eurocentric appetite for so-called functional, picturesque ideals that later guide the planning of the Capital City. As profiled here, the stones and styles that fill halls and line walls of buildings of the 'Hill,' designed by Chief Dominion Architect Thomas Fuller, not surprisingly originate from some of the most violent empires in the world, from Italy and France, Belgium to Britain. While quarries in Quebec and Manitoba may have served as hosts for the construction and architecture of Parliament, the Parliament stands both as a cornerstone of colonialism and as a microcosm of the *Extractive State*. Its political geology stands tall as a mine site turned upside down, ordered by federal, administrative design. Atop a once vibrant beech and hemlock forest, the Haudenosaunee lands upon which the capital of Canada rests on, serve as an important reminder of deep histories and underlying futures of land to be reclaimed. ΔΔΔ

Power Inc.

From Petro-Canada to Power Corporation of Canada, Before and Beyond

The nation's largest resource companies have served as testbeds for training the nation's most influential power brokers. From the earliest forms of political Indigenous resistance, this graphic timeline redraws the geopolitical and continental eras of power dynamics by proposing a deeper history of land and life. Here, the space of Indigenous peoples and Indigenous Nations are visualized and expressed as contemporary, political cultures contesting corporate and governmental structures for more than eight centuries at various stages of contact with Europeans. Together, they are leading to the inevitable weakening of state structures and colonial control through affirmation of anti-imperial freedoms, reconstruction of anti-patriarchal community relations, and resurgence of

General James Wolfe
Major General, British Army, 1740-1759; Known for victory over the French at the Battle of the Plains of Abraham

William Johnson
Superintendant of
Affairs, 1755-1774;
r-General, British
my, 1755-1774

Chief Pontiac
Odawa War Chief, Pontiac War, 1763-1766

Chief Thayendanegea
"Joseph Brant"
742-1807; Mohawk Chief
nber of Six Nations Iroquois
federacy; Early Advocate of Self-Determination

Tomah Denys
Mi'kmaq Grand Chief, ca. 1759
Cumberland district, pre-1759
Escasoni, post-1759

Chief Tecumseh
1768-1813; Shawnee Indian Leader
Formed tribal confederacy to unite against USA

General Louis-Joseph,
Marquis de Montcalm
Lieutenant General, French Army, 1721-1759; French Commander at Battle of the Plains of Abraham

King George III
King of Great Britain and Ireland, 1760-1820; Issued Royal Proclamation, 1763

Kahkewaquonaby (Peter Jones)
Ojibwa Methodist Minister, 1802-1856

Alexander Macdonell
First Roman Catholic Bishop of Kingston, Upper Canada, 1819

William Lyon Mackenzie
First Mayor of Toronto, 1834-1835; Member of Legislative Assembly, Upper Canada, 1829-1834; President of the Republic of Canada, 1837-1838; Leader during 1837 Upper Canada Rebellion

Joseph Onasakenrat
Mohawk Chief of Kanesatake, 1868
Methodist Minister, 1880

Shanawdithit
Last Beothuk Woman, 1801-1829

Isobel "John Fubbister" Gunn
Scottish woman masqueraded as a man to work for Hundon's Bay Company, 1806-1807; First European Woman to travel to Rupert's Land

Egerton Ryerson
Superintendent of Schools for Upper Canada, 1844; Released a report on Indian education (preceding residential school system), 1847

Queen Victoria
Queen of the United Kingdom
Empress of India; Reign: May 1, 1876 - January 22, 1901

John A. MacDonald
First Prime Minister of Canada 1867-1873; Leader of the Opposition 1873-1878; Prime Minister of Canada 1878-1891

Sandford Fleming
Railway Engineer, Northern Railway, 1852-1862; Engineer-in-Chief, Intercolonial Railway 1867-1876; Chief Engineer, Canadian Pacific Railway, 1871-1880; Director, Canadian Pacific Railway, 1884; Inventor of Time Zones

Hector-Louis Langevin
Public Works Minister, 1869-1873 and 1879-1891; One of the Fathers of Confederation; Proponent of residential school system, advocated for the separation of indigenous children from their families

Edgar Dewdney
Lieutenant Governor of the Northwest Territories, 1881-1888; Lieutenant Governor of British Columbia, 1892-1897; Member of Canadian Parliament for Assiniboia East, 1888-1892; Member of Canadian Parliament for Yale, 1872-1879

American Progress (1872)

Chief Big Bear
Cree chief involved
with the signing of Treaty 6, 1876

Louis Riel
President of the Provisional Government
of Saskatchewan, 1885; Member of
Parliament, 1873-1874

Gabriel Sylliboy
Mi'kmaq Grand Chief, 1919;
First Chief to fight for treaty
recognition, 1929; Treaty of
1752

Chief Sitting Bull
Hunkpapa Lakota Holy Man
Performer, Buffalo Bill's Wild West show,
1885

Buffalo Bill
American Scout, US Army, c. 1870;
Performer, Buffalo Bill's Wild West show,
1872-1901

Alexander Morris
Member of Parliament for Lanark South,
1867-1872; Lieutenant Governor of the
Northwest Territories, 1872-1876;
Lieutenant Governor of Manitoba,
1872-1877; Treaty Commissioner,
Treaties, 3, 4, 5, and 6

Gabriel Dumont
President of the Republic of St. Laurent,
1873; Metis Leader in North-West
Rebellion 1885

James Douglas
Governor of Vancouver Island,
1851-1864; Governor of British Columbia,
1858-1864

Lawrence Vankoughnet
Deputy Superintendent General of Indian
Affairs, 1874

Belinda Mulrooney
Entrepreneuer of during the Klondike Gold
Rush, 1896-1899

Agnes Deans Cameron
Teacher, Writer, Journalist, Lecturer,
Adventurer; First White Woman to reach
Arctic Ocean, 1908

John Stoughton Dennis
Deputy Minister of the Department of the
Interior, 1878-1881; Surveyor General of
Canada, 1871

Residential School Children
Fort Simpson Indian Residential School,
1922

Thomas Moore
Student at Regina Industrial Residential
School, 1896

Duncan Campbell Scott
Deputy Superintendent of the
Department of Indian Affairs,
1913-1932

Dr. Peter Bryce
Indian Affairs Chief Medical Officer, 1906

Tahltan Chiefs
Declaration of the Tahltan Tribe

Frank Oliver
Minister of the Interio
1905-1911;
Responsible for havi
Papaschase Cree remove
Treaty 6 Land

Queen Victoria (1900)

Sir Wilfrid Laurier (1922)

← Power Inc.

Deskaheh
Cayuga Chief of
enosaunee Confederacy
1917-1925

Viola MacMillan
President of Prospectors and Developers Association of Canada, 1941

CD Howe
Chief Engineer, Board of Canadian Grain Commissioners, 1913-1916; MP, Minister of Railways and Canals & Minister of Marine, 1935-40; Minister of Munitions and Supply, 1940-44; Minister of Trade and Commerce, 1948-51

Hayter Reed
er of the Council of the
st Territories, 1883-1885;
n Commissioner of the
west Territories Deputy
endent General of Indian
Affairs

William Morris Graham
Indian Commissioner, 1918-1932

Alice Tye Fonds
Cousin of William Graham
Writer; Collector of Letters, Articles and Correspondence pertaining to Long Lance, William Graham, and the Department of Indian Affairs (Glenbow Museum)

Robert Alexander Hoey
Indian Affairs Superintendent of Welfare and Training, Department of Mines and Resources, 1936-1945

Marshall McLuhan
Professor of English, philosophy, and communication theory; Author of Understanding Media, 1964; The Medium is the Massage, 1967

Lester B Pearson
Prime Minister of Canada, 1963-68; Nobel Peace Prize, 1957

John Diefenbaker
Prime Minister of Canada, 1957-63

Queen Elizabeth II
Queen of the United Kingdom and the other Commonwealth realms Reign: February 6, 1952 - present

Robert Winters
MP, Minister of Reconstruction and Supply, 1948-50; MP, Minister of Resources and Development, 1950-53; MP, Minister of Public Works, 1953-57; President and CEO, Rio Algom Mines, 1963-65; MP, Minister of Trade and Commerce, 1966-68

Dan George
Chief of Tslei-Waututh Nation, 1951-1963; Actor, 1960-1981

Walter Gordon
Bank of Canada, 1939-45; President of the Privy Council, 1942-44; Chair, Royal Commission on Canada's Economic Prospects, 1955-57; MP, Minister of Finance, 1963-65; Co-founder, Walter and Duncan Gordon Foundation, 1965; Co-Founder, Committee for an Independent Canada, 1970

Harold Innis
Professor of political economy at
niversity of Toronto; Author of Empire and Communications, 1950

Walter E. Duffett
Wartime Prices and Trade Board, 1942-44; Researcher, Bank of Canada, 1944-54; Director of Economics and Research,
Department of Labour, 1954-1957; Dominion Statistician, Dominion Bureau of Statistics, 1957-1972

Chief Yellowbird
The Indian Chiefs of Alberta
Citizen's Plus "The Red Paper", 1970

Pierre Trudeau
Board Member, Power Corporation of Canada; Prime Minster of Canada, 1968-79, 1980-84

Jean Chrétien
Minister of National Revenue, 1968; Minister of Indian Affairs and Northern Development, 1968-1974; President of the Treasury Board, 1974-1976; Minister of Industry, Trade and Commerce, 1976-1977; Minister of Finance, 1977-1979; Minister of Justice, 1980-1982; Minister of Energy, Mines and Resources, 1982-1984; Leader of the Opposition, 1990-1994; Prime Minister of Canada, 1993-2003

Phebe Nahanni
Dene geographer
The Dene Mapping Project, 1974; Dene Women in the Traditional & Modern Northern Economy in Denendeh, Northwest Territories, Canada, 1992; Research Director, National Indian Brotherhood

Justice Thomas Berger
Supreme Court of British Columbia, 1972-1983; Mackenzie Valley Pipeline Inquiry, 1977; Calder Case, 1973

Georges Erasmus
President, Dene Nation 1974; Opponent of Mackenzie Pipeline Project, 1974; Order of Canada, 1987
Chief of the Assembly of First Nations, 1985-1991

The Indian Chiefs of Alberta
Authors of Citizen's Plus "The Red Paper" 1970

Harold Cardinal
Citizen's Plus, Indian Association of Alberta, 1970; Author of The Unjust Society, 1969

Elijah Harper
First Treaty Indian elected to provincial government, 1981; Chief for Red Sucker Lake Band, 1978-1982; MLA for Rupertsland, 1981-1992; MP for Churchill, 1993-1997

Paul Martin
Executive Assistant to Maurice Strong, 1970-73; CEO, Canada Steamship Lines (subsidiary of PCC) 1973-76;
Prime Minister of Canada, 2003-06
Founder, Martin Aboriginal Education Initiative, 2008; Co-Founder, CAPEFund
Chair, Congo Basin Forest Fund

Maurice Strong
President, Power Corporation of Canada, 1964-66; President, Canadian International Development Agency, 1966
CEO, Petro-Canada, 1976-78;
Secretary General, UNECD, 1992
CEO, Ontario Hydro, 1992-95

Paul Desmarais
Trans Canada Corporation Fund, 1965-1969; Chairman and CEO, Power Corporation of Canada, 1968-96; Founder, Business Council on National Issues, 1976; Founder, Canada-China Business Council, 1978

Peter Munk
Founder, Barrick Gold, 1984-2014; President, PM Capital, 1992-Present; Founder and Chairman, Trizec Properties, 2002-06; Porto Montenegro, 2006-Present

Francois Paulette
Dene Suline Elder, Smith's Landing Chief, Indian Brotherhood, 1971; Opponent of Mackenzie Pipeline Project, 1972; (Paulette Case)

Art Manual
Chair, Shuswap Nation Tribal C
1997-2003; Co-Chair, Delgam
Implementation Strategic Com
Spokesperson, Defenders of th
Activist, Idle No More Mover
Unsettling Canada: A National V
Call, 2016

Paul Tellier
Deputy Minister, Indian Affairs and Northern Development, 1972-82
Deputy Minister, Energy, Mines and Resources, 1982-85; Board of Directors, Bell Canada Enterprises 1999-2010; President and CEO, Bombardier, 2003-04; Board of Directors, Rio Tinto, 2007-Present

Queen Elizabeth II (1977)

John George Diefenbaker

chefferville Mine Workers
Company of Canada announces
ıre of Schefferville Mine, 1982

Brian Mulroney
nt, Iron Ore Company of Canada, 1983; Prime Minister of Canada, 993; Chairman of Board, Barrick 993-2013; Senior Partner, Norton ulbright, 1993-Present; Chairman rd, Quebecor Inc., 2014-Present

Pierre Lassonde
Founder of Franco-Nevada, 1982-2002; Author of The Gold Book, 1990; President of Newmont Mining Corp. 2002-2007; Chairman of World Gold Counvil, 2005-2009; Chairman of Franco-Nevada, 2008-present

Conrad Black
CEO, Ravelston Corp (voting control over Argus), 1978; Hollinger Mines consolodated into Hollinger-Argus, 1985

Eric Sprott
Founder, Sprott Resource Corp., 1994 Chairman, Sprott Inc., 2008-Present Founder, Sprott Physical Gold Trust, 2009; Founder, Sprott Physical Silver Trust, 2010; Chairman of the Board, Kirkland Lake Gold Inc, 2015

Lloyd Axworthy
MP, Minister of Transport, 1983-84 MP, Minister of Labour, 1993-95 MP, Minister of Employment and Immigration, 1993-96; MP, Minister of Foreign Affairs, 1996-2000; Board of Directors, MacArthur Foundation, 2000-01; Board of Directors, Hudbay Minerals Inc., 2006-09

Lester B. Pearson (1989)

Robert Skidders "Mad Jap"
Mohawk Warrior, Oka Crisis, 1990

Brad "Freddie Krueger" Larocque
Mohawk Warrior, Oka Crisis, 1990

Patrick Cloutier
Private, Canadian Forces, 1990

John Ciaccia
Member of National Assembly for Mont-Royal, 1973-1998; Minister of Energy and Natural Resources, 1985-1994; Key Player in Oka Crisis, 1990

Ronald "Lasagna" Cross
Mohawk Warrior, Oka Crisis, 1990

Gordon "Noriega" Lazore
Mohawk Warrior, Oka Crisis, 1990

Roger "20-20" Lazore
Mohawk Warrior, Oka Crisis, 1990

Alanis Obomsawin
Filmmaker, Kanehsatake: 270 Years of Resistance, 1993

Bob Lovelace
Ardoch Algonquin First Nation Chief, 2008; Instructor, Sir Sandford Fleming Community College

Marcel Coutu
President and CEO, Canadian Oil Sands Ltd., 2001-14; CFO and Senior Vice President, Gulf Canada Resources Ltd., 1999-2001; Board of Directors, Power Corporation of Canada, 2011-Present; Board of Directors, Enbridge, 2014-Present; Chairman, Syncrude Canada

Michael Ignatieff
Leader of the Liberal Party of Canada, 2009-2011; Member of Parliament, Etobicoke-Lakeshore, 2006-2011

Frank Iacobucci
Pusine Judge of the Suprime Court of Canada, 1991-2004

Phil Fontaine
National Chief, Assembly of First Nations, 1997-2000, 2003-09; Special Advisor, Royal Bank of Canada, 2009-Present; Board of Directors, Avalon Rare Metals, 2009-Present; Board of Directors, Chieftan Metals Corp, 2010-Present; Board of Directors, New Brunswick Power, 2013-14

Ellen Gabriel
Mohawk Activist, Kanehsatà:ke Nation (Turtle Clan); Spokesperson, People of the Longhouse, Oka Crisis, 1990; Art Teacher, Mohawk Immersion School; President, Quebec Native Women's Association, 2004; Golden Eagle Award, 2005; Jigonsaseh Women of Peace Award, 2008

Clement Chartier
President of the World Council of Indigenous Peoples, 1984-1987
President of Metis Nation - Saskatchewan, 1998-2003; President of Metis National Council, 2003-present

Oren Lyons
Haudenosaunee Faithkeeper, b. 1930; Activist, Speaker, Artist; Former Lacrosse Player

David Tuccaro
President and General Manager, Neegan Development Corporation Ltd., 1991-2001
Director, Aboriginal People's Television Network, 1999; President and CEO, Tuccaro Inc., 2001-Present Investment Committee, CAPEFund

Mary Simon
Canadian Ambassador for Circumpolar Affairs, 1994-2003; President of the Inuit Tapiriit Kanatami, 2006
Executive Council Member and President, Inuit Circumpolar Conference, 1980-1994; Member of Nunavut Implementatiion Commission, 1993

Anita Zucker
Governor, Hudson's Bay Company, 2008

Michaëlle Jean
Governor General of Canada, 2005-2010

Edward John
Minister of Children and Families of British Columbia, 2000-2001
Elected Councillor of Tl'azt'en Nation, 1974-1990; Chief of Tl'azt'en Nation, 1990-1992; Chief of Carrier-Sekani Tribal

Patrick Brazeau
National Chief of the Congress of Aboriginal Peoples, 2006-2009
Expelled from Conservative caucus following arrest for domestic and sexual assault, 2013
Lost a celebrity boxing match to Justin Trudeau, 2012

John Baird
MP, Minister of Foreign Affairs, 2011-15
International Advisor, Barrick Gold, 2015; Board of Directors, Canada Pacific Railway, 2015
Senior Advisor, Bennett Jones LLP, 2015;

Amanda Polchies
Mi'kmaq Anti-fracking protester, 2013

David Johnston
Governor General of Canada, 2010-prese

Elder Stephen Augustine
Principal of Unama'ki College; 2012-present, Mi'kmaq Elder

Clayton Thomas Mulle
Activist, writer, public speak and facilitator, for Indigeno self-determination and environmental justice
Co-Director; Indigenous T Sands Campaign of the Po Institute
Organizer; Defenders of the Idle No More

Inside the Economy of Appearances

Anna Lowenhaupt Tsing

"the unemployed become murderers
with uniforms and badges of rank
vast forests are torn apart

It is necessary that I emphasize
the problem of power
that tends to turn people into bandits"[1]

—Pramoedya Ananta Toer, as reworked
by Peter Dale Scott in *Minding the Darkness*,
[1996] 2000

Scale. Relative or proportionate size or extent.[2]

Indonesia's profile in the international imagination has completely changed. From the top of what was called a "miracle," Indonesia fell to the bottom of a "crisis." In the middle of what was portrayed as a timeless political regime, students demonstrated, and, suddenly, the regime was gone. So recently an exemplar of the promise of globalization, overnight Indonesia became the case study of globalization's failures.

The speed of these changes takes one's breath away—and raises important questions about globalization. Under what circumstances are boom and bust intimately related to each other? If the same economic policies can produce both in quick succession, might *deregulation* and *cronyism* sometimes name the same thing—but from different moments of investor confidence? Such questions run against the grain of economic expertise about globalization, with its discrimination between good and bad kinds of capitalism and policy. Yet the whiggish acrobatics necessary to show how those very economies celebrated as miracles were simultaneously lurking crises hardly seem to tell the whole story. A less pious attitude toward the market may be necessary to consider the specificities of those political economies, like that of Suharto's Indonesia, brought into being together with international finance.[3]

This essay brings us back to the months just before Indonesia so drastically changed, to canoe at the running edge of what turned out to be a waterfall, and thus to think about a set of incidents that can be imagined as a rehearsal for the Asian financial crisis as well as a minor participant in the international disillusion that led to the Suharto regime's downfall. In 1994, a small Canadian gold prospecting company announced a major find in the forests of Kalimantan, Indonesian Borneo. Over the months, the find got bigger and bigger, until it was the biggest gold strike in the world, conjuring memories of the Alaskan Klondike and South Africa's Witwatersrand. Thousands of North American investors put their savings in the company, called Bre-X. First-time investors and retired people joined financial wizards. Whole towns in western Canada invested.[4] The new world of Internet investment blossomed with Bre-X. Meanwhile, Bre-X received continuous coverage in North American newspapers, especially after huge Canadian mining companies and Indonesian officials entered the fray, fighting over the rights to mine Busang, Bre-X's find.[5] The scandal of Indonesian business-as-usual, opened to public scrutiny as corruption, heightened international attention and garnered support for Bre-X. But, in 1997, just when expectation had reached a fevered pitch, Busang was exposed as barren: There was nothing there. Gasps, cries, and law suits rose from every corner. Even now, as I write two years later, the drama rumbles on. The Toronto Stock Exchange is changing its rules to avoid more Bre-Xs.[6] Bre-X law suits set new international standards.[7] Bre-X investors still hope and complain across the Internet, as they peddle the remains of their experiences: jokes, songs, and stock certificates (as wallpaper, historical document, or irreplaceable art, ready to hang).[8] Meanwhile, Indonesian mining officials and copycat prospecting companies scramble to free themselves from the Bre-X story, even as they endlessly reenact its scenes, hoping to rekindle investor enthusiasm.[9] Hope's ashes are inflamed even by ridiculous claims; recently the Bre-X chief geologist, named in many lawsuits, says there *is* gold at Busang. Who is to prove him wrong?[10]

The Bre-X story exemplifies popular thinking about the pleasures and dangers of international finance and associated dreams of globalization. The story dramatizes North-South inequalities in the new capitalisms; it celebrates the North's excitement about international investment, and the blight of the South's so-called crony capitalisms: business imagined not quite/not white. Painting Southern leaders as rats fighting for garbage, the story also promises new genres of justice for the Northern investor who dares to sue. Finance looks like democracy: The Internet, they say, opens foreign investment to the North American everyman. But the Bre-X story also narrates the perils of the downsized, overcompetitive economy: the sad entrepreneurship of selling worthless stock certificates on-line. As one writer put it, mixing metaphors, "The Bre-X saga will come to be known as the demarcation of the Internet as the weapon of choice for investors."[11]

Most salient to my concerns about the specificity of global economic promises is the genre convention with which Bre-X started its own story, and by which it was finished off. Bre-X was always a performance, a drama, a conjuring trick, an illusion, regardless of whether real gold or only dreams of gold ever existed at Busang. Journalists compared Busang, with its lines of false drilling samples, to a Hollywood set.[12] But it was not just Busang; it was the whole investment process. No one would ever have invested in Bre-X if it had not created a performance, a dramatic exposition of the possibilities of gold.

Performance here is simultaneously economic performance and dramatic performance. The "economy of appearances" I describe depends on the relevance of this pun; the self-conscious making of a spectacle is a necessary aid to gathering investment funds. The dependence on spectacle is not peculiar to Bre-X and other mining scams: It is a regular feature of the search for financial capital. Start-up companies must dramatize their dreams in order to attract the capital they need to operate and expand. Junior prospecting companies must exaggerate the possibilities of their mineral finds in order to attract investors so that they might, at some point, find something. This is a requirement of investment-oriented entrepreneurship, and it takes the limelight in those historical moments when capital seeks creativity rather than stable reproduction. In speculative enterprises, profit must be imagined before it can be extracted; the possibility of economic performance must be conjured

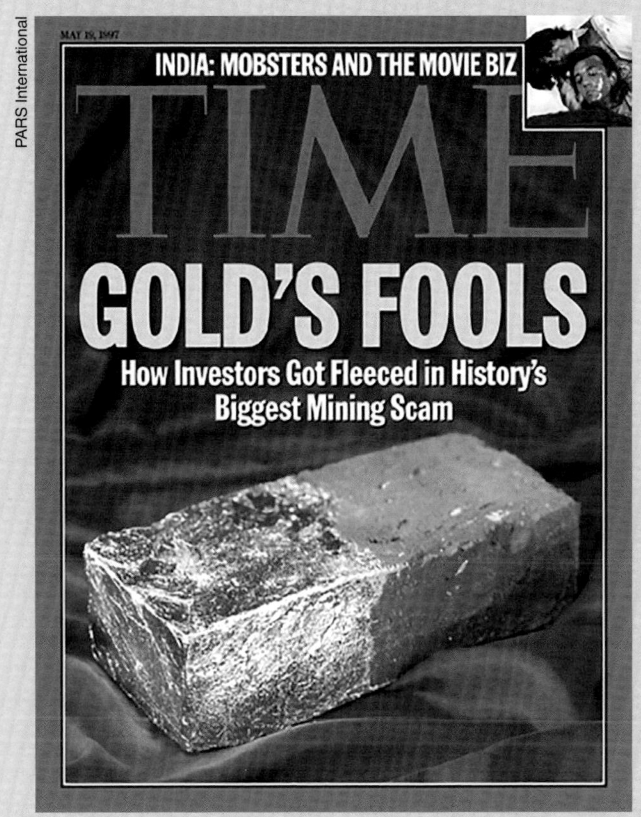

MAY 19, 1997
INDIA: MOBSTERS AND THE MOVIE BIZ
TIME
GOLD'S FOOLS
How Investors Got Fleeced in History's
Biggest Mining Scam

Covering the Cover-Up of Canadian Corporation Bre-X
Cover of TIME Magazine, 1997

like a spirit to draw an audience of potential investors. The more spectacular the conjuring, the more possible an investment frenzy. Drama itself can be worth summoning forth.[13] Nor are companies alone in the conjuring business in these times. In order to attract companies, countries, regions, and towns must dramatize their potential as places for investment. Dramatic performance is the prerequisite of their economic performance.

Yet conjuring is always culturally specific, creating a magic show of peculiar meanings, symbols, and practices. The conjuring aspect of finance interrupts our expectations that finance can and has spread everywhere, for it can only spread as far as its own magic. In its dramatic performances, circulating finance reveals itself as both empowered and limited by its cultural specificity.

Contemporary masters of finance claim not only universal appeal but also a global scale of deployment. What are we to make of these globalist claims, with their millennial whispers of a more total and hegemonic world-making than we have ever known? Neither false ideology nor obvious truth, it seems to me that the globalist claims of finance are also a kind of conjuring of a dramatic performance. In these times of heightened attention to the space and scale of human undertakings, economic projects cannot limit themselves to conjuring at different scales—they must conjure the scales themselves. In this sense, a project that makes us imagine globality in order to see how it might succeed is one kind of "scale-making project"; similarly, projects that make us imagine locality, or the space of regions or nations, in order to see their success are also scale-making projects. The scales they conjure come into being in part through the contingent articulations into which they are pushed or stumble. In a world of multiple divergent claims about scales, including multiple divergent globalisms, those global worlds that most affect us are those that manage tentatively productive linkages with other scale-making projects.

Social scientists have been thrilled to discover the excitement about the global that has taken corporate planners, bureaucrats, and political activists by storm over the last decade. Anthropologists have hoped that by attaching ourselves to this excitement we might propel our discipline beyond a heritage of studying obscure villages to reposition it in the middle of all kinds of world-making projects, big and small. Yet our tools for thinking about the big picture are still rudimentary. Holding on, as I think we should, to a disciplinary heritage of attention to up-close detail, we find ourselves with data about how a few people somewhere react, resist, translate, and consume. From here it is an easy step to invoke distinctions between local reactions and global forces, local consumption and global circulation, local resistance and global structures of capitalism, local translation and the global imagination. I find myself doing it. Yet we know that these dichotomies are unhelpful. They draw us into an imagery in which the global is homogeneous precisely because we oppose it to the heterogeneity we identify as locality. By letting the global appear homogeneous, we open the door to its predictability and evolutionary status as the latest stage in macronarratives. We know the dichotomy between the global blob and local detail isn't helping us.

We long to find cultural specificity and contingency within the blob, but we can't figure out how to find it without, once again, picking out locality.

A number of scholars are making important suggestions about how to find the specificity and contingency of the global by attention to spreading "scapes," shifting "routes," cosmopolitanisms, transnationalisms, and the like.[14] This essay suggests further that we pay attention to the making of scale. Scale is the spatial dimensionality necessary for a particular kind of view, whether up close or from a distance, microscopic or planetary. I argue that scale is not just a neutral frame for viewing the world; scale must be brought into being: proposed, practiced, and evaded, as well as taken for granted. Scales are claimed and contested in cultural and political projects. A "globalism" is a commitment to the global, and there are multiple, overlapping, and somewhat contradictory globalisms; a "regionalism" is commitment to the region, and so on. Not all claims and commitments about scale are particularly effective. Links among varied scale-making projects can bring each project vitality and power. The specificity of these articulations and collaborations also limits the spread and play of scale-making projects, promising them only a tentative moment in a particular history. Neoliberalism, as an economic doctrine promoting the free circulation of goods, currencies, labor, and more across national and regional borders, is a set of scale-making projects. These are powerful projects, and some of us have begun to imagine a world remade entirely according to neoliberalism's dreams. Before conceding the hegemony of their global claims, however, it seems useful to take a look at the limitations of their reach. The performative dramas of financial conjuring offer one perspective from which to appreciate the specificity and contingency of particular niches within the neoliberal project.

One of the chief puzzles of globalist financial conjuring is why it works. We've all seen advertisements for hamburgers, express mail, or computers bridging cultures across the globe. But it's one thing to offer a stylish picture of diversity, and another thing to figure out how entrepreneurial projects actually manage to affect people who may not pay them any mind. Conjuring is supposed to call up a world more dreamlike and sweeter than anything that exists; magic, rather than strict description, calls capital. The puzzle seems deeper the more the material and social worlds to be reshaped and exploited are geographically, culturally, and politically remote from financial conjuring centers. How do the self-consciously glossy and exaggerated virtual worlds conjured by eager collectors of finance become shapers of radically different peoples and places? My frame highlights contingent articulations in which the globalist financial conjuring links itself with regional and national scale-making projects, making each succeed wildly—if also partially and tentatively. It seems likely that successfully conjuring the globe is possible, at least now, only in thick collaboration with regional and national conjurings; certainly financial conjuring has been deeply implicated in promises of making regional and national dreams come true.

Often, globalist financial conjuring supports the most bizarre and terrible of national and regional dreams. Certainly this was the case in the Bre-X story. Finance capital

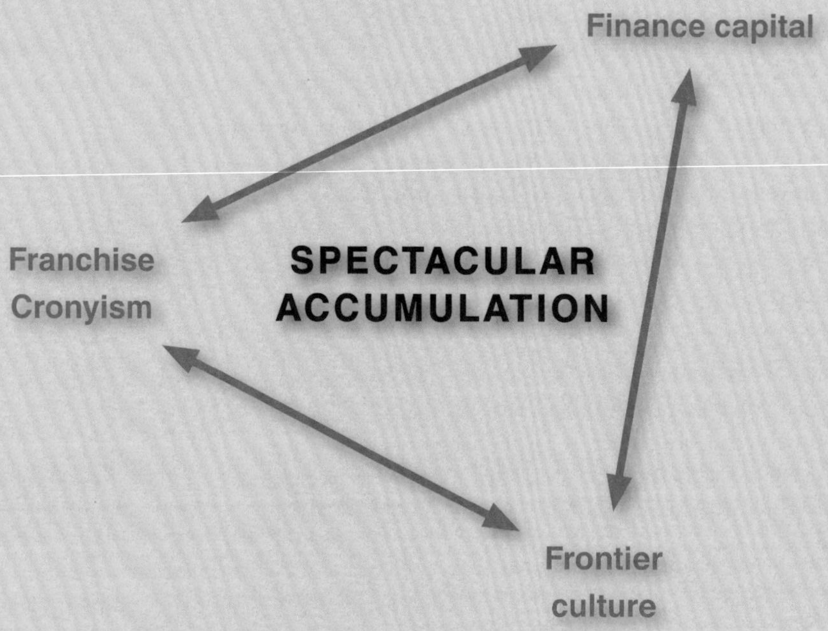

Finance capital

Franchise
Cronyism

**SPECTACULAR
ACCUMULATION**

Frontier
culture

F-C Articulations in the Economy of Appearances
This diagram is both serious and a joke

became linked with greedy elite dreams of an authoritarian nation-state supported by foreign funds and enterprises; this is a nation-making project I call franchise cronyism to mark the interdependence of corruption and foreign investment. These in turn became linked with migrant dreams of a regional frontier culture in which the rights of previous rural residents could be wiped out entirely to create a Wild West scene of rapid and lawless resource extraction: quick profits, quick exits. To present this rather complicated set of links, I offer a diagram. Diagrams by their nature are oversimplifications, and this one is certainly no exception. Indeed, I have named each of the three scale-making projects I discuss in a self-consciously joking manner. Yet the playfulness is also a serious attempt to focus attention on the specificity and process of articulation. Finance capital is a program for *global* hegemony; franchise cronyism is one particular *nation*-making project; frontier culture is an articulation of a *region*. Each is a scale-making project with its sights set on a different scale: global, national, and regional. The links among them cross scales and strengthen each project's ability to remake the world. At the same time, not one of these three projects is predictable or ubiquitous in the world. Coming together as they did for a moment in Suharto's Indonesia, they created a great fire.[15] Looking back, we see that they didn't create an evolutionary ladder to the stars. Isn't this sense of specificity and contingency in globalist claims what scholars and social commentators need most to bring into view?

"Yes, We Are Still in Business"

Bre-X was the brainchild of a Canadian stock promoter named David Walsh. Walsh dropped out of high school at the end of tenth grade and soon joined a Montreal trust company, rising quickly to become the head of the investment department. After thirteen years, he left to try to form his own trust company. Three unsuccessful years later, he agreed to start an office in Calgary for another firm, only to quit the next year. From then on, Walsh worked to set up his own companies: first, the oil-oriented Bresea Resources Ltd. (named after his sons Brett and Sean) and then, in 1985, Bre-X Minerals Ltd., which from the first aimed to find gold.[16]

Gold mining had become a profitable industry in the 1970s when the United States ended the gold standard, and the price of gold, which had been held constant at $35 per ounce for many years, skyrocketed, hitting $850 per ounce in 1980. Canadian companies rushed to take advantage of the new gold prices by exploring not only in the Canadian West but around the world. Junior mining exploration companies, whose goal is to find the minerals that can be exploited by major companies, sprouted by the dozens. Toronto became the world's mining finance capital. In 1997, there were 1,225 publicly traded mining companies in Canada, and mining stocks represented 21.5 percent of all trades on the Toronto Stock Exchange.[17] In this industry, the line between various kinds of expertise is thin: geologists (with salaries supplemented by stock options) must be promoters to raise the money to finance their mineral finds, and stock promoters must explain their offerings in geologically convincing terms. Canadian preeminence in mining depended on both its mining history and its position as a center of mining finance.

For a stock promoter like David Walsh to become president of a gold exploration company was not unusual in this climate. Consider the trajectory of the president of Barrick Gold, Canada's biggest gold-mining company. Peter Munk is a high-flying but not always successful entrepreneur. In the 1950s, he founded a television and hi-fi company that crashed, leaving the government of Nova Scotia in deep debt; he went on to build hotels on South Sea islands funded by Saudi Arabian princes. Nothing in his background gave him expertise in minerals. In 1986, however, he bought a worked-over mine in Nevada. It turned out to be the most profitable gold mine in the world, pushing Munk's company into a leading position.[18] Peter Munk was "a dreamer who became a king."[19] In this context, David Walsh's little experience made sense.

In 1988, Walsh listed Bre-X on the Alberta Stock Exchange at 30 Canadian cents per share. His wife Jeanette supported the household by working as a secretary. The family bought on credit, and, over C$200,000 in debt, both David and Jeanette Walsh declared personal bankruptcy in 1992. Bre-X shares sometimes fell as low as two cents; in his 1991 annual report, David Walsh wrote, "Yes, we are still in business." In 1993, however, Walsh pulled together some money for a trip to Indonesia. There he met with Dutch-born geologist John Felderhof, who had achieved some fame in identifying the Ok Tedi copper and gold mine in Papua New Guinea in 1967, but had suffered hard times in the 1980s. Felderhof agreed to help Walsh find gold in Kalimantan and contacted Filipino geologist Michael de Guzman for the project. Filipino geologists had been in much demand in Indonesia because of their experience, education, and regional savvy.[20] De Guzman brought several Filipino associates to the team.[21]

Mining properties were cheap and available in Indonesia in the early 1990s because the Australians, who had come in some ten years before in their own wave of national mining speculation, were trying to get out. Felderhof had worked for some Australian companies and had witnessed the financial boom and bust in which mineral exploration was begun and then abandoned, promising or not. He convinced Walsh to form a partnership around the creek called Busang in East Kalimantan, and Walsh raised the money to drill some holes. The results were disappointing, and by December 1993 they were about to close the property. Then, early in 1994, de Guzman struck gold. Walsh was quick and effective in informing investor newsletters and brokerage firms. Felderhof's estimates grew bigger and bigger. In 1993, Bre-X was trading at 51 Canadian cents a share; by May 1996, stocks were trading at C$286.50, accounting for a ten-to-one split. In April, Bre-X had been listed on the Toronto Stock Exchange; in August, the stock was listed in the United States on NASDAQ; in September, it was also listed on the Montreal Stock Exchange. By then the company's market capitalization was over 6 billion Canadian dollars.[22] Awards started to roll in: Mining Man of the Year for Bre-X president David Walsh; Explorer of the Year for chief geologist John Felderhof. On March 9th and 10th 1997, Bre-X officers and geologists were feted in awards dinners and ceremonies at the Prospectors and Developers Association of Canada meetings. They were at the height of their success.

Conjuring

On 19 March 1997, Michael de Guzman fell eight hundred feet from a helicopter into the rainforests of Kalimantan. Although up to that point he had been considered little more than the Filipino sidekick, at the moment of his death de Guzman became the company, his face displayed everywhere in the news media over charts of the company's finds and stock prices. If Bre-X had been a big story before, it was truly dramatic now.

Mysteries abounded. A suicide note was found in which de Guzman wrote that he couldn't stand the pain of hepatitis. But he was an optimistic man and in quite good health. What happened to the third man the other Filipino geologists had seen enter the helicopter? Rumors circulated like wildfire. One Philippines scholar confided to me, "When I heard about the watch found in the helicopter, I had to find out what kind it was. When they said 'Rolex,' I knew he was murdered. No Filipino gangster would dispose of his victim without first removing his Rolex." A sign, a trophy. The trouble was, the scene was cluttered with signs, clues, false leads. Wives of de Guzman who knew nothing of each other's existence cropped up everywhere: in Manila, Jakarta, Manado, Samarinda.[23] Rumors circulated that de Guzman had parachuted into Zurich.[24] When a corpse was finally found, the face and much of the body had been devoured by wild pigs. Multiple autopsies failed to establish the identity of the body beyond controversy: Could Bre-X have changed the fingerprint on his employee identity card? Did the dental records match? And where were his geologist friends when it was time for the funeral?[25]

It was at this point, too, that the gold deposit at Busang came to seem just as mysterious. Bre-X had been drilling core samples at Busang since 1993; by 1997 some sites looked like "Swiss cheese."[26] As Busang became famous, industry professionals came to visit. Bre-X president Walsh later complained, "Virtually every mining geologist, analyst went to the site, but I never received one letter or phone call during that whole period that something was amiss over in Indonesia."[27] The "analysts" Walsh refers to were mining stock analysts, and, indeed, dozens visited the site, each fueling investors' attraction with more glowing reports. But in March 1997, the American company Freeport-McMoRan sent their assayers as the "due diligence" element of their agreement to become partners with Bre-X at Busang. Freeport found nothing. Furthermore, they claimed that the kind of gold in Bre-X's samples was inappropriate for the site: It was stream-rounded alluvial gold instead of igneous gold. Bre-X's assay methods were now open to question. Rumors flew of plots and cover-ups, and the price of Bre-X stocks roller-coastered. Perhaps Freeport was making false claims to take over the property. Perhaps Bob Hasan, an Indonesian partner, was buying up cheap stocks at a bargain-basement price. Why else had he taken out a bank loan to log just in this area, just at this time?[28] Perhaps New York investors were trying to beat out Canadians.[29] The gamble drew stock speculators into the fray, and on April 2nd trading was so intense that it closed down the computer system at the Toronto Stock Exchange.[30] On May 3rd, however, the independent test report arrived. Its finding: *no* economic gold deposit. The ballooning

Prinçen Stol.
Oude Myn.
Zuyder Schacht.
Oude Maleysche werken.
Zuyd Zuyd W. eingang.

F. Doct: Grims Schacht.
G. Ingang in de Myn agter de Smits Winckel.
H. Een Stol die na de Zuyder Schacht werd gedreven.

I. Dwars slagen om daar door de Mal. werken op te soek.
K. Doorgang van de Z. Schacht.
L. Zuyder Redout.
M. Huys der Berg-lieden.

laven wooning.
ngang in de Myn agter het
Probeer huys.
ulver Kelder.
Huys waar in de dreck legt.

R. Een Oord.
S. Groote Schacht is 195. voet diep.
T. Onderste Schacht.
V. Staat vol water.
W. Een Oord.

X. Dagh Trom.
Y. Het Maleysche Oord.
Z. Berglieden Huys.
✱. Een Blaasbalg in de Princen

stock swap immediately deflated; Bre-X stocks were officially worthless.[31] Yet what are we to make of the mysteries?

I am not a journalist, and my concern does not involve just which gold miners and which Indonesian officials and which stock market participants knew about or participated in various conjuring acts. I'm more interested in the art of conjuring itself, as practiced not only by Bre-X officers and employees but also by the analysts, reporters, investors, and regulators who formed their retinue. I am struck by two counterintuitive observations. First, mystery, rumor, and drama did not come to Bre-X just at the tail end of its ride; these qualities marked the Bre-X story from its beginnings. Rather than closing Bre-X down, mystery and drama kept Bre-X alive and growing; it was only when an official report stopped the show that the company died. Second, Bre-X is not the only company that has required spectacle to grow. Bre-X seems typical of the "junior" Canadian mineral exploration companies that it has helped usher into the international spotlight—except, of course, it was more successful at first and later more despised. Junior companies don't have the equipment or capital to take their mining ventures very far. They must make a big splash, first, to attract enough investors to keep prospecting and, second, to bring in big mining companies to buy out their finds.

One can draw the net wider. The mystery and spectacle Bre-X cultivated is representative of many kinds of companies in which finance capital is the ruling edge of accumulation. Such companies draw investments through drama. And the importance of drama guarantees that it is very difficult to discern companies that have long-term production potential from those that are merely good at being on stage. The charismatic and dramatic attraction of international finance capital was a key feature of Southeast Asian development strategies during the so-called economic miracle. After the 1997 financial crisis, we were told to distinguish between the real and the fake, but does not the whole design of these accumulation strategies work against our ability to draw this line? As in a beauty contest, artistry and drama are necessary to compete; spectacle and mystery, playing equally across the line of the real and the fake, establish the winning reality of performance.[32]

Bre-X initially attracted investors because of the excitement of the reports coming out about Busang. From the very first, Bre-X was in the news, and journalists constantly wrote about Busang. The success with which Bre-X attracted investors depended on these reports and particularly on the ways they used and elaborated tropes that brought the Bre-X find into other circulating stories of wealth, power, and fulfillment. Some of these stories were colonial adventure tales: the search for hidden, uncounted riches in remote places. *Maclean's*, a Canadian magazine, wrote: "Two, four, and then maybe six million ounces will be pulled from Busang annually. There has never been an El Dorado like this."[33] Other stories told of frontier independence and the promise of wealth "at the end of a miner's rainbow," as the *New York Times* put it, where "independent-mindedness" led miners to "forbidding jungles" in search of the "century's greatest gold strike."[34] There were stories of science in the

Calgary Herald / Postmedia Archives

Duplicitous Duo
Bre-X President David Walsh (left) with Senior Vice-President and Chief Geologist John Felderhof (right), ca. 1993

service of human innovation. There were stories of war and conquest—"the battle for Busang"—recalling the French and the United States in Vietnam, and Bre-X became, repeatedly, a "rumble in the jungle," and then, eventually, a "bungle" or a "jumble" in the jungle.[35]

There were also pervasive stories of underdog charisma. After announcing the Busang find, Bre-X has to fight for the rights to mine it. The story of little Bre-X's up against the big North American mining companies and the big Indonesian establishment gathered an overwhelming response, ushering in Bre-X's greatest period of popular investment. When U.S. ex-president George Bush and Canadian ex-prime minister Brian Mulroney put pressure on Jakarta at the request of a big company, Bre-X's David Walsh, with his high school education, his beer belly, and his ineptness, looked like David up against Goliath. He made such a convincing "little guy" that after the scam was exposed many refused to believe him responsible. As *Fortune* magazine's reporter wrote, "Even now I have trouble believing that Walsh participated. … Walsh looked more like some poor schlemiel who had just won the lottery and couldn't locate his ticket."[36]

Stockholders, too, contributed to the stories swirling around Bre-X.[37] Bre-X established an Internet presence early on, with stories posted on their web site; meanwhile, investors' chat lines buzzed with Bre-X news.[38] The more controversy swirled around Bre-X, the more investors talked and exchanged rumors, extending, too, dreams of wealth, conspiracy theories, reinterpretations of the mining geology and engineering, and romances of unexpected underdog advantage. On the Internet, dramatic presentation was often clearly the point. As one *Silicon Investor* contributor wrote about the Bre-X Internet thread, "The theater is open. The stage is set 24 hrs. a day. There is always an audience."[39]

One Internet contributor signed herself "Ole49er," reminding readers that the Wild West is never far from discussion of Bre-X. A Canadian shareholder with whom I spoke explained that Canadians were excited about the chance to invest in minerals in Indonesia because of the symbolic importance of mining in Canada as well as a national anxiety about the closing of the frontier. Environmental regulations in Canada, he explained, made it difficult to mine profitably in the last wide open spaces of the Canadian West. Yet those open spaces might be pursued abroad in foreign lands. As a substitute frontier, Bre-X's Kalimantan continued the excitement of the frontier story of Canada's development.

But what of Kalimantan in this story? In the moment of frontier making, financial conjuring runs up against the landscape, but not quite as in its dreams. At this trick in the magic show, an opportunity presents itself to ask about how the magic works, and how it doesn't.

Frontiers

Let me return to the beginning of the Bre-X story, or at least *one* beginning, in Kalimantan. A brave man is hacking his way through the jungle, alone and surrounded by disease and danger. There is nothing there but mud, malaria, leeches, hepatitis, and the pervasive loneliness of the jungle trek. But one day … he discovers gold!

This is an old story, told not only about Borneo, but about many a jungle or lonely rock cliff. But it is also what the Bre-X miners told the press. According to the *Far Eastern Economic Review*:

> "The story of how Felderhof and de Guzman unearthed Busang's golden secret is the stuff that fables are made of. Mostly it is a human saga—about two quiet, very ordinary men from opposite sides of the world who persevered for years in the face of tropical illnesses and some of the harshest terrain in the world. …
>
> Felderhof first landed in Indonesia in 1980. … he worked his way from west to east across the centre of the vast Borneo island. … Those were hard times. He lived in jungle villages, eating whatever was available, carrying what he needed on his back, and hacking his way through some of the remotest rain forest in Asia.
>
> De Guzman has gone through similar hardships. … He's had malaria 14 times since he met up with Felderhof in 1986. … Travelling alone, de Guzman took a seven-hour boat ride up the Mahakam River, then trekked 32 kilometers. … During a week of old-fashioned prospecting, he recognized a geological setting he had come to know well in the Philippines."[40]

The story has come close to the moment of discovery. The New York Times tells that moment succinctly: "1994. Michael de Guzman, a mining geologist, is trekking through the Busang site doing work for Bre-X when a bit of yellow rock on a river bank catches his eye. 'Check it out,' he writes on a plastic strip and tacks it to the rock. His assistant follows with a more detailed analysis and writes: 'Checkmate.'"[41]

One might be suspicious of this sudden appearance from nowhere of an assistant, but let's let that pass and move on to the second act in the discovery, the scientific theory that lets the miners know that a glitter of gold might be the tip of an underground hoard. Without science, de Guzman is just a Filipino guy playing with rocks; with science, he is the translator between nature and North American industry. The *Far Eastern Economic Review* story continues: "Working alone at 3 o'clock one morning, it suddenly becomes clear [to de Guzman]. Bounding upstairs, he woke Puspos [the assistant] and explained his theory: that

TONY O'REILLY OF HEINZ: HIS DAY HAS 57 VARIETIES PAGE 72

BusinessWeek

ECEMBER 17, 1984 A McGRAW-HILL PUBLICATION $2.00

THE DEATH OF MINING

PAGE 64

51
743675

"The Death of Mining," *BusinessWeek* Special Issue, December 17, 1982 (Bloomberg / YGS Group)

Busang, with its dome-like geological structure, lay at a fault-line crossroads. Together on that January morning, the two spent eight hours in 'nonstop technical brainstorming.'"[42]

It's a lovely story, but it bears a very odd relationship to everything else one might want to say about mining in Kalimantan. By the time de Guzman arrived at Busang, foreign prospectors, migrant miners, and local residents had all combed the landscape for gold. Yet his "discovery" story does not reflect the interdependence of his knowledge with that of other miners. Local residents, government regulations, mining camps, churches, markets, bus schedules, army officers, village heads, property dispute: All are missing. The frontier story requires that de Guzman wander alone on an empty landscape.

The story of lonely prospectors making independent discoveries in a remote jungle moved North American investors and stimulated the capital flow that made Bre-X rich. These images are repeated in the portfolio of every North American prospecting company working in Kalimantan. Consider the exploration of a Canadian prospecting company called International Pursuit, as reported in Gold Newsletter's Mining Share Focus. International Pursuit was prospecting in the wake of the Bre-X drama and needed to distance itself from Bre-X. Yet the frontier story is precisely the same. The company has sent its prospectors into the empty, wild landscape of Kalimantan, where, through a combination of luck and science, they stumble on precious metals. They are in particular luck, because in late 1997 the devastation brought about by El Niño drought and forest fires had given the landscape some of the wild loneliness of which they dreamed. El Niño, they say, was a godsend because the streams dried up, revealing hidden minerals. "It was as if Mother Nature had lifted a curtain, exposing the secrets below," the industry reporter wrote. Ignoring local residents' hunger, fire damage, smoke inhalation, and displacement, he bragged, "International Pursuit's exploration teams were among the relative few left on the ground."[43]

I first began to think about gold mining because of the spread of desolation and wildness across large swaths of the Kalimantan landscape. I had been working with people in this area for some time and knew them not so long ago when traditional resource claims were strong.[44] Suddenly, a new landscape began to develop, and I couldn't help but think of it as a "frontier," because so many themes and images of frontiers, from the U.S. Wild West to the Brazilian Amazon, were self-consciously overlaid on the Kalimantan landscape. The mad rush for gold joined and stimulated mad rushes for logs, birds' nests, incense woods, marble, and even sand, to transform the landscape in the 1990s from a quiet scene of forests, fields, and houses to a wild terrain of danger, urgency, and destruction.[45]

Three developments seemed essential to the transformation of the terrain. First, big and small entrepreneurs were interdependent. Transnational companies and transregional migrants from all over Indonesia complemented each other. In mining, big companies have official permits; small-time miners are illegal. Small miners lead company prospectors to the best spots. The companies displace them but then employ them for prospecting. The companies complain about the illegals, blaming them for environmental problems, thus protecting their own reputations. The illegals supply the everyday services for the companies. The companies follow the small miners and the small miners cluster around the companies; they are codependents living off common loot.

Second, nature had to be made into loot, free for all. In the 1980s, logging companies worked hard to extinguish local rights to resources. "This place belongs to Indonesia, not to you," the logging bosses said when residents complained. The military followed and supported their claims, creating an authoritarian lawlessness that made resources free to those who could take them. By the mid 1990s, local residents told me, "We still know our customary legal rights, but no one cares about them." The roads invited migrants, violence became key to ownership, and swarming miners escalated the terror, the risk, and the urgency of taking everything out, right away.

Third, frontier migrants arrived at the end of long chains of culture and capital. Small miners rarely own the hydraulic pumps with which they removed topsoil; they contract them from urban entrepreneurs in profit-sharing arrangements. Often, they subcontract with local residents, dividing imagined profits even further. Migrants and local residents learn to share cultural practices of entrepreneurship that reach from distant cities deep into the forest. Local residents bring mining activities up the rivers, through their familiar forests, close to home, spreading frontier standards of rapid resource extraction. They are drawn into the competition and a diet of coffee and canned sardines. In these entrepreneurial chains, a spreading frontier culture is created. It is a culture dedicated to the obliteration of local places, local land and resource rights, local knowledges of flora and fauna. The makeshift camps of the miners proliferate across the landscape, mixing migrants and local residents in an antilocal regionality in which commitment to the local landscape is as useless as the gravel residue left after gold has been picked out and taken away.

Frontier culture is a conjuring act because it creates the wild and spreading regionality of its imagination. It conjures a self-conscious translocalism, committed to the obliteration of local places. Such commitments are themselves distinctive and limited—and thus "local" from another perspective. Yet they break with past localisms in self-conscious regionalism. This is a conjuring of scale, and frontier resource extraction relies on it.

A distinctive feature of this frontier regionality is its magical vision; it asks participants to see a landscape that doesn't exist, at least not yet. It must continually erase old residents' rights to create its wild and empty spaces where *discovering* resources, not stealing them, is possible. To do so, too, it must cover up the conditions of its own production. Consider the contrasts between the features of the story of the frontier that must be told and the frontier conditions I observed: The lone prospector replaces swarming migrants and residents, searching the landscape. The excitement of scientific discovery replaces the violence of expropriation as local resource rights are extinguished and armed gangs enforce their preeminence. The autonomy of the prospector's find replaces the interdependent negotiations of big companies and illegal miners, each leading each other to new sites and trading political and material assets as they form complementary players.[46]

Why does the frontier story have any power at all, considering what it erases? How can it imagine the Kalimantan landscape so wrongly? These trompe l'oeils are possible only because of national discipline: the violence of the military, which spreads a regional lawlessness; the legal regulations that privilege company rights and profits yet allow illegal migrants to accumulate in the spreading wildness; the confusion between private entrepreneurship and public office that forged the national government. This is not the only kind of nation-making that can exist. To explore its specificity, it is useful to turn our attention to CoWs.

CoWs

CoWs are Contracts of Work. No mining company can extract minerals in Indonesia without one. Like animals, CoWs come in generations. The first CoW was a much revered and singular ancestor, granted to Louisiana's Freeport-McMoRan to mine in Irian Jaya.[47] Succeeding generations of CoWs have become more differentiated, more limited, more finely detailed.[48] Yet as they develop, they continue to be icons, even fetishes, held up to show the relationship between the Indonesian nation and the world. Ideally, they guarantee that resource extraction activities work in the interests of the Indonesian nation as well as the mining company. They specify the conditions that create mutual benefits shared between the nation and its foreign investors.

CoWs are magical tools of the national elite. Although merely paper and ink, they conjure a regular income for the Indonesian nation-state. Their terms must be secure and attractive by international standards, or they will not draw capital. But if they meet these standards, they can also conjure the funds that allow the nation-state to produce itself as what one might call a "miracle nation": a nation in which foreign funds support the authoritarian rule that keeps the funds safe. I earlier called this "franchise cronyism." In exchange for supplying the money to support the national leaders who can make the state secure, investors are offered the certainties of the contract, which ensures title to mineral deposits, fixes taxation rates, and permits export of profit.

The CoW guarantees that investors are not working with "dictators." As one Canadian Bre-X investor explained to me, in investing his modest funds he always avoided the countries of dictators. This was the reason the Bre-X investment seemed reasonable. I was confused: What is a dictator? As we talked, I realized that a dictator is a foreign ruler who interferes with Canadian investment. Indonesia's President Suharto was not a dictator, at least before Bre-X. As Bre-X president David Walsh put it, his company—like other Australian and Canadian mining companies in the 1980s and 1990s—targeted Indonesia "by virtue of its geological setting, favorable investment climate and political stability."[49]

Other nations could be and were imagined in Indonesia. Suharto's New Order regime emerged violently in 1966 from an earlier scene of diverse and competitive programs for making the nation, and from the start the regime depended on the repression of other Indonesian national visions through censorship and militarization. Political quietude was nurtured, too, through internationally sponsored "development," which came to refer to programs of state expansion dedicated to convincing diverse local people of the unified national standards of state power. Through "development," the state conceived a legal framework to claim the nation's resources and make them available for foreign expropriation, thus amassing the materials through which its version of the nation could prosper.

When investment capital began to circulate wildly across national boundaries in the 1980s, the Indonesian elite was ready for it. They beckoned to the mining sector, saying: "There are still vast tracts of unexplored land in Indonesia. For those who dare to venture, Indonesia offers immense possibilities."[50] The Australians came. One Indonesian ex-mining official candidly admitted that these were "irresponsible investments" with "rubbish technology." When the Canadian companies followed in the 1990s, it was more of the same: "It's a repetition of history. It's not really a gold rush. It's rather a stock market rush."[51] Yet the government transformed the stock market rush into a gold rush by offering it regional frontiers in the making. The regime gave the companies Contracts of Work, despite their irresponsibility. The CoWs wrote away local rights. Military men deployed to enforce CoWs felt encouraged to start their own entrepreneurial schemes, creating a model of government in which administrators by definition doubled as entrepreneurs who, supported by kickbacks, freed up resources for investors, including themselves. Civil servants became franchise entrepreneurs, too, learning to conjure the miracle nation locally. It was they who sanctioned the mass migration of illegal small loggers and miners that kept the regional economy afloat while bigger investors took out bigger resources and profits. In this escalating mobility and lawlessness, the mysteries of the search for buried treasure became possible. Whether gold was there or not, the economy could grow, spurred on by fabulous dreams.

By the 1990s the Suharto regime began to take for granted its domestic stability and international support. The work of disguising official kickbacks as sound investment policy seemed already complete. Perhaps this is how to understand how the set performance of the miracle nation could have been allowed to deteriorate into dramatic excess. Drama ultimately embarrassed the Suharto regime, allowing investors to label it corrupt. At the same time, it provided a moment of opportunity for investors, who could maneuver within the new embarrassments of national performance to gain a better position for themselves. No investors did better than Bre-X and its rivals in opening up those dramatic cracks and making them visible on international news screens. However unselfconscious the manipulation, it seems clear that the over-the-top drama of franchise cronyism set off around Bre-X allowed the Bre-X investment bubble to last far longer than it could have otherwise, drawing out the drama from the 1994 announcement of gold to the 1997 death of de Guzman. This drama popularized the Bre-X story, vastly enlarging investment.

Calgary Herald / Postmedia Archives

High-Flying Geologist

Michael de Guzman, Bre-X Geologist, preparing to board the Alouette III helicopter with Indonesian Air Transport, 1997

In January 1994 Barrick Gold offered to buy a stake in Bre-X. Barrick was conducting an aggressive campaign of acquisitions in an attempt to become the world's biggest gold-mining company.[52] As mentioned above, Barrick's CEO, Peter Munk, was a risk taker who made his fortune by a chance buy of a Nevada mine, which turned out to be a fabulous mother lode. Barrick had targeted Indonesia as a possible site for high-profit, low-cost mining, and Bre-X's Busang was just Barrick's kind of buy.

When Bre-X turned him down, Barrick moved into the cracks of the Indonesian regime, working the connection between greed and vulnerability that their nation-making performances had produced. The high-powered politicians on Barrick's advisory board pressured the minister of mines and energy, and even President Suharto.[53] Barrick then approached the president's daughter, Tutut, a tycoon in road construction contracts. He offered her Busang's construction contracts, and she pushed the minister of mines to negotiate a Barrick–Bre-X split in which the government would control 10 percent of the mine, Barrick 70 percent, and Bre-X 20 percent. Bre-X was vulnerable because they had begun drilling without a CoW; now the ministry pulled out even their temporary exploration permit to clinch the deal.[54]

Bre-X stockholders went wild with anger. In the glare of their dissatisfaction, the drama escalated. Placer Dome, another Canadian company, made a better bid.[55] Meanwhile, Bre-X made a spectacular play by approaching the president's son, Sigit, and offering him 10 percent of the mine plus an eye-catching one million dollars per month to push their case. Whether naive, as read at the time, or unimaginably clever, this move took the performance of franchise cronyism to its extreme limits, offering the investment drama a new life. Now Suharto's children were pitted against each other publicly.

Barrick continued to have government support until it gained a new opponent: Indonesian entrepreneur Bob Hasan, the longtime friend and golfing buddy of the president, and the man who best knew how to mimic foreign investors' ploys to enlarge his own empire. Hasan played the patriot of the miracle nation, arguing passionately that the enrichment and empowerment of national elites is the first principle of national interest. Barrick irritated him by sending in North American politicians, and he railed against their "cowboyisms," which made Indonesia look helpless under the American thumb.[56] Meanwhile, he managed to acquire a 50 percent share in Bre-X's Indonesian partner.[57]

The game was almost over when President Suharto asked Bob Hasan to work out what to do with Bre-X, and yet this, too, was a moment of momentous drama. Bre-X stockholders were at the edge of their seats. Activist stockholder George Chorny wrote a public letter to Hasan, reminding him that it was Bre-X that had discovered the gold. "It's not any of the guys at Barrick or Placer. It's not the Indonesian government. All the Indonesian government did was to welcome the people of Bre-X to come into their country with open arms to explore this jungle, this desolate jungle in the middle of nowhere."[58] For Chorny, the frontier is always already empty. But Hasan had a different perspective; making the frontier was a national responsibility. Hasan dismissed both Barrick and Placer Dome and brought in the regime's favorite company, Freeport-McMoRan. In Hasan's solution, Freeport-McMoRan would take 15 percent of the mine and become sole operator; the Indonesian government would take 10 percent; Hasan's companies would take 30 percent; and the remaining 45 percent would remain with Bre-X. Backed to the wall, Bre-X signed gracefully.[59]

Freeport, unlike Bre-X and even Barrick, was not in the business of spectacular accumulation, the economy of appearances. Freeport worked with its own cultural logic of investment and development, which, at least in this period, differed from that of Bre-X. Freeport was no mining "junior," amassing capital to finance further exploration. Instead, it had established itself as a big, solid outpost of "American civilization" in Indonesia. As CEO Jim Bob Moffett put it, "We are thrusting a spear of economic development into the heartland of Irian Jaya."[60] Freeport built residential neighborhoods in Irian Jaya reminiscent of U.S. suburbs; Moffett performed Elvis Presley imitations during Christmas visits. Freeport's culture of business, then, offered Americanization rather than "frontier discovery" as a model of profitability.

Freeport had long since gained its miracle deal from the Suharto regime. Its personal Contract of Work far exceeded the benefits of all other investors. In turn, Freeport was the largest source of investor tax revenue for the Indonesian government. It had spent an enormous amount of money developing its Grasberg mine in Irian Jaya, where it depended on the support of the army to keep local residents properly behaved. In 1995, however, riots closed the mine; in 1996, Irian tribal leaders sued the company for environmental destruction and human rights abuses, and the Overseas Private Investment Corporation (an agency of the U.S. government) canceled the company's risk insurance because of its environmental policies. In 1997, then, Freeport was busy living down international accusations of environmental and social responsibility in Irian Jaya. It needed a green profile and solid production results, not an economic miracle.[61] In this spirit, Freeport sent in a sober team to assess the gold at Busang. There was nothing there. "It makes me sick every time I think about it," said Jim Bob Moffett, Freeport's CEO.[62] Following a few impressive last gasps, the spectacle wound down and collapsed. Said Bre-X president David Walsh, "Four and a half years of hard work and the pot at the end of the rainbow is a bucket of slop."[63]

On Spectacular Accumulation

What does this story tell us about transnational finance and its globalist aspirations? In the midst of their dramatic roles, the major players usefully remind us of the stage they have laid. Like Bob Hasan, I am struck by the North American character of the dreams and schemes of investment that swirled around Bre-X. With Bre-X stockholders, I marvel at the ability of Indonesian nation makers to usurp an economic process that has been imagined as

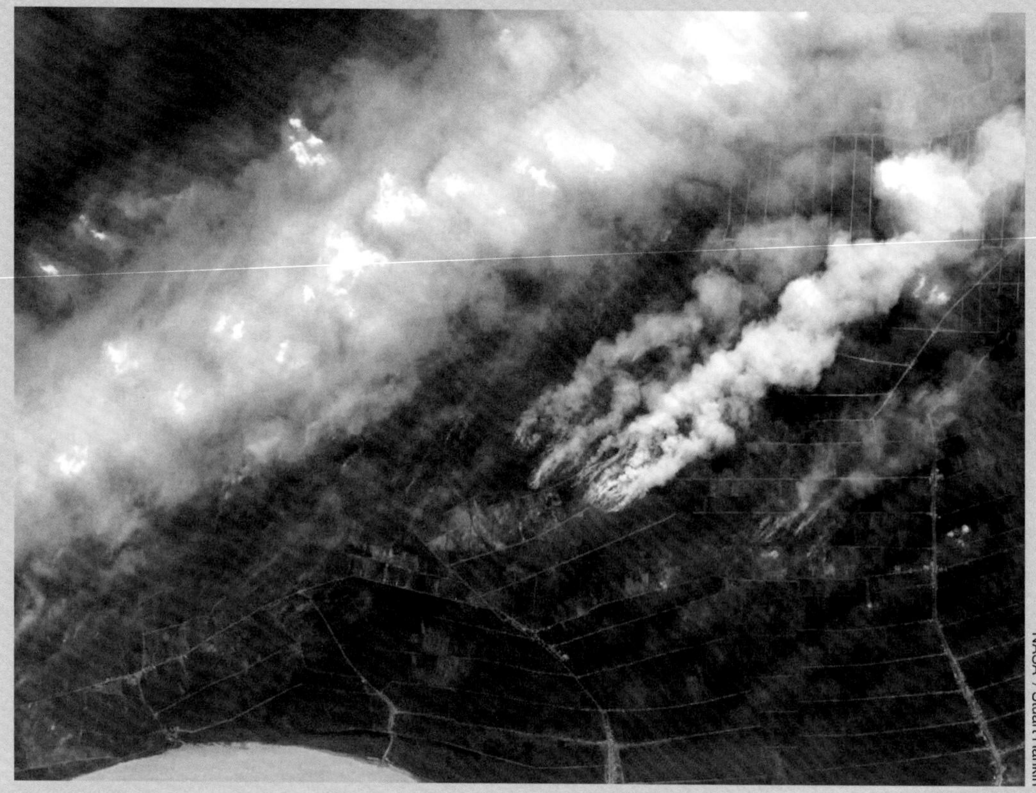

NASA / Stuart Rankin

Disastrous Deforestation
Forest clearing in the Borneo Region
for rampant development of palm oil plantations, 2016

fiercely independent from national controls. And as for the Kalimantan landscape, it is hard not to mourn: The pot at the end of the rainbow is a bucket of eroding mud, damaged forests, and mercury-poisoned rivers. Slop indeed.

It was the Canadian imagination of the combined frontier of investment and mining that made this drama possible. The mining industry has been historically important to Canada's economy and identity. By the 1980s, its locus had shifted from mining *in* Canada to mining *for* Canada. It represented opportunity, potential prosperity, and the sense of initiative in national character. Bre-X's run from the bottom of the Alberta stock exchange to the top of the Toronto exchange and into the world was a source of pride for many Canadians. As much as for profit, Canadians invested for reasons of national pride.[64]

Yet the national specificity of attraction to investments disappears in the excitement of commitments to globalism in the financial world. When one thinks about finance in the Bre-X case, there was nothing worldwide about it at all; it was Canadian and U.S. investment in Indonesia. Yet it is easy to assimilate this specific trajectory of investment to an imagined globalism to the extent that the global is defined as the opening-up process in which remote places submit to foreign finance. Every time finance finds a new site of engagement, we think the world is getting more global. In this act of conjuring, *global* becomes the process of finding new sites. In the force field of this particular globalism, Canadian national dreams are reimagined as transcendent, circulating, beyond culture.

Despite the enormous coercions and seductions of financiers, which aim to make the whole world ready for investment, there is great particularity not only in the reasons a Canadian might want to invest but also in the places where he or she can invest. In the 1990s, when the dreams of the Indonesian elite linked with those of Canadians to jointly conjure the promise of gold, Indonesia became one of those places. Images of remote wild places that could make independent-minded Canadians rich and free touched Indonesian visions of a miracle nation, a nation that could come into being in the arms of foreign finance. Flying in the face of financiers' fantasies of making the nation disappear before the greater mobility of capital, the magic of the miracle nation, waving its CoWs, asserted itself as the only door to North American investment. CoWs, as I have argued, are not merely mechanical adjustments of economic affairs. They are fetish objects, charged with conjuring the miracle nation in the face of competing, threatening alternative visions that unless warded off might come to control the apparatus of the state. From investors' perspectives, they are charged, too, with the security of profit and property. As a gift, they remake the identities of both giver and receiver, vitalizing the miracle nation and its globalist speculators.

For the aspirations of international investors and national elites to emerge as more than a moment's daydream, however, they must be made tangible on a regional landscape. They must engage people, places, and environments. The antilocal culture of Kalimantan frontier regionalism nurtured and raised up both the miracle nation and Canadian speculation. Here is a truly cosmopolitan scene, where varied dreams are jumbled together, naming and renaming creeks, valleys, routes, and towns. The dreamers jostle, fight bitterly, and patronize each other. As they make their own new places, these, too, are knocked away. Even old residents become aliens, as the familiar landscape is transfigured by trauma, danger, and the anxiety of the unknown. Here mystery can flourish, and unexpected discoveries can be made. Unimagined riches can be found because the layout of wealth and poverty is unsettled, un-imagined. Impossible promises cannot be ignored. On this landscape, the economy of appearances seems so real that it must be true.

When the spectacle passes on, what is left is rubble and mud, the residues of success and failure. People with other stakes and stories will have to pick up the pieces.

⁝⁝⁝

At the intersection of projects for making globes, nations, and regions, new kinds of economies can emerge. In the Bre-X drama, globalist commitments to opening up fresh sites for Canadian mining investments enabled Indonesian visions of a miracle nation at the same time as they stimulated the search for mining frontiers. The program of the miracle nation offered speculators security as it forced potential frontier regions into lawless violence and abolished customary tenure. When Kalimantan responded by developing a wild frontier, its regional reformation confirmed the proprietary rights of the miracle nation. The Kalimantan frontier could then appeal to globalist speculation, offering a landscape where both discovery and loss were possible. Three scale-making projects came together here: the globe-making aspirations of finance capital, the nation-making coercions of franchise cronyism, and the region-making claims of frontier culture (fig. 1). Globalist, nationalist, and regionalist dreams linked to enunciate a distinctive economic program, the program of spectacular accumulation.

Spectacular accumulation occurs when investors speculate on a product that may or may not exist. Investors are looking for the appearance of success. They cannot afford to find out if the product is solid; by then their chances for profit will be gone. To invest in software development requires this kind of leap: Software developers sell their potential, not their product. Biotechnology requires a related if distinctive leap of faith to trust the processes of innovation and patenting to yield as-yet-unknown property rights and royalties. Real estate development requires an assessment of desirability and growth, not demonstrated occupancy; it sells investors attractiveness. In each of these cases, economic performance is conjured dramatically.[65]

I use the term *spectacular accumulation* mainly to argue with evolutionary assumptions in popular theories of the ever changing world economy. According to regulation theorists, "flexible accumulation" is the latest stage of capitalism. Flexible accumulation follows Fordist production as barbarism follows savagery, that is, up a singular political-economic ladder. David Harvey's writing has made this conceptualization influential among anthropologists, who suggest correlated changes in culture, spatiality, and scale to go along with this evolutionary progression.[66] Thus, too, we imagine evolutionary changes in the making of space and time. With Harvey, anthropologists have begun to imagine a worldwide condensation of space and time in which spaces grow smaller and times more instantaneous and effortless. Consider, however, the space-time requirements of Bre-X's spectacular accumulation: Space is hugely enlarged; far from miniature and easy, it becomes expansive, labored, and wild, spreading muddy, malarial frontiers. Time is quickened, but into the rush of acceleration, not the efficiency of quick transfers. It is not effortless; if you can't feel the rush and the intensity, you are missing the point, and you'll keep your money at home. Moreover, this spectacular accumulation does not call out to be imagined as new. It is self-consciously old, drawing us back to the South Sea bubble and every gold rush in history. In contrast with flexible accumulation, its power is not its rejection of the past but its ability to keep this old legacy untarnished.

I have no desire to add yet another classificatory device to the annals of capitalism. Instead, my point is to show the heterogeneity of capitalism at every moment in time. Capitalist forms and processes are continually made and unmade; if we offer singular predictions we allow ourselves to be caught by them as ideologies. This seems especially pressing when considering the analysis of scale. At the end of this century, every ambitious world-making project wants to show itself able to forge new scales. NGOs, ethnic groups and coalitions, initiatives for human rights and social justice: We all want to be creative and self-conscious about our scale-making. We want to claim the globe as ours. In this context, rather than ally myself with globalist financiers to tell of *their* globe, I want to trace how that globe comes into being both as a culturally specific set of commitments and as a set of practices. The investment drama of the Bre-X story shows how articulations among globalist, nationalist, and regionalist projects bring each project to life. In the spirit of serious but joking diagrams, I offer an acronym to refocus your attention (fig. 2). The particularity of globalist projects, I am arguing, is best seen in the contingent articulations that make them possible and bring them to life: These are APHIDS, Articulations among Partially Hegemonic Imagined Different Scales.

Often we turn to capitalism to understand how what seem to be surface developments form part of an underlying pattern of exploitation and class formation. Yet before we succumb to the capitalist monolith called up in these analyses, it is useful to look at the continual emergence of new capitalist niches, cultures, and forms of agency. For this task, Stuart Hall's idea about the role of articulation in the formation of new political subjects is helpful. New political subjects form, he argues, as preexisting groups link and, through linking, enunciate new identities and interests.[67] Social processes and categories also can develop in this way. I have used this insight to trace the spectacular accumulation brought into being by the articulation of finance capital, franchise cronyism, and frontier culture. While each of these linked projects achieved only a moment of partial hegemony, this was also a moment

Globalist projects
come into being
as
APHIDS

APHIDS =
Articulations
among
Partially **H**egemonic
Imagined **D**ifferent **S**cales

APHIDS
This acronym is both serious and a joke

of dramatic success. Soon after the events I have narrated, Indonesia's economy precipitously collapsed; the miracle nation was discredited as a site for investment, and the articulation fell apart.

Afterward, analysts scrambled to describe the difference between good and bad investments. They recognized that the Busang saga had contributed in a small way to the Indonesian crash. But they ignored or refused its allegorical quality: Bre-X offered a dramatic rendition of the promises and perils of the economic miracle attributed, in Indonesia and beyond, to globalization.[68]

As the century turns, the field of anthropology has taken on the challenge of freeing critical imaginations from the specter of neoliberal conquest—singular, universal, global. Attention to contingency and articulation can help us describe both the cultural specificity and the fragility of capitalist—and globalist—success stories. In this shifting heterogeneity there are new sources of hope, and, of course, new nightmares.

Notes

This essay was first presented as a talk at the Society for Cultural Anthropology meetings in San Francisco in May 1999. A number of colleagues encouraged the author to write about Bre-X: Rheana Parrenas offered invaluable research assistance; Steven Feld and Tania Li contributed their newspaper clippings on the Bre-X saga; Arjun Appadurai, Kathryn Chetkovich, Paulla A. Ebron, Celia Lowe, and Lisa Rofel offered their comments on earlier drafts. This essay was originally published as "Inside the Economy of Appearances," *Public Culture* 12, no. 1 (2000); and re-published as "The Economy of Appearances," chap. 2 in *Friction: An Ethnography of Global Connection*, ed. Anna Lowenhaupt Tsing (Princeton, NJ: Princeton University Press, 2005), reproduced with permission from Anna Lowenhaupt Tsing.

1 Pramoedya Ananta Toer, "My Apologies, in the Name of Experience," trans. Alex G. Bardsley, *Indonesia* 61 (April 1996): 1–14, as quoted in Peter Dale Scott, *Minding the Darkness: a poem for the year 2000* (New York, NY: New Directions, 2000), 212–14.

2 *Oxford English Dictionary*, 2nd ed., s.v. "scale."

3 As much as any place in the world, Indonesia rode high on the wave of enthusiasm for mobile, finance-driven international investment in the 1980s and early 1990s. Between the late 1980s and 1997, economic growth averaged about 8 percent annually, and in early 1997 economists saw "little sign of the turmoil that was to emerge" when the economy crashed later that year and the Suharto regime followed in 1998 (Ross McLeod, "Indonesia's Crisis and Future Prospects," in *Asian Contagion: The Causes and Consequences of a Financial Crisis*, ed. Karl Jackson (Boulder, CO: Westview Press, 1999), 209). Looking back, many once enthusiastic analysts blamed the crisis on "cronyism," residual protectionism, and bad national regulatory practices. See, for example, Karl Jackson, "Introduction: The Roots of the Crisis," in *Asian Contagion*, 1–27.

4 Dale Eisler, "Sorrow in St. Paul," *Maclean's* 100, no. 14, April 7, 1997, 55.

5 From the first, the Bre-X story caught the popular imagination in Canada, and thousands of articles have been written about it. Most major Canadian newspapers covered the story in detail; the *Calgary Sun, Calgary Herald*, and *Ottawa Citizen* have had many installments of Bre-X news. Only rather late in the game did U.S. newspapers cover the Bre-X story regularly, but at the height of its fame and infamy much of the media across North America covered the story. Indonesian newspapers and news magazines also offered considerable coverage. Indonesian nongovernmental organizations added their distinctive perspectives in flyers and newsletters. Several books have been published about Bre-X, mainly by journalists. Diane Francis, *Bre-X: The Inside Story* (Toronto, ON: Key Porter Books, 1997); and Douglas Goold and Andrew Willis, *The Bre-X Fraud* (Toronto, ON: McClelland and Stewart Inc., 1997) tell the story with lively excitement. Vivian Danielson and James Whyte (editor and staff writer for the *Northern Miner*, respectively) offer their expertise on the people and politics of the mining scene in *Bre-X: Gold Today, Gone Tomorrow* (Toronto, ON: The Northern Miner, 1997). Bondan Winarno's *Bre-X: Sebungkah emas di kaki pelangi* (Jakarta: Inspirasi Indonesia, 1997) includes useful detail on Indonesian politics and texts of a number of Bre-X documents. Jennifer Wells's coverage for the Canadian news magazine *Maclean's* and John McBeth's coverage for the *Far Eastern Economic Review* have also been very informative. John Behar's "Jungle Fever," (*Fortune*, June 9, 1997, 116–28) offers a useful description of Bre-X's operations. Internet investor chat lines with Bre-X "threads" offer a wealth of both technical information and personal views on the drama.

6 "TSE Raises Listing Standards for Mining and Exploration Companies," and "TSE Tightens Rules for Junior Miners," *CFRA News Talk Radio* website, August 20, 1998.

7 Bre-X lawsuits have their own website: the Bre-X/Bresea Shareholder Class Action Information website. Key issues have involved the liability of stock exchanges as well as Bre-X officials (Diane Francis, "Brokers Must Pay for Their Role in Bre-X," *National Post Online*, May 20, 1999); the ability of Canadians to participate in U.S. class-action lawsuits ("Will Canadians Be Allowed in American Bre-X Suit?," *Daily Mining News*, March 31, 1999); and the nationwide (vs. provincial) scope of Canadian class action lawsuits (Sandra Rubin, "Let All Canadian Bre-X Shareholders in Class-Action Suit, Court Urged," *Financial Post*, February 11, 1999).

8 The Bre-Xscam.com website peddled jokes, news, and art about the Bre-X saga on their "Bungle in the Jungle" web page. A Canadian stockholder named Ross Graham recorded a song called "The Bre-X Blues" and made the CD available for sale on the *Red Deer Advocate* web page. He claims to have made back his losses on Bre-X investment through selling these CDs ("Bre-X Investor Gets Last Laugh with Song," *Breaking News*, December 22, 1998).

9 Junior companies prospecting for gold in Kalimantan with "post–Bre-X" advertising of their claims include Kalimantan Gold Corporation (Vancouver, BC); Twin Gold Corporation (Toronto, ON); and Nevada Manhattan Mining Inc. (Calabasas, CA). See "Indonesia Trying to Recover Reputation after Bre-X, Says Specialist," *CFRA News Talk Radio*, March 20, 1999.

10 "Felderhof Still Insists Bre-X Site Has Gold," *Financial Post*, March 1999.

11 David Zgodzinski, "Bre-X: The Battle Between Bulls and Bears on SI," *Silicon Investor* website, 1997.

12 Behar, "Jungle Fever," 121.

13 To "conjure" is both to call forth spirits and to perform magical tricks; in each case, the term highlights the intentionality of the performance, the studied charisma of the performer, and the hope of moving the audience beyond the limits of rational calculation. These features characterize the economic strategies I discuss here, in which everyday performance requirements—for contracts, marketing, investor reports, and the like—are made into dramatic shows of potential.

14 For example, Arjun Appadurai, *Modernity At Large* (Minneapolis, MN: University of Minnesota Press, 1996); James Clifford, *Routes* (Cambridge, MA: Harvard University Press, 1997); Pheng Cheah and Bruce Robbins, ed., *Cosmopolitics* (Minneapolis, MN: University of Minnesota Press, 1998); Roger Rouse, "Thinking through Transnationalism: Notes on the Cultural Politics of Class Relations in the Contemporary United States," *Public Culture* 7 (1995): 353–402.

15 It is hard to forget that the capitalist development of the late Suharto regime did, literally, create a great fire. Government subsidies supported the conversion of "degraded forest" into oil palm and pulp and paper plantations; eager plantation managers used the 1997 droughts to burn their way into plantation conversions, sometimes burning local residents out of farms, orchards, and homes. The fires created a regional smoke haze so great as to provoke transnational alarm. Many accounts, however, refused to discuss the alliance of government and business in this kind of environmentally destructive entrepreneurship. Indeed, some fires were still smoldering in 1998 when the IMF demanded that Indonesia expand the making of plantations by admitting a greater range of foreign investors. See WALHI/Redaksi, "Ramai-ramai bakar hutan dan lahan perkebunan," *Tanah Air* 18, no. 5 (1998): 2–14.

16 This version of the much told story is taken from Goold and Willis, "The Bre-X Fraud."

17 *The Privateer Gold Pages* (http://www.the-privateer.com/gold.html) reviews the recent history of government, interstate, and private uses and prices of gold.

18 Jennifer Wells, "King of Gold," *Maclean's*, December 9, 1996, 32–40.

19 Peter Newman, "Peter Munk: A Dreamer Who Became a King," *Maclean's*, December 9, 1996, 42.

20 The importance of Filipino geologists in Indonesian mining highlights the national and regional cultural demands of the industry which, like other industries, operates through stereotypes about the appropriate cultural specifications of labor power. Goold and Willis ("The Bre-X Fraud," 170) write: "The Filipinos have their own culture in the mining world. They are well-educated and often trained at big U.S.-owned mines in the Philippines. They are fun-loving, attuned to Western tastes and sensibilities, yet Asian." Meanwhile, the resentment of Indonesian professionals toward the access of Filipinos to Indonesian projects makes them easy scapegoats when things go wrong. Bondan Winarno says Indonesian geologists use the term "Filipino Mafiosi" (*Bre-X: Sebungkah emas di kaki pelangi*, 164).

21 This version draws on Goold and Willis, "The Bre-X Fraud."

22 Goold and Willis, "The Bre-X Fraud," 64–65.

23 Peter Waldman and Jay Solomon, "Geologist's Death May Lie at Heart of Busang Mystery," *Wall Street Journal*, April 9, 1997, A10.

24 "Rumors Swirl around Bre-X," *Ottawa Citizen*, March 25, 1997.

25 The autopsy results are discussed in the following news items: "Buried Body Not Geologist: Report," *Ottawa Citizen*, April 10, 1997; Michael Platt, "Dead or Alive?" *Calgary Sun*, April, 11 1997; "Foul Play Fears Haunt Geologist," *Calgary Sun*, April 19, 1997; "Family Accepts Autopsy," *Calgary Sun*, April 21, 1997; "ID Challenged," *Calgary Sun*, August 5, 1997; "Print Identified as De Guzman's," *Calgary Sun*, April 24, 1997. Bondan Winarno (*Bre-X: Sebungkah emas di kaki pelangi*, 134–35) discusses the theory that although de Guzman wore false teeth, the corpse had natural teeth.

26 Nisid Hajari, "Is the Pot at the End of the Rainbow Empty?" *Time* 149, no. 14, April 7, 1997.

27 "Ultimate Betrayal: Bre-X Boss Says Pair Ruined Dream," *Calgary Sun*, October 12, 1997.

28 Joe Warmington, "Bre-X Takeover Claim," *Calgary Sun*, April 6, 1997, and "Bank on More Intrigue," April 8, 1997.

29 Joe Warmington, "Yanks Waiting: Americans Ready to Gobble up Bre-X Shares," *Calgary Sun*, April 13, 1997.

30 David Jala, "Frenzy Stuns Market," *Calgary Sun*, April 2, 1997.

31 Strathcona Mineral Services conducted the independent technical audit. In the words of the report: "We very much regret having to express the firm opinion that an economic gold deposit has not been identified in the Southeast zone of the Busang property, and is unlikely to be. We realize that the conclusions reached in this interim report will be a great disappointment to the many investors, employees, suppliers, and the joint-venture partners associated with Bre-X, to the Government of Indonesia, and to the mining industry elsewhere. However, the magnitude of the tampering with core samples that we believe has occurred and the resulting falsification of assay values at Busang, is of a scale and over a period of time and with a precision that, to our knowledge, is without precedent in the history of mining anywhere in the world." Graham Farquarson [Strathcona Mineral Services], "Busang Technical Audit: Interim Report," reproduced in *Gatra*, May 17, 1997, 27.

32 McLeod gives considerable weight to the Bre-X affair in bringing on the Indonesian financial crisis by undermining investor confidence. He writes, "Perhaps the most significant recent event to crystallize attitudes on the part of the general public, the intellectual and business elite, and the foreign investment community regarding the direction in which government had been heading was the so-called Busang saga" ("Indonesia's Crisis and Future Propsects," 215). In my view, although the Bre-X saga did not in itself shake the Indonesian economy very much, it *dramatized* issues of what came to be called, following Philippine precedent from the 1980s, "crony capitalism."

33 Jennifer Wells, "Greed, Graft, Gold." *Maclean's*, March 3, 1997, 40.

34 Anthony DePalma, "At End of a Miner's Rainbow, A Cloud of Confusion Lingers," *New York Times*, March 31, 1997, A1.

35 Jennifer Wells, "Rumble in the Jungle," *Maclean's*, February 3, 1997, 38–39; Bre-Xscam.com, "The Bungle in the Jungle"; Michael Platt, "Rush Hour in the Jungle," *Calgary Sun*, May 23, 1997.

36 Behar, "Jungle Fever," 123.

37 Goold and Willis ("The Bre-X Fraud," 207) write about the Canadian scene: "Because so many small investors held Bre-X stock, virtually everyone knew someone who had won or lost money. Their stories, often exaggerated, played out across the country. In this environment, any rumor had legs."

38 "On Silicon's popular net forum Techstox, Bre-X dominated for months. More than 4,000 new people a day were searching for Bre-X information, and more than 700 items were being posted about the company for all to read" (Francis, *Bre-X: The Inside Story*, 153).

39 David Zgodzinski, "Bre-X: The Battle between Bulls and Bears on SI," *Silicon Investor*, May 4, 1997.

40 John McBeth, "The Golden Boys," *Far Eastern Economic Review* 10, March 6, 1997, 42–43.

41 DePalma, "At End of a Miner's Rainbow," C10.

42 McBeth, "The Golden Boys," 44.

43 Brien Lundin, "International Pursuit: Turning World-Class Potential into World-Class Reality," *Gold Newsletter's Mining Share Focus*, 2, no. 1 (1998): 4.

44 Anna Tsing, *In the Realm of the Diamond Queen: Marginality in an Out-of-the-Way Place* (Princeton, NJ: Princeton University Press, 1993).

45 In the early 1990s, the first small-time migrant gold miners came to my research areas in South Kalimantan; in other parts of Kalimantan, they arrived somewhat earlier. They came together with two technical innovations: the extension into the forest of logging roads, which opened the terrain to motor transport; and the introduction of a hand-held hydraulic pump, small enough to carry, yet powerful enough to hose away a hillside to expose its gold-bearing gravel. Each of these technologies carried social and political conventions and practices and dreams, as described in the text below. I have written elsewhere in more detail about the self-conscious making of a "frontier" in Kalimantan. See Anna Tsing, "How to Make Resources in Order to Destroy Them (and Then Save Them?) on the Salvage Frontier," chap. 3 in *Histories of the Future*, ed. Susan Harding and Daniel Rosenberg (Durham, NC: Duke University Press, 2005), 51–73. The importance of logging roads for the mining industry in Kalimantan is discussed in Danielson and Whyte, *Bre-X: Gold Today, Gone Tomorrow*, 43.

46 Personal narratives of frontier gold discovery often give more clues to the interdependence of company miners and independent gold seekers than they admit in their official "discovery tales." Bre-X geologists were quite willing to admit, informally, that they chose their prospecting sites through the advice of small-scale Kalimantan miners, who led them to gold-rich sites. Consider, for example, the narrative Bre-X chief geologist Felderhof told reporters about how he first found Busang: "Felderhof had first heard of Busang in the early 1980s, when he and his Australian colleague Mike Bird were exploring the logging roads of Borneo on rented motorcycles. They roared from village to village, asking the locals to point out where they had found gold nuggets in stream beds" (Goold and Willis, "The Bre-X Fraud," 36). Indeed, one of the earliest clouds of suspicion about the rich "southeast zone" of Busang was generated by the fact that small-scale miners did not pan for gold in this area (Danielson and Whyte, *Bre-X: Gold Today, Gone Tomorrow*, 197). Despite acknowledging this interdependence, it never occurred to the geologists that it might interfere with their personal rights—or their official stories—of "discovery."

47 In April 1967, Freeport Indonesia was granted a tax holiday, concessions on normal levies, exemption from royalties, freedom in the use of foreign personnel and goods, and exemption from the requirement for Indonesian equity. The terms were changed slightly in 1976, canceling the remaining eighteen months of the tax holiday and allowing the Indonesian government to purchase an 8.5 percent share (Hadi Soesastro and Budi Sudarsono, "Mineral and Energy Development in Indonesia," in *The Minerals Industries of ASEAN and Australia: Problems and Prospects*, ed. Bruce McKern and Praipol Koomsup (Sydney: Allen and Unwin, 1988), 161–208).

48 More restrictive second generation and less restrictive third generation CoWs were introduced in 1968 and 1976, respectively. 1986–87 marked the Australian "gold rush" in Indonesia and introduced fourth generation CoWs, with "one year of general survey ending with 25% relinquishment of concession area; three years of exploration with 75% relinquishment by the end of the fourth year; an Indonesian partner in the CoW; and equity divestment after five years of operation so that ideally after 10 years of production the local partner holds 51%" (Carolyn Marr, *Digging Deep: The Hidden Costs of Mining in Indonesia* (London, UK: Down to Earth, 1988), 16). Fifth generation contracts began in 1990, with tax incentives and low tariff property taxes. A 1992 law required foreign investors in frontier areas to reduce their equity shares to a maximum of 95 percent within five years and to 80 percent within twenty years (Marr, *Digging Deep*, 17). Sixth generation CoWs, requiring environmental impact assessments, were offered in 1997 (Bondan Winarno, *Bre-X: Sebungkah emas di kaki pelangi*, 28).

49 Wells, "Greed, Graft, Gold," 42.

50 Dr. Soetaryo Sigit, ex-Director General of Mines, "Current Mining Developments in Indonesia," quoted in Marr, *Digging Deep*, 26.

51 Rachman Wiriosudarmo, quoted by Wells, "Greed, Graft, Gold," 41.

52 Goold and Willis, "The Bre-X Fraud," 99–100.

53 U.S. ex-president Bush wrote Suharto to express his "highly favorable" impression of Barrick (Wells, "Rumble in the Jungle").

54 Wells, "Greed, Graft, and Gold."

55 By this time, it was clear to all the players that the Indonesian government was calling the shots. Placer Dome sent their bid directly to the president (John McBeth and Jay Solomon, "First Friend," *Far Eastern Economic Review* 8 (February 20, 1997): 52–54).

56 Hasan's perspective was developed in a context in which other Indonesians were calling for greater national control of Busang. Islamic leader Amien Rais, for example, argued that the gold "should be kept for our grandchildren in the 21st century" (cited in Francis, *Bre-X: The Inside Story*, 130). Bondan Winarno (*Bre-X: Sebungkah emas di kaki pelangi*, 84–94) details nationalist claims.

57 Hasan's investments during this period are detailed in McBeth and Solomon, "First Friend."

58 Quoted in Jennifer Wells, "Gunning for Gold," *Maclean's*, February 17, 1997, 52.

59 Bre-X president David Walsh said he was settling with an arrangement that reflected "Indonesia's political, economic, and social environment" (Richard Borsuk, "Bre-X Minerals Defends Pact with Indonesia," *Wall Street Journal*, February 2, 1997, B3a).

60 Cited in Marr, *Digging Deep*, 71.

61 The history of Freeport-McMoRan is the subject of an unpublished book titled *Corporate Power and Civil Society: The Story of Freeport-McMoRan at Home and Abroad* by journalist Robert Bryce and anthropologist Steven Feld. For Freeport's arrangements with the Indonesian government, see n46; Francis, *Bre-X: The Inside Story*, 129; and Goold and Willis, "The Bre-X Fraud," 113–14. Marr, *Digging Deep*, details Freeport's Irian operations, with special attention to the mine's history of environmental problems and human rights abuses.

62 Quoted in Behar, "Jungle Fever," 128.

63 David Walsh died of a stroke on 4 June 1998. He spent the last months of his life fighting class action suits and trying to clear his name (Sandra Rubin, "Obituary: David Walsh," *Financial Post*, June 5, 1998). See also "Ultimate Betrayal: Bre-X Boss Says Pair Ruined Dream," *Calgary Sun*, October 12, 1997.

64 The importance of small, popular investment in Bre-X highlights the importance of this national agenda. "At its peak in May 1996, 70 percent of Bre-X's 240 million shares were in the hands of individual investors" (Goold and Willis, "The Bre-X Fraud," 239). This contrasts with a more ordinary Canadian company, which might have 30 percent of its shares owned by individuals. According to Goold and Willis (105), in 1996 Bre-X had 13,000 shareholders, including pension funds and insurance companies. According to Francis (*Bre-X: The Inside Story*, 199), about 5 percent of Bre-X trading was from outside Canada and 90 percent of all Bre-X trading was conducted on the Toronto Stock Exchange (197).

65 It is possible to make a great deal of money out of speculation even if the product comes to nothing. Bre-X shareholders made money merely by selling their shares while the price was still high. The outspoken investors Greg and Kathy Chorny, for example, sold two thirds of their stock for C$40 million and lost a comparatively minor sum on remaining shares (Francis, *Bre-X: The Inside Story*, 196). At the end, smart investors made money by "short selling," that is, borrowing Bre-X shares from brokers, selling them, and returning them by buying them back at a lower price. Goold and Willis ("The Bre-X Fraud," 221) report that 5.5 million Bre-X shares were sold short. Francis (*Bre-X: The Inside Story*, 203) reports that the investment bank Oppenheimer and Co. made C$100 million shorting Bre-X stock. Meanwhile, other firms and individuals, including Quebec's public pension fund and the Ontario Teachers Pension Plan Board, lost major amounts of money (Goold and Willis, "The Bre-X Fraud," 248).

66 David Harvey, *The Condition of Postmodernity* (Oxford, UK: Basil Blackwell, 1989).

67 Stuart Hall, "On Postmodernism and Articulation: An Interview with Stuart Hall," edited by Lawrence Grossberg, in *Stuart Hall: Critical Dialogues in Cultural Studies*, ed. David Morley and Kuan-Hsing Chen (London: Routledge, 1996), 131–150.

68 Other Bre-X allegories have been suggested—for example, that greed blinds everyone's eyes (Goold and Willis, "The Bre-X Fraud," 267), or that the international flow of money means business must deal with "exotic and troublesome regimes" (Francis, *Bre-X: The Inside Story*, 232). Outside of Canada, the allegorical reading arose that this was just the way of Canadian business, where stock exchanges are a "regulatory Wild West" (quotation attributed to the U.S. mass media in Bondan Winarno, *Bre-X: Sebungkah emas di kaki pelangi*, 208). My reading refuses the distinction between the seeing and the blind to point to the money being made even in a scam. I also emphasize the exotic and troublesome nature of capitalism itself—both in and beyond Canada and Indonesia.

Empire Lite

A Conversation with Michael Ignatieff

Michael Ignatieff is a Canadian politician, writer, broadcaster, and educator. He is currently serving as Rector and President of Central European University in Bucharest after teaching at the Harvard Kennedy School (2013–16). As Member of Parliament (2006–11), he served as Leader of the Liberal Party of Canada (2009–11). His books include *Empire Lite: Nation Building in Bosnia, Kosovo, Afghanistan* (2003), *Virtual War: Kosovo and Beyond* (2000), and *Blood and Belonging* (1993).

MI: You're a Canadian studying risk and resilience

X: I am.

MI: I'm interested in risk and resilience.

I spent two weeks in Japan in the nuclear evacuation zone talking to people who lost everything as a result of the tsunami, the earthquake, and the nuclear fallout.

I became struck by the perfectly obvious things. Resilience is not an individual thing. It's tremendously important what community you're a part of. Your own capacity to leverage strength is dependent on the capacity of your community to leverage strength.

What everybody was impressed with in Japan was the small town character of these places and their very local and regional values. They are tough folks with a deep consciousness of traditions of endurance that go back a thousand years.

When you ask a guy who's got half of his fields polluted by nuclear stuff, and he's still hanging on, he hasn't given up… And you ask him why he is staying on, and he says, "well, in 1695, my family came to this region because they were fleeing famine in another part of Japan. My ancestors have been through a disaster and survived it. I wish to do the same."

The second example of that was in a place called Minamisoma on the North East coast, right by the sea. It was just devastated by the tsunami. I was very struck by how important, an apparently irrelevant thing—or I thought it was irrelevant—was the Horse Festival. There is a Horse Festival on the beach in Minamisoma that has been going on for a thousand years. It relates to Samurai warrior traditions in that area.

I was struck by the number of men. I didn't even ask how they were doing, how they were hanging on. They just started talking about what was meaningful to them as men. They said they had just heard that they are going to bring back to the Festival this year. "I go to the Horse Festival every year and it's who I am."

You got the strong sense that they thought of themselves as Samurais. All of this historical stuff is very important, I think.

Resilience is a rooted connection in time to meaningful past, but also a deep conviction that there is a future. Resilience involves a metaphysical, existential commitment to a future that you cannot see. I am hanging on because I want the past that I value to continue into the future. It is a commitment to time. And the commitment to time is not just a commitment to *your* time, it's a commitment to the time of your community, of your people, of the people who you regard as significant.

I think that as a British Columbian, you will know immediately that your fellow Aboriginal citizens have a deep commitment to rooted historical traditions through time as the proof that we have survived everything. We are still here. We will be here in the future *because* we have been here for so long.

"Empires survive only by understanding their limits... At the beginning of the first volume of 'The Decline and Fall of the Roman Empire,' published in 1776, Edward Gibbon remarked that empires endure only so long as their rulers take care not to overextend their borders. Augustus bequeathed his successors an empire 'within those limits which nature seemed to have placed as its permanent bulwarks and boundaries: on the west the Atlantic Ocean; the Rhine and Danube on the north; the Euphrates on the east; and towards the south the sandy deserts of Arabia and Africa.' Beyond these boundaries lay the barbarians. But the 'vanity or ignorance' of the Romans, Gibbon went on, led them to 'despise and sometimes to forget the outlying countries that had been left in the enjoyment of a barbarous independence.' As a result, the proud Romans were lulled into making the fatal mistake of 'confounding the Roman monarchy with the globe of the earth.'
This characteristic delusion of imperial power is to confuse global power with global domination."

—Michael Ignatieff,
New York Times, 2003

A lot of the language of resilience is tremendously diminished by being psychologized and individualized and stripped of these crucial historical dimensions.

From a design perspective, I also think of physical buildings; physical shrines. In the Japanese case, it was so significant that the Shinto shrines that are built up on the high ground all survived the earthquake and tsunamis; and *all* of the modern buildings built near the coast were swept away. A lot of people took some pretty significant messages from that.

Chief "Highest Peak in Mountain Range" in front of "House Where People Always Want to Go," Haina, Haida Gwaii archipelago, 1888 (Richard Maynard, courtesy of Royal BC Museum, BC Archives AA-00073)

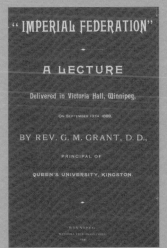

We didn't listen to our ancestors. And not in some metaphysical abstract way, but because their ancestors in 1964 had been through *exactly* the same thing. This is not vague. The ancestors built at 25 meters above sea level because they knew damn well that a tsunami could take them out. So the placement of these Shinto shrines up on the hills was kind of a warning. A reproach to the present that they had forgotten.

There is a lot written about this, but I did think it was fascinating that in the previous tsunami that happened in 1896, people were putting stone markers that basically say, "this is where the water reached." You can still see them. So Japan literally didn't listen to the markers.

The risk it took was not listening to the physical environment.

X: I appreciate the cultural connection to resilience.

You coined the term "empire lite," and at one point proposed that the US creates a "humanitarian empire." How do you see Canada's relationship with empire?

MI: A humanitarian empire… I was involved in the International Commission on Intervention and State Sovereignty. It was less about a humanitarian empire. It was, in fact, a Security Council-mandated Chapter 7 authorization for the use of military force in locations like Syria where the regime is massacring its own people, or a situation like Rwanda where there is a civil war, where the State engages in a preemptive act of genocide to prevail in the war.

If you want to use force to stop atrocities, there are not many combat-capable militaries that can do it. So you end up—whether you like it or not—having to deal with Washington.

It's not that I want the Americans to extend any kind of empire, I just want people to stop being massacred. The lead example of that being Bosnia. I spent a lot of time in the Balkans between '91 and '95. The Europeans said they would stop all of this killing. They didn't really stop until the Americans came in and said: "we're going to make it stop."

It doesn't make me an imperialist; it just makes me a realist.

"IMPERIAL FEDERATION"

A LECTURE

Delivered in Victoria Hall, Winnipeg,

On September 13th 1889,

BY REV. G. M. GRANT, D. D.,

PRINCIPAL OF

QUEEN'S UNIVERSITY, KINGSTON.

WINNIPEG

George Monro Grant, "Imperial Federation," 1890 (Manitoba Free Press)

The other thing I would say is of course you don't want to use military force to stop this stuff. Prevention, prevention, prevention, prevention. Conflict avoidance, conflict negotiation. Really any set of tools that you can use to prevent mass atrocities you should deploy. But there may come a point at which—when Assad is barrel bombing Aleppo—you want this to stop. There is only one way to do it. That's where I come from.

X: Do you see Canada having a different kind of empire?

I recently saw you speak at the Harvard Law School where you stated that, "The ceiling is too low in Canada. We are not nearly ambitious enough. The Canada standard is not good enough." Is Canada forgoing its empire?

MI: If by empire you mean soft power influence in the world, I think the story Canada tells itself—and has told itself since the peacekeeping era—is that we do things differently than the Americans. We are the peacekeepers and they are the war makers.

I don't think you can do peacekeeping with blueberries and side arms.

The thing that's burned into my soul is that in 1993, it was a Canadian detachment that was sent to Srebrenica under the Safe Havens Mandate to protect the civilians. We did not have the combat capability to repel the Serbs had they chosen to come in. We were then replaced by the Dutch. The Dutch sat and watched while 10,000 men were taken away and exterminated by the Serbs because they didn't have the capacity to defend these people. It destroyed the military prestige of the Dutch. It might well have done the same to us, hadn't we changed places.

I think that Canada has to have a combat-capable military. The NATO military spending is two percent. We never get close to it, but we need to spend enough on our military to defend ourselves. We will always do things differently than the Americans because we're not American. We have neither their capabilities nor do we have the same set of values. I think it's appropriate for us to have a very different foreign policy.

If you want to do human rights work in the world—that is if you want to protect civilians who are at risk of being massacred—you simply have to have the capacity to do so. You can't do it without robust capabilities.

But we should go our own way. The current Liberal government has committed to take F-16s out of the fight against ISIS. Fine. We should define our own objectives. If we choose to put trainers in as opposed to combat aircraft, it's fine with me.

What I don't buy is the Canadian narrative that we're all just soft power. Development. Human rights promotion. Civil society promotion.

All of that stuff is great. I think that Canada does some of that pretty well. I like those analyses that show that we have spent very little on those things in the past twenty years. If you want to be a country that is really international, you've got to spend much more on foreign aid and development than we do, and you've got to spend much more on defense.

These things divide Canadians. I understand all that. I lived it in politics.

I'm not a Harperite in foreign policy. I *loathed* his foreign policy. But the thing I don't want liberal progressive Canadians to do is to think that you don't need combat capability. You absolutely do. This is a very dangerous and unpleasant world.

The other thing is that we don't want to be defended by the Americans. It's a nationalist thing. We want to be able to defend ourselves. We want to be able to patrol our own airspace and be able to rescue Canadians when they get into trouble.

X: You went on to say that we can't separate the policy and the messenger. "We think of policy as separate from the abstraction of making it possible. You can have the best idea but if you can't sell it, it won't happen. So then you start to push policy you can sell. So then the message drives the policy."

You were absent for the Bill C-300 vote (a bill that supported the principle of Corporate Social Responsibility for Canadian mining in developing countries, proposed by Liberal John McKay). What was your message in absentia?

MI: I thought it was a bad bill. I got huge flak for it. I just thought it was a bad bill. My absence turned out to be electorally devastating.

Jack Layton made a huge deal of it in the English language debate. The fact is that it was standard practice. This will convince nobody, but it happens to be true. It was the practice of Liberal Prime Ministers and Leaders of the Opposition to absent themselves from free votes. Because a free vote has to be free. A leader—in principle—shouldn't be voting, because it appears to express the view of the caucus. A free vote is a free vote.

I thought Bill C-300 just hadn't thought through what it wanted to do. It had Export Development Canada riders attached to it that didn't make any sense. It didn't have enforcement mechanisms that seemed credible.

I think it would be terrific for the government of Canada to have the capacity to inspect Canadian mining operations overseas, make sure they are CSR (corporate social responsibility) compliant.

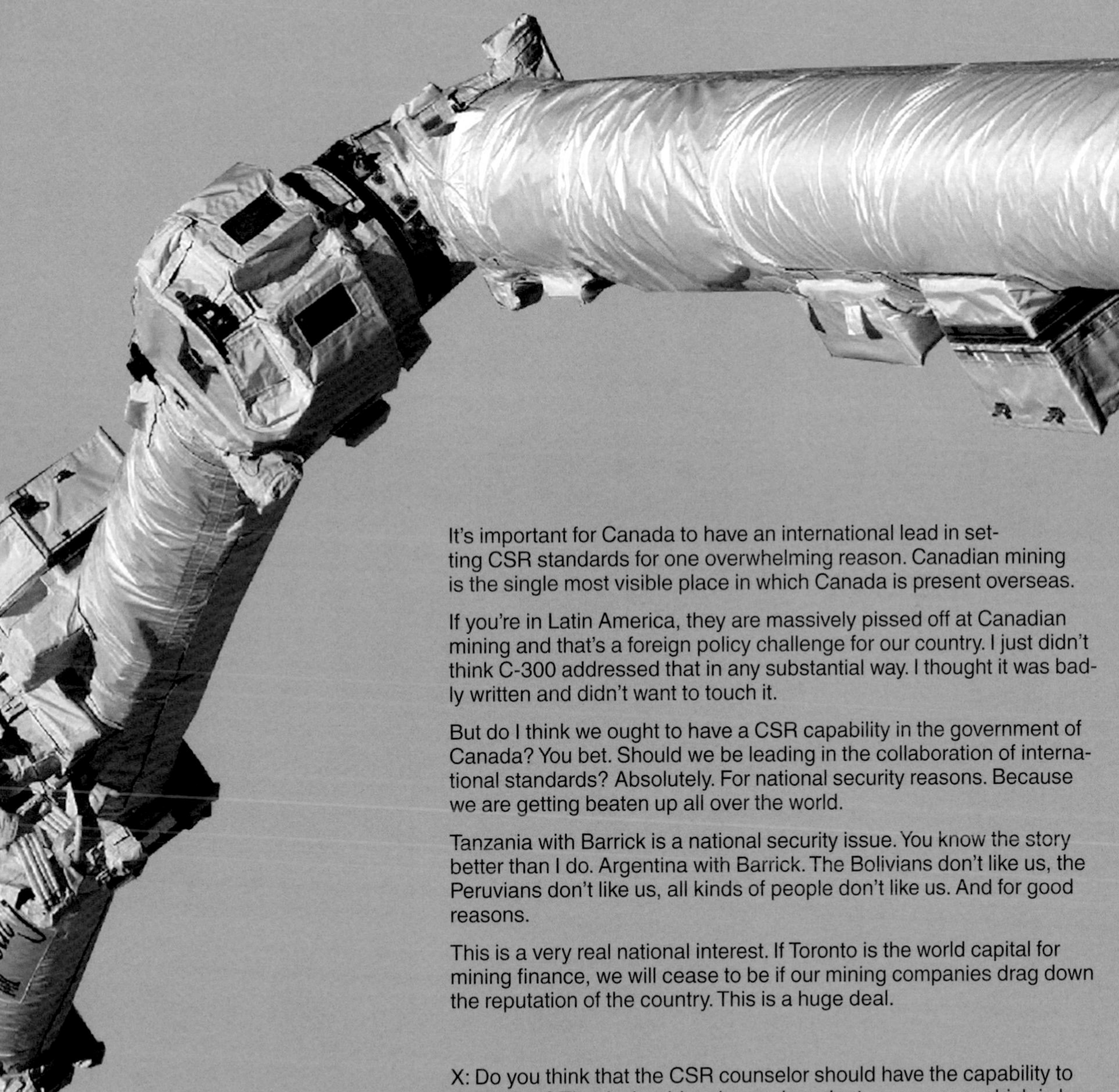

It's important for Canada to have an international lead in setting CSR standards for one overwhelming reason. Canadian mining is the single most visible place in which Canada is present overseas.

If you're in Latin America, they are massively pissed off at Canadian mining and that's a foreign policy challenge for our country. I just didn't think C-300 addressed that in any substantial way. I thought it was badly written and didn't want to touch it.

But do I think we ought to have a CSR capability in the government of Canada? You bet. Should we be leading in the collaboration of international standards? Absolutely. For national security reasons. Because we are getting beaten up all over the world.

Tanzania with Barrick is a national security issue. You know the story better than I do. Argentina with Barrick. The Bolivians don't like us, the Peruvians don't like us, all kinds of people don't like us. And for good reasons.

This is a very real national interest. If Toronto is the world capital for mining finance, we will cease to be if our mining companies drag down the reputation of the country. This is a huge deal.

X: Do you think that the CSR counselor should have the capability to prosecute? That it should go beyond a voluntary process, which is how it currently stands?

MI: I don't know enough about what's in place. I think if the government of Canada is involved, and the reputation of Canada is involved, you've got to have some capacity to inspect. You've then got to have some

CSA / NASA

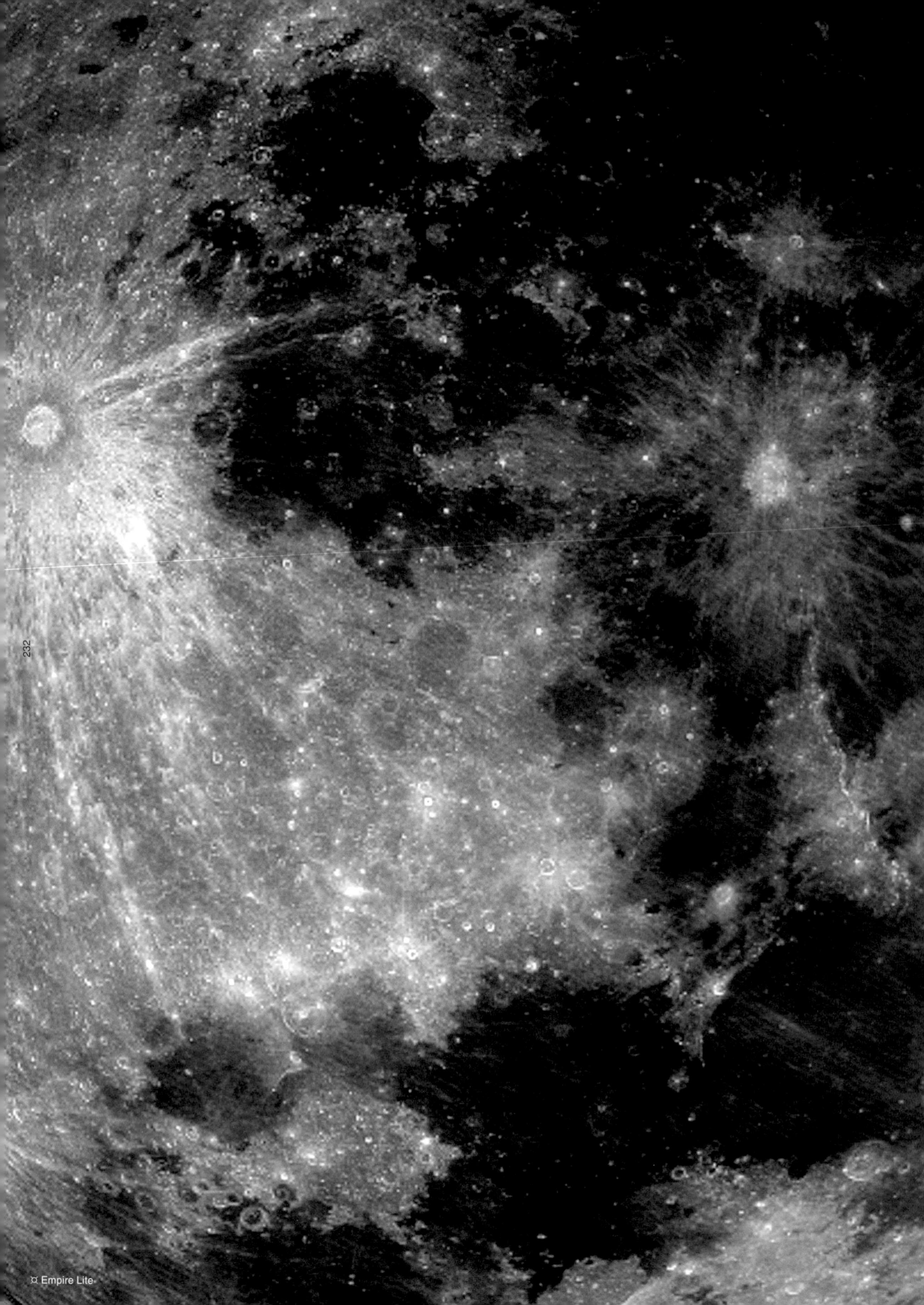

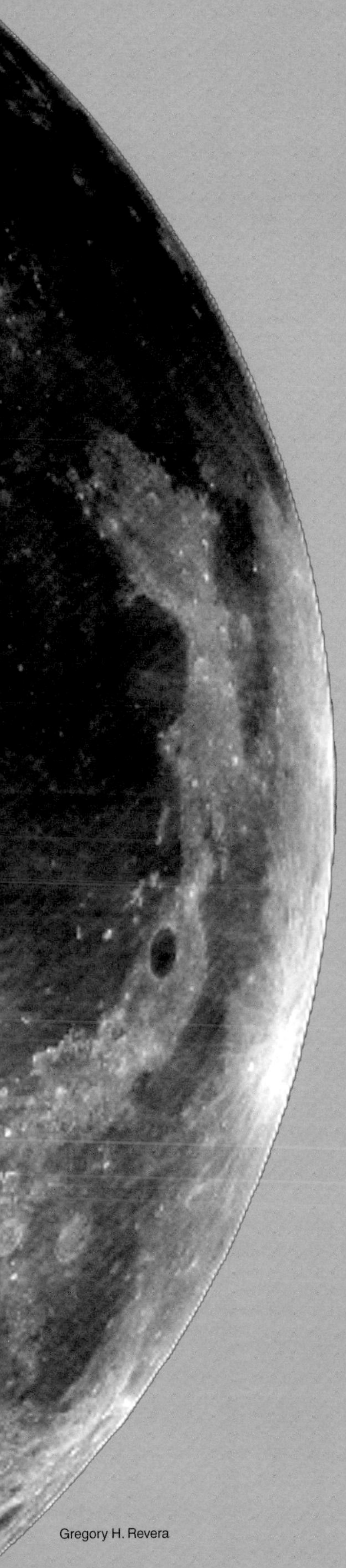

Gregory H. Revera

capacity for effected citizens in mining operations to make representations to the government.

There are some very complicated issues of extraterritoriality here.

If there are some Chileans complaining about a Canadian operation in Chile, you've got to work this out with Chile as to what recourse a Chilean citizen has.

I don't think a purely voluntary system is going to work.

But none of this was thought through in Bill C-300. I got beaten up by everybody about this. They said, "you've just caved to the mining industry." *No!* Part of the problem is that I actually know something about this stuff. I don't like voting for stuff that I just don't think is any bloody good.

X: Canada is starting to move into space extraction, such as lunar mining.

The Liberals have historically supported aerospace leadership in Canada. For example, in 2008, the Government of Canada intervened in the sale of MacDonald, Dettwiler and Associates, Ltd. The Liberals supported this decision to protect aerospace industry from foreign takeover, and subsequently called for an increase in investment and the development of a coherent strategy to make Canada a leader in space. Do you think that is an important role for Canada to take on?

MI: Well, We've got Marc Garneau, don't we? And the Canadarm.

We've got a big node of high tech capability in Vancouver and Montreal in these areas. Absolutely. Whether we should be extracting from the moon is a question I have never been asked in a long and happy life.

Should we be in aerospace? Should we be in deep space? I *absolutely* think so. But in consortium with others. We can't do it alone. But it is a hugely significant technology for the future and we've got some pretty good folks in that area.

But pulling coal out of the moon? I don't know. ¤

This conversation took place between Michael Ignatieff and Genevieve Ennis Hume in 2015 and 2016.

Geology's Extractive Impulse

The Case of Victorian Exploration and Empire

Suzanne Zeller

"They say we lack audacity...
But I say to you, they do not know where to look,
        and have not the eyes to see.
For audacity is all around us,
Boldness sits in the highest places,
We are riddled with insolence."[1]

—F.R. Scott, "Audacity ('Audacity Is Missing in Canada,' *The Times*, 30/11/59)," 1964

Unlike the Puritans' goal of religious freedom in what became the Thirteen American Colonies, Canada's first European visitors sought a northern *El Dorado*, whether in the mythic Kingdom of the Saguenay or in an overland route to Asia's storied riches. They memorialized these efforts in their naming of Lachine (near Montreal) in 1667, after neither dream panned out.[2] Hope persisted, however, in a relentless extractive impulse that has focused ever since on a sequence of natural resources, from fish and furs, timber, minerals, and wheat, to oil and natural gas. Canada, according to the Toronto political economist Harold Innis's famous staples thesis, took its current form as a by-product of this transcontinental pursuit—"not in spite of geography," Innis famously declared in *The Fur Trade in Canada* (1930), "but because of it."[3]

A generation later, the Montreal poet F.R. Scott instead saw "bank robbers … helping themselves to the wealth of the land like the French and the English before them, *coureur-de-bois* and fur trader rolled into one." "Do you want audacity?" Scott exclaimed in 1964, "You may marvel at the boldness of promoters of oil and natural gas … Getting their hands on concessions and rights, access to underground treasures awaiting man's use in the womb of our northland."[4]

The extractive impulse, as glimpsed from these diverging perspectives, had been cemented in place during the nineteenth century by the rise of modern science—especially in the case of geology—in a Victorian context that sharpened exploration's focus, purpose, and significance. Britain's burgeoning Industrial Revolution demanded overseas markets and devoured new sources of mineral wealth apace—no longer gold so much as coal, iron, and other metallic ores—empowered by unprecedented new partnerships between science and technology.[5] It also revived in Britain's North Atlantic culture a powerful Baconian[6] philosophical heritage that valued large-scale, collaborative, and empirical approaches to nature and the land, through firsthand observations and inductive reasoning, as the best way to ensure modern civilization's rapid material and moral progress. Scottish culture in particular extolled, as two sides of the same coin, science's production of useful as well as theoretical knowledge—and its diffusion through public education—in hopes of emulating, not only among Scots at home but also among waves of Scottish immigrants to British North America, England's remarkable industrial take-off.

Nor was it only the Mother Country's sojourning Gullivers who led the way in intensified geological investigations of British North America: the Crusoes who sought to build a New-World life for themselves adopted the extractive impulse's imperialist discourses, adapting its expansionist goals, methods, and institutions to their own particular circumstances. The resulting cultural transfers ran in both directions as New World evidence challenged and transformed Old World assumptions. These transatlantic exchanges encouraged the rise of secondary imperialisms that were reshaped, in their turn, by both collaboration and resistance. While a francophone majority in what is now the province of Quebec, for example, recoiled from the British regime's increasing scientific scrutiny of their homeland, some leading bourgeois *Canadiens* nevertheless came to admire and participate not only in Victorian culture's gentlemanly amateur naturalist tradition, but also in extractive processes that welcomed foreign investment, first in railways and then in far more intrusive branch-plant mining enterprises.[7] And while Indigenous nations found far greater reason to suspect European imperialism in all of its manifestations, they routinely assisted geological surveys across British North America in untold ways—earning scant, if any, recognition of their presence, let alone of their indispensable contributions, in the resulting scientific reports, at least before twentieth-century plans for a northern pipeline threatened to destroy their homelands altogether.[8]

Geology's extractive impulse overlapped British North American political developments in three phases: as explorations moved inland along developing transatlantic scientific networks after 1815; institutionalized colonial government surveys during the 1830s and 40s; and consolidating its transcontinental foothold from the 1850s.

Neptune's Notebook

The end of the Napoleonic Wars in Europe in 1815 opened a new era for British overseas exploration, as an overabundance of military officers, many of whom were also amateur naturalists, sought to escape redundancy through useful peacetime applications of their advanced scientific training. A long-term exploratory collaboration between the Royal Society of London (founded in 1660) and the Hudson's Bay Company (HBC, founded in 1670) gained more specialized stakeholders with the establishment of the Geological Society of London (GSL) in 1807. It also gained new traction when the Royal Navy's Second Secretary to the Admiralty, Sir John Barrow, interpreted growing evidence of polar warming in 1816 as an unprecedented—and therefore urgent—opportunity to revisit the centuries-old dream of opening a North-West Passage.[9] From 1818, the Royal Navy accordingly supplemented a series of polar voyages with Sir John Franklin's assignment to approach the Arctic shore from inland, down the Mackenzie and Coppermine Rivers. Franklin's reports duly noted apparent outcroppings of coal deposits that could, he tantalized, supply future fueling stations for steamships along an Arctic route.

While traditional areal exploration generated such welcome mineralogical information, the emergence of geology's modern theoretical paradigm during those years added powerful new stratigraphical and palaeontological tools that keyed the Earth's formations to ordered layers and their characteristic minerals and fossils.[10] Followers of A.G. Werner of the Freiberg School of Mines adopted "Neptunist" interpretations that emphasized water's historic role in depositing sedimentary formations over lengthy periods of time.[11] Among Werner's students, the University of Edinburgh's Regius Professor of Natural History, Robert Jameson, in turn forged an extraordinary corps of overseas observers—British explorers, surveyors, military officers, whaling captains, and colonists—to report their findings in Jameson's widely subscribed *Edinburgh Philosophical Journal*. British North America, with its vast interlocking systems of rivers and lakes linking three oceans, appeared to these Wernerian agents to reveal clearly inscribed pages from Neptune's notebook. Their enthusiasm in turn recruited new observers among HBC officers stationed in Rupert's Land, the vast Hudson Bay watershed over which the Company enjoyed a 200-year monopoly.

Farther east, the reports of Royal Navy hydrographers on the Great Lakes, Royal Engineers stationed at Quebec, the Canada Land Company's Warden of the Woods and Forests in Upper Canada's Huron Tract, and various other geological explorers in Nova Scotia, New Brunswick, and Newfoundland—many of them with Scottish connections—reflected Jameson's Neptunist influence. J.J. Bigsby in particular, the University of Edinburgh-educated medical officer for the International Boundary Commission, is credited as the author of British North America's first geological publications.[12] Bigsby's work during the 1820s inspired his editor, David Chisholme, a Scottish member of the Montreal business community, to demand colonial mineralogical surveys as British subjects' rightful due.[13] While several American states were setting the precedent during those years, a government survey even of the Mother Country was not forthcoming until 1835. Yet enough was already known of Nova Scotia's lucrative coalbeds for Britain to grant private monopoly rights in that colony to the General Mining Association in 1826, precluding a public survey there, at least, for the foreseeable future.

For the time being, interesting anomalies that strained Neptunist assumptions riveted geological explorers' attention across British North America. They grew fascinated especially with widespread scatterings of "rolling" rocks ("erratics") among lighter debris ("drift") deposited at considerable distances from their apparent places of origin. Nor could water's slow sedimentary action explain the unstratified granite formations of the Canadian Shield that comprised the geological backbone of northern North America. One way to a clearer understanding, inspired by another of Werner's students, the Prussian scientific traveler Alexander von Humboldt, was to think in continental terms, seeking out global—even "cosmic"—patterns of distribution for as many mappable natural phenomena as possible, thereby to illuminate their more fundamental interconnections.[14]

Humboldt's multi-volume *Personal Narrative of Travels to the Equinoctial Regions of the New Continent, During the Years 1799–1804*, which became available during the 1820s, impassioned observers as remotely situated as HBC fur traders were, to contribute to imperial synoptic projects data about the natural world to which their location gave them exclusive access. They eagerly anticipated Humboldt's magnum opus, *Cosmos: Sketch of a Physical Description of the Universe*, the first volume of which appeared in 1845, supplemented by an atlas replete with infinite possibilities for further synoptic inquiries.[15]

In response, the Scottish naturalist (Sir) John Richardson, Franklin's medical officer during the 1820s Arctic expeditions and a long-time Humboldt admirer, collated decades of accumulated information to attempt a transcontinental geological overview of British North America (1851). Richardson's map envisioned the northern half-continent as framed by a right angle between its Rocky Mountain and Laurentian axes, an apparent structural pattern that Richardson hoped one day to investigate more fully.[16] That same year, and in a similar mode, J.J. Bigsby mapped all that was known of North American boulder erratics, with vectors marking the paths of their "scratches" in an otherwise impenetrable Canadian Shield.[17] Bigsby's cartographical synopsis formed an important database for the new glacial understanding of the Earth's geological history that was taking shape in the hands of the Swiss-American geologist Louis Agassiz.

Laurentian Extensions

To Humboldt's spatial outlook, the Scottish geologist (Sir) Charles Lyell's foundational *Principles of Geology* (3 vols., 1830–33) added a temporal dimension, in his *Attempt to Explain the Former Changes of the Earth's Surface by Reference to Causes Now in Operation*.[18] Lyell's "uniformitarian" approach recalled the insights of his forebear James Hutton, a major scientific contributor to Scotland's rich eighteenth-century Enlightenment culture. Hutton had recognized both the "Vulcanist" power of fire in shaping the Earth's crust and the necessity of accepting "deep" time—far longer than the Bible had dictated—in any theory of Earth's geological history.[19] Lyell's subsequent synthesis of Werner's Neptunist and Hutton's Vulcanist insights into geological processes allowed for floating icebergs as possible transporters of boulder erratics. His general insistence on uniformitarian over "catastrophist" explanations however, delayed for decades a fuller British—and British North American—acceptance of Agassiz's ice-age theory.[20]

Despite its controversial aspects, Lyell's uniformitarianism promoted confidence in geology's ability, more accurately, to pinpoint workable sources of industrial minerals. He borrowed Hutton's term "metamorphism" to denote the effects of heat and pressure in forming contorted, unstratified masses like those of the Canadian Shield, with a focus on the metallic ores they were known to contain. Equally important, Lyell's social position and wide-ranging professional contacts offered colonial geologists a welcome pipeline to the Geological Society of London through his active sponsorship, as he declared their work "destined to surprise us yet" with invaluable geological and mineralogical data.[21] During a North American tour for this purpose in 1841, Lyell became a mentor to the Nova Scotian geologist (Sir) J.W. Dawson, to whom he introduced a new colleague, the Canadian-born (Sir) William Logan: both Dawson and Logan had studied natural history at the University of Edinburgh, and both would contribute important revisions to future editions of Lyell's *Principles*.

Like Lyell, Logan was preoccupied with the coalfields of Nova Scotia and Pennsylvania.[22] A professional knowledge of coal-seams, acquired while working at his uncle's copper works in South Wales, had distinguished him in British scientific as well as mining circles for explaining coal's in situ origins, in places where plants during the Carboniferous Age had lived, died, and become carbonized by metamorphic processes. Unemployed since his uncle's recent death, Logan was also visiting his brother, James Logan, a member of the Montreal business community, in hopes of securing the directorship of the newly United Province of Canada's newly-funded geological survey. Despite justifiable concerns from some scientific quarters, that Logan lacked sufficient palaeontological expertise to carry out the full range of these responsibilities, he nevertheless won the appointment on the basis of his personal connections, as a native Montrealer, to leading Canadian commercial interests poised to industrialize their growing enterprises. To that particular end, his supporters insisted, Logan's record as a self-styled "practical coal miner of education" qualified him exceptionally well.[23]

Indeed, the Geological Survey of Canada's (GSC) survival turned almost entirely on its founding director's keen grasp of the various forces that drove Victorian industrial culture. The far-reaching vision of former Governor General Lord Durham—himself "an English landed gentleman whose crop was coal"—in his prescriptive *Report on the Affairs of British North America* (1839) had postulated a modern economic future for the colonies based on rail and steam, fueled by Durham's estimation of "extensive regions of the most valuable minerals" and other natural resources as key to British North America's promise.[24] Logan and his one assistant, Alexander Murray, now traversed the province of Canada from Lake Huron to Gaspé in Durham's intellectual footsteps, shouldering the burden of his—and many others'—great expectations. The GSC trod a fine line not only between theoretical contributions to scientific knowledge and practical mineral discoveries, but also between the danger of promising too much and that of predicting too little. Logan's counterparts, Abraham Gesner in New Brunswick and J.B. Jukes in Newfoundland, had recently learned this hard lesson from experience, as their own respective reports on their searches for coal cost each his public funding after only one season in the field.[25]

Logan also had to face down his own private suspicions. Preliminary inquiries in the strata near Montreal had already revealed "a coal field, with the coal left out"; it was not long before the edges of the great North American coalfields to the south and east showed their adjacent strata dipping downward in Canada's direction.[26] Just as Logan feared, the entire province lay geologically too far below the coal-rich Carboniferous strata for workable deposits, at least according to "present geological experience," ever to be found there.[27] News of the dreaded conclusion, which called into question Canada's modern economic future, met with bitter resistance even from ardent GSC supporters. William Dunlop, the Canada

Lands Company's former Warden of the Woods and Forests and now a Member of Canada's Legislative Assembly, echoed public disillusionment in a contemptuous dismissal of Logan's report as the "statement of theoretical reasoners which he thought had no foundation": Dunlap, for one, "could not see because the coal vein in Pennsylvania dipped upward that therefore there should be coal up in the moon."[28] Like many others, he countered instead that he had himself seen coal scattered about in several districts of the colony. Canadians who insisted that the province simply must contain coal, and who did not wish to hear otherwise, repeatedly tested Logan's skill and veracity—not to mention his patience—over the years, by surreptitiously salting alleged mines. In the most notorious example, in Bowmanville as late as 1858, a cheese sandwich emerged with a bucket of coal planted in one such potential 'mine.' Logan's Nova Scotian colleague J.W. Dawson, now principal of McGill College in Montreal, entered the fray in his defense, chiding that:

> "The thing that we cannot have, is always that which we most desire, and the more richly we are endowed otherwise, the more earnestly do we long for the one object that may have been withheld. So it would seem to be with the Canadian public in the matter of coal. … Like the child whose toys are all valueless because mamma cannot give it the moon to play with in its own hand, it turns its eyes away from all its other treasures, and cries for coal."[29]

Logan understood the problem only too well as no simple matter of childish pique. His painful judgment threatened not only the GSC's funding renewal, but also Canadians' faith in modern science more generally. In the first of a series of countermeasures, he redirected public attention to the north shore of Lake Superior, where the Canadian Shield might just yield more of the copper already reported on the Michigan side of the border—and which he knew existed in the strata below the limestone beds that underlay the settled parts of Canada.[30] His previous experience in South Wales reinforced Logan's understanding that, besides its traditional applications in construction, coinage, and the military, copper stood next to iron and zinc as indispensable to the Victorian industrial age. The only metal found in both its native form as well as in various ores, copper conducts heat and electricity exceptionally well, supporting a nascent electrical industry at the time especially through its use in the wiring for electric telegraphs. A run on limited copies of GSC reports among political and commercial interests in 1846 launched a corresponding rush for Lake Superior mining claims, as Logan himself fed into heightened public anxieties over the Oregon Boundary Dispute between Britain and the United States: "When the British Government gave up the Michigan territory at the end of the last American war, with as little concern as if it had been so much bare granite," he warned,

> "I dare say they were not aware that 12,000 square miles of a coal-field existed in the heart of it … ready to supply American steamers with fuel on the lakes, while ours on the same waters, in case of war, must depend on

wood, or coal expensively transported from Nova Scotia or Cape Breton Island, or across the Atlantic from the United Kingdom."[31]

Even while this focus on the lucrative combination of copper and coal formed the matrix of Logan's initial approach to the GSC, he knew better than to share the widespread eagerness for Canadian investment in Lake Superior's mineral resources. New technologies that had permitted the rise of industrial coal smelting during his time in South Wales had elevated Swansea (with its nearby coal deposits) over Cornwall (where the copper originated) to become the copper-producing capital of the world, because copper was less expensive to transport than coal was. By the same token, Logan reluctantly recognized, without either the discovery of coal farther west—as was rumored about Saskatchewan—or the development of new electrical technologies, the laws of economics "naturally destined" Michigan or Ohio to process Lake Superior copper from both sides of the international border.[32]

Logan's time in South Wales also taught him the importance of public display: the larger the mineralogical specimens, he believed, the more valuable their deposits would appear to be, "especially in the minds of the unlearned." In preparing a proposed Canadian Museum of Economic Geology during the 1840s, he accordingly requested of his assistant "a thundering piece of gypsum … as white as possible," along with "a huge slab of lithographic stone."[33] In this same spirit, Logan delved wholeheartedly into several opportunities during the 1850s to curate Canada's contributions to international exhibitions. The Great Exhibition of the Works of Industry of All Nations in London's remarkable iron and glass Crystal Palace—itself a visual paean to the age of iron—in 1851 saw Logan expertly courting potential British investors.[34] Linking Canadian mineral specimens conceptually to their means of extraction and processing, he played down Canada's problems with coal and copper, highlighting instead the availability, in particular, of workable iron ore.

In a trifecta of first prizes and high praise earned first in London, then at New York's Industrial Exhibition (1853) and the Paris Universal Exposition (1855), Logan emerged from his efforts a member of the Royal Society of London, a *Chevalier* in France's Legion of Honour, and the first ever British knight among Canadian *enfants du sol*. For the quality of his extensive scientific analyses, including his "Laurentian" series of metamorphic formations comprising the greater part of the Precambrian Canadian Shield—which he recognized as "the oldest known rocks, not only of North America, but of the Globe"—he furthermore took home the GSL's coveted Wollaston Medal,[35] successfully fighting off the best efforts of the British Geological Survey's archetypal "scientist of empire," Sir Roderick Murchison, to subsume the Laurentian series under his own system and to rename it "Lewisian."[36] Indeed, whereas Logan had originally envisaged the GSC and its British counterpart to be "mutually serviceable," with Logan planning to situate Canada as "the measure of a correct geological comparison" between Europe and America, he had instead felt frustrated early on by erroneous GSL classifications of Canadian specimens;[37] and "mortified" years later at finding GSC specimen boxes in the British Survey's basement, still unopened.[38] As a result, consulting with the New York Geological Survey's nearby collections at Albany meant adopting American nomenclature after all—and necessitated that much more diligence when it became necessary to challenge Murchison's stratigraphical declensions.[39]

Territorial Extractions

Repeated waves of international recognition reflected back on a Canadian population newly aware, along with the rest of the world, that the colony just might, after all, find the wherewithal to stand tall on Victorian culture's ordered scale of industrializing nations. In exchange for further extensions to both the GSC's funding and its staff in 1855, Logan assured a Canadian parliamentary inquiry of the GSC's continued commitment to "economic researches carried on in a scientific way." In so doing, he signaled his determination to walk the political tightrope between theory and practice, possibilities and limitations.

Logan's international success inspired even *The* [London] *Times* in 1855 to concede that while Canada did lack coal,

> "who shall say for what purpose? Perhaps to stimulate their industry in clearing away those interminable forests, interposed between Western civilization and the Rocky Mountains. Certainly we may hope to enable Canada to compete with Sweden in supplying our iron trade with an abundance of the finest quality of iron smelted with wood charcoal."[40]

The Times thus obligingly connected the dots in Logan's purposeful strategy. Following Charles Lyell's uniformitarian tendency to look to geological more than political units as an analytical guide—especially while technology continued to favor coal in industrial smelting—Logan's map for the Paris Exposition drew public attention outward, not just to the limits of settlement in Canada, but well beyond the colony's political horizons. Just as J.W. Dawson's masterwork *Acadian Geology* (1855) covered not only Nova Scotia, but also New Brunswick and Prince Edward Island, so Logan's map of Canada transcended contiguous formations in New Brunswick and New York to embrace—with Dawson's full agreement—all of Acadia, as well as Logan's *pièce de resistance*, his Laurentian series.[41] In thus laying out the distribution of Canada's mineral resources, he emphasized, he was effectively mapping its manufacturing potential, affording nothing less than a scientific glimpse of the colony's future settlement patterns. As Dawson elaborated in 1858:

> "Physically considered, British North America is a noble territory, grand in its natural features, rich in its varied resources. Politically, it is a loosely united aggregate of petty states, separated by barriers of race, creed, local interest, distance, and insufficient means of communication. As naturalists, we hold to its natural features as fixing its future destiny, and indicating its present interests, and regard its local subdivisions as arbitrary and artificial."[42]

Logan had disclosed his own version of this expansive vision to his friend and mentor, the founding director of the Geological Survey of Great Britain, Sir Henry De la Beche, as early as 1845: "Just look at Arrowsmith's little map of British North America," he had confided:

> "You will see that Canada comprises but a small part of it. Then examine the great rivers and lakes which water the interior between that American Baltic, Hudson's Bay, and the Pacific Ocean … It will become a great country hereafter. But who knows anything of its geology? Well, I have a sort of presentiment that I shall yet, if I live long enough, be employed by the British Government … to examine as much of it as I can, and that I am here in Canada only learning my lesson, as it were, in preparation."[43]

On this expansionist front, Logan once again tacked his sails to prevailing historical winds. Among the most immediate, the Annual Report of Canada's Commissioner of Crown Lands, Joseph Cauchon, announced in 1856 that settlement in the province had finally bumped up against the Canadian Shield, so denuded of soil as to be impervious to the plough.[44] Whereas this harsh reality had not prevented the government's removal of Indigenous peoples to "23,000 rocks of granite, dignified with the name of Manitoulin Islands" during the 1830s,[45] it now appeared directly to threaten the future of the colony. The rush for mining claims in those same Laurentian formations on the upper Great Lakes had also seen the Canadian government scramble to "negotiate" legal rights to these traditional Indigenous hunting grounds, resulting in the famous Robinson Treaties, in a process that formed a template for the numbered treaties that were to follow farther west.[46] Moreover, the impending expiry in 1869 of the Royal Charter that continued to hold the crumbling HBC fur-trading empire out of reach, ignited an increasingly vocal territorial expansionist movement centered mainly in Toronto and along the Ottawa Valley, expecting Canada to inherit the Great Northwest as its British birthright. Growing fears of an increasingly aggressive American Manifest Destiny along the 49th parallel served only to heighten the overall sense of urgency.

Practitioners of Victorian science, including geology, contributed actively to these discussions. In 1855, the Métis and former HBC employee Alexander Kennedy Isbister, who had been born at the Company's Cumberland House and educated at the Universities of Aberdeen and Edinburgh, published an important paper "On the Geology of the Hudson's Bay Territories, and of Portions of the Arctic and North-Western Regions of America" in the GSL's *Quarterly Journal*. A sharp critic of the HBC's secretive hold on its vast territories, Isbister touted the Mackenzie River valley as a "mass of minerals," and the entire region's mining potential as far exceeding the fur trade in value. His accompanying map assembled evidence to highlight in particular "a vast coalfield, skirting the base of the Rocky Mountains for a great extent, and continued probably far into the Arctic Sea," inflating quite considerably John Franklin's report a generation earlier.[47]

The purported coalfield then re-appeared as a key feature of other important expansionist documents. By 1857, both the British and the Canadian governments were actively seeking more evidence of the future settlement potential of the HBC territories. Dual parliamentary inquiries, supplemented by dual exploring expeditions,[48] found themselves facing a deeply interested third party as the Smithsonian Institution published Lorin Blodget's massive *Climatology of the United States, and of the Temperate Latitudes of the North American Continent* that same year.[49] An accumulation of climatological records had inspired the construction of Humboldtian isothermal maps of North America, their north-westward-leaning lines of equal temperature persuading competing Canadian and American expansionists of the HBC territories' unexpectedly rich agricultural and settlement potential as ripe for the picking. Canada's Chief Justice, W.H. Draper, testified before the British committee of inquiry by brandishing his Department of Crown Lands' remarkable "Map of the North-West Part of Canada, Indian Territories, and Hudson's Bay," complete with both the relevant isotherms as well as Isbister's coal projections. A simplified version of this same map appeared almost simultaneously in the Toronto *Globe* newspaper, owned and edited by the outspokenly expansionist Reform Party leader George Brown.[50]

Keeping the GSC discreetly out of the public eye in these highly political matters, William Logan scored further international success in 1857 by joining J.W. Dawson in hosting the annual meeting of the American Association for the Advancement of Science in Montreal. He selected this important occasion to refine his hard-won Laurentian formations by introducing a separate Huronian series to his scientific analysis.[51] As westward territorial matters worked their way through the political system at Britain's highest levels, the GSC sent "consulting" geologists eastward, to crack the codes of both New Brunswick's highly contorted strata and Newfoundland's "harsh and forbidding" repository of "the chippings of the world." With the General Mining Association's 30-year monopoly finally broken in 1858, Dawson expressed his preference for a similar arrangement with the GSC for Nova Scotia, with "the whole brought into one great work at the close."[52] Logan thus positioned his GSC to hit the ground running in each of the new Dominion of Canada's four new provinces—and beyond, since Logan's Newfoundland envoy, his original assistant Alexander Murray, remained there at his post even after the colony declined to join Confederation—when the institution finally attained permanent status under new constitutional arrangements in 1867.

By that time, Logan's sterling reputation had culminated in his original mandate fulfilled. His magnum opus, *Geology of Canada* (1863), a cumulative report supported by a state-of-the-art geological map, reconfirmed the painful irony that the province, surrounded by coal, was nevertheless bereft of it. On the surface of things, it seemed hardly to matter. When the craggy retrenchment-minded Liberal premier J.S. Macdonald dared to suggest—despite the work's dazzling reviews—that the GSC had in fact failed to discover any of Canada's working mines, his own ministers lampooned his "gross and ludicrous ignorance": Macdonald objected to Logan's studies of Canadian fossils, his critics chuckled, "because that was carrying personalities too far."[53] In 1867, the British North America Act assigned natural resources to provincial jurisdiction, effectively releasing the GSC, as a federal institution, from Macdonald's pointed charge.

Coal's Extraction

Logan retired in 1869, before the fledgling Dominion of Canada extended its reach in rapid succession to include Rupert's Land (1870), British Columbia (1871), Prince Edward Island (1873), and Britain's claims to the Arctic Archipelago (1880); leaving to his successor, A.R.C. Selwyn, a rapidly distending GSC on the verge of a worldwide economic slowdown. British fears of impending "coal exhaustion" not only redoubled geological interest in assessing Newfoundland's potential and led Nova Scotians to anticipate their destiny as Britain's coal-and-iron heir—"a capitalist among nations"—as "simply a question of time, and … inevitable."[54] While they soon learned the hard way that it took more than the presence of coal and iron ore to accede to industrial leadership, the race was also on to identify new sources of petroleum and electricity, alternative fuels that would ultimately launch a Second Industrial Revolution in which the resource cards would be reshuffled, and stacked entirely differently.

For the time being, the GSC struggled to balance the immediate need for an expansive, broad-stroke reconnaissance—including basic topographical maps—of Canada's vastly enlarged territories, with continuing demand for more finely detailed observations and analyses that would identify new mineral deposits. It did so under the leadership of a British-born director whose Geological Survey of Victoria, Australia had recently lost its funding. Under these enormously challenging conditions, Selwyn shrewdly appointed to his senior staff a remarkable new generation of geologists including J.W. Dawson's son George Mercer Dawson, Robert Bell, J.B. Tyrell, and A.P. Low, whose responsibilities took them west to British Columbia and north as far as the high Arctic, where the effects of glacial ice sheets came once more to the fore.[55] The global sweep of Dawson's geological insights, in particular, correlated evidence of metamorphism in Canada to the central axis of the Rocky Mountains and, farther afield, to volcanic activity in Chile and elsewhere, vindicating deep interconnections that Alexander von Humboldt and John Richardson, before him, could only suggest. While Dawson remained one of the last uniformitarian holdouts who rejected ice-age theory, his keen observations of the powerful effects of continental ice-sheet dynamics in sculpting the landscape also held out practical mining possibilities. During the 1890s, placer gold occurrences, this time in the Klondike region of the Yukon, incited wild public expectations all over again—once again inflicting on Indigenous peoples the damaging territorial and cultural fallout that resulted.[56]

Moreover, Dawson infused his geological surveys with systematic ethnological inventories intended to advise governments in their formulation of policy toward Indigenous peoples. He cast their declining numbers in the harsh light of apparent inevitability in the face of the systematic onslaught whose vanguard his work served so effectively. The moral dilemma inherent in his position did not escape Dawson, even as he promoted a social evolutionary paradigm that preferred assimilation to segregation. His emphasis on collecting over other forms of anthropological fieldwork, widely admired and supported by both the British Association for the Advancement of Science and the Royal Society of Canada, advanced the cause of national repositories and encouraged international rivals to compete for cultural artifacts essentially looted from their owners.

In all these ways, Victorian geology lent both physical backbone, daring imaginative vision, and scientific justification to a colonial world undergoing profound reconceptualization. The extractive impulse, exposed and transformed by the heady brew of empire-building, infused Canadians' understanding of their historical situation with a nagging source of tension: what was the country actually capable of becoming in an industrializing world? In so doing, it focused—sometimes frantically—as much on territory as it did on the mineral wealth that British North America did (or did not) contain. Harold Innis, it seems, lent us an early twentieth-century political economist's translation of a uniformitarian worldview inherited from Victorian geology—which, it appears, continues in powerful and even surreptitious ways to shape our assumptions and define our expectations of Canada.[57] It should come as no surprise, then, that as it pushed its self-serving metaphors and myths[58] ever northward in the twentieth century, the extractive impulse and its imperial power structures replicated themselves recursively, replaying their cycles of dissemination and domination over and over again. Recent critical scholarship shows how even the best of intentions can have a dark side that needs to be acknowledged; there is all the more reason to do so as the extractive impulse in turn drives Canadians abroad.[59] We might be well advised—as F.R. Scott insisted—to break the cycle by considering how we would wish history to remember *our* cumulative audacities in the world.

Notes

1 This poem refers to an article written years before it, which stated "Audacity is missing in Canada," *The Times*, November 30, 1959. See F.R. Scott, "Audacity," in *Signature* (Vancouver, BC: Klanak Press, 1964), 46. See also "Audacity," in *Leaving the Shade of the Middle Ground: The Poetry of F.R. Scott*, rev. ed. (Waterloo, ON: Wilfrid Laurier University Press, 2013); and Eric Ormsby, *Facsimiles of Time: Essays on Poetry and Translation* (Erin, ON: The Porcupine's Quill, 2001), 22.

2 Timothy Brook, *Vermeer's Hat: The Seventeenth Century and the Dawn of the Global World* (New York, NY: Bloomsbury Press, 2008), 46.

3 Harold A. Innis, *The Fur Trade in Canada: An Introduction to Canadian Economic History*, rev. ed. (Toronto, ON: University of Toronto Press, [1930] 1956), 393.

4 Scott, "Audacity."

5 See Crosbie Smith, *The Science of Energy: A Cultural History of Energy Physics in Victorian Britain* (Chicago, IL: University of Chicago Press, 1998).

6 The Victorian version re-envisioned the work of the English philosopher, statesman, and scientist Sir Francis Bacon (1561–1626).

7 For a critical assessment, see Pierre Elliott Trudeau, ed., *The Asbestos Strike* [Quebec 1949], trans. James Boake (Toronto, ON: James Lewis & Samuel, 1974).

8 See Thomas R. Berger, *Northern Frontier, Northern Homeland: The Report of the Mackenzie Valley Pipeline Inquiry*, rev. ed. (Vancouver, BC: Douglas & McIntyre, 1988).

9 Polar warming from 1816 seemed all the more remarkable as it coincided with the so-called "year without a summer" farther south; both have since been explained by the global effects of several major volcanic eruptions across the Pacific Ocean during those years. See Brian Fagan, "The Year without a Summer," chap. 10 in *The Little Ice Age: How Climate Made History, 1300–1850* (New York, NY: Basic Books, 2000), 167–80; and Gillen D'Arcy Wood, "The Polar Garden," chap. 6 in *Tambora: The Eruption That Changed the World* (Princeton, NJ: Princeton University Press, 2014), 121–49.

10 Martin J. S. Rudwick, *Bursting the Limits of Time: The Reconstruction of Geohistory in the Age of Revolution* (Chicago, IL: University of Chicago Press, 2005).

11 See Rachel Laudan, *From Mineralogy to Geology: The Foundations of a Science, 1650-1830* (Chicago, IL: University of Chicago Press, 1987); Hugh S. Torrens, "Geology in Peace Time: An English Visit to Study German Mineralogy and Geology (and visit Goethe, Werner and Raumer) in 1816," *Algorismus* 23 (1998): 147–75; and Alexander M. Ospovat, "Romanticism and German Geology: Five Students of Abraham Gottlob Werner," *Eighteenth-Century Life* 7, no. 2 (1982): 105–17.

12 See John J. Bigsby, "Notes on the Geography and Geology of Lake Huron," *Transactions of the Geological Society of London* 2, no. 1 (1824): 175–209; John J. Bigsby, "Notes on the Geography and Geology of Lake Superior," *Quarterly Journal of Science, Literature, and the Arts* 18, no. 25 (October 1824): 1–34; and G. Morey, "Early Geologic Studies in the Lake Superior Region: The Contributions of H.R. Schoolcraft, J.J. Bigsby, and H.W. Bayfield," *Earth Sciences History* 8, no. 1 (1989): 36–42.

13 David Chisholme, "Editorial," *The Canadian Review and Magazine* 2, no. 4 (February 1826): 319–21, 332–33. See also William E. Cormack, "Account of a Journey Across the Island of Newfoundland, by W.E. Cormack, Esq. in a letter addressed to the Right Hon. Earl Bathrust, Secretary of State for the Colonies," *The Edinburgh New Philosophical Journal* 10, no. 19 (1824): 156–62; and John J. Bigsby, "On the Utility and Design of the Science of Geology," *Canadian Review and Literary and Historical Journal* 2, no. 2 (December 1824): 377–95.

14 See Alexander Von Humboldt, *Cosmos: A Sketch of a Physical Description of the Universe*, trans. E.C. Otté, 5 vols. (New York, NY: Harper & Brothers, 1845–61).

15 Ibid.

16 John Richardson, "British North America," map in *Arctic Searching Expedition: A Journal of a Boat-voyage through Rupert's Land and the Arctic Sea, in Search of the Discovery Ships under Command of Sir John Franklin. With an Appendix on the Physical Geography of North America*, 2 vols. (London, UK: Longman, Brown, Green, and Longmans, 1851), 1:x; reproduced in Suzanne Zeller, "The Colonial World as Geological Metaphor: Strata(gems) of Empire in Victorian Canada," *Osiris* 15 (2000): 90.

17 See "A Map of the Canadas and Adjacent Part of the United States, to Illustrate Dr. Bigsby's Paper on the Canadian Erratics," map in John J. Bigsby, "On the Erratics of Canada," *Quarterly Journal of the Geological Society of London* 7 (1851): 215–38, print; reproduced in Zeller, "The Colonial World as Geological Metaphor," 92.

18 Charles Lyell, subtitle in *Principles of Geology: Being an Attempt to Explain the Former Changes of the Earth's Surface by Reference to Causes Now in Operation*, 3 vols. (London, UK: John Murray, 1830–33).

19 See Stephen Jay Gould, *Time's Arrow, Time's Cycle: Myth and Metaphor in the Discovery of Geological Time* (Cambridge, MA: Harvard University Press, 1987).

20 Robert Silliman, "Agassiz Vs. Lyell: Authority in the Assessment of the Diluvium-Drift Problem by North American Geologists, with Particular Reference to Edward Hitchcock," *Earth Sciences History* 13, no. 2 (1994): 180–86.

21 Charles Lyell to J. W. Dawson, 2 Feb. and 2 May 1843, 30 May 1854; Dawson to Lyell, 5 Sept. 1845, 17 Feb. 1849; R. Brown to Dawson, 18 Nov., 9 Dec. 1845, 10 Mar, 1846; S. Cunard to Dawson, 28 Nov., 8 Dec. 1845, 19 Feb 1846; Dawson to Hon. G. R. Young, n.d. 1846, 20 Apr. 1848, 17 Jan. 1849, John William Dawson Papers, McGill Univ. Archives. W. E. Logan to James Logan, 16 Aug. 1841, William Edmond Logan Papers, McGill University Archives.

22 William E. Logan, "On the Coal-Fields of Pennsylvania and Nova Scotia," *Proceedings of The Geological Society of London* 3, Part 2, no. 88 (1842): 707–12.

23 Contrast Robert A. Stafford, *Scientist of Empire: Sir Roderick Murchison, Scientific Exploration and Victorian Imperialism* (Cambridge, MA: Cambridge University Press, 1989), 65–68; for details of campaigns for a Geological Survey of Canada and Logan's appointment, see Suzanne Zeller, *Inventing Canada: Early Victorian Science and the Idea of a Transcontinental Nation*, chaps. 1–2, 2nd ed., Carleton Library Series #214 (Montreal, QC: McGill-Queen's University Press, 2009); on Logan's mining background, see Hugh Torrens, "How, When and Where did William E. Logan Learn his Geology in Britain 1831-1841?" forthcoming in *Geoscience Canada*. William E. Logan to James Logan, 14 Aug. 1841, Logan Papers.

24 See John Lambton, Earl of Durham, *Report on the Affairs of British North America* (London, UK: J.W. Southgate, 1839).

25 See Abraham Gesner, *Remarks on the Geology and Mineralogy of Nova Scotia* (Halifax, NS: Gossip and Coade, 1836), print; Abraham Gesner, *First [to Fourth] Report on the Geology of New Brunswick* (Halifax, NS: n.p., 1839–42); James F. W.

Johnston, *Report on the Agricultural Capabilities of the Province of New Brunswick* (Fredericton, NB: J. Simpson, 1850); J. Beete Jukes, *Report on the Geology of Newfoundland* (St. John's, NL: Ryan, 1839); J. Beete Jukes, *General Report of the Geological Survey of Newfoundland, Executed under the Direction of the Government and Legislature of the Colony during the Years 1839 and 1840* (London, UK: J. Murray, 1843); James W. Ross to John William Dawson, 1 May 1846, Dawson Papers.

26 See Frederick Henry Baddeley, "On the Geognosy of a Part of the Saguenay Country," *Transactions of the Literary and Historical Society of Quebec* 1 (1829): 163–64; and Frederick Henry Baddeley, "Additional Notes on the Geognosy of Saint Paul's Bay," *Transactions of the Literary and Historical Society of Quebec* 2 (1831): 91–93.

27 See Logan to De la Beche, 5 Oct. 1840, 31 May 1843, 12 May 1845, De la Beche Papers, National Museum of Wales; Geological Survey of Canada, *Preliminary Report*, 6 Dec. 1842; and "Report of Progress for the Year 1843," in *Journals of the Legislative Assembly of Province of Canada* (House of Assembly, Province of Canada, 1844–45).

28 Province of Canada, *Statutes*, 1843, 7 Vict., c. 45; *Montreal Gazette*, January 23, 1845.

29 John William Dawson, "Coal in Canada: The Bowmanville Discovery," *Canadian Naturalist and Geologist* 3, no. 3 (June 1858), 212–13. See also "To Our Reviewers," *Canadian Naturalist and Geologist* 3, no. 5 (October 1858), 400.

30 Proceedings of Geological Society of Canada, *Report of Proceedings* (1843), print.

31 William E. Logan as quoted in Bernard J. Harrington, *Life of Sir William E. Logan: First Director of the Geological Survey of Canada. Chiefly Compiled from His Letters, Journals and Reports* (Montreal, QC: Dawson Brothers, 1883), 234–35.

32 *Montreal Gazette*, March 11, 1846; Proceedings of Geological Society of Canada, *Report of Proceedings* (1846–47).

33 Harrington, *Life of Sir William E. Logan*, 180–81.

34 See Geological Society of Canada, "Report of Progress for the Year 1850–51," in *Journals of the Legislative Assembly of Province of Canada* (House of Assembly, Province of Canada, 1852); and Audrey Short, "Canada Exhibited: 1851–1867," *Canadian Historical Review* 48, no. 4 (December 1967): 353–64.

35 House of Assembly of the Province of Canada, *Journals of the House of Assembly of the Province of Canada* (House of Assembly, Province of Canada, 1847); *Montreal Gazette*, May 15, 1845; May 26, 1847; Nov 20, 1846; Civil Secretary to James Logan et al., 4 Apr. 1846, McGill University Archives.

36 Edward John Chapman to Logan, 13 Feb., 4 Mar., 10 Mar., 9 Apr. 1855; 22 Mar. 1856, Logan Papers.

37 Logan to Sir Henry De la Beche, 19 Oct., 3 Dec. 1841; 24 Apr. 1843; 12 May 1845, Sir Henry De la Beche Papers.

38 Logan to Dawson, 4 June 1843. 10 Jan. 1853, Dawson Papers; Logan to De la Beche, 20 Apr., 11 Nov. 1844; 12 May, 27 Dec. 1845, De la Beche Papers.

39 Murchison named the Silurian system, ultimately dedicating to Logan a revised edition of his major work *Siluria* (4th ed., 1867). See Sir Roderick Impey Murchison, *Siluria: A History of the Oldest Rocks in the British Isles and Other Countries; With sketches of the Origin and Distribution of Native Gold, the General Succession of Geological Formations, and Changes in the Earth's Surface*, 4th ed. (London, UK: John Murray, 1867).

40 *The* [London] *Times*, September 7, 1855, in Logan Scrapbook, LP, 1207/16; Logan to Francis Hinks, 11 May 1855, Logan Papers.

41 Letterbooks, Logan to Carter, 21 May 1866, 115. Murray to Logan, 23 Apr. 1864; Edward Morris to Logan, 16 Oct. 1866; Logan to Morris, 5. Nov. 1866, Logan Papers.

42 J.W. Dawson, "Reviews and Notices of Books: Pamphlets on British America," *Canadian Naturalist and Geologist* 3, no. 5 (October 1858): 392–93. Dawson was reviewing, among other similar items, a like-minded Alexander Morris, *Nova Britannia: British North America, Its Extent and Future* (Montreal, QC, 1858). Morris went on to become Lieutenant Governor of Manitoba, where he negotiated many of the so-called Numbered Treaties with Indigenous peoples.

43 Logan as quoted in Harrington, *Life of Sir William E. Logan*, 234–35.

44 See Joseph Cauchon, *Étude sur l'union projetée des provinces britanniques de l'Amérique du Nord* (Quebec, QC: Typographie d'Agustin Coté et Co., 1858).

45 Aborigines Protection Society, *Report of the Indians of Upper Canada* (London, UK: William Ball, Arnold, and Company, 1838), 26, as quoted in Michael D. Blackstock, "Trust Us: A Case Study in Colonial Social Relations Based on Documents Prepared by the Aborigines Protection Society, 1836–1912," chap. 2 in *With Good Intentions: Euro-Canadian and Aboriginal Relations in Colonial Canada*, ed. Celia Haig-Brown and David A. Nock (Vancouver, BC: UBC Press, 2006), 59, 70n65, adding that "The United States' Indian Policy was similar to that advocated by Sir Frances Bond Head: 'removal of entire tribes to more isolated locations west of the Mississippi River where they could pursue their own cultures and develop their political institutions according to their aspirations and capacities.' Royal Commission on Aboriginal Peoples, *Final Report*, vol. 1, chap. 9, sec. 6."

46 See Julia Jarvis, "Robinson, William Benjamin," in *Dictionary of Canadian Biography* [DCB], vol. 10, *1871–80* (Toronto, ON: University of Toronto Press, 1972), 622–24; and Jean Friesen, "Morris, Alexander," *DCB*, vol. 11, *1881–90* (Toronto, ON: University of Toronto Press, 1982), 608–15. See also Haig-Brown and Nock, eds., *With Good Intentions*.

47 A.K. Isbister, " Mr. Alexander Isbister, called in; and further examined," [June 23, 1857] in *Report from the Select Committee on the Hudson's Bay Company; Together with the Proceedings of the Committee, Minutes of Evidence, Appendix and Index* (London, UK: The House of Commons, 1857), 355; A.K. Isbister, "On the Geology of the Hudson's Bay Territories, and of Portions of the Arctic and North-Western Regions of America; with a Coloured Geological Map," *Quarterly Journal of the Geological Society of London* 11 (1855): 513. See also W.O. Kupsch, "The History of Canadian Geology: Métis and Proud," *Geoscience Canada* 4, no. 3 (1977): 147–48.

48 See Henry Youle Hind, *Narrative of the Canadian Red River Exploring Expedition of 1858 and of the Assinniboine and Saskatchewan Exploring Expedition of 1858*, 2 vols. (London, UK: Longman, Green, Longman, and Roberts, 1860).

49 Lorin Blodget, *Climatology of the United States, and of the Temperate Latitudes of the North American Continent* (Philadelphia, PA: J.B. Lippincott and Co., 1857).

50 Crown Lands Department, "Map of the North-west Part of Canada, Indian Territories, and Hudson's Bay," (Toronto, ON: Public Archives of Ontario, March 1857), map B-24. See Zeller, *Inventing Canada*, xxxii. See also J.E. Hodgetts, *Pioneer Public Service: An Administrative History* of the United Canadas, 1841–1867 (Toronto, ON: University of Toronto Press, 1955), 118; *Report from the Select Committee on the Hudson's Bay Company*; and *The* [Toronto] *Globe*, March 23, 1857.

51 See A.N. Rennie, "Natural History Society of Montreal: Report for 1857," *Canadian Naturalist and Geologist* 2, no. 3 (July 1857): 233–40; William E. Logan, "On the Division of the Azoic Rocks of Canada into Huronian and Laurentian," *Canadian Journal of Industry, Science, and Art* 2, no. 12 (November 1857): 439-42; and William E. Logan, "On the Probable Subdivision of the Laurentian Rocks of Canada," *Canadian Journal of Industry, Science, and Art*, Vol. 3, no 13 (January 1858): 1–5.

52 Dawson to Logan, 18 Mar. 1858; 10 Oct. 1861, Logan Papers.

53 See Dawson Papers, "1863," 66; *Montreal Gazette*, 25 Sept. 1863; *Quebec Gazette*, 28 Sept. 1863, 10 June 1864; and Harrington, *Life of Sir William E. Logan*, 352.

54 See William Stanley Jevons, *The Coal Question; An Inquiry concerning the Progress of the Nation, and the Probable Exhaustion of our Coal-mines* 2nd rev. ed. (London, UK: Macmillan and Co., 1866); and Robert G. Haliburton, *The Coal Trade of the New Dominion* (Halifax, NS: Printed by T. Chamberlain, 1868).

55 See Morris Zaslow, chap. 7 in *Reading the Rocks: The Story of the Geological Survey of Canada, 1842–1972* (Toronto, ON: Macmillan Company of Canada, 1975).

56 See Julie Cruikshank, "Images of Society in Klondike Gold Rush Narratives: Skookum Jim and the Discovery of Gold," *Ethnohistory* 39, no. 1 (1992): 20–41.

57 See, for example, Paul Kellogg, *Escape from the Staple Trap: Canadian Political Economy after Left Nationalism* (Toronto, ON: University of Toronto Press, 2015).

58 See, for example, Daniel Francis, *National Dreams: Myth, Memory, and Canadian History* (Vancouver, BC: Arsenal Pulp Press, 1997).

59 See, for example, Paula Butler, *Colonial Extractions: Race and Canadian Mining in Contemporary Africa* (Toronto, ON: University of Toronto Press, 2015).

The Ballad of Frog Plain

Would you like to hear me sing
Of a true and recent thing?
It was June nineteen, the band of Bois-Brûlés
Arrived that day,
Oh the brave warriors they

We took three foreigners prisoners when
We came to the place called Frog, Frog Plain.
They were men who'd come from Orkney,
　　Who'd come, you see,
To rob our country.

Well we were just about to unhorse
When we heard two of us give, give voice.
Two of our men cried, "Hey! Look back, look back!
　　The Anglo-Sack
　　Coming for to attack."

Right away smartly we veered about
Galloping at them with a shout!
You know we did trap all, all those Grenadiers!
　　They could not move
　　Those horseless cavaliers.

Now we like honourable men did act,
Sent an ambassador—yes, in fact!
"Monsieur Governor! Would you like to stay?
　　A moment spare—
　　There's something we'd like to say.

Governor, Governor, full of ire.
"Soldiers!" he cries, "Fire! Fire!"
So they fire the first and their muskets roar!
　　They almost kill
　　Our ambassador!

Le gouverneur qui se croit empereur,
Il veut agir avec rigueur;
Le gouverneur qui se croit empereur
À son malheur, agit trop de rigueur.

Ayant vu passer tous ces Bois-Brulés
Ill a parti pour les épouvanter;
Étant parti pour les épouvanter;
Il s'est trompé, il s'est bien fait tuer.

Il s'est bien fait tuer.
Quantité de grenadiers.
J'avons tué Presque tout son armée,
Sur la band' quatre ou cinq s'sont sauvés.

Si vous aviez vu tous ces Anglais
Et tous ces Bois-Brûlés après
De butte en butte les Anglais culbutaient.
Les Bois-Brûlés jetaient des cris de joie.

Qui en a composé la chanson?
C'est Pierre Falcon, poète du canton.
Elle a été faite et composée.
Sur la victoire que nous avons gagnée.
Elle a été faite et composée
Chantons la gloire de tous les Bois-Brûlés.

La Bataille des Sept Chênes

(The Battle of Seven Oaks)

Pierre Falcon

ca. 1816

Chanson de la Grenouillère

Voulez-vous écouter chanter
Une chanson de vérité?
Le dix-neuf de juin, la band' des Bois-Brûlés
Sont arrivés comm' des braves guerriers.

En arrivant à la Grenouillère
Nous avons pris trois prisonniers;
Trois prisonniers des Arkanays
Qui sont ici pour piller not' pays.

Étant sur le point de débarquer
Deux de nos gens se sont mis écriés
Deux de nos gens se sont mis écriés
Voilà l'Anglais qui vient nous attaquer.

Tout aussitôt bous avons déviré
Avons été les rencontrer
J'avons cerné le band' des Grenadiers,
Ils sont immobiles, ils sont tout démontés.

J'avons agi comme des gens d'honneur,
J'avons envoyé un ambassadeur,
Le gouverneur, voulez-vous arrêter
Un p'tit moment, nous voulons vous parler?"

Le Gouverneur qui était enragé
Il dit à ses soldats: Tirez!
Le premier coup c'est l' Anglais qu'a tiré,
L'ambassadeur ils ont manqué de tuer.

Governor thought himself a king.
He wished an iron rod to swing.
Like a lofty lord he tries to act.
Back luck, old chap!
A bit too hard you whacked!

When we went galloping, galloping by
Governor thought that he would try
For to chase and frighten us Bois-Brûlés.

Catastrophe!
Dead on the ground he lay.

Dead on the ground lots of grenadiers too.
Plenty of grenadiers, a whole slew.
We've almost stamped out his whole army.
Of so many.
Five or four left there be.

You should have seen those Englishmen—
Bois-Brûlés chasing them, chasing them.
From bluff to bluff they stumbled that day
While the Bois-Brûlés
Shouted "Hurray!

Tell, oh tell me who made up this song?
Why it's our own poet, Pierre Falcon
Yes, she has written this song of praise
For the victory
We won this day
Yes, she was written this song of praise—
Come sing the glory
Of the Bois-Brûlés.

You and What Army?

Indigenous Rights and the Power of Keeping Our Word

Naomi Klein

"I never thought I would ever see the day that we would come together. Relationships are changing, stereotypes are disappearing, there's more respect for one another. If anything, this Enbridge Northern Gateway has unified British Columbia."[1]

—Geraldine Thomas-Flurer, Coordinator of the Yinka Dene Alliance, a First Nations coalition opposing the Enbridge Northern Gateway pipeline, 2013

The guy from Standard & Poor's was leafing through the fat binder on the round table in the meeting room, brow furrowed, skimming and nodding.

It was 2004 and I found myself sitting in on a private meeting between two important First Nations leaders and a representative of one of the three most powerful credit rating agencies in the world. The meeting had been requested by Arthur Manuel, a former Neskonlith chief in the interior of British Columbia, now spokesperson for the Indigenous Network on Economies and Trade.

Arthur Manuel, who comes from a long line of respected Native leaders, is an internationally recognized thinker on the question of how to force belligerent governments to respect Indigenous land rights, though you might not guess it from his plainspoken manner or his tendency to chuckle mid-sentence. His theory is that nothing will change until there is a credible threat that continuing to violate Native rights will carry serious financial costs, whether for governments or investors. So he has been looking for different ways to inflict those costs.

That's why he had initiated a correspondence with Standard & Poor's, which routinely blesses Canada with a AAA credit rating, a much coveted indicator to investors that the country is a safe and secure place in which to sink their money. In letters to the agency, Manuel had argued that Canada did not deserve such a high rating because it was failing to report a very important liability: a massive unpaid debt that takes the form of all the wealth that had been extracted from unceded Indigenous land, without consent—since 1846.[2] He further explained the various Supreme Court cases that had affirmed that Aboriginal and Treaty Rights were still very much alive.

After much back-and-forth, Manuel had managed to get a meeting with Joydeep Mukherji, director of the Sovereign Ratings Group, and the man responsible for issuing Canada's credit rating. The meeting took place at S&P's headquarters, a towering building just off Wall Street. Manuel had invited Guujaaw, the charismatic president of the Haida Nation, to help him make the case about those unpaid debts, and at the last minute had asked me to come along as a witness. Unaware that, post-9/11, official ID is required to get into all major Manhattan office buildings, the Haida leader had left his passport in his hotel room; dressed in a short-sleeved checked shirt and with a long braid down his back,

Lyle Stafford

Mary Jack at Idle No More Highway Blockade
Patricia Bay Highway, BC, 2013
Vancouver Island Douglas Treaties
(WSÁNEĆ Territory)

Guujaaw almost didn't make it past security. But after some negotiation with security (and intervention from Manuel's contact upstairs), we made it in.

At the meeting, Manuel presented the Okanagan writ of summons, and explained that similar writs had been filed by many other First Nations. These simple documents, asserting land title to large swaths of territory, put the Canadian government on notice that these bands had every intention of taking legal action to get the economic benefits of lands being used by resource companies without their consent. These writs, Manuel explained, represented trillions of dollars' worth of unacknowledged liability being carried by the Canadian state.

Guujaaw then solemnly presented Mukherji with the Haida Nation's registered statement of claim, a seven-page legal document that had been filed before the Supreme Court of British Columbia seeking damages and reparations from the provincial government for unlawfully exploiting and degrading lands and waters that are rightfully controlled by the Haida. Indeed, at that moment, the case was being argued before the Supreme Court of Canada, challenging both the logging giant Weyerhaeuser and the provincial government of British Columbia over a failure to consult before logging the forests on the Pacific island of Haida Gwaii. "Right now the Canadian and British Columbia governments are using our land and our resources—Aboriginal and Treaty Rights—as collateral for all the loans they get from Wall Street," Manuel said. "We are in fact subsidizing the wealth of Canada and British Columbia with our impoverishment."[3]

Mukherji and an S&P colleague listened and silently skimmed Manuel's documents. A polite question was asked about Canada's recent federal elections and whether the new government was expected to change the enforcement of Indigenous land rights. It was clear that none of this was new to them—not the claims, not the court rulings, not the constitutional language. They did not dispute any of the facts. But Mukherji explained as nicely as he possibly could that the agency had come to the conclusion that Canada's First Nations did not have the power to enforce their rights and therefore to collect on their enormous debts. Which meant, from S&P's perspective, that those debts shouldn't affect Canada's stellar credit rating. The company would, however, continue to monitor the situation to see if the dynamics changed.

And with that we were back on the street, surrounded by New Yorkers clutching iced lattes and barking into cell phones. Manuel snapped a few pictures of Guujaaw underneath the Standard & Poor's sign, flanked by security guards in body armor. The two men seemed undaunted by what had transpired; I, on the other hand, was reeling. Because what the men from S&P were really saying to these two representatives of my country's original inhabitants was: "We know you never sold your land. But how are you going to make the Canadian government keep its word? You and what army?"

At the time, there did not seem to be a good answer to that question. Indigenous rights in North America did not have powerful forces marshaled behind them and they had plenty of powerful forces standing in opposition. Not just government, industry, and police, but also corporate-owned media that cast them as living in the past and enjoying undeserved special rights, while those same media outlets usually failed to do basic public education about the nature of the treaties our governments (or rather their British predecessors) had signed. Even most intelligent, progressive thinkers paid little heed: sure they supported Indigenous rights in theory, but usually as part of the broader multicultural mosaic, not as something they needed to actively defend.

However, in perhaps the most politically significant development of the rise of Blockadia-style resistance,[4] this dynamic is changing rapidly—and an army of sorts is beginning to coalesce around the fight to turn Indigenous land rights into hard economic realities that neither government nor industry can ignore.

The Last Line of Defense

As we have seen, the exercise of Indigenous rights has played a central role in the rise of the current wave of fossil fuel resistance. The Nez Perce were the ones who were ultimately able to stop the big rigs on Highway 12 in Idaho and Montana; the Northern Cheyenne continue to be the biggest barrier to coal development in southeastern Montana; the Lummi present the greatest legal obstacle to the construction of the biggest proposed coal export terminal in the Pacific Northwest; the Elsipogtog First Nation managed to substantially interfere with seismic testing for fracking in New Brunswick; and so on. Going back further, it's worth remembering that the struggles of the Ogoni and Ijaw in Nigeria included a broad demand for self-determination and resource control over land that both groups claimed was illegitimately taken from them during the colonial formation of Nigeria. In short, Indigenous land and treaty rights have proved a major barrier for the extractive industries in many of the key Blockadia struggles.

And through these victories, a great many non-Natives are beginning to understand that these rights represent some of the most robust tools available to prevent ecological crisis. Even more critically, many non-Natives are also beginning to see that the ways of life that Indigenous groups are protecting have a great deal to teach about how to relate to the land in ways that are not purely extractive. This represents a true sea change over a very short period of time. My own country offers a glimpse into the speed of this shift.

The Canadian Constitution and the Canadian Charter of Rights and Freedoms acknowledge and offer protection to "aboriginal rights," including treaty rights, the right to self-government, and the right to practice traditional culture and customs. There was, however, a widespread perception among Canadians that treaties represented agreements to fully surrender large portions of lands in exchange for the provision of public services and desig-

Road Block
La Loche, SK, 2014
Northern Trappers Alliance , Fur Block N19
(Dene Territory)

Mykel Riley / Twitter

Anti-Fracking Protest
Elsipogtog, NB, 2013
(Mi'kmaq Territory)

nated rights on much smaller reserves. Many Canadians also assumed that in the lands not covered by any treaty (which is a great deal of the country, 80 percent of British Columbia alone), non-Natives could pretty much do what they wished with the natural resources. First Nations had rights on their reserves, but if they once had rights off them as well, they had surely lost them by attrition over the years. Finders keepers sort of thing, or so the thinking went.[5]

All of this was turned upside down in the late 1990s when the Supreme Court of Canada handed down a series of landmark decisions in cases designed to test the limits of Aboriginal title and treaty rights. First came *Delgamuukw v. British Columbia* in 1997, which ruled that in those large parts of B.C. that were not covered by any treaty, Aboriginal title over that land had never been extinguished and still needed to be settled. This was interpreted by many First Nations as an assertion that they still had full rights to that land, including the right to fish, hunt, and gather there. Chelsea Vowel, a Montreal-based Métis educator and Indigenous legal scholar, explains the shockwave caused by the decision. "One day, Canadians woke up to a legal reality in which millions of acres of land were recognized as never having been acquired by the Crown," which would have "immediate implications for other areas of the country where no treaties ceding land ownership were ever signed."[6]

Two years later, in 1999, the ruling known as the *Marshall* decision affirmed that when the Mi'kmaq, Maliseet, and Passamaquoddy First Nations, largely based in New Brunswick and Nova Scotia, signed "peace and friendship" treaties with the British Crown in 1760 and 1761, they did not—as so many Canadians then assumed—agree to give up rights to their ancestral lands. Rather they were agreeing to *share* them with settlers on the condition that the First Nations could continue to use those lands for traditional activities like fishing, trading, and ceremony. The case was sparked by a single fisherman, Donald Marshall Jr., catching eels out of season and without a license; the court ruled that it was within the rights of the Mi'kmaq and Maliseet to fish year-round enough to earn a "moderate livelihood" where their ancestors had fished, exempting them from many of the rules set by the Federal Government for the non-Native fishing fleet.[7]

Many other North American treaties contained similar resource-sharing provisions. Treaty 6, for instance, which covers large parts of the Alberta tar sands region, contains clear language stating that "Indians, shall have right to pursue their avocations of hunting and fishing throughout the tract surrendered"—in other words, they surrendered only their *exclusive* rights to the territory and agreed that the land would be used by both parties, with settlers and Indigenous peoples pursuing their interests in parallel.[8]

But any parallel, peaceful coexistence is plainly impossible if one party is irrevocably altering and poisoning that shared land. And indeed, though it is not written in the text of the treaty, First Nations elders living in this region contend that Indigenous negotiators gave permission for the land to be used by settlers only "to the depth of a plow"—considerably less than the cavernous holes being dug there today. In the agreements that created modern-day North America such land-sharing provisions form the basis of most major treaties.

In Canada, the period after the Supreme Court decisions was a tumultuous one. federal and provincial governments did little or nothing to protect the rights that the judges had affirmed, so it fell to Indigenous people to go out on the land and water and assert them—to fish, hunt, log, and build ceremonial structures, often without state permission. The backlash was swift. Across the country non-Native fishers and hunters complained that the "Indians" were above the law, that they were going to empty the oceans and rivers of fish, take all the good game, destroy the woods, and on and on. (Never mind the uninterrupted record of reckless resource mismanagement by all levels of the Canadian government.)

Tensions came to a head in the Mi'kmaq community of Burnt Church, New Brunswick. Enraged that the *Marshall* decision had empowered Mi'kmaq people to exercise their treaty rights and fish outside of government-approved seasons, mobs of non-Native fishermen launched a series of violent attacks on their Native neighbors. In what became known as the Burnt Church Crisis, thousands of Mi'kmaq lobster traps were destroyed, three fish-processing plants were ransacked, a ceremonial arbor was burned to the ground, and several Indigenous people were hospitalized after their truck was attacked. And it wasn't just vigilante violence. As the months-long crisis wore on, government boats staffed with officials in riot gear rammed into Native fishing boats, sinking two vessels and forcing their crews to jump to safety in the water. The Mi'kmaq fishers did their best to defend themselves, with the help of the Mi'kmaq Warrior Society, but they were vastly outnumbered and an atmosphere of fear prevailed for years. The racism was so severe that at one point a non-Native fisherman put on a long-haired wig and performed a cartoonish "war dance" on the deck of his boat in front of delighted television crews.

That was 2000. In 2013, a little more than an hour's drive down the coast from Burnt Church, the same Mi'kmaq Warrior Society was once again in the news, this time because it had joined with the Elsipogtog First Nation to fend off the Texas company at the center of the province's fracking showdown. But the mood and underlying dynamics could not have been more different. This time, over months of protest, the warriors helped to light a series of ceremonial sacred fires and explicitly invited the non-Native community to join them on the barricades "to ensure that the company cannot resume work to extract shale gas via fracking." A statement explained, "This comes as part of a larger campaign that reunites Indigenous, Acadian & Anglo people." (New Brunswick has a large French-speaking Acadian population, with its own historical tensions with the English-speaking majority.)[9]

Many heeded the call and it was frequently noted that protests led by the Elsipogtog First Nation were remarkably diverse, drawing participants from all of the province's ethnic groups, as well as from First Nations across the country. As one non-Native partici-

Tom Hanson / The Canadian Press

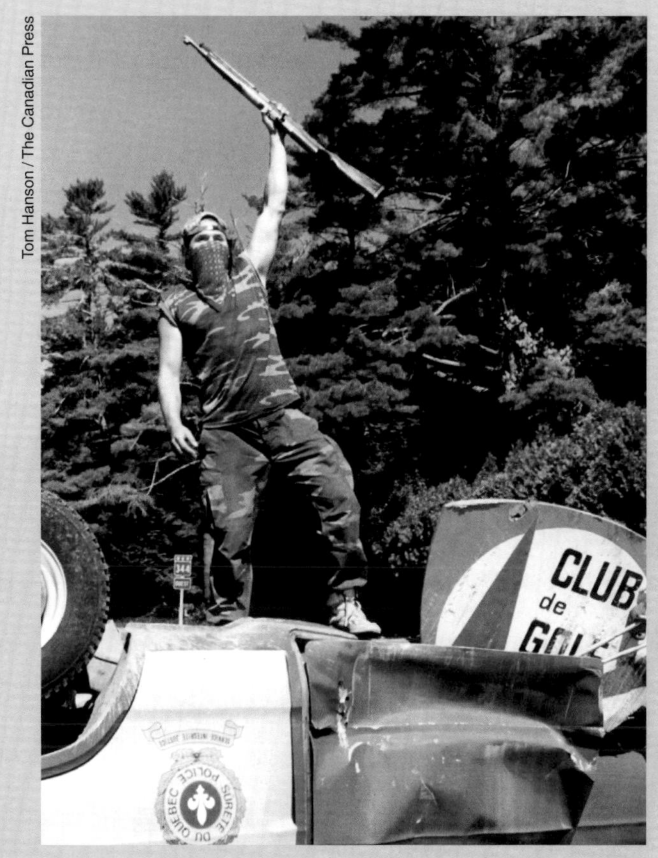

Road Barricade & Standoff
Kanesatake Lands, Oka, QC, 1990
(Mohawk Territory)

Dave Chidley

CN Rail Blockade
St. Clair Spur Line, Sarnia, ON, 2012
(Aamjiwnaang First Nation, Anishinaabeg Territory)

pant, Debbi Hauper, told a video crew, "It's just a real sense of togetherness. We are united in what is most important. And I think we're seeing more and more of government and industries' methods of trying to separate us. And let's face it, these methods have worked for decades. But I think we're waking up."[10]

There were attempts to revive the old hatreds, to be sure. A police officer was overheard saying "Crown land belongs to the government, not to fucking Natives." And after the conflict with police turned violent, New Brunswick premier David Alward observed, "Clearly, there are those who do not have the same values we share as New Brunswickers." But the community stuck together and there were solidarity protests in dozens of cities and towns across the country: "This is not just a First Nations campaign. It's actually quite a historic moment where all the major peoples of this province—English, French and Aboriginal—come together for a common cause," said David Coon, head of the Green Party in New Brunswick. "This is really a question of justice. They want to protect their common lands, water and air from destruction."[11]

By then many in the province had come to understand that the Mi'kmaq's rights to use their traditional lands and waters to hunt and fish—the same rights that had sparked race riots a dozen years earlier—represented the best hope for the majority of New Brunswickers who opposed fracking.[12] And new tools were clearly required. Premier Alward had been a fracking skeptic before he was elected in 2010, but once in office, he promptly changed his tune, saying the revenue was needed to pay for social programs and to create jobs—the sort of flip flop that breeds cynicism about representative democracy the world over.

Indigenous rights, in contrast, are not dependent on the whims of politicians. The position of the Elsipogtog First Nation was that no treaty gave the Canadian government the authority to radically alter their ancestral lands. The right to hunt and fish, affirmed by the *Marshall* decision, was violated by industrial activity that threatened the fundamental health of the lands and waters (since what good is having the right to fish, for instance, when the water is polluted?). Gary Simon of the Elsipogtog First Nation explains, "I believe our treaties are the last line of defense to save the clean water for future generations."[13]

It's the same position the Lummi have taken against the coal export terminal near Bellingham, Washington, arguing that the vast increase in tanker traffic in the Strait of Georgia, as well as the polluting impacts of coal dust, violates their treaty-protected right to fish those waters. (The Lower Elwha Klallam tribe in Washington State made similar points when its leaders fought to remove two dams on the Elwha River. They argued, successfully, that by interfering with salmon runs the dam violated their treaty rights to fish.) And when the U.S. State Department indicated, in February 2014, that it might soon be offering its blessing to the Keystone XL pipeline, members of the Lakota Nation immediately announced that they considered the pipeline construction illegal. As Paula Antoine, an employee of the Rosebud tribe's land office, explained, because the pipeline passes through Lakota treaty-protected traditional territory, and very close to reservation land, "They aren't recognizing our treaties, they are violating our treaty rights and our boundaries by going through there. Any ground disturbance around that proposed line will affect us."[14]

These rights are real and they are powerful, all the more so because many of the planet's largest and most dangerous unexploded carbon bombs lie beneath lands and waters to which Indigenous peoples have legitimate legal claims. No one has more legal power to halt the reckless expansion of the tar sands than the First Nations living downstream whose treaty-protected hunting, fishing, and trapping grounds have already been fouled, just as no one has more legal power to halt the rush to drill under the Arctic's melting ice than Inuit, Sámi, and other northern Indigenous tribes whose livelihoods would be jeopardized by an offshore oil spill. Whether they are able to exercise those rights is another matter.

This power was on display in January 2014 when a coalition of Alaskan Native tribes, who had joined forces with several large green groups, won a major court victory against Shell's already scandal-plagued Arctic drilling adventures. Led by the Native village of Point Hope, the coalition argued that when the U.S. Interior Department handed out drilling permits to Shell and others in the Chukchi Sea, it failed to take into account the full risks, including the risks to Indigenous Inupiat ways of life, which are inextricably entwined with a healthy ocean. As Port Hope mayor Steve Oomittuk explained when the lawsuit was launched, his people "have hunted and depended on the animals that migrate through the Chukchi Sea for thousands of years. This is our garden, our identity, our livelihood. Without it we would not be who we are today. … We oppose any activity that will endanger our way of life and the animals that we greatly depend on." Faith Gemmill, executive director of Resisting Environmental Destruction on Indigenous Lands, one of the groups behind the lawsuit, notes that for the Inupiat who rely on the Chukchi Sea, "you cannot separate environmental impacts from subsistence impacts, for they are the same."[15]

A federal appeals court ruled in the coalition's favor, finding that the Department of the Interior's risk assessments were based on estimates that were "arbitrary and capricious," or presented "only the *best* case scenario for environmental harm."[16] Rather like the shoddy risk assessments that set the stage for BP's Deepwater Horizon disaster.

John Sauven, executive director of Greenpeace U.K., described the ruling as "a massive blow to Shell's Arctic ambitions." Indeed just days later, the company announced that it was putting its Arctic plans on indefinite hold. "This is a disappointing outcome, but the lack of a clear path forward means that I am not prepared to commit further resources for drilling in Alaska in 2014," said Shell CEO Ben van Beurden. "We will look to relevant agencies and the Court to resolve their open legal issues as quickly as possible." Without Indigenous groups raising the human rights stakes in this battle, it's a victory that might never have taken place.[17]

Canadian Dimension

Occupation of Anicinabe Park
Kenora, ON, 1974
(Anishinaabeg Territory)

Crossing the Border
US-Canada Border at Niagara Falls, July 14, 1928
(Haudenosaunee and Anishinaabeg Territory)

National Museum of the American Indian, Smithsonian Institution (P14081)

Worldwide, companies pushing for vast new coal mines and coal export terminals are increasingly being forced to similarly reckon with the unique legal powers held by Indigenous peoples. For instance, in Western Australia in 2013, the prospect of legal battles over native title was an important factor in derailing a planned $45 billion LNG (liquefied natural gas) processing plant and port, and though the state government remains determined to force gas infrastructure and fracking on the area, Indigenous groups are threatening to assert their traditional ownership and procedural rights in court. The same is true of communities facing coal bed methane development in New South Wales.[18]

Meanwhile, several Indigenous groups in the Amazon have been steadfastly holding back the oil interests determined to sacrifice new swaths of the great forests, protecting both the carbon beneath the ground and the carbon-capturing trees and soil above those oil and gas deposits. They have asserted their land rights with increasing success at the Inter-American Court of Human Rights, which has sided with Indigenous groups against governments in cases involving natural resource and territorial rights.[19] And the U'wa, an isolated tribe in Colombia's Andean cloud forests—where the tree canopy is perpetually shrouded in mist—have made history by resisting repeated attempts by oil giants to drill in their territory, insisting that stealing the oil beneath the earth would bring about the tribe's destruction. (Though some limited drilling has taken place.)

As the Indigenous rights movement gains strength globally, huge advances are being made in recognizing the legitimacy of these claims. Most significant was the United Nations Declaration on the Rights of Indigenous Peoples, adopted by the General Assembly in September 2007 after 143 member states voted in its favor (the four opposing votes—United States, Canada, Australia, and New Zealand—would each, under domestic pressure, eventually endorse it as well). The declaration states that, "Indigenous peoples have the right to the conservation and protection of the environment and the productive capacity of their lands or territories and resources." And further that they have "the right to redress" for the lands that "have been confiscated, taken, occupied, used or damaged without their free, prior and informed consent." Some countries have even taken the step of recognizing these rights in revised constitutions. Bolivia's constitution, approved by voters in 2009, states that Indigenous peoples "are guaranteed the right to prior consent: obligatory consultation by the government, acting in good faith and in agreement, prior to the exploitation of non-renewable natural resources in the territory they inhabit." A huge, hard-won legal victory.[20]

Might vs. Rights

And yet despite growing recognition of these rights, there remains a tremendous gap between what governments say (and sign) and what they do—and there is no guarantee of winning when these rights are tested in court. Even in countries with enlightened laws as in Bolivia and Ecuador, the state still pushes ahead with extractive projects without the consent of the Indigenous people who rely on those lands.[21] And in Canada, the United States, and

Australia, these rights are not only ignored, but Indigenous people know that if they try to physically stop extractive projects that are clearly illegal, they will in all likelihood find themselves on the wrong side of a can of pepper spray—or the barrel of a gun. And while the lawyers argue the intricacies of land title in court, buzzing chainsaws proceed to topple trees that are four times as old as our countries, and toxic fracking fluids seep into the groundwater.

The reason industry can get away with this has little to do with what is legal and everything to do with raw political power: isolated, often impoverished Indigenous peoples generally lack the monetary resources and social clout to enforce their rights, and anyway, the police are controlled by the state. Moreover the costs of taking on multinational extractive companies in court are enormous. For instance in the landmark "Rainforest Chernobyl" case in which Ecuador's highest court ordered Chevron to pay $9.5 billion in damages, a company spokesman famously said: "We're going to fight this until hell freezes over—and then we'll fight it out on the ice." (And indeed, the fight still drags on.)[22]

I was struck by this profound imbalance when I traveled to the territory of the Beaver Lake Cree Nation in northern Alberta, a community that is in the midst of one of the highest-stakes legal battles in the tar sands. In 2008, the band filed a historic lawsuit charging that by allowing its traditional territories to be turned into a latticework of oil and gas infrastructure, and by poisoning and driving away the local wildlife, the provincial and federal governments, as well as the British Crown, had infringed no fewer than fifteen thousand times on the First Nation's treaty rights to continue to hunt, fish, and trap on their territory.[23] What set the case apart was that it was not about one particular infringement, but an entire model of poisonous, extractive development, essentially arguing that this model itself constituted a grave treaty violation.

> "The Governments of Canada and Alberta have made a lot of promises to our people and we intend to see those promises kept," said Al Lameman, the formidable chief of the Beaver Lake Cree Nation at the time the lawsuit was filed (Lameman had made history before, filing some of the first Indigenous human rights challenges against the Canadian government). Against the odds, the case has proceeded through the Canadian court system, and in March 2012 an Alberta court flatly rejected government efforts to have the case dismissed as "frivolous," an "abuse of the Court's process," and "unmanageable."[24]

A year after that ruling, I met Al Lameman, now retired, and his cousin Germaine Anderson, an elected band councilor, as well as the former chief's niece, Crystal Lameman, who has emerged as one of the most compelling voices against the tar sands on the international stage. These are three of the people most responsible for moving the lawsuit forward, and Germaine Anderson had invited me to a family barbecue to discuss the case.

Courtesy of Ground Zero Occupy INAC

Occupying the INAC Office
Toronto, ON, 2016
(Wendat, Anishinaabeg, Haudenosaunee, Métis, and Mississauga Territories)

Michael Toledano / VICE

Pipeline Standoff & Blockade
Unist'ot'en Camp, Widzin Kwah/Morice River, BC, 2014
(Wet'suwet'en Territory)

It was early July and after a long dark winter it was as if a veil had lifted: the sun was still bright at 10 p.m. and the northern air had a thin, baked quality. Al Lameman had aged considerably in recent years and slipped in and out of the conversation. Anderson, almost painfully shy, had also struggled with her health. The spot where the family met for this gathering was where she spent the summer months: a small trailer in a clearing in the woods, without running water or electricity, entirely off the grid. I knew the Beaver Lake Cree were in a David and Goliath struggle. But on that endless summer evening, I suddenly understood what this actually meant: some of the most marginalized people in my country—many of them, like all the senior members of the Lameman clan, survivors of the intergenerational trauma of abusive residential schools—are taking on some of the wealthiest and most powerful forces on the planet. Their heroic battles are not just their people's best chance of a healthy future; if court challenges like Beaver Lake's can succeed in halting tar sands expansion, they could very well be the best chance for the rest of us to continue enjoying a climate that is hospitable to human life. That is a huge burden to bear and that these communities are bearing it with shockingly little support from the rest of us is an unspeakable social injustice. A few hours north, a different Indigenous community, the Athabasca Chipewyan First Nation (ACFN), recently launched another landmark lawsuit, this one taking on Shell and the Canadian government over the approval of a huge tar sands mine expansion. The band is also challenging another Shell project, the proposed Pierre River Mine, which it says "would significantly impact lands, water, wildlife and the First Nation's ability to utilize their traditional territory." Once again the mismatch is staggering. The ACFN, with just over one thousand members and an operating budget of about $5 million, is battling both the Canadian government and Shell, with its 92,000 employees across more than seventy countries and 2013 global revenues of $451.2 billion. Many communities see odds like these and, understandably, never even get in the ring.[25]

It is this gap between rights and resources—between what the law says and what impoverished people are able to force vastly more powerful entities to do—that government and industry have banked on for years.

"Honour the Treaties"

What is changing is that many non-Native people are starting to realize that Indigenous rights—if aggressively backed by court challenges, direct action, and mass movements demanding that they be respected—may now represent the most powerful barriers protecting all of us from a future of climate chaos. Which is why, in many cases, the movements against extreme energy extraction are becoming more than just battles against specific oil, gas, and coal companies and more, even, than pro-democracy movements. They are opening up spaces for a historical reconciliation between Indigenous peoples and non-Natives, who are finally understanding that, at a time when elected officials have open disdain for basic democratic principles, Indigenous rights are not a threat, but a tremendous gift. Because the original Indigenous treaty negotiators in much of North America had the foresight to include language protecting their right to continue living off their traditional lands, they bequeathed to all residents of these and many other countries the legal tools to demand that our governments refrain from finishing the job of flaying the planet. And so, in communities where there was once only anger, jealousy, and thinly veiled racism, there is now something new and unfamiliar. "We're really thankful for our First Nations partners in this struggle," said Lionel Conant, a property manager whose home in Fort St. James, British Columbia, is within sight of the proposed Northern Gateway pipeline. "[They've] got the legal weight to deal with [the pipeline] … because this is all unceded land." In Washington State, anti-coal activists talk about the treaty rights of the Lummi as their "ace in the hole" should all other methods of blocking the export terminals fail. In Montana, the Sierra Club's Mike Scott told me bluntly, "I don't think people understand the political power Natives have as sovereign nations, often because they lack the resources to exercise that power. They can stop energy projects in a way we can't."[26]

In New Brunswick, Suzanne Patles, a Mi'kmaq woman involved in the anti-fracking movement, described how non-Natives "have reached out to the Indigenous people to say 'we need help.'"[27] Which is something of a turnaround from the saviorism and pitying charity that have poisoned relationships between Indigenous peoples and well-meaning liberals for far too long. It was in the context of this gradual shift in awareness that Idle No More burst onto the political scene in Canada at the end of 2012 and then spread quickly south of the border. North American shopping centers—from the enormous West Edmonton Mall to Minnesota's Mall of America—were suddenly alive with the sounds of hand drums and jingle dresses as Indigenous people held flash mob round dances across the continent at the peak of the Christmas shopping season. In Canada, Native leaders went on hunger strikes, and youths embarked on months-long spiritual walks and blockaded roads and railways.

The movement was originally sparked by a series of attacks by the Canadian government on Indigenous sovereignty, as well as its all-out assault on existing environmental protections, particularly for water, to pave the way for rapid tar sands expansion, more mega-mines, and projects like Enbridge's Northern Gateway pipeline. The attacks came in the form of two omnibus budget bills passed in 2012 that gutted large parts of the country's environmental regulatory framework. As a result, a great many industrial activities were suddenly exempt from federal environmental reviews, which along with other changes, greatly reduced opportunities for community input and gave the intractable right-wing government of Stephen Harper a virtual free hand to ram through unpopular energy and development projects. The omnibus bills also overhauled key provisions of the Navigable Waters Protection Act that protect species and ecosystems from damage. Previously, virtually 100 percent of the country's water bodies had been covered by these protections; under the new order, that was slashed to less than 1 percent, with pipelines simply exempted. (Documents later revealed that the latter change had been specifically requested by the pipeline industry.)[28]

Ossie Michelin

Anti-Fracking Protest
Amanda Polchies holding feather on Highway 11, Elsipogtog, NB, 2013
(Mi’kmaq Territory)

Ossie Michelin

Muskrat Falls Protest
13-year-old Allyson Gear, of Postville, Nunatsiavut, drum dancing at the encampment near Happy Valley-Goose Bay, NL, 2016 (Inuit Territory)

Canadians were in shock at the extent and speed of the regulatory overhaul. Most felt powerless, and with good reason: despite winning only 39.6 percent of the popular vote, the Harper government had a majority in Parliament and could apparently do as it pleased.[29] But the First Nations' response was not to despair; it was to launch the Idle No More movement from coast to coast. These laws, movement leaders said, were an attack on Indigenous rights to clean water and to maintain traditional ways of life. Suddenly, the arguments that had been made in local battles were being taken to the national level, now used against sweeping federal laws. And for a time Idle No More seemed to change the game, attracting support from across Canadian society, from trade unions to university students, to the opinion pages of mainstream newspapers.

These coalitions of rights-rich-but-cash-poor people teaming up with (relatively) cash-rich-but-rights-poor people carry tremendous political potential. If enough people demand that governments honor the legal commitments made to the people on whose land colonial nations were founded, and do so with sufficient force, politicians interested in re-election won't be able to ignore them forever. And the courts, too—however much they may claim to be above such influences—are inevitably shaped by the values of the societies in which they function. A handful of courageous rulings notwithstanding, if an obscure land right or treaty appears to be systematically ignored by the culture as a whole, it will generally be treated tentatively by the courts. If, however, the broader society takes those commitments seriously, then there is a far greater chance that the courts will follow.[30]

As Idle No More gained steam, many investors took notice. "For the first time in six years, Canadian provinces failed to top the list of the best mining jurisdictions in the world in a 2012/13 survey," Reuters reported in March 2013. "Companies that participated in the survey said they were concerned about land claims." The article quoted Ewan Downie, chief executive of Premier Gold Mines, which owns several projects in Ontario: "I would say one of the big things that is weighing on mining investment in Canada right now is First Nations issues."[31]

Writing in *The Guardian*, journalist and activist Martin Lukacs observed that Canadians seemed finally to be grasping that,

> "Implementing Indigenous rights on the ground, starting with the United Nations Declaration on the Rights of Indigenous Peoples, could tilt the balance of stewardship over a vast geography: giving Indigenous peoples much more control, and corporations much less. Which means that finally honoring Indigenous rights is not simply about paying off Canada's enormous legal debt to First Nations: it is also our best chance to save entire territories from endless extraction and destruction. In no small way, the actions of Indigenous peoples—and the decision of Canadians to stand alongside them—will determine the fate of the planet. This new under-

> standing is dawning on more Canadians. Thousands are signing onto educational campaigns to become allies to First Nations. … Sustained action that puts real clout behind Indigenous claims is what will force a reckoning with the true nature of Canada's economy—and the possibility of a transformed country. That is the promise of a growing mass protest movement, an army of untold power and numbers."[32]

In short, the muscle able to turn rights into might that Standard & Poor's had been looking for in that meeting with Arthur Manuel and Guujaaw back in 2004 may have finally developed. The power of this collaboration received another boost in January 2014 when the rock legend Neil Young kicked off a cross-Canada tour called "Honour the Treaties." He had visited the tar sands several months earlier and been devastated by what he saw, saying (to much controversy) that the region "looks like Hiroshima." While in the region, he had met with Chief Allan Adam of the Athabasca Chipewyan and heard about the lawsuits opposing Shell's tar sands expansions, as well as the health impacts current levels of oil production are already having on the community. "I was sitting with the chief in the teepee, on the reserve. I was hearing the stories. I saw that the cancer rate was up among all the tribes. This is not a myth. This is true," Young said.[33]

And he concluded that the best way he could contribute to the fight against the tar sands was to help the Athabasca Chipewyan First Nation exercise its rights in court. So he went on a concert tour, donating 100 percent of the proceeds to the court challenges. In addition to raising $600,000 for their legal battles within two months, the tour attracted unprecedented national attention to both the local and global impacts of runaway tar sands development. The prime minister's office fought back by attacking one of Canada's most beloved icons, but it was a losing battle. Prominent Canadians spoke up to support the campaign, and polls showed that even in Alberta a majority were taking Young's side in the dispute.[34]

Most importantly, the Honour the Treaties tour sparked a national discussion about the duty to respect First Nation legal rights. "It's up to Canadians all across Canada to make up their own minds about whether their integrity is threatened by a government that won't live up to the treaties that this country is founded on," Young said. And the country heard directly from Chief Allan Adam, who described the treaties his ancestors signed as "not just pieces of paper but a last line of defense against encroaching reckless tar sands development that my people don't want and that we are already suffering from."[35]

Oka Crisis Mercier Bridge Road Fire, QC (1990)

Road Blockade & Tire Fire
Argyle Street, Caledonia, ON, 2006
(Six Nations of Grand River, Haudenosaunee Territory)

Wayne Glowacki / The Canadian Press

Opposing Meech Lake Accord
Elijah Harper, Manitoba Legislative Assembly, 1990

Making the most of that last line of defense is a complex challenge involving much more than rock concerts and having cash in hand to pay lawyers. The deeper reason why more First Nations communities aren't taking on companies like Shell has to do with the systematic economic and social disenfranchisement that makes doing business with heavily polluting oil or mining companies seem like the only way to cover basic human needs. Yes, there is a desire to protect the rivers, streams, and oceans for traditional fishing. But in Canada, according to a 2011 government report, the water systems in 25 percent of First Nations communities are so neglected and underfunded that they pose a "high overall risk" to health, while thousands of residents of Native reserves are living without sewage or running water at all. If you are the leader of one such community, getting those basic services taken care of, no matter the cost, is very likely going to supersede all other priorities.[36]

And ironically, in many cases, climate change is further increasing the economic pressure on Indigenous communities to make quick-and-dirty deals with extractive industries. That's because disruptive weather changes, particularly in northern regions, are making it much harder to hunt and fish (for example when the ice is almost never solid, communities in the far north become virtually trapped, unable to harvest food for months on end). All this makes it extremely hard to say no to offers of job training and resource sharing when companies like Shell come to town. Members of these communities know that the drilling will only make it harder to engage in subsistence activities—there are real concerns about the effects of oil development on the migration of whales, walruses, and caribou—and that's without the inevitable spills. But precisely because the ecology is already so disrupted by climate change, there often seems no other option.

The paucity of good choices is perhaps best on display in Greenland, where receding glaciers and melting ice are revealing a vast potential for new mines and offshore oil exploration. The former Danish colony gained home rule in 1979, but the Inuit nation still relies on an annual infusion of more than $600 million (amounting to a full third of the economy) from Denmark. A 2008 self-governance referendum gave Greenland still more control over its own affairs, but also put it firmly on the path of drilling and mining its way to full independence. "We're very aware that we'll cause more climate change by drilling for oil," a top Greenlandic official, then heading the Office of Self-Governance, said in 2008. "But should we not? Should we not when it can buy us our independence?" Currently, Greenland's largest industry is fishing, which of course would be devastated by a major spill. And it doesn't bode well that one of the companies selected to begin developing Greenland's estimated fifty billion barrels of offshore oil and gas is none other than BP.[37] Indeed the melancholy dynamic strongly recalls BP's "vessels of opportunity" program launched in the midst of the Deepwater Horizon disaster. For months, virtually the entire Louisiana fishing fleet was docked, unable to make a living for fear that the seafood was unsafe. That's when BP offered to convert any fishing vessel into a cleanup boat, providing it with booms to (rather uselessly) mop up some oil. It was tremendously difficult for local shrimpers and oystermen to take work from the company that had just robbed them of their livelihood—but what choice did they have? No one else was offering to help pay the bills. This is the way the oil and gas industry holds on to power: by tossing temporary life rafts to the people it is drowning.

That many Indigenous people would view the extractive industries as their best of a series of bad options should not be surprising. There has been almost no other economic development in most Native communities, no one else offering jobs or skills training in any quantity. So in virtually every community on the front lines of extractive battles, some faction invariably makes the argument that it's not up to Indigenous people to sacrifice to save the rest of the world from climate change, that they should concentrate instead on getting better deals from the mining and oil companies so that they can pay for basic services and train their young people in marketable skills. Jim Boucher, chief of the Fort McKay First Nation, whose lands have been decimated by the Alberta tar sands, told an oil-industry-sponsored conference in 2014, "There is no more opportunity for our people to be employed or have some benefits except the oil sands"—going so far as to call the mines the "new trap line," a reference to the fur trade that once drove the economics of the region.[38] Sadly, this argument has created rancorous divisions and families are often torn apart over whether to accept industry deals or to uphold traditional teachings. And as the offers from industry become richer (itself a sign of Blockadia's growing power), those who are trying to hold the line too often feel they have nothing to offer their people but continued impoverishment. As Phillip Whiteman Jr., a traditional Northern Cheyenne storyteller and longtime opponent of coal development, told me, "I can't keep asking my people to suffer with me."[39]

These circumstances raise troubling moral questions for the rising Blockadia movement, which is increasingly relying on Indigenous people to be the legal barrier to new, high-carbon projects. It's fine and well to laud treaty and title rights as the "last line of defense" against fossil fuel extraction. But if non-Native people are going to ask some of the poorest, most systematically disenfranchised people on the planet to be humanity's climate saviors, then, to put it crassly, what are we going to do for them? How can this relationship not be yet another extractive one, in which non-Natives use hard-won Indigenous rights but give nothing or too little in return? As the experience with carbon offsets shows, there are plenty of examples of new "green" relationships replicating old patterns. Large NGOs often use Indigenous groups for their legal standing, picking up some of the costs for expensive legal battles but not doing much about the underlying issues that force so many Indigenous communities to take these deals in the first place. Unemployment stays sky high. Options, for the most part, stay bleak.

If this situation is going to change, then the call to Honour the Treaties needs to go a whole lot further than raising money for legal battles. Non-Natives will have to become the treaty and land-sharing partners that our ancestors failed to be, making good on the

Samantha Eldridge

Water Protectors Isabella & Alyssa Klain at Dakota Access Pipeline Protest
Standing Rock Sioux Reservation, ND, 2016
(Sioux Territory)

full panoply of promises they made, from providing health care and education to creating economic opportunities that do not jeopardize the right to engage in traditional ways of life. Because the only people who will be truly empowered to say no to dirty development over the long term are people who see real, hopeful alternatives. And this is true not just within wealthy countries but between the countries of the wealthy postindustrial North and the fast-industrializing South.

☼ ☼ ☼

Notes

This essay was originally published as "You and What Army? Indigenous Rights and the Power of Keeping Our Word," chap. 11 in *This Changes Everything: Capitalism vs. The Climate* (Toronto, ON: Alfred A. Knopf Canada, 2014), reproduced with permission from Naomi Klein.

1 Geraldine Thomas-Flurer, Yinka Dene Alliance coordinator, as quoted in Melanie Jae Martin and Jesse Fruhwirth, "Welcome to Blockadia!" *YES!*, January 11, 2013.

2 Gurston Dacks, "British Columbia After the Delgamuukw Decision: Land Claims and Other Processes," *Canadian Public Policy* 28 (2002): 239–55.

3 "Statement of Claim between Council of the Haida Nation and Guujaaw suing on his own behalf and on behalf of all members of the Haida Nation (plaintiffs) and Her Majesty the Queen in Right of the Province of British Columbia and the Attorney General of Canada (defendants)," Action No. L020662, Vancouver Registry, November 14, 2002, http://www.haidanation.ca; *Haida Nation v. British Columbia* (Minister of Forests) 3 SCR 511 (SCC 2004); "Government Must Consult First Nations on Disputed Land, Top Court Rules," CBC News, November 18, 2004; personal interview with Arthur Manuel, August 25, 2004.

4 For a deeper understanding of Naomi Klein's use of the term "Blockadia," see her essay "Blockadia: The New Climate Warriors," chap. 9 in *This Changes Everything*, 253–90.

5 Personal email communication with Tyler McCreary, PhD candidate, York University, January 30, 2014.

6 *Delgamuukw v. British Columbia*, [1997], 3 SCR 1010; British Columbia Treaty Commission, "A Lay Person's Guide to Delgamuukw v. British Columbia," November 1999, http://www.bctreaty.net; Chelsea Vowel, "The Often-Ignored Facts About Elsipogtog," *Toronto Star*, November 14, 2013.

7 Melanie G. Wiber and Julia Kennedy, "Impossible Dreams: Reforming Fisheries Management in the Canadian Maritimes After the Marshall Decision," in *Law and Anthropology: International Yearbook for Legal Anthropology*, Vol. 2, ed. René Kuppe and Richard Potz (The Hague: Martinus Nijhoff Publishers, 2001), 282–97; William Wicken, "Treaty of Peace and Friendship 1760," *Aboriginal Affairs and Northern Development Canada*, https://www.aadnc-aandc.gc.ca; R. v. Marshall, 3 SCR 456 (1999); "Supreme Court Decisions: R. v. Marshall," *Aboriginal Affairs and Northern Development Canada*.

8 "Map of Treaty-Making in Canada," Aboriginal Affairs and Northern Development Canada, https://www.aadnc-aandc.gc.ca; "Alberta Oil Sands," Alberta Geological Survey, last modified June 12, 2013, http://www.ags.gov.ab.ca; "Treaty Texts—Treaty No. 6; Copy of Treaty No. 6 Between Her Majesty the Queen and the Plain and Wood Cree Indians and Other Tribes of Indians at Fort Carlton, Fort Pitt, and Battle River with Adhesions," Aboriginal Affairs and Northern Development Canada, https://www.aadnc-aandc.gc.ca.

9 "Emergency Advisory: Mi'kmaq say, 'We Are Still Here, and SWN Will Not Be Allowed to Frack,'" press release, Halifax Media Co-op, November 3, 2013.

10 Martha Stiegman and Miles Howe, "Summer of Solidarity—A View from the Sacred Fire Encampment in Elsipogtog," (video), Halifax Media Co-op, July 3, 2013.

11 "'Crown Land Belongs to the Government, Not to F*cking Natives,' " APTN, October 17, 2013; Martin Lukacs, "New Brunswick Fracking Protests Are the Frontline of a Democratic Fight," *Guardian*, October 21, 2013; Renee Lewis, "Shale Gas Company Loses Bid to Halt Canada Protests," Al Jazeera America, October 21, 2013.

12 "FORUMe Research Results," PowerPoint, MQO Research, presented at FORUMe conference, New Brunswick, June 2012, http://www.amiando.com; Kevin Bissett, "Alward Facing Opposition from N.B. Citizens over Fracking," The Canadian Press, August 30, 2011.

13 Stiegman and Howe, "Summer of Solidarity."

14 Richard Walker, "In Washington, Demolishing Two Dams So That the Salmon May Go Home," *Indian Country Today*, September 22, 2011; "Press Release 02/26/2014," Shield the People, press release, February 26, 2014; "Keystone XL Pipeline Project Compliance Follow-up Review: The Department of State's Choice of Environmental Resources Management, Inc., To Assist in Preparing the Supplemental Environmental Impact Statement," United States Department of State and the Broadcasting Board of Governors, February 2014; Jorge Barrera, "Keystone XL 'Black Snake' Pipeline to Face 'Epic' Opposition from Native American Alliance," APTN, January 31, 2014.

15 Steve Quinn, "U.S. Appeals Court Throws Arctic Drilling into Further Doubt," Reuters, January 23, 2014; *Native Village of Point Hope v. Jewell*, 44 ELR 20016, No. 12-35287 (9th Cir., 01/22/2014); "Native and Conservation Groups Voice Opposition to Lease Sale 193 in the Chukchi Sea"; World Wildlife Fund, press release, February 6, 2008; Faith Gemmill, "Shell Cancels 2014 Arctic Drilling—Arctic Ocean and Inupiat Rights Reality Check," Platform, January 30, 2014.

16 *Native Village of Point Hope v. Jewell.*

17 Terry Macalister, "Shell's Arctic Drilling Set Back by US Court Ruling," *Guardian*, January 23, 2014; "New Shell CEO Ben van Beurden Sets Agenda for Sharper Performance and Rigorous Capital Discipline," Shell, press release, January 30, 2014.

18 Erin Parke, "Gas Hub Future Unclear After Native Title Dispute," ABC (Australia), February 7, 2013; "Environmentalists Welcome Scrapping of LNG Project," ABC (Australia), April 12, 2013; Andrew Burrell, "Gas Fracking Wars to Open Up on a New Front," *Australian*, December 30, 2013; "Native Title Challenge to Canning Gas Bill," Australian Associated Press, June 20, 2013; Vicky Validakis, "Native Title Claimants Want to Ban Mining," Australian Mining, May 14, 2013.

19 "Ecuador: Inter-American Court Ruling Marks Key Victory for Indigenous People," Amnesty International, press release, July 27, 2012.

20 ORIGINAL VOTE: United Nations News Centre, "United Nations Adopts Declaration on Rights of Indigenous Peoples," United Nations press release, September 13, 2007; LATER ENDORSEMENTS: "Indigenous Rights Declaration Endorsed by States," Office of the United Nations High Commissioner for Human Rights, press release, December 23, 2010; "HAVE THE RIGHT," "REDRESS": *United Nations Declaration on the Rights of Indigenous Peoples*, G.A. Res. 61/295, U.N. Doc. A/Res/61/295 September 13, 2007, 10–11, http://www.un.org; CONSTITUTION (ORIGINAL SPANISH): República de Bolivia, Constitución de 2009, Capítulo IV: Derechos de las Naciones y Pueblos Indígena Originario Campesinos, art. 30, sec. 2; CONSTITUTION (ENGLISH TRANSLATION): Leah Temper et al., "Towards a Post-Oil Civilization: Yasunization and Other Initiatives to Leave Fossil Fuels in the Soil," EJOLT Report No. 6, May 2013, p. 71.

21 Alexandra Valencia, "Ecuador Congress Approves Yasuni Basin Oil Drilling in Amazon," Reuters, October 3, 2013; Amnesty International, "Annual Report 2013: Bolivia," May 23, 2013, http://www.amnesty.org.

22 John Otis, "Chevron vs. Ecuadorean Activists," *Global Post*, May 3, 2009.

23 "Beaver Lake Cree Sue over Oil and Gas Dev't," *Edmonton Journal*, May 14, 2008; "Beaver Lake Cree Nation Draws a Line in the (Oil) Sand," Beaver Lake Cree Nation, press release, May 14, 2008.

24 Ibid.; Court of the Queen's Bench, Government of Alberta, 2012 ABQB 195, Memorandum of Decision of the Honourable Madam Justice B. A. Browne, March 28, 2012.

25 Bob Weber, "Athabasca Chipewyan File Lawsuit Against Shell's Jackpine Oil Sands Expansion," The Canadian Press, January 16, 2014; Chief Allan Adam, "Why I'm on Tour with Neil Young and Diana Krall," Huffington Post Canada, January 14, 2014; "Administration and Finance," Athabasca Chipewyan First Nation, http://www.acfn.com; "Shell at a Glance," Shell Global, http://www.shell.com/global.

26 Emma Gilchrist, "Countdown Is On: British Columbians Anxiously Await Enbridge Recommendation," *DesmogCanada*, December 17, 2013; personal interview with Mike Scott, October 21, 2010.

27 Benjamin Shingler, "Fracking Protest Leads to Bigger Debate over Indigenous Rights in Canada," *Al Jazeera America*, December 10, 2013.

28 OMNIBUS BILLS: Bill C-38, Jobs, Growth and Long-Term Prosperity Act, 41st Parliament, 2012, S.C. 2012, c. 19, http://laws-lois.justice.gc.ca; Bill C- 45, Jobs and Growth Act 2012, 41st Parliament, 2012, S.C. 2012, c. 31, http://laws-lois.justice.gc.ca; REVIEWS: Tonda MacCharles, "Tories Have Cancelled Almost 600 Environmental Assessments in Ontario," *Toronto Star*, August 29, 2012; COMMUNITY INPUT: Andrea Janus, "Activists Sue Feds over Rules That 'Block' Canadians from Taking Part in Hearings," *CTV News*, August 15, 2013; ACT: Navigable Waters Protection Act, Revised Statutes of Canada 1985, c. N-22, http://laws-lois.justice.gc.ca; FROM PRACTICALLY 100 PERCENT: "Omnibus Bill Changes Anger Water Keepers," CBC News, October 19, 2012; TO LESS THAN 1 PERCENT: "Legal Backgrounder: Bill C-45 and the Navigable Waters Protection Act" (RSC 1985, C N-22), EcoJustice, October 2012; "Hundreds of N.S. Waterways Taken off Protected List; Nova Scotia First Nation Joins Idle No More Protest," CBC News, December 27, 2012; PIPELINES: See amendments 349(5) and 349(9) of Bill C-45, Jobs and Growth Act 2012, 41st Parliament, 2012, S.C. 2012, c. 31; DOCUMENTS REVEALED: Heather Scoffield, "Documents Reveal Pipeline Industry Drove Changes to 'Navigable Waters' Act," The Canadian Press, February 20, 2013.

29 "Electoral Results by Party: 41st General Election (2011.05.02)," Parliament of Canada, http://www.parl.gc.ca; Ian Austen, "Conservatives in Canada Expand Party's Hold," *New York Times*, May 2, 2011.

30 Indeed, it may be no coincidence that in June 2014, the Supreme Court of Canada issued what may be its most significant indigenous rights ruling to date when it granted the Tsilhqot'in Nation a declaration of Aboriginal title to 1,750 square kilometers of land in British Columbia. The unanimous decision laid out that ownership rights included the right to use the land, to decide how the land should be used by others, and to derive economic benefit from the land. Government, it also stated, must meet certain standards before stepping in, and seek not only consultation with First Nations, but consent from them. Many commented that it would make the construction of controversial projects like tar sands pipelines—rejected by local First Nations—significantly more difficult.

31 Julie Gordon and Allison Martell, "Canada Aboriginal Movement Poses New Threat to Miners," Reuters, March 17, 2013.

32 Martin Lukacs, "Indigenous Rights Are the Best Defence Against Canada's Resource Rush," *Guardian*, April 26, 2013.

33 "Neil Young at National Farmers Union Press Conference" (video), YouTube, Thrasher Wheat, September 9, 2013; Jian Ghomeshi, "Q exclusive: Neil Young Says 'Canada Trading Integrity for Money'"(video), CBC News, January 13, 2014.

34 Personal interview with Eriel Deranger, communications manager, Athabasca Chipewyan First Nation January 30, 2014; "Poll: How Do You Feel About Neil Young Attacking the Oilsands?" *Edmonton Journal*, January 12, 2014.

35 Ghomeshi, "Q exclusive: Neil Young Says 'Canada Trading Integrity for Money'"; Adam, "Why I'm on Tour with Neil Young and Diana Krall."

36 "National Assessment of First Nations Water and Wastewater Systems," prepared by Neegan Burnside for Department of Indian and Northern Affairs, Canada, April 2011, 16, http://www.aadnc-aandc.gc.ca.

37 In 2012, Greenland's subsidy from Denmark was about 3.6 billion Danish kroner, equal to 31 percent of its GDP that year. The subsidy was also about 3.6 billion Danish kroner in 2013: "Greenland in Figures: 2014," Statistics Greenland, 2014, 7–8; Jan. M. Olsen, "No Economic Independence for Greenland in Sight," Associated Press, January 24, 2014; "OUR INDEPENDENCE": McKenzie Funk, *Windfall: The Booming Business of Global Warming* (New York, NY: Penguin, 2014), 78.

38 Angela Sterritt, "Industry and Aboriginal Leaders Examine Benefits of the Oilsands," CBC News, January 24, 2014.

39 Personal interview with Phillip Whiteman Jr., October 21, 2010.

Counter-Development

A Conversation with John Van Nostrand

John Van Nostrand is an architect, planner, and founding principal of SvN in Toronto. His practice focuses on planning and design of affordable housing and community infrastructure, including a number of major mine-related housing projects in Africa, Latin America, and Canada.

X: We've been speaking over the past couple of years about your experience, your research, and your practice. It's become extremely clear that you have a unique, territorial understanding of urbanization, and of the North in particular, through the lens of the metropolis and the hinterland. How did you arrive at this geographic perspective?

JVN: There a lot of issues that pervade the North, ranging from territory and territoriality, to Indigenous sovereignty, resource extraction, immigration, infrastructure inequality, water quality, colonization, and the rights of First Nations. I've just tried to formulate a different urban discourse in Canada that, as I've mentioned before, draws from a deeper, historical understanding of territory and its influence on patterns of urbanization across the country. I have also been working simultaneously in former colonies in Africa for over 35 years, most often with Indigenous peoples, many of whom were drawn to large metropolitan areas.

"The problem with radical and reactionary forms of development is that what they miss is the nearly continuous, informal, undeclared, disguised forms of autonomous resistance by lower classes: forms of politics that I call 'everyday resistance.'"

—James C. Scott, "Everyday Forms of Resistance," 1989

X: Does this dichotomy between the metropolis and the hinterland still exist?

JVN: Yes, very much so. Let me begin with the Century Initiative (CI), a group that I have been advising, whose Mission is: "A competitive global nation of 100 Million Canadians unified in 2100 by diversity and prosperity." Basically, it was founded by a think-tank later creating the Liberal government's economic policy. The process is simple: it's based on immigration. First, they look at population growth. Canada is already the fastest-growing country in the G8. If currently we have a population of about 35 million, that would be 55 to 65 million by 2050, and well over 100 million by 2100. Second, immigration will form 60-80 percent of this projected growth and of course, we need to think about it. Third, the population of First Nations is growing. They're the fastest growing part of the population, and in total, we are the fastest growing country in the G7. The Greater Golden Horseshoe (which stretches from Niagara, to Hamilton, Toronto, and on to Oshawa) itself is like the second fastest growing region in North America, if not the first. This is really unheard of.

X: With all the territorial and ecological complexities of development, is this about a geographic approach to urbanism?

JVN: We're turning our attention to urban development across a range of scales. We need to figure out two main things. One, what to do about fly-in/fly-out workers and man camps—there are 700,000 of them—taking local and regional economies away to other parts of Canada, who take it away with them. Those economies need to stay *in place*; two, foreign students, who have become a great source of revenue for the universities. Although, Canadian universities charge half the tuition of private universities such as Harvard, there is a concern that when those graduates leave, they take with them the knowledge and what they could potentially do with it. It all comes down to urban development.

"We know that the Creator did not give the settlers the right to exclusively benefit from our natural wealth and resources. It is colonialism that gave settlers the power to economically exploit our lands, crush our culture, and dominate our peoples."

—Arthur Manuel, Unsettling Canada, 2015

Mid-Canada Plan Development Corridor by Richard Rohmer, 1967 (Acres Research & Planning)

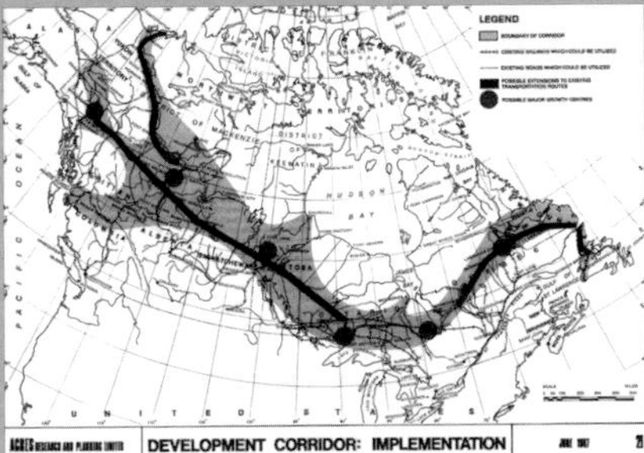

Mid-Canada Corridor Planning Project, John Van Nostrand, 2004 (Chris Brackley, adapted from Pamela Ritchot. Courtesy of John van Nostrand)

X: How has that geographic lens on urbanism been received?

JVN: In a recent presentation at the Neptis Foundation, I've said, "look, the cities are growing and we have these megaregions (including Toronto, Vancouver, Montreal, etc.), they're very important. However, in Canada, we also need to recognize that there are a series of emerging new mega-economies—the whole West Coast which is seeing massive immigration especially from China, the Arctic Ocean Coast as a result of the opening of the Northwest Passage, and even the East Coast which will experience growth resulting from ocean shipping and oil and gas refineries—all of which Canada is ignoring, except from an environmental position, where the first steps have been taken on the West and Arctic fronts to identify areas requiring protection. Then there's the Mid-Canada (also known as Northern or Boreal) Corridor which contains 75% of the country's mineral and forestry wealth."

It's only the economists that see these regions emerging over the remainder of the century—in spite of their potentially profound impact on settlement. By the way, I think it's difficult to describe what is already beginning to happen on these ocean fronts and across the Middle of Canada, as some form of 'urbanism.' The words I would use are the "resettlement" of these mega-regions, which requires an 'unsettling' of the current, largely colonial, and exploitive planning and development controls, and its replacement by the overlay of a new economic, social, and environmental framework which first, embraces and grows the Indigenous (formerly hunter-gatherer or resource-based around fishing, forestry, hunting, etc.) economies what existed—and continue to exist—along and across these economic fronts. It is guided by underlying and overlying landscape conditions rather than by theoretical grids planned and laid-out to define 'Crown land,' which is subsequently excluded from economic development without the mysterious permission of the Crown ('Her Majesty'). I find it fascinating and ingenious that governments are wanting to purchase Churchill—which is going to re-emerge as one of Canada's largest and most important ports once the North-West passage opens—the port that over 300 years exported the equivalent of hundreds of millions of dollars of Indigenously-created furs at "below-market" prices, when the Hudson watershed was the most productive economic zone in the British Empire.

X: You're pushing city-centric planners and urban designers to see beyond the limits of urban megaregions—or metropolitan regions as they're defined now—to really come up with strategies that project on a series of different coasts; new territorialities. This goes beyond your earlier work when you wrote about Rohmer's Plan in the 1970s for the development of a Mid-Development Boreal Corridor?

JVN: I'm trying to revive a conversation about planning the nation—new and present nations, through an understanding of territoriality. Richard Rohmer was probably one of the only ones that was thinking on that scale; but I've grown a lot since that 2014 article, "If We Build It, They Will Stay," published in *The Walrus*.[1] The Northern Corridor is very important, but so to are the coasts, because of major pipeline projects.

1 John Van Nostrand, "If We Build It, They Will Stay," *The Walrus*, September 8, 2014, https://thewalrus.ca/if-we-build-it-they-will-stay/.

X: So we have to monitor global forces that potentially affect territory and infrastructure?

JVN: I've been mentally tracking Chinese development in Canada. I first heard that they were building a giant port in Jamaica, and I wondered, why would you put a port in Jamaica of this size? Then suddenly you realize, Jamaica's at the eastern mouth of the Panama Canal. It's one of the first islands you hit when you come out of the Panama Canal. China was also trying to develop a major port in Iceland two years ago, at the eastern end of the Northwest Passage. From there, Chinese cargo ships could get to that port from mainland China in two days, and therefore control the entire shipping business in the Atlantic Ocean.

X: Is that the end of it? How far does China's infrastructure span?

JVN: Well, that's part of China's strategy and they are affecting many other countries and regions. They are planning to build a major corridor from China to the Mediterranean, just like the old Silk Road. It's called the Mid-Asia Corridor. Paying attention to those forces is pretty important given that the Chinese National Offshore Oil Company (CNOOC) has, also, huge interests in Canadian resources. CNOOC also bought up Nexen in the Tar Sands in 2013.

X: Does this also have to do with your territorial outlook? And how to break the myth of the North as a remote, removed, or isolated area?

JVN: Yes, and then of course, there's no grid there. So we're talking about economies that depend on landscape, ground, geology, waters, and fish. This hinterland is set against a settler colonial landscape. Remember, the Department of the Interior had been putting down the early forms of settlement in Canada, in favor of mining and extraction sites, during the early twentieth century. Before Britain withdrew from Canada to place its force entirely in India at the end of the 1850s and early 1860s, immigration policy of Canada tried to use the mining sector to coax Britain into staying here.

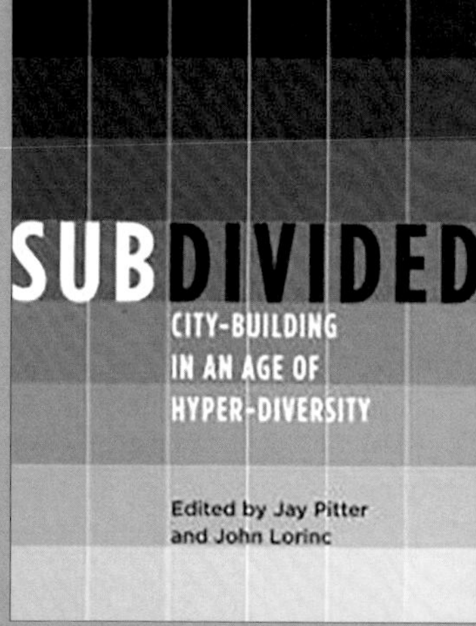

Jay Pitter and John Lorinc, eds., *Subdivided: City Building in an Age of Hyper-Diversity*, 2016

X: So immigration, resource extraction, and territory are bound together?

JVN: This is important, and so is the conflict with territory, sovereignty, history, and extraction. Recently, a fossil was found in central Canada (Nuvvuagittuq Supracrustal Belt) that claims and proves that Northwestern Ontario was the oldest civilization in the world. At the pragmatic level, if you have that number of people coming here, you need to create a lot of work, and the whole process is related to education and land. If we don't watch out, and don't deliver, you push immigrants into cities, into high-rise towers, as renters. At a certain point, the momentum will grow, people are going to rise up and say, "we're not taking this anymore. We're taking over."

X: Why is that massive change inevitable?

JVN: It's the movement and location of this major shift of population. I've been in touch with people like Jay Pitter, a planner and author. She grew up in one of the tall towers in Toronto, and she's been writing about that experience. Her book, which you'd really love, is called *Subdivided: City-Building in an Age of Hyper-Diversity*. Her book is not explicitly about the spatial subdivision of land—although there are important parallels—it's actually about the subdivision of races within places like Regent Park or Jane & Finch (in Toronto) that just aren't being picked up on, understood, or acknowledged. Regent Park's so-called success is all about making a developer a hell of a lot of money for replacing people into smaller units compared to what they had before, and placing them on rent. On the safety and security side, people have no idea what this all means.

X: So is it this colonial interpretation of urbanization that essentially brought you to rethinking ideas of incremental housing that are based also on territory?

JVN: Yes.

X: Is this the result of your very early work, informed by practice in different uncolonized or decolonized parts of the world?

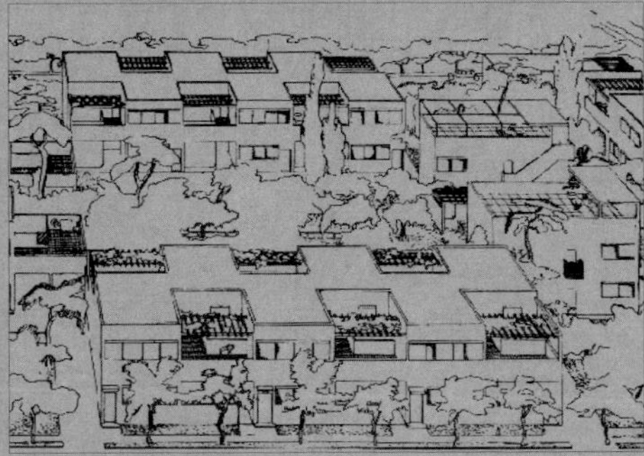

Workers' Housing, Pessac (France) by architect Le Corbusier ca. 1920 (Philippe Boudon, *Pessac de le Corbusier,* 1969)

JVN: Yes, that's why in the past year, I decided to create a development company. I'm really not doing too much work within the spheres of urban planning and design anymore, because I'm actually interested in the design of development. The devil is in the details. Right now, we've bought two sites, and we're developing buildings that lower the costs of purchase.

X: Is this is a form of counter-development?

JVN: We're trying to change how development is done; first of all, they're condos, which I love and hate. I don't understand condo tenure, it seems so sick to me, but it's the norm now. So we're working with the condo format, which, it turns out, was actually invented for at-grade subdivisions as a way of governing or regulating subdivisions. And so, we'd start with a block, for instance in Hamilton; a normal block, a street and lots on the side, and we show how the block has been developed from 1890 until now, and it's always changing. Then we say: if we don't have enough space—now we've run out of those blocks—so we need to build another one above that one, and another one above that one, and another one above that one, and so on. Now, we're building a six-storey building, that holds only the core infrastructure—the structure to hold it up, but also the water and wastewater systems, and so on—just exactly what we did in Africa, to be quite honest, but in three dimensions. That's the whole thing we are at the moment, even the skin of the building is malleable. Basically, you get to choose your lot, which delineates your place within the structure—we don't use shear walls, by the way, which, in effect, facilitates change while maintaining order. You can buy a lot of 225 square feet—which is the size of a building that we have already built at Evangel Hall for the Presbyterian Church of Canada—as a basic, self-contained unit, two of which make a one-bedroom, three of which make a two-bedroom, and four of which make a three-bedroom. And now we're actually even saying to some people that come to us, if they can't find land for example, we say, "we'll sell you a floor." We lower the price for entry to make it work for them, even if it is all on paper so far. We lower the price to the threshold of affordability for a household income of 30,000 dollars per annum.

X: Wow.

JVN: And we've done it by saying that you can actually take the lot then, or we can complete it for you as a finished condo. Or, you just move in and we offer the added little perk of working with you on the process of designing the condo. Or, we give it to you in a legally occupiable condition, but you finish it yourself or we help you finish it. Or, we give you the halfway, whatever the in-between is. And then, once you have at least two lots, you could also rent out one of them. So you could buy two, invest some money into one, fix it up in the part you want to rent. Or you fix up your place, until you get married or for any reason, you decide you want it back. All of this is based on the urban patterns of 1900-1950.

This model is just about the reintroduction of what is happening in the rest of the world. This is what the rest of the world is doing. They're doing it here too, but illegally and behind closed doors; nobody wants to talk about it.

X: Can you talk about your concept of incremental development?

JVN: It's the opposite of master planning. The idea of incremental development goes back to the work of architect Le Corbusier. I don't really talk about that stuff too much, but I've always been very, very interested in one of his early projects, Pessac.

It's Le Corbusier's first built project in a suburb of Bordeaux, Southern France. The architect built 70 houses for, actually, sugar workers. For their small, square shapes they became known in town as the "Sugar Cubes." They included the five principles of modern architecture—the open plan, open façade, raised ground floor, long horizontal windows, open roof area. Sociologist Philippe Boudon was going through the town of Pessac in the 40s, 20 years after the construction, looked at these cubic buildings to find out they were all changed and modified over time. They had shutters and all kinds of decoration. They all looked different. They looked like Bordeaux. Anyway, over two years, he talked to the 72 families. Without having a clue about the five rules of Le Corbusier's architecture, the houses really spoke to them all. They talked about the walls. The walls were not structural so you can make holes in them. They loved the roof terrace because you can rent it out. You could go up at night and look over the city. They loved the fact that the ground floor was open. In the end, the best thing, the residents said, "we can make them our own." Boudon went back to Le Corbusier, who was still alive, and he showed him the project; twenty years later with photographs. Boudon said, "do you realize what's going on with these homes?" And Corb responded with his famous quote, "it's always life that's right and architecture that's wrong."

The Republic of Botswana, 1978

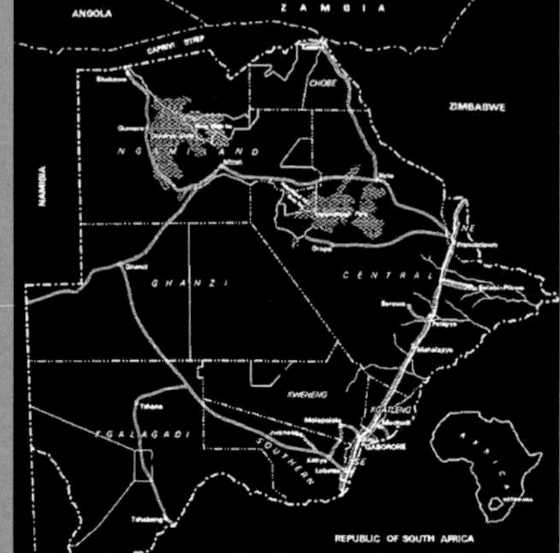

X: What I think is unsettling about what you're mentioning also, is the fact that you're kind of putting into question the whole idea of master planning, and the idea of fixing design permanently down to the structural nature of the walls to, essentially, modifications of the interior as much as the exterior. You're also turning the idea of the vertical condominium skyscraper on its side and reclaiming this relationship through and by land. Then all of a sudden, it grounds urban life and the diversity that it enables. It generates racial diversity, class diversity, bringing then together, associated. Is that fair?

JVN: Yes. This practice coming together slowly, but it was critical being in Africa and acknowledging that life *is* right. If life wasn't right or tied to land, we would really be fucked, because they'd be invading us for our land—so, they've figured out how to live with land.

X: Is this related to one of your first projects in Botswana?

JVN: One thing about the Botswana project was that there were no shared boundaries between properties; no shared walls, and no shared fences. The aim was to make this so-called "squatter settlement"—the first time they'd ever heard the word—a part of the city. And, it was actually the camp originally for the people who built the capital of the country of Botswana, Gabarone. When the British left, and 'gave' Botswana its independence in 1968, Britain said, "we're going to give you a present." The present was a Garden City for the new capital, designed by third or fourth generation British planners. It wasn't the Welwyn Garden City, but something similar, a semicircular town that was meant to become a circle over time, with a railway passing through the middle and a mall placed on the railway station, with government buildings beside it. But, by the time I arrived there ten years later, 25 percent of the population was living in the construction camp of 300 acres, outside, on the edge of town.

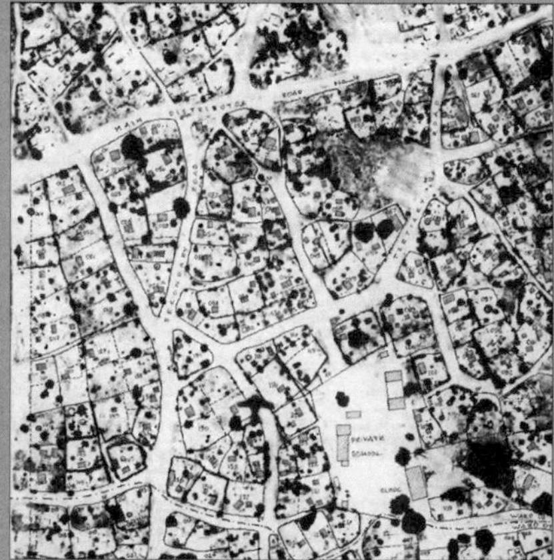

Proposed Plot Layout for Agisanyang, Botswana (John Van Nostrand, *Old Naledi*, 1982)

X: Living space for the service population was unaccounted for?

JVN: Yes, because there was no place for the laborers in the town. I mean, everybody needed them desperately because they did all the hard labor of the city—the gardening, the shops, the taxis, but nobody had built them into the fabric of the town in any way. The project emerged by accident from Botswana President Sir Seretse Khama who

I eventually met. He tried to bulldoze the suburban slums, because in 1978 that's what you did in southern Africa, as far as he knew. You just bulldoze them and they go away. And when he started bulldozing they said, "what's going on here? We just elected you president. Why are you doing this?" And basically he said, "because I don't know what else to do."

Morula Tree as Central Market, Old Naledi (John Van Nostrand, *Old Naledi*, 1982)

X: So the project was about trying to find out what else could you do based on your observations and different ways of life?

JVN: Yes, interestingly the town they were laying out was their version of the villages they came from. It was not disorderly or a slum in any respect. It was actually a copy of the villages that they came from, into the city. They didn't come to stay there, they just came to the capital when they ran out of money, cattle, or crops, to earn a bit of money and go back. They built these temporary settlements using the same patterns as their villages.

There, just by fluke, I ran into an anthropologist, wandering around in one of those villages, and I said, "what are you up to?" He said, "I don't know if you realize this, but the Botswanans are probably the most sophisticated urban people in Africa. *What?* And he took me and showed me these villages, which are the largest in Africa. In fact, ten years later, notable journalist Patrick Martin from the CBC did a documentary on democracies of the world, and he started with Botswana. It was the purest democracy he'd ever been to. And then, here were these planners, copying the Garden City, saying that "these squatter settlements are a mess, 'slums'; put them all out somewhere else where we don't have to look at them."

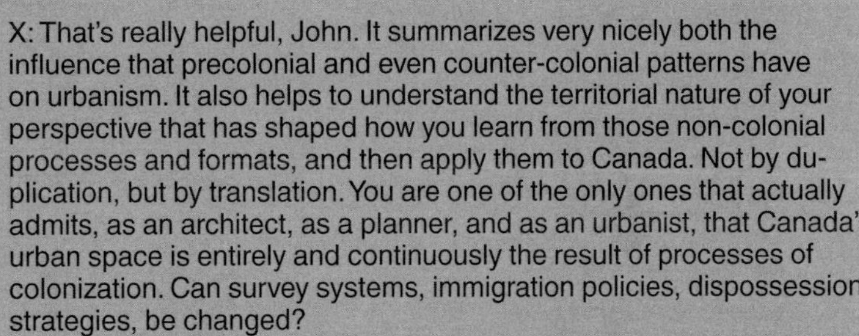

X: That's really helpful, John. It summarizes very nicely both the influence that precolonial and even counter-colonial patterns have on urbanism. It also helps to understand the territorial nature of your perspective that has shaped how you learn from those non-colonial processes and formats, and then apply them to Canada. Not by duplication, but by translation. You are one of the only ones that actually admits, as an architect, as a planner, and as an urbanist, that Canada's urban space is entirely and continuously the result of processes of colonization. Can survey systems, immigration policies, dispossession strategies, be changed?

JVN: Yes, absolutely. And meanwhile, I'm going mad because the architecture of buildings increasingly perpetuate ongoing processes of colonization, class segregation, and spatial injustices. There are many people who think I'm nuts, when I speak about colonization. They think I've spent too much time in the North, when, in fact, the first ones who have observed that colonization is not only oppressive, but that its oppression goes across multiple generations, are the First Nations.

Incremental shed building, Hay River, 2017 (Pierre Bélanger)

X: Is there anyone else who has particularly shaped or influenced your thinking?

JVN: Yes, the young Dënesųłiné communications coordinator Eriel Deranger, from the Athabasca Chipewyan First Nation (ACFN) has been an important voice from the rapidly urbanizing North. She's brilliant. And to think that Indigenous people were not sophisticated is horrendous. They made a place—a territory, out of ways of life. I noticed this when I looked at some very early survey maps but didn't really read them properly. Hugh Brody was also influential. His *Maps and Dreams* was important too. Ed Chamberlin, who taught me English at the University of Toronto when I was there as a student, was important. I took Scandinavian literature with him. He was amazing and it turns out that he was very involved in the Mackenzie Valley. He had written several books, but one of them, the first one was important for me: *Come Back to Me My Language* (1993).

X: Was it about the difference between cultures of hunter-gatherers and agriculturalists?

JVN: Brody's book is called *Hunters, Farmers, and the Shaping of the World*. Chamberlin and Brody both say the same thing, that for the hunter-gatherers, their place was their home, their world. Gods and beliefs all came from there. They buried all their people there. They moved through it and looked after the whole place very intimately. They weren't looking to expand. They had all they needed. The theory is that agriculturalists destroyed it. They were always pushing, pushing, and pushing for more. And it's interesting to read colonial strategies for assimilating and enfranchising Indigenous peoples—the first one was an attempt to make them into farmers.

AN ACT

LAND SURVEYORS

SURVEY OF LANDS.

QUEBEC:
PRINTED BY STEWART DERBISHIRE & GEORGE DESBARATS,

1860.

An Act respecting Land Surveyors and the Survey of Lands, 1859

X: With no recognition of their life as hunter-gatherers?

JVN: Correct. And then Brody put a value on that. By saying that if you put a value on a pound of caribou meat, Indigenous peoples have an enormous economy. We already know that because we're not letting them get the caribou meat, so we're having to subsidize them, or charge them 14 dollars for a pound of chuck steak in their original territory. Utterly bizarre. It's the whole idea of economy and territoriality, so intimately related.

X: It's ironic that the colonial project of settlement was entirely based on agricultural development. That was the image and iconography of colonial settlement. The technology of territorialization was the survey grid for parcels of agricultural development. And now you're identifying that with the lack of an immigration strategy at the moment and a plan divorced from territory and land, it forces people into the service industry of cities. Immigration without land, immigration without territory, in a sense, is a form of enslavement.

JVN: Yes, exactly.

X: Lastly. With your deep profiling of territorialization, does the survey and the grid lie at the origins of expression of the colonial project?

JVN: Look, the main implementing agency of the colonial government in Canada and across the British Empire was the Land Office. When I went to Botswana, it still existed there. It was called the Department of Lands, Surveys, and Physical Planning. That department existed in every British colony. Here it's called the Ministry of National Resources (MNR) now. But it was Lands and Forests, and at one point it was Lands and Surveys, and Lands, Surveys, and Planning. And if you want to go into any planning history before 1900, you don't go to the archives, you go to the MNR records. Because they're all there, all of the towns and everything else.

X: Because it was the Department of Interior?

JVN: Yes, the Department of Lands, Surveys, and Physical Planning was at the district level, under the District Commissioner. In Canada, it is the Premier of the Province. It's like a big district. But they have, just like in Africa, the District Commissioner, who I barely would ever meet in Africa, and always lived outside of town. If there was a District Commissioner of Toronto, their office would probably be in Newmarket. But they would actually control the city and the territory around the city, the hinterland. They had power, while the city, Newmarket itself, had no leverage. Toronto had no power. No city in Canada had any power. It was always the Province. As I've said in interviews and lectures, every fucking drawing I do in the architecture and planning offices does not become law until it goes through the Province.

X: The Provincial Crown?

JVN: Yes, so we go to the Province, through the Ontario Municipal Board, for approvals—zoning, building, planning. Everybody goes there to get the final ruling on development.

X: And the ultimate irony is that it's the district representative of the Federal Crown. It's a subdivision of the colonial project, back to the same old settler-state structure?

JVN: Yes, it is. And so, the *district* is hugely important—the next scale of importance to the British. Surveying and mapping were extraordinarily important. To do it, first of all, you had an excuse to say to an Indigenous nation, or a country, that you're about to take over their land without their knowledge or consent. They're there to help because they're actually going to make these maps, and need help to get around. Then with maps you'll be able to tax people and generate income. But what they were really doing was getting to understand, intimately, the physical, anthropological, sociological makeup of that place. In mapping it, you had to know all those things. You even made names for new places which was part of the colonizing process. And throughout the process, though, they found out how territories were defined by people who were there, and then you could map it. But then, those maps were used for settlement, infrastructure, displacement, and dispossession.

Indigenous Peoples of the Arctic (W.K. Dallmann ©Norwegian Polar Institute)

X: Did other international or transnational experiences shape your thinking about territory?

JVN: I remember going into the office of the Government of Palestine when I was working nearby in Amman, Jordan. Somebody explained land tenure to me about Palestine. The land office would draw up a list and say, here are the territories we've mapped, numbered 1 to 45. We just would like you to sign up in the next 90 days. If you own one of these, which ones do you own. And then on the bottom in small type it says: *If you don't sign up, they revert to us*.

X: Similar to the extinguishment clause?

JVN: Basically, across India, which is one of the worst examples of extinguishment, woke up to the fact that they had lost control of their entire country by the British. The British government did this to people who weren't even allowed to leave England. If you went to India before 1937, you had to speak 3 to 4 dialects, at least, and you had to know the history and the language. If you didn't pass it, you didn't get into the colonial service. You needed to communicate in their language.

X: Is this how Hugh Brody worked?

JVN: Hugh Brody, author of *Maps and Dreams* (1981), learned Cree and 14 dialects. He learned Inuit and Cree for his work in Canada, and he was far ahead of any other Southerner, to even talk about ways of life. Think of the Inuit, who have 35 words for "snow." So, surveying is not a way of imposing order, it's actually about learning intimately the place yourself through community, and its wealth, its resources, and its economy. Drawing the line of the survey was about describing where all those things were. Unfortunately, and disastrously, their lines took resources away from the people who owned them already.

X: Is this because Indigenous society is not considered politically organized, to become subjects of the state?

JVN: Of course. The language of land is so important.

X: How did this relate to the Canadian example?

JVN: There's another perfect example; I didn't even realize it until we started working in Nunavut, that there are three tribal areas. I honestly thought it was just one big blob that they turned from the eastern Arctic into Nunavut, but it's actually three Inuit Nations, with the Nation that occupied Baffin Island, because they weren't First Nations. They were never pinned down as being subject to the Indian Act. In the 1980s, they negotiated the simple ownership of all their traditional lands, which, they had them all mapped out and knew exactly where they were and here they are. They gave the government the map and said, "ok, we're in, but we own all these lands." And now they are the largest territorial landowners in the world.

Just as importantly, they're sitting on the Northwest Passage, which brings us full circle to global pressures and the question of territorial infrastructures. They're running fishing businesses and resource mining—and that was just one big negotiation where they were politically and economically organized. But now the problem is the people who run the administration. It's all run by managers from outside, who are just getting at them in different ways. The last place you would go if you were trying to sort that out would be to the Department of Lands, Surveys, and Physical Planning, or the Ministry of Natural Resources.

X: Wow. I don't know who would ever go there?

JVN: At the end of all this, think about the current entitlement of land. What do you think the proportion of Crown Land is in Canada? If you had to guess, what would you think the portion of the area of Canada is Crown Land?

X: I'm not sure, it's quite large?

JVN: Based on different calculations, one is just under 92 percent owned by the Crown; but it's actually more like 99 percent. How can we call ourselves a country? It's unbelievable. One sixth of the landmass of the world is owned by the Queen of England. It's an incredible, practically unbelievable statistic. Moving into the future, the project of urbanization needs to necessarily be one of decolonization, decoupling the centralized power of the Crown from Indigenous lands. We have to come to grips with the fact that we're literally floating on what appears to be Crown Land. But in fact, we're sitting on top of stolen ground. These lands need to be unsettled. ●

This conversation took place between John Van Nostrand and Pierre Bélanger in 2017.

John Van Nostrand, *Old Naledi*, 1982

REUTERS / Toru Hanai

Confronting Empire

"The political claims that are most urgent in decolonized space are tacitly recognized as coded within the legacy of imperialism: nationhood, constitutionality, citizenship, democracy, socialism, even culturalism. In the historical frame of exploration, colonization, decolonization—what is being *effectively* reclaimed is a series of regulative political concepts, the supposedly authoritative narrative of the production of which was written [and conceived] elsewhere, in the social formations of Western Europe."[1]

—Gayatri Chakravorty Spivak, 1993

f extraction is the process and practice that defines Canada at home, then it is also the policy and dynasty that shapes its image abroad. With over 60 minerals and metals, Canada is the most productive mining nation in the world. Of the nearly 20,000 mining projects worldwide from Africa to Latin America, more than half are Canadian-operated. Not only does the mining economy employ close to 400,000 people in Canada, it contributed $52.6 billion to Canada's GDP in 2012 alone. Globally, more than 75% of the world's mining firms are based in Canada.[2] Importantly, 57% of the world's public mining companies are listed on the TSX—Toronto's Stock Exchange,[3] accounting for 48% of global mining equity transactions in 2013.[4]

Seemingly impossible to conceive, the scale of these statistics naturally extends the state logic of Canada's historical legacy, not as a middle power, but now as a center of power. From Confederation to Commonwealth, plotting its sources and profiling its systems requires new representations of its historical foundations in order to rewrite its past and reproject its future once as province, as nation, and now, as global resource empire.

Originally developed to propel the Victorian expansion of the British Empire, the country's constitutions and institutions are built today on a foundation of primary resource mining and distribution—from explorations to extractions, all the way to exports. As geographic and spatial representations of its political, cultural, and industrial histories, it is the ecological profiling of its manifold processes that open a world wide web of industries, policies, and operations. If, as the political economist Harold Innis argued in 1930, "Canada emerged as a political entity with boundaries largely determined by the fur trade … from the Atlantic to the Pacific [then] the present Dominion emerged not in spite of geography but because of it."[5] As profiled here, it is the spatial and geographic re-representation of the uneven, contested, and complex ecologies of extraction that render so compelling the political, cultural, and industrial histories of resource mining. But fundamentally, at its core, are the subliminal realities of land and land rights that are categorically assumed and fundamentally overlooked. This brief introduction opens and images this vast landscape of extraction from its territorial worlds to its paper worlds, where future lands and future lives will be decided—Indigenous and non-Indigenous.

Innis' geopolitical regionalism stood in stark contrast to the form of cultural continentalism that was promoted early on in the construction of the Canadian Pacific Railway by the Scottish émigré and engineer Sandford Fleming. The royally-appointed secretary George M. Grant captured and crystallized the moral inter-coastal impulse of Canadian imagination in the country's nineteenth-century declaration, *A Mari Usque Ad Mare*, "From Sea to Sea" (*D'un Océan à l'Autre*).[6] Those royal continental sentiments echoed again and again, in nearly every generation thereafter: "If some countries have too much history," as Prime Minister Mackenzie King proposed in 1936,

"Canada has too much geography."[7]

Whereas Innis largely saw geography, and sub-disciplines of geology and topography, as deterministic media of both economic and communicative substance, the royally-appointed surveyors saw no existing myth or material meaning in the surface of terrain or topography—mere obstacles to be conquered, really. Seen as subjects on the surface of the State, Indigenous populations could then be successively displaced and replaced, assimilated as subjects rooted in foundations of British Law, unwillingly tethered as movable demographic obstacles.[8]

Canada's constitution after all, was entirely based in the inheritance of a British legal system that irreversibly guaranteed rights and freedoms for settler-citizens, but resulted in a legal form of encirclement. As a psychological (read cybernetic) system, this form of territorial enclosure also relied on three strategies: the annexation of a few through assimilation, the eradication of many through attrition and extermination, and the displacement of many through segregation, relocation, and concentration. The supreme law of the nation—the early and present Constitution of Canada—was itself based on one of the most significant historical documents in the English-speaking world, the *Magna Carta*:

Bearing Tree
Fig. 1

Department of the Interior, *Description of Boundary Monuments Erected on Surveys of Dominion Lands*, 1917

Live Monument

Bearing tree as basic survey monument of the Dominion Lands Survey for government to sell and settle lands

1763

Government of Saskatchewan, Land Titles Office, 1915

PROVINCE

LAND TITLES O

INFORMATION REGA

USED IN T

DC

The original surveys of lands into Townships and sections in the Province of Manitoba, Saskatchewan and Alberta were made by the Dominion Government under the direction of the Surveyor General of Dominion Lands.

These directions were embodied in Manuals of Survey issued at different times and the monuments used in marking these surveys have been changed in character from time to time, these changes being illustrated in the different editions of the Manual.

All subsequent surveys of these lands under the authority of the various Acts passed by the Provincial Legislature are based on the Original Surveys.

As the exact position in which the original corner was established must in a large number of cases be determined by evidence and as this evidence is in many cases still to be found in the shape of remains of Mounds, Pits, Bearing Trees and other objects which were intended to be a witness to the corner as originally established, it is of the utmost importance that Surveyors be acquainted in every detail with the nature of the monuments used at different times and the periods of time during which the same were used in order that they may intelligently interpret the evidence afforded by whatever remains of these Mounds, Pits, Bearing trees &c when determining therefrom the exact position of a Corner as originally established.

With this in view I have compiled the following information with the hope that it will be directly of use to Surveyors and indirectly bring them to a realization of the necessity of obtaining a thorough knowledge of these details which is necessary to the conduct of a successful survey practice.

The information and illustrations contained herein have been obtained from the Manuals of Survey issued by the Surveyor General of Dominion Lands in the following years, viz 1871, 1881, 1883, 1890 (2nd 4th edit) 1903, 1905, 1910 (?)

Regina 15th April 1915, Land Titles Office — Sgd. E H Phillips, Acting Chief Surveyor

The Surveyor General at Ottawa has been kind enough to proof read this information and through his interest considerable valuable matter has been added which would not otherwise be available.

TOWNSHIP CO

TREE (1871–1881)

In a timbered country should a tree be found at the precise spot indicated it is to be squared and marked for the corner.

STONE (1871–1881)

In a region where stone abounds the township corner will be a single stone planted and marked with a small pyramid of stones beside it. Stones to be referenced with bearing trees where possible.

WOODEN POST (1871–1881)

In a timbered country a wood en post as shown was used. Post to be referenced by one or more bearing trees.

WOODEN POST AND MOUND (1871–1881)

POST AND MOUND ON CORRECTION LINE ON PRAIRIE

In Prairie country Township corners were marked with Post and Mound.

SECTION CO

QUARTER SECTION

WOODEN POST IN MOUND (1871–1881)

WOODEN POST AND M (1881–1887)

WITNESS

WITNESS MOUND (1871–1881)

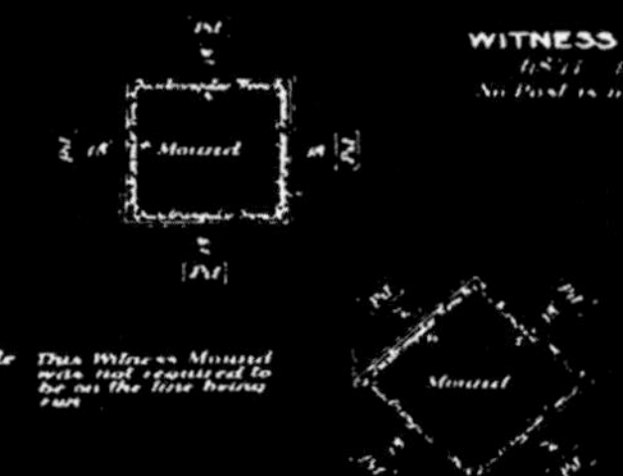

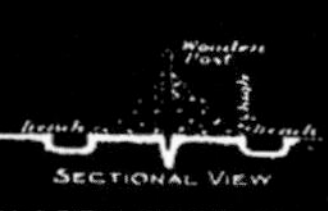

WOODEN POST IN WITNESS MOUND

302

TCHEWAN

SURVEYS BRANCH

SURVEY MONUMENTS

L SURVEYS

NDS

MONUMENTS

MONUMENTS

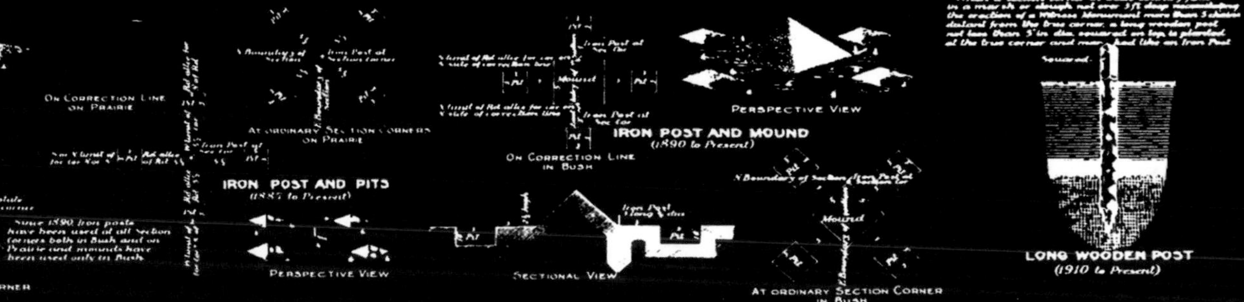

MONUMENTS

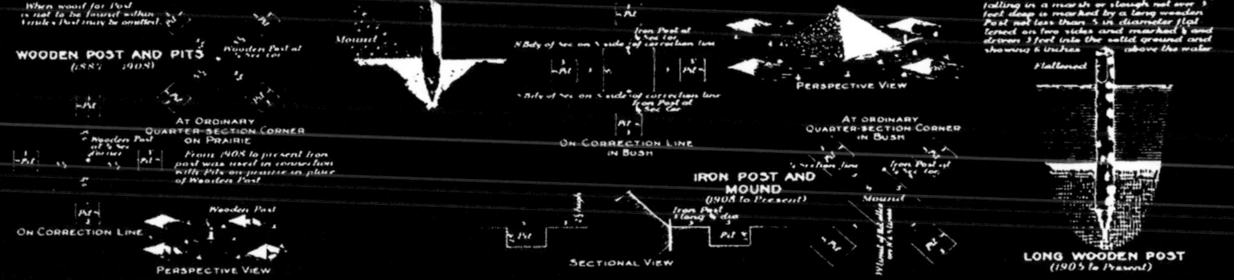

MONUMENTS

WITNESS IRON POST AND TRENCH

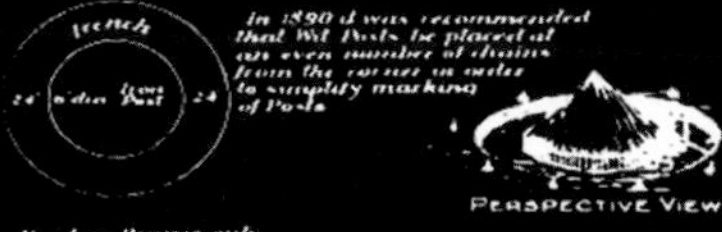

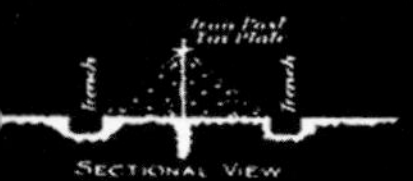

WITNESS IRON POST AND MOUND

Used on Prairie only

WITNESS IRON POST AND MOUND

1903 to Present in Bush

Sectional View

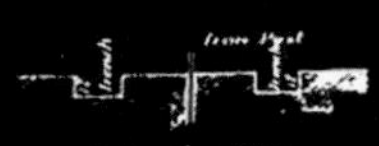

Sectional View

Sectional View

303

The World as Empire

The extents of British colonies throughout the world with a tenfold increase in territorial conquest during the 19th century under the reign of Queen Victoria

1886

> "No free man shall be seized or imprisoned, or stripped of his rights or possessions, or outlawed or exiled, or deprived of his standing in any way, nor will we proceed with force against him, or send others to do so, except by the lawful judgment of his equals or by the law of the land."[9]

From the *Freedom Charter* to the *Forest Charter*, this meta-law of the land rose between 1215 and 1217, underwrote the 1670 *Charter of the Hudson's Bay Company*, annexed with the *1763 Royal Proclamation*, entrenched into Canadian law of the *British North American Act of 1867* (Confederation), then later patriated in 1982—while the Queen, then still the Sovereign, and now, Head of State.[10]

With ground penetrating foresight, surveyors saw below the legally-encircled surface of the State, a vast subsurface of resources to be territorialized (through survey), financialized (through measure), and incorporated (by state) where the Crown served as, and was served to, the supreme and central objective of the royal domain—its dominion.

TERRITORIALIZATION

Tools for Jewels

At the core of the Dominion's power was, and still is, the separation of rights of the surface, from the subsurface; rights to be later known as mineral rights. And it is through the political, technological, and economic decoupling of the ground from the surface, and subsurface, that lies the core of the production of foreign colonial power as force, disempowerment, and dispossession of preexisting Indigenous population as effect.

If the underground served a body politic of extraction, then jewels of the Crown were served up with the tools of Confederation: maps, stats, and acts. Across the vast Canadian Shield—from temperate to tundra climates—the geological projection of Canada as *extractum* served the

Trans-Canadian myth of nation building that stood in for imperial ambitions of British expansionism and Victorian *extractivism*; a mythological project that, albeit realized, remains unfinished to this day, according to scientific historian Suzanne Zeller:

> "Here not palm and pine, but coal and gold—and the tensions induced by their absence—redefined territories claimed and defended by geology's sultans of science. Not only do imperial colonial power structures in the history of science work both ways, they also breed new empires, replaying their cycles of dissemination and domination over and over again."[11]

Through the specialization of mapping, the scientific discipline of geology was a strategic tool and technique of British expansionism. Founded in 1807, the mandate of the (Royal) Geological Society of London was already, by the 1820s, one of collecting and categorizing mineral deposits across the reaches of the Empire;[12] a categorization that implied the erasure of territory in favor of its techno-scientific representation. At a time when it had depleted all of its own energy sources (from forests to other fuels like coal), scientific advancement on the forefront of geological science served the compounded role of imperial acquisition on new spatial frontlines:

> "The conceptual development of British geology was accompanied by an institutional one … the Geological Society was recognized as an active scientific arm of the expanding British Empire."[13]

As this extractive project moved through the middle of the nineteenth century towards Confederation, the influence of the Geological Society registered within the geoscientific community and broadly within society where exploration—extraction's prerequisite—was afforded. If extraction was the ultimate strategy of colonization, then settlement served as the façade of geology, the scientific tool for political territorial acquisition under the veil of nation building:

> "Already during the 1830s, the use of geology to facilitate and rationalize such an inventory of Canadian resources helped to

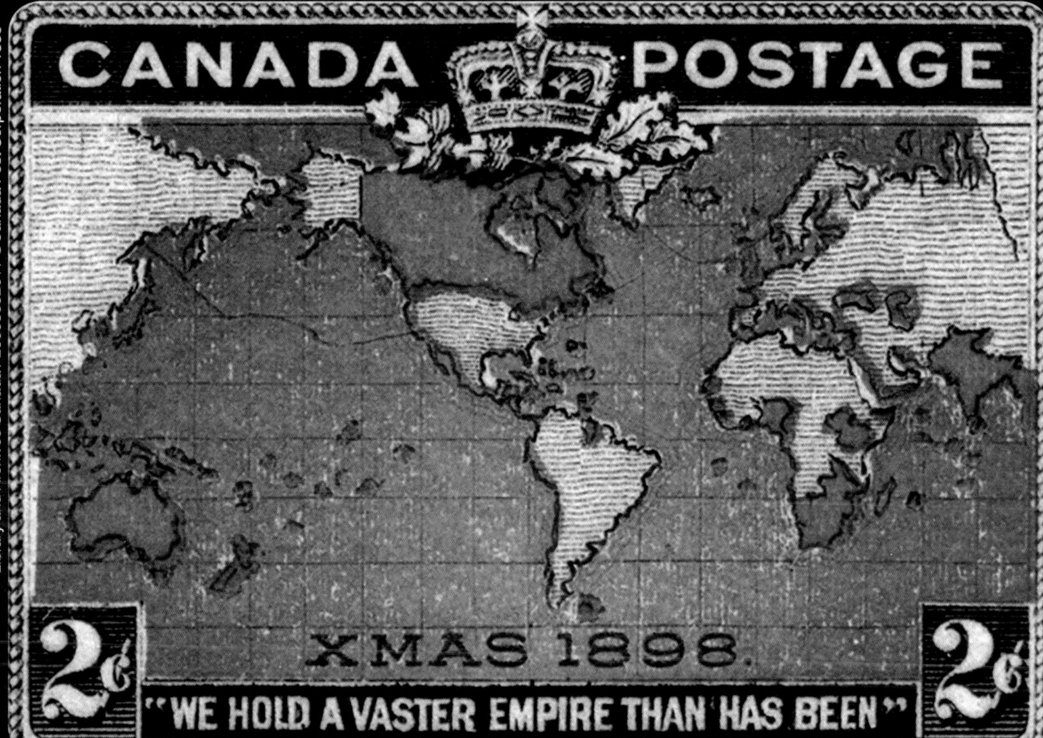

Britain's Bitch

The vast conquest of British Empire using Canada as its mouthpiece for its supreme spoken words, "We Hold a Vaster Empire Than Has Been"

Semiotic Substitute

Sandford Fleming's design for the first ever Canadian Postage Stamp replacing the growing head of Queen Victoria Regina with the slim trim body of the beaver looking west

1851

reshape the science from a systematic method of classification to an ideology forged in the image of British—and especially Scottish—industrial civilization."[14]

Preempting the scramble for Africa with the extractive march of British imperialists such as Cecil Rhodes, the institutions that facilitated colonialism were transformed as geologists devised means to interpret, between map and territory, specific Victorian ideals of the most powerful woman to roam the surface of the Earth—the Sovereign: Queen Victoria Regina.

Precedent to discovery and progenitor to the extractive state, surveying was paramount to territorial demarcation and to the production of *territory* itself. For the Empire, the territory was the map. Informed by and embedded with a litany of monarchist legislations—ranging from the Settlement Act of 1701 to the Royal Land Proclamation of 1763, through the numerous treaties signed into the twentieth century—the imperial process of exploration served as the cloak of settlement strategies in many ways: by encircling, enclosing, and ultimately containing, within carefully defined limits, the radial extents of the Crown. The expanding line of territorial delineation ensured by the early premises of the *Magna Carta*—the law of and underlying the *Land Proclamation*—served to keep *in* what could be acquired and accumulated through enclosure (i.e. resource) and in turn, keep out what was deemed unnecessary through exteriorization (i.e. outsource). Invoking terra nullius—the perceived uninhabited ground—this process of exteriorization and exclusion defined expedient ideologies of extraction, its speed of operations, and ground-penetrating systems of state building by pulverizing the diversity found in its path.[15] From this enclosure, the formation of the notions "homelands" and "frontiers" became pivotal.[16] A new territory is created, an old one abandoned, without condition or consent.[17]

In the depleting circumstances of the home country's sources of wealth within the motherland's own *reserves*, extended exploration of new reserves (through colonization) and development of new *sources* (through imperialism) inadvertently yet irreversibly formed what the seventeenth century produced: the idea of wealth from the etymological invention of *resources* serving to secure power through authority by replenishing origi-

The Geological Society

Everything Under the Earth
Quicquid Sub Terra Est as the subliminal signifier, ideological agenda, and imperial insignia of the Geological Society of London in 19th Century England
1807

Point of Demarcation

Standard rock post of the Canada Lands Survey as the basic marker and point that lies in between territory and map, land and law, people and paper, Crown and culture

1874

nal reserves and original sources of wealth for the homeland. That moving boundary line of the frontier fueled the frictions, distinctions, and growing distances between the center and periphery, bringing map and territory closer together. The frontier dramatized the differences between what was civilized and what was wild (i.e. uncivilized), what was in-law or out-law, what was resource (fish, fur, forest, fauna, farmland, fact, ferrous minerals) and what was not (tribal populations, Aboriginal agreements, seasonal cycles).

Once the military-political complex that bound First Nations and Settlers together were no longer needed, technological nationalism and Indigenous apartheid went into overdrive.

In short, the litigious line between what was inside and what was outside of the Homeland (i.e. the United Kingdom) and what could be extracted, drew a greater divide between what was fixed and unfixed. In other words, the control of resources and all its infrastructures of mobilization formed an epistemology of Empire. Not only did it create and perpetuate metaphors of frontier and mythologies of wilderness, the nationalization of nature through the Victorian mythologization of the North, in the constructed imagination of the Empire, continues today.

While it served many Victorian aspirations, the nationalization of nature however was deeply contradictory and contested. Political geographies and divine mythologies were embedded forms of Christian naturalism and creationism that obscured pre-state ecologies and often obliterated Indigenous pre-histories.[18] If Victorian nature served as an imperial construction and tool of the British State for what stood beyond settlements, but within declared territorial boundaries, then,

> "Nationalism [was] based upon a belief in the moral [and cultural] superiority of the English over the lesser breed of men … The conviction that the English were a chosen people [by God's Destiny], elected to enjoy the fruits of virtue at home and to rule over palm and pine abroad, was peculiar to the Victorians."[19]

The Victorian production of nature and the perceived superiority of the protestant Calvinist work ethic, entailed subjugation of what fell outside notions of the natural—as unnatural (non-British, non-original, non-civilized) or unordered (deformed, abnormal, savage) in order to reinforce and reinstate supremacy of what formed a near-divine pre-destiny of what Nature was in the eye of the British, and especially the Royal Monarchy. If they were denoted as natural, the nationalization of the notion of Nature simply served the assumed endowment of source materials to replenish original stocks—the fuels and feedstocks of Empire. Thus mythologized, the royal connotation of resources placed, positioned, and subjugated nature in service of the British State through what came to be viewed as *natural resources* in *natural exchange* for its self-righteous *civilizing mission*.

The ongoing process of exploration, extraction, and exporting of mineral resources through science and policy was program and protocol of the extractive state leaving in its wake a new territory where the built, urban, and infrastructural artifacts that are required to support the process constitute the hardware of territorialization. Innis, a political economist later turned resource theorist, proposed "the development of mining," but the advancement of extractive ideology (i.e. industry) it fueled, provided a new geological axis between "Precambrian and Cordilleran regions." "Because geology fed and grew upon new data, the science was particularly concerned with exploration," as the Geological Society of London claimed and reflected in their motto *Quicquid Sub Terra Est*, so would the totalizing agenda of Dominion be interested in "whatever is under the Earth."[20] Not only did the construction of the transcontinental railway in the late nineteenth century attempt to tighten this resource regionalism and mineral territorialism of the subsurface, it sealed the Victorian utopia and appearance of Canadian unity from above (the Crown) and from below (the Confederation).[21]

DOMINION & DIVINITY

Development through Displacement

If surveying and exploration were the techniques for imaging and reimagining the *surface* of State, then extraction provided tools for *projecting* and *protecting* the power of State. Supporting and upholding these powers for more than 300 years was the invention of mineral rights, made possible by the separation of surface rights—on the ground, and mineral rights—below the ground. As early as 1763, King George's Royal Proclamation claimed "dominion" over much of North America; no one individual could buy land directly from Aboriginal people—then derisively called "Indians,"[22] the lands could only be sold to the British Monarch, through the Crown. The right to purchase, and thus the question of land and life turned into property of the monarchy, was reserved for the King and his heirs, a rule that has been in place at least since the fourteenth-century Settlement Act of England.[23]

Although the 1763 Proclamation was largely limited, if not futile, in its fight against unjust and unlawful settlement of Native lands, the decision (as if divine) to separate what was on the surface from what was below surface, was exclusive:

> "And We do further declare it to be Our Royal Will and Pleasure, for the present as afore-said, to reserve under our Sovereignty, Protection, and Dominion, for the use of the said Indians, all the Lands and Territories not included within the Limits of Our said Three new Governments, or within the Limits of the Territory granted to the Hudson's Bay Company, as also all the Lands and Territories lying to the Westward of the Sources of the Rivers which fall into the Sea from the West and North West as aforesaid.
>
> And We do hereby strictly forbid, on Pain of our Displeasure, all our loving Subjects from making any Purchases or

Settlements whatever, or taking Possession of any of the Lands above reserved. Without our especial leave and Licence for that Purpose first obtained.

And We do further strictly enjoin and require all Persons whatever who have either willfully or inadvertently seated themselves upon any Lands within the Countries above described. Or upon any other Lands which, not having been ceded to or purchased by Us, are still reserved to the said Indians as aforesaid, forthwith to remove themselves from such Settlements.

And whereas great Frauds and Abuses have been committed in purchasing Lands of the Indians, to the great Prejudice of our Interests. And to the great dissatisfaction of the said Indians: In order, therefore, to prevent such Irregularities for the future, and to the end that the Indians may be convinced of our Justice and determined Resolution to remove all reasonable Cause of Discontent, We do. With the Advice of our Privy Council strictly enjoin and require. that no private Person do presume to make any purchase from the said Indians of any Lands reserved to the said Indians, within those parts of our Colonies where, We have thought proper to allow Settlement: but that. if at any Time any of the Said Indians should be inclined to dispose of the said Lands, the same shall be Purchased only for Us, in our Name, at some public Meeting or Assembly of the said Indians, to be held for that Purpose by the Governor or Commander in Chief of our Colony respectively within which they shall lie: and in case they shall lie within the limits of any Proprietary Government."[24]

Signed on the 7th of Day of October of 1763, in the third year of its Reign, the Proclamation was appropriately undersigned—instituting the rise of the supremacy of the monarchy above divinity:

"GOD SAVE THE KING."

1842 Geological Survey of Canada

1907 Department of Mines

established under Wilfred Laurier

1936 Department of Mines & Resources

includes: Department of Indian Affairs, Department of the Interior

1950 Department of Mines & Technical Surveys

includes: Department of Reconstruction and Supply

1953 Department of Northern Affairs and National Resources

formerly: Department of Resources & Development

1966 Department of Energy, Mines, Resources, & Forestry

1995 Department of Natural Resources

includes: Department of Energy, Mines & Resources, Department of Forestry

2001 Natural Resources Canada

known as NRCan

Colonial Name Game I

The institutional & administrative histories of the nationalization of nature and of the naturalization of resources 1842–Present

1755 British Indian Department
established as a branch of the military

1830 British Indian Department
becomes independent from the military

1860 Commissioner of Indian Affairs
Crown Lands Department responsible for Indian Affairs

1867 Superintendent General of Indian Affairs
position established following Confederation

1880 Minister of Interior
holds position of Superintendent General of Indian Affairs

1936 Minister of Mines & Resources
becomes responsible for Indian Affairs

1950 Minister of Citizenship & Immigration
becomes responsible for Indian Affairs

1966 Department of Indian Affairs
and Northern Development

2011 Aboriginal Affairs &
Northern Development Canada

2015 Indigenous and Northern Affairs Canada

Colonial Name Game II
The strategic re-naming and re-branding
of the Department of Indian Affairs
1755–Present

By the KING,

A PROCLAMATION.

GEORGE R.

WHEREAS We have taken into Our Royal Consideration the extensive and valuable Acquisitions in *America*, secured to Our Crown by the late Definitive Treaty of Peace, concluded at *Paris* the Tenth Day of *February* last; and being desirous, that all Our loving Subjects, as well of Our Kingdoms as of Our Colonies in *America*, may avail themselves, with all convenient Speed, of the great Benefits and Advantages which must accrue therefrom to their Commerce, Manufactures, and Navigation; We have thought fit, with the Advice of Our Privy Council, to issue this Our Royal Proclamation, hereby to publish and declare to all Our loving Subjects, that We have, with the Advice of Our said Privy Council, granted Our Letters Patent under Our Great Seal of *Great Britain*, to erect within the Countries and Islands ceded and confirmed to Us by the said Treaty, Four distinct and separate Governments, stiled and called by the Names of *Quebec*, *East Florida*, *West Florida*, and *Grenada*, and limited and bounded as follows; viz.

First. The Government of *Quebec*, bounded on the *Labrador* Coast by the River *St. John*, and from thence by a Line drawn from the Head of that River through the Lake *St. John* to the South End of the Lake nigh *Pissin*; from whence the said Line crossing the River *St. Lawrence* and the Lake *Champlain* in Forty five Degrees of North Latitude, passes along the High Lands which divide the Rivers that empty themselves into the said River *St. Lawrence*, from those which fall into the Sea; and also along the North Coast of the *Baye des Chaleurs*, and the Coast of the Gulph of *St. Lawrence* to Cape *Rosieres*, and from thence crossing the Mouth of the River *St. Lawrence* by the West End of the Island of *Anticosti*, terminates at the aforesaid River of *St. John*.

Secondly. The Government of *East Florida*, bounded to the Westward by the Gulph of *Mexico*, and the *Apalachicola* River; to the Northward, by a Line drawn from that Part of the said River where the *Chatahouchee* and *Flint* Rivers meet, to the Source of *St. Mary's* River, and by the Course of the said River to the *Atlantick* Ocean; and to the Eastward and Southward, by the *Atlantick* Ocean, and the Gulph of *Florida*, including all Islands within Six Leagues of the Sea Coast.

Thirdly. The Government of *West Florida*, bounded to the Southward by the Gulph of *Mexico*, including all Islands within Six Leagues of the Coast from the River *Apalachicola* to Lake *Pontchartrain*; to the Westward by the said Lake, the Lake *Mauripas*, and the River *Mississippi*; to the Northward, by a Line drawn due East from that Part of the River *Mississippi* which lies in Thirty one Degrees North Latitude, to the River *Apalachicola* or *Chatahouchee*; and to the Eastward by the said River.

Fourthly. The Government of *Grenada*, comprehending the Island of that Name, together with the *Grenadines*, and the Islands of *Dominico*, *St. Vincents*, and *Tobago*.

And, to the End that the open and free Fishery of Our Subjects may be extended to and carried on upon the Coast of *Labrador* and the adjacent Islands, We have thought fit, with the Advice of Our said Privy Council, to put all that Coast, from the River *St. John's* to *Hudson's Streights*, together with the Islands of *Anticosti* and *Madelaine*, and all other smaller Islands lying upon the said Coast, under the Care and Inspection of Our Governor of *Newfoundland*.

We have also, with the Advice of Our Privy Council, thought fit to annex the Islands of *St. John's*, and *Cape Breton* or *Isle Royale*, with the lesser Islands adjacent thereto, to Our Government of *Nova Scotia*.

We have also, with the Advice of Our Privy Council aforesaid, annexed to Our Province of *Georgia* all the Lands lying between the Rivers *Attamaha* and *St. Mary's*.

And whereas it will greatly contribute to the speedy settling Our said new Governments, that Our loving Subjects should be informed of Our Paternal Care for the Security of the Liberties and Properties of those who are and shall become Inhabitants thereof; We have thought fit to publish and declare, by this Our Proclamation, that We have, in the Letters Patent under Our Great Seal of *Great Britain*, by which the said Governments are constituted, given express Power and Direction to Our Governors of Our said Colonies respectively, that so soon as the State and Circumstances of the said Colonies will admit thereof, they shall, with the Advice and Consent of the Members of Our Council, summon and call General Assemblies within the said Governments respectively, in such Manner and Form as is used and directed in those Colonies and Provinces in *America*, which are under Our immediate Government; and We have also given Power to the said Governors, with the Consent of Our said Councils, and the Representatives of the People, so to be summoned as aforesaid, to make, constitute, and ordain Laws, Statutes, and Ordinances for the Publick Peace, Welfare, and Good Government of Our said Colonies, and of the People and Inhabitants thereof, as near as may be agreeable to the Laws of *England*, and under such Regulations and Restrictions as are used in other Colonies: And in the mean Time, and until such Assemblies can be called as aforesaid, all Persons inhabiting in, or resorting to Our said Colonies, may confide in Our Royal Protection for the Enjoyment of the Benefit of the Laws of Our Realm of *England*; for which Purpose, We have given Power under Our Great Seal to the Governors of Our said Colonies respectively, to erect and constitute, with the Advice of Our said Councils respectively, Courts of Judicature and Publick Justice, within Our said Colonies, for the hearing and determining all Causes, as well Criminal as Civil, according to Law and Equity, and as near as may be agreeable to the Laws of *England*, with Liberty to all Persons who may think themselves aggrieved by the Sentences of such Courts, in all Civil Cases, to appeal, under the usual Limitations and Restrictions, to Us in Our Privy Council.

We have also thought fit, with the Advice of Our Privy Council as aforesaid, to give unto the Governors and Councils of Our said Three New Colonies upon the Continent, full Power and Authority to settle and agree with the Inhabitants of Our said New Colonies, or with any other Persons who shall resort thereto, for such Lands, Tenements, and Hereditaments, as are now, or hereafter shall be in Our Power to dispose of, and them to grant to any such Person or Persons, upon such Terms, and under such moderate Quit-Rents, Services, and Acknowledgements as have been appointed and settled in Our other Colonies, and under such other Conditions as shall appear to Us to be necessary and expedient for the Advantage of the Grantees, and the Improvement and Settlement of Our said Colonies.

And whereas We are desirous, upon all Occasions, to testify Our Royal Sense and Approbation of the Conduct and Bravery of the Officers and Soldiers of Our Armies, and to reward the same, We do hereby command and impower Our Governors of Our said Three New Colonies, and all other Our Governors of Our several Provinces on the Continent of *North America*, to grant, without Fee or Reward, to such Reduced Officers as have served in *North America* during the late War, and to such Private Soldiers as have been or shall be disbanded in *America*, and are [illegible]ually residing there, and shall personally apply for the same, the following Quantities of Lands, subject at the Expiration of Ten Years to the same Quit-Rents as other Lands are subject to in the Province within which they are granted, as also subject to the same Conditions of Cultivation and Improvement; viz.

To every Person having the Rank of a Field Officer, Five thousand Acres.—To every Captain, Three thousand Acres[illegible]To every Subaltern or [illegible] Officer, Two thousand Acres.—To every Non-Commission Officer, Two hundred A[illegible]—To every Private Man, Fifty Acres.

We do lik[illegible] authorize and require the Governors and Commanders in Chief of all Our said Colonies upon the Continent [illegible] *North America*, to grant the like Quantities of Land, and upon the same Conditions, to such Reduced Officers of Our Navy, of like Rank, as served on Board Our Ships of War in *North America* at the Times of the Reduction of *Louisbourg* and *Quebec* in the late War, and who shall personally apply to Our repective Governors for such Grants.

And whereas it is just and reasonable, and essential to Our Interest and the Security of Our Colonies, that the several Nations or Tribes of *Indians*, with whom We are connected, and who live under Our Protection, should not be molested or disturbed in the Possession of such Parts of Our Dominions and Territories as, not having been ceded to, or purchased by Us, are reserved to them, or any of them, as their Hunting Grounds; We do therefore, with the Advice of Our Privy Council, declare it to be Our Royal Will and Pleasure, that no Governor or Commander in Chief in any of Our Colonies of *Quebec*, *East Florida*, or *West Florida*, do presume, upon any Pretence whatever, to grant Warrants of Survey, or pass any Patents for Lands beyond the Bounds of their respective Governments, as described in their Commissions; as also, that no Governor or Commander in Chief in any of Our other Colonies or Plantations in *America*, do presume, for the present, and until Our further Pleasure be known, to grant Warrants of Survey, or pass Patents for any Lands beyond the Heads or Sources of any of the Rivers which fall into the *Atlantick* Ocean from the West and North West, or upon any Lands whatever, which, not having been ceded to, or purchased by Us as aforesaid, are reserved to the said *Indians*, or any of them.

And We do further declare it to be Our Royal Will and Pleasure, for the present as aforesaid, to reserve under Our Sovereignty, Protection, and Dominion, for the Use of the said *Indians*, all the Lands and Territories not included within the Limits of Our said Three New Governments, or within the Limits of the Territory granted to the *Hudson's Bay* Company, as also all the Lands and Territories lying to the Westward of the Sources of the Rivers which fall into the Sea from the West and North West, as aforesaid; and We do hereby strictly forbid, on Pain of Our Displeasure, all Our loving Subjects from making any Purchases or Settlements whatever, or taking Possession of any of the Lands above reserved, without Our especial Leave and Licence for that Purpose first obtained.

And We do further strictly enjoin and require all Persons whatever, who have either wilfully or inadvertently seated themselves upon any Lands within the Countries above described, or upon any other Lands, which, not having been ceded to, or purchased by Us, are still reserved to the said *Indians* as aforesaid, forthwith to remove themselves from such Settlements.

And whereas great Frauds and Abuses have been committed in the purchasing Lands of the *Indians*, to the great Prejudice of Our Interests, and to the great Dissatisfaction of the said *Indians*; in order therefore to prevent such Irregularities for the future, and to the End that the *Indians* may be convinced of Our Justice, and determined Resolution to remove all reasonable Cause of Discontent, We do, with the Advice of Our Privy Council, strictly enjoin and re[illegible]ire, that no private Person do presume to make any Purchase from the said *Indians* of any Lands reserved to t[illegible] [illegible]id *Indians*, within those Parts of Our Colonies where We have thought proper to allow Settlement; but that if, at any Time, any of the said *Indians* should be inclined to dispose of the said Lands, the same shall be purchased only for Us, in Our Name, at some Publick Meeting or Assembly of the said *Indians* to be held for that Purpose by the Governor or Commander in Chief of Our Colonies respectively, within which they shall lie; and in case they shall lie within the Limits of any Proprietary Government, they shall be purchased only for the Use and in the Name of such Proprietaries, conformable to such Directions and Instructions as We or they shall think proper to give for that Purpose: And We do, by the Advice of Our Privy Council, declare and enjoin, that the Trade with the said *Indians* shall be free and open to all Our Subjects whatever; provided that every Person, who may incline to trade with the said *Indians*, do take out a Licence for carrying on such Trade from the Governor or Commander in Chief of any of Our Colonies respectively, where such Person shall reside; and also give Security to observe such Regulations as We shall at any Time think fit, by Ourselves or by Our Commissaries to be appointed for this Purpose, to direct and appoint for the Benefit of the said Trade; and We do hereby authorize, enjoin, and require the Governors and Commanders in Chief of all Our Colonies respectively, as well Those under Our immediate Government as Those under the Government and Direction of Proprietaries, to grant such Licences without Fee or Reward, taking especial Care to insert therein a Condition, that such Licence shall be void, and the Security forfeited, in case the Person, to whom the same is granted, shall refuse or neglect to observe such Regulations as We shall think proper to prescribe as aforesaid.

And We do further expresly enjoin and require all Officers whatever, as well Military as Those employed in the Management and Direction of *Indian* Affairs within the Territories reserved as aforesaid for the Use of the said *Indians*, to seize and apprehend all Persons whatever, who, standing charged with Treasons, Misprisions of Treason, Murders, or other Felonies or Misdemeanors, shall fly from Justice, and take Refuge in the said Territory, and to send them under a proper Guard to the Colony where the Crime was committed of which they stand accused, in order to take their Tryal for the same.

Given at Our Court at *Saint James's*, the Seventh Day of *October*, One thousand seven hundred and sixty three, in the Third Year of Our Reign.

GOD save the KING.

LONDON:

Printed by *Mark Baskett*, Printer to the King's most Excellent Majesty; and by the Assigns of *Robert Baskett*. 1763.

King Kong

The royal territorial proclamation of King George III, after the global Seven Years' War, prior to the Treaty of Paris

1763

That sovereign right to land—a patriarchal duty and hegemonic entitlement—was akin to the divine right of (to) territorial acquisition in the name of God: "Christians have an obligation, a mandate, a commission, a holy responsibility to reclaim the land for Jesus Christ—to have dominion in the civil structures, just as in every other aspect of life and godliness."[25]

That doctrine, the Doctrine of the Divinity of Kings, was an assumed model and unquestionable mainstay of monarchy, "aimed at instilling obedience by explaining why all social ranks were religiously and morally obliged to obey their government." It was the foundation of future doctrines of discovery and colonization. Although consequences of disobedience and un-Christian behavior would result in the "public spectacle of execution" (burning, decapitation, hanging, drawing, quartering), "the main way of instilling obedience, however, was propaganda."[26]

Crystallized by the motto of the monarch of the United Kingdom—the Sovereign, "Dieu et mon Droit," when King of England Richard I (*Richard Coeur de Lion* never spoke English) employed the expression in 1198 as territorial passcode, and later adopted by Henry V.[27]

As a declaration of Sovereignty—*God Save the King*, or today, *God Save the Queen*, the motto established a political-religious bond applicable in either the right of passage or a rule of law, between no one else but the Monarch—the King and God, and God alone. Christendom essentially defined this divine right—the *Corpus Christianorum*, delineated in the Middle Ages "as an imagined present and future for a world united by its beliefs and aspirations," a belief that "emerged along with the fall of the Roman empire."[28]

What captures this celestial presence of the monarchy is the orientation of the world, whose center on Jerusalem at the source of the Mediterranean and center of the world in the Middle Ages placed Europe to the left, Africa to the right, and Asia to the top. The East, as the True North, where the sun rose, where Gods lay and lived, and sometimes even Kings (think of Louis XIV, the Sun King). Not only did this world-view provide justification for imperial alignments and divine parallels, it placed the king of the queen in a higher position with the Gods, side-by-side with divinity, above the people and the populace. More importantly, this sovereign celestial alignment, up in the sky, placed its power above land.[29]

Befittingly, to prevent anarchy during times of civil war, royal philosopher and essayist of the seventeenth century Thomas Hobbes, embodied this elevated position and alignment of the monarchy in his 1615 *Leviathan*—the embodiment of the all-encompassing power of the State—that is "The Matter, Forme, and Power of a Common-wealth Ecclesiasticall and Civill ,"[30] with an enduring image, "a crowned giant emerging from the landscape, clutching a sword (a symbol of earthly power) and a crosier (a symbol of church power) … Facing the colossus [away from the viewer], three hundred humans compose the corpus, showing how they are represented by their contracted leader, who draws his strength from their collective agreement"—a veritable body politik. Below the giant, is the quote from the Book of Job: "Non est potestas Super Terram quae Comparetur ei" (There is no power on earth to be compared with him), linking the king, the biblical monster, and god, as presiding above society at large. The security of the monarchy, as it is today, equal, akin, and above the celestial body of divinity, far beyond the people or the populated ground, would irreversibly seal its supremacy as it still does today, in contemporary imagination.[31]

As settlement moved westward, this royal geo-logic oriented the nation's gaze in that direction and reflected back eastward. But that gaze was not oriented towards the political navel of the country, between Lower Canada (Quebec, and the French) and Upper Canada (Ontario, and the English) but towards the center of control in England. Aiming chiefly at the original power center, as Emily Carr proposed, empire was iterated by a return glance towards the Motherland, a cultural bow to the Crown[32]:

> "Nearly all the people in Victoria were English and smiled at how they tried to be more English than the English themselves, just to prove to themselves and the world how loyal they were being to the Old Land."[33]

Oriens

MARE OCEANUM

ASIA
Sem

Mare magnum sive

mediterraneum

EUROPA
Iafeth

AFRICA
Cham

Meridies

Occidens

LEVIATHAN
Or
THE MATTER, FORME
and POWER of A COMMON-
WEALTH ECCLESIASTICALL
and CIVIL.
By THOMAS HOBBES
of MALMESBURY.
London
Printed for Andrew Crooke
1651.

Government of the United Kingdom

Doctrine of Discovery
Motto of the British Monarchy, *Dieu et Mon Droit*,
(between King and God) to explore, discover, conquer
1785

Tom Thomson, *The West Wind*, 1917

Exclusive Natures

The explorative, exploitive, abstract, and depoliticized rendering of virgin wilderness by the Group of Seven crafted a national identity of nature with broad strokes, void of Indigenous presence

1917

When Confederation came a century after the British North American Act in 1867, the powers over the Dominion were securitized by Victoria Regina—Queen Victoria herself, at just 30 years of age:

> "We, therefore, by and with the advice of Our Privy council, have thought fit to issue this Our Royal Proclamation, and, We do ordain, declare, and command that on and after the First day of July, One Thousands Eight Hundred and Sixty-seven, the Provinces of Canada, Nova Scotia, and New Brunswick, shall form and be One Dominion, under the name of CANADA,
>
> GOD SAVE THE QUEEN."[34]

While Victorian geologists mapped the mineral regions and rock resources, the Victorian *joie-de-vivre* of newly minted Western settlers then reaffirmed the political bedrock of the Crown. If "you could pretend you weren't where you were" as essayist John Ralston Saul proposes, then the psychological separation of the imperial state from colonial states as evidenced in the reverence for the far distant motherland from *British Columbia* naturally followed, and flowed from the unhindered amalgamation of political provinces from above and the uncontested legislative separation of surface and subsurface mineral rights, from the side.[35]

Easily confused and conflated with Crown Land today, little if anything has changed in this legislative landscape that supports the reality of the Dominion since the enactment of Confederation, save for a few minor land acts built into previous declarations.[36] As it were, over 90% of Canada today is titled, as it has been, as Crown Land. While the notion of *public land* in Canada is widely upheld and understood today as meaning open and accessible to all, that widespread Canadian concept is not only erroneous and ironically naïve, but rife with contradictory and discriminatory meaning.

Save for a few meters of municipal utilities today, the subsurface of the 10-million square-kilometer nation is practically, for all intents and purposes, entirely owned and regulated by the Crown. That power is extended by the provinces, the state's arms. Like the complex layering and geometric entanglement of crosses and fields on the flags of the United Kingdom since the twelfth century—from St. Andrew to St. George—the idea of dominion is inscribed within the righteous lines traced across the Kingdom's land. Masking the separation between what is above from what is below, these lines mirror the symbolic Saints memorialized in the flags of England—once known itself as the *Land of Angles*. Like the dignified complexity of warrior and martyr, lines both connect and contain as much as they divide and exclude. Indicative of the contradictions of empire, and that of the deeper running history of Canada as extractive and ideological its historical ecologies are both unified, partitioned, segmented, and simultaneously entangled.

CROWN VS. CORPORATION

Upholding the power of the Crown is therefore the decoupling of the power inherently embedded in the inseparability and complexity of surface ecologies and subsurface environments—the complex interface of the ground recognized as *land*. Yet, the roots of the historic legacy of the Crown's influence on the configuration of the country, runs deeper than the Confederation that created the nation in 1867 through the British North America Act, and even deeper than the Royal Proclamation of 1763. In fact, the country existed, and more specifically, the territory operated for well over two centuries prior to its royal confederation and national constitution, as a vibrant corporation: the Hudson's Bay Company.[37]

Merely one hundred and fifty years old, the Confederation of what the Nation and the Proclamation of the Dominion as it is today, was vastly preceded by the incarnation of another previous empire. For well over two centuries prior to its official incarnation as a single dominion—far more than its current existence as a so-called modern State—the country, it seems, existed with an entirely different form and function. With over 600 outposts distributed across is entire superficies—its acronym transliterated

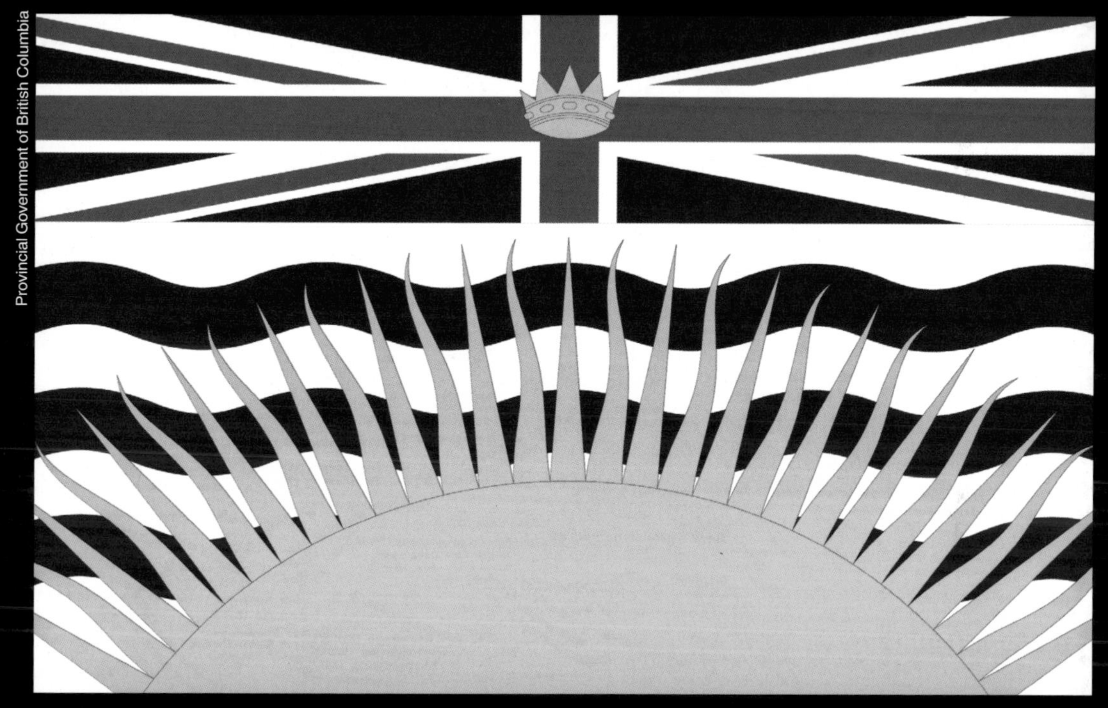

Provincial Government of British Columbia

Crowning Achievement
The annexation of the Colony of British Columbia as territorial linchpin for Queen Victoria's plan for Confederation whose motto shines forever, "Splendour without Diminishment"
1960

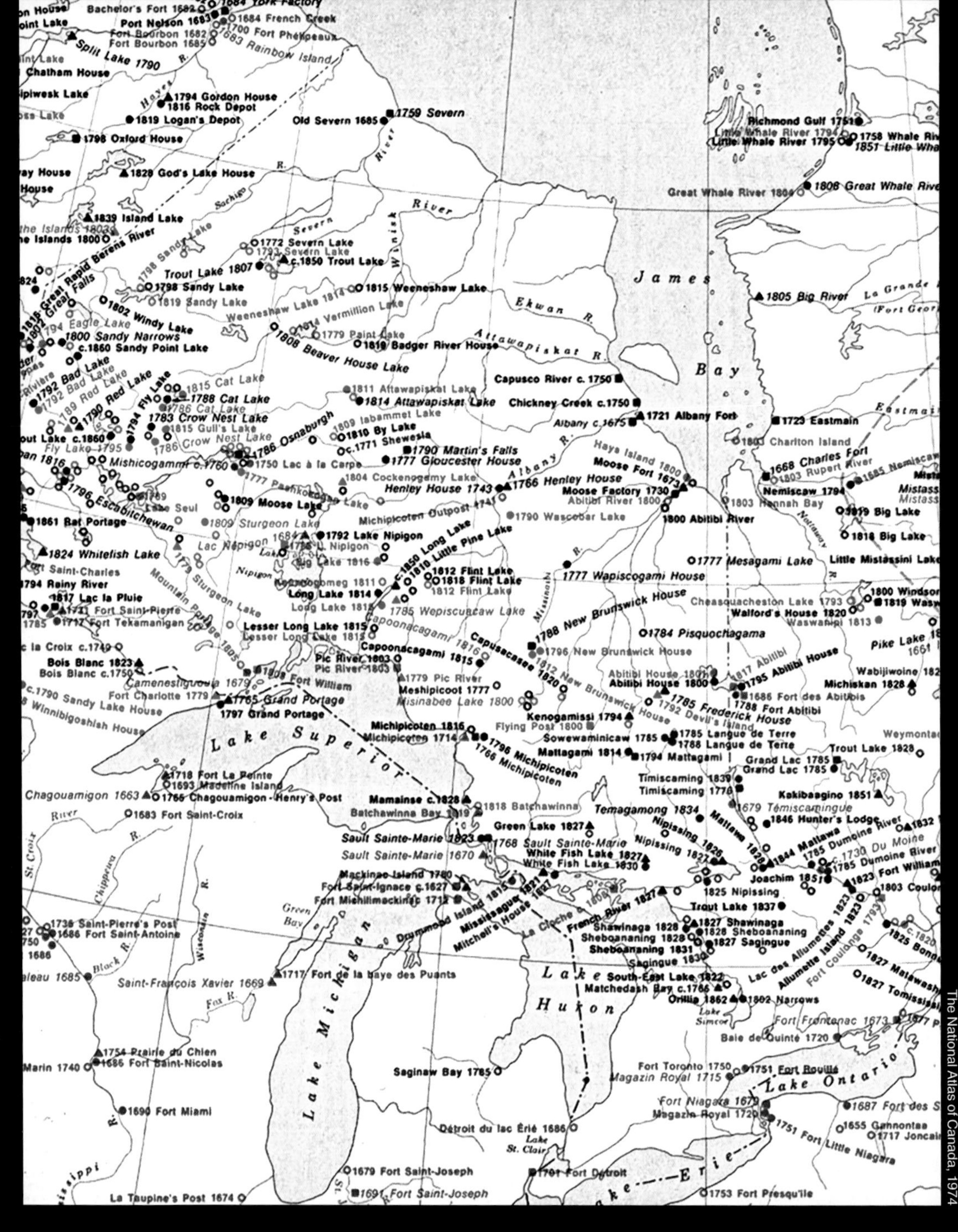

The National Atlas of Canada, 1974

Trap-Line Nation
Map of the hundreds of outposts of the Hudson's Bay Company
1670–1870

"Here Before Christ,"[38] the Hudson's Bay drainage basin served as its own infrastructure. Its territory was its diagram, and the company, its coat-of-arms.[39]

Like *terra nullius*, the country was delineated as land by incorporation as early as 1670 in King James II's Royal Charter, in honor of (Belgian) cousin Prince Rupert. Lodged in between the lines of King James II's words, the origins of these powers found themselves in the rights relinquished across 3 million square kilometers of the watershed that surrounded Hudson's Bay—the perfect transportation system to access beaver laden forests and Indigenous trap-lines.

Across this vast territory (nearly two-thirds of modern Canada today), two of the world's most important woodsman-cum-tradesman—the now famed coureurs-des-bois Pierre-Esprit Radisson and Médard des Groseilliers—would naturally lead to the formation of the "oldest continuously running corporation in the world," the Hudson's Bay Company.[40] Its Charter was not only a joint-stock, commercial, and corporate endeavor, but the map of a fluid, hydrological enterprise whose very constitution—the *x-factor of land*—depended on the navigability of its land and the fluidity of its territorial waters:

> "We have given, granted and confirmed, and by these Presents, for Us, Our Heirs and Successors, Do give, grant, and confirm, unto the said Governor and Company, and their Successors, the sole Trade and Commerce of all those Seas, Streights, Bays, Rivers, Lakes, Creeks, and Sounds, ... with the Fishing of all Sorts of Fish, Whales, Sturgeons, and all other Royal Fishes, in the Seas, Bays, Inlets, and Rivers within the Premisses, and the Fish therein taken, together with the Royalty of the Sea upon the Coasts within the Limits aforesaid, and all Mines Royal, as well discovered as not discovered, of Gold, Silver, Gems, and precious Stones, to be found or discovered within the Territories, Limits, and Places aforesaid, and that the said Land be from henceforth reckoned and reputed as one of our Plantations or Colonies in America, called Rupert's Land."[41]

E.A. Hegg-Cantwell / Library and Archives Canada / C-005142

Rush
Miners and Prospectors climb the Chilkoot Trail during the Klondike Gold Rush
1897

As geopolitical fuel, fur was the original gold that fashioned the building and the northwestern expansion of British Empire:

> "Friend, once 'twas Fame that led thee forth
>
> To brave the Tropic Heat, the Frozen North,
>
> Late it was Gold, then Beauty was the Spur;
>
> But now our Gallants venture but for Fur."

Unlocked by the territorial Charter of 1670, Rupert's Land proclaimed the lending of land while ensuring an underlying vision of pelt trade which stands in stark opposition to mineral extraction, where the drainage basin formed the basis, not the bottom, of empire and the building of a new hydrological civilization, whose territory was formed by its tributaries:

> "AND WHEREAS the said Undertakers, for their further Encouragement in the said Design, have humbly besought Us to incorporate them, and grant unto them, and their Successors, the sole Trade and Commerce of all those Seas, Streights, Bays, Rivers, Lakes, Creeks, and Sounds, in whatsoever Latitude they shall be, that lie within the entrance of the Streights commonly called Hudson's Streights, together with all the Lands, Countries and Territories, upon the Coasts and Confines of the Seas, Streights, Bays, Lakes, Rivers, Creeks and Sounds, aforesaid, which are not now actually possessed by any of our Subjects, or by the Subjects of any other Christian Prince or State."[43]

The incorporation of territory for well over two hundred years was premised on the leasing of land and the rights of passages. Under the auspices of trade, and arguably that of capitalism through incorporation, the originally Indigenous, pre-Canadian, or French movement of Radisson and Groseillers across the waters of the Hudson's Bay drainage basin and beyond that formed the Indigenous and Métis world—from the Red River, to the Athabasca, to the Rupert River—were irreversibly locked out of the imminent inter-coastal and continental ideology of the empire's railway. The very

Edward Wright

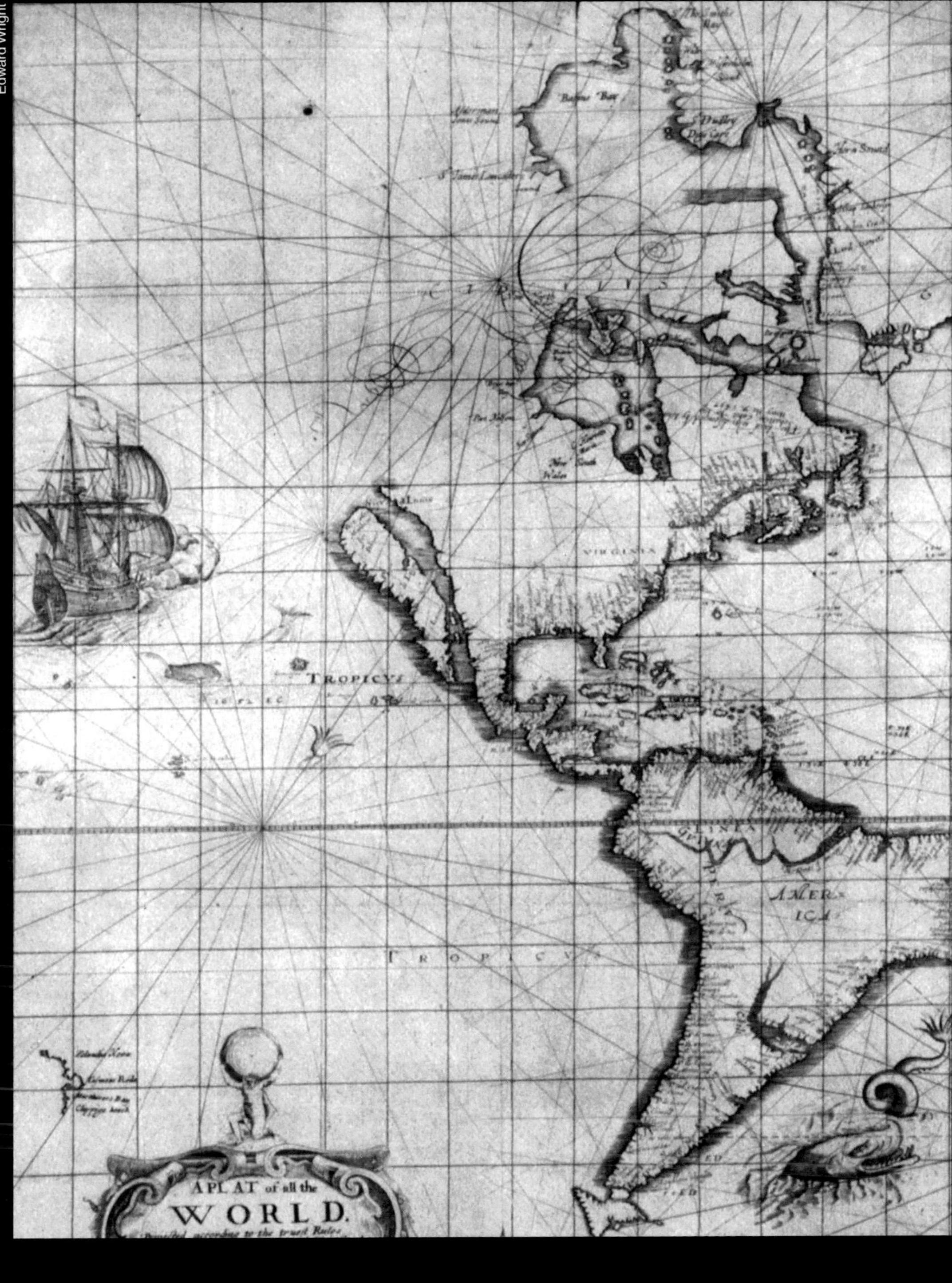

Terra Nullius

The early depiction of the ‘undiscovered land’ as white space for discovery and exploration in Wright-Molyneux’s Map of the World

Hudson's Bay Company Archives / Archives of Manitoba

Territorial Monopolization

The 8-foot scroll of the Hudson's Bay Company Charter incorporating nearly two-thirds of North America in a rush to dominate the beaver fur trade

1670

BY THE QUEEN!

A PROCLAMATION

For Uniting the Provinces of Canada, Nova Scotia, and New Brunswick, into one Dominion, under the name of CANADA.

VICTORIA R.

WHEREAS by an Act of Parliament, passed on the Twenty-ninth day of March, One Thousand Eight Hundred and Sixty-seven, in the Thirtieth year of Our reign, intituled, "An Act for the Union of Canada, Nova Scotia, and New Brunswick, and the Government thereof, and for purposes connected therewith," after divers recitals it is enacted that "it shall "be lawful for the Queen, by and with the advice of Her Majesty's "Most Honorable Privy Council, to declare, by Proclamation, that "on and after a day therein appointed, not being more than six "months after the passing of this Act, the Provinces of Canada, "Nova Scotia, and New Brunswick, shall form and be One Domi- "nion under the name of Canada, and on and after that day those "Three Provinces shall form and be One Dominion under that "Name accordingly;" and it is thereby further enacted, that "Such Persons shall be first summoned to the Senate as the Queen "by Warrant, under Her Majesty's Royal Sign Manual, thinks fit "to approve, and their Names shall be inserted in the Queen's "Proclamation of Union:"

We, therefore, by and with the advice of Our Privy Council, have thought fit to issue this Our Royal Proclamation, and We do ordain, declare, and command that on and after the First day of July, One Thousand Eight Hundred and Sixty-seven, the Provinces of Canada, Nova Scotia, and New Brunswick, shall form and be One Dominion, under the name of CANADA.

And we do further ordain and declare that the persons whose names are herein inserted and set forth are the persons of whom we have by Warrant under Our Royal Sign Manual thought fit to approve as the persons who shall be first summoned to the Senate of Canada.

Given at our Court, at Windsor Castle, this Twenty-second day of May, in the year of our Lord One Thousand Eight Hundred and Sixty-seven, and in the Thirtieth year of our reign.

GOD SAVE THE QUEEN.

Strategy of Encirclement

Crown coronation and forced union of the provinces and territories of Canada through the Declaration of Confederation by Her Majesty Queen Victoria

1867

"For the Hide, a Skin"
Chartered in London (England), the Hudson's Bay Company is the oldest and largest transnational corporation in the world
1670

E.E. Rich, "Pro Pelle Cutem," *The Beaver* (Autumn 1980): 55-59; R. Watson (ed.), "The Coat of Arms" *The Beaver* (1929/1945): 16-18; provided by HBC Corporate Historian Joan Murray. For additional historic trademark information, see Canadian Intellectual Property Office (CIPO), Trademark SPECIAL & LABEL DESIGN, Expunged, 0484347, TMDA034446 (1904) characteristics of fur and navigation produced a very particular relationship between exploration and exploitation—live, fluid, interactive—and in it, produced a culture of exchange, both transactional and transformative. As the medium and matter of exchange, fur was not only the basis for the earliest economies, but was arguably one of the earliest manifestations of Modernity and of the most modern civilizations: the underrecognized half-breeds and overlooked bastards of the Métis whose relationships were essential to the formation of Empire.[44] Métis modes of communications were vital in the sustainability of *exchange*:

> "As many observers have also noted, the relatively greater harmony of the Indian-European exchange in this territory and the emergence of mediating communities—the Métis—also were by-products of the European adaptation to Aboriginal culture. In many parts of northern North America, especially in the west and the north, Europeans accepted and worked within Aboriginal cultural conventions. In this way, Aboriginal people guided the creation of a separate country."[45]

If fur became the basis for growing Métis communities and a transforming Indigenous culture of exchange, then the fluid navigability of its waters was the basis for its body politic.

GEOLOGY AS MEDIUM, EMPIRE AS MESSAGE

The process of cataloging the plants and minerals through botany and geology in Victorian Canada was intimately tied to the process of asserting economic power over the hinterland and the building of a transcontinental nation. As Suzanne Zeller writes:

> "A common inventorial purpose linked early Victorian scientific pursuits in British North America, and they are here collectively called 'inventory science' to highlight the mapping

avec Privilege du
DES
CASTORS DU CANADA
Leur Industrie a Bâtir des Chaussées pour
est tout a fait Merveilleuse.
Bucherons qui avec leurs Dents coupent de
Gros Arbres, quils font tomber a travers le Ruis
seau, pour Servir de fondement a leurs Chaussées
Charpentiers qui Coupent les Branches
Porteurs de Bois pour la Construire
Ceux qui font du Mortier
Commendant ou l'Architecte
Inspecteur des Invalides
Ceux qui Trainent le Mortier sur leurs Queües
Castor Incommodé de la Queüe pour avoir trop

Nicolas Guérard, ca. 1698

Colonial Imagination

Nicolas Guérard's false depiction of the beaver as perfect colony of industrialized laborers—lumberjacks to carpenters—commanded and overseen by a chief architect 1698–1715

Imperial Accumulation

The élaborate inventorying, accounting, packing of fur pelts at Fort Chipewyan, Alberta, one of the oldest outposts of the British Empire prior to shipping to the fashion market in London

1890

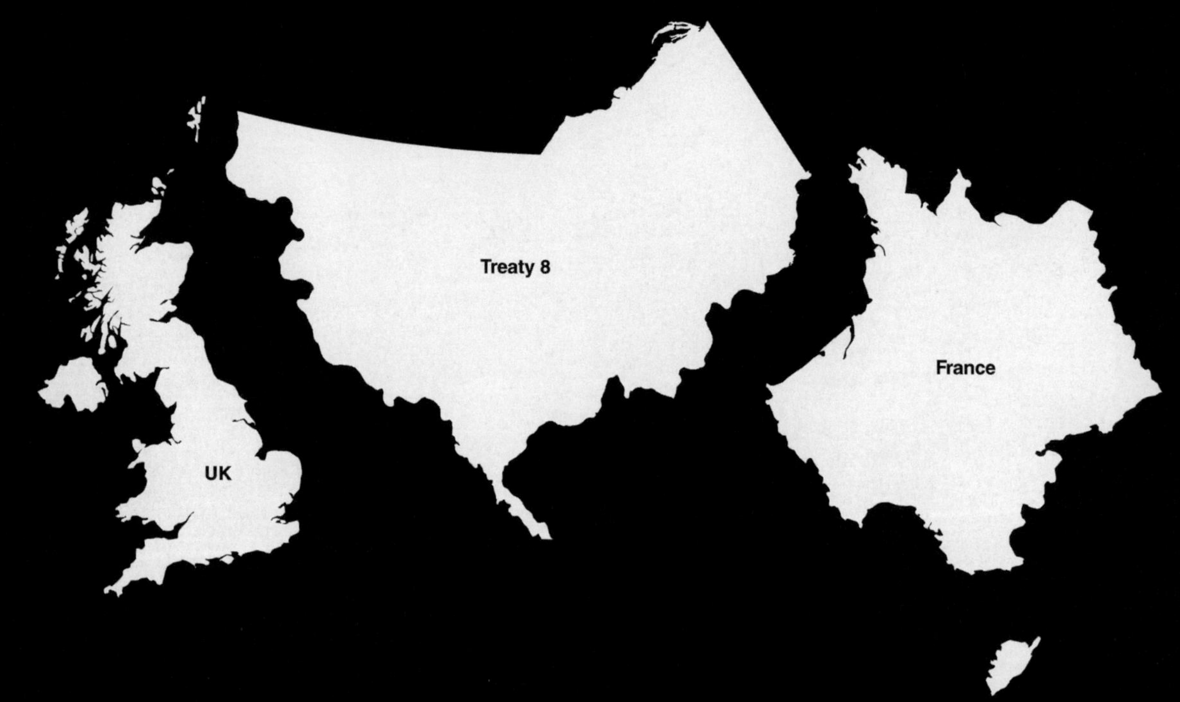

Empires, in Comparison
UK and France dwarfed by the size of Treaty 8 Lands
2016

> and cataloguing of resources and other natural phenomena which preoccupied the colonists. Inventory science in early Victorian Canada included geology, terrestrial magnetism, meteorology, botany, and to a lesser extent entomology, zoology, and anthropology."[46]

The formation of this "visible empire" was essential and it was carried and communicated in the representational media of knowledge: illustrations, inventories, charts.

> "Botanical illustrations helped to discipline the eye of both naturalists and artists; they served as entry point, instrument, and result of natural historical investigation; they were essential within a culture of gift-exchange and patronage; and they provided a tool with which to think. If an imperial eye brought certain objects into sharp focus, however, ... it did so by a process of selective blindness. 'Images preserved the impermanent and transported the distant,' but they did so by excising precisely what made the objects of illustration so important: the place of their production. The 'indigenous' (people, plants, soil) were removed even as indigeneity itself became a global commodity."[47]

From exploring and surveying to measuring and valuing since the birth of the Canadian Geological Survey in 1842, the discerning yet discriminating territorial work of the geologist can also be seen in the statistician's work today. The colonial statistical mind defined the categories and the metrics through which culture could be tabulated and evaluated. In their words, this territorial body of knowledge became their *body bureaucratic*.

If, as Walter L. Duffett, Chief Statistician for the Dominion Bureau of Statistics, claimed in 1967, "there is no history without context,"[48] then the task of the Dominion statistician would quantitatively reformulate the context of the country to qualitatively reframe and rewrite its history.

By design, the drawing of tables and the drafting of treaties served to quantify, numerically designate, disorganize, displace, or dispossess In-

Cultural Appropriation
The decadent display of Britain's imperial acquisitions in Canada's Exhibition at the self-indulgent World Fair in London
1851

TO THE STAIRCASES.
CANADA

Mining as Victorian Extraction
Celebrating the stolen mineral wealth from Indigenous territories of the colonies to serve the white supremacy of the imperial center
1851

MINING

digenous populations through the fracking of surface and subsurface rights. Together, they formed the basic formation of state powers and the externalization of Indigenous populations from cultural and economic production:

> "The Canadian government commenced the treaty-making process with the indigenous populations of the Athabasca region in 1870, motivated by the Geological Survey of Canada's reports that petroleum existed in the area."[49]

Assimilation remained a central objective effectively operationalized through the destructive "residential and industrial school system" from 1876 onwards, for well over a century.[50] In a very short period of time, two successive stages of retrocession of lands took place following the Declaration of Confederation in 1867: the exit strategy and royal alibi, that would transfer the jurisdictions and powers over to the Canadian Crown, as the Charter of the Hudson's Bay Company on Rupert's Land would soon expire three years later in 1870. Providing the Crown with the jurisprudence to move into and settle with farmland in order to move resources out, Confederation, with its intent towards federation through forced consent, was more an action of conquest and extinction than integration or incorporation.

One of the earliest Superintendents of Indian Affairs in 1901, Duncan Campbell Scott, confirms the racist supremacy that was entrenched in the Residential School System which provided a pathway to indoctrination and dispossession, or sometimes even death:

> "I want to get rid of the Indian problem. I do not think, as a matter of fact, that the country ought to continuously protect a class of people who are able to stand alone. Our object is to continue until there is not a single Indian in Canada that has not been absorbed into the body politic, and there is no Indian question, no Indian department."[51]

Later, from 1912 and 1932, where he saw assimilation through "education and intermarriage as central elements in the policy of enfranchisement,"[52] Scott reiterated:

Retrocession
The geography of Treaties across Indigenous lands following the Declaration of Confederation by Queen Victoria
1871–1921

David Chancellor, *Intruders*, 2012 / Kiosk

Resource Risk
Barrick Gold's reinforced security vehicle patrolling
North Mara Gold Mine in Tanzania
2012

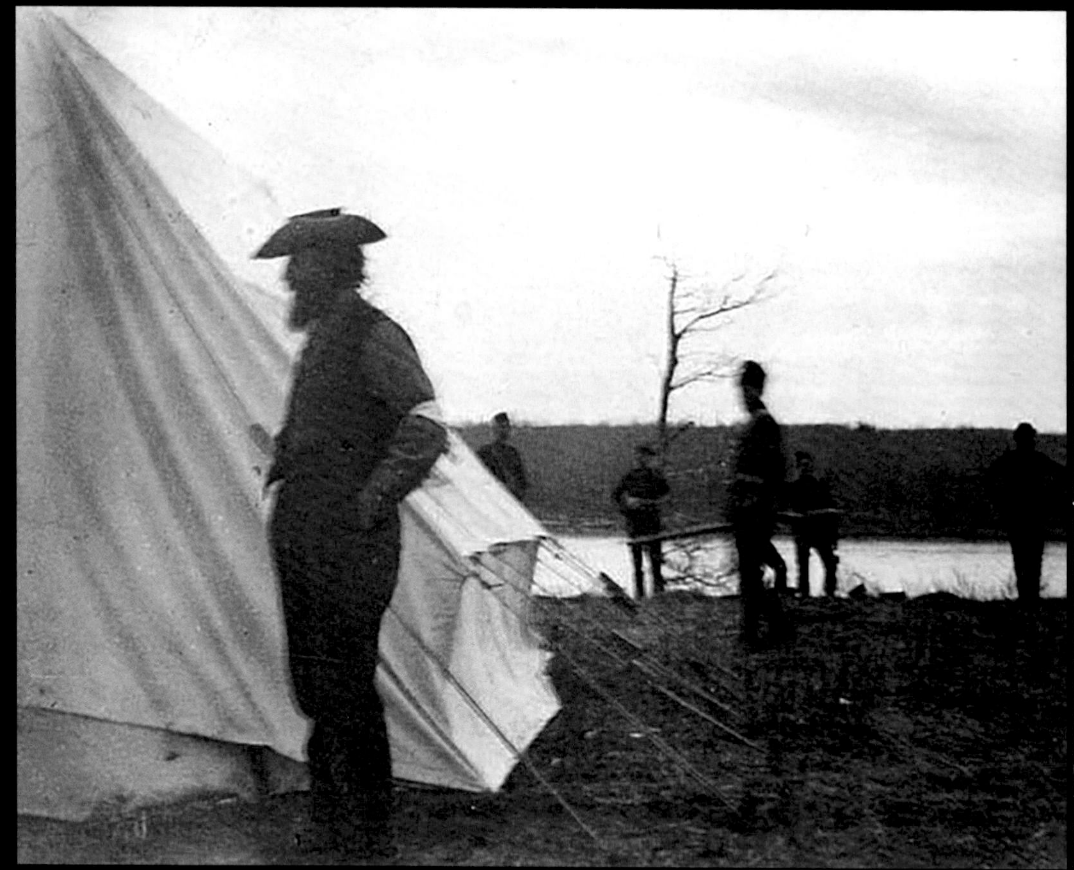

James Peters / Library and Archives Canada / e011156619_s3 ; C-003450

Surface Resistance
The capture of Métis Nation Leader Louis Riel following the 1869–70 Red River Rebellion
1885

> "The happiest future for the Indian race is absorption into the general population, and this is the object of the policy of our government ... The great forces of intermarriage and education will finally overcome the lingering traces of native custom and tradition."[53]

Along with the swift consolidation of pre-Confederate laws affecting Aboriginal populations, "the great aim of our legislation" as the first Prime Minister John A. Macdonald proclaimed in 1887 in reference to the inception of the Indian Act of 1880, among others, "has been to do away with the tribal system and assimilate the Indian people in all respects with the other inhabitants of the Dominion as speedily as they are fit to change."[54] Although retrocession provided an opportunity for Queen Victoria to backtrack on Confederation through the numbered treaties of the West (No.1–No.11), they were not only preempted by a series of regional rebellions; retrocession also lead to wide sweeping cultural dismemberments, including drastic restrictions on the roles and matriarchal rights of Indigenous women and families, deprivation of tribal languages and Indigenous names, bans on social and political organization (potlatch, dance, voting), and trade (alcohol, ammunition, farm products, pemmican, land), to name a few.

By design, the most violent and destructive constitutional instrument of the *Indian Act* was, and still is, the Reserve System. In its name, lay its predispositions and prejudices. The French translation, *Loi sur les Indiens*, was more operative and inherently instrumental. Although vested in Her Majesty, the "Law *on* the Indians" (as opposed to the Anglo-Saxon 'of'), referenced very different Franco-Catholic undertones in addition to influences that implied superimposition and overlay of the dominant white power structure: its segregationist reserve system would be its final solution. Instituting geographic seclusion and cultural sequestration, the reserve system whose legal designation was (and still is),

> "a tract of land, the legal title to which is vested in Her Majesty, that has been set apart by Her Majesty for the use and benefit of a band."[55]

Provisions and conditions of the Indian Act were many, but they also found another, more extreme application in another part of the British Empire. By the 1930s, Canada's cheaper and more economic Indian Reserve System became the template and blueprint for apartheid-era systems of segregation in South Africa.

Seeking guidance and parallel instruction from one continent to another, just a few years after Canada's Confederation, the Indian Act, and the Dominion's Land Act, "South African leaders turned to the Canadian experience with 'Indian reservations' for guidance in crafting and forming up their homeland structures in the 1920s." Together with the South African Land Settlement Acts of 1912 and 1913,[56] these parallels could frame and functionalize apartheid:

> "It is significant that South Africa came to Canada at different times since the Boer War asking and [obtaining] permission to study the Canadian system by which Indian people were controlled and managed separately from the politically dominant white population. South Africa took what it needed and applied it to its own situation: first to segregation, and after the Second World War to apartheid. The fundamental difference between Canada and South Africa was that Canada was interested in segregating and managing, as cheaply as possible, a population it did not want as an important source of labour. South Africa was interested in the same type of relationship, but for a people whose labour it needed and wanted cheaply."[57]

South Africa's own reserve system was extremely timely. By then, the mining industry was in full development now that the British imperialist Cecil Rhodes, fully financed by banking financier N.M. Rothschild & Sons, was able to create a monopolistic world market in the early 1880s when diamonds were discovered in the Central Region of the country at the junction of the Orange and Vaal rivers. The small ratio of Whites to Blacks and Browns in South Africa practically required this, while it was inversely proportional to the large proportion of Whites to Aboriginal populations in Canada who clearly dominated in terms of numbers.[58] But, more striking

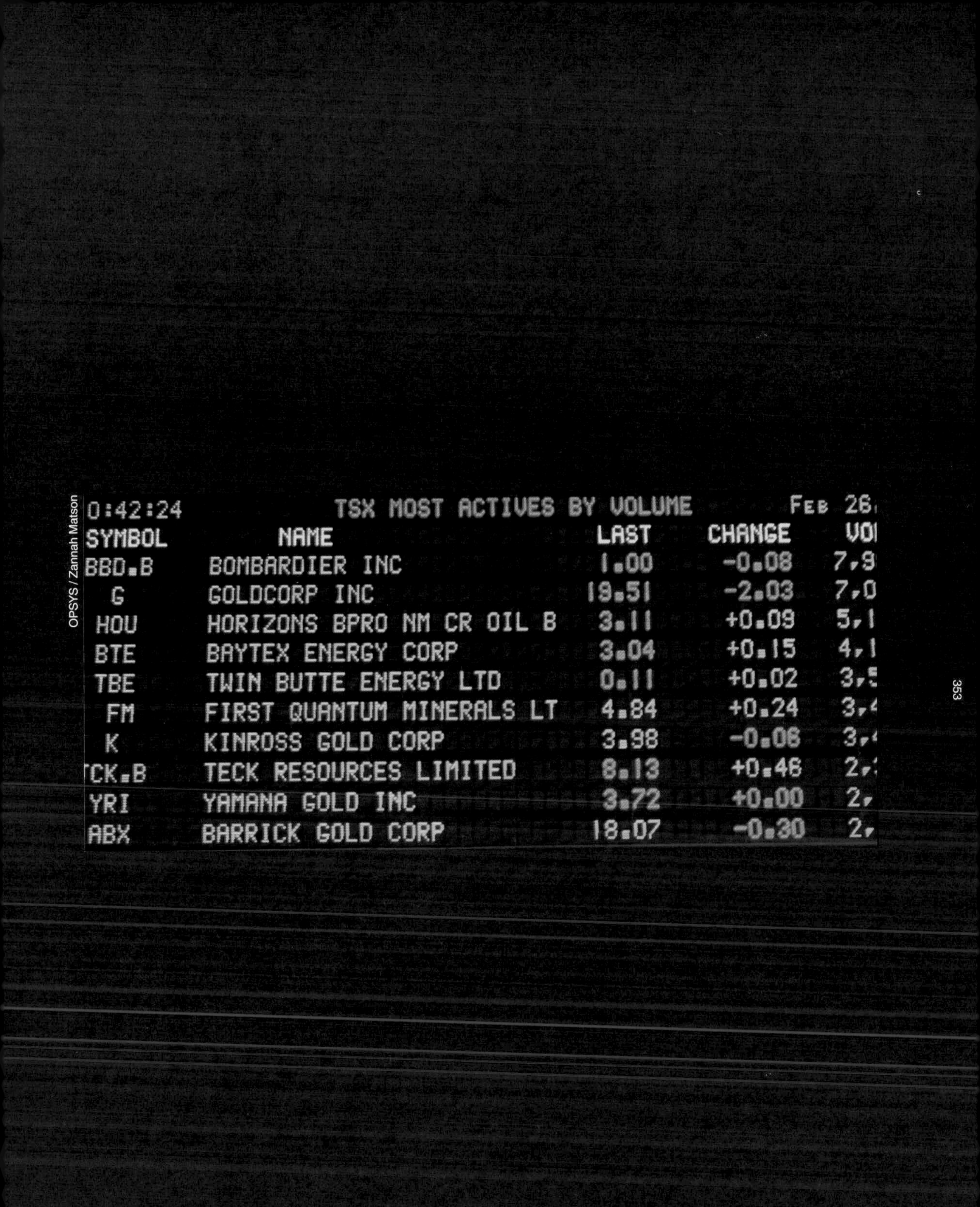

OPSYS / Zannah Matson

0:42:24 TSX MOST ACTIVES BY VOLUME FEB 26,

SYMBOL	NAME	LAST	CHANGE	VO
BBD.B	BOMBARDIER INC	1.00	-0.08	7,9
G	GOLDCORP INC	19.51	-2.03	7,0
HOU	HORIZONS BPRO NM CR OIL B	3.11	+0.09	5,1
BTE	BAYTEX ENERGY CORP	3.04	+0.15	4,1
TBE	TWIN BUTTE ENERGY LTD	0.11	+0.02	3,5
FM	FIRST QUANTUM MINERALS LT	4.84	+0.24	3,4
K	KINROSS GOLD CORP	3.98	-0.06	3,4
TCK.B	TECK RESOURCES LIMITED	8.13	+0.46	2,3
YRI	YAMANA GOLD INC	3.72	+0.00	2,
ABX	BARRICK GOLD CORP	18.07	-0.30	2,

Commodities & Stocks
Prices, Toronto Stock Exchange
2016

REUTERS / Miraflores Palace / Handout

Golden Touch
Nationalization of Crystallex Gold Mine by Hugo Chavez
in Las Cristinas, Venezuela
2008

U.S. Central Intelligence Agency, 1979

Subjugation through Segregation
The Homelands Reserve System of South Africa developed during the regime of apartheid, modeled on Canada's Indian Reserve System 1948–1994

ly, the system of Aboriginal subjugation, concentration, and extermination served another sinister purpose.

In his book, *Adolf Hitler: The Definitive Biography*, Pulitzer author John Toland, notes:

> "Hitler's concept of concentration camps as well as the practicality of genocide owed much, so he claimed, to his studies of English and United States history. He admired the camps for Boer prisoners in South Africa and for the Indians in the Wild West; and often gave praise to his inner circle the efficiency of America's extermination—by starvation and uneven combat—of the red savages who could not be tamed by captivity.
>
> He was very interested in the way the Indian population had rapidly declined due to epidemics and starvation when the United States government forced them to live on the reservations. He thought the American government's forced migrations of the Indians over great distances to barren reservation land was a deliberate policy of extermination. Just how much Hitler took from the American example of the destruction of the Indian nations is hard to say; however, frightening parallels can be drawn. For some time, Hitler considered deporting the Jews to a large 'reservation' in the Lubin area where their numbers would be reduced through starvation and disease."[59]

In the evolution of the North American model of segregation systems and through the advent of Canadian methods of retrocession (ceding lands formerly colonized), at once isolating and containing First Nations, the numerical statistical and scientific accounting of value and volume of land served the Victorian transformation of virgin nature into valuable resource. Later, they would be served by oblique, misleading, and racist ideologies of environmental preservation and land conservation in the nationalization project of nature itself; blindly perpetuating the conflation and near total incorporation of Victorian ideals of nature.[60] As early as 1908:

> "That this region is stored with a substance of great economic value is beyond all doubt, and, when the hour of development comes, it will, I believe, prove to be one of the wonders of Northern Canada. We were all deeply impressed by this scene of Nature's chemistry, and realized what a vast storehouse of not only hidden but exposed resources we possess in this enormous country."[61]

Through the edification of *land* as *nature*, the power of governance that is upheld today in the designation of Canadian territory as 'Crown Land' was produced and reproduced across the geological characterization of the surface and subsurface of the State, by the bureaucratic systems of the State.

Closer to Confederation, "as the materialization of belief and of political will, the railroad was then perceived as the consequence of political rhetoric, of discourse which constitutes power." Set against the background of this imagined nature and constructed wilderness, the Canadian Pacific Railway (CPR) was an expression of the technological nationalism required to both unite and contain the country, while at the same time extracting resource, as the first Prime Minister John A. MacDonald explained:

> "The road will be constructed … and the fate of Canada, will then as a Dominion, be realized. Then will the fate of Canada, as one great body be fixed."[62]

But, as rhetorical expert Maurice Charland argues much like Innis, the transcontinental railway was a technological media of political communications and ideological bias:

> "The CPR's existence is discursive as well as material, for it stands as an articulation of political will. While the CPR proved economically profitable for its backers, the linking of Montreal to Vancouver was not a happenstance or the result of a private entrepreneurial venture, rather the road was built under the auspices of Canada's federal government for the explicit purpose of extending spatial control over a territory. That is to say, the determination of Canada to remain British in char-

Canada One Hundred 1867–1967, Dominion Bureau of Statistics, Handbook and Library Division, 1967

CANADA

ONE HUNDRED

1867-1967

Magnum Opus

Canada's last printed yearbook issued by preeminent statistician Walter E. Duffett for the Centennial Anniversary of Confederation

1967

CANADIAN EXPLOSIVES
Royal Exchange, Cobalt, Ont.

acter rather than be absorbed by the United States preceded the railway's construction. Furthermore, the construction of the Pacific Railway was not even a necessary condition to British Columbia's entry into Confederation: That Pacific colony had demanded only that Ottawa build a wagon road. Thus, the CPR was part of a rhetorical ploy. Cartier and MacDonald offered more than was necessary, a rail link to the west coast within ten years of British Columbia's joining the Dominion. Consequently, the CPR cannot be viewed as the product or manifestation only of economy. The construction of the railroad was more than an overdetermined response to material and political exigencies; a will to statehood preceded it. It was an element of a strategy based in the belief that a nation could be built by binding space."[63]

The solidification of the surface of the State through *technological nationalism* also required a strong binding agent: a nationalist group with affiliations to the British Empire. Short lived yet prematurely effective, the Canada First Movement saw one of its leaders and most loyal bureaucrats, Charles Mair, quickly rise while simultaneously fueling an adversarial relationship with none other than Louis Riel.

In the hotspot of pre-established systems of trade and exchange, then came confrontation with policies of extraction. A successive series of resistance movements and Red River rebellions emerged in Manitoba before and after the 1867 period of Confederation.[64] Although surface resistance led to the era of legal land treaties in the Prairies (successively numbered 1 to 11), the political technology of the treaty itself as territorial agreement did not correspond to the long-term transgenerational needs of Indigenous populations related to housing, schooling, servicing, and provisioning. Nor did it succeed in leveraging the power of its original providence.

Treaty 9, for example, was negotiated during a period of two summers (1905–1906) across 14 different outposts for nearly 100 signatories, from chief and leaders of Indigenous groups during different periods of hunting and foraging seasons.[65] Speaking a foreign language, the angst and anxiety of foreign colonial representatives made practically impossible any form of just and equitable agreement, as preordained forms were presented in the media of Empire. Recognizably foreign but binding through unconditioned consent, the quintessential "X," or the Cree Syllabic, that are recognized today as assumed marks of contractual approval, are the very source of unconditional contestations. For cultures that did not practice law through 'property' nor recognized the paper space of the State, the imperial medium itself raises today fundamental questions of validity, trickery, coercion, deceit, theft, and dispossession.

Here, the medium of Empire speaks for itself: English calligraphy, language and style of monarchy; black India ink on white British calfskin, the materials of imperial, empirical record.

As witnessed in the ongoing tensions of oil extraction in the Tar Sands of Alberta, between the Crown and First Nations in the heart of the Treaty 8 Lands of the Athabasca River Delta, or in the Treaty 9 Lands of James Bay, the paper space of the treaties extinguish surface resistance and exclusively annex the Chipewyans in the State search of the subsurface, "[rendering] the Indian … virtually powerless to resist the white civilization."[66]

The technological nationalism that was required to both build and bind the surface of the State did meet considerable resistance well before and after Confederation; both in its clearing the plains or transitioning the continent from fur and timber, then towards wheat and oil. A more recent iteration of technological nationalism emerged during Canada's Centennial in 1967. Created by reconnaissance pilot and military veteran Richard Rohmer, the *Mid-Canada Development Corridor* inherited ideologies of colonial extraction towards a proposal for the mostly "undeveloped" and "unsettled" Mid-North: "a treasure house of natural resources … that produces most of the world's nickel, zinc, iron, asbestos, potash, uranium, newsprint, hydro-electric power, oil and gas."[67] It essentially iterated the image of the state created and curated by the British Empire at its World Fair in 1851 and in Paris, 1858, where the wealth of resources from across the Empire were displayed as geological gifts to the world.

Ruler of the Earth

Queen Victoria II, Ruler of the British Empire, Empress of India, Queen of Canada, State Sovereign

Treaties, Surrenders and Agreements RG10, Vol. 1851, IT 436 ©Government of Canada. Reproduced with permission of Library and Archives Canada (2017)

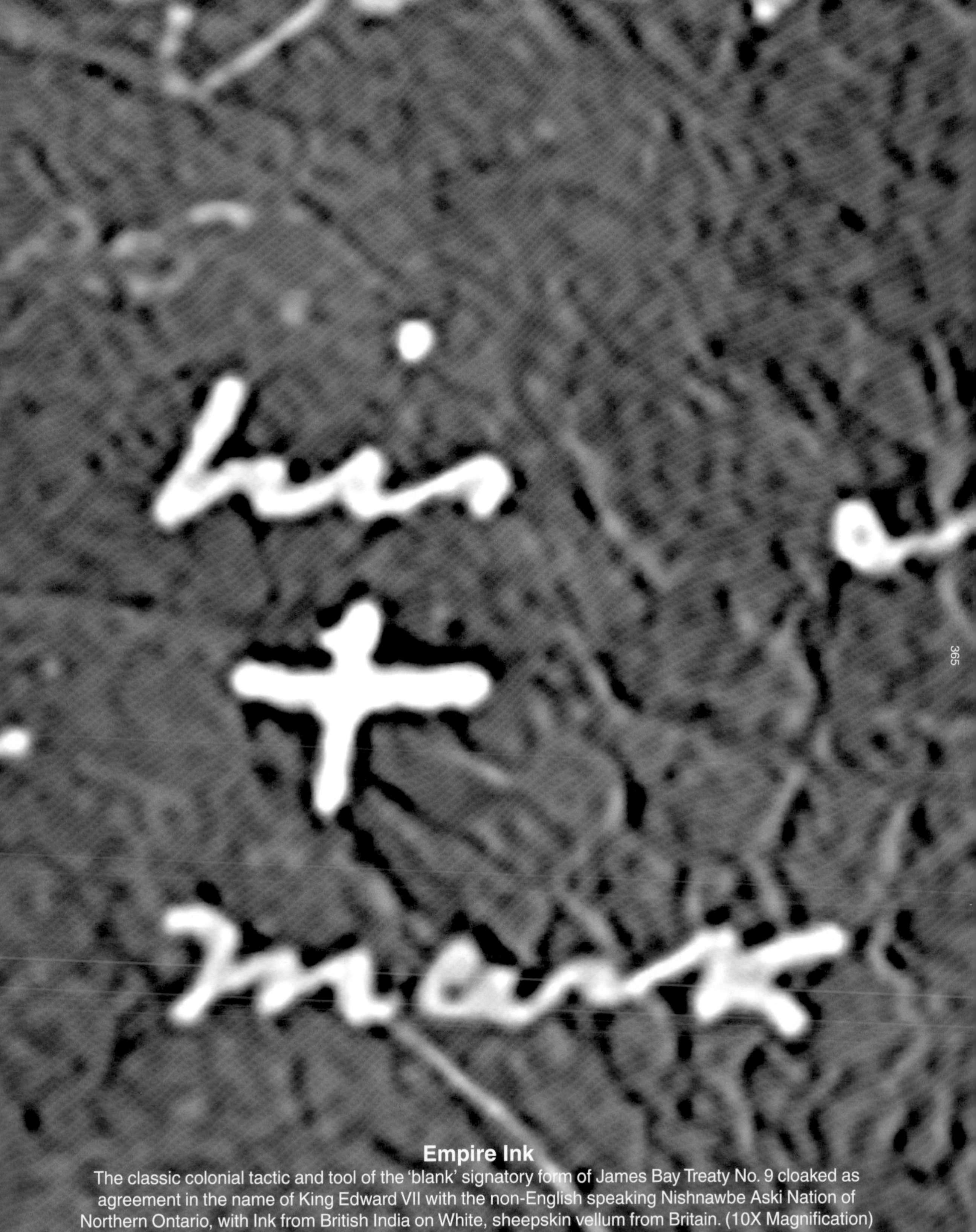

Empire Ink

The classic colonial tactic and tool of the ‘blank’ signatory form of James Bay Treaty No. 9 cloaked as agreement in the name of King Edward VII with the non-English speaking Nishnawbe Aski Nation of Northern Ontario, with Ink from British India on White, sheepskin vellum from Britain. (10X Magnification)
1905–1906

Library and Archives Canada / R12590-2206

Petro-State
Petro-Canada founded as Crown Corporation under Pierre-Elliott Trudeau's National Energy Plan and Maurice Strong as CEO
1975

While the plan for 'Mid-Canada' remains unrealized, and unrealizable, according to a former Royal Commission in 1957, it nevertheless completely overlooked and obviated the presence of pre-Confederation populations:

> "You businessmen, you mining people, you labour people, where are your Indian foremen, your Indian stewards, your Indian field bosses?"[68]

Yet, amidst the extensive explanation of geological potential in Rohmer's Plan, there is not a single mention of the *pre-occupation* and extensive inhabitation of the Mid-Canada region already by the majority of First Nation populations. The only reference to *reserves* in the 1967 document is to *known reserves,* not of people, but of *mines and minerals*. In the case of settlements, inhabitation functionalized by growth zones, and a multitude of typological '*centres*' (service *centres*, resource *centres*, administrative *centres*, medical *centres*). The Elizabethan blanket statement of natural resources once again emerges as the *tabula rasa* uncovering the Victorian extractive project.

The nascent projection of imperialism through geological knowledge—maps that often advanced the imperial world[69]—could then be only understood by an emerging school of imperial thought and its *geological medium* or *colonial metaphors*. Unparalleled and unmatched in their conflation of empire, technology, and communications, a new school of thinkers surfaced from resource economics and media theory. The message of empire was found in the medium of geological envy, as media theorist and staunch Catholic Marshall McLuhan proposed:

> "Today, even natural resources have an informational aspect. They exist by virtue of the culture and skill of some community. The reverse, however, is also true. All media—or extension of man—are natural resources that exist by virtue of the shared knowledge and skill of a community."[70]

So, in the first stage of state simplification, natural resources were media, and in the imperial eyes of cartographic science, the environment is data-

REUTERS / Hereward Holland

Unprotected Labor
The machine-less excavations and gravity-fed slurries of gold mining in Ghana, West Africa
2013

technology of Empire building. As media, these resources se
s of the state in construction of its power, and as a result, becar
nic staples and royal resources. Thus, the expression and impres
alties, levies from extractive industries paid directly to the Crown
npelling analysis of staple economies and its counterintuitive c
n to communications theory, resources were media of empire: th
thods of communication and systems of organization that later
extremely sophisticated stock market, fiscal structure, royalty rem
em, and insurance industry.

As Harold Innis described, the tools and techniques of imperi
action were a product of its materials and resources—its *media*:

> "In the organization of large areas, communication occupies a vital place … The effective government of large areas depends to a very important extent on the efficiency of communication. The concepts of time and space reflect the significance of media to civilization. Media that emphasize time are those that are durable in character, such as parchment, clay and stone. The heavy materials are suited to the development of architecture and sculpture. Media that emphasize space are ap to be less durable and light in character, such as papyrus and paper. The latter are suited to side areas in administration and trade. The conquest of Egypt by Rome gave access to supplies of papyrus, which became the basis of a large administration empire. Materials that emphasize time favour decentralization and hierarchical types of institutions, while those that emphasize space favour centralization and systems of governments less hierarchical in character."[72]

the British Empire, Queen Victoria has both light and heavy m
mobilize its power: maps, mines, materials, and now, money. Inni
of resource dependency would then lead to the emergence and ex
w decades later of an advanced 'staples theory' in the 1960s. Re
e not only material commodities or contributors to economic

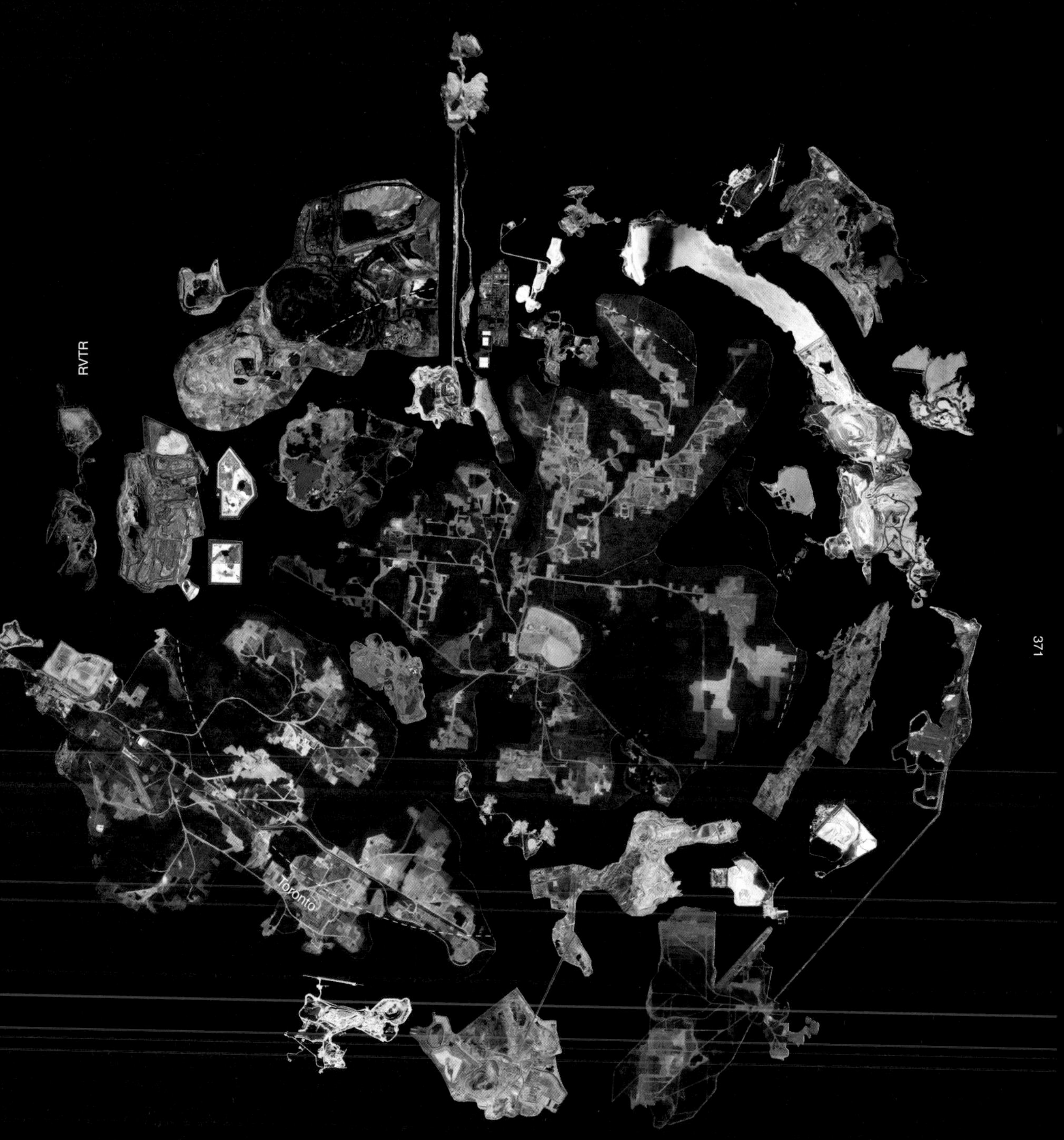

Canada's Glory Holes
The largest Canadian-owned, Canadian-operated sites
of mineral resource extraction in the world
2016

Bronze Cap

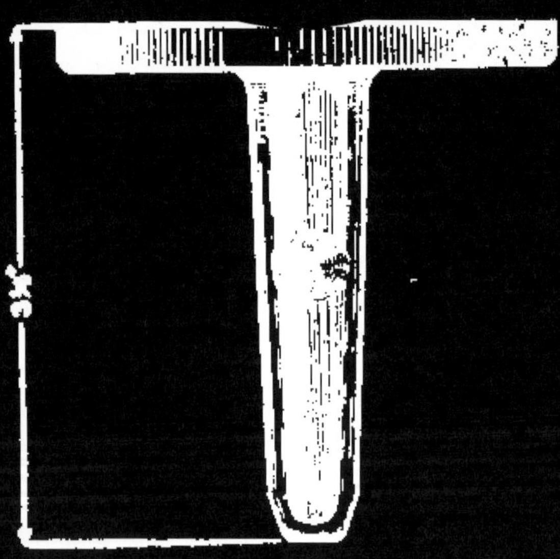

Short Iron Post and Stone Mound
Fig. 5

Department of Mines and Resources, Manual of Instructions for the Survey of Dominion Lands

Monument
Benchmark for the Dominion Land Survey initiated after Confederation
by the Government of Canada for land settlement
1869–1917

they communicated the politic, sociologic importance of empire, that is t British imperialism. Later, it became Canadian capitalism, where Canada's most distinctive contribution to the world of political economy was, as it is today, the extraction of raw, unprocessed resources[73]:

> "Staple theory became central to the creation of the 'New Canadian Political Economy' in the past fifty years, which sees the Canadian capitalist class, which manages our relationship with the global economy, as having a 'staples fraction' with the muscle to maintain the status quo of resource dependency while masking that behind the rhetoric of the inherent goodness of the market."[74]

In other words, if propaganda served the Church as a form of soft, social political coercion in the submission of society to the power of the Crown then the hard, scientific exploration and extraction of resources would prove to be a violent, destructive system of economic production and settlement concentration, Aboriginal alienation, and colonial power projection of the Monarchy. In other words, not only was the territory a *map*, the territory was a *mine*.[75] The geologists—and their discoveries—lead to an increase in prospectors, which led to an increase in miners, which lead to an increase in investment capital. The early days of Confederation found this rush to capitalize on mineral wealth essentially buttressed by its political elite.[76]

FINANCIALIZATION

Stats, States, Stocks, & Staples

If the Queen's Declaration on the formation of "one dominion" a century prior was the first strategic act of spatial encirclement and distancing population further away from territorial rights and access to land, then the creation of the Toronto Venture Stock Exchange in 1999 may have been the second. With 70% of the globe's mining equity flowing through the TSX (the Toronto Stock Exchange), Canada by far out-finances the rest of the

Shaney Komulainen / The Canadian Press

Tête-à-Tête
Face-off at Oka and Kanesatake between
Private Patrick Cloutier and Brad 'Freddy Krueger' Larocque
1990

world. It is now the financial-industrial complex that blanches the mineral medium and political territories of extraction. Between 2009 and 2013, the TSX provided $157 billion in equity for the mining industry, more than any other trading floor on the planet.[77] In 2012 alone, capital flowing through Bay Street contributed over $3 billion towards mining projects in Latin America and Africa. Canada's institutions not only adapted to a growing world demand, they have also capitalized on it. The development organization for the gold market seen in the membership of the World Gold Council of predominantly Canadian-based companies.[78]

With the increasing role of the TSX in mediating the resource industry, Toronto has not surprisingly become a resource capital and the financial operating platform for the world's mining industry. Located around the Stock Exchange on Bay Street, one finds the financial hallmarks of the big five banks: TD, BMO, RBC, Scotia, CIBC. Each incipient devices of royal, monarchist, and imperial beginnings cloaked in catchy acronyms: RBC for Royal Bank of Canada, BMO for Bank of Montreal, Scotiabank, CIBC for Canadian Imperial Bank of Commerce, TD for Toronto Dominion.

Through the financialization of resources and mathematization of their volume through the scientific sector, the symbolic relationship between geology and nationhood also created a symbiotic axis between minerals and power. Thanks to the science of resource speculation and the rise of the risk insurance industry, this powerful cocktail of exploration and extraction is leveraged through the force of finance and the flow of credit. For the future State, statistics were a means to account for itself, and more specifically to the Crown. The image of the state relied on a level of abstraction found in its oldest department, the Dominion Bureau of Statistics:

> "It is therefore not surprising that social statistics—stats about people—should be a constant preoccupation of society and of government, since programs of this kind and magnitude [welfare programs including health, education, and social assistance] have to be developed in the light of statistical measures, and such measures are also necessary at a later stage to monitor and control."[79]

Paul Chiasson / The Canadian Press

Territorial Health Hazard
Rail oil disaster near Lac-Mégantic, Quebec
2013

Being a statistician for the State was the bureaucrat's supreme dream. Walter E. Duffett was Chief Statistician for the Bureau, who was central in the era of Pierre Trudeau and Confederation's Centennial when, fresh off EXPO '67, Canada projected itself as the evolved, internationalist middle power. Statistics were leveraged for the rhetoric of the market and mobilized for the science of demographics. The modern extractive State—its economy and enterprise—was quantifiable and calculable through the clear categorization of culture and conversion of living matter from land into resources above, and below the surface, from public power, to political power, to economic power.

Therefore, its extractive economy could be planned in advance, as to minimize hazards, avoid emergence, and avert risk. Requiring a deeper level of resource financialization, the extractive State could not only be described as numerical and mathematical, but highly if not purely ideological. Profoundly territorial, the financialization process of resource extraction then carried deep spatial and symbolic implications:

> "It is a process whereby a set of narrative, metaphoric and procedural resources imported from the financial world come to help explain and reproduce everyday life and the capitalist totality of which we are a part. But, in so doing, they also transform that reality more broadly. To the extent that we see ourselves as miniature financiers, investing in and renting out our human capital, we act, behave, cooperate, and reproduce social life differently. To the extent that we see health, education, government programming, relationships, games, shopping and work as investments and see our lives as fields of paranoid securitization, we build up an ideological armature which occludes certain aspects of social reality and precludes certain futures. But, further (and this is crucial), financialization also means a moment when the financial system, and the capitalist economy of which it is a part, is dependent on and invested in the ideologies, practices and fictions of daily life, as never before."[80]

If oil, gold, and uranium in the twentieth century have now replaced fish, fur, and forests of the eighteenth century as financial resources (no longer natural), then political economist Mel Watkins proposes a geopolitical reclamation of Innis' staples theory:

> "It is striking the extent to which the Canadian economy remains driven by commodity exports, with the Canadian dollar classified as a petrocurrency, while the petro-politics of Alberta pushed us into free trade with the U.S., with its gluttonous appetite for resources, and continues to transform national politics."[81]

With the increasing role of the TSX in expanding the resource industry, Canada's underground landscape has not surprisingly become a resource frontier: a global operating platform for the world's mining industry. In other words, the Taylorization of territory for production goes hand-in-hand with the extraction of resources for fastest gain. Highly complex, and largely opaque, the TSX blurs distinctions between promoter and speculator: by not only creating a climate for the financialization of resource extraction, but also through the encouragement of greater economies of scale with larger levels of risk associated with resource speculation.

> Back in 1993,
>
> "Bre-X's run from the bottom of the Alberta stock exchange to the top of the Toronto exchange and into the world was a source of pride for many Canadians. As much as for profits, Canadians invested for reasons of national pride.[82] ... In this industry, the line between various kinds of expertise is thin: geologists (with salaries supplemented by stock options) must be promoters to raise the money to finance their mineral funds, market analysts must explain their offerings in geologically convincing terms. Canadian preeminence in mining depended on both its mining history and its position as a center of mining finance."[83]

Fred Chartrand / The Canadian Press

The reliance on self-regulation across the industry and remarkably low royalties both promoted these affiliations while attracting considerable capital from elsewhere. The lack of consequences for white-collar crime (especially in theaters of operation across borders and boundaries of different states) entrenched these practices and helped to ensure Toronto, and the TSX, became the mining capital of Canada.[84]

It first took though a resource boom 400 kilometers to the north of its downtown to sow the seeds of Toronto's dominance: the Royal Exchange. Before the big banks that we recognize today as financial icons of the country, there was an earlier financial prototype. Built in 1910, the Royal Exchange in Cobalt, Ontario, "at the corner of Silver Street and Prospect," was "a milestone, pointing the way to new and better Cobalt. Missive [sic] in size, the block is built with a visible structure of ferro-concrete floors and wall piers, infilled with brick ... a handsome and dominant element of the Cobalt landscape."[85] Besides the Stock Exchange itself, the same block contained the Canadian Explosives Office and the General Electric Office, as well as Cobalt's branch of the Bank of Toronto and the Ontario Surveyor's Office. The Northern Miner Press—to this day the industry paper of record—was housed here and, in the event the workday's end was met with either good or bad fortune, a parlor and restaurant completed the complex. Although the Royal Exchange in Cobalt is now an important part of the Cobalt Mining District National Historic Site of Canada, it is celebrated for developing economies of extraction and production.[86]

Cobalt expressed the bipolarity of a resource town. When the Prince of Wales arrived to Cobalt in 1919, "Silver Street was a sea of Union Jacks and loyal subjects. But flags and crowds couldn't throw off the dismal cloak of industrial grime that hung over the town. It was relentless." Since the advent of the exchange, in such a short period, several silver mines had shut down and influenza had hit the town hard. After visiting the town, which included a trip past the Royal Exchange Building, the Prince of Wales remarked, far from the bright blue stone the city represented, "Ah, it's a gray, wee town."[87] But as impactful as Cobalt's mineral resources were at the be-

MONEY MATTERS

A CRITICAL LOOK AT BANK ARCHITECTURE

Canadian Centre for Architecture

CENTRE CANADIEN D'ARCHITECTURE/CANADIAN CENTRE FOR ARCHITECTURE

Capitalisme, à l'Anglaise

'Golden Tower' of the Royal Bank of Canada on the cover of the Bank Architecture Exhibition at the Canadian Centre for Architecture 1990–1991

L'OR ET LA PIERRE

UN REGARD CRITIQUE SUR L'ARCHITECTURE DES BANQUES

CENTRE CANADIEN D'ARCHITECTURE/CANADIAN CENTRE FOR ARCHITECTURE

Matérialisme, à la Française

'Gold to Gravel' on the translated cover of the Bank Architecture Exhibition at the Canadian Centre for Architecture 1990–1991

DeBeers

Diamonds, De-Territorialized
De Beers' postwar marketing slogan and propaganda
1958

Diamonds, Re-Territorialized
Map of Ottawa to Attawapiskat, downstream from De Beers' Victor Diamond Mine
2017

ginning of the twentieth century, its influence on financialization of mining and exchange is still felt and ingrained in the twenty-first.

For the first time, royalties were generated for the province of Ontario from the mined resources around Cobalt, along with the profits from financial stocks for the companies that invested in nearby Cobalt deposits. In *Harvest from the Rock*, Philip Smith writes:

> "It was at Cobalt that Canadian financiers overcame their traditional reluctance to invest in mines, and the money that they made there, encouraged them to finance the great expansion of mining that would occur in Ontario over the next half-century. Bankers too—after that initial reluctance to bankroll Noah Timmins—gained valuable experience in the financing of mining ventures."[88]

The Toronto Stock Exchange, and by extension Canada's global mining footprint, rests on the back of this royal heritage of regional risk. Extending Victoria Regina's political bosom, the financial framework of the 145-year old TSX today captures more than cross-border capital, it epitomizes cross-continental ambitions. As the merger of stock exchanges from Vancouver, Alberta, Toronto, and Montreal, the TSX has grown to become the ultimate financial extension of Sandford Fleming's vision. The TSX ignites a new global beginning.

Canada's continentalist motto from "sea to sea" as the original King James Bible proclaimed, is a much vaster and broader project, with potentially more expansive implications whose scope is planetary and imperialist in proportion:

> "He shall have dominion also from sea to sea, and from the river unto the ends of the earth."[89]

Not surprisingly then, this coat of arms—the ensign inscribed on every Canadian passport, and as such—is the first legal document (*le passe-partout Canadien*) that has entrenched, beyond the world of financialization, the legalization of rights and self-entitled jurisprudence associated with the Vic-

Edward Burtynsky

Terrifying Beauty
Silver Lake Operations, Lake Lefroy, Western Australia
2008

TREATY OF UTRECHT
Dated 31 - 11 day of
March - April 1713
Article XV

TREATY OF GHENT
(1814)
Restoring Indians' Sovereignty by Article 9.

SILVER COVENANT CHAIN

TWO ROW WAMPUM
Self Explanatory

HALDIMAND PLEDGE
(7th April, 1779)

HALDIMAND TREATY
(October 25th, 1784)

Indentification of the Six Nations Iroquois Confederacy of North America.

Under our Council's Seal.

Two Row Times, courtesy of Jim Windle

torian colonial project of extraction, generated and relied exclusively upon paper worlds, written in the 1791 law, for example, of Lower-Canada's *Assemblée Nationale*:

> "*Les mines sont à la disposition de la nation.*"[90]

This paper space, in which the Empire built itself, leaves a paper trail that reveals a long legacy of class separation and racial subjugation by an English intellectual elite. Through language and law of maps, stats, and acts, this paper world controlled the surface of the State, further increasing the distance between Indigenous populations and the land they depended on, and alienating cross-regional relationships. By imposed rule of law, centralized authority, and exclusionary history, this marginalization has grown to a level beyond contestation. In this paper space, the territory of the Crown could not only be surveyed and mapped, it could be translated with plans on paper, it could be researched and quantified in scientific reports and negotiated in treaties, on the imperial parchment of white British calfskin, literally and figuratively. Moving beyond the mere manifest destiny and utopic dream of cross-continentalism that was typical of American ambitions, the quiet revolution of this verse inscribed as coat-of-arms maps out, with religious fervor, the geographic aspirations to travel up the rivers of resource regions and across coasts of mineral rich countries world-wide, across an ever expanding empire on "which the sun never sets."[91] Penetrating deep into its territory, it comes to represent an infinite and depersonalized space of resource potential without apparent variation or without horizon, but with deep cultural biases and bases on territorial dispossession and territorial exploration and expansion without limit, approximating the divinity of extraction akin to the Sun God:

> "His name shall endure for ever: his name shall be continued as long as the sun."[92]

If, as Innis described in 1930, "Canada supplied the British and American economies through the exploitation of its considerable bounty," then it has now become Empire in its own right. Thanks to the Mining Act dating back to thirteenth-century English law, and in some cases to fifth-century

Getty Images / Eddie Mulholland

Tour of Duty
State Sovereign, Queen Elizabeth II, visits the gold vaults
at the Bank of England during an annual tour and inspection
2012

Roman law, Canada is now home to a legion of its own surface mining firms whose practices reflect Canadians everywhere on the surface of the planet. Beyond a mere industrial storehouse for Great Britain, it not only supplies the industry as financial system, it systematizes the ideology of extraction as economic policy across the surfaces of the State and subsurfaces of other states.

The Extractive State thus thrives on legalization; through the translation of conflicted territorial and ecological conditions in the form of documents—from white calfskin to white paper—as a medium of communication (in mind), of control (in form), and of containment (in time). It is in between the lines of the law, in the transfers of internal governmental powers, in the renaming of its institutional organizations, that the Extractive State has learned to grow without consent. It is therefore in the maps, stats and acts that it both builds and preserves power and suppresses emergent errors and flaws.

Indeed, if we can trace new historical periods through the mathematical financialization of the mining industry, then we can also open new territorial scales of time through its statistical quantification. In the eyes of the Bureau of Statistics and of the pens of Dominion Statisticians, the country's statistical geography also describes a quantitative historiography reflective of another deeper level of extractive and exclusive ideologies.

Often associated with extraction technology, the awe-inspiring sites with the large machines, big mines, powerful modeling systems, and complex algorithms of extraction that are the subject of much interest and opposition, it is the *State* itself—Canada—that was, and still is, *a priori,* the preeminent extractive technology. In the process and project of colonization, the extractive State knows no end. After all, its underground is practically limitless. It simply changes in name, sometimes in form, mutating in institutions, while relations and networks of power simply grow as they have since the beginning of the last century:

> "If Canadians choose, in perfecting their mining laws, they have at their disposal a marvelous wealth of experience, rich

The Central Bank of the Bahamas

Bank of Canada

The Central Bank of England

Money, Monarchy, Mimesis
The 3-dollar, 2-dollar, and 1-pound bills as currencies of the Crown & Commonwealth
1974

Franco Nevada

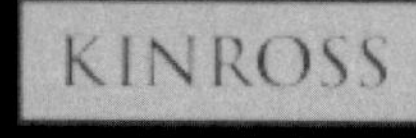

SILVER WHEATON

YAMANA GOLD

World Gold Council

Golden Touch
14 of the original 19 member companies
of the World Gold Council are based in Canada
2015

> with the spoils of time and with the reasoned conclusions of the great systems of jurisprudence which have contributed most to civilization and to human progress."[93]

Whereas territory is usually seen as technology of the State, it is the State that mobilizes territory as a system of political empowerment, cultural settlement, and then as international lever. In fact, its statistical assessment confirms this fact. In the 100-year inventory of the country by the Bureau (now Department) of Statistics, between 1867 and 1967, in what was largely composed as a report to the Crown. As its chief statistician proudly claimed in the last printed yearbook in 1967, celebrating Canada's Centennial:

> "'Canada 1967,' portrays the changing 'Face of Canada' during a *century of progress*; the land and its people, the Canadian manner of nation-building, the spread of settlement, and the development of the pioneer colonial community into a modern industrial state."[94]

The revenge and resurgence of political theory and political economy in Canada then lies in the reformulation of the colonized foundations of land, or the rethinking of colonization of land altogether, whose imperial base has now grown larger and further than ever. These processes of expansion have become telescopic: tracing global vectors of capital flow, as much as localized material demand proves to be as temporal as it is territorial. The interrelationships between staples, communications, and empire all imply that the ongoing actions of the State as 'extractive technology,' requires continued observance and testing of the validity of the theories of staples, communication, and empire building. Though Mel Watkins, and others, have resurrected a materialist perspective of territory, largely as a result of a perceived, premature, yet clear "decline of British [power] and rise of American political influence,"[95] any reassessment of staple theory demands the re-understanding of Canadian empire-building in its own right—this is where the project of decolonization must begin.

BUSINESS WEEK

June 6, 1964

Fifty cents

A McGraw-Hill publication

Below: There's almost nothing that Chmn. F. William Nicks' Bank of Nova Scotia won't try — even trading in gold bars [Finance]

Bloomberg / YGS Group

Money Man

Chairman F. William Nicks on the cover of Business Week in his new gold vault in the basement of the Bank of Nova Scotia

1964

While mining may only appear to account for 10% of the country's GDP, the significance of the mining sector on Canada's GDP has been compared to the financial industry's influence in the United States. As far as the diaspora of global mining is concerned, Canada is now its undisputed homeland and hinterland; both frontier and front-line. With the TSX, Canada is base camp to more than 1,500 of the world's resource extraction firms thanks to increased regulatory freedom and remarkably low royalty rates. "Canada has always been more of a hinterland than a colony,"[96] and now it produces hinterlands.

Canada has experienced various iterations of infrastructural and economic development, but they were not consecutive, nor symmetrical. They were synergistic, made manifest in the territorialization, deterritorialization, and re-territorialization of Indigenous land. In its first iteration, placer mining (for gold mainly) was impetus for the confederation of British Columbia, while the rebellions in the Prairies hastened the building of the transcontinental railway.[97] As the materialization of colonial desire, the railroad was the consequence of political rhetoric and territorial discourse which constituted power.

As John A. MacDonald put it, speaking in the House of Commons:

> "The road will be constructed … and the fate of Canada, will then as a Dominion, be realized. Then will the fate of Canada, as one great body be fixed."[98]

Primarily, the CPR enmeshed Canada within a series of networks of domination. As Innis observes and the suppression of the Métis uprising of 1885 makes manifestly clear, spacebinding technologies extend power as they fos-

David Chancellor, *Intruders*, 2012 / Kiosk

Border Patrol
Security Guard on periphery of
North Mara Gold Mine in Tanzania
2012

While Prairie Provinces were in revolt against Confederation, and would be subjects of future retrocession, the extractive state of Canada required the instituted Britishness of the most western province along the Pacific. In a somewhat preemptive strategy, Queen Victoria named the colony of British Columbia as her territorial linchpin in 1858, one that mobilized power but immobilized resources, less for Canada—the country—but more in the name of the Empire, as worn on its 1895 coat-of-arms of its westernmost arm of the Empire:

"*Splendor sine occasu.*"

"Splendor without diminishment."[100]

Apparently irreversible, that power is now inscribed in every single passport and oath of citizenship for 35 million Canadians—*jus soli*, as birth right:

> "Over the course of the first half of the twentieth century all states gradually developed an increasing range of policy instruments and tools to regulate the movement of people. During this period, passport, border posts, and laws limiting access to employment, investment opportunities, and government services became the core features of the way states regulated international mobility. To a large extent, these policy instruments were geared towards the territorial exclusion of foreigners form national jurisdictions. As part of these developments, the right and capacity of states to refuse entry to unwanted migrants at their external borders—alongside their right to tax goods and defend territory—became a lynchpin of sovereign authority. Simultaneously, the unequivocal right of national citizens to enter their state's territory had become a core privilege of modern citizenship."[101]

With the legislation of Confederation spanning from coast-to-coast, the continental strategy of encirclement of the Crown was complete in 1867, and so was the forging of a new rhetorical, paper-based power at the height of its territorial conquest:

Government of Canada; Public Sector Pension Investment Board; Defence Construction Canada; Standards Council of Canada; Atomic Energy of Canada Limited; Bank of Canada; Business Development Bank of Canada; Canada Lands Company; Canadian Broadcasting Corporation; Canadian Commercial Corporation; Export Development Canada; Royal Canadian Mint

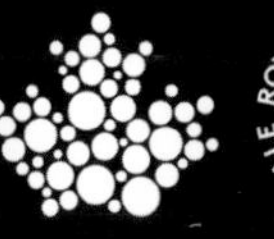

Crown Corporations

Conceived over a century ago, these hybrid entities operate somewhere between the space of a government agency for industries of extraction, export, and energy

1867–2017

"The Crown is not a person, it is a concept. The Crown is legitimacy. With the arrival of responsible government, the Crown could no longer be presented, as the legitimate will of any individual, even if ministers were doing the talking. It could no longer be represented as an expression of power, legitimate or not. Instead, it gradually became an expression of legitimate authority built upon an abstract representation of the land, place and people. So the Honour of the Crown is not simply the obligation to respect formal commitments. It is the responsibility of the civilization to respect its reality."[102]

The pro-British figure, John A. Macdonald, emerged as a local, loyal Victorian hero who not only respected this rhetoric, but persistently represented its political image of superiority. Prioritizing ties to the monarchy, "a new school of historians emerged from 1867 determined to treat Confederation as a brand-new beginning designed to make up for past failures."[103] On paper, Confederation was as much an appeal to the theory of transcontinentalism as it was a denial of the model of navigation-based Métis civilization, that had been, until then, cultivated over two centuries.

From Macdonald, to Mackenzie King, to Martin, this legislative legacy has been a continuous and ongoing one. A statement made by former Prime Minister Paul Martin in 2012 may gather much controversy today, but in retrospect, is less astonishing for its near routine banality:

"We have never admitted to ourselves that we were, and still are, a colonial power."[104]

Under the Crown—in between the lines of the laws, the charters and the agreements—the ontological notion of 'land' disappeared, and instead, appeared a new power that relied on the strict separation of surface rights and subsurface rights. As a result, defenders of land, that have existed in contraposition to this Empire of paper and parchment, have consequently been forced to erect the blockades, stage interventions, and reveal the political agencies of land, often in ways made illegal by the same state structure that has been constantly re-ordering and re-territorializing the state of affairs through waves of re-surveying, re-financing, re-legalizing, relicensing, resampling, re-exploring. Now that Crown Land assumes a value greater than its monetary potential, a value greater than its market-worth. This elevated posture of the Crown is an enclosure of the imagination as much as an enclosure on the map. It assumes its own rightfulness that eventually meets resistance and revolution.

Increasingly transnational, the perception of the Victorian project of extraction remains arguably industrial, but not yet imperial. It refuses to openly acknowledge its colonial underpinnings. Although its colonial role as resource warehouse may have substituted the demand from the UK for the US—its natural economic next-of-kin—the magnitude of Canada's operations worldwide makes it what it never sought to imagine; a global resource empire whose territorial influence has become a colossal geological map of the world.

In the context of future change, rarely, as Frantz Fanon observed in his 1961 *The Wretched of the Earth*, is the process of change, unperturbed:

> "National liberation, national renaissance, the restoration of nationhood to the people, commonwealth: whatever may be the headings used or the new formulas introduced, decolonization is always a violent phenomenon."[105]

But what underlies the extractive state—under and in between the white space of legislation, are the living populations of pre-State First Nations, including nearly 1.5 million people, distributed across different urban areas and over 800 reserves across the country. Between exploration and exploitation, extraction arose from the lands and cultures of First Nations: Cree, Mohawk, Dënesųłiné, Huron, Iroquois, and hundreds more Indigenous populations. The project of extraction required resource acquisition, and to acquire resources, the project of extraction not only requires access to the subsurface. It also requires access across lands of Treaties and Territories. The partitioning of Canada into provinces and territories followed patterns of nineteenth-century settlement and resource development, giving rise to post-confederation rebellions throughout contested land claims in the Praiies and the North. From Louis Riel in the Red River Rebellions in 1884-85 to Elijah Harper's refusal to sign on to the Meech Lake Accord in the Manitoba Legislature in 1990, evidence demonstrates that the Victorian era of consolidation of the country over-reached beyond not only the Royal Proclamation of 1763, but over time, saw the gradual erosion of Indigenous rights, while seeing it more recently reemerge.

But they are preceded by earlier extractive actions of displacement through a litany of measures of dispossession. The case of the Mohawk Nation moving from Hochelaga in 1717 (now Downtown Montreal) to Kanesatake and parallel displacements of the Seven Nations—Mohawk of Akwesasne, Mohawk of Kahnawake, Mohawk and Anishinaabeg of Kenesatake, Abenaki of Odanak, Abenaki of Becancour (Wôlinak), Huron of Jeune-Lorette (Wendake), Onondaga (of Oswegatchie)—known as the Oka crisis of 1990, or better understood as the confrontation of the State on the Mohawks of the Akwesasne Nation, ought to be considered in this history as a global turning point.

As a conflict over land, it is the most pivotal case in Canadian history and arguably, in the world. The standoff that later took place during 79 long days of the summer of 1990, between two nations—the Mohawk Nation and the Crown Nation—illustrated the depth and extent to which five centuries of displacement attempted to dismiss, dislodge, and debase the actual potential of pre-Canadian nations, as well as their territorial rights. Triggered by the badgering and condescending proposal of town Mayor Jean Ouellette for the extension of a 9-hole golf course into an 18-holer, with an accompanying residential development that was largely opposed by local residents, the Mohawk nation refused to allow access to lands that not only featured sacred and beautiful riverside forests of Eastern White Pines; the lands also hold an ancient burial ground of great cultural Aboriginal value. The disrespectful and unapproved actions of a single mayor, then exacerbating reactions from the Mohawks who barricaded a major route into the area, regionally caught the attention of all Montrealers, further when two major bridges were blocked.[106] Stopping flow through territory was both a

OPSYS / Zannah Matson & Jacob Moginot, 2017

Crown & Country Architecture

The Royal Bank Plaza Building in Toronto designed by WZMH Architects, after the CN Tower (1976) and before the Scotia Plaza (1989)

1976–1979

Crown & Corporation
The classic imperial logo and heraldic lion head
of the Royal Bank of Canada looking leftward—to the west, before the turn of the millennium
1962–2001

Royal Bank of Canada

Crownless Corporation

Looking rightward—to the east, the abbreviated, truncated, globalized logo of the "new" Royal Bank of Canada, mining bank to the

2001–Present

technique of delay placed on the common citizen and also a frontal assault on the surface of the State, a plea for an open international discourse.

The tumultuous standoff in 1990 also drew an active body of thousands of supporters from as far as Mexico and Guatemala, Japan to Indonesia, in peace camps located in Quebec and Manitoba demonstrating the power of Indigenous solidarity and its resurgence. The importance of the confrontation at Oka is not only the result of a nation declaring war on its own people through the deployment of its Federal Army, it was essentially one of many in a long line and litany of oppressive acts of subjugation and oppression. Colonial power could read through nature, development, and gender. Importantly it is read through the value of land itself. “This story is also broader, pointing far beyond this particular conflict to a much larger ecological and political crisis in which colonizers regarded common, undeveloped lands as unappropriated. The lands were thus available for expropriation unlike ‘Property.’”[107] Through the racist extensions of its provincial police (*La Sûreté du Québec*), the unapologetic aggression of a local mayoral office and the gross negligence and dangerous obliviousness of a federal government state, events at Oka in the hot summer of 1990 epitomized the image of its own domination that transcends over 5 generations.

If Oka erupted in the summer of 1990 as the nation’s and the world’s most explosive land claim clash in contemporary history,[108] then the state irreversibly was marked with a renewed threat revealing the thinness of the paper on which rests the fundamentals of its self-proclaimed power.

Furthermore, Oka revealed to many Canadians a deeper history of resistance that has been ongoing and growing in kind with the pervasive pattern of global resource urbanism. From the Dene Autonomous Movement of the 1970s, to Ipperwash and Gustafsen Lake in the 1990s, to Idle No More, Lac Mégantic, and #OccupyINAC today, these events are no longer isolated acts of resistance. They are movements from the ground up that have taken different forms in response to unreported contraventions, undisclosed assaults, and unfulfilled responsibilities of the Crown in Indigenous populations. Yet, they remain an underlying dimension of the complexities of resource extraction, industrial development policies, envionmental impacts, unpaid royalties, and territorial claims associated with and rights. While the Government of Canada has sought to redress injustices and inequities as a result of these complexities through emerging Land Claims Agreements at home (the Territory of Nunavut marked a major turning point in the political geography of Canada in 1999), the presence of Canadian mining operations and investment interests abroad are taking unprecedented scales of operations with international governments and Indigenous populations in developing regions worldwide.

In recent legal cases, the court has become the mediator of story and empire, land and communication. Recently, pre-confederate conceptions of land and communication have become more explicitly linked through these court cases whereby the medium of communication comes to represent different appeals to sovereignty. Through these cases we begin to note a renewed validation of oral culture. After the 1969 White Paper, which sought to annihilate Indian title, attempted to erase Aboriginality and act as a final solution in the struggle to assimilate, these cases are an important renegotiation of Aboriginal [Land] Title in Canadian courts.

Since the White Paper, land claims have passed uncomfortably through the courts, from the *Calder v. British Columbia* case of 1973 reaffirming Aboriginal Title with reference to the Royal Proclamation of 1763, to the important case of the KI First Nation of the James Bay, who were compelled to insist their treaty rights were not overruled by the Ontario Mining Act.[109] *Chartrand v. British Columbia* continues now to sit in this tradition of interpretation, in this case that the Douglas Treaties established for the Hudson's Bay Company in the middle of the nineteenth century.

But two cases are particularly important in their treatment of the Oral World, not the Paper World: *Delgamuukw v. British Columbia* (1997) and *T'silhqot'in v. British Columbia* (2014). The landmark Delgamuukw land claims of 1997 has been read as an attempt to legislate title out of law and as a return to the unsuccessful 1969 White Paper. Importantly, the case was one that set precedence in determining that oral evidence can be submitted during a land claim case. In 2014, this precedent would prove vital and helped secure victory for the Tsilhqot'in Nation. In that sense, the Tsilhqot'in case,

> "made new law in the areas of the duty to consult and accommodate, governments' justification of infringements of Aboriginal title, and federalism ... the decision is extremely important for at least two reasons. First, as part of its return to principles set out in the Court's 1997 decision in *Delgamuukw v. British Columbia*, [1997] 3 S.C.R. 1010, Tsilhqot'in Nation includes a return to an equal role for Aboriginal perspectives that includes Aboriginal laws, instead of the exclusive focus on Aboriginal practices."[110]

The other important aspect of Tsilhqot'in is that it "accords with a territorial approach to aboriginal title, one that does not require and piece together intensive use of well-defined tracts of land."[111] It is fluid, as it were.

When, at Delgamuukw, Chief Justice Lamer ruled oral history as admissible evidence, and that oral evidence to that point had been dealt with in an inadequate way, he found laws of evidence must adapt to accommodate oral ways of knowing which are equal with other forms of historical evidence. Lawyer P. Michael Jerch has laid out the consequences of this case and found that despite the weight of the decision, oral history has still not been interpreted in the manner Lamer intended. There has been a shift to hear oral evidence, but this change has been applied inconsistently. "This means that First Nations with Aboriginal rights claims have been faced with the uncertainty of putting forward their oral history evidence without knowing how, or even if, it will be heard."[112]

During the Tsilhqot'in case, the First Nation relied on testimony from their band's elders. "In Justice Vickers' decision he gave careful consideration, not only to the oral history evidence, but also to the issue of oral history as evidence. As a result, the decision in the Tsilhqot'in case is one of the most clear and thorough treatments of oral history as evidence that any court in Canada has yet provided."[113]

> The classic modern monolithic nation-state has been built around the written word. This is the tool for defining meaning, narrowing meaning and asserting power … Part of the atypical nature of Canada lies in the persistence of the oral."[114]

It is no wonder communications theory emerged from such a society and landscape as Canada's. Saul would argue that this is tied directly to our origins as an Aboriginal culture, a Métis Nation. He would remind us that though it is commonplace to think that Aboriginal populations for centuries after contact were illiterate, so were most of both Francophone and Anglophone communities. These Canadians would have had important legal texts read to them, as part of a communal custom. Indeed for Saul, this had a direct effect on how Canadian texts have been composed since: "The meaning of Canadian texts was digested within an oral tradition."[115]

Despite the Delgamuukw and Tsilhqot'in cases, there is still no protocol among the courts to deal with oral testimony or how oral evidence should be weighed against different types of evidence. Until now it has resided next to heresy, not as actual evidence that is regarded equally alongside written records. In the past, oral evidence has required corroboration by written records. Yet, the Tsilhqot'in decision demonstrates a return to the oral tradition from which Canada was founded as a colonial, multi-national society, echoed by Judge Vickers' statement:

> "Courts that have favored written modes of transmission over oral accounts have been criticized for taking an ethnocentric view of the evidence. Certainly the early decisions in this area did little to foster Aboriginal litigants' trust in the court's ability to view the evidence from an Aboriginal perspective. In order to truly hear the oral history and oral tradition evidence presented in these cases, courts must undergo their own process of decolonization."[116]

The entire tension between the oral and written, between the textual bias of the western legal system, as it is expressed through conflict and negotiation over land rights and Aboriginal consultation, demands a constant reinterpretation of the forms, relations, and translations of Empire. The confluence of Canadian state power, like Saul describes, is the combination of a bilingual colonial regime, "meaning" becomes contested out of the gate, though seen in the light of Innis:

> "An oral tradition implies freshness and elasticity ... the binding character of custom in primitive cultures."[117]

When we begin to acknowledge the nature of pre-confederation Canada as oral, as fluid and interpreted, we can more easily trace the colonial construction of Confederation as a loose leaf federation whose entire foundation is built on the parchment of propaganda and the vellum of dichotomous divinity; Confederation as a skewed, if not grossly simplified myth.

AID TO EXTRACTION

From People to Paper to Policy

In order to continue projecting power in the world, the extractive state must continually produce new frontiers and new frontlines. The paper trails left by the resource authorities, the federal land departments, the mine operators, the investors, and the landowners' associations reveal the importance not only of the players involved across processes of development, but also the layers of the legal landscape materialized by the thumbprint signatures of local, illiterate citizenry, in the complex and sometimes uneven world of land agreements, laws, licenses, and leases.

If the legacy of this paper trail left by the extractive state is one built on a history of land leases, captured by the Confederation, the Proclamation, the HBC Charter, the Ensign, the Treaty, and beyond, then its future export to the world should reflect a reformulation of this power.

If Canada has become the managerial wing of the world's resource extractive industry, then it comes as no surprise that as the quintessential extractive state, Canada has grown into a heavily systemic, bureaucratic, and administrative entity, in the image of the British Empire. It is the territorialized, financialized, and incorporated state apparatus for the world's mining industry. It relies on the creation of new frontiers and new financial hinterlands to colonize untapped mineral grounds as the vehicle for normalizing its economic supremacy. Like its former parent, the extractive State colonizes, it colonizes the surface, displaces any and all living things from it, in order to penetrate and extract the financial ore below.

Capitalizing on Canada's outdated do-good, humanitarian image, scholars have struggled to describe the true nature of Canada's resource economy.

> "Concerns about whether or not Canada is becoming a 'petro-state' fold into the reality that it is already the quintessential extractive state, superintending the global mining industry while simultaneously pursuing fossil fuel projects and related enabling policies."[118]

While the status of Canada as a petro-state may seem at odds with its conventional national image, its effects extend to the rights and freedoms of the individual. As imperial ideology, 'extractivism,' as Naomi Klein describes, is a,

> "nonreciprocal, dominance-based relationship with the earth, one purely of taking. It is the opposite of stewardship, which involves taking but also taking care that regeneration and future life continue."[119]

Within this definition, the extractive and extractivist state of Canada has been at work far before Confederation: in its early machinations of the Hudson's Bay Company, and much earlier to the Magna Carta that enabled conquest. It is in the historically-rooted fiscal innovation and legal organization of Canada' extractive sector that the financialization of resources becomes important to recognize as instruments a priori of incorporation.

Government of Canada

Coded Arms
Canada's Colonial Coat-of-Arms "A Mari Usque Ad Mare"
(from Sea to Sea) adopted by Proclamation of King George V
1921

With these extractive co-dependencies, Canada is more aptly described as a 'geo-state,' whose legacy is as much geographic as it is geological. Resources have permeated the economic activity of Canada ever since the Hudson's Bay Company was incorporated with a 1670 Charter, formally entrenching land as colonial enterprise in service of extraction, well before Confederation in the late nineteenth century.

Essential to this colonial arrangement has been the construction of 'the frontier,' described by anthropologist Anna Tsing as:

> "An edge of space and time: a zone of not yet—not yet mapped, not yet regulated. It is a zone of unmapping: even in its planning, a frontier is imagined as unplanned. Frontiers aren't just discovered at the edge; they are projects in making geographical and temporal experience."[120]

Now Canada reinforces and exports that imaginary through its national corporate branches. Of exploration in the Amazon, Tsing writes: "Giant mining conglomerates were licensed to save the land from the depredations of wild miners, yet legal and illegal prospectors were inseparable. 'They go where we go,' a Canadian engineer explained, 'and sometimes we follow them.'"[121]

The myth of the twenty-first-century voyageur transforms into a real-time transnational conquistador at the moment of global incorporation. Both acting as corporate entities transcending the laws and boundaries of nations. The myth of frontier became mentality, supported by an incorporated construction of state. Now, it unifies extractive power more than ever, uniquely relying on systems and speculation, in the sands, steppes, and seafloors of the globe.

If this foreign policy is reflective of not only "mining in Canada" but also of "mining for Canada,"[122] then it expresses the export of extractive ideologies aided and abetted by a global geological intelligence across the Commonwealth. This ideology is made manifest through development generated by geological associations and mining institutions cultivated since Confederation. The extractive state is irrespective of political regimes, it is established within the code of state bureaucracy, transcending superficial political shifts.

Now, the international presence of institutions perform a formative role in Canada's mining industry. As an arm of the Department of Foreign Affairs, Trade and Development (DFATD), the Canadian International Development Agency (CIDA) has also become a subsidizing branch of mining operations overseas, "increasingly partnering with some of the largest Canadian mining companies, such as Barrick Gold and IAMGold."[123] As the industry grows abroad, so does the demand for technology and expertise at home. At the cost of $24.6 million, CIDA is funding the Canadian International Institute for Extractive Industries and Development (CIIEID) that began in 2013 and will be completed by 2018 with the goal to collaborate "with developing country leaders and institutions in support of their efforts to properly develop and manage their extractive sectors."[124] The Canadian state and Canadian investment capital are in service of each other's imperialist aims.

The Crown mobilizes these efforts through its arm's-length corporations, structured to extract resources from developing nations as a Canadian parallel of Washington Consensus foreign policy.[125] The rise of the Canadian mining giants since the 1890s has been the result of diligent work by the Canadian state. This holds especially true through the massive creation of Canadian crown corporations and institutions, especially Export Development Canada (EDC) and (until 2013) CIDA. The restructuring of these entities has led to a quantitative jump in the scale of Canadian operations over the past two-decades. Case studies abound, but perhaps no anecdote is better suited to demonstrate Canada's foreign development imperialism than its engagement in Colombia.

Canada has spearheaded a reorganization of Colombia's regulatory regime and a restructuring of mining policy that has resulted in a Canadian hegemony of the country's economy, beginning with a 1996 direct payment from CIDA to the Colombian Ministry of Mines and Energy of $11.3 million CAD. Colombia has been a model for the liberalizing approach Canada has promoted across the Global South:

The Canadian approach consists of a 'foreign policy tool-kit' of institutional reforms, expanded land markets, stability measures, lip service to human rights, aid financing, non-governmental organizations (NGOs) and CSR [corporate social responsibility]."[126]

By 2011, Canadian Mining Firms would make up 65% of exploration companies in Colombia, and 75% of those exploring oil and gas.[127] As an acknowledgement of success in this regard, the Canadian Minister of International Cooperation has diverted funding towards NGOs working in collaboration with mining operations. Canadian aid, as the extension of CIDA, has become the exported face of empire.

As the language of development takes on a "post-Washington consensus" tone, CSR becomes a carrot for transnationals, eager to take advantage of CIDA's funding packages that promote the friendly face of empire. This Ottawa Method of Empire has as its prerequisite the restructuring of state institutions to benefit imperial extractivist ambitions. Further to this tethered funding, CIDA has been instrumental in restructuring the mining codes of developing countries. As Political Scientist Todd Gordon observed in his book *Imperialist Canada*, "The result has been a field day for Canadian Mining corporations."[128] And while Canadian cultural idiosyncrasies may not resound as forcefully as other powers, the Canadian state's imperial intentions are as forceful as any competitor. "Canada's pockets may not be as deep as those of the United States or some of the European powers, but its commitment in this regard is not weaker."[129]

Other arm's length institutions prop-up Canadian mining capital following from this model. Regionally, the Canadian Investment Fund for Africa (CIFA) is set up with public funds and the CIDA-INC (Industrial Cooperation Program) held a mandate "to provide financial and technical assistance to Canadian industrial start-ups in developing countries or countries in transition to a market economy." Amazingly, CIDA-INC as of 2012 was the developing world's largest single source creditor for the private sector, with a bulk of the $12 billion share going to extractive industries.[130] EDC invests money from private interests in companies looking to operate

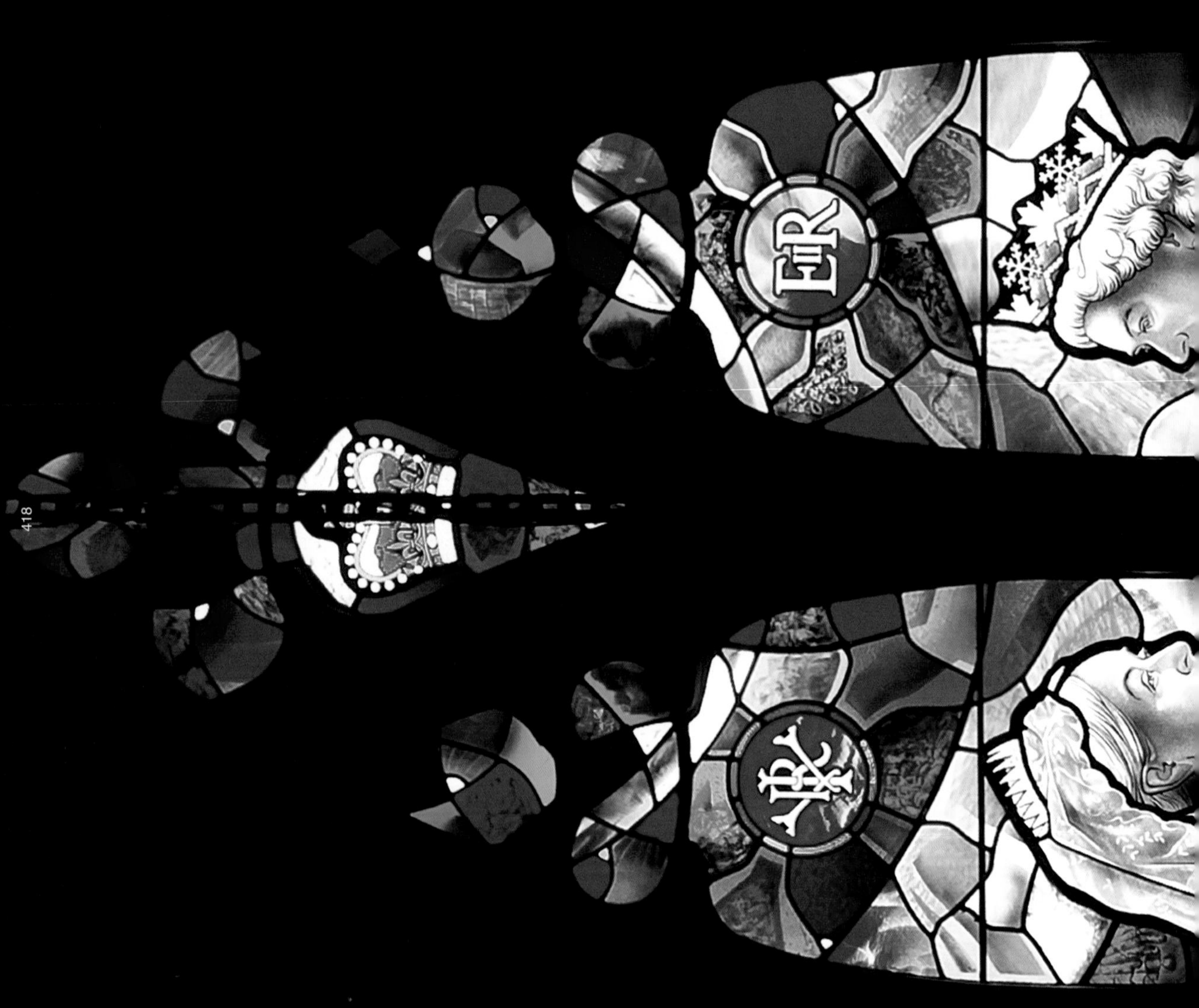

418

Mirage of Monarchy
Portraying a boastful Canada in the image of Britain with an oversize window in the Canadian Senate, gifted by Stephen Harper to the Queen on the Diamond Jubilee
2012

Saffron Blaze, via http://www.mackenzie.co

overseas. As Alain Deneault and William Sacher explain in their book *Imperial Canada Inc.*:

> "The companies' capital, and hence, their solvency, is ensured by the government of Canada. Mining companies also benefit from tied-aid Canadian cooperation programs with weak industrialized countries for major transportation infrastructure projects or hydroelectric dams."[131]

Loans and investment guarantees are siphoned through DFATD and EDC. Canadian state investments implicate citizens in these imperialist ambitions.

> "Whenever we contribute to retirement funds, buy insurance, invest in RRSPs, or put money into savings accounts at banks or credit unions, we are sending money to be invested on the TSX, and a good part of this money will probably end up in the mining industry."[132]

The Canadian state has been actively promoting a liberalization of markets in the Global South by way of the liberalization of its own development branches. The mandate for the Canadian Empire is thus self-fulfilling. As EDC takes care of most of the financing for the industry abroad, the Halifax Initiative reported as far back as 2001:

> "EDC is an important financier of large infrastructure and resource extractive projects, which by nature are more likely than others to have adverse impacts on the poor, labor, human rights and the environment."[133]

The EDC further fulfills its role within the Empire by providing Political Risk Insurance (PRI) to the extractive industry for international operations. PRI can cover investments when operations are threatened by risks as wide-ranging as political violence or threats of repossession.[134] PRI can therefore allow a corporation to begin an operation in a region and destabilize it through uneven development with the assurance that it will recoup maximum financial gains. The collusion between development and industrial strategy has become more entrenched as these Crown and arm's-length institutions cooperatively strategize global conquests.

Operating in support of the Crown, the Canadian banking industry has provided these institutions with leverage and inspiration. As Gordon writes:

> "RBC did not build up its international empire over the last century by seeking out so-so returns. Predators never go in half-hearted."[135]

This ruthlessness is mutually assured and exhibited even in the heart of conflict. In Iraq, at the height of the Global War on Terror, the EDC offered risk protection to RBC and their investments in that company's infrastructural programming.

> "And so it goes. A Canadian bank exploring naked imperialism, and in the process helping other Canadian companies get their piece of the Iraqi pie."[136]

Since the fall of Communism, the largest mass uprising in the country of Romania has been in protest of the proposed Rosia Montană gold mine by Canadian company Gabriel Resources. Thought to be a routine industrial intervention for Canadian industry, instead Romanians turned out in the tens of thousands, over several months, to protest the proposed operation. Through an "underwriting syndicate," RBC raised $254 million for Gabriel. CIBC and the Bank of Montreal provided financing. If built, Rosia Montană would become the largest open pit cyanide-leech gold mine in Europe.

In 2013, the Romanian government attempted to pass a bill that would, among other things, declare the project as a public utility, overriding the public interest. Fought hard by opposition groups in Romania, the proposed "Special Draft Law" would have given unprecedented powers to Gabriel. The bill, which would have been in conflict with EU law, did not pass into legislation. As of July 2015, Gabriel is seeking compensation from the Romanian state based on Bilateral Investment Treaties (BITs) between

Canada and Romania, despite its own risk insurance. The Canadian state has become oriented to ensure whether or not mineral extraction occurs, financial extraction is guaranteed.

This new strategy dovetails a series of emerging agreements between Canada and dozens of countries—such as Free Trade Agreements (FTAs) and Foreign Investment Protection Agreements (FIPAs). This imperial-industrial complex continues to grow with an extractive logic across powerful African states with a contemporary confederationalist and colonial logic that is forming a new Commonwealth. But, as Deneault and Sacher caution:

> "From the establishment of its own Dominion from Sea to Sea, to its participation in other country's civil wars, to its all-out defense of the interests of the extractive industry in international institutions, Canada has reinvented itself time and again only to better serve the same interests: those of the speculators and exploiters of the resources of the world's land."[137]

LAND AS RESISTANCE

Displacing & Decentering the Settler-State

The bureaucratic apparatus at the State's Service—from accountants and lawyers in government, to the financial analysts, geomatic experts, corporate social responsibility officers, the face of surveyors and explorers of today may have changed, but its predominantly white power structure remains the same, with very few exceptions. Between the rows of financial information and in between the lines of the leases and agreements, Eugene Forsey's 1967 centennial provocation resonates:

> "The British North America Act was designed by British overlords; from which it follows, of course, that we must now scrap it and give ourselves a homemade one."

Fool's Gold

1.5kg gold chalice by Canadian mining company Buffalo Gold
gifted to Pope Benedict XVI as gratitude for mining in Sardinia, Italy
2008

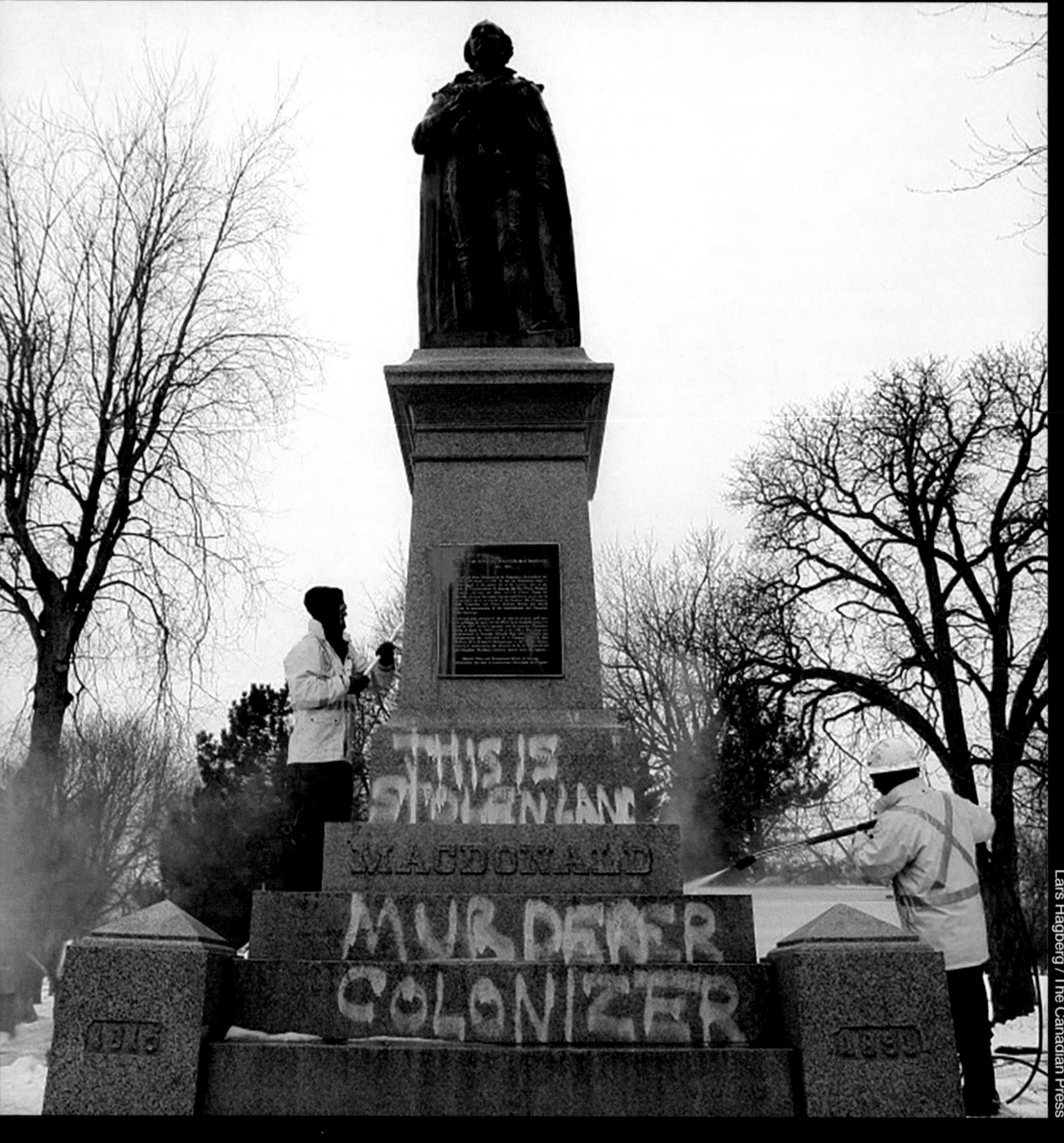

Lars Hagberg / The Canadian Press

On Stolen Land

Statue of John A. Macdonald, Kingston, Ontario
on the 198th birthday of Canada's first Prime Minister
2013

Be it Guatemalan gold, Congolese coltan or Albertan oil, resources are required to support contemporary urban life; but dependency on extraction and exploitation is not a foregone nor inevitable conclusion. Either in the assembly of consumer goods like smartphones or the construction of concrete highways, Canadian life may be mediated, currently, through mineral extraction, but is it really the permanent urban ore in which the future is cast?[138] Can we not imagine a world in which enough is enough? Where we can live on top of the water that we drink? Where materials are made from infinitely renewable materials that are consistently replenished, as opposed to depleted, through our ways of life?

Particular to Canada, the so-called marriages of geography and geology, distance and the extractive economy, Indigenous territorialities and the disturbances of Empire, reveal the contradictory elements and paradoxical inequities of the modern extractive media. As Marshall McLuhan proclaimed:

> "When one of these staples becomes dominant as a social bond, it serves, also, as a store of value, and as a translator or exchanger of skills and tasks. The classic curse of Midas, his power of translating all he touched into gold, is in some degree the character of any medium, including language. This myth draws attention to a magic aspect of all extensions of human sense and body; that is, to all technology whatever. All technology has the Midas touch. When a community develops some extension of itself, it tends to allow all other functions to be altered to accommodate that form."[139]

Could a re-orientation towards alternative material media and territorial justice be expressions of these new communities that confront current regimes of extraction and to current formations of Empire? It was the Hudson's Bay Company, incorporation and territorialization together *in utero*, that developed into contemporary Canada amidst Victorian representation: the corporate extension of ourselves, our Confederation. Could new representations alter and bring an end to these colonial and corporate extensions of empire? For this, we must not only imagine but also unbuild concepts and constructs of Empire in order to dismantle the colonial world of oppression and destruction it has created.[140]

As Arthur Manuel concludes with proclaiming "the end of colonialism" in his 2015 book *Unsettling Canada: A National Wake-Up Call*:

> "We know that the creator did not give the settlers the right to exclusively benefit from our natural wealth and resources. It is colonialism that gave settlers the power to economically exploit our lands, crush our cultures, and dominate our peoples. It is our responsibility to move Canada beyond this exploitation and help the global community move one step closer to peace and security for all peoples."[141]

Moving into the twenty-first century, the opposing process of extraction is a counter-project that clearly requires a different kind of imagination, new ways of engagement, and new forms of representation. This book aims to open a lens on these new horizons and develop a deeper discourse on the ecologies of extraction, nationally, transnationally, and territorially. If it is to do so justly, and intelligently, it will have to grapple with the geographic advantages as much as the social reactions of transnational operations, the environmental realities of resource extraction as much as the economic myths of mining cultures. Canada needs to re-examine and re-imagine its imperial role throughout the world, as much as it does at home, for the foreseeable future and for the current generation. But it will have to re-educate itself by listening to the stories of atrocities, inhumanities, and injustices it has committed over the past two to three centuries and act swiftly and justly on the 800-year old legacies of the past and of the present through reflective cognition and reflexive action.

As histories of domination continue to grow over and over again the varying scales of domination that the Extractive State has inflicted and continues to impose will require transparency of its actions, disclosure of its management and bureaucracy, as well as a gradual weakening of the Crown's control on the surface and superficies of the State to avoid histories of colonial domination and extractive oppression from repeating themselves. That process of deconstruction is built from rage as much as it needs to be built with love for land, and for the Indigenous peoples that have protected it, cherished it, and nurtured it unconditionally, for over ten thousand years and beyond.[142] In this vein, the legacy of the next generation can be rebuilt and regrown—one seed and one skin at a time, across shorelines of conflict and collaboration, from one nation to the next.

× × ×

Notes

1 Gayatri Chakravorty Spivak, *Outside the Teaching Machine* (New York, NY: Routledge, 1993), 48.

2 This statistic has been quoted in various publications and from different perspectives, from Prospectors and Developers Association of Canada (PDAC) and The Mining Association of Canada, "Canada's Mining Industry: Socially Responsible Global Leader," September 1, 2010, http://www.pdac.ca/docs/default-source/public-affairs/fact-sheet-mining.pdf?sfvrsn=6; to Dave Dean, "75% of the World's Mining Companies Are Based in Canada," *VICE News*, July 9, 2013, http://www.vice.com/en_ca/read/75-of-the-worlds-mining-companies-are-based-in-canada. It assumes 1,800 Canadian-based companies operating globally as of 2009.

3 Including the Toronto Venture Exchange, otherwise known today as the TMX. For simplicity purposes, the more commonly known TSX is used here.

4 Mining Association of Canada, "Facts and Figures 2013," January 29, 2014, http://mining.ca/sites/default/files/documents/FactsandFigures2013.pdf.

5 Harold A. Innis, *The Fur Trade in Canada: An Introduction to Canadian Economic History*, rev. ed. (Toronto, ON: University of Toronto Press, [1930] 1956), 393.

6 David Jackel, "Ocean to Ocean: G.M. Grant's 'Round Unvarnish'd Tale,'" *Canadian Literature* 81 (1979): 7; See also George Monro Grant, *Ocean to Ocean: Sandford Fleming's Expedition through Canada in 1872*, Enl. and rev. ed. (Toronto, ON: Rose Belford Pub. Co., 1879).

7 The actual quote was "we have too much geography," which was adapted into this text from Prime Minister Mackenzie King, Speech to the House of Commons, June 18, 1936, *House of Commons Debates, 18th Parliament*, 1st session: vol. 4 (1936), 742, accessed at Library of Parliament, http://parl.canadiana.ca/view/oop.debates_HOC1801_04/742?r=0&s=3.

8 See Harold A. Innis, *Empire and Communications* (Toronto, ON: University of Toronto Press, [1950] 1972).

9 Magna Carta, 1215, clause 39, English translation by *British Library*, accessed June 5, 2015, https://www.bl.uk/magna-carta/articles/magna-carta-english-translation.

10 Canada Department of Justice, "The Canadian Constitution," http://www.justice.gc.ca/eng/csj-sjc/just/05.html. Those principles actually resulted in the exclusion of those not considered to be citizens of the royal realm or the kingdom, further excluding those considered to be savages, or uncivilized. See Ralph V. Turner, "The Meaning of Magna Carta since 1215," *History Today* 53, no.10 (October 2003): 29–35. The Centre for Constitutional Studies at the University of Alberta reports that the "Magna Carta is not formally part of Canada's Constitution, but it played an important role in its creation. Concepts such as the rule of law and many common law principles derive from it, and these do form part of our Constitution." Although is "not formally our constitution, but influences it," law student Mark Moore reports in a lucid straightforward manner that "Constitutional expert Peter Hogg and court cases such as R v. Dobell and R v. Jebbett argue that Magna Carta is not a "constitutional document" in Canada. That said, Canada's founding constitutional document at Confederation was the British North America Act, 1867 (now known as the Constitution Act, 1867). The Preamble of the Constitution Act, 1867 states that Canada adopts a constitution "similar in Principle to the United Kingdom," which does include principles from Magna Carta. Much of the actual text of Magna Carta was formally removed from the statute books by the British Parliament a few years before Confederation, but Magna Carta principles such as the rule of law shaped our Constitution at Confederation. See Centre for Constitutional Studies, "Part II: Magna Carta and Canada's Constitution," *The Constitution* (web publication), November 23, 2015, http://ualawccsprod.srv.ualberta.ca/ccs/index.php/constitutional-issues/democratic-governance/884-part-ii-magna-carta-and-canada-s-constitution#_edn2. See also R v. Dobell, [1978] B.C.J. no. 1041 (SC); R v. Jebbett, [2003] B.C.C.A. 69 (par. 4); *Constitution Act*, 1867 (UK), 30 & 31 Vict., c 3, reprinted in RSC 1985, App II, no. 5, Preamble; and Carolyn Harris, *Magna Carta and Its Gifts to Canada: Democracy, Law and Human Rights* (Toronto, ON: Dundurn, 2015), 91–95.

11 Suzanne Zeller, "The Colonial World as Geological Metaphor: Strata(Gems) of Empire in Victorian Canada," *Osiris* 15, no.1 (2010): 101.

12 During the rise of the British Empire, the discipline of geology was exclusively focused on subsurface resources, in sufficient quantities identified for extraction. This objective became more and more important into the mid to late nineteenth century, with depleting reserves of coal. See C.L.E. Lewis and S.J. Knell, eds., *The Making of the Geological Society of London* (London, UK: The Geological Society, 2009).

13 Suzanne Zeller, *Inventing Canada: Early Victorian Science and the Idea of a Transcontinental Nation* (Montreal, QC: McGill-Queen's University Press, 2009), 15–16.

14 Ibid., 34.

15 Charles Geisler, "Disowned by the Ownership Society: How Native Americans Lost their Land," *Rural Sociology* 79, no. 1 (2014): 56–78.

16 Ibid., 74.

17 See Henri Lefebvre, *The Production of Space* (Oxford, UK: Blackwell, 1991), quoted in Geisler, "Disowned by the Ownership Society," 74, ref. "a new geographical space has been produced and an old one retired, and not by mutual agreement."

18 Richard White, "The Nationalization of Nature," *The Journal of American History* 86, no. 3 (December 1999): 976–86.

19 Maurice Quinlan, *Victorian Prelude: A History of English Manners 1700–1830* (Hamden, CT: Archon Books, 1965), 253.

20 John M. Mackenzie, ed. *Imperialism and the Natural World* (Manchester, UK: Manchester University Press, 1990), 69–70; see also, Francis Bacon, "Aphorisms on the Interpretation of Nature and the Empire Of Man," in *Novum Organum* (1620): 345.

21 Harold A. Innis, *Settlement and the Mining Frontier*, vol. 9, part 2 of *Canadian Frontiers of Settlement*, ed. W.A. Mackintosh and W.L.G. Joerg (Toronto, ON: Macmillan, 1936), 397.

22 John Cabot (Giovanni Cabotto) was actually searching for China and instead found Canada mistaking it for India, which led to the entrenched and erroneous formulation of Aboriginal populations as Indians.

23 George Grant, "Reclaiming the Land," in *The Changing of the Guard: The Vital Role Christians Must Play in America's Unfolding Political and Cultural Drama* (Nashville, TN: Broadman & Holman, 1995).

24 *Royal Proclamation*, King George III, 1763, http://www.solon.org/Constitutions/Canada/English/PreConfederation/rp_1763.html.

25 George Grant, *The Changing of the Guard: Biblical Principles for Political Action* (Ft. Worth, TX: Dominion Press, 1987), 50.

26 J.P. Somerville, "The Divine Right of Kings," 367-04, https://faculty.history.wisc.edu/sommerville/367/367-04.htm.

27 Juliet Barker, *Agincourt: The King, The Campaign, The Battle* (London, UK: Little, Brown, 2005).

28 Mark Greengrass, *Christendom Destroyed: Europe 1517–1648* (New York, NY: Viking, 2014), 3.

29 Wesley M. Stevens, "The Figure of the Earth in Isidore's 'De natura rerum,'" *Isis* 71, no. 2 (June 1980): 268–77.

30 Thomas Hobbes of Malmesbury, *Leviathan: Or The Matter, Forme, and Power of a Common-wealth Ecclesiasticall and Civill* (London, UK: Andrew Crooke, 1651), subtitle.

31 "Hobbes's Leviathan," British Library, http://www.bl.uk/onlinegallery/takingliberties/staritems/55hobbesleviathan.html.

32 The explorative, abstract rendering of virgin wilderness by the Group of Seven crafted a national identity of nature with broad strokes, void of Indigenous presence. See Linda Morra,

"Canadian Art According to Emily Carr: The Search for Indigenous Expression," *Canadian Literature* 185 (Summer 2005): 43–57. Bordo and others have argued that the Group of Seven, of which Carr would become an influential member, perpetuated a homogeneous national identity—either appropriating or erasing aboriginal culture from their art, despite (or perhaps because of) a worldview informed by nature as the mystical replacement for Anglicanism. Thus hybrid identity became Canada's ruling representative language of modernity.

33 Emily Carr, *The Book of Small* (Toronto, ON: Oxford University Press, 1942), 112.

34 *A Proclamation*, (March 26, 1867), Queen Victoria Regina, https://upload.wikimedia.org/wikipedia/commons/b/b0/Proclamation_Canadian_Confederation.jpg.

35 John Ralston Saul, *A Fair Country: Telling Truths about Canada* (Toronto, ON: Viking Canada, 2008), 14–15.

36 The Dominion Land Acts were a form of state standardization as a tool of settlement. Demanding and deploying standards of survey (cadastral), eventual homesteaders would then be required to demonstrate a value-added element by way of construction or cultivation. While this was meant to encourage European settlement, it explicitly did not extend rights to the subsurface from the Crown to the landowner. "While the availability of land helped to entice settlers, the Dominion Lands Act also ensured that inhabitants would have no control over the land and other resources: The Dominion would hold this control until 1930, when the Federal Government transferred responsibility of land issues to the provinces." Elizabeth Mooney "Dominion Lands Act / Homestead Act," (Canadian Plains Research Centre, 2006), http://esask.uregina.ca/entry/dominion_lands_act__homestead_act.html.

37 See John S. Galbraith, *Hudson's Bay Company, as an Imperial Factor 1821–1869* (Los Angeles, CA: University of California Press, 1957).

38 Emerson Hough, "Bucking the Hudson Bay: What modern Competition is Doing to the Earth's Oldest Monopoly," *The Saturday Evening Post* 183, no.19 (November 5, 1910): 3.

39 Chartered in London (England), the Hudson's Bay Company is the oldest and largest transnational corporation in the world. See E.E. Rich, "Pro Pelle Cutem," *The Beaver Magazine* (Autumn 1980): 65–59; R. Watson (ed.), "The Coat of Arms" *The Beaver Magazine* (1929/1945): 16–18; provided by HBC Corporate Historian Joan Murray. For additional historic trademark information, see Canadian Intellectual Property Office (CIPO), Trademark SPECIAL & LABEL DESIGN, Expunged, 0484347, TMDA034446 (1904).

40 The Hudson's Bay Company was originally called the *Governor and Company of Adventurers of England Trading into Hudson's Bay*; see Siân Madsen, "The Evolution of Recordkeeping at the Hudson's Bay Company," *Archivaria* 66 (Fall 2008): 25.

41 This poem is attributed to John Dryden, 1672. See Sir James Schooling, K.B.E., *The Hudson's Bay Company 1670–1920: The Governor and Company of Adventurers of England Trading into Hudson's Bay During Two Hundred and Fifty Years 1670–1920* (London, UK: The Hudson's Bay Company, 1920), 68.

42 In contrast to Karl Wittfogel concept of hydraulic civilization where form and power hinged on the control of water for irrigation; see Karl August Wittfogel, *Oriental Despotism: A Comparative Study of Total Power* (New Haven, CT: Yale University Press, 1957).

43 Hudson's Bay Company, *The Royal Charter, for Incorporating the Hudson's Bay Company, Granted by His Majesty King Charles II in the Twenty-Second Year of His Reign, A.D. 1670* (London, UK: Printed by J. Brettell, 1819).

44 See Emmanuelle Saada and Arthur Goldhamme's discussion of hybrids and bastards in "An Imperial Question," in *Empire's Children: Race, Filiation, and Citizenship in the French Colonies* (Chicago, IL: University of Chicago Press, 2012), 17.

45 Friesen Gerald, *Citizens and Nation: An Essay on History, Communication, and Canada* (Toronto, ON: University of Toronto Press, 2000), 54. See also "The Coat of Arms," in the Hudson's Bay Company newsletter *The Beaver* (1945): 16–18.

46 Zeller, *Inventing Canada*, 4.

47 Suman Seth, "Putting Knowledge in its Place: Science, Colonialism, and the Postcolonial" in *Postcolonial Studies* 12, no.4 (2009): 383; see also Daniela Bleichmar, *Botanical Expeditions and Visual Culture in the Hispanic Enlightenment* (Chicago, IL: University of Chicago Press, 2012).

48 Statistics Canada, "Summary Chronologies," March 31, 2008, http://www65.statcan.gc.ca/acyb_r016-eng.htm.

49 Jen Preston, "Neoliberal Settler Colonialism, Canada and the Tar Sands," *Race & Class* 55, no. 2 (2013): 42.

50 Robert J. Carney, "The Hawthorn Survey (1966–1967), Indians and Oblates and Integrated Schooling," CCHA, *Study Sessions* 50 (1983): 609–30.

51 Public Archives of Canada (PAC), RG10, vol. 6810, file 470–2–3, vol. 7: Evidence of Duncan Campbell Scott to the Special Committee of the House of Commons examining the Indian Act amendments of 1920, pp. 55 (L–3) and 63 (N–3) respectively; see Memorandum for the Special Committee of the House of Commons re. Bill 14 (An Act to amend the Indian Act) from the Secretary, Six Nations Council, March 30, 1920, as quoted in Kahn-Tineta Miller, George Lerchs, and Robert G. Moore, "The Impact of Immigration and WWI: 1906–1927," chap. 7 in *The Historical Development of the Indian Act*, 2nd ed. John Leslie and Ron Maguire (Ottawa, ON: Treaties and Historical Research Centre, P.R.E. Group, Indian and Northern Affairs, 1978), 114, 176n57, http://www.kitselas.com/images/uploads/docs/The_Historical_Development_of_the_Indian_Act_Aug_1978.pdf. See also *I am a Boy*, directed by Louise BigEagle (October 28, 2015, RIIS Media), 11 min 53 sec, http://www.riismediaproject.com/i-am-a-boy on Thomas Moore Keesicksik, the "face of assimilation" in Canada who died at the age of 12 while he attended Regina Indian Industrial School in 1912.

52 Carney, "The Hawthorn Survey," 612.

53 See Duncan Campbell Scott, "Indian Affairs, 1867–1912" in *Canada and its Provinces, Volume VII, Section IV, The Dominion Political Evolution Part II*, gen. eds. Adam Shortt and Arthur G. Doughty (Toronto, ON: Brook & Co., 1914), 622–23.

54 John A Macdonald, "Memorandum," *Sessional Papers Volume 16. First Session of the Sixteenth Parliament of the Dominion of Canada*, January 3, 1887, 37.

55 *Indian Act*, R.S.C., 1985, c. I-5, accessed 30 September 2015, http://laws-lois.justice.gc.ca/eng/acts/i-5/, p.3.

56 "The Act's most catastrophic provision for Africans was the prohibition from buying or hiring land in 93% of South Africa. In essence, Africans, despite being more in number, were confined to ownership of 7% of South Africa's land." For more information on the "Laws on Land Dispossession and Segregation" in South Africa, see http://www.sahistory.org.za/topic/list-laws-land-dispossession-and-segregation.

57 Ron Bourgeault, "Canada Indians: The South African Connection," *Canadian Dimension* 21, no. 8 (1988), 7–8. For further information about flows of ideas and transfers of ideologies between Canada, South Africa, India, the United States, and Germany, see Pierre Bélanger with Kate Yoon, "Footnote 57, Canada's Apartheid: The Duplicitous Diffusion of Canadian Strategies of Indigenous Segregation, Assimilation, and Extermination," in this volume.

58 For a comprehensive comparison of the debate and discourse on the exchange and fluidity of ideas on Canada's Indian Reserve System and the Townships of South Africa's Apartheid, see Maria-Carolina Cambre, "Terminologies of Control: Tracing the Canadian-South African Connection in a Word," *Politikon* 34, no. 1 (2007): 19–34.

59 John Toland, *Adolf Hitler: The Definitive Biography* (New York, NY: Anchor Books, [1976] 1992), 702; and James Pool, *Hitler and His Secret Partners: Contributions, Loot and Rewards, 1933–1945* (New York, NY: Pocket Books, 1997), 254–55,

273–74, as quoted in Anthony J. Hall, *Earth into Property: Colonization, Decolonization, and Capitalism* (Montreal, QC: McGill-Queen's University Press, 2010), 244.

60 See Margaret Atwood's unfinished PhD dissertation, "Nature and Power in the English Metaphysical Romance of the Nineteenth and Twentieth Centuries" (PhD diss. [unfinished] Harvard University, Faculty of Arts and Sciences, 1973). For example, as Preston, "Neoliberal Settler Colonialism," describes, the Treaty making process itself was foreseen in the documentation of land by geologists; see also John Sandlos, "Where the Scientists Roam: Ecology, Management and Bison in Northern Canada," *Journal of Canadian Studies* 37, no. 2 (2002): 93.

61 Charles Mair, *Through the Mackenzie Basin: A Narrative of the Athabasca and Peace River Treaty Expedition of 1899*, ed. Roderick MacFarlane (Toronto, ON: W. Briggs, 1908), 121.

62 Canada House of Commons, Debates, January 17, 1881, 488, as quoted in Maurice Charland, "Technological Nationalism," *Canadian Journal of Political and Social Theory* 10, no. 1–2 (1986): 201.

63 Charland, "Technological Nationalism," 200–1.

64 W.L. Morton, "Two Young Men, 1869: Charles Mair and Louis Riel," *MHS Transactions* 3, no. 30 (1973–74 season), http://www.mhs.mb.ca/docs/transactions/3/twoyoungmen.shtml.

65 See "Treaty Texts - Treaty No. 9: The James Bay Treaty - Treaty No. 9 (Made in 1905 and 1906) and Adhesions Made in 1929 and 1930," *Indigenous and Northern Affairs Canada*, last modified August 30, 2013, http://www.aadnc-aandc.gc.ca/eng/1100100028863/1100100028864.

66 Thomas Berger, "The B.C. Indian Land Question and the Rights of the Indian People" (speech, 9th annual convention of the Nishga Tribal Council, Port Edwards, BC, November 1, 1966), 3. See also Michèle DuCharme, "The Segregation of Native People in Canada: Voluntary or Compulsory?" *Currents* (Summer 1996); Daniel Raunet, *Without Surrender, Without Consent: A History of the Nishga Land Claims* (Toronto, ON: Douglas & Mclntyre, 1984), 167.

67 Richard Rohmer, *Mid-Canada Development Corridor: A Concept*, 5th ed. (Niagara Falls, ON: Acres Research and Planning Limited, 1960), 4.

68 Walter Currie, President of the Indian-Eskimo Association of Canada, in a speech to the opening conference for the Mid-Canada Development Corridor in 1970. See Tristin Hopper, "The failed plan to develop the Canadian North" *Yukon News*, October 25, 2008, http://www.yukon-news.com/life/the-failed-plan-to-develop-the-canadian-north.

69 See William Smith's 1815 geological map of England, Wales and Scotland Royal Geological Society in Simon Winchester, *The Map that Changed the World: William Smith and the Birth of Modern Geology* (New York, NY: HarperCollins, 2001).

70 Marshall McLuhan, *Understanding Media: The Extensions of Man*, 1st ed. (New York, NY: McGraw-Hill, 1964), 142.

71 This systemic reformulation of environment as amalgamation and inventory of information, draws from the work of Robert G. Pietrusko, "The Organizational A Priori: History of Land Classification" (lecture, #decoding Doctor of Design Conference, Harvard Graduate School of Design, Cambridge, MA, March 11, 2016); Benjamin H. Bratton "The Aesthetics of Logistics: Architecture, Subjectivity & the Ambient Database," in *Language Systems: After Prague Structuralism*, ed. Louis Armand with Pavel Cernovsky (Prague: Litteraria Pragensia, 2007), 114–48; and McLuhan, *Understanding Media*.

72 Innis, *Empire and Communications*, 7.

73 Melville H. Watkins, "A Staple Theory of Economic Growth," *The Canadian Journal of Economics and Political Science* 29, no. 2 (1963): 141–58.

74 Jim Stanford, ed. *The Staple Theory @ 50: Reflections on the Lasting Significance of Mel Watkins' "A Staple Theory of* Economic *Growth"* (Ottawa, ON: Canadian Centre for Policy Alternatives, 2014), 129, https://www.policyalternatives.ca/sites/default/files/uploads/publications/National%20Office/2014/03/Staple_Theory_at_50.pdf.

75 Alfred Korzybski, *Science and Sanity: An Introduction to Non-Aristotelian Systems and General Semantics* (Brooklyn, NY: Institute of General Semantics, 1933), 750; and Alessandra Ponte, "Maps and Territories," Log 30 (Winter 2014): 61–65.

76 See Innis, *Settlement and the Mining Frontier*.

77 Toronto Stock Exchange, *A Capital Opportunity: A Global Market for Mining Companies* (2014 presentation), 10, http://www.global-mining-finance.com/gmf-autumn/pdfs/Presentations2014/TSX-TSXV.pdf.

78 Made up of the world's leading gold producers, 14 of the 19 members of the World Gold Council are Canadian-based companies.

79 Statistics Canada, *1971 Census of Canada = Recensement Du Canada 1971*, ed. Canada Dominion Bureau of Statistics (Ottawa, ON: Information Canada, 1972), 217–18.

80 Max Haiven, *Cultures of Financialization: Fictitious Capital in Popular Culture and Everyday Life* (Basingstoke, UK: Palgrave Macmillan, 2014), 14.

81 Melville H. Watkins, "Harold Innis: An Intellectual at the Edge of Empire" *Canadian Dimension* 41, no. 3 (2006): 45.

82 Anna Lowenhaupt Tsing, *Friction: An Ethnography of Global Connection* (Princeton, NJ: Princeton University Press, 2005), 73.

83 Ibid., 60.

84 Alain Deneault and William Sacher, eds., *Imperial Canada Inc.: Legal Haven of Choice for the World's Mining Industries*, trans. Robin Philpot and Fred A. Reed, (Vancouver, BC: Talonbooks, 2012), 184.

85 Historic Cobalt, "Historic Buildings, the Royal Exchange Building," *The Corporation of the Town of Cobalt*, August 26, 2015, http://cobalt.ca/visitors/historic-buildings/.

86 Canada's Historic Places, "Cobalt Mining District National Historic Site of Canada," *Parks Canada*, accessed August 26, 2015, http://www.historicplaces.ca/en/rep-reg/place-lieu.aspx?id=4183.

87 Charlie Angus and Brit Griffin, *We Lived a Life and Then Some: The Life, Death, and Life of a Mining Town* (Toronto, ON: Between The Lines, 1996), 49.

88 Philip Smith, *Harvest from The Rock: A History of Mining in Ontario*, ed. Ontario Ministry of Northern Development and Mines (Toronto, ON: MacMillan of Canada, 1986), 152.

89 Psalms 72:8 (KJV).

90 "The mines are the service of the nation" in J. M. Clark, "Mining Legislation in Canada," *The Annals of the American Academy of Political and Social Science* 45 (January 1913): 157.

91 George Macartney, *An Account of Ireland in 1773: By a Late Chief Secretary of that Kingdom* (London, UK: 1773).

92 Psalms 72:17 (KJV).

93 Clark, "Mining Legislation in Canada," 157.

94 Walter E. Duffet, *1867–1967 Canada Centennial Yearbook* (Ottawa, ON: Queen's Printer & Controller of Stationery, 1967): vii.

95 Melville H. Watkins, "The Dismal State of Economics in Canada" in *Close the 49th Parallel etc.; The Americanization of Canada*, ed. Ian Lumsden for the University League for Social Reform (Toronto, ON: University of Toronto Press, 1970), 205; Melville H. Watkins, *Staples and Beyond: Selected Writings of Mel Watkins*, eds. Hugh Murray Grant and David A. Wolfe (Montreal, QC: McGill-Queen's University Press, 2006), 217.

96 Watkins, *Staples and Beyond*, 223; and Zeller, *Inventing Canada*.

97 Innis, *Settlement and the Mining Frontier*, 397–98.

98 John A. Macdonald, *Canada House of Commons, Debates* (17 January 1881), 488, cited in Charland, "Technological Nationalism," 201.

99 Charland, “Technological Nationalism,” 202.

100 B.C.’s first Coat of Arms was adopted on July 19, 1895, and created by Rector, Canon, and Sub-Dean of Christ Church Cathedral, Canon Arthur Beanlands of Victoria; see “British Columbia’s Coat of Arms,” British Columbia, http://www2.gov.bc.ca/gov/content/governments/celebrating-british-columbia/symbols-of-bc/coat-of-arms.

101 Darshan Vigneswaran, *Territory, Migration and the Evolution of the International System* (London, UK: Palmgrave Macmillan, 2013), 96.

102 Saul, *A Fair Country*, 79.

103 Ibid., 158.

104 Mark Kennedy, “Colonial Legacy Haunts Canada, Paul Martin says” *Ottawa Citizen*, November 3, 2012, http://o.canada.com/news/national/colonial-legacy-haunts-canada-paul-martin-says.

105 Frantz Fanon, “Concerning Violence,” in *The Wretched of the Earth [Les damnés de la terre]*, trans. Constance Farrington (New York, NY: Grove Press, [1961] 1968), 35.

106 The cost of the standoff was evaluated at $155 million; see *Kanehsatake: 275 Years of Resistance*, directed by Alanis Obomsawin (National Film Board of Canada, 1993), DVD, 1 h 59 min, https://www.nfb.ca/film/kanehsatake_270_years_of_resistance/.

107 Amelia Kalant, *National Identity and the Conflict at Oka: Native Belonging and Myths of Postcolonial Nationhood in Canada* (New York, NY: Routledge, 2004), 202. See also Glen Coulthard, “The Colonialism of the Present” interviewed by Brad Epstein, *Jacobin*, January 13, 2015, https://www.jacobinmag.com/2015/01/indigenous-left-glen-coulthard-interview/.

108 The Magna Carta remains iconic for its assumed inclination towards individual liberty, but its effects on territory were equally consequential, drawing English Land Law into maturation in the form of a lease between commoner and state, allowing access to common land through a companion document titled the *1217 Charter of the Forest*. In this way, the Magna Carta informed statehood as much as Westphalia, and, “While the Treaty of Westphalia is considered the Magna Carta of the state system, its historical purpose was to keep the Empire functioning politically”; in Mathias Risse, “What to Say about the State,” *Social Theory & Practice* 32, no.4 (2006): 671. A western interpretation of land and law (and Empire) has been enshrined through these paper documents, and these decrees have preserved for years where a pre-settler system has been forgotten. At Oka, we can clearly see pre-settler nations resisting the conception and assumption inherent in these core Eurocentric conceptions of land and title. Francis Fukuyama’s famous thesis was exactly this: a statist interpretation of a shift in state relations with the state. If “The End of History?” was in 1989 intended to predict the demise of anything other than the neo liberal state, it was also therefore the rhetorical extension of NAFTA; the clarion call of the fall of the Berlin Wall. Yet at Oka the following year, Indigenous populations from around the world de-bordered conflict, arrived to fight and act in solidarity with Mohawk Warriors. Movements aligned themselves in a globalism of resistances. Conflict since Oka has been read through Zapatismo and Bolivian water wars, anti-globalization, anti-apartheid, and racial solidarity organizing from Soweto to Seattle to Ferguson, MO. Recently, even the radical wing of the Kurdish resistance has abandoned the project of statehood in favor of libertarian municipalism and the militant anarchism of the Rojava revolution. This worlding, understood as universal history, can be read—if we are indeed to be Voltaire’s Bastards—through Susan Buck-Morss, “The critical writing of history is a continuous struggle to liberate the past from within the unconscious collective that forgets the conditions of its own existence” in *Hegel, Haiti, and Universal History*, ed. Susan Buck-Morss (Pittsburgh, PA: University of Pittsburgh Press, 2009), 85.

109 For both cases see Christopher Moore, “The Calamity of Caledonia: What B.C. can Teach Ontario about Native Land Claims,” *Literary Review of Canada* 18, no. 3 (2010), http://reviewcanada.ca/magazine/2010/04/the-calamity-of-caledonia/.

110 Jonnette Watson Hamilton, “Establishing Aboriginal Title: A Return to Delgamuukw,” *University of Calgary Faculty of Law Blog*, July 2, 2014, http://ablawg.ca/2014/07/02/establishing-aboriginal-title-a-return-to-delgamuukw/.

111 Ibid.

112 P. Michael Jerch, “The Changing Stature of Oral History as Evidence,” *Jerch Law Office*, April 23, 2009, http://jerchlaw.com/Oral%20history_T%27silhqot%27in.pdf.

113 Ibid.

114 Saul, *A Fair Country*, 125.

115 Ibid., 126.

116 Justice D.H. Vickers statement in Tsilhqot’in Nation v. British Columbia, [2007] B.C.S.C. 1700, 36–37, http://www.courts.gov.bc.ca/Jdb-txt/SC/07/17/2007BCSC1700.pdf.

117 Harold A. Innis, *The Bias of Communication* (Toronto, ON: Universty of Toronto Press, 1951), 4.

118 Stephen Collis and Samir Gandesha, “The State of Extraction: A Conference Primer” (SFU Vancouver - Coast Salish Territories: SFU Institute for the Humanities, March 27–29, 2015).

119 Naomi Klein, *This Changes Everything: Capitalism vs. the Climate* (Toronto, ON: Alfred A. Knopf Canada, 2014), 148; see also Henry Veltmeyer, *The New Extractivism: A Post-Neoliberal Development Model Or Imperialism of the Twenty-First Century?*, eds. James F. Petras and Verónica Albuja (London, UK: Zed Books, 2014), 2: “At issue in these developments–the ‘new extractivism’ as it is termed in Latin America—are the dynamics of a system in crisis. This capitalist development dynamic, based on an expanding extractive frontier with intensifying social conflicts over territorial rights, land, water and associated natural resources, can be viewed through the prism of class struggle, political conflict and the resource wars that have accompanied the extraction process.”

120 Tsing, *Friction*, 28–29.

121 Ibid, 32.

122 Anna Lowenhaupt Tsing, “Inside the Economy of Appearances,” *Public Culture* 12, no. 1 (2000): 139.

123 Jen Moore, “Canadian Development Aid: No Longer Tied—just Shackled to Corporate Mining Interests,” MINING issue, *Canadian Dimension* 47, no. 6 (2013): 31.

124 “About,” *Canadian International Institute for Extractive Industries and Development*, accessed September 15, 2014, http://cirdi.ca/about/.

125 In her December 6, 2011 speech at the “Coalition for Health, Ethics and Society (CHES) Annual Lecture,” World Health Organization Director-General, Dr. Margaret Chan, described the 1989 Washington Consensus as “largely failed, [setting] out economic policy prescriptions as a standard ‘reform package’ for debt-ridden developing countries,” http://www.who.int/dg/speeches/2011/ches_20111205/en/; *Encyclopaedia Britannica Online* defines the term as “a set of economic policy recommendations for developing countries, and Latin America in particular, that became popular during the 1980s. The term Washington Consensus usually refers to the level of agreement between the International Monetary Fund (IMF), World Bank, and U.S. Department of the Treasury on those policy recommendations. All shared the view, typically labelled neoliberal, that the operation of the free market and the reduction of state involvement were crucial to development in the global South,” s.v. “Washington Consensus,” https://www.britannica.com/topic/Washington-consensus.

126 Veltmeyer, *The New Extractivism*, 132.

127 Viviane Weitzner, *Holding Extractive Companies to Account in Colombia: An Evaluation of CSR Instruments through the Lens of Indigenous and Afro-Descendent Rights* (Riosucio: RICL, PCN and The North-South Institute, July 2012), 57.

128 Todd Gordon, *Imperialist Canada* (Winnipeg, MB: Arbeiter Ring Pub, 2010), 168.

129 Ibid, 169.

130 Deneault and Sacher, *Imperial Canada Inc.*, 42.

131 Ibid., 36.

132 Ibid., 36–37.

133 "Canada's Export Development Corporation - Financing Disaster," *Halifax Initiative*, last updated May 25, 2009, http://www.halifaxinitiative.org/content/canadas-export-development-corporation-financing-disaster-0.

134 Peter Diekmeyer, "Managing Political Risk in the Global Mining Sector. Proactive and Protective Measures Key to Overseas Ventures," *CIM Magazine*, December 2009–January 2010, http://www.cim.org/en/Publications-and-Technical-Resources/Publications/CIM-Magazine/2009/December/upfront/Managing-political-risks.

135 Gordon, *Imperialist Canada*, 260.

136 Ibid.

137 Deneault and Sacher, *Imperial Canada Inc.*, 184.

138 According to the Ontario Stone, Sand & Gravel Corporation (OSSGA), the first year of a baby's life will require 2,000 diapers, 225 litres of milk and 14 tonnes of aggregate. Ontario consumes 170 million tonnes of aggregate each year. "A technologically mediated Canadian culture, based in the experience of media commodities, would contribute little to a Canadian self-understanding. Rather than interpreting some supposedly Canadian experience, and offering 'a sense of balance and proportion,' technological nationalism can only offer itself in a constantly mutating form. We must develop new rhetorics about and for ourselves, and create our cultures otherwise and elsewhere." Charland, "Technological Nationalism," 217.

139 McLuhan, *Understanding Media*, 139.

140 This definition of empire is based on Michael Hardt & Antonio Negri's threefold definition of empire as concept (not metaphor), boundless and limitless rule (as opposed to defined by nation-state boundaries), and as total regime (as world building). See, Michael Hardt and Antonio Negri, *Empire* (Cambridge, MA: Harvard University Press, 2000), xiv-xv.

141 Arthur Manuel, *Unsettling Canada: A National Wake-Up Call* (Toronto, ON: Between the Lines, 2015), 227.

142 The concept of 'love' as counter-position and counter-strategy to the imperial and industrial techniques of 'extraction' is informed by the life-work of Leanne Betasamosake Simpson, namely her book *Islands of Decolonial Love: Stories & Songs* (Winnipeg, MB: ARP, 2013), as well as by the pedagogical work of Matthew Wildcat, Mandee McDonald, Stephanie Irlbacher-Fox, and Glen Coulthard collected in a special of *Decolonization: Indigeneity, Education & Society* 3, no.3 (2014) outlined in their essay "Learning from the land: Indigenous land based pedagogy and decolonization" (i-xv).

King.

hey promise and engage that they will, in all respects, obey and abide by the law; that will maintain peace between each other, and between themselves and other tribes of Indians, and en themselves and others of His Majesty's subjects, whether Indians, Half-breeds or Whites, this inhabiting and hereafter to inhabit any part of the said ceded territory; and that they will not st the person or property of any inhabitant of such ceded tract, or of any other district or country, or re with or trouble any person passing or travelling through the said tract or any part thereof, and they will assist the officers of His Majesty in bringing to justice and punishment any Indian ding against the stipulations of this Treaty or infringing the law in force in the country so ceded.

nd it is further understood that this Treaty is made and entered into subject to an agreement dated third day of July between the Dominion of Canada and Province of Ontario, —— which is attached.

Witness Whereof His Majesty's said Commissioners and the said Chiefs and Headmen hereunto set their hands at the places and times set forth in the year herein first above written.

ned at Osnaburgh on the twelfth day of July 1905 by Majesty's Commissioners and the Chiefs & headmen in the presence undersigned witnesses after having been first interpreted & explained

Witnesses

Thomas Clouston Rae C.T.
Alex. George Meindl M.D.
Jabez Williams Clerk H.B.Co.

Duncan Campbell Scott
Samuel Stewart
Daniel George MacMartin

Missabay his + mark
Thomas his + mark Missabay
George his + mark Wahoaskung
Kweash his + mark
Nawokesic his + mark
Sombash? his + mark
David his + mark Sainnawap? / Skunk
John his + mark Skunk
Thomas his + mark Panacheese

Signed at Fort Hope on the nineteenth day of July 1905 by His Majesty's Commissioners and the Chiefs and headmen in the presence of the undersigned witnesses after having been first interpreted & explained

Witnesses

C. F. A. Fafard O.M.I.
Thomas Clouston Rae
Alex. George Meindl M.D.
Chas. H. M. Gordon H.B.Coy.

Duncan Campbell Scott
Samuel Stewart
Daniel George MacMartin

Yesno his + mark
Moonias his + mark
John his + mark Wesley?
George his + mark Namay
Joe his + mark Goodwin
Jacob?
Wenangsie his + mark
Abraham his + mark Atlook
George his + mark Quisess
Harry his + mark Oakinee
Katchang his + mark
Noah his + mark Neshinopais

not colonial powers. 5. On Oral Communication: As a vehicle, the voice of communication can control and also liberate. If oral knowledge transcends *time* through an inheritance of lived histories and transfer of transgenerational techniques, the spoken word not only communicates history, but as a forensic form of evidence, it also preserves and produces previously excluded histories. Like a map, the medium of the voice not only challenges the written foundations of Victorian jurisprudence, it challenges the legal origins of the *Magna Carta Libertatum* of the 13th century altogether. Always mean what you say, and do what you say. No more empty promises. 6. On the Planning Paradox: Resistance to industrialization is hard, negotiation is complicated. Institutional infiltration and systemic unplanning are tenuous and often contradictory proposals, but eventually they will become necessary strategies. Like a Trojan horse, the role of the counter-planner is to identify and fuel the emergence of counter-economies within new political ecologies and bodies. Life, not greed, is the future. 7. On Production: If it's not grown or harvested, it will apparently be mined, according to the resource sector. Thanks to new markets and forms of exchange, there is a huge potential in designing means of co-production where sites of extraction are increasingly closer to the spaces of consumption. If a resource is not left in or on the ground, refine it where you harvest it. 8. On Consumption: Resource dependency is not waning. If two-thirds of the world is currently under development, then new patterns of resource cultivation and distribution will be required for the 21st and 22nd centuries. Existing minerals and metals can be recycled infinitely. Mining raw virgin resources is obsolete. Every single material should be renewable, making wood as well as proximity to forests, essential to an infinitely renewable future. Learn to love the trees for the generations they embody. 9. On Metropolis: The world is not, and will never be, a city. Concentrated cities will have to weaken their stronghold on the hinterland by rethinking their exclusionary footprint. Consumption simply should not, cannot continue to drain and starve what it considers to be its hinterland. Otherwise, the city, as we know it, will become the backwaters of the territories it chooses to ignore and dispossess. We will never be urban. 10. On Energy: To rethink extraction is to re-think energy. If we are to rethink forms of power, we must conquer technologies of combustion and destroy the British order of the

source culture is emerging across the more than 80% of lands throughout the world beyond cities and where almost 1.5 billion Indigenous peoples live or belong to. This cultural resurgence marks a territorial turning point in transnational politics and political geographies with the imminent dissolution of the hegemony of the Crown over resources, and of the States over cities. The banking systems that hold them together will have to weaken. 16. On Land: Territorial claims and ecological politics of pre-State populations—whose reclamation of surface and subsurface rights, as well as the live engagement of the 'non-anthropogenic'—will form the basis and bones for the independence of individual nations and eventually lead to the birth of new constituencies and reclaimed freedoms. Moving towards the 22nd century, it is the freedom that this living landscape and dynamic ground provides that we will all eventually have to fight for, in this lifetime. 17. On Water: There is no greater freedom than living on top of the water you drink. There is no greater freedom than living off the life that depends on that same water. Since waters are indivisible—rivers, lakes, oceans, estuaries, groundwaters, rains, clouds, dews, droughts, snows, ice, are all connected *in place* and *in time*—all forms of transformation, like families, need to guarantee the longevity of this interconnected system. Love the infinity that water represents as part of your family. 18. On Sovereignty: Since boundary delineation has inflicted unspoken oppression on Indigenous cultures that existed long before the Crown—lands must be unmapped, unsourced, and unsurveyed by the Crown to be reinstituted to the originals: First Nations, Inuit, Métis. Canada's Constitution needs to be deconstituted, in order to be reconstituted. Unsettling the lands and resources of the State, the Crown will eventually have to cede and surrender its power. 19. On Deterritorialization: The Doctrines of Discovery issued in the Papal Bulls of the 15th and 16th centuries that promote violence, uphold dispossession, and illicit racism through beliefs of *terra incognita* and *white supremacy*, must be unconditionally rejected by the Church, State, and Crown. Their repudiation will affirm the political and territorial sovereignty of Indigenous peoples and bring down the perpetuation of racial inequities, political injustices, and environmental inhumanities that Christian, Eurocentric, and heteropatriarchal power has upheld. The reclamation of land is thus non-negotiable, for life and self-determination, if there is to be hope for the people, and possibly for the Pope. ×

Undermining Empire

A Landscape Manifesto for this Generation.

0. On Displacement: All development, like all forms of colonization, displaces. If displacement follows extractive development, then more just, consensual, equitable, and negotiated forms of transformation and organization are needed to avoid further dispossession. If displacement—of peoples, lands, animals, plants, or waters—cannot be avoided, rethink your development and your design. Transformation takes time. 1. On Borders: All States encircle and enclose. If their purpose is to control and contain, then their territorial techniques and technologies produce borderlines that exclude as much as they connect. Since lines separate, counter-boundaries and buffer-zones need to be created for the indivisible, Indigenous systems tying land, life, and language together. Water, intrinsically, knows no borders, and river valleys provide the most natural way of organizing change. Overthrow the grid. 2. On Frontiers: Remoteness, like wilderness, is an imperial myth and a colonial lie. There is no frontier—no *terra nullius*—there never was. Every single square kilometer of Canada, no matter how apparently remote, has live boundaries, edges, peripheries, thresholds, and lived histories. Today, there are only frontlines. Doctrines of (mineral) discovery latent in the ideologies of remote (resource) exploration are coming to an abrupt end on new fronts of action and retroaction. Resist the picturesque. 3. On Paper: Property in Canada is built on the back of land stolen from First Nations, Inuit, and Métis. Land has been subdivided by a colonial bureaucracy leading to territorial agreements and retroactive treaties—from the Charter of the Hudson's Bay Company in 1670, to the Confederation of 1867, to the 11 Post-Confederation Treaties between 1871 and 1921. Future legal revisions to colonial laws and imperial maps lie in the medium, media, and means in which they were originally drafted. Subvert the survey and the property map. 4. On the Crown: If the assumed and unquestioned power of the Crown lies in the separation of surface rights above and mineral rights below, then the supremacy of the paper world of extraction—where power lies in between the lines of laws and leases—needs to be broken. The system of royalties for the 95% of land in Canada controlled by the Crown needs to be radically redesigned to reinforce the integral and indivisible rights of Indigenous peoples whose livelihoods are based *on*, *in*, and *through* land. Serve lands,

combustion engine to render fossil fuels obsolete and propose alternative carbon flows and live forms of energy. The power of human energy is the greatest and most precious renewable resource on the planet. 11. On Work: Labor, like knowledge, is a precious commodity and a rare resource. Unlike numbers or units, cross-generational labor—however mobile it may seem—is not easily moved nor is it easily pushed around, as if people were data, digits, or statistics. Economies of labor can only be achieved by unlocking the energy and creativity of its workforce, whose intelligence is rooted in land and community itself. A labor of love is always about uncoerced collaboration. 12. On Mining: Megasize mines should be banned. Current forms of extraction can result in the uneven stratification of wealth. Friction and confrontation often occur in the self-regulation of an industry that benefits from the under-estimation of environmental and security risks. Greater levels of economic exchange and cultural interaction across multiple generations can eventually eliminate the damaging practices of high-grade mining and recalibrate the disparities of over-exploitation. Harvesting and foraging will make mining obsolete. 13. On Geography: Mines radically redefine regions far beyond the limits of their boundaries. They draw new borders and inscribe new geographies. Along with the waste they create, mines radically alter and damage water systems and the health of its inhabitants—downstream and upstream. Obscured by the quintessential image of the dry open pit, every mine is located inside a watershed drainage area, above an aquifer, and upstream from a larger body: a lake, a lagoon, a river, a bay, a delta, a gulf, an ocean, an atmosphere. If the risks of effluents and emissions cannot be eliminated, than that resource is not a resource, it's a liability. Leave it in the ground. 14. On Exchange: Resource industry is not an industry, it is a colonial pattern of urbanization in the age of petrocapitalism. In lieu of extraction and its deleterious effects, cities are the gold mines of the future. Exchange will soon usurp and supplant extraction where new, unborn cultures can begin to decolonize industrialized territories. To avoid the violence that usually comes with decolonization, those new spaces of exchange need to be opened and imagined to extinguish the hegemony of the Victorian extractive sector and beyond the imperial staple-based economy. Markets should never be virtual, they should always be material. 15. On Territory: Beyond the relatively small and concentrated footprints of cities, a new re-

give momentary access to
the landscape behind or under
the future cracks in the plaster

when the houses, capsized, will slide
obliquely into the clay seas, gradual as glaciers
that right now nobody notices.

That is where the City Planners
with the insane faces of political conspirators
are scattered over unsurveyed
territories, concealed from each other,
each in his own private blizzard;

guessing directions, they sketch
transitory lines rigid as wooden borders
on a wall in the white vanishing air
tracing the panic of suburb
order in a bland madness of snows

The City Planners
Margaret Atwood
1965-1975

Cruising these residential Sunday
streets in dry August sunlight:
what offends us is
the sanities:
the houses in pedantic rows, the planted
sanitary trees, assert
levelness of surface like a rebuke
to the dent in our car door.
No shouting here, or
shatter of glass; nothing more abrupt
than the rational whine of a power mower
cutting a straight swath in the discouraged grass.

But though the driveways neatly
sidestep hysteria
by being even, the roofs all display
the same slant of avoidance to the hot sky,
certain things:
the smell of spilled oil a faint
sickness lingering in the garages,
a splash of paint on brick surprising as a bruise,
a plastic hose poised in a vicious
coil; even the too-fixed stare of the wide windows

Decolonization of Planning

Pierre Bélanger, Christopher Alton, Nina-Marie Lister

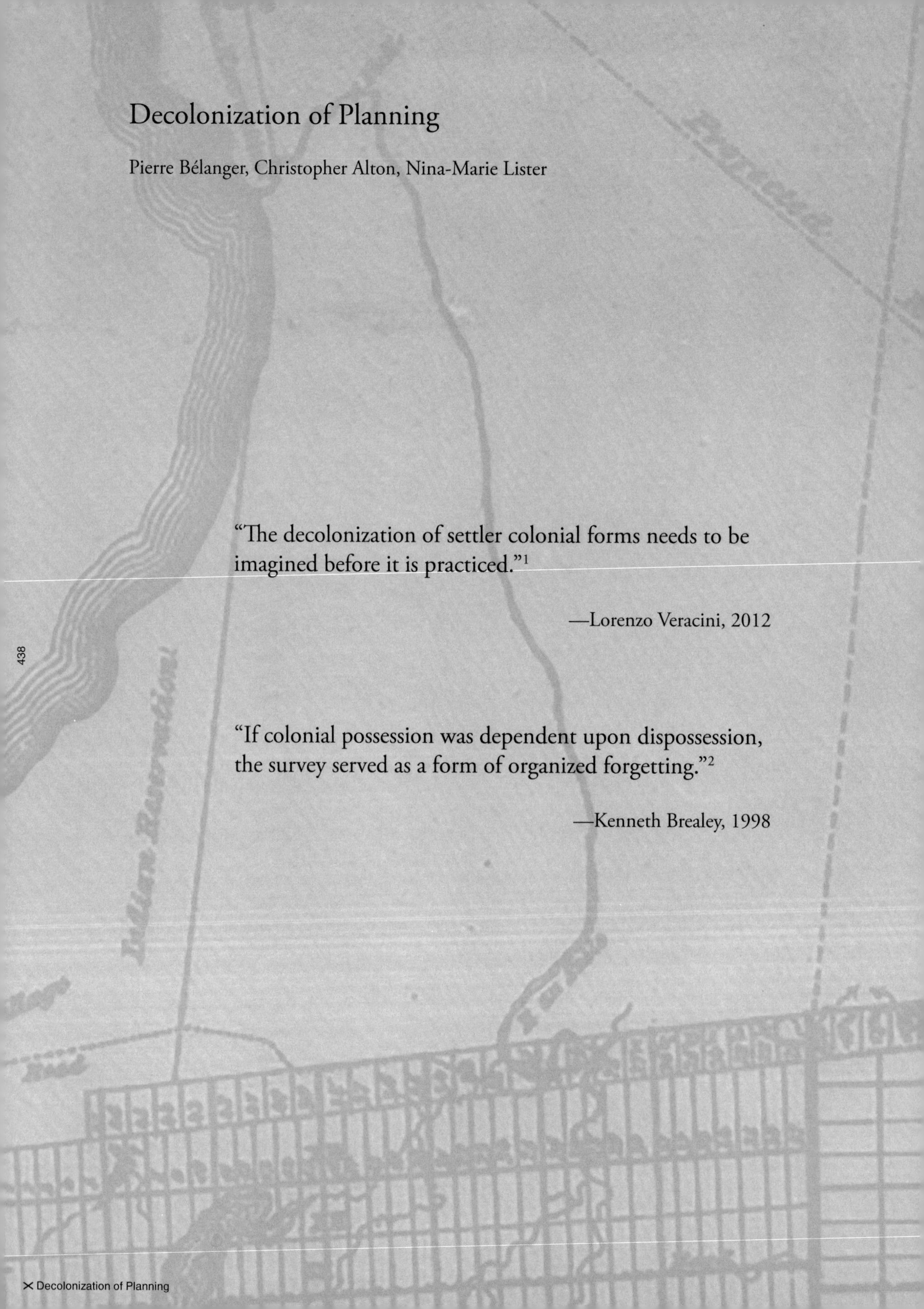

"The decolonization of settler colonial forms needs to be imagined before it is practiced."[1]

—Lorenzo Veracini, 2012

"If colonial possession was dependent upon dispossession, the survey served as a form of organized forgetting."[2]

—Kenneth Brealey, 1998

Planning in Canada is a colonial institution. As legislative framework established by the state and deployed by the academy of the university, the professional practice of planning both exudes and embodies the logic of settler-state colonialism.

If "the early set of practices that constituted a nascent planning practice were," according to what Australian urbanists Ed Wensing and Libby Porter stated in 2015, "centrally important to the colonial dispossession of lands through surveying, naming and town-building,"[3] then those past practices embed and entrench administrative logics in what constitutes today *the colonial present* and its re-figuration. Anthropologist and historian Nicholas Thomas further observes:

> "Colonialism is not best understood primarily as a political or economic relationship that is legitimized or justified through ideologies of racism or progress. Rather, colonialism has always, equally importantly and deeply, been a cultural process; its discoveries and trespasses are imagined and energized through signs, metaphors and narratives; even what would seem its purest moments of profit and violence have been mediated by and enmeshed in structures of meaning. Colonial cultures are not simply ideologies that mask, mystify or rationalize forms of oppression that are external to them; they are also expressive and constitutive of colonial relationships in themselves."[4]

Embedded with colonial motivations of land domination through territorial acquisitions, the practice of planning is thus captured in the organizational geometries of the grid, the uncontested hegemony of the plan, the political erasure of the ground, and the simplification of complex ecological processes through land use controls. More precisely, its attendant roles were located in property development at one end and resource management at the other, between the city and the countryside. Often touted under the benevolent intentions of economy, health, or conservation, the practice of planning—whether by land survey, land use zoning, or property development—was originally conceived and today remains, a politicized practice consisting in resource management, hierarchical bureaucracy, and population control with ideological foundations deeply and firmly rooted in class, gender, race, and religion. Surveying, mapping, and platting *were* and still *are* policy-making in space and in time, and for all practical purposes of the settler-colonial context of Canada, they have become synonymous *with* planning.

Courtesy of Glenbow Archives

Trail Blazing

Cutting lines across the boreal forest of the 49th parallel, 1860

Indeed, it is through the contemporary history of "Canada" as administration of "colonial power"[5] and as construction of "colonial space"[6] from where the institution of planning has formed. An inquiry into the tools and technologies of territorial control, in tandem with the bureaucratic systems and scales of spatial administration, traces critical contours of the colonial origins and serves to ground the imperial antecedents of the planning profession in Canada in histories of racial violence, Indigenous displacement, and land dispossession.

Surveying Settler-State Space

Drawing from colonial 'lessons' of Great Britain prior to 1919, Scottish-born and British-trained architect Thomas Adams founded what was first known as the *Town Planning Institute of Canada*, and later, the *Canadian Institute of Planning—l'Institut Canadien des Urbanistes*. With the objectives "to advance the study of town planning" and "civic design,"[7] Adams' focus was greatly influenced, among others, by the work of Scottish town planner Patrick Geddes. His study of *Civics: As Applied Sociology* (1905) and *Cities in Evolution: An Introduction to the Town Planning Movement* (1915) traced the formation of colonial thinking of other planners across the British Empire. As Geddes made his way east to India in the early twentieth century, following the founding of the British Town Planning Institute in 1910, so too did Adams, heading west across the Atlantic Ocean to Canada.

Applying his seasoned British perspective to the young Canadian context, Adams' basis for the *Town Planning Institute* in Canada was "to promote the scientific and artistic development of land in urban and rural districts."[8] In a series of iterations between 1910 and 1917, "town-planning schemes" and "municipal town planning acts"[9] were proposed by the Victorian views that Adams traveled with from England to Canada, and later onto the United States. Adams was a reformist thinker of imperial-era industrialism of Great Britain. Here, at the beginning of his professional sojourn in "the Dominion,"[10] Adams drew heavily on the British example of town planning and the rising movement of garden cities in response to industrial pollution, overcrowding, and traffic congestion. He was immediately concerned with the tight and tidy delineation of the "urban problem" at the scale of the "city," to be administered with policy of the State, applied to "everything connected with the city ... and well-being of the citizens."[11] Precluding how citizenship was defined (through the bias of race, class, gender) Adams drew from clear instruments of housing, sanitation, and transportation, as important (albeit not exhaustive) infrastructures to shape the Victorian colonial project of urbanism as spatial reform. Practically ad-libbing Geddes' methods, Adams saw the mapping and surveying of existing conditions as prerequisites for the development of town-planning schemes. Together with the alignment of arterial roads, streets, and heights of buildings, the layout of zones of industrial, residential, and open spaces could then be made.[12]

Seeing beyond the aesthetic predispositions of the turn-of-the-century City Beautiful Movement, Adams largely advocated for the value of planning and preemptive measures based on rational, scientific, and sanitary paradigms of urban reform. Most notably, there were three inter-related but unchecked ills and evils in relation to "city life and growth" where "prevention," according to Adams, "is better than cure."[13] Bad housing (slums, over-crowding, building heights, speculation), poor sanitation (sewage, disease, waste), and poor transportation (congestion, commuting, hauling) were often reported concerns.[14] But Adams' all-encompassing perspective was quick to totalize the complexities of housing policy, traffic congestion, sewage, and sanitation by problematizing them as the comprehensive purview of town planning expertise. Adams view was comprehensive in the sense that it called for addressing several issues simultaneously, but it was totalizing through the top-down authority *a priori* of the state-based planner. However, it was largely a graphic and technical exercise, a paper-based process, not unlike surveying to establish a skeleton of what the settler colonial settler-state could achieve.

Although membership was multidisciplinary, the Institute's composition and clientele were also hierarchical and patriarchal; granted first to registered Architects with the Royal Architecture Institute of Canada (ca. 1907), then enrolled Engineers from the Engineering Institute of Canada (ca. 1887), Dominion Land Surveyors, Landscape Architects, Lawyers, and finally members from the Town Planning Institute of Great Britain.[15] Its emphasis was on "development" across a spectrum of settlements, between extents of the urban and the rural as it was perceived in Britain, consistently relying on a range of scale-based states, from the federal, to the provincial, to the municipal. These nested hierarchies of policy established dominant scales of decision-making that would express, entrench, and encode colonial ideologies of cultural and racial domination in the spatial patterns of development.

Together with a well-established base of professionally-qualified practitioners, Adams crafted a profession with tools of spatial intervention: the survey map and the town plan—with technologies of policy—by-laws, regulations, zones, and codes. Together, these tools and technologies could first be pre-conceived, pre-calculated, pre-determined, pre-developed, and pre-detailed by engineers and architects. Second, execution and construction could be outsourced to developers and builders while specified, checked, and inspected by surveyors. Third, and last, following implementation and occupancy, they then could be ruled, regulated, and policed by the state, in the form of the municipality or the province.

Through the crafting of state-based projects of spatial reform and development schemes using the English Garden City model, the colonial instruments of surveying existing conditions—the grid, the line, the orientation, the density, the layout—could thus be naturally transferred and applied for the planning of future scenarios. Articulated in the pre-eminent tool, technology, and dispositif of the *plan*, the invisible hand of the settler-state whose arm was the professional practice of *planning* could therefore be made visible on the surface of the state.

"CHRISTIAN STATESMANSHIP."
Sir John: Indians starving? Oh, well, they're not "friends of Dewdney,"
you know. I'll see that *you* don't come to want, though, Mr. Contractor.
[Note: Edgar Dewdney was the Lieutenant-Governor of the North West Territories.]

Starvation as Federal Strategy
Sir John A. Macdonald in conversation with
Edgar Dewdney, Lieutenant Governor of the Northwest Territories, 1888

Steeped in the British view of civics, Adams saw the practice of town planning as charting the path towards the modern mission of development; an evolutionary, Darwinian project of progress, production, and civilization. From his firmly rooted imperialist, and late-Victorian yet technocratic point of view that saw him become the first president of the British Town Planning Institute in 1910, Adams' new *Institute* in Canada mirrored the "legal, institutional, and professional structure" of the English parent institution.[16] Combining parallel interests in individual liberty and private property with fierce rationality and oversight, his ideals and ideologies from the British Garden City Movement were also combined with German development ideas of scientific rationalism, administrative centralization, and spatial control, firmly rooted in central place theory.[17]

Planning, in the Briton's mind, charted the path towards modern progress and civilization,

> "... if you realize your responsibilities and do your duty, the power of this country of yours as a factor in the progress of civilization will be largely increased and made still more manifest in the future than in the past."[18]

Planning the advent of progress made the civilizing project "virtually synonymous with culture,"[19] referring to it as "an interconnected web of the arts, sciences, politics, religion, social relations, and material goods of a people."[20] But unlike Britain, the late Victorian period of Canada saw 'civilization' as co-dependent and co-constituted with its opposite, "barbarism."[21] Important to note that the "civilization" project, or more precisely the project of "civilizing," largely depended on an advocacy of the virtue of a sophisticated material culture and purity of race. It catered to a growing bourgeoisie and the civilized culture that "the Indigenous lacked," according to political theorist Mark Francis in his 1998 essay "The 'Civilizing' of Indigenous People in Nineteenth-Century Canada":

> "[civilization and culture] had a continuing impact in structuring European views of aboriginal peoples."[22]

This perception was so embedded in Britain's institutional tradition and colonial language of planning that it was transferred to Canada practically seamlessly. Demarcation of cultural differences and cultural superiority were inscribed in the form, location, and layout of urban form, drawing a distinction between what was 'civilized' and what was not, steeped in victory of the British and the English speaker, over the French, and all other subjugated nations:

> "The fortress of Quebec with its frowning citadel was taken, and soon all Canada belonged to the British. Ever since then the Union Jack has been Canada's flag and the Canadians are as proud of it as the British, and just as ready to die for it, because it stands now and has always stood for all that is greatest and noblest in the life of Britain."[23]

More importantly, as what could be considered a puppet planner or state prop, Adams' early and formative role in a branch organization, the Commission on Conservation, would serve one of the utopic aims of its parent agency, the Department of the Interior. Together, the Department and the Commission promoted civilization by settlement, under the authority of Clifford Sifton, then Chairman of the Commission. With conservation as trope for nationalism, Sifton was lauded as "one of the greatest constructive statesman that Canada … produced."[24] As Head of the Department of the Interior, his primary mandate was driven by "the conservation and better utilization of the natural resources of Canada," supported by secondary top-down bureaucratic tasks "to make … inventories, collect and disseminate such information, conduct such investigations inside and outside Canada, and frame such recommendations as seem conducive to the accomplishment of that end."[25]

Overseeing the Department of the Interior[26] between 1896 and 1905, Sifton was also Superintendent General of Indian Affairs, overseeing an intense period of administration and transformation of Crown Lands especially in the West. Referred to as the "Sphinx of Canada,"[27] Sifton tightly coupled the process of settlement (especially in the Prairies) and resource management across the administration of Indian Affairs under a racially-coded regime. Sifton privileged white, Anglo-Saxon immigration[28] as a cover for increased Indigenous population assimilation. In these eliminationist and extractivist contexts,[29] conservation had less to do with material or environmental preservation as it would be assumed today, than with race-based power domination from territorial dispossession and resource extraction. Here, colonization could take the form of ordered administration, protracted exploitation, and regulated extraction of so-called human resources (white settlers, immigrant labor, Indians) together with natural resources (furs, fisheries, forests, farms, minerals).

Created in 1873, the Department of the Interior that presided over the Commission of Conservation was the third of three instruments created by John A. Macdonald, Prime Minister of Canada between 1867 to 1873, as well as 1878 to 1891. The purpose was to manage lands and reserves, promote settlement, and exploit natural resources. Largely to legitimize the survey and settlement of the Prairies, the two instruments—first, the Dominion Lands Act of 1872, and second, the North-West Mounted Police in 1873—were administered by the powerful position of the Minister of the Department of the Interior. Shifting across the 63-year lifespan of the Department to secure, survey, and settle Crown Lands, the Interior's mandate was both fluid and expansive. According to federal archivist and scholar Terry Cook, the list of programs within the Department's purview grew considerably throughout that period,

> "to explore the western region; remove the natives from the open plains; settle outstanding grievances with the Métis; survey and subdivide the area; establish land reserves for natives, schools, the Hudson's Bay Company,

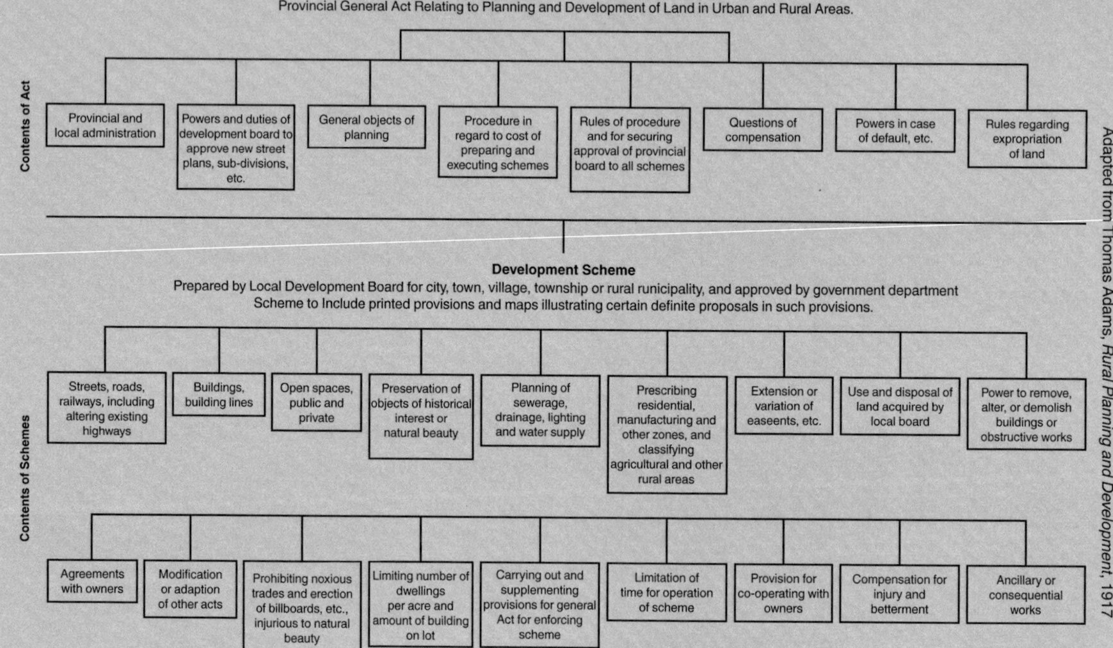
Planning Legislature and Procedure
Provincial General Act Relating to Planning and Development of Land in Urban and Rural Areas.
Contents of Act
Provincial and local administration
Powers and duties of development board to approve new street plans, sub-divisions, etc.
General objects of planning
Procedure in regard to cost of preparing and executing schemes
Rules of procedure and for securing approval of provincial board to all schemes
Questions of compensation
Powers in case of default, etc.
Rules regarding expropriation of land
Development Scheme
Prepared by Local Development Board for city, town, village, township or rural runicipality, and approved by government department
Scheme to Include printed provisions and maps illustrating certain definite proposals in such provisions.
Contents of Schemes
Streets, roads, railways, including altering existing highways
Buildings, building lines
Open spaces, public and private
Preservation of objects of historical interest or natural beauty
Planning of sewerage, drainage, lighting and water supply
Prescribing residential, manufacturing and other zones, and classifying agricultural and other rural areas
Extension or variation of easeents, etc.
Use and disposal of land acquired by local board
Power to remove, alter, or demolish buildings or obstructive works
Agreements with owners
Modification or adaption of other acts
Prohibiting noxious trades and erection of billboards, etc., injurious to natural beauty
Limiting number of dwellings per acre and amount of building on lot
Carrying out and supplementing provisions for general Act for enforcing scheme
Limitation of time for operation of scheme
Provision for co-operating with owners
Compensation for injury and betterment
Ancillary or consequential works
Adapted from Thomas Adams, *Rural Planning and Development*, 1917

Colonial Procedure

The bureaucratic hierarchy of the Department of the Interior and its Planning Hierarchies, 1873 railways, towns and swamp lands; grant or sell millions of acres of homestead lands; encourage immigration; lease lands for timber, grazing, mining and water rights; create the national park system; protect wildlife; and administer and conduct scientific research on a whole range of natural resources. At one time or another over its sixty-three-year existence, [the Department of the] Interior had, in addition to its central administrative core, more than twenty-five distinct branches and agencies under its umbrella, from Dominion Lands to Tourism, without counting the several units that administered the Canadian North."[30]

Underpinning these three federal instruments of the mid 1870s, was a fourth instrument of the State. In 1876, in an effort to amalgamate all other previous legislation on Aboriginal Affairs, the Indian Act was created. Contested ever since, its hierarchical, restrictive, and prohibitive foundations would irreversibly underpin future forms of spatial and territorial planning between the "White Settlers" and the "Indians" within a massive and powerful bureaucracy of the Department of the Interior whose principles emulated those of the British Indian Office of 1858. But, by "amending and consolidating [passed] laws respecting Indians," the Indian Act racialized those functional relations, "well meant but inoperative statutes," towards imperial aims of colonial assimilation and extermination through industrial means such as education, agriculture, and towns.[31] According to then Department of the Interior Minister David Laird in 1876:

> "Our Indian legislation generally rests on the principle that the aborigines are to be kept in a condition of tutelage and treated as wards or children of the State. The soundness of the principle I cannot admit. On the contrary, I am firmly persuaded that true interests of the aborigines and of the State alike require that every effort should be made to aid the Red man in lifting himself out of his condition of tutelage and dependence, and that is clearly our wisdom and our duty, through education and every other means, to prepare him for a higher civilization by encouraging him to assume the privileges and responsibilities of full citizenship. In this spirit and with this object the enfranchisement clauses in the proposed Indian Bill have been framed."[32]

If rule over land was created by the power of planning lodged in between the reality of the territory on the ground and the abstraction of the map on paper; then the relations of power were created in the formation of a geographical imagination built on the scaffold of colonial aspirations and imperial desires. When seen from above, through the eye of the state, those power relations relied on the depoliticization and deterritorialization of Indigenous peoples[33]:

> "In the partitioning of North America, itself 'part of a vast European process and experiment, an ongoing development of worldwide imperialism,'

Settlement as Provincial Strategy of Dispossession I
Manitoba Advertisement, 1883

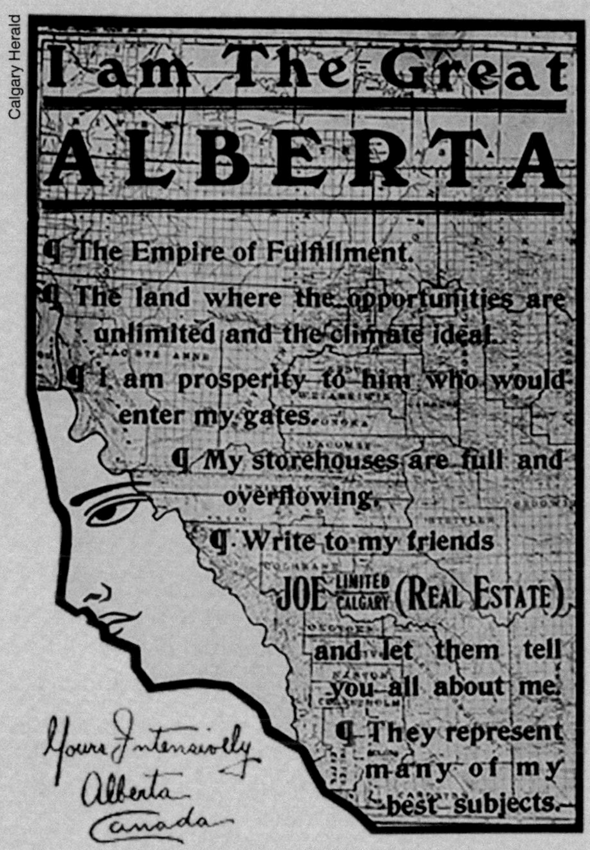

Settlement as Provincial Strategy of Dispossession II
Alberta Advertisement, 1910

> the 'very lines on the map exhibited this imperial power and process because they have been imposed on the continent with very little reference to indigenous peoples, and indeed in many places with little reference to the land itself. The invaders parceled the continent among themselves in designs reflective of their own complex rivalries and relative power.'"[34]

In parallel with practices of planning, Sifton's dual executive positions underscore the spatial translation of power through bureaucracies. It is important to note that as Head of the Department of the Interior, overseer of the Commission of Conservation, and simultaneously as Superintendent General of Indian Affairs, Sifton oversaw both the administration of lands of settler immigrants as well as the administration of Indians and their reserve lands. Combined, this powerful bureaucratic position allowed him to functionally slip and slide interchangeably between roles and goals of settlement, assimilation, dispossession, and exploitation, under the seemingly apolitical guise of conservation. Not surprisingly, over the course of his tenure and afterwards with his successors, the Department's imperially-motivated and racially-biased practices later saw a scathing report on the inhuman treatment, "criminal"[35] government neglect, and "injustice"[36] towards the "welfare of the Indian wards of the nation." Authored in 1922 by then retired Dr. Henderson P. Bryce, the report titled *The Story of a National Crime*, provided "a record for the health conditions of the Indians of Canada from 1904 to 1921"[37] in addressing, among many others, thirty-five Indian schools in three Prairie Provinces. The report was sounding a national alarm for more than a decade with Chief Medical Officer Bryce making an urgent plea that was directed beyond the Department of the Interior, and directly to King Edward I, for what Bryce called:

> "An Appeal for Justice to the Indians of Canada—the Wards of the Nation, our Allies in the Revolutionary War, Our Brothers-in-Arms in the Great War."[38]

While the Department of the Interior saw its budget rise tenfold for purposes of settlement and immigration over the course of Sifton's tenure, the budget of the Branch of Indian Affairs barely increased by thirty percent during the same period of time.[39] In subsequent years, changes in health administration were barely perceptible, seeing the dwindling health, poor living conditions, and increasing abuse, as well as violence in the treatment and assimilation of Indians in the industrial reservation and boarding school programs across Canada.[40]

Here, Australia's example and experience with the inhuman marginalization and racist segregation of non-whites is instructive, showing that colonization entailed more than just mere settlement by foreign immigrants:

> "[Colonization] was based on establishing a white settler colony in a land previously occupied by an Indigenous peoples, the Aborigines. The desire

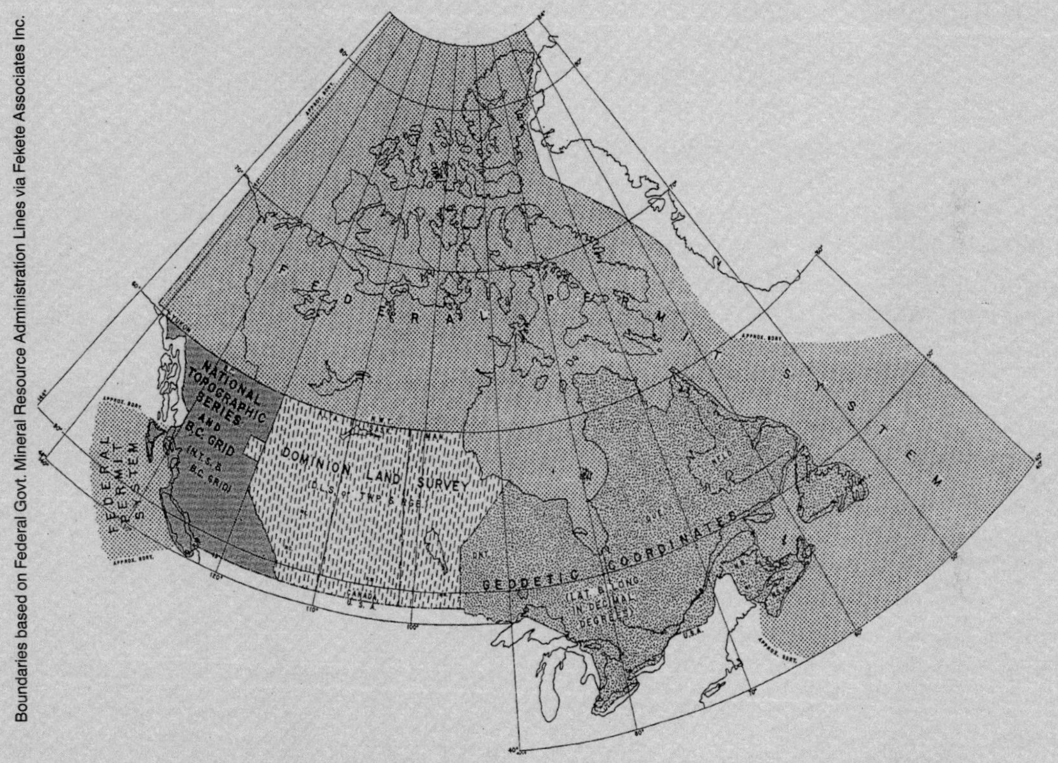

Boundaries based on Federal Govt. Mineral Resource Administration Lines via Fekete Associates Inc.

Super Survey
The Western Meridians of the Dominion Lands Survey, 1869

to establish settler colonies depended upon the will of erasure or, when this failed, systematic containment of Indigenous peoples. … This 'erasure' was inaugurated by the notion of *terra nullius*, land unoccupied, which became the foundational fantasy of the Australian colonies. The justice of *terra nullius* was debated in the British Parliament and its truth was daily challenged by the undeniable presence of Aborigines in the colonies. From the outset it has been a most unstable foundation for the nation. Its tenuous and debated reality was necessarily shored up by a whole range of spatial technologies of power such as the laws of private property, the practices of surveying, naming, and mapping the procedures of urban and regional planning."[41]

Crafting the Uneven Cartographies of Colonialization

Adams was working under the authority of the Crown. The power structure of that hierarchy was already embedded in the institutional organization of the Department of the Interior founded on Sifton's nationalist intentions and racist undercurrents. At the onset of the early decades of the twentieth century, Adams was entering a rapidly and potentially aggressive settlement period of the newly-forming country during one the fastest periods of immigration that Canada had ever experienced. Between the three decennial censuses of 1891 and 1921, the nation saw over three million immigrants, mostly from Europe, settle in less than three decades.[42]

As a necessary prerequisite and operational background for westward immigration, was the largest survey to ever take place in the history of the world: The Dominion Land Survey. Nearly one million square kilometers of land (250 million acres) were surveyed, a region almost the size of the Roman Empire, in less than thirty years. It was a confirmation of Britain's grand design of Canada's "terra incognita."[43] In his massive report for the Commission of Conservation in 1917 titled *Rural Planning and Development*,[44] Adams conveyed Sifton's multiscalar predispositions towards Victorian ideals of scientific rationality, sanitary living, and bourgeois elitism, from the national territorial scale to the municipal town level:

> "The Chairman of the Commission of Conservation expresses himself thus, as to the ideal we should seek to attain in building up the Canadian democracy in the future: 'The ideal state is that in which all the citizens, without exception, have the opportunity of living a sane, clean and civilized life, partaking of at least all the necessary comforts provided by modern science, and enjoying the opportunity of spiritual and intellectual improvement.'"[45]

A few decades after Queen Victoria took dominion over Canada by the royal proclamation of Confederation in 1867, Adams observed the importance of imperial standards and spec-

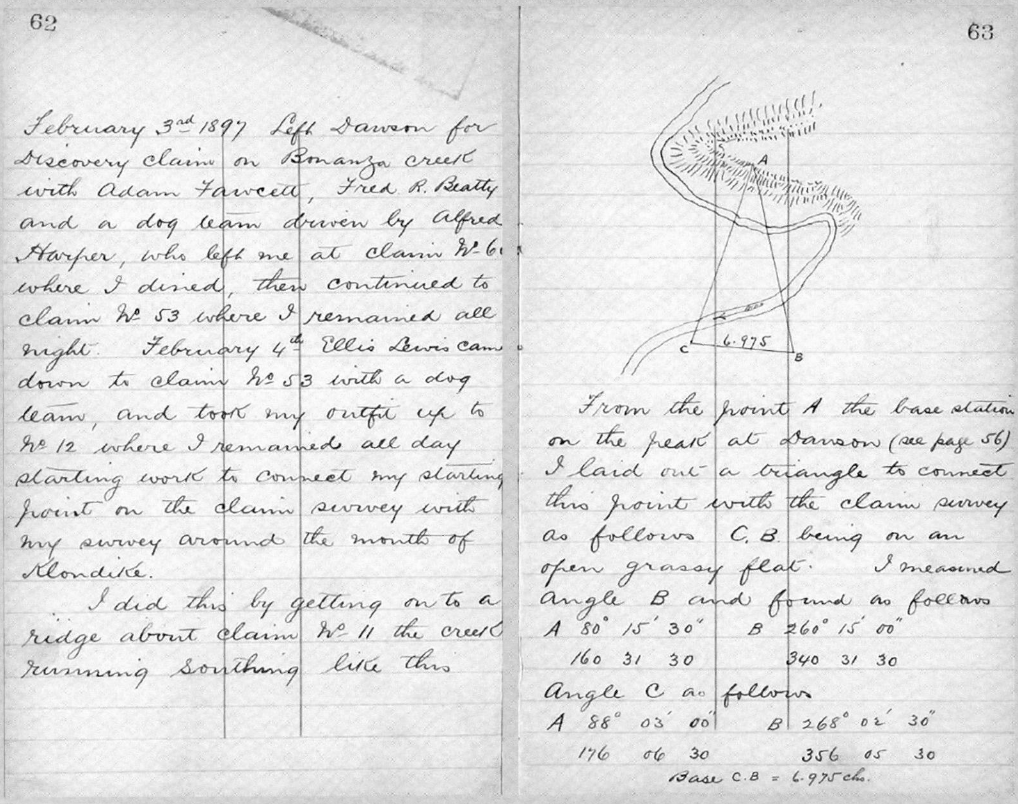

62

February 3rd 1897 Left Dawson for Discovery claim on Bonanza creek with Adam Fawcett, Fred R. Beatty and a dog team driven by Alfred Harper, who left me at claim No 6[illegible] where I dined, then continued to claim No 53 where I remained all night. February 4th Ellis Lewis came down to claim No 53 with a dog team, and took my outfit up to No 12 where I remained all day starting work to connect my starting point on the claim survey with my survey around the mouth of Klondike.

I did this by getting on to a ridge about claim No 11 the creek running southing like this

63

A

6.975

C

B

From the point A the base station on the peak at Dawson (see page 56) I laid out a triangle to connect this point with the claim survey as follows C. B. being on an open grassy flat. I measured angle B and found as follows

A 80° 15' 30" B 260° 15' 00"
160 31 30 340 31 30

Angle C as follows

A 88° 03' 00" B 268° 02' 30"
176 06 30 356 05 30

Base C.B = 6.975 chs.

Lines of Abstraction
Field notes of Dominion Land Surveyor William Ogilvie, 1897
(note pre-printed center lines on notebook pages to follow survey line on ground)

ifications for the planning profession across the continent, drawing from geometries—the grid—of the Dominion Land Survey of 1871:

> "*The Manual of Instructions for the Survey of Dominion Lands* states [that] 'The directions for the survey of township and section lines may in the mountains have to be departed from, but must be adhered to as closely as the nature of the ground will allow.'"[46]

To be mapped and settled, either for provincial farms or municipal towns or Indian reserves, land first had to be surveyed, and then drawn, placed on a map to draft and craft the imagination of settler-space, on paper. While historical grids and geometries may have varied, the Dominion Land Survey remained consistent as state planning project,

> "the settlement of the vast territory of Canada, acquired from the French in the Seven Years' War, created new opportunities for the land surveyor, most associated with the Dutch-born Samuel Holland. ... Governors such as Dorchester and Simcoe tried to standardize the size and design of townships, but a variety of approaches remained until the 1867 British North America Act established the Dominion of Canada, and nine-square mile townships (23 km^2) and 600-acre (243 ha) sections became the norm."[47]

As such, the surveyor was precursor to the planner. Adams observed the increasing sophistication of the department-sanctioned, surveying process across a range of survey scales. As the State took on more and more territory, at greater and greater pace, complexity in its resolution and definition lacked:

> "Under the Dominion Lands Surveys Act of 1908 land is required to be laid out in quadrilateral townships, each containing 36 sections, and each section divided into quarter sections of 160 acres. ... Surveys are now made more accurate and elaborate than formerly, and a good beginning has been made in recent years in collecting information regarding the situation and character of townships. This survey, however, is not sufficient and is too general in character, to provide a basis for classification or a proper system of settlement."[48]

But in spite of the pace and apparent precision lauded by state officials regarding the survey, Adams critically reiterated and strongly criticized that the creation of the survey, or the survey plan for that matter, was not a process of planning nor projecting into the future, but merely a systematic method of cartography, and inventory:

> "As a general rule there has been no proper planning of rural and urban areas in Canada—merely adherence to a rectangular system of survey. Land

has been divided according to certain principles laid down by land surveyors, to whom have been assigned greater responsibilities in defining boundaries of municipal areas and land divisions than in older countries."[49]

Furthermore, Adams critically mentioned the absence of terrain and topographic information, perhaps as a justification for its expediency and economy. Less a reconnaissance project of geological resources, timberlands, or terrain, the forests and water bodies were often seen as obstacle and impediment to the procedure of surveying, rather than assets. Unlike the US, the northern boreal forest was not the midwestern prairie. Groundless, the white space of the survey map flattened the ground in the imaginary of the settler-state, where forests and waters impeded the course of settlement:

> "From the surveyor's point of view [the inception and development of the survey plan] appears to have been influenced in its growth by two main considerations. These were, first, the necessity of mapping out the territory on a geometrical plan, without regard to the physical features of the surface of the land owing to the vastness of the areas to be dealt with, and, second, the need for accuracy and simplicity in defining boundaries of the different units of a geometrical system."[50]

But in this so-called frontier space of the Prairies, the Dominion Land Surveyor was one of the first and most common disciplines represented within the institution of planning. As the basis of colonial settlement and therefore of Indigenous displacement, surveyors were more than merely charting lines and inventorying resources as 'field workers' for the Church or the Crown. Blazing trails, building monuments, and mapping townships, they were inscribing on the land a grid of spatial settlement, tracing territorializing power structures of future institutions. After all, there was nothing that threatening of the surveyor nor of their equipment *per se* (a theodolite, a chain, a notebook). Conquest was therefore achieved in the *execution* of the survey, which then opened up settler-colonial space that was then fulfilled by the pace of immigration and filled in with settlers.

From the Geological Survey in 1841 to the Dominion Land Survey in 1871, the cartography of planning had to be conceived as "an intricate, controlled fiction" from the level of the state (or formerly the colony of the British and the French).[51] Since "boundaries and systems of property rights" are intertwined—according to the former Department of the Interior, now renamed Natural Resources Canada—survey maps were largely touted as tools of efficiency for enabling those rights.[52] In the earlier words of Clifford Sifton, at the Department of the Interior and Aboriginal Affairs, "the survey" was, a priori, "the practical means of developing the country and planning of the nation."[53] From the survey followed the grid. If, "logically speaking," according to historian and art critic Rosalind Krauss, "the grid extends, in all directions, to infinity,"[54] it would also be lent the appearance of permanence and quickly naturalized, as Krauss emphasized, "once the grid appears it seems quite resistant to change."[55]

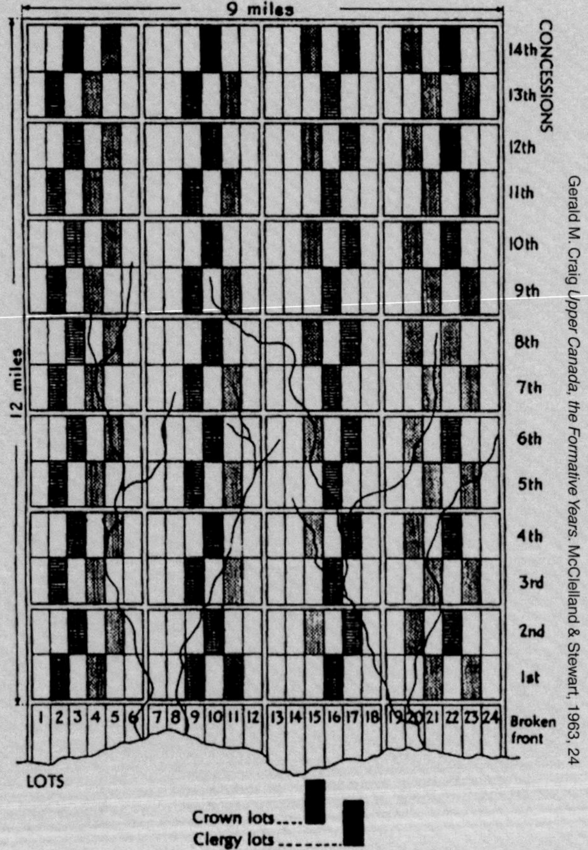

Gerald M. Craig *Upper Canada, the Formative Years*. McClelland & Stewart, 1963, 24

Grid as Colonial Device I
Ontario Township Plan, 1793

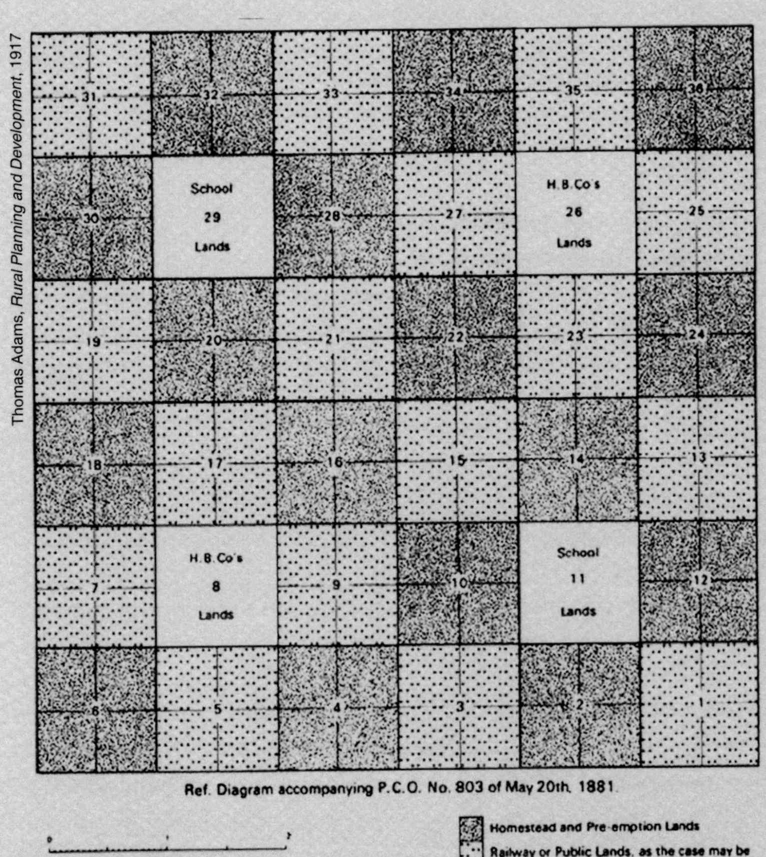

Thomas Adams, *Rural Planning and Development*, 1917

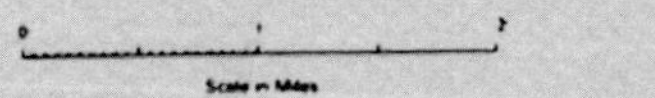

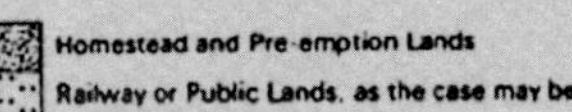

Grid as Colonial Device II
Manitoba Township Plan, 1881

Planning, Plotting & Patenting Systems of Erasure

In the paper space of survey plans and their transfer to colonial town maps and municipal development schemes, the grid was simply the preeminent state tool of spatial delineation, domination, and displacement par excellence. With near infinite geometric divisibility and unlimited levels of geographic deployability, soon after and even long after, treaties were incorporated or overlaid onto the two-dimensional frame of settler-colonialism.

The survey plan—a record of the ground in the form of a view from above—was means and media by which "colonization itself, … an organized action by an external country" could take form. As Thomas Adams further justifies:

> "In spite of their defects, however, it is probable that no better series of systems of surveying lands could have been devised when we have regard to the object of making the surveys. That object—important in itself—was to secure accurate measurements and divisions of the land for rapid settlement; and the fact that it did not include a topographical survey, a scheme of classification of the land and planning of the roads was no fault of the surveyor."[56]

After all, the survey grid was a tried and tested instrument of empire, dating back centuries if not millennia. As federal historian Don W. Thomson affirms:

> "The chief interest for Canadians in the Roman surveys lies in the reflection that more than [two thousand years] ago, a system of surveying land in rectangular blocks was in effect in Europe and Africa and that its influence persists today … In a very real sense the vast Canadian land measurements stand with those of many states of the United States as a New World tribute to achievement to surveyors of Roman times."[57]

Today, the grid and the plan—the paper-based record of the survey as spatial geographic referent and the forward-looking projection of the state as legislative map—secured itself a position as the preeminent technology of the Crown. Like the treaty and the land patent, the paper plan of the survey was always the privileged and paternalistic medium of the State. It locked in the rules of State law in the privileged territorial medium of graphic, visual, legal space of the plan.

Manitoba planner Kent Gerecke confirms that "the relationship between a colony and mother country [could then be] governed by a set of regulations enacted into law. … One main function of colonial government in Canada was to administer policies regarding alienation of land from the Crown and the extraction of revenue from [resources of that] land. Under the British, land policy was seen as a means to develop a society with the right kind of social, political, and religious institutions to serve British interests."[58]

The role of the survey grid then was more than an instrument, its basis was a colonizing system of imperial motivations with monarchist and clerical origins. A century earlier than the Dominion Land Survey, "the chequered plan" for example, "by Surveyor General D.W. Smyth in 1792, [which had its origins in rectangular Roman survey systems from the first three centuries BCE] stands as an example of how land policy could mold the social structure. … Seven lots on each concession were reserved for the Crown and Clergy … consciously done to 'tie a government, a church and a people to the land.'"[59] However, concessions of the survey plan served to exclude, marginalize, and displace others (namely pre-state Indigenous populations to native reserves and immigrant populations including settler labor to town peripheries), from church and state reserves set aside for settlers, beyond the boundaries of the map.

Possession necessarily entailed dispossession. More important than any single example underlying the development of all Canadian cities were the impositions of the colonial survey grids and imperial titles that would serve to erase the history, memory, and premise of Indigenous foundations. From east to west, every Canadian city was, and still is, a story of stolen land and Indigenous dispossession masked by a middle history of trading outposts, remnants of the imperial-sanctioned corporation of the Hudson's Bay Company that carpeted the north of the nation, from Fort Vancouver, to Fort Chipewyan, to Fort Pitt, to Fort Timiskaming, to Fort Montreal.

This surrogate history of stolen land remains the contentious ground upon which all Canadian cities lie on: Halifax from the Mi'kmaq, Quebec City and Montreal from the St. Lawrence Iroquois and relocated groups from the Haudenosaunee Confederacy and Huron-Wendat, Ottawa from the Anishinabeg, Toronto (including Kingston to Hamilton) from the Mississauga, Winnipeg and Regina from the Plains Cree and Ojibwe, and Nakoda Oyadebi, Edmonton from the Papaschase Cree, Calgary from the Niitsítapi (or Blackfoot), Tsuut'ina (or Sarcee), and Stoney, as well as Vancouver and Victoria from the Coast and Straits Salish.[60]

According to Cole Harris, "one might say that imperialism entails an ideology of land on which colonialism (the actual taking up of land and dispossession of its formers owners) depends."[61] One might actually say that imperial constructs generate particular kinds of knowledge and representation of land by means of which colonial dispossessions take place, where the other ones (Indigenous peoples or immigrants) are out-of-place or displaced. The paper space of state planning clearly was, and still is, part of this spatial supremacy, imperial representation and misrepresentation.

"In spite of their defects however," state planner Thomas Adams acknowledged surveyor's skill and significance where technical observations often served to erase and depoliticize the land that was being surveyed:

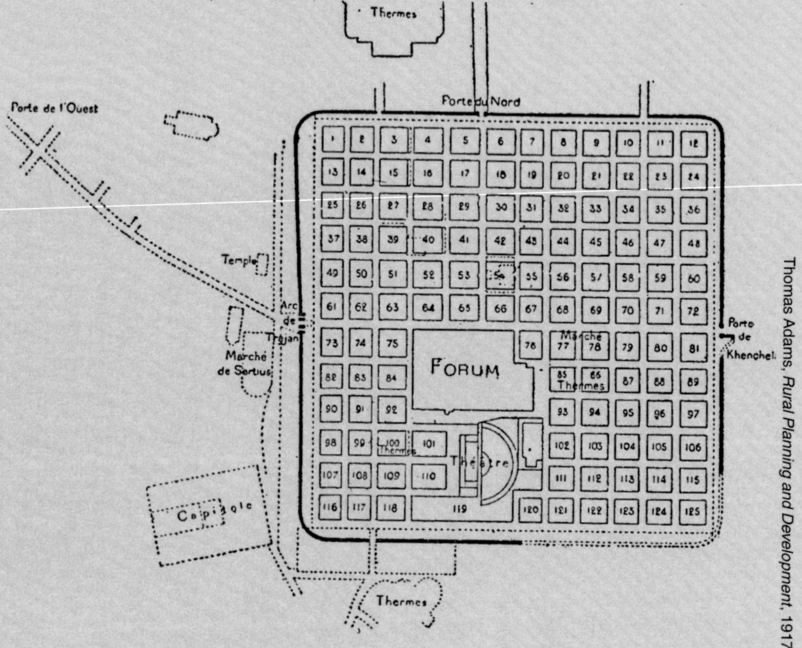

Thomas Adams, *Rural Planning and Development*, 1917

Imperial Origins of the Grid I
Timgad Ancient Rectangular Plan, AD 100

Thomas Adams, *Rural Planning and Development*, 1917

Imperial Origins of the Grid II
Irregular Plan of Medieval Period, 1550

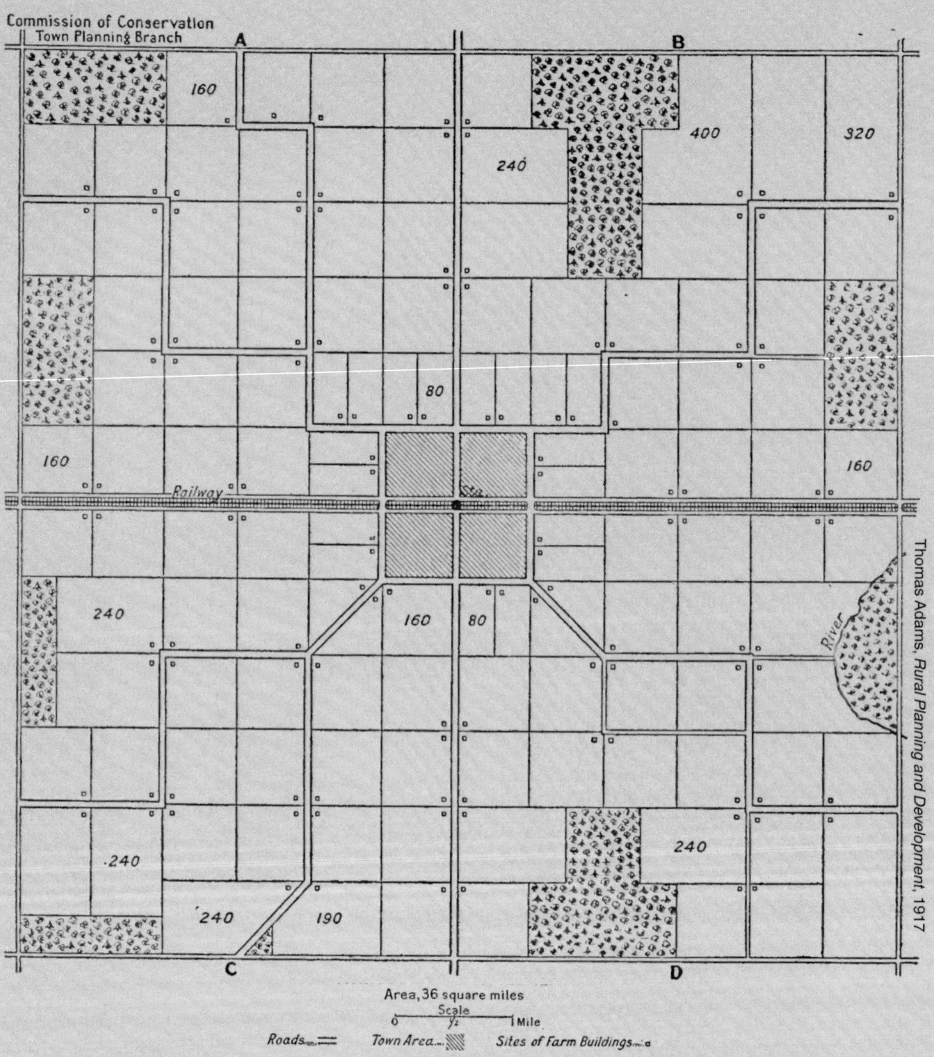

Thomas Adams, *Rural Planning and Development*, 1917

Grid as Settlement Strategy
Township Layout for Agricultural Settlement, 1897

"His duties have been circumscribed within a narrow radius and within that radius he has performed his task with great skill and energy. Under the leadership of Dr. Deville, the Surveyor General, the surveyors of Canada have given able and devoted service to the country, and it is necessary to make it quite clear that what is objected to is not the work or methods of the surveyor, nor the rectangular system as a means to secure accurate measurement, but to the scope of the surveyor's duties having been too limited, and to the rectangular system as a plan for land settlement. The surveyor should not only measure the land, but make a survey of its conditions in the real sense; and the rectangular survey should not be the plan for settlement, but only provide the basis on which a proper development plan for each township should be prepared."[62]

With the legal survey on paper and the pace at which it was executed however, the grid and administrative plan delineated very precisely "regulations" that "established the system of government, monopolies, trading laws of tariffs and preferential treatment of goods, settlement policies, and in some cases plans for urban settlement," as Gerecke explains, "in the context of Canadian colonial development, the rules of colonization represented a far more planned approach to land use and development than had been true in France or, later, in England."[63]

In the paper space of territorial colonialization, the scales and extents of the survey represented a consolidation of economic, political, and spatial power. Not only did the survey describe or delineate extents of territory, it further served to inscribe and encode forms of development to come as a form of future planning at the scale of empire, either through infrastructures or institutions. In the creation of a cadastre and translation of sovereignty then, the survey plan was not only a cartographic tool of the surveyor but it was a symbolic, semiotic, and biased declaration of the state,

"the cadastral map is an instrument of control which both reflects and consolidates the power of those who commission it. … The cadastral map is partisan: where knowledge is power, it provides comprehensive information to be used to the advantage of some and the detriment of others, as rulers and ruled were well aware in the tax struggles of the eighteenth and nineteenth centuries. Finally, the cadastral map is active: in portraying one reality, as in the settlement of the new world … it helps obliterate the old."[64]

James C. Scott echoes these reflections on gross simplifications and abstractions of state mapping projects that depoliticized the ground, quantified its surface area, and romanticized its history:

"The value of the cadastral map to the state lies in its abstraction and universality. In principle, at least, the same objective standard can be applied

> throughout the nation, regardless of local context, to produce a complete and unambiguous map of all landed property. The completeness of the cadastral map depends, in a curious way, on its abstract sketchiness, its lack of detail—its thinness. Taken alone, it is essentially a geometric representation of the borders or frontiers between parcels of land. What lies inside the parcel is left blank—unspecified—since it is not germane to the map plotting itself … The most significant instance of myopia, of course, was that the cadastral map and assessment system considered only the dimensions of the land and its value as a productive asset or as a commodity for sale. Any value that the land might have for subsistence purposes or for the local ecology was bracketed as aesthetic, ritual, or sentimental values."[65]

Infrastructural Apartheid & Territorializing Toponymies

Grid lines not only served as center lines to section land for settlement and partitioned townships, they served thick and robust functions of infrastructure building. Lines designated the drainage of swamps, the clearing of forests, the burning of fields, and the alignment of roads. Grid lines also played an especially important role in the surveying of lands and layout of settlement by plotting positions and geometries of different infrastructures. Forest clearing, boundary marking, township parceling, and road building were techniques used as part of a vast linear and legible system of infrastructure necessary for the colonial project of empire building.

Divisions on paper marked endless inscriptions of self-assigned access and forced entry on the ground. They were also projective alignments and guidelines, literally, for future infrastructure and institutions as outgrowths of the colonial outposts. In the case of the concession or township survey of the eighteenth and nineteenth centuries,

> "individuals provided the buildings and fixtures required by their occupations … government provided the normal institutions of the state, the courthouses, the jails, registry offices, post offices, and most important, roads."[66]

The interrelationships of these functions, between the settler and the state, naturally made way for the distribution and decentralization of the Crown's functions tightly woven into a now thick, three-dimensional matrix of socio-spatial structures. Between the cities, towns, and villages in which they were planned, to the territories, townships, and districts in which they were nested, all these infrastructures, substructures, and superstructures carried with them surrogates. Swamps as setbacks, lakes as reservoirs, valleys as dams, rivers as power, reserves as islands or prisons, forests as timberlands, bedrock as mines, nature as parks, railways as resource trains, streets as borders, roads as clearings, and settlement zones as fortresses, as if the country—the largest of land superficies in the world—could itself be engineered

Tabula Rasa
Cutting out the right of way of Yonge St.
by the Queen's Rangers under Lt. Governor John Graves Simcoe, 1795

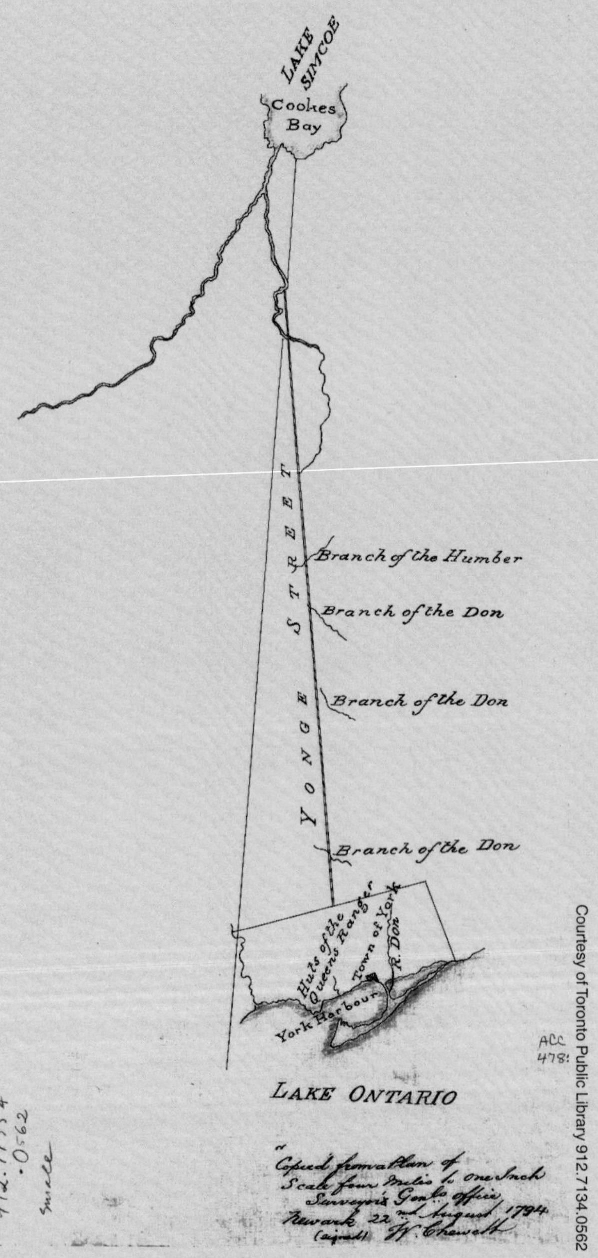

Imperial Axis

Partial Survey Plan of the 1,895 km Yonge St. from Lake Ontario to Lake Simcoe and far beyond, 1794 and administered as a supersize planning project. It is as if the state saw settlement of the country as one big, homogeneous city.

While overtly portrayed as enabling physical and economic mobility, many of their functions and services covertly served as state-sanctioned strategies of expansion, exclusion, or expropriation much like zoning served purposes of class or racial segregation by way of subdivision design versus multi-family high-rises, mortgaged estates versus rental properties, while embedding violent gender-biases and dangerous labor divisions within the structure of a paternalistic economy of wartime industrial production.[67]

By virtue of cutting through the forests and so-called 'unoccupied' lands, the technological and linear infrastructure of roads naturalized the colonial settler-imagination. Roads tamed lands that were perceived as either unknown, unseen, wild, wasted, or savage. Roads then displace the focus and centrality of the city for the periphery of systems of infrastructure. If cities characterized settlements as spaces of containment, then infrastructure conveyed systems of privileged communication, testifying to the equal significance of roads to settlements.

Rather than exclusively focus on the legible identity of cities then, an alternative focus on "infrastructures as dynamic relational forms," offers much potential. "Roads," more specifically, as anthropologists Penny Harvey and Hannah Knox observe, when "approached as infrastructural technologies" might provide new perspectives in the politics of contemporary social relations."[68] Albeit "a mundane material structure [that registers] histories and expectations of state presence and of state neglect,"[69] "the development in the study of infrastructures," and that of roads is a complex and entangled endeavor. As anthropologist Claude Lévi-Strauss mentioned, the study of infrastructure "is a task which must be left to history—with the aid of demography, technology, historical geography, and ethnography."[70]

As historian Jo Guldi further explains, "roads were a mechanism for government."[71] They enabled a project of deep penetration into already occupied Indigenous lands to ultimately facilitate conquest by extraction of land-based resources for a centralized imperial economy.

"Under the name of Colonization Roads [since 1853],"[72] lines of movement were opened "through wholly new territory" for and by the Crown "to render unoccupied parts of the country accessible."[73] Also called 'government roads' or 'highways' as they were later known, the thick space of the grid lines served as primary channels and preemptive infrastructures of the colonial settler-state well before the advent of town planning in the early twentieth century.

While "roads [that] opened through unoccupied lands of the Crown for the purpose of promoting their settlement," reports of their use were fairly technical and quantita-

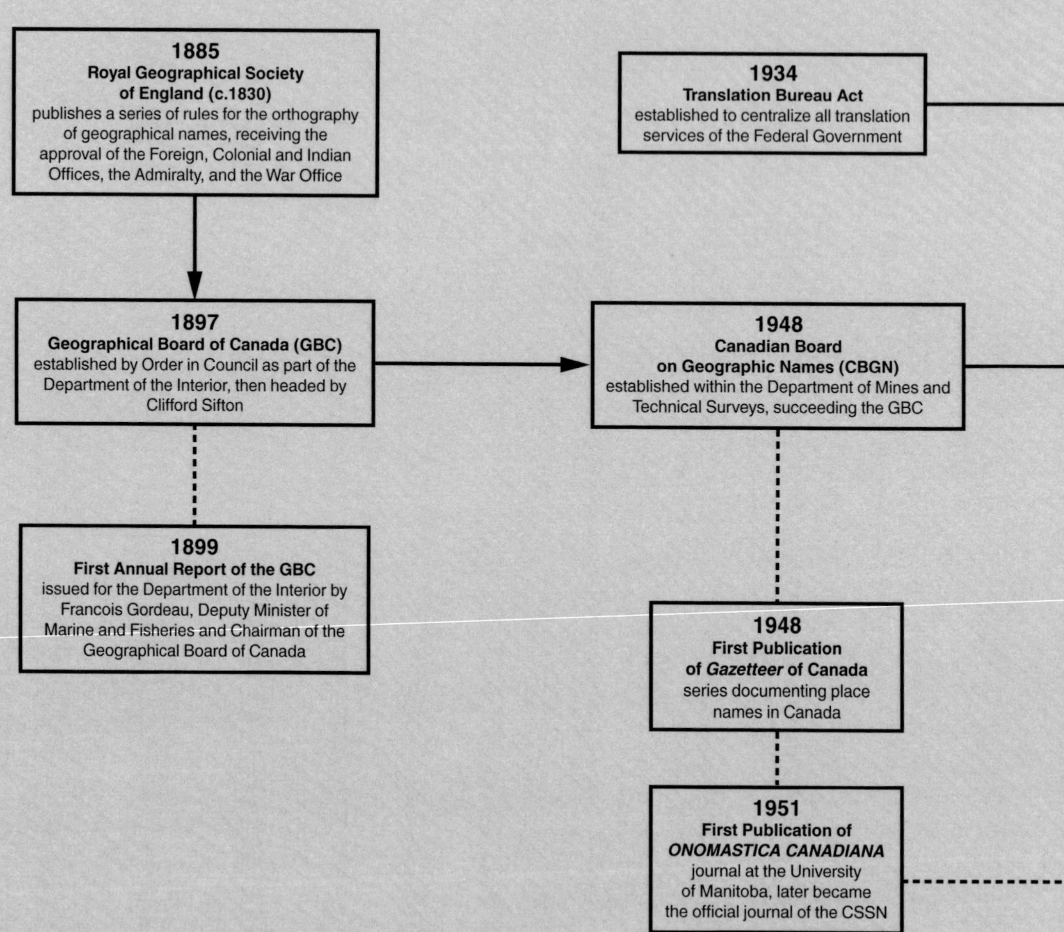

1885
Royal Geographical Society of England (c.1830)
publishes a series of rules for the orthography of geographical names, receiving the approval of the Foreign, Colonial and Indian Offices, the Admiralty, and the War Office
1934
Translation Bureau Act
established to centralize all translation services of the Federal Government
1897
Geographical Board of Canada (GBC)
established by Order in Council as part of the Department of the Interior, then headed by Clifford Sifton
1948
Canadian Board on Geographic Names (CBGN)
established within the Department of Mines and Technical Surveys, succeeding the GBC
1899
First Annual Report of the GBC
issued for the Department of the Interior by Francois Gordeau, Deputy Minister of Marine and Fisheries and Chairman of the Geographical Board of Canada
1948
First Publication of Gazetteer of Canada
series documenting place names in Canada
1951
First Publication of ONOMASTICA CANADIANA
journal at the University of Manitoba, later became the official journal of the CSSN

The Colonial Toponymical Procedures of Indigenous Erasure

As seen through a brief timeline of the Geographical Names Board of Canada

Today, the GNBC is comprised of 31 members. Its Chair is appointed by the Minister of Natural Resources Canada (NRCan). Each of the provinces and territories are represented, also including various federal departments concerned with separate entities of Mapping, Archives, Defense, Translation, First Nations Reserves, National Parks, and Statistics.

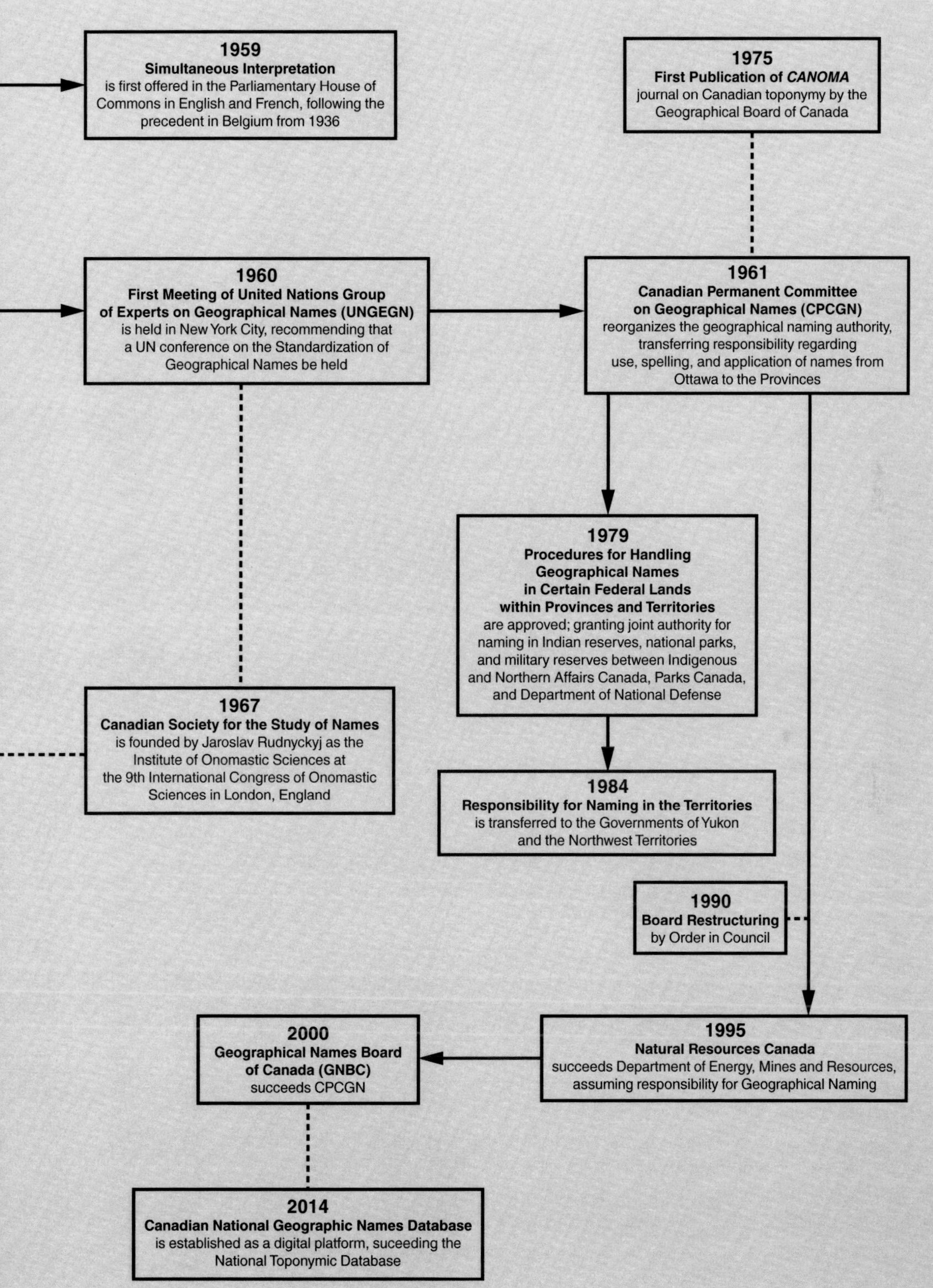
1959
Simultaneous Interpretation
is first offered in the Parliamentary House of Commons in English and French, following the precedent in Belgium from 1936
1975
First Publication of CANOMA
journal on Canadian toponymy by the Geographical Board of Canada
1960
First Meeting of United Nations Group of Experts on Geographical Names (UNGEGN)
is held in New York City, recommending that a UN conference on the Standardization of Geographical Names be held
1961
Canadian Permanent Committee on Geographical Names (CPCGN)
reorganizes the geographical naming authority, transferring responsibility regarding use, spelling, and application of names from Ottawa to the Provinces
1979
Procedures for Handling Geographical Names in Certain Federal Lands within Provinces and Territories
are approved; granting joint authority for naming in Indian reserves, national parks, and military reserves between Indigenous and Northern Affairs Canada, Parks Canada, and Department of National Defense
1967
Canadian Society for the Study of Names
is founded by Jaroslav Rudnyckyj as the Institute of Onomastic Sciences at the 9th International Congress of Onomastic Sciences in London, England
1984
Responsibility for Naming in the Territories
is transferred to the Governments of Yukon and the Northwest Territories
1990
Board Restructuring
by Order in Council
2000
Geographical Names Board of Canada (GNBC)
succeeds CPCGN
1995
Natural Resources Canada
succeeds Department of Energy, Mines and Resources, assuming responsibility for Geographical Naming
2014
Canadian National Geographic Names Database
is established as a digital platform, suceeding the National Toponymic Database

tive as if to say their primary purpose was assumed, if not already obvious. Serving to locate other institutional functions like post offices and banks,

> "these roads were intended to have been made practicable for travel, in the first instance at the cost of Government, by wagons loaded with, at least, half a ton weight, and drawn by one span of horses or a yoke of oxen. They have cost from $300 to $800 per mile, according to the character of the surface over which they have been carried, and other difficulties attending their construction."[74]

As preemptive basis for the project of colonization, the inception and conception of roads in the colonial imagination stemmed from road building techniques in seventeenth and eighteenth-century England pioneered by the likes of pavement engineers such as Thomas Telford. In 1833, Sir Henry Parnell of the British Institution of Civil Engineers, proclaimed in his 1833 *Treatise on Roads* that would later be used as a basis for the colonial project in Canada:

> "The making of roads, in point of fact, is fundamentally essential to bring about the first change that every rude country must undergo in emerging from a condition of poverty and barbarism."[75]

Political invasion was hastened by material and spatial infrastructures. Colonial representatives from England, among others, would often translate British principles and apply them towards imperial aims as invasive infrastructures in the colonies. Six years after the publication of Parnell's *Treatise on Roads*, Colonial Governor of Nova Scotia Colin Campbell would order its reproduction for the Crown in 1839 "chiefly for the use of Road Commissioners in Nova Scotia" during the advances on the eastern edge of Canada as state infrastructure. Colonization Roads were therefore also the primary vehicles for creating notions of a mythologized frontier and perpetuating the courage of the pioneer. Land clearing for road-building served as a means and metaphor for nation-building as Parnell explains in his original *Treatise* of 1883:

> "It is, therefore, one of the most important duties of every government to take care that such laws be enacted, and such means provided, as are requisite for the making and maintaining of well-constructed roads into and throughout every portion of the territory under its authority."[76]

Later, as French and British advances were being made in Lower (now Quebec) and Upper Canada (now Ontario) into the late eighteenth and mid-nineteenth centuries, the central colonizing tenets of road building (inherited from French or British Engineering) would underlie the surfacing of the colonial settler-state by the Commission on Crown Lands and precursor to the Department of the Interior. As underlying and imperative infrastructural

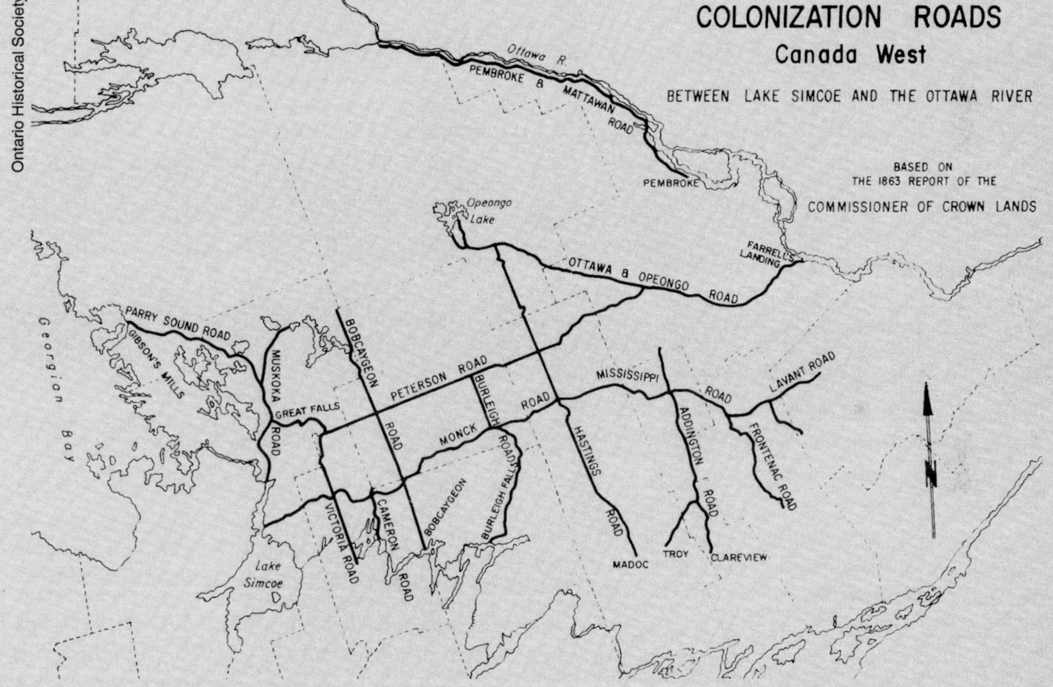

Ontario Historical Society

Imperialism Through Infrastructure
Nineteenth-Century Map of Colonial Roads in Central Ontario, 1841

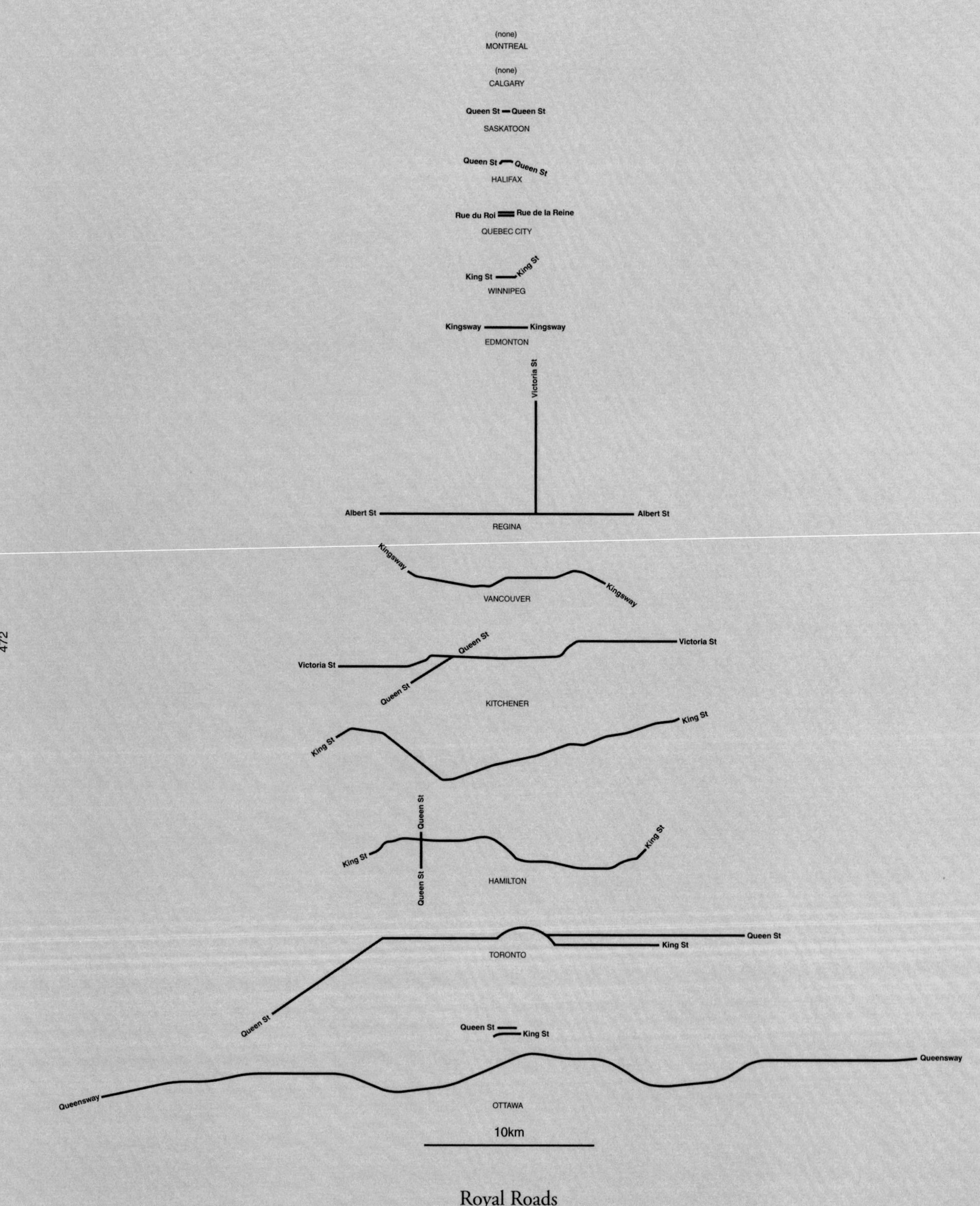

Royal Roads
Queen and King Streets cutting across Canadian Cities

investments by the Crown, Agent in charge of Colonization Roads Boucher de la Bruère and Commissioner of Crown Lands Philip Michael VanKoughnet reaffirmed in the 1862 *Report on Colonization Roads in Lower Canada*, that the roads served the primary purpose of settling land 'deemed' open and unoccupied:

> "The progress of colonization has also been greatly favored by the Legislative reforms in this matter since 1854. Before that period, wealthy land proprietors of the Townships, for the most part unknown and many of them foreigners, had proven the principal obstacles to the settlement of our lands, by demanding too high a price for their property, and by not contributing in any way towards local improvements. The Legislature wisely provided against these disadvantages, by compelling these persons to contribute to the municipal fund, and to assist in the opening of the roads."[77]

Further attesting to the institutional effects of investments into colonization roads, de la Bruère and VanKoughnet underscored that:

> "Another great proof of the daily increasing importance of the Townships, owing to the opening of Colonization Roads, lies in the fact that in 1850, two years only after the creation of this office, the ecclesiastical authorities had already provided for the erecting of churches in the vicinity of the newly-opened roads."[78]

And, these settlement roads would simultaneously serve as vectors of state access and production:

> "For some years past, the public has been so fully convinced of the real value of land in the Townships, that many farmers from the old parishes have been known to sell their comfortable homesteads in order to come and settle upon these new lands. Formerly such localities were selected by none but the man of straitened means, who devoted all his powers to the hard task of clearing his land. If sufficient sums are voted by the Legislature for the opening of new roads, there can be no doubt but that in a few years the Townships will become a mine of wealth to the country; for the prosperity of the United States, and even of Upper Canada, may be traced to their newly cleared land; flour, wheat and corn are principally derived from the West, in other words, from land more or less new."[79]

In other words, the microgrids of roads were *a priori* ideological in nature, for imperial penetration, territorial conquest, resource mining, and Indigenous displacement:

> "The efforts which the Government is making to favor the colonization of our wild lands will, it is to be hoped, be crowned with success, and the sending of agents to Europe will largely contribute to bring the resources of this country under the notice of the population of those countries. A new band of brethren has come to make part of the Canadian population, and the numerous subscriptions that have been made throughout the country for the unfortunate Acadians bear witness.
>
> … For whatever may be the strength, vigor, and courage of the foreign settler, none can surpass the Canadian as a pioneer.
>
> … The friends of the country trust that a vigorous impulse will be given to colonization by a grant proportionate to its actual wants; for, as regards the question of colonization, the judicious money grant is a gain and not a loss. The settlers, for their part, are anxiously expecting the opening of new roads; for it must not be forgotten that if courage has been their guide to the midst of the forest, it is hope that keeps them there."[80]

These extensions of the rigid, colonial steel frame of the survey grid, on the surface, below the surface, and above the surface. Road clearings and alignments could be understood as important extensions and subdivisions of the work of the settler-state based on different altitudes of colonization to build a three-dimensional frame out of the two-dimensional survey. Roads were not just lines on paper, they were spatialized and topographic corridors of movement and communication. *Zoning* remains its most significant contribution in the conditioning of development form with control of segregation of population through roads, buildings, and land uses. Persistent within the administrative and colonial frame of the grid-iron of the city to come, roads could be regulated and trafficked through speed limits. Buildings could be regulated through density regulations and height restrictions. Land uses could be pre-established to divide labor and industry, trade, and commerce from different types of housing and residential construction. Across all these different formats of zoning and spatial regulation, the use of setbacks, buffers, and distances would maintain and strictly separate what was deemed incompatible through eyes of the settler-state, thanks to the pre-emptive construction of the roads as foundation to the colonial project.

Like invisible walls and enclosures, these zones of extrusions created interiors and exteriors. They created barely perceptible boundaries in many forms, with exclusionary effects: from buildings, to blocks, to parks, to districts, to highways, to cities, to regions, or to provinces.

In reference to imperial antecedents and "the invention of the British Infrastructure State," Guldi notes the attendant expansion of a government apparatus and a centralized road system in England as a foreshadowing example:

Thomas Adams, *Rural Planning and Development*, 1917

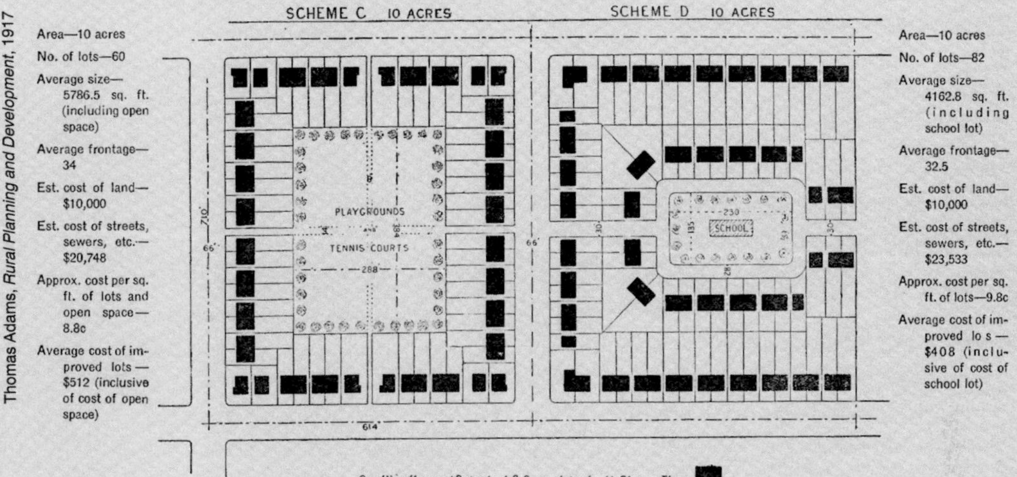

The saving in road space in Schemes C and D increases the size of the average lot over what is obtained by mere reduction in the number of lots. Scheme D differs from the other three schemes in that it represents the development of one-half of a 20-acre block and has no street frontage on one side. The interior streets in schemes B and D are deliberately designed to hinder through traffic.

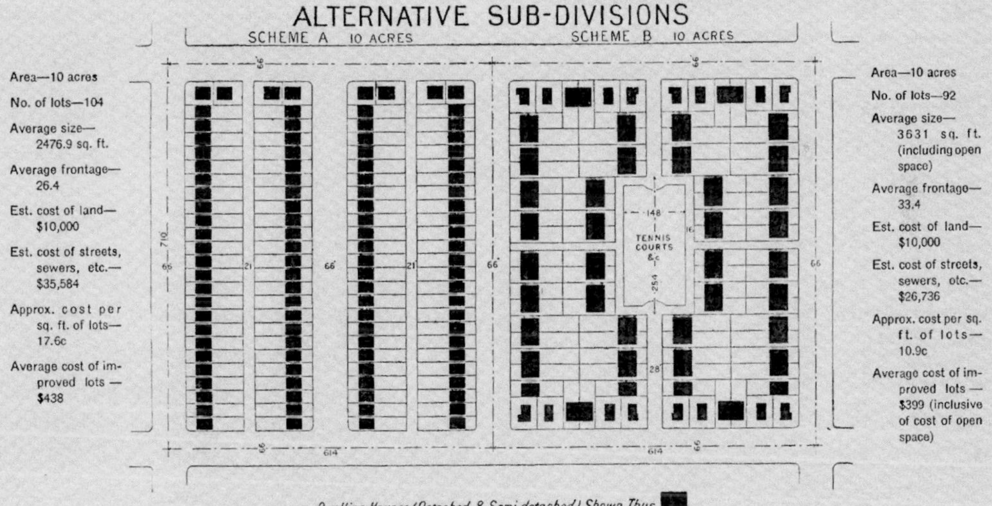

The cost of the land is the same in each scheme, A, B, C and D. The difference in the figures is in the cost of the local improvements, due to different systems of planning. Detached houses are shown in Scheme A as being typical of development in Canadian cities. In the other schemes most of the houses are shown in pairs, which is a more desirable arrangement where practicable. Whether the houses are single or in pairs, however, has no bearing on the cost of development of the land.

Colonial Subdivisions
Internal Partitioning of the Metropolitan Grid, ca. 1911

> "Expanding infrastructure constituted an unprecedented expansion of bureaucracy, foreshadowing later government undertakings in the name of welfare and public health. Infrastructure united the nation in terms of mail, commerce, and travel, and it promised to do more. Promoting the highway system in Parliament, the Scottish landlords urged the power of roads to transform a nation, to unite separate ethnicities (like Celts and Englishmen) into a single people, to promote intermarriage, to overcome linguistic divides, and to quell the risk of military rebellion. The roads seemed to promise even greater forms of unity for the nation, a valuable tool for governments in the service of peace. The origins of modern infrastructure have been frequently misunderstood in ways that mask the role of the state."[81]

Technological historian of American and British Empires, Rosalind Williams, iterates the "divisive," "geographic space" of "large technological systems" inherent to the "cultural origins" of modern state building. As the infrastructural underbelly of the democratic "concept of connective systems and pathways," Williams illustrates the dualistic nature of infrastructure as state technological system[82]:

> "The outstanding feature of the modern cultural landscape is the dominance of pathways over settlements ... The pathways of modern life are also corridors of power, with power being understood in both its technological and political senses. By channeling the circulation of people, goods, and messages, they have transformed spatial relations by establishing lines of force that are privileged over the places and people left outside those lines."[83]

Williams further adds on the spatial, organizational power of these infrastructural systems:

> "In the creation of [these tangible infra]structures, nature is understood primarily as *space*, and the system as a *means of organizing space* [of the state]."[84]

Centering on the Capital of the Colonial Settler-State

While the mid-1950s would see the advent of major postwar master-planning projects following a prolonged period of planning dormancy during the Depression, many schemes failed by sheer bureaucracy of administration unable to keep pace with immigration, housing pressure, infrastructure requirements, and labor demands in town. The much-lauded work of the planning of the city of Ottawa for example, is probably one of planning's most illustrative yet overlooked failures. Designated in 1857 as the Nation's Capital by State Sovereign and Monarch Queen Victoria, the royal location in the backwater of Bytown was the product of imperial, geographic, and political concessions in what was nothing more than a lumber town. Lodged between two historic empires, the Capital was sited at the location of a trading post at the confluence of the Ottawa, Rideau, and Gatineau rivers. The Capital was in the center of the 'nation' at the time, on the border where respective French and English territories met between Lower and Upper Canada. Later named Ottawa—an English derivative of the Anishinabeg *Odawa* [part of this encompassing Indigenous Nation] who were displaced to make way for the city—the Capital City was the supreme example and application of central place theory. Ottawa was a national-imperial idea applied locally.

As poster child of the Garden City and the City Beautiful models, the Capital underwent a major planning exercise between 1948 and 1950 by Parisian architect Jacques Gréber, commissioned by Canada's longest serving Prime Minister, Garden City advocate William Lyon Mackenzie King. Beyond the inception of Ottawa as "Washington of the North,"[85] the Capital would become the national model of a Prime Minister's desire to express sovereignty and independence through town planning; following decades of planning processes and masterplans since the foundation of the Ottawa Town Planning Commission by engineer Noulan Cauchon, Thomas Adams' longtime colleague and collaborator. Ironically, it was through the British utopia of the Garden City that the urban declaration of the Capital City was conceived. It was an attempt to distinguish it from imperial neighbors to the east (London, England) and south (Washington DC, US). Rehearsing Thomas Adams' well-established British planning ideals, Mackenzie King even cited word-for-word garden-city tenets and core periphery theories in his 1918 manifesto on *Industry and Humanity*:

> "The garden city movement was founded in England in 1899, and has spread to different countries throughout the world. It recognizes the slum as the product of bad means of transit and high land values, combined with the necessity of men living near their work. By providing cheap and rapid transit and controlling land values, it has been able 'to provide a maximum of comfort, convenience, and happiness at the minimum of financial and personal cost.' It marks a widening of community rights and an enlargement of community services, the building of the city by the city itself, from the foundations upward and from centre to circumference."[86]

Furthermore, in an iteration of utopic ideologies of the colonial settler-state, Mackenzie King lauded the healthy, sanitary social environment that the garden-city could and would offer:

> "The garden city is in effect its own landlord. Indirectly it is a house builder and house owner. The ordinary city is left to the unrestrained license of speculators, builders, and owners. It is a community of unrelated, and, for the most part, uncontrolled, property rights. The garden city, whether promoted by cities, co-operative companies, or private individuals, and

UK Archives, scanned by Cannon Ivers

Capital Concession
Location Map of the Capital City of Ottawa between Upper and Lower Canada, 1857

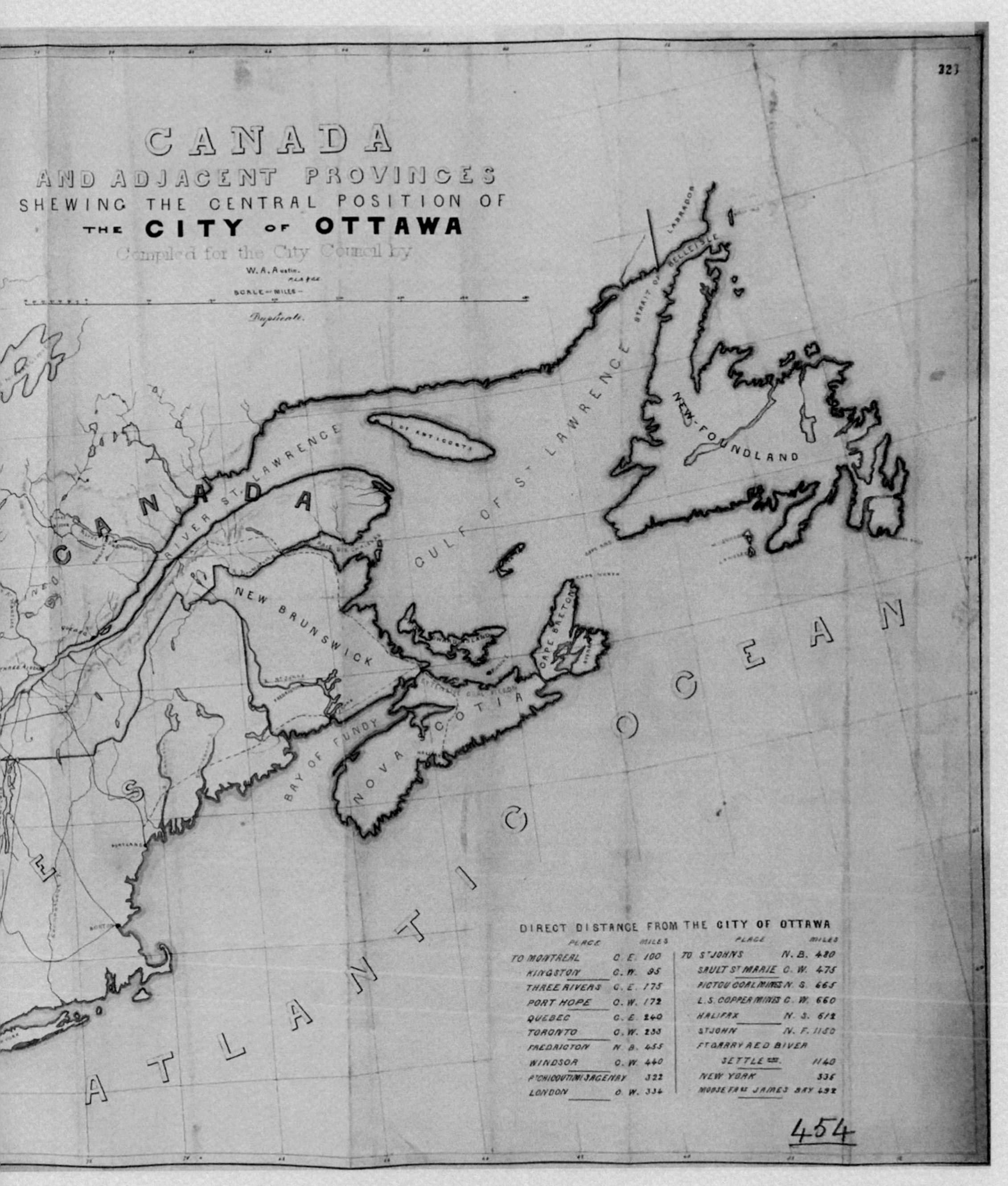
323
CANADA
AND ADJACENT PROVINCES
SHEWING THE CENTRAL POSITION OF
THE CITY OF OTTAWA
Compiled for the City Council by
W. A. Austin.
SCALE—MILES—
Duplicate.
LABRADOR
STRAIT OF BELLEISLE
NEW-FOUNDLAND
GULF OF ST LAWRENCE
RIVER ST LAWRENCE
CANADA
NEW BRUNSWICK
CAPE BRETON
NOVA SCOTIA
BAY OF FUNDY
ATLANTIC OCEAN
DIRECT DISTANCE FROM THE CITY OF OTTAWA
PLACE MILES
TO MONTREAL C. E. 100
KINGSTON C. W. 95
THREE RIVERS C. E. 175
PORT HOPE C. W. 172
QUEBEC C. E. 240
TORONTO C. W. 233
FREDRICTON N. B. 455
WINDSOR C. W. 440
PT CHICOUTIMI SAGENAY 322
LONDON C. W. 334
PLACE MILES
TO ST JOHNS N. B. 480
SAULT ST MARIE C. W. 475
PICTOU COAL MINES N. S. 665
L. S. COPPER MINES C. W. 660
HALIFAX N. S. 612
ST JOHN N. F. 1150
FT GARRY RED RIVER SETTLEMT. 1140
NEW YORK 336
MOOSE FALLS JAMES BAY 492
454

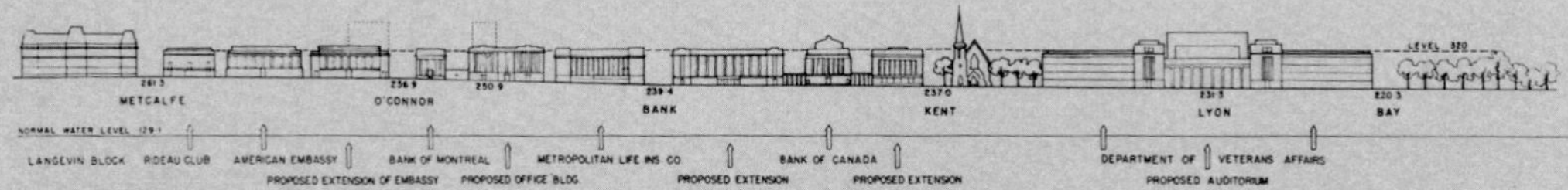

Jacques Gréber, National Capital Plan, 1948

Capital Expansion & Extension
The Administrative Territory of the Capital after Ottawa
framed by Township Grid (below) and Building Envelopes (above), 1948

whether it be a self-contained industrial community, a garden suburb, or a factory village built about a manufacturing plant, is a community intelligently planned and with the emphasis always on the rights of the community rather than on the rights of the individual property owner. The aim of the garden city is to bring dividends in human health and happiness as well as a return on property investment. It has plenty of places of rest, recreation, and play. Building restrictions are imposed, and the maximum number of houses to the acre is fixed. The improved health and condition of employees due to better homes and the open air, yield a return that pays for the investment. The co-ordination of garden cities with rural life, and of agriculture with the city, keeps down the cost of living. Existing garden communities have demonstrated that clean, wholesome, comfortable cottages are possible at a low rental, and that life is lengthened, the death and infant mortality rate reduced, and Labor [sic] in these open-air communities rendered more efficient than in the cities."[87]

Creating the *Central Mortgage and Housing Corporation* to finance planning schools and housing projects, as well as the *Community Planning Association of Canada* in 1947, to revive the Town Planning Institute of Canada that had fallen into disfavor during the Depression, Mackenzie King was a strong advocate of town planning and community building, one of the basic tenets of Adams' Town Planning Institute and early advisor to the Capital in the 1920s. Co-founder of the Town Planning Institute, the surveyor-railway-engineer-turned-town-planner and collaborator of Adams, Canadian-born Noulan Cauchon served as Chairman of the Ottawa Town Planning Commission from 1921 to 1935, iterating similar objectives under the City Scientific perspective that came with it[88]:

> "Scientific town planning, that which provides for proper arrangement and widths of streets, for height and proportion of buildings, so controls the value of land, the calls upon public services, so keeps down civic capitalization, as to enable wholesome and happy conditions within the reach of all."[89]

According to Ottawa historian David Gordon, Noulan Cauchon's role in Ottawa was not only foundational or technical in Ottawa but also influential in the national revival of town planning across the country in the early 1950s:

> "[Cauchon's] 'City Scientific' approach to planning would become the dominant mode of operation for the first generation of Canadian planners, and would influence the profession's resurgence after World War II."[90]

But in the background of these developments were views of the dominant "white supremacist state within the context of a settler colony in North America"[91] that Mackenzie King,[92] including his government and as a member of the white Anglo-Saxon elite, represented and consolidated during and after World War II, a period that saw extensive immigration. After all, as was often repeated and claimed in meetings between Canadian Prime Minister Mackenzie King and US President Roosevelt:

"This continent must belong to the white races."[93]

As most likely the country's most over-studied and most over-planned city, the Capital was riddled from the beginning with uncontrollable population pressure, awash with "undiluted" and "selective borrowing" of imperial forms from both French and British Architecture, with a penchant for "modern buildings" in "pre-modern civic spaces," according to French Architect and Master Planner Jacques Gréber.[94] Again, historian David Gordon explains these appropriations and borrowings of the Capital City:

> "The Beaux Arts architecture, landscape and urban design in Philadelphia evolved into a multi-layered approach to urbanism for Canada's capital, working at several scales. Gréber's architecture evolved from the classical to the modern, his landscape repertoire expanded from the *jardin à la française* to incorporate wilderness parks, and his planning embraced the CIAM without adopting their urban design manifesto. Gréber's urban public spaces typically avoided the tabula rasa, super-block and free plan. He preferred formal streets, blocks and public plazas."[95]

As pointed out "a closer reading of [Gréber's promiscuous] plan reveals a remarkable montage of themes"[96] which can be read as symptomatic of the subservient colonial mind and complex of architectural inferiority felt across the young nation state. Urban form was appropriated and borrowed as if it was depoliticized media. While layers of imperial appropriations from British and American sources plagued the 1950 Plan for the Capital—from Geddes' "Civic Survey," Olmsted's "Park System," Burnham's "City Beautiful," Howard's "Greenbelt," Abercrombie's "Satellite Towns," Cauchon's "City Scientific," to CIAM's "Land Use Planning"[97]—the major failure of town planning across the country, like in Ottawa, was an underestimation of its population dynamics with the pronounced pace of immigration, from east *and* west, that would take place in the following decades, along with the erasure of Indigenous peoples that would re-emerge later.[98]

While Ottawa—and its carefully planned and unobstructed silhouette—may have provided a picture-perfect precedent and model of master-planning for burgeoning and sprawling postwar Canadian cities, it barely succeeded in showing the power of planning, mainly an exercise of surveying and engineering. The Master Plan failed to prove how the Crown could instill colonial-style spatial control and territorial containment, in spite of the now clichéd uses of greenbelts and parkways, with other development controls such as buffers and setbacks, that simply would not, could not succeed. With this spatial militariza-

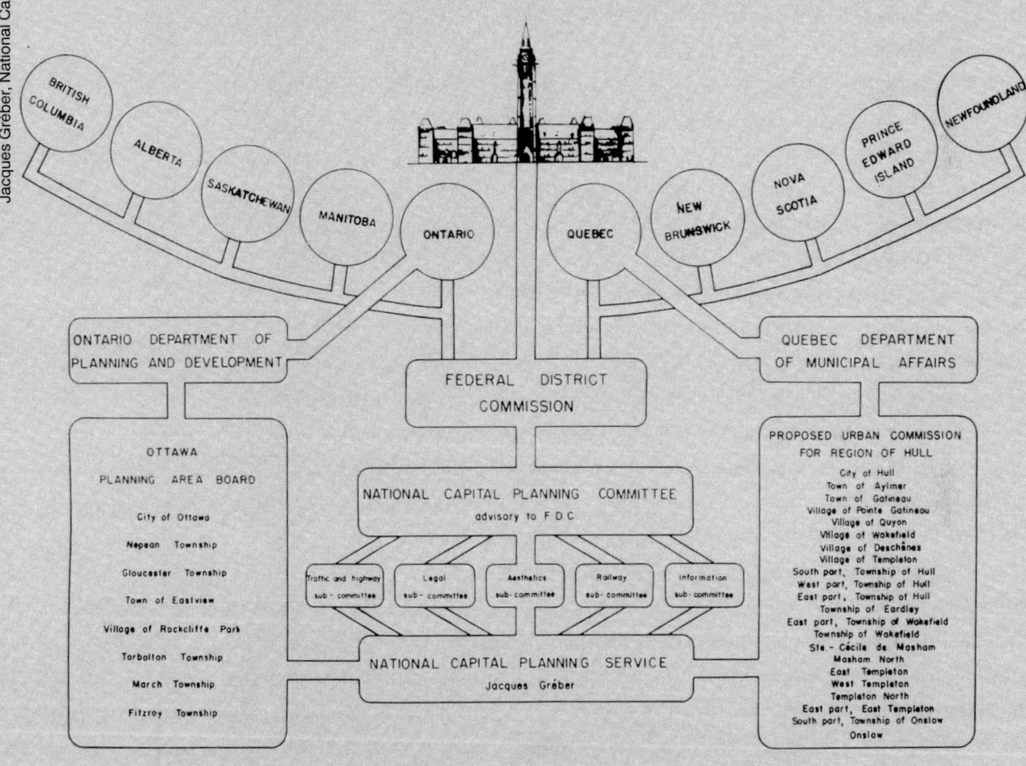

Jacques Gréber, National Capital Plan, 1948

From the Top Down

Planning model of cities using Ottawa as example and planning template for Canadian Cities, 1948

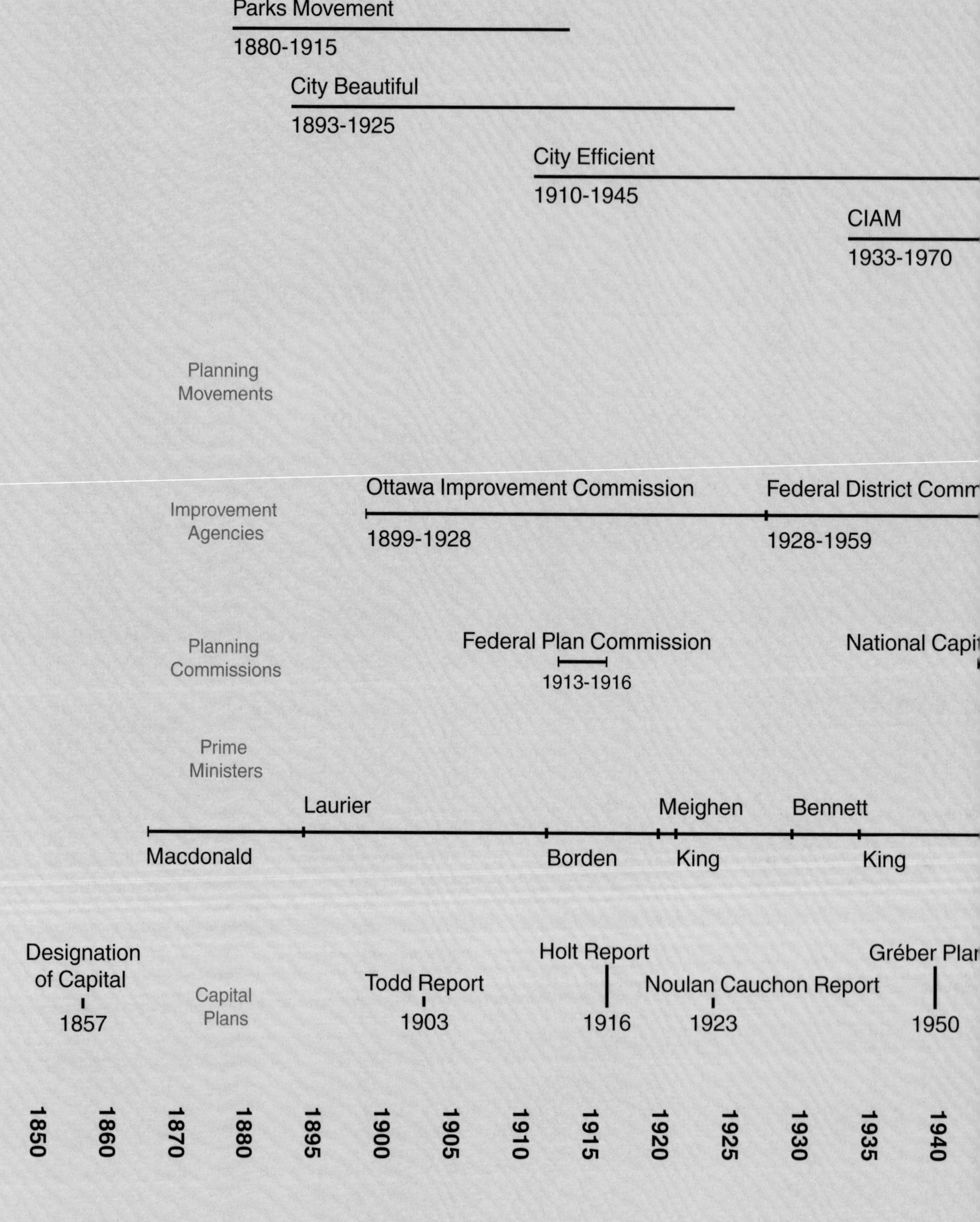
Parks Movement
1880-1915
City Beautiful
1893-1925
City Efficient
1910-1945
CIAM
1933-1970
Planning Movements
Improvement Agencies
Ottawa Improvement Commission
1899-1928
Federal District Comm
1928-1959
Planning Commissions
Federal Plan Commission
1913-1916
National Capi
Prime Ministers
Laurier
Meighen
Bennett
Macdonald
Borden
King
King
Designation of Capital
1857
Capital Plans
Todd Report
1903
Holt Report
1916
Noulan Cauchon Report
1923
Gréber Plar
1950
1850
1860
1870
1880
1895
1900
1905
1910
1915
1920
1925
1930
1935
1940

Capital Planning Paradigms

Historical Evolution of Planning Concepts of Ottawa, 1857–2020

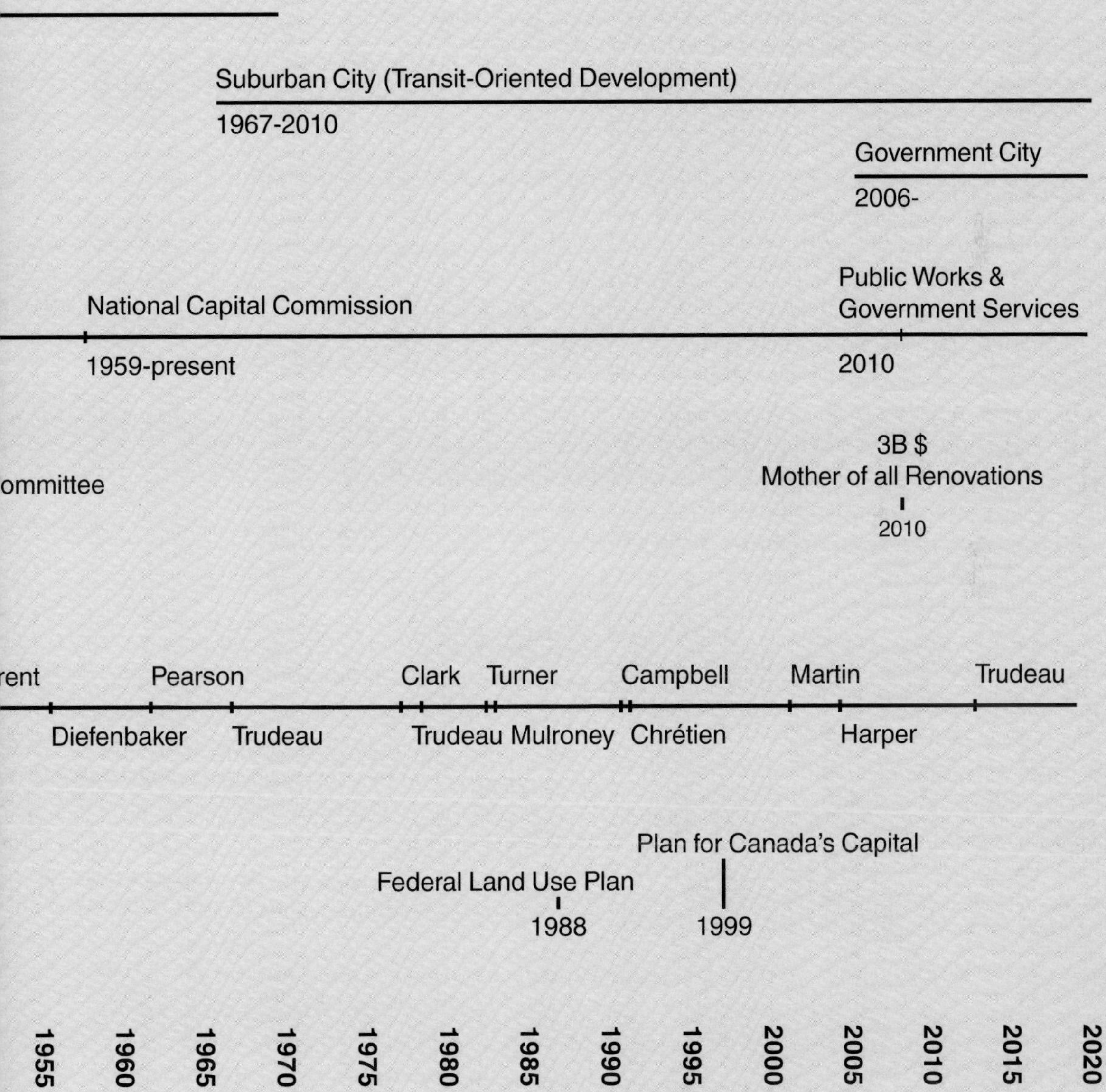

tion of the Garden City, the nostalgic Romanesque and Picturesque of the State Architect from France could not compete with either the political dynamics of growing immigration, transportation demands, and Indigenous territorial claims from which it originated or that continue to challenge it today, like most cities across Canada. For the foreseeable future, these challenges continue to confront and contest the plan of the city as state space for what it really is, the colonial metropolis.

Metropolitan Monopolies & the Colonization of Hinterlands

Underlying these transformations, between town and territory, was a geographic toponomy rooted in settler-colonial imagination. As the language of the British Empire, the English language produced an Anglo-Saxon system of appellations, spatial nomenclatures, and cognitive representations of the colonizing state. English became the predominant linguistic frame through which place could be named and land could be territorialized. Thus subjugated by representatives of the Crown, references to the motherland of the British Empire could be found anywhere, from a street corner and road intersection to the name of a post office, school, or hospital. Streets, blocks, buildings, districts, regions, counties, townships, tracts, would all put planning in a different place in reference to a far distant motherland lodged within the space of the grid, imprinted at every intersection and overlap of the grid lines: the imperial hosts creating territorial ghosts.

Territory could thus be conquered through a taxonomy of verbal referents and place designations that would dislocate any Indigenous appellations or immigrant influences. It centralized the historical narratives through erasure and ablation. From all the King streets to the Queensways, the naming of places, streets, districts in the grid thus entrenched and encoded colonial memes that would be accreting as fast as the grid could be surveyed and developments laid out. The colonizing effect of this territorial and administrative language was ubiquitous, hegemonic, and homogenizing.[99]

While the metropolis today may be portrayed in global culture as the deliverance and liberation from colonization, it is the presence of the centralizing forces of cities, towns, and villages that reproduce three-dimensional power structures of the colonial settler-state. The reclamation of a critical discourse in the rise of the colonial metropolis, and of its hinterland, is not insignificant as part of the setter-colonial construct, spatially and historically. Cities were best understood as the burgeoning outposts of financial, fiscal, regulatory, and administrative power, choice explosions at the intersection points of the settler-state survey grid.

While the concept of the metropolis-hinterland dichotomy has most often been understood as an outward-looking spatial concept, it has rarely been conceived as an inward-looking one. Classically, when understood as an axis, the metropolis-hinterland dialectic has stretched between empire and colony. Invoking relations between master and servant,[100] the metropolis was at the center of a motherland and core of an empire, while the hinterland spread far beyond in distant, remote colonies abroad. The further the colony, the potentially more significant the bureaucracy. Several decades after Adams into the early 1970s, and far from the political center of Ottawa, radical Marxist and editor of socialist journal *Canadian Dimension* based in Winnipeg, Cy Gonick, began to revive the issues and topics of colonialism and capitalism beyond the predominant discourse on provincialism that was pervasive across the country. Of relevance in this revival, in 1972, Gonick explained the theme of upcoming issues through emerging conflicts inherent between the colonial metropolis and its hinterlands:

> "Conflict in Canada has economic, racial, regional, and international dimensions. Any attempt at writing Canadian history which ignores conflict and struggle … deal only with what lies at the periphery of Canadian historical development. … To understand the economy of a country, you must place it within the larger economic system of which it is a part. Capitalism is viewed as a global economic system with one or a few metropolitan centres. These can shift from century to century—Holland, Spain, France, England, Germany Japan, the U.S.A. In chain-like fashion, it extends the metropolis through its many satellites."[101]

In this historical, urban, and territorial context, the works of Harold Innis and his theory of staple-based development from the late 1920s, as well as the later work with Mel Watkins in the 1960s, are groundbreaking. As a major proponent of Innis' staple theory thesis, Gonick focused on expanding the discourse of the metropolis-hinterland axis to reveal its internal territorializing tendencies and structures. Profiled in several issues of *Canadian Dimension* between 1972 and 1974, these internal processes of colonization could not only explain the growing spatial divide between the metropolis and the hinterland, they were actually contributing to the widening economic divide between classes:

> "The concept of metropolis-hinterland may also be applied within countries. The underdevelopment of some regions is usually linked with the overdevelopment of other regions. 'Probably the most interesting way to write the history of Canada,' writes Mel Watkins, 'is to write the history of Ontario: look up the chain and you see New York—at least if you sit in Toronto that is what you see. Then look down the chain and you see the Atlantic Provinces, the Prairie provinces (and further down, the Caribbean and Brazil.)' The overdevelopment of Ontario is the other side of the coin to the underdevelopment of rural, northern and fishing communities in Canada's Prairie, Atlantic and Northern regions."[102]

Again, Gerecke also echoed this reexamination of the metropolitan-hinterland thesis in the late 1970s:

> "Early Canadian cities were situated to protect colonial monopolies and to act as transportation points for the staples being exploited. Halifax, Quebec City, Montreal, Toronto and Winnipeg all owe their existence to these imperatives of colonialism. They also owed their respective places in the hierarchy of cities to these imperatives. As staples were exhausted in the East, the next westerly city became the gateway city to the West. This enhanced its competitive position and placed it in a position of dominance over the other Canadian cities. The westward movement of the dominant Canadian city should have moved from Toronto to Winnipeg but stopped at Toronto primarily due to its linkages with New York. This established a permanent dominance by Toronto. The relationships among Canadian cities is well-explained by the concept of metropolis and hinterland. This idea has primarily been used to understand the economy of different countries by identifying their place in the global economic system.[103]

Borrowing from Gerecke, if "the metropolis-hinterland theory explains the size and prosperity of Canada's cities historically," then planning's complicity in "Canadian colonization"[104] can be interpreted by reviewing planning's tools and techniques. As Gerecke further explains:

> "The 'gridiron plan' (rectangular blocks with straight streets and an intersection at each corner) was applied to the expansion of all Canadian cities. Its dominance is due to its preference by land speculators because it was simple to layout and survey. ... In short, colonization established the location, rate of growth, land system, and pattern of Canadian cities. This was done as a system of planning, to meet the needs of colonization, and not by chance. There was extensive government involvement from England and by the colony-based governments in order to implement colonialism in a formal way. Government worked for the interests of major economic forces. Any understanding of Canadian city planning [and that of westward expansion] must recognize this part of the history of Canadian development and its relevance to today."[105]

As the double end of the metropolis-hinterland axis was therefore a colonial-settler sword drawn between the frontier on the periphery, and the empire at the core. Exploitation and extraction of resources served the imperial center. According to resource economist Harold Innis, the development of Canada was lodged in between and merged from the exploitation and movement of raw commodities between "the centre and the margin,"[106] with the imperial metropolis as the center and the colonial hinterland as the margin. From pelt to fur, fish to fodder, forest to timber, pulp to paper, mine to ore, "the development of Canada has always depended on the production and export of one of a few key resource-based commodities. ... The nature of the particular staple product had much to do with the definition and the content of the imperial relationship."[107]

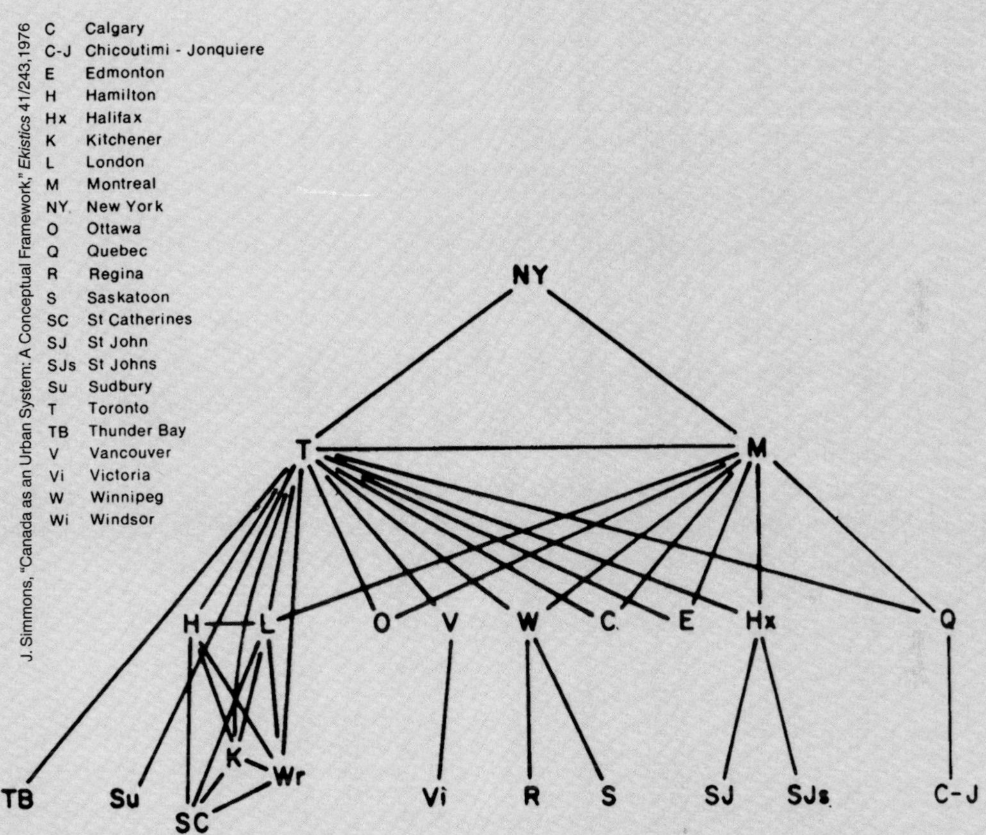

Colonial Metropolises
Urban Networks and the Erasure of Territorial Hinterlands, 1973

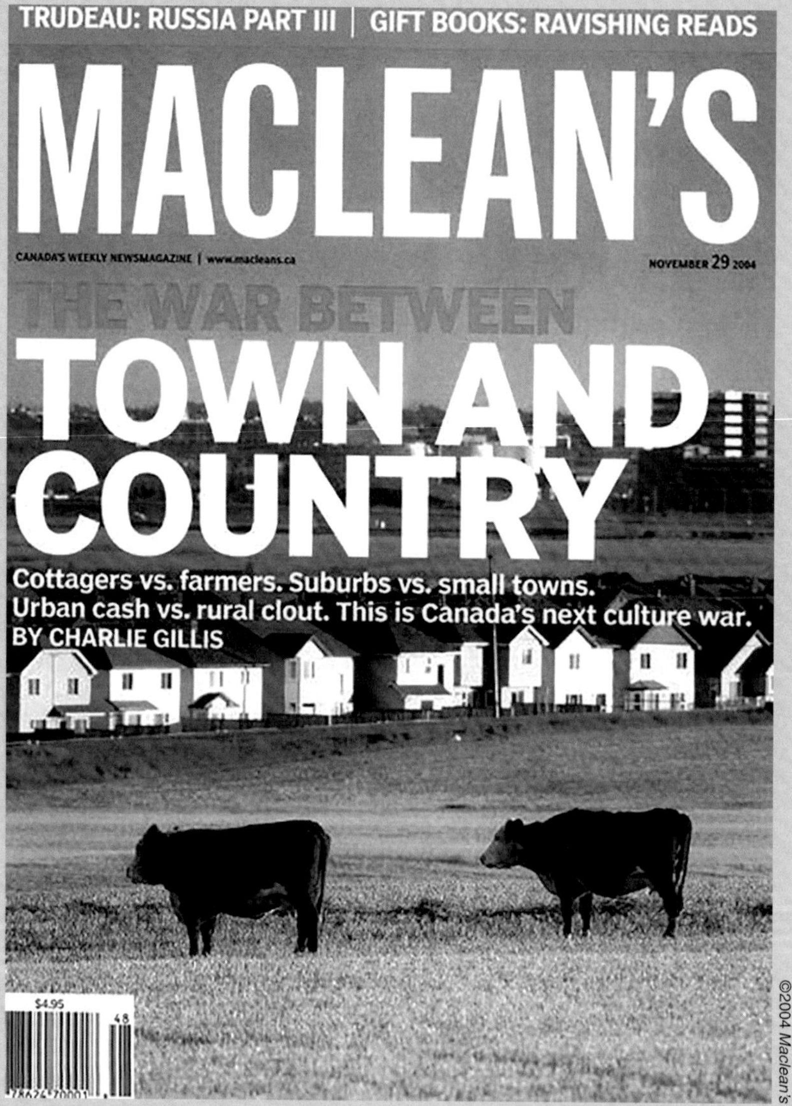

The Metropolitan Hinterland Axis
Cover of *Maclean's* "Town and County" Issue, 2004

Surpassing Adams' objections to the tendencies of the survey towards tabula rasa and scrubbing the ground clean, the system of the rectangular grid served as the template for settlement and speculation in spite of his aversion against the grid. For Adams, the rational and mathematical form of the grid was also "a crime against both nature and society and an economic blunder of the worst kind."[108] From the obvious influence of civic planner Patrick Geddes, Adams' city-centric view may have been simplistic, but it was also dogmatic and inflexible, tethered to weighty British-style state bureaucracy in the administration of internalized colonization:

> "Proper planning should follow no hard and fast rule. It should intelligently dispose the boundaries of at least the smaller divisions of land to suit industrial requirements, to conform to the natural conditions and physical features of the locality, and to provide for the most economical, convenient and healthy development."[109]

Adams revealed perhaps a misreading or selective understanding of Geddes' oeuvre. He overlooked the often-cited territorial nature of the influential Scottish planner calling for an "awakening to a more vital interest in civic problems and in the science of land development and industrial organization."[110] Topography, the valley, and the watershed were essential to Geddes' perspective.[111]

The city, at this juncture, was more than a settler-colonial space. The city, as colonial-metropolis, was a territorial technological infrastructure of segregation, exclusion, and exploitation. It would be an intellectual mistake to historicize the concept since it has grown. As historian Jane M. Jacobs observes:

> "In the areas where cities grew, those of first settlement, quick and comprehensive colonization was necessary to create the material preconditions for the realization of a 'settler' colony. Towns provided the home for settler 'authority'; they were administrative hubs of colonial governmentality and orchestrated the acquisition and redistribution of material resources. From these secure administrative centres, settler expansion could move outward into those where the fantasy of terra nullius was less surely inscribed."[112]

Any singular focus on the city obviated the dynamics that extended either beyond the simplistic dualities between town and country as discussed in Great Britain, or between the urban and the rural as characterized in the US. The 'grid,' as it still does today, instilled a political and historical geometry. 'Settlement,' 'speculation,' and 'development' were code words for colonial development and Indigenous dispossession. Beyond the failed development schemes and town planning acts that Adams proposed across Canada between 1910 and 1930—nationally, provincially, and municipally—Adams' 'master plans' also overlooked the political geographies and historical dynamics in which settlement and development were already framed and taking place. As historian Neville Morley explains of the Eurocentric biases:

> "The unequivocal view of cities and the Western bourgeoisie as centres of economic progress as well as political virtue and social emancipation from tyranny is very much a product of the nineteenth century industrial capitalist transformation of Europe."[113]

"While Canadian society and history may be viewed," according to Arthur K. Davis, "as a series of hinterland reactions to foreign metropolitan imperialisms," there was a need to understand the complexities of internal colonization and the unprecedented growth of domestic metropolises.[114] Not unlike criticisms of its British counterparts, the importance of costs in development schemes associated with classes of inhabitants were systematically overlooked. Evident in the criticism of a masterplan for a suburb of the Capital City of Ottawa by state planner Thomas Adams, costs of class segregation, labor mobility, and social stratification were obviated:

> "The original purpose … to provide decent, low-cost housing to those in need, had become obscured by a complex combination of aesthetic and social preconceptions. The rationalism and scientific management which was hailed by the reformers as the solution to Canada's urban social and moral woes became submerged into a style of new traditionalism."[115]

The continuous subdivision of land into property was practically automated by its simplicity and near infinite applicability in a range of micro-grids. It was a plight that drew very few observations by the transplanted planner Thomas Adams, who by now was ignoring, if not erasing, any forms inequity, injustice, or inhumanity associated with the colonialism of the Department of the Interior and the Commission of Conservation.[116] While his 1917 report focused on the work of conservation of resources as one frontier, and of planning of towns as another, this traditional Eurocentric perspective reified the "town" as "archetypal, generative, economically progressive city" mistook the city "as an independent, cross-cultural variable. … 'The town as a physical object is turned into a taken-for-granted social object,'" and much like in contemporary city planning today, "'a captivating focus of analysis in its own right.'"[117]

Proposing the abandonment of the concept of the town and therefore the exclusive study of settlements altogether across fields of urban history and sociology, historian Philip Abrams has proposed an understanding "of towns as sites in which the history of larger systems—states, societies, modes of production, world economies—is partially, but crucially worked out."[118] Therefore, in "Notes on the Difficulty of Studying the State (1977)," Abrams' substituted the safer and more simplistic categories of town (urban) and country (rural), in lieu of seeing the state of colonialism itself as the planning project, or the settler-state as object of study across a range of scales. With due cause of scale and complexity, Abrams re-iterated that:

> "The state is not the reality which stands behind the mask of political practice. It is itself the mask which prevents our seeing political practice as it is. There is a state-system: a palpable nexus of practice and institutional structure centred [sic] in government and more or less extensive, unified and dominant in any given society. There is, too, a state-idea, projected, purveyed and variously believed in different societies at different times."[119]

Put otherwise, in the settler-colonial state, the synthetic and seemingly apolitical nature of the grid masked, obscured, and erased the political embodiment of the ground and its history. It served to generate an imagination that separated occupation from extraction, settlement from migration, surface from subsurface, urban from rural, core from the periphery.

Indigenous Dispossession by Urban Displacement

As the Dominion Land Survey neared completion in the late 1920s, soon the project of settlement took on structural changes with the growth of small towns into larger cities, taking on more complex financial, legal, administrative functions for the growing bourgeois elite and its service population. As dominant forms, these cities were taking on new levels of in-migration as depopulation drained failing farmlands and centralization of resource exploitation took place. In 1927, Innis was instructive:

> "The economic history of Canada has been dominated by the discrepancy between the centre and the margin of western civilization. … Agriculture, industry, transportation, trade, finance, and governmental activities tend to become subordinate to the production of the staple for a more highly specialized manufacturing community."[120]

Yet, in the production of new frontiers and the perpetuation of new hinterlands, Innis further wrote,

> "the importance of metropolitan centres in which luxury goods were in most demand was crucial to the development of colonial North America. In these centres goods were manufactured for the consumption of colonials and in these centres goods produced in the colonies sold at the highest price."[121]

Extending Innis' staple theory of resource development as a powerful centralizing force with territorial implications, 'frontierism' entailed a condition of 'metropolitanism.' As Canadian historian James Careless explained in 1954, this relation was not only structural but it was historical and spatial:

"It will come as no surprise that my approach to urban development in Canada may be labelled metropolitanism. That really makes me both less and more than an urban historian—but more or less an urban historian anyway. On the one hand, many major themes in urban history do not come into a metropolitan analysis, except in a subsidiary way. On the other, much of its concern lies outside the city, and in the area of the hinterland. For a metropolitan approach essentially seeks to relate town and country, or better, to study the complex of reciprocal relationships between the concentrated population centre and the extended community beyond it. Any 'reciprocal' surely should suggest that both sides need equal billing. Put another way, the metropolitan concern is not primarily how 'the city' affects those within it—its internal patterns, processes and problems—but how 'the city' affects and is affected by those outside it."[122]

In other words, the city was not so much an urban frontier, as it was the periphery of a double-ended axis between the metropolis and hinterland. Pushing of the frontier (under the Sifton régime of fast-paced immigration) towards the West and mapping of its hinterland (through the Dominion Lands Survey), were therefore the first part of the colonial extractivist project. Once lands were settled and property secured, mass extraction could begin. In an iteration of imperial space, the frontier marked the periphery of empire on one end of the exploitation axis by establishing an axis between hinterland and metropolis, a much more complex iteration of the town and country dialectic, or the urban-rural dichotomy.[123]

While the "metropolis-hinterland relationship with the British Empire is seen as a mutually beneficial rather than exploitative arrangement," its "inter-relatedness" was more than two-dimensional, it was three-dimensional with a hierarchical, internal and external structure. Institutionally, it required an internalized structure of state administration.[124] The class structure of the metropolitan-hinterland relationship thus, was changed; no longer only administrative or political, it was sociological. By polarizing economic classes, it produced new social structures and reproduced old ones as sociologist and anthropologist Arthur K. Davis explained in the 1972 edition of *Canadian Dimension*:

"Roughly speaking, *metropolis* refers to centres of economic and political control, usually in the larger cities. *Hinterland* means any comparatively underdeveloped or colonial area which exports for the most part raw or semi-processed materials—including people migrating to better opportunities elsewhere. In the broad sense, it seems convenient to use *metropolis* and *overclass* interchangeably, likewise *hinterland* and *underclass* ... The dialectical premise is that major changes in the institutional structure of a particular society stem from internal and external oppositions in that society and its milieu. These oppositions or confrontations of interests and values eventually work into a new institutional pattern not like either of the original opposing complexes but usually including important elements of each. … In the Canadian context, the dialectical approach may be seen in the conflict of interests between metropolis and hinterland and in the inherent tendency of hinterlands to fight back—though these conflicts at different times may be hidden or counteracted by other factors, as conditions vary."[125]

These internalizing social structures would then be exacerbated by changes and directions in the power of the Crown from the national-federal to the provincial-regional. Institutionally, the Natural Resources Transfer Acts of 1930 led to increased control of resources and lands in the hands of the Provinces. Divisions in the powers of the Crown, between the federal and provincial governments, strategically left the complex administration of Indian Affairs (formerly within the Department of the Interior) under the newly formed Department of Mines and Resources in 1936. Under the staple theory of extractive development, provincial resources could now form and take control of distant spaces of production in albeit remote hinterlands through the power of legislation, royalty levy, and licensing, set against a counter-prevailing space of consumption as towns were expanding beyond the original surveyed limits.

"Urbanizing frontiers and outlying frontiers were highly interrelated," as Australian historian Penelope Edmonds adds in her unpacking of the colonial settler city of Victoria, British Columbia:

"As well as lobbying for the sale of the reserve, and for the extinguishment of Native title in outlying lands, Yates also made an address to the legislature 'praying for the immediate removal of the Northern Indians from the Towns of Victoria.' He, like many colonists, was deeply concerned about creating property and preserving its value by expropriating the local Lekwungen and other First Nations groups from the town and from outlying areas, only highlighting settler colonialism's most idealized geography, the state of commerce and the telos of best use embodied by the idea of the settler city, which rests upon the precondition of Indigenous absence."[126]

Since the "legendary," "steel frame of colonial bureaucracy" was now well developed, if not "over-developed" in some instances thanks to the Department of the Interior and the Town Planning Institute, its intent was "to control colonized peoples and ensure their submission to colonial imperatives, specifically the extraction and transfer of surplus to the metropolis."[127]

Periphery to Property: The Colonial Geographies of Capitalism

As divisions within the planning in the administration of resource extraction and Indian affairs grew, so too did town planning take shape as colonial template for the contemporary city and its infrastructure:

> "Adams' aim in Canada was to establish planning as a central function of government, buttressed by an integrated structure of legislation, administration, public support and professional expertise, organization and education."[128]

With the 1915 Civic Improvement League, and later, the Canadian Planning Act of 1919, Adams emulated the 1909 British Town Planning Act by establishing a relation between the state and citizen through the disembodiment of land based in property: "The act included zoning, subdivision control, public space reservation and the encouragement of co-operation between landowners and municipalities."[129]

Frequently citing the model of town planning of his English homeland,[130] Adams' imagination of planning was strictly utilitarian but Eurocentric: "the scientific and orderly disposition of land and buildings in use and development with a view to obviating congestion and securing economic and social efficiency, health and well-being in urban and rural communities."[131] But amidst an "enmity and indifference of a 'frontier' society to resource management and environmental planning,"[132] since "Adams had developed his basic concept of metropolitan regional planning in pre-war Britain, he was aware that the cultural and economic hinterland of a metropolitan city extended far beyond its boundaries and felt that it was essential to secure centralized planning control over this territory [by the State], effectively subjugating it to the interests of the core city."[133 134]

At the scale of the city, this centralizing approach by way of urban planning as civil institution would then, as post-colonial political theorist Tariq Amin-Khan would claim, "reinscribe" and possibly *extend* the process of colonialization "by dependency, subordination, and underdevelopment" in an urban setting.[135]

In the imagination of the British-trained planner then, and of the Canadian planner today, land could be assumed as a resource (not a culture, nor an economy, nor a politic) entrenched and embedded in the colonial concept of property that would subsume it. As capital, commodity, and consumable, land was a resource to be surveyed, partitioned, and developed *en masse*.

In short, land was industrialized by processes of draining, clearing, and leveling for settlement. Clearly, as the economist John Kenneth Galbraith claimed, "in the last century capital became more important than land. Power associated itself with this [new reality]."[136]

1. *Military*
2. *Geodetic*
3. *Geological*
4. *Hydrographic*
5. *Archaeological*
6. *Topographic*
7. *Engineering*
8. *Minerals*
9. *Parliamentary*
10. *Sanitation*
11. *Divisions*
12. *Jurisdiction*
13. *Census*
14. *Waste*
15. *Title*
16. *Property*
17. *Miscellaneous*

Surveying System
The 17 Good Reasons for a Survey according to
The Ordnance Survey of the United Kingdom, 1886

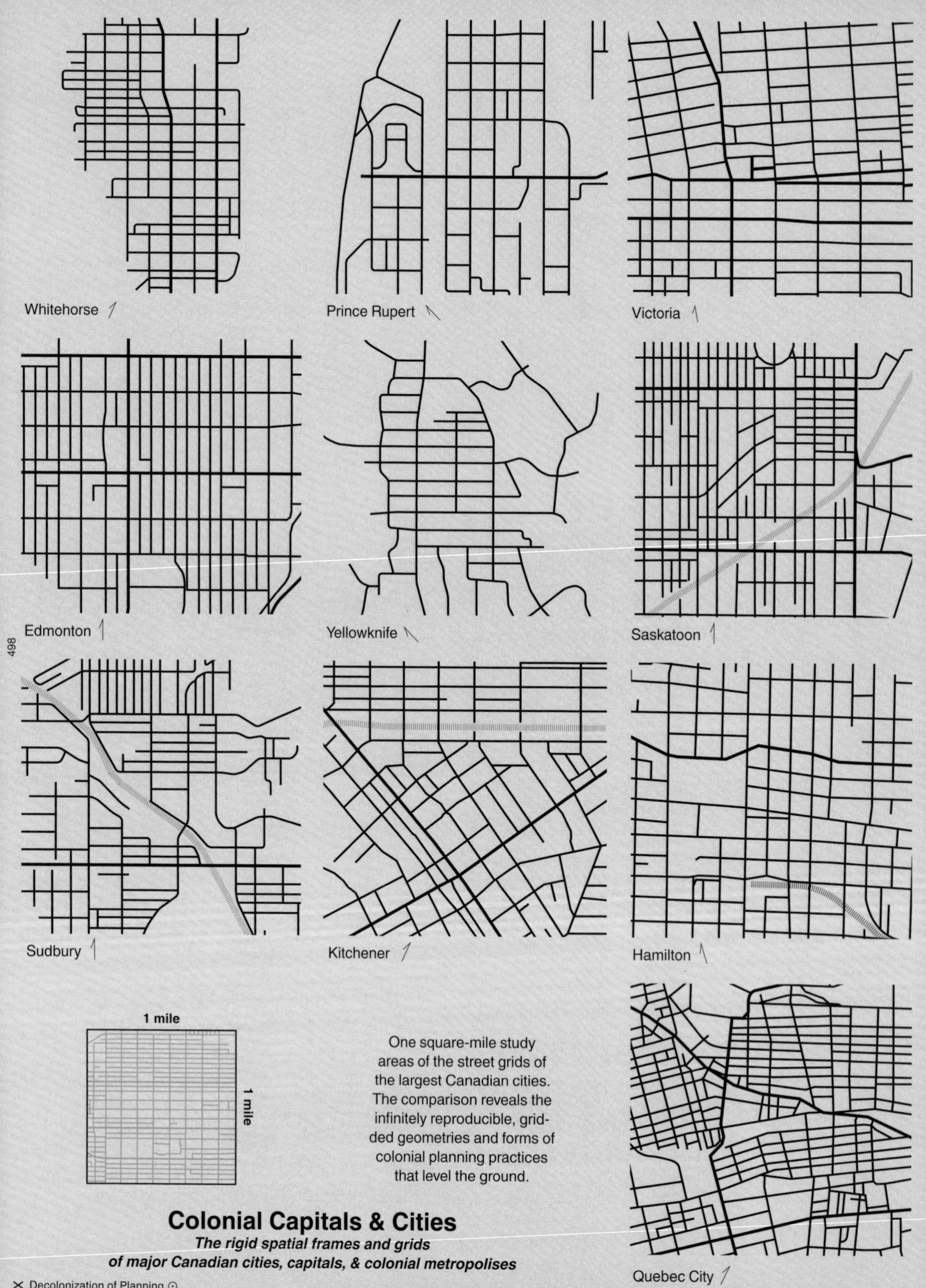

One square-mile study areas of the street grids of the largest Canadian cities. The comparison reveals the infinitely reproducible, gridded geometries and forms of colonial planning practices that level the ground.

Colonial Capitals & Cities

The rigid spatial frames and grids of major Canadian cities, capitals, & colonial metropolises

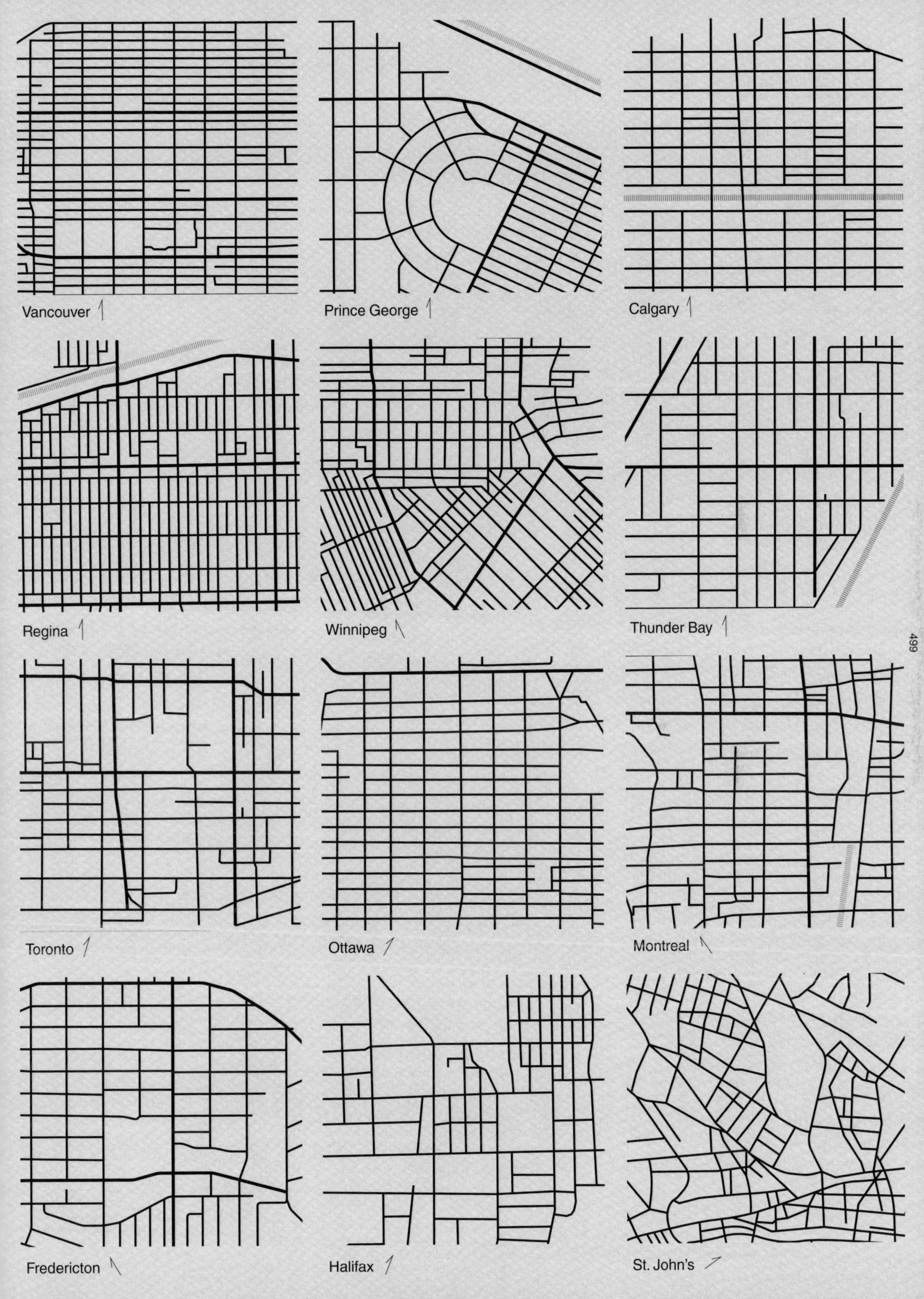
Vancouver
Prince George
Calgary
Regina
Winnipeg
Thunder Bay
Toronto
Ottawa
Montreal
Fredericton
Halifax
St. John's

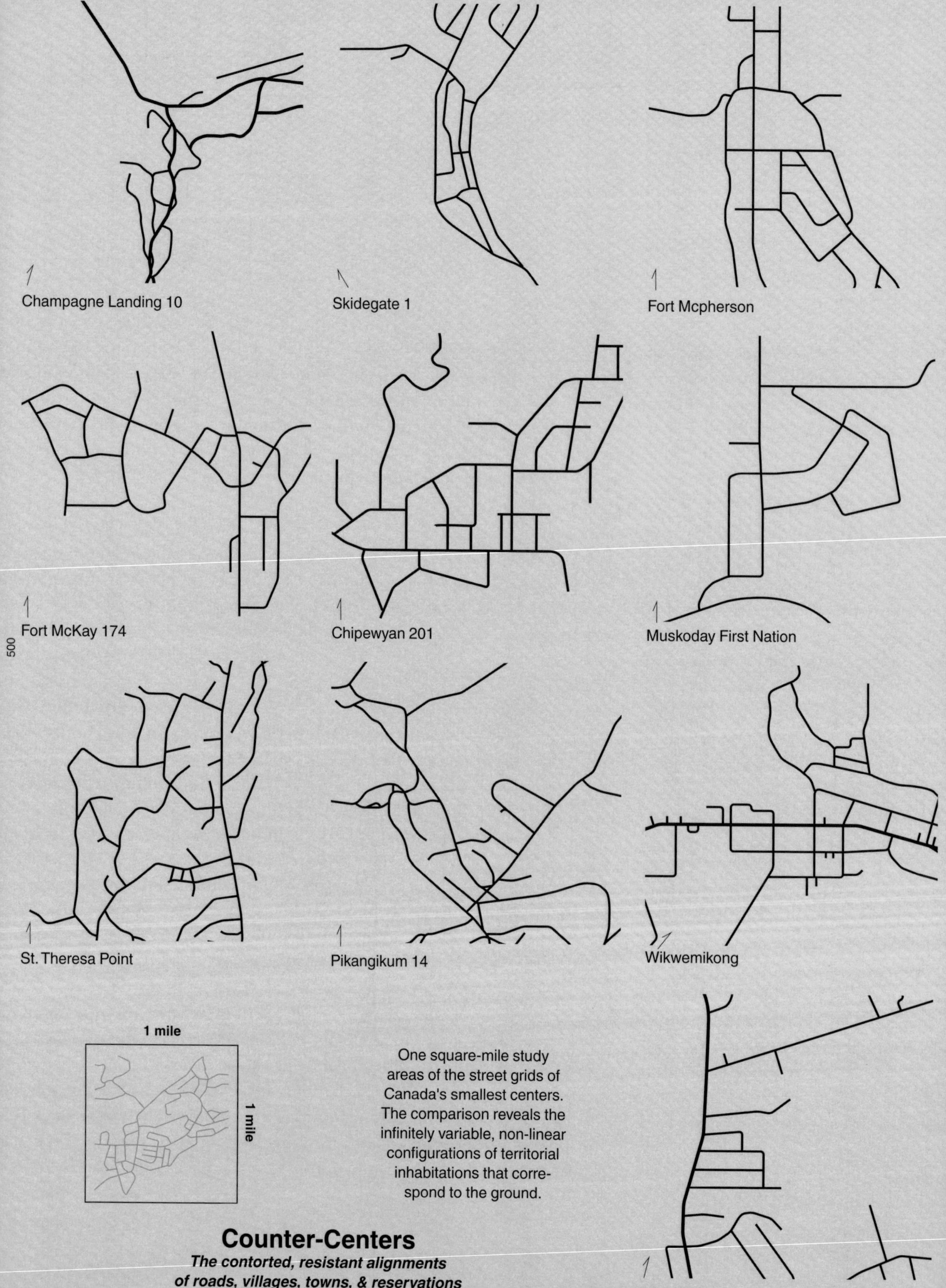

One square-mile study areas of the street grids of Canada's smallest centers. The comparison reveals the infinitely variable, non-linear configurations of territorial inhabitations that correspond to the ground.

Counter-Centers

The contorted, resistant alignments of roads, villages, towns, & reservations

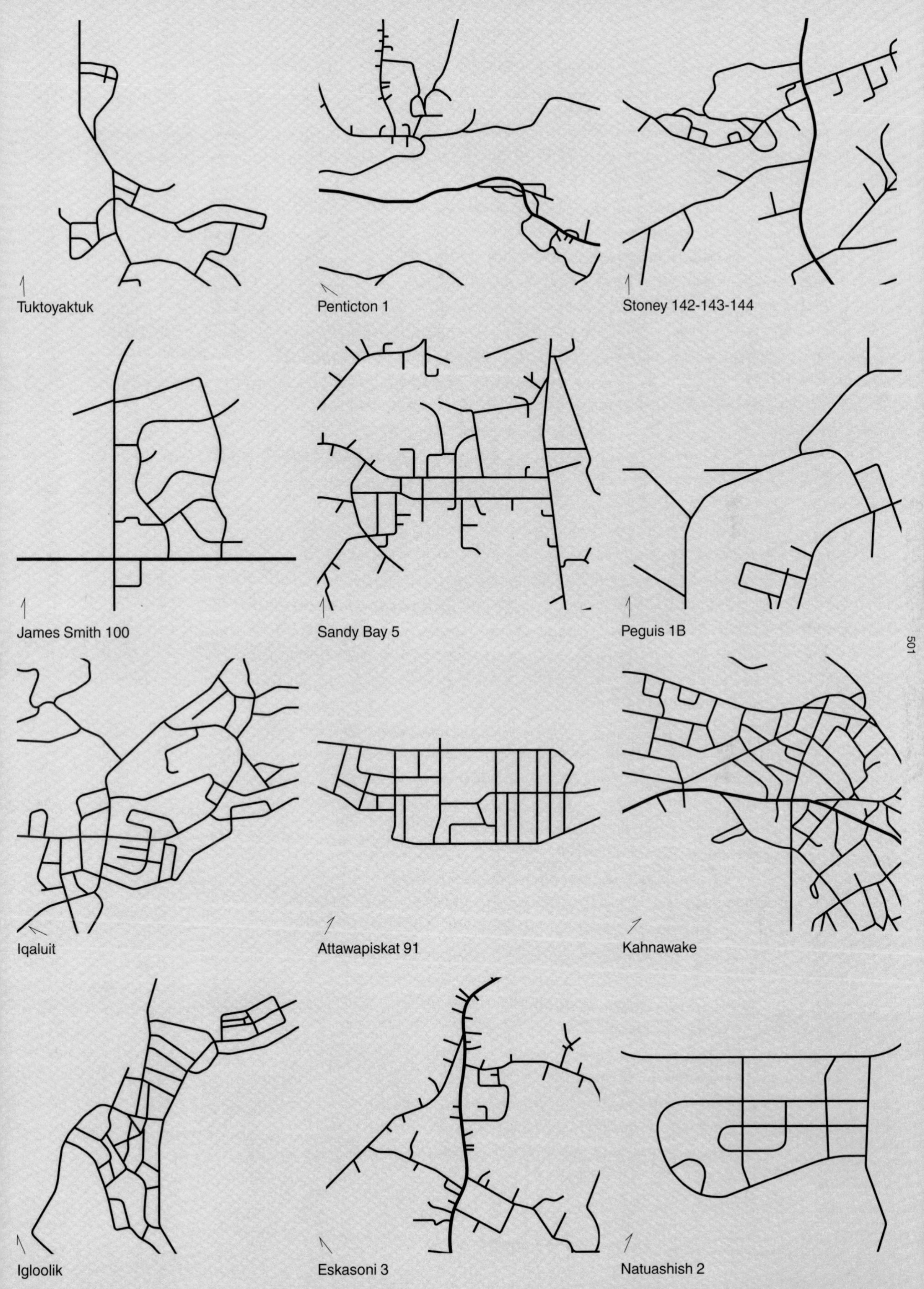
Tuktoyaktuk
Penticton 1
Stoney 142-143-144
James Smith 100
Sandy Bay 5
Peguis 1B
Iqaluit
Attawapiskat 91
Kahnawake
Igloolik
Eskasoni 3
Natuashish 2

As urban lawyer Charles Abrams proposed in the late 1930s, this so-called *revolution in land* involving "the patterns of … land use and ownership" in the entrenchment of "possession" and property,

> "flowed necessarily from the geographic, economic, and social backgrounds [and outcomes] of colonization."[137]

This was (and still is) the power of property as privilege and possession, territory as technology, survey as system, plan as tool. "[Signifying] ownership and the rights of private property," as Cole Harris observes in his sharp study of "colonialism, resistance, and reserves in British Columbia," titled *Making Native Space*, was especially true for the English in, "understanding of the relationship between property rights and land use"[138]:

> "Ownership was secured by action rather than word, action that made use of the land in ways that English people could appreciate—planting and tilling, gardening, building a house, bounding a space, and in so doing all of this, creating a visible presence on the landscape, that signified possession. No other colonists … made such use of surveyors, or wrote so much about gardens. For English colonists, planting a garden, and in the process, subduing the land, was much more than horticultural experience. A properly fenced garden was property."[139]

In both its development through disposition and disposal, the "landed power" of property as Galbraith called it, would legitimize and legitimate the power of the State—*comme l'État*, and control of property—*as estate*. When viewed historically and geographically, the twofold effect of territorialization—capital fixing and authority projecting—reveals significant environmental consequences, social trauma, and bodily effects.

Geographer Nicholas Blomley notes,

> "Physical violence, whether realized or implied, is important to the legitimation, foundation, and operation of a Western property regime. Certain spatializations notably those of the frontier, the survey, and the grid play a practical and ideological role at all these moments. Both property and space … are reproduced through various enactments. While those enactments can be symbolic, they must also be acknowledged as practical, material, and corporeal."[140]

As environmental lawyer Joseph Sax clarifies in Sydney Plotkin's *Keep Out: The Struggle for Land Use Control*[141]:

> "Property does not exist in isolation. Particular parcels are tied to one another in complex ways, and property is more accurately described as being

National Topographic System, NRCan

National Topographic System · Le Système national de référence cartographique

Dual highway, hard surface	**Route à 2 chaussées séparées,** revêtement dur
Road, hard surface, more than 2 lanes	**Route,** revêtement dur, plus de 2 voies
Road, hard surface, 2 lanes	**Route,** revêtement dur, 2 voies
Road, hard surface, less than 2 lanes	**Route,** revêtement dur, moins de 2 voies
Street	**Rue**
Road, loose or stabilized surface, all season, 2 lanes or more	**Route** de gravier, aggloméré, toute saison, 2 voies ou plus
Road, loose or stabilized surface, all season, less than 2 lanes	**Route** de gravier, aggloméré, toute saison, moins de 2 voies
Road, loose surface, dry weather	**Route** de gravier, temps sec
Unclassified road, street	**Route non classée, rue**
Vehicle track or winter road; gate	**Chemin de terre ou d'hiver; barrière**
Trail, cut line or portage; portage	**Sentier, percée, portage; portage**
Road, under construction	**Route,** en construction
Highway interchange with number; **traffic circle**	**Échangeur routier** avec numéro; rond-point
Highway route number	**Numéro de route**
Built-up area; street; park/sports field	**Agglomération; rue; parc ou terrain de sports**
Indian reserve, small	**Réserve indienne,** petite
Railway, single track; **railway station; turntable**	**Chemin de fer,** voie unique; **gare ferroviaire; plaque tournante**
Railway, multiple tracks	**Chemin de fer,** voies multiples
Railway, under construction	**Chemin de fer,** en construction
Railway, abandoned	**Chemin de fer,** abandonné
Railway on road; special track railway	**Voie ferrée sur route; chemin de fer à voie spéciale**
Rapid transit route: rail; road	**Transport rapide:** voie ferrée; route
Bridge; footbridge; snowshed	**Pont; passerelle; paraneige**
Bridge: swing, draw, lift; **tunnel**	**Pont:** tournant, basculant, levant; **tunnel**
Cut; embankment, causeway	**Déblai; remblai, chaussée**
Dyke/levee; carrying road/railway	**Digue/levée; portant une route/chemin de fer**
Ferry Route	**Traverse**
Ford	**Gué**
Submarine cable	**Câble sous-marin**
Navigation light; navigation beacon	**Feu de navigation; balise de navigation**
Coast Guard station; exposed shipwreck	**Station de la garde-côtière; épave émergée**
Seaplane base; seaplane anchorage	**Hydrobase; mouillage d'hydravions**
Crib, abandoned bridge pier	**Caisson, pilier de pont abandonné**
Airfield, position approximate; **heliport**	**Terrain d'aviation,** position approximative; **Héliport**
Building(s)	**Bâtiment(s)**
Church; non-Christian place of worship; shrine	**Église; lieu de culte non chrétien; lieu de pèlerinage**
Educational building; grain elevator; fire station	**Bâtiment d'enseignement; élévateur à grains; caserne de pompiers**
Sports/race track; stadium	**Piste de course; stade**
Silo; kiln; dome	**Silo; séchoir; dôme**
Cemetery; historic site, point of interest	**Cimetière; lieu historique, lieu d'intérêt**
Landmark object; on building; with height	**Élément-repère; sur bâtiment; avec hauteur**
Campground; picnic site; service centre	**Terrain de camping; terrain de pique-nique; air de service**
Golf course; golf driving range; drive-in theatre	**Terrain de golf; champ d'exercice-golf; ciné-parc**
Wind-operated device; ruins; greenhouse	**Éolienne; ruines; serre**
Aerial cableway, ski lift, conveyor	**Téléphérique, remontée mécanique, convoyeur**
Ski area, ski jump	**Station de ski, tremplin de ski**
Wall; fence	**Mur; clôture**
Tank(s): vertical; horizontal	**Réservoir(s):** vertical; horizontal
Warden, ranger station; Customs	**Poste de garde forestier; poste de douane**
Well: oil, natural gas	**Puits** de pétrole de gaz naturel
Crane: vertical; horizontal	**Grue:** verticale; horizontale
Rifle range with butts	**Champ de tir** avec buttes
Power transmission line, multiple lines	**Ligne de transport d'énergie électrique,** lignes multiples
Telephone line; firebreak	**Ligne téléphonique; coupe-feu**
Pipeline with control valve	**Pipeline** avec un valve de contrôle
Multiple pipelines underground	**Pipelines multiples** souterrains
Electric facility; oil or natural gas facility	**Installation éléctrique; installation pétrolière ou gazière**
Pit: sand, gravel, clay; quarry	**Sablière, gravière, glaisière; carrière**
Mine; cave	**Mine; caverne**

Natural Resources Canada · Ressources naturelles Canada

Canada

Colonial Symbology

Legend Marks of Settler Colonial Land Uses & Designations, 2014

City Centrism
The anti-territorial agenda of the Canadian Institute of Planners, 2017

> inextricably part of a network of relationships that is neither limited to, nor usefully defined by, the property boundaries with which the legal system is accustomed to dealing."[142]

In short, the contested acquisition of land—either by treaty or trickery,[143] by force or by edict—resulted in the production of a settler-state space through territory. In the hands of the State—either the Crown, the Clergy, or the Corporation—land was a spatial, material resource, a political tool for settlement, a capital commodity for taxation, and an environmental technology for resource extraction with a host of externalities. Together, this complex and protracted process of territorial production, projected colonial control and maintained imperial power.

So, if territorialization consisted and relied in the pushing of a boundary and the creation of a frontier, then the grid enabled spaces to be surveyed, then partitioned, subdivided, and settled into micro-structures of itself. That spatialization of this process aspired to the irreversible and permanent transition from land—as sovereign, political, Indigenous space (which was altogether marginalized, suppressed, atomized, and erased)—to territory as jurisdictional, industrial, and settler-state space. Political scientist Sydney Plotkin cites controversial Canadian economist John Kenneth Galbraith in this process of mass-industrialization:

> "For the production system at large, land remains a vital, if subordinate, ingredient in the work of society. The social industrial workshop cannot expand without space. The drive of capital to 'nestle everywhere, settler everywhere, establish connections everywhere' rarely abates. Indeed, the growth of capitalism as an integrated industrial system, elbowing out across huge interconnected metropolitan centers, places high premiums on well-located sites. Paradoxically, whereas land no longer focuses the politics of production and class, but remains the one industrially relevant resource—aside from labor itself—still in the hands of a population otherwise disconnected from the means to control industry. John Kenneth Galbraith is right: 'the control of labor or land accords no reciprocal power to command capital. But like labor, which can strike, the controllers of land restrict capital by refusing its cravings for space.'"[144]

As Jane M. Jacobs further confirms, "[these processes of] colonial aspirations of territorialisation depended upon fine-grained spatial technologies of power such as town planning regulations [control structures like the plan and the grid] and policing [of that system]."[145] The prompting of the creation of the North-West Mounted Police in 1873 (precursor to the Royal Canadian Mounted Police today) following the Dominion Lands Act in 1871 and the Department of the Interior, all three by institutions created by John A. Macdonald, attest to the securitization of territory.

The significance of these three technological constructs—the survey, the grid, the plan—can be found in the connections they draw between tools of representation, instruments of exclusion, and techniques of control. Jacobs further reiterates that, "imperial expansions established specific spatial arrangements in which imaginative geographies of desire hardened into material spatialities of political connection, economic dependency, architectural imposition, and landscape transformation."[146] As spatial technologies of power then, the survey, the grid, and the plan are now embedded as invisible constructs and therefore accepted as the naturalizing bureaucratic background of contemporary development control and land use zoning. While their ends may seem modest or well-intended, these tools violently and dangerously structure the practice and perception of urban planning today with their attendant spatialities of the colonial settler-state.

Planning without Planners: Subverting the Settler-State

The epistemological inquiry into the colonial origins of planning's tools and techniques, however mundane or banal they may be, reveal the range of a surprisingly large arsenal of technologies where spatial power and territorial force have been exercised and exerted. Through its institutionalization and its professionalization, the practice of planning has contributed to these exertions leaving indelible marks on the ground that have further stratified the colonial-industrial complex of the settler-state. Through the prism of planning, engineering, and architecture today, colonial statecraft is now entrenched in the reproduction of settler-space. With its codes and protocols, the design of urban spaces and engineering of territorial infrastructures have remained historically and critically unquestioned for far too long. Ironically, the invocation of British-based colonial antecedents and Roman-era imperial standards have barely raised any self-conscious form of responsibility or complicity. The survey grid and town plan, still, remain the default techniques of both the aspiring and practicing planner. In fact, colonialism seems to both elude and evade most curricula of planning schools either for young urban scholars or for established practitioners. Yet, amidst the recent historical context of the various programs, pedagogies, and practices of planning—either at regional, city, community, or neighborhood scales—its modern constitution continues to represent itself as the baseline, if not necessary, intermediary practice across the urban disciplines of city building. Across its lifespan of just over 100 years, the profession of planning has learned how to establish, uphold, and even innovate the systems of the colonial settler-state. Today, planners are notoriously known as go-betweens for the state and the developer (the surrogate settler), while leaving out overlooked, or worse, marginalized others that are lesser organized or represented.

However, as profiled above, these practices clearly stem from much longer histories of colonial statecraft, far beyond the conventional and short-sighted discourses of post-war modernity in the US and post-industrial reform in Europe. The systems of the colonial settler-state of Canada—and thereby of the Crown, Clergy, and Corporation—run so much deeper than modern renditions of subdivision plans and development schemes of the past century. Ironically, planning has been dwarfed and caricaturized by sub-municipal scales of community development, under the heavy hand of provincial and federal regulations. Through a range of techniques and technologies, this miniaturization of planning's scope has now become merely the micro-infill of the macro-survey grid; where settler-colonial values of class separation, racial segregation, gender discrimination, and economic apartheid remain in place. The proliferation of these overt injustices is deployed by covert, and often subliminal structures that included physical architectures and spatial infrastructures of state power. That transfer and fixing of spatial power in the leisurely banality of contemporary life, however distorted or aberrant, is made possible by the underlying foundation and reality of the contemporary metropolis. This is the democratic community that Adams, Sifton, and Mackenzie King wished for, and sought after, at the beginning of the twentieth century. As geographer Derek Gregory notes, "while they may be displaced, distorted, and (most often) denied, the capacities that inhere within the colonial past are routinely reaffirmed and reactivated in the colonial present."[147]

So, what alternatives, if any, exist? Can the displacement, dispossession, and disempowering effects of the settler-state, across centuries and generations, be redressed? Can the totalizing territorial logic of the grid be subverted or subdued? Can the all-encompassing, hierarchical institution of planning, weaken?

Not surprisingly, Australian urbanist Libby Porter asks an even more profound question on the planning of the settler-state:

> "Can we bring ourselves to a point of meaningful recognition of Indigenous coexistence in cities? If we are to do so, cities must be seen as unsettled places where Indigenous title, connection, and contemporary culture rightfully belong."[148]

These questions, and many other contestations, are ultimately matters of representation. They are rooted in the deep-seeded ideological politics and power of the settler-state standing against the sovereignties of Indigenous land and rights of immigrants. To ground these matters and questions of representation is dimension and depth of land, beyond the power of imagination. Representation brings into the question of the dispossession of land by way of the production of state territory and property. As Unangax scholar Eve Tuck and postcolonial scientist Wayne Yang suggest, "decolonization in the settler colonial context must involve the repatriation of land simultaneous to the recognition of how land and relations to land have always already been differently understood and enacted." For Tuck and Yang, "decolonization is not a metaphor," but rather, it entails "all of the land, and not just symbolically."[149] [150]

This re-placing of Indigenous land necessarily entails the dis-placing of planning's assumed state and the planner's perceived role. Within the settler-space of the colonial me-

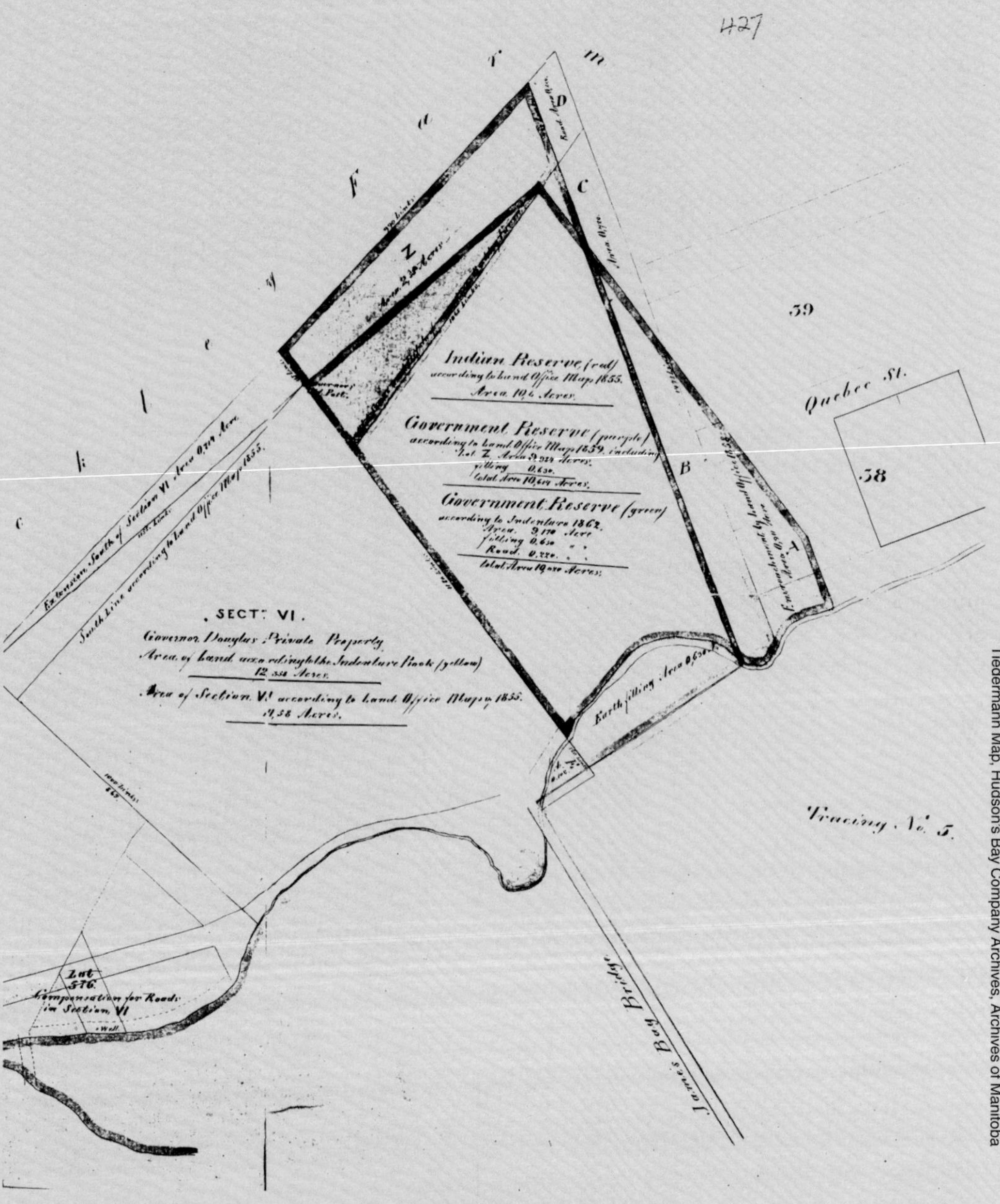

Tiedermann Map, Hudson's Bay Company Archives, Archives of Manitoba

Reclamation
Historical location of Songhees Reserve in the context of Victoria, ca. 1855

tropolis, a representative example in the assertion and affirmation of historical land titles and Indigenous rights can be found in the recent case involving the city of Victoria. In the Capital of the Province of British Columbia nicknamed the "Garden City," claims were made by Songhees and Esquimalt First Nations in 2001 to lands of the BC Government Legislative Buildings, putting into question the Crown's assumed rule of law and right to property. Reportedly in breach of the 1850 Douglas Treaty, the land claim was not only located at the center of a city historically known for "aboriginal erasure," but according to Penelope Edmonds, set an important precedent for retroactive land cession:

> "The successful claim [by the Songhees and Esquimalt First Nations in inner Victoria] contested and overturned the pervasive Western historicizing narrative structured upon a stadial evolutionary sequence of progress, one that has been both temporal and spatial, operating on the replacement of Indigenous spaces and peoples by immigrants. The claim disrupted the driving syntax of empire, showing that the city is a post-colonial and syncretic entity, and that Indigenous people with their own histories of emplacement remain active participants in the urban polity. Likewise the claim reminds us of the longstanding, if changing, Indigenized landscape of our cities and asserts the sovereignties staked in them."[151]

The representation of the politics, sovereignties, and histories of the erased and of the marginalized is therefore central to the un-planning of the city, and as in the case of the provincial capital City of Victoria, of the un-settling of the State. This transfiguration of the urban environment towards a political landscape and politicized ground, beyond the state, brings new spatial possibilities in view, based on non-settler narratives, immigrant realities, and importantly, Indigenous sovereignties.

If these changes are spatial, they must necessarily be made legible. As the economy of the metropolis opens and loosens its industrial stronghold, new visible forms of change can thrive and transcend its socio-economic strata. Simultaneously, this de-industrial change enables other non-anthropogenic, territorial systems, to flourish. Here, the space of production is not marginalized deindustrialization of the colonial settler-state of affairs. Rather it is enlivened by exchange, difference, and experience, acting as cultural generators, social powers, and material forces.

If then, "of all the ways space is produced, the creation of a settler-colonial city in the midst of an Indigenous homeland is perhaps one of the most visible,"[152] then the displacement of the city, that is the rigid steel frame of the colonial metropolis, will ultimately result in the reconfiguration of the city, its spaces, its surfaces. In other words, the ground on which the city rests. *Sous le pavé, le territoire*.

Not surprisingly then, the Nation's Capital itself has recently witnessed the largest land claim this century between the Federal Government and the Algonquin Anishinabeg

First Nation. Initiated by Algonquins of Pikwàkanagàn First Nation in 1983, the legal case sought aboriginal title to a significant territory, nearly three times the size of Lake Ontario, ironically encompassing much of downtown Ottawa and the Gréber Plan, including the grounds of Parliament Hill:

> "The [Algonquin Land] claim covers a territory of 36,000 square kilometres in eastern Ontario that is populated by more than 1.2 million people. If successful, the negotiations will produce the province's first modern-day constitutionally protected treaty. The Algonquins of Ontario assert that they have Aboriginal rights and title that have never been extinguished, and have continuing ownership of the Ontario portions of the Ottawa and Mattawa River watersheds and their natural resources. The boundaries of the claim are based largely on the watershed, which was historically used and occupied by the Algonquin people."[153]

Dating as far back as 1772, the claims of Indigenous land and rights have been expressed by the Algonquins and impressed upon the Crown with the existence of Algonquin peoples in the region of the Ottawa River Valley, a presence that has been documented to span 8,000 years. Among others, hunting and fishing rights were affirmed as part of the Treaty of 1760 with the King of England and his Superintendent General of Indian Affairs Sir William Johnston. Instituted as part of the Royal Proclamation of 1763 with King George III, those rights have, according to Pikwàkanagàn First Nation, been contravened, hindered, obstructed, infringed, and attacked.[154] Underlying these assertions, is the State's 300 year-old legal responsibility and fiduciary duty. "Canada and/or its agent (the National Capital Commission) have never compensated the Algonquin Anishinabe Nation for any of the federal interferences, either fairly or at all."[155] Adams, Sifton, and Mackenzie King would all balk at the greatest shortcoming of their masterplans that they fought so hard to erase: underlying, Indigenous sovereignty of land. Below property, below the State, below the Crown, is land.

So, finally then, if the hard core of capitalism and the building blocks of colonialism are lodged in between the imperial nature of Crown property and of State rule, then the key to unlock its stronghold may be in the dis-placement and dis-location of the foundations of property itself. As geographer Nicholas Blomley notes on the replacement of Indigenous sovereignties and emplacements of alternative futures:

> "Property relations can be configured as exclusionary, violent, and marginalizing. They can also be a means by which people find meaning in the world, anchor themselves to communities, and contest dominant power relations. To get at these multiple possibilities, we need to unsettle dominant treatments and recognize property's diverse meanings and often unsettled politics."[156]

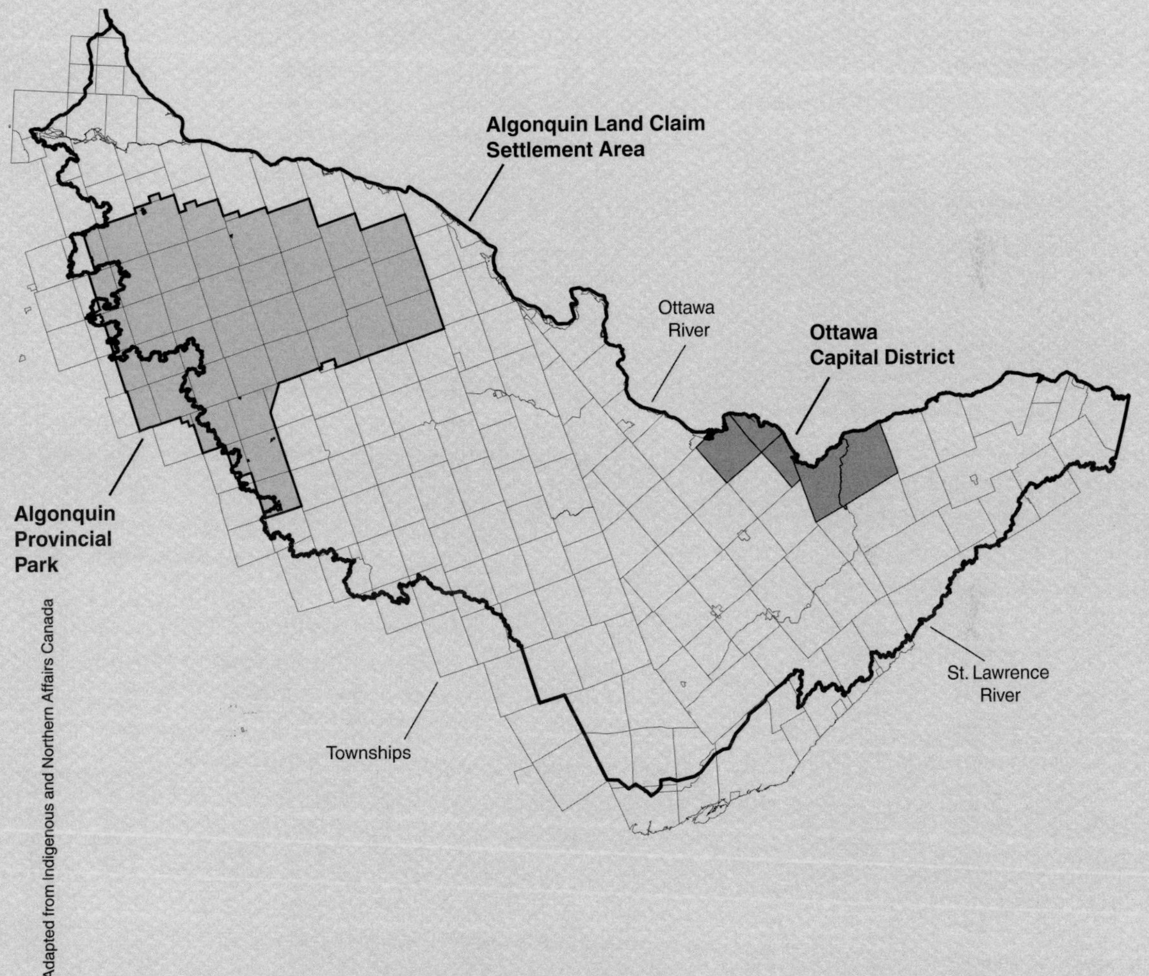

Reclaiming Territory & Colonial Displacement
The Algonquin Land Claim Encompassing Capital City of Ottawa, 2017

Once negotiated, the Algonquin land claim of the Pikwàkanagàn First Nation will therefore open new, alternative, and potentially unsettling modes of practice. Cessions, supercessions, and surrenders of land *a priori* will certainly continue to grow and be required in the decentering of the colonial settler-state. And, as a result, the displacement of this center of power and the replacement with new spatial sovereignties, these practices will have to find new legible and identifiable representations. The spatialities will be found in between the space of the survey grid and the town plan, between the territory and the city, between the ground and the property. These spatialities will be both inside and outside of the profession… within and without the colonial frame of the established institution. Across these spatialities lie the dynamics of watersheds, their waters, and their peoples.

Legal and constitutional transfiguration is therefore on the horizon. Common assumptions about the spatial constitutions of democracy that formed the visions of nation building in the work of Adams, Sifton, and Mackenzie King obviously need to be turned over. Either in the slackening of Adams' rational ideals of land use control (dezoning), the dismantling of Sifton's segregationist and preferential immigration policies (deracialization), or the weakening of Mackenzie King's ideals of the nation as single community (denationalization), will not only become central to the unsettling of colonial rule and settler centrality, but for the emancipation of Indigenous peoples, and new publics. For the next generations of planners and policy makers, the shift is jarring.

Consequently, these rights and sovereignties confront the orthodoxy of the planning discipline in two ways. First, they put into question the authority of the planner, and second, they put into question the centrality of the city and of the colonial metropolis in the imagination of the colonial settler-state. For this next generation, transfiguration of the institution will follow a series of historical shifts and declarations: the decoupling of the profession of planning from the exclusive agency of the state, the displacement of the city as core economic center, and finally, the dismantling of the planning establishment. First, in this break with colonial traditions of Canadian planning, the profession of planning can no longer serve exclusively as representative of the state. Second, in the retreat of its primary focus on cities as homogenous settler-state space, living forms of interventions will eventually subdue the colonial structures of racial, linguistic, gender, class-based superiority, that have marginalized the non-whites and the non-European.

Finally, moving forward, the proposed dismantling of planning's establishment must be based on a reconciliation and reckoning of its own violent and oppressive roots. The re-placing of academic curricula must assess, expose, and engage the inequalities, injustices, and inhumanities of the white, colonial settler-state, whose privilege and welfare has been constructed on. At all costs, it must avoid the often self-promotional nature and thematic platitudes of conciliatory diversity, nationalist multiculturalism, or participatory tokenism that only serve to perpetuate the hegemony of establishment. Should these alternative pathways towards re-placed possibilities entail the de-professionalization and de-institutionalization of planning as a whole, then the planner may have to accept weaker roles in favor of wider, more central representations, along with different methods and new timelines for change, across generations.

As a design process, 'unplanning' must then be transformative, transitive, and retroactive. Rethinking its educational and historical foundations that pre-date the colonial settler-state by several millennia, an imperative. To do this requires a fundamental shift in the pedagogies of planning programs and the paper worlds in which it is premised on. This transfigured curriculum and media must break with Modernist traditions and Victorian tendencies towards deepening knowledge of Indigenous land rights, lived experiences, immigrant realities, and territorial engagements. This transfiguration not only privileges fluid identities and substantiates spatial sovereignties, but demonstratively opens a wide array of ways to see, listen, hear, and experience as a basis for learning beyond the colonized spaces and hermetic campuses of university institutions.

Untethered from the State, the pluralization of planning practices can therefore rethink the institution's exclusivity, by replacing the state and the representation of its authority on more level ground. Unplanning of the establishment affirms a multiplicity of identities, beings, citizens, peoples, nations, of other pre-settler and post-settler states. As Tuck and Yang propose "settler colonialism and its decolonization implicates and unsettles everyone."[157] Here, Libby Porter also suggests calls for a model of relational action, albeit conflicted and potentially antagonistic, where planning must be "comfortable with conflict and the possibility of incommensurability. It does not seek to resolve away the tensions inherent in any post-colonial relationship, but sees that as constitutive of this particular social domain."[158]

It is perhaps *here*—in the decoupling between planning and state, as well as the decentralization of the discourse on the city and the country—that new strategies can emerge. This multiplicity of method and media can grow and recombine material practices, lived experiences, systems of productions with (not against) land-based constituencies and in relation (not in opposition) to territorial sovereignties. It is perhaps here where the project of the decolonization of planning can begin. After all, the colonial settler-state must be un-designed and un-planned, not simply overwritten.

For not only does the institution of planning need to replace and reposition the binary divide between town and country, city and territory, urban and rural, country and citizen, or metropolis and hinterland, but perhaps the more important reclamation and disposition of land rests in the transitional process and imagination of new relational and temporal infrastructures. Grounded, land-based living systems go hand in hand with territorial sovereignties. This relation is asserted by the figures of water bodies, watersheds, water cycles, water systems, water stories. They can help transfigure the oppressive and violent hierarchies of planning's grid; its structure is rooted in the imagined permanence of the survey, the cultural benchmarks, and paper-based monuments, literally and figuratively. This means that old states can be dismantled, suppressed states can be revived and represented, new states can be sown. States working and living across other states: trans-states.

Albeit seemingly monumental in scope, the retroactive process of unplanning the profession across new states does however require inception that spans lifecycles, lifetimes, and living generations *on* the ground, *in* place, today.

In other words, the works of reconceiving the territory, between the town and the country, lies in refiguring the difference between the image and reality, between map and territory, between oppressor and oppressed, wherein lies the questions and powers of representation beyond the grid. As an inquiry and survey then, the future lies not only beyond the limits of the top-down, hegemonic view of the city-state or the nation-state, but beyond the hegemony of the plan itself.

×××

Notes

1 Lorenzo Veracini, *Settler Colonialism: A Theoretical Overview* (New York, NY: Palgrave MacMillan, 2012), 108.

2 Kenneth Brealey, "Travels from Point Ellice: Peter O'Reilly and the Indian reserve system in British Columbia," *BC Studies* 115/116 (Autumn/Winter 1997/1998), 181–236, as referenced in Nicholas Blomley, "Law, Property, and the Geography of Violence: The Frontier, the Survey, and the Grid," *Annals of the Association of American Geographers* 93, no. 1 (2003): 128.

3 Ed Wensing and Libby Porter, "Unsettling planning's paradigms: towards a just accommodation of Indigenous rights and interests in Australian urban planning?" *Australian Planner* 53, no. 2 (December 2015): 90. Research experiences across the Commonwealth, on the parallel histories of settler-colonialism and spatial histories of different colonies across the British Empire are significant throughout scholarship on settler-colonialism. Their collaboration was also influenced by Libby Porter and Janice Barry's "Bounded recognition: Urban planning and the mediation of Indigenous rights in Canada and Australia," *Critical Policy Studies* 9, no. 1 (2015): 22–40.

4 Nicholas Thomas, *Colonialism's Culture: Anthropology, Travel and Government* (Cambridge, UK: Polity Press, 1994), 2. A significant number of Indigenous (mainly) and non-Indigenous scholars have been leading the discourse on decolonization and territorial resurgence by arguing that colonization is an ongoing process: Glen S. Coulthard, Kim Tallbear, Chelsea Vowel, Eve Tuck, K. Wayne Yang, Leanne Betasamosake Simpson, Patrick Wolfe, Edward Said, Derek Gregory, and Cy Gonick to name a few.

5 Madelaine Drohan, "Canada as Colonial Power: Not quite the way we like to think of ourselves," *Literary Review of Canada*, January–February 2011, http://reviewcanada.ca/magazine/2011/01/canada-as-colonial-power/.

6 Cole Harris profiles a brilliant understanding of colonization and colonial order through the historical narrative of geographical change and the colonial construction of space in Western Canada, in his *Making Native Space: Colonialism, Resistance, and Reserves in British Columbia* (Vancouver, BC: UBC Press, 2002), xxi.

7 "First Meeting of Town Planning Institute," *Construction: A Journal for the Architectural Engineering and Contracting Interests of Canada* 12, no.1 (January 1919): 207–8.

8 Town Planning Institute of Canada, *Constitution and By-Laws* (July 5, 1920), art. 3b.

9 Thomas Adams, "Town Planning and Housing in Canada," *Conservation of Life* [Issued under direction of the Commission of Conservation of Canada, Ottawa] 1, no.3 (January 1915): 62. See also Thomas Adams, "Town Planning Housing and Public Health," in *Commission of Conservation Annual Report* (Montreal, QC: The Federated Press, 1916), 116–17.

10 Adams, "Town Planning and Housing in Canada," back cover.

11 Ibid, 53–54.

12 Thomas Adams, ed., "Town Planning: What Municipalities Can Do," *Conservation of Life* 1, no.3 (January 1915): 71.

13 Adams, "Town Planning and Housing in Canada," 53.

14 See Adams, "Town Planning and Housing in Canada."

15 Town Planning Institute of Canada, *Constitution and By-Laws*, 3.

16 See Town Planning Institute of Canada, *Constitution and By-Laws*.

17 See David Lewis Stein, "Thomas Adams, 1871–1940," *PLAN Canada* [75th Anniversary Special Edition] 34, no.3 (1994): 14–15; and Michael Simpson, "Thomas Adams in Canada, 1914–1930," *Urban History Review* 11, no. 2 (October 1982): 1–16.

18 Thomas Adams, "The British Point of View," in *Proceedings of the Third National Conference on City Planning. Philadelphia, Pennsylvania: May 15–17, 1911* (Boston, MA: The university Press, 1911), 37.

19 Mark Francis, "The 'Civilizing' of Indigenous People in Nineteenth-Century Canada," *Journal of World History* 9, no.1 (1998): 56.

20 Ibid, 52.

21 "The great interest excited throughout the British Empire by the display at the Colonial and Indian Exhibition of 1886, which illustrated the vast wealth in natural products and the commercial, industrial, artistic and educational achievements of the various Colonies and of India, led His Royal Highness the Prince of Wales to suggest that a permanent Institution, designed to afford a thorough and living representation of the progress made in the development of their resources and elaborated upon a scale commensurate with the importance of their relations to the prosperity of the Empire, might constitute a fitting national memorial marking the fiftieth year of the reign of Her Majesty—an epoch within which some of the most important and thriving Colonies passed from insignificance, and even comparative barbarism, to exalted positions in the commercial and civilised world." See Frederick Abel, "The Imperial Institute and the Colonies" in *Canada: An Encylopædia of the Country - The Canadian Dominion Considered in its Historic Relations, its Natural Resources, its Material Progress, and its National Development*, ed. J. Castell Hopkins (Ottawa, ON: The Linscott Publishing Company, 1900), 6:41.

22 Francis, "The 'Civilizing' of Indigenous People," 56.

23 Clifford Sifton and Edwin R. Peacock, *Canada: A Descriptive Text-Book* (Toronto, ON: Warwick, Bros. & Rutter, 1900), 3.

24 Augustus Bridle, *Sons of Canada* (Toronto, ON: J.M. Dent & Sons, 1916), 79, as quoted in James White, *Conservation in 1918* (Ottawa, ON: Commission of Conservation, 1918), 5.

25 Clifford Sifton, *Commission of Conservation Annual Report* (Ottawa, ON: The Morimer Co., 1910), viii.

26 Terry Cook, "The Canadian West: An Archival Odyssey through the Records of the Department of the Interior," *The Archivist* 12, no. 4 (July–August 1985): 1–4.

27 Bridle, *Sons of Canada*, 79.

28 See Clifford Sifton's notes on "British and Continental Immigration" in *Annual Report of the Department of the Interior for the Year 1900* (Ottawa, ON: S.E. Dawson, 1900): xvi. On his exclusive, discriminatory, and racist views of 'desirable' immigrants, Sifton observed in a 1922 issue of *Maclean's* 35, no. 7 "The Immigrants Canada Wants": "When I speak of quality … I think of a stalwart peasant in a sheepskin coat, born on the soil. Whose forefathers have been farmers for ten generations, with a stout wife and a half-dozen children … It is quite clear that we received a considerable portion of the off-scourings and dregs of society," April 1, 16, 32–34.

29 For a greater discussion of the politics of cultural assimilation, Indigenous elimination, and genocidal policies, see Patrick Wolfe, "Settler Colonialism and the Elimination of the Native," *Journal of Genocide Research* 8, no. 4 (December 2006): 387–409.

30 Cook, "The Canadian West."

31 Department of Interior Minister David Mills, *Annual Report of the Department of the Interior for the year ended 30th June, 1876* (Ottawa, ON: Maclean, Roger & Co, 1877), xiv.

32 Governor of the Northwest Territories David Laird, as quoted by successor Mills, *Report Department of Interior 1876*, xiv.

33 J. Brian Harley, "Maps, Knowledge, and Power," in *The Iconography of Landscape*, ed. Denis Cosgrove and Stephen Daniels (Cambridge, UK: Cambridge University Press, 1988), 277.

34 Donald W. Meinig, *The Shaping of America: A Geographic Perspective on 500 Years of History*, vol.1, *Atlantic America, 1492–1800* (New Haven, CT: Yale University press, 1986), 232, as quoted in Harley, "Maps, Knowledge, and Power," 282.

35 Henderson P. Bryce, *The Story of a National Crime* (Ottawa, ON: James Hope & Sons, 1922), 14.

36 Ibid., 17.

37 Ibid., 3.

38 Ibid., 1.

39 "Indian Affairs had long been closely associated with the Department of Interior, which was the principal instrument through which the Federal Government attempted to implement its development policies for the prairie West. The Dominion authorities were charged with responsibility for all of Canada's Indians, but it was the prairie Indians that created the greatest problems for the government, and to whom the government had the most obligations. Indian Affairs was still a branch of the Department of the Interior when most of the numbered treaties were signed in the 1870s. Although it created a separate department in 1880, it thereafter normally retained its association with the Department of the Interior by coming under the aegis of the Minister of the Interior until 1936. Thus the Indians were viewed always in the context of western development, their interests, while not ignored, only rarely commanded the full attention of the responsible minister. Sifton illustrates these problems well. There is plenty of evidence of his desire to serve what he believed to be the best interests of the Indians. Yet he shared some pretty conventional prejudices and misconceptions about them, was heavily influenced by his officials and always had an eye on the political repercussions of his policies. He further obscured the already hazy separate identity of Indian Affairs by placing it in the Interior department under a single deputy minister. During Sifton's tenure, furthermore, the national budget more than doubled, the Department of the Interior budget nearly quintupled, but that of Indian Affairs increased by less than 30%. The fact was that the government—and, indeed, Parliament—had an unvaryingly parsimonious attitude towards the Indians." See Gregory P. Marchildon, *Immigration and Settlement, 1870-1939* (Regina, SK: University of Regina Press, 2009), 183–84.

40 These issues are still currently in discussion and subject of major reform, including, for example, the Royal Commission on Truth and Reconciliation, the Missing and Murdered Indigenous Women Inquiry, Drinking Water Advisories for First Nations, and the Indigenous Youth Suicide Epidemic.

41 Jane M. Jacobs, *Edge of Empire: Postcolonialism and the City* (London, UK: Routledge, 1996), 105.

42 See K.G. Basavarajappa and Bali Ram, "Section A: Population & Migration," Statistics Canada, last modified July 2, 2014, http://www.statcan.gc.ca/pub/11-516-x/section-a/4147436-eng.htm.

43 Joseph Bouchette, *The British Dominions in North America; or a Topographical and Statistical Description of the Provinces of Lower and Upper Canada, New Brunswick, Nova Scotia, the Islands of Newfoundland, Prince Edward, and Cape Breton* (London, UK: Longman, Rees, Orme, Brown, and Green, 1831), 1:1.

44 According to Michael Simpson, the 1917 Report was the last of three reports, two of which were never produced: "Adams came from the Chadwick tradition of scientific social investigation and had an academic cast of mind. These influences, combined with the investigatory responsibilities of the Commission of Conservation, led him to initiate three research projects, two of which—on housing and urban conditions—were never completed owing to the pressure of other work. The third was *Rural Planning and Development*." See Simpson, "Thomas Adams in Canada," 5.

45 Thomas Adams, *Rural Planning and Development: A Study of Rural Conditions and Problems in Canada* (Ottawa, ON: Commission of Conservation Canada, 1917), 249–250, citing Clifford Sifton's essay "The Foundations of a New Era" in *The New Era in Canada: Essays dealing with the Upbuilding of the Canadian Commonwealth*, ed. J.O. Miller (Toronto, ON: J.M. Dent & Sons, 1917), 57.

46 *Manual of Instructions for the Survey of Dominion Lands, 1871*, quoted in Adams, *Rural Planning and Development*, 258.

47 Robert K. Home, *Of Planting and Planning: The Making of British Colonial Cities* (London, UK: Taylor & Francis, 1996), 40.

48 Adams, *Rural Planning and Development*, 47.

49 Ibid., 45.

50 Ibid.

51 Harley, "Maps, Knowledge, and Power," 287.

52 "Canada Lands Survey System (CLSS)," Natural Resources Canada, last modified June 22, 2017, http://www.nrcan.gc.ca/earth-sciences/geomatics/canada-lands-surveys/canada-lands-survey-system/10870.

53 For Sifton, Indians were "Wards of the State." See David John Hall, *Clifford Sifton*, vol. 2, *A Lonely Eminence: 1901–1929* (Vancouver, BC: UBC Press, 1981), 49. In the Manitoba Free Press, a paper which Sifton owned, he further stated in 1900, "that the tendency of the Indian race within the limits of the Dominion is not in the direction of becoming extinct." See *Clifford Sifton*, vol. 1, *The Young Napoleon: 1861–1900* (Vancouver, BC: UBC Press, 1981), 269.

54 Rosalind Krauss, "Grids," *October* 9 (Summer 1979): 60.

55 Ibid., 64.

56 Adams, *Rural Planning and Development*, 52.

57 Don W. Thomson, *Men and Meridians: The History of Surveying and Mapping in Canada*, vol. 1, *Prior to 1867* (Ottawa, ON: Queen's Printer, 1966), 11.

58 Kent Gerecke, "The History of Canadian City Planning," in *The Second City Book: Studies of Urban and Suburban Canada*, ed. James Lorimer and Evelyn Ross (Toronto, ON: Charlottetown Group Publishing, 1977), 155–56.

59 Ibid., 156; and Gary Teeple, "Land, Labour and Capital in Pre-Federation Canada," in *Capitalism and the National question in Canada*, ed. Gary Teeple (Toronto, ON: University of Toronto Press, 1972), 47, as quoted by Gerecke.

60 This list of displacement and enumeration of First Nations is not exhaustive, nor complete. Given the contentious histories, conflicting accounts, and incomplete histories of these regions, sources have been compiled as a preliminary set of associations that require further research. The information reported here regarding First Nations and their displacement from metropolitan areas were compiled and compared from a variety of archival, encyclopedic, press, and Indigenous sources, such as the "Treaties, Surrenders and Agreements Database" from Library and Archives Canada, last modified March 18, 2008, http://www.collectionscanada.gc.ca/data-

bases/treaties/001040-100.01-e.php; "The Canadian Encyclopedia" from the Historica Canada organization, http://www.thecanadianencyclopedia.ca/en/; the Native Council of Nova Scotia, http://ncns.ca/; the Haudenosaunee Confederacy, http://www.haudenosauneeconfederacy.com; the Algonquin Anishinabeg Nation Tribal Council, http://www.anishinabenation.ca; the Papaschase First Nation, http://www.papaschase.ca; and the Songhees First Nation, http://www.songheesnation.ca/. It is also important to note that a considerable amount of land claims remain in progress, and are soon to come for the foreseeable future as pre-state territoriality is asserted by First Nations, Métis, and Inuit.

61 Harris, *Making Native Space*, 48. Harris' statement applies and extends the equally important work of Edward Said in his reference to *Culture and Imperialism* (New York, NY: Vintage Books, [1993] 1994), 78: "the actual geographical possession of land is what empire in the final analysis is all about," as quoted by Harris, *Making Native Space*, 47.

62 Adams, *Rural Planning and Development*, 52–53.

63 Gerecke, "History of Canadian City Planning," 155.

64 Roger J.P. Kain and Elizabeth Baigent, *The Cadastral Map in the Service of the State: A History of Property Mapping* (Chicago, IL: University of Chicago Press, 1992), 344.

65 James C. Scott, *Seeing like a State: How Certain Schemes to Improve the Human Condition Have Failed* (New Haven, CT: Yale University Press, 1998), 44, 47.

66 John Clarke, *Land, Power, and Economics on the Frontier of Upper Canada* (Montreal, QC: McGill-Queen's University Press, 2001), 87.

67 See Stuart Mill, *The Subjection of Women* (London, UK: Longman, Greens, Reader, and Dyer, 1869), 1.

68 Penny Harvey and Hannah Knox, *An Anthropology of Infrastructure and Expertise* (Ithaca, NY: Cornell University Press, 2015), 3–4.

69 Ibid.

70 Claude Lévi-Strauss, *The Savage Mind* (Chicago, IL: University of Chicago Press, [1962] 1966), 130.

71 Jo Guldi, *Roads to Power: Britain invents the Infrastructure State* (Cambridge, MA: Harvard University Press, 2012), 5.

72 Alexander T. Galt, "Report of the Minister of Finance of Canada" in *Public Accounts of the Province of Canada for the 1860* (Quebec, QC: Thompson, Hunter & Co., 1861), 24.

73 Ryan McMahon, an Anishinabe filmmaker, activist, and comedian, recently produced the film *Colonization Road*, profiling how roads built by the Crown serve as "startling reminders of the colonization of Indigenous territories and the displacement of First Nations people … [the roads were built] across the province, direct reminders of the Public Lands Act of 1853 and its severe impact on First Nations, their treaties, and their land in the name of "Canadian settlement." https://www.colonizationroad.com/.

74 William M. McDougall, *Report of the Commissioner of Crown Lands of Canada, for the year 1862* (Quebec, QC: Hunter, Rose, & Co., 1863), xiv.

75 Henry Parnell, *A Treatise on Roads: Wherein the principles on which roads should be made are explained and illustrated, by the plans, specifications, and contracts made use of by Thomas Telford, Esq. on the Holyhead Road* (London, UK: Longman, Rees, Orme, Brown, Green & Longman, 1833), 2. Six years after its publication, Parnell's Treatise would be reproduced and republished for the Crown at the onslaught of road building in Nova Scotia, on the edge of the British Empire, prior to the formation of Eastern Canada. See Henry Parnell, *Extracts of a Treatise on Roads published by order of His Excellency The Lieutenant Governor chiefly for the use of Road Commissioners in Nova Scotia* (Halifax, NS: Gossip & Coade, 1839).

76 Parnell, *Extracts of a Treatise on Roads*, iii.

77 Boucher de la Bruére, *Report on Colonization Roads in Lower Canada, for the year 1861* (Quebec, QC: Hunter, Rose & Lemieux, 1862), 4.

78 Ibid.

79 Ibid.

80 Ibid., 5–6.

81 Jo Guldi, *Roads to Power*, 15.

82 Rosalind Williams, "Cultural Origins and Environmental Implications of Large Technological Systems," *Science in Context* 6, no.2 (1993): 380.

83 Ibid., 381, 395.

84 Ibid., 380.

85 As Wilfrid Laurier (one of Mackenzie King's predecessors) referred to it during the 1899–1925 Ottawa Improvement Commission: "it shall be my pleasure … to make the city of Ottawa the centre of the intellectual development of this country," in Carleton Ketchum, *Federal District Capital* (Ottawa, ON: 1939), cited in Ken Hillis, "A History of Commissions: Threads of An Ottawa Planning History," *Urban History Review* 21, no.1 (October 1992): 46. See also David Gordon, *Planning Twentieth Century Capital Cities* (New York, NY: Routledge, 2006), 151.

86 William Lyon Mackenzie King, *Industry and Humanity: A Study In the Principles underlying Industrial Reconstruction* (Toronto, ON: Thomas Allen, 1918), 359–60, citing Frederic C. Howe, "The Garden Cities of England," *Scribner's Magazine* 52 (July 1912): 1–19.

87 Mackenzie King, *Industry and Humanity*, 360–61.

88 See David L.A. Gordon, "'Agitating People's Brains': Noulan Cauchon and the City Scientific in Canada's Capital," *Planning Perspectives* 23, no.3 (2008): 349–79.

89 Noulan Cauchon, *Hamilton* (Noulan Cauchon papers, LAC, Ottawa, MG 30 C105), 46–7, cited in Gordon, "Agitating People's Brains," 356.

90 Gordon, "Agitating People's Brains," 351. See also Sara E. Coutts, *Science and Sentiment: The Planning Career of Noulan Cauchon* (Graduate Research Essay, Institute of Canadian Studies, Carleton University, 1982).

91 John Price, "'Orienting' the Empire: Mackenzie King and the Aftermath of the 1907 Race Riots," *BC Studies* 156 (Winter 2007/08): 55.

92 William Lyon Mackenzie King, "International Relations: Mission to Great Britain, United States - Japan, January–March 1908, Volume 1, Private Diary of W.L. Mackenzie King," *Diaries of William Lyon Mackenzie King* (Toronto, ON: 1971), manuscript fiche 98–105, as quoted in Kirt Niergarth, "'This Continent must belong to the White Races': William Lyon Mackenzie King, Canadian Diplomacy and Immigration Law, 1908," *The International History Review* 32, no. 4 (2010): 605. In another report in 1908, Mackenzie King also mentions: "That Canada should desire to restrict immigration from the Orient is regarded as natural, that Canada should remain a white man's country is to be not only desirable for economic and social reasons … is necessary on political and national grounds." *Report by W.L. Mackenzie King, C.M.G., Deputy Minister of Labour, on Mission to England to confer with the British Authorities on the subject of Immigration to Canada from the Orient and Immigration from India in Particular* (Ottawa, ON: S.E. Dawson, 1908), 12, as quoted in Price, "'Orienting' the Empire," 70.

93 See Niergarth, "This Continent must belong to the White Races," 605.

94 David Gordon, "Weaving a Modern Plan for Canada's Capital: Jacques Gréber and the 1950 Plan for the National Capital Region," *Urban History Review* 29, no.2 (March 2001): 56–57.

95 Ibid., 57.

96 Ibid., 56.

97 This comparative genealogy was adapted from Gordon's formulation in "Weaving a Modern Plan," 56. The abbreviation CIAM denotes the *Congrès Internationaux d'Architecture Moderne* which had an important influence in the world of architecture, urban design, and planning during its lifespan between 1928 and 1956.

98 See The Canadian Press, "Ontario First Nation lays claim to downtown Ottawa, including Parliament Hill," *The Globe and Mail*, December 8, 2016, https://www.theglobeandmail.com/news/national/ontario-first-nation-lays-claim-to-downtown-ottawa-including-parliament-hill/article33275568/.

99 "The imposition of permanent surnames on colonial populations offers us a chance to observe a process, telescoped into a decade or less, that in the West might have taken several generations. Many of the same state objectives animate both the European and the colonial exercises, but in the colonial case, the state is at once more bureaucratized and less tolerant of popular resistance. The very brusqueness of colonial naming casts the purposes and paradoxes of the process in sharp relief." See Scott, *Seeing like a State*, 68–69.

100 See Michel Foucault, "Space, Knowledge, and Power," in *The Foucault Reader*, ed. Paul Rabinow (New York, NY: Pantheon Books, 1984), 239–55.

101 Cy Gonick, "Metropolis/Hinterland Themes," *Canadian Dimension* 8, no.6 (March-April 1972): 24.

102 Ibid., 25.

103 Gerecke, *The History of Canadian City Planning*, 155.

104 Ibid., 156.

105 Ibid., 157.

106 Harold A. Innis, *The Fur Trade in Canada: An Introduction to Canadian Economic History*, rev. ed. (Toronto, ON: University of Toronto Press, [1930] 1956), 385.

107 Gonick, "Metropolis/Hinterland Themes," 26–27.

108 Simpson, "Thomas Adams in Canada," 5.

109 Adams, *Rural Planning and Development*, 46.

110 Ibid., 241.

111 Patrick Geddes, "The Valley Plan of Civilization," *The Survey* (June 1925): 288–90, 322–23, 235.

112 Jacobs, *Edge of Empire*, 105.

113 Neville Morley, *Metropolis and Hinterland: The City of Rome and the Italian Economy 200 B.C.–A.D. 200* (Cambridge, UK: Cambridge University Press, 1996), 22–23.

114 Arthur K. Davis, "Metropolis/Overclass, Hinterland/Underclass: A New Sociology," *Canadian Dimension* 8, no. 6 (March-April 1972): 38.

115 Jill Delaney, "The Garden Suburb of Lindenlea, Ottawa: A Model Project for the First Federal Housing Policy, 1918–24," *Urban History Review* 19, no. 3 (February 1991): 164.

116 "Adams cannot be faulted on strategic grounds, for he had little choice but to try to advance on several fronts at once. His principal error was in persisting with a sophisticated British legislative and institutional framework which was far too complex for a relatively primitive society. However, what principally defeated the first attempt to establish planning as part of the governmental and developmental process in Canada was simply the enmity and indifference of a 'frontier' society to resource management and environmental planning." See Simpson, "Thomas Adams in Canada," 11.

117 Philip Abrams, "Towns and Economic Growth: Some Theories and Problems," in *Towns in Societies: Essays in Economic History and Historical Sociology*, ed. Philip Abrams and Edward Anthony Wrigley (Cambridge, UK: Cambridge University Press, 1978), 9, as quoted in Morley, *Metropolis and Hinterland*, 21–22n58.

118 Abrams, "Towns and Economic Growth," 3, 10, as quoted in Morley, *Metropolis and Hinterland*, 22n59.

119 Philip Abrams, "Notes on the Difficulty of Studying the State (1977)," *Journal of Historical Sociology* 1, no.1 (March 1988): 58.

120 Innis, *The Fur Trade in Canada*, 385.

121 Ibid., 384.

122 James M.S. Careless, "Urban Development in Canada," *Urban History Review* 3, no.1 (June 1974): 9.

123 For a greater discussion of spatiality of frontiers, hinterlands, territories, and staples development, see John Van Nostrand, "Counter-Development," interview by Pierre Bélanger, in this volume.

124 Melville H. Watkins, "The Innis Tradition in Canadian Political Economy," *Canadian Journal of Political and Social Theory* 6, no. 1–2 (Spring 1982): 22–23.

125 Davis, "Metropolis/Overclass, Hinterland/Underclass," 37.

126 Penelope Edmonds, "Unpacking Settler Colonialism's Urban Strategies: Indigenous Peoples in Victoria, British Columbia, and the Transition to a Settler-Colonial City," *Urban History Review* 38, no.2 (Spring 2010): 12.

127 Tariq Amin-Khan, *The Post-Colonial State in the Era of Capitalist Globalization: Historical, Political, and Theoretical Approaches to State Formation* (New York, NY: Routledge, 2012), 111.

128 Simpson, "Thomas Adams in Canada," 4.

129 Ibid.

130 "What has been done … in England points the way to a very real and very substantial success." See Thomas Adams, "Town Planning and Housing Reform in Canada," *Conservation of Life* 1, no.3 (January 1915): 53. It is worth noting that the Canadian Commission of Conservation formed in 1909, was largely based on the U.S. Commission of Conservation formed by President Theodore Roosevelt in 1908. Roosevelt's issued a declaration on the principles of conservation addressing forests, waters, lands, and soils. Critical to the understanding of the conservation movement and the conservation of resources is its historical basis rooted in the displacement of Indigenous peoples and the de-politicization of lands beyond cities in the aesthetic creation of the picturesque movement and the ideological perception of wilderness, formerly occupied, inhabited, and governed territories (like the early historical example of Yosemite Valley before the Indian War of 1850-51) later turned into "national assets" like national parks.

131 Town Planning Institute of Canada, as quoted in Simpson, "Thomas Adams in Canada," 7, 13n52.

132 Simpson, "Thomas Adams in Canada," 11.

133 Ibid., 6.

134 Dr. Charles Hodgetts, a medical advisor with the Canadian Commission of Conservation of Natural Resources, was responsible for proposing Thomas Adams' appointment to Canada for the creation of the Town Planning Institute. Working closely with Clifford Sifton, then Chair of the Canadian Commission of Conservation from 1909 to 1918, the coupling of natural and human resources was particularly significant as noted in the 1991 issue of *The Survey* (October 21): "In Canada official recognition is given to the fact that the health and lives of the people are as well worth conservation as are natural resources." ("Housing in Canada," 1030). Given the strong ties that were forming between the so-called conservation of national resources and the health of the public, an important racial sentiment began

to focus on the exclusive health of metropolitan populations, as noted in a 1910 note on the Canadian Commission of Conservation and Public Health in the Journal of the American Public Health Association, "The building [of a public health superstructure] must be upon the higher lines of public health, embracing all that tends to prevent disease, to improve environment, to propagate and build up a healthy race, prolong life, and lessen suffering" (405). The legislation establishing Yellowstone in 1872, *An Act to set apart a certain Tract of Land lying near the Head-waters of the Yellowstone River as a Public Park*, says: "Be it enacted by the Senate and House of Representatives of the United States of America in Congress assembled, that the tract of land in the Territories of Montana and Wyoming, lying near the head-waters of the Yellowstone river, … is hereby reserved and withdrawn from settlement, occupancy, or sale under the laws of the United States, and dedicated and set apart as a public park or pleasuring-ground for the benefit and enjoyment of the people."

135 Amin-Khan, *The Post-Colonial State*, 1.

136 John Kenneth Galbraith, *The New Industrial State*, 2nd rev. ed. (New York, NY: Mentor, [1967] 1972), 388.

137 Charles Abrams, *Revolution in Land* (New York, NY: Harper & Brothers, 1939), 10.

138 Harris, *Making Native Space*, 48–49.

139 Ibid., 48.

140 Blomley, "Law, Property, and the Geography of Violence," 121.

141 Sidney Plotkin, *Keep Out: The Struggle for Land Use Control* (Berkeley, CA: University of California Press, 1987), 66.

142 Joseph L. Sax, "Takings, Private Property and Public Rights," *The Yale Law Journal* 81, no.2 (December 1971): 152.

143 See *Trick or Treaty*, directed by Alanis Obomsawin (Ottawa, ON: National Film Board of Canada, 2014), 84 minutes, https://www.nfb.ca/film/trick_or_treaty/.

144 Galbraith, *The New Industrial State* (Boston, MA: Houghton-Mifflin, 1967), 55, as quoted in Plotkin, *Keep Out*, 57–58, 269n33.

145 Jacobs, *Edge of Empire*, 21.

146 Ibid, 19.

147 Derek Gregory, *The Colonial Present: Afghanistan. Palestine. Iraq* (Oxford, UK: Blackwell Publishing, 2004), 7.

148 Libby Porter, "Coexistence in Cities: The Challenge of Indigenous Urban Planning in the Twenty-First Century," chap. 12 in *Reclaiming Indigenous Planning*, ed. Ryan Walker, Ted Jojola, and David Natcher (Montreal, QC: McGill-Queen's University Press, 2013), 304.

149 Eve Tuck and K. Wayne Yang, "Decolonizing is not a metaphor," *Decolonization: Indigeneity, Education & Society* 1, no. 1 (2012): 7.

150 Internal colonization also includes the current tendency towards treating aboriginal land claims and Indigenous sovereignty as matters to be incorporated into 'the plan,' which otherwise go unchanged in their tools, techniques, and procedures along with the laws, legislations, and legal systems. See the professional journal of the Canadian Institute of Planning, *PLAN Canada* 53, no. 2 (Summer 2013) "Indigenizing Planning / Planning to Indigenize." In a recent 2016 issue of *PLAN Canada*: "Research shows that in bands that are still very close to the land, member's home attachment is to their hunting territory and camps, rather than to the houses they inhabit in their reserve. Therefore, it is not surprising that while public consultation on territorial development and land claims are very popular, consultations on reserve expansion plans can look like rubber-stamping exercises … With very few exceptions, reserves were planned or widely extended after the Second World War, when the Canada Mortgage and Housing Corporation was created and began to spread suburban housing development models and housing development models and when municipalities adopted urban regulations for their expansion that led to the reproduction of these urban forms. Therefore, it is not surprising that many First Nations communities were developing along the same lines, considering these models dominated the views and practices of professional surveyors, planners, engineers, and construction firms, whether they worked for federal agencies or as consultants to band councils. Although these suburban developments have been the object of criticism in planning circles for the past 40 years, and their inadequacy for Indigenous communities voiced by their members and numerous observers, they are still reproduced today, with all their well-known problems such as the discontinuity between urban fabrics, the infrastructure, and environmental costs, and the lack of permeability. On the one hand, those same consultants are hired again and again by many Indigenous communities because they deliver what the bands expect in terms of timelines; they know how to respond to the specific conditions imposed by INAC's financial programs to maximize housing production and limit immediate infrastructure costs; and they produce the plans and regulations that are prioritized by financial institutions as well as by INAC." See Denise Piché, "Quebec Indigenous Communities and Urban Planning: A World Apart?" *PLAN Canada* [Indigenous Planning / Urbanisme Authochtone] 56, no. 4 (Winter 2016): 55–57. Of noteworthy importance, INAC is the abbreviation for *Indigenous and North Affairs Canada*, a federal department renamed under Canada's "Federal Identity Program" in the late 1970s. Officially, legally, and historically, INAC is authorized as the "Department of Indian Affairs and Northern Development," whose identities and designations may have changed over the past century and a half, but its surrogate form, paternalistic powers, and colonial structure have remained relatively the same. Ironically, its responsibilities have been carried under various federal departments including the "Minister of Citizen and Immigration," and the "Department of Mines and Resources," but INAC's origins are found in the Department of Interior, created in 1873, following Confederation.

151 Edmonds, "Unpacking Settler Colonialism's Urban Strategies," 17.

152 Patrick A. Dunae, John S. Lutz, Donald J. Lafreniere, and Jason A. Gilliland, "Making the Inscrutable, Scrutable: Race and Space in Victoria's Chinatown, 1891," *BC Studies* 69 (Spring 2011): 53.

153 See "The Algonquin Land Claim," Government of Ontario, last modified April 7, 2017, https://www.ontario.ca/page/algonquin-land-claim.

154 See the Petition of the Chief Council, and People of the Algonquins of Golden Lake, on Behalf of the Algonquin Golden Lake First Nation, addressed To His Excellency Edward Shreyer, Governor General of Canada and Personal Representative of her Majesty the Queen Elizabeth the Second, March, 1983, http://www.tanakiwin.com/wp-system/uploads/2013/10/j-Algonquin-Petition-of-1983.pdf.

155 The Canadian Press, "Ontario First Nation lays claim," par. 9.

156 Nicholas Blomley, *Unsettling the City: Urban Land and the Politics of Property* (New York, NY: Routledge, 2005), 156.

157 Tuck and Yang, "Decolonization is not a metaphor," 7.

158 Porter, "Coexistence in Cities," 291.

A Glacial Pace

Delineating the Contours of Colonization in Canada's National Parks System

Tiffany Kaewen Dang

"Landscapes organize the creation and dissemination of national myths, which are naturalized over time."[1]

—Rod Barnett, "Designing Indian Country," 2016

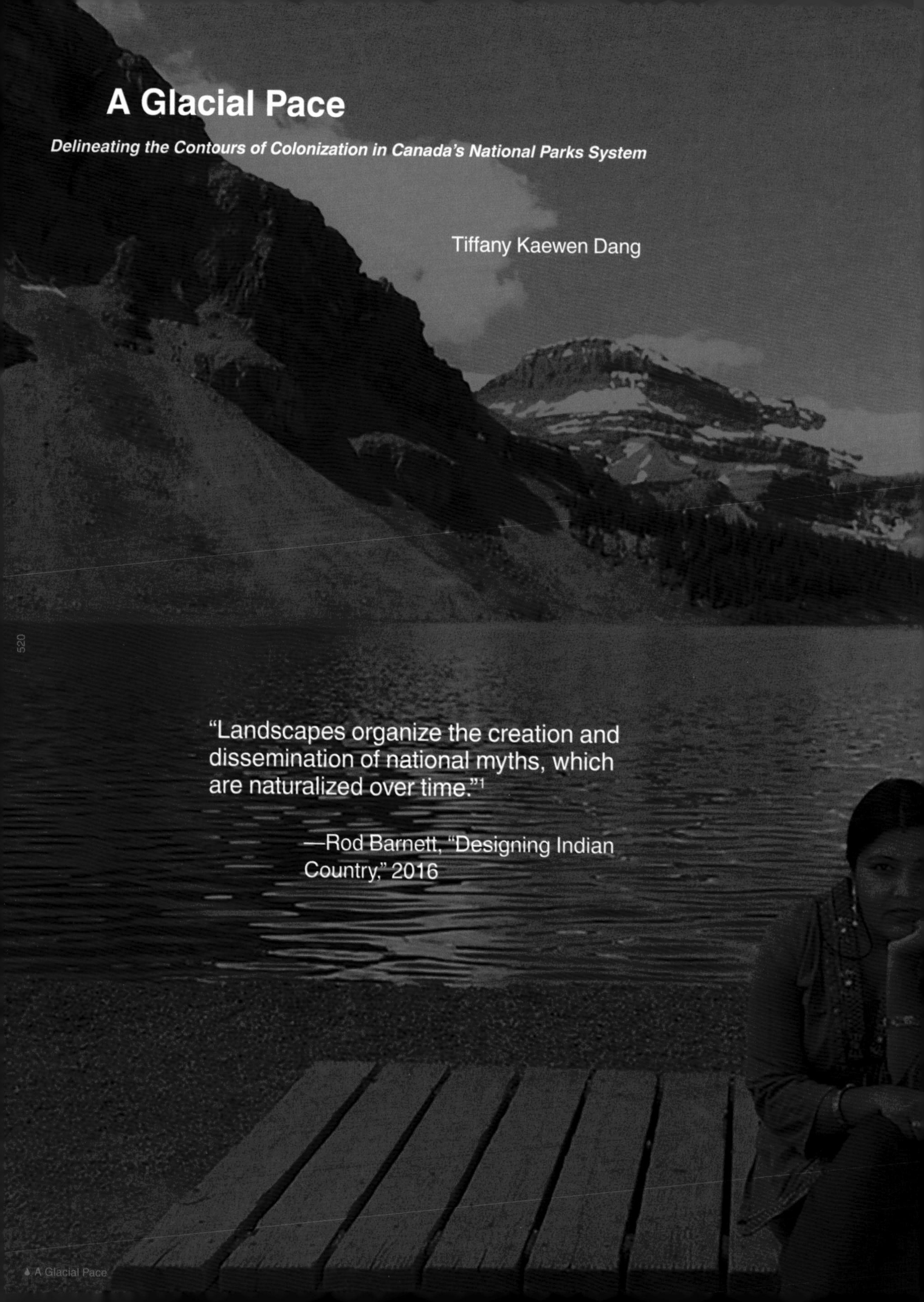

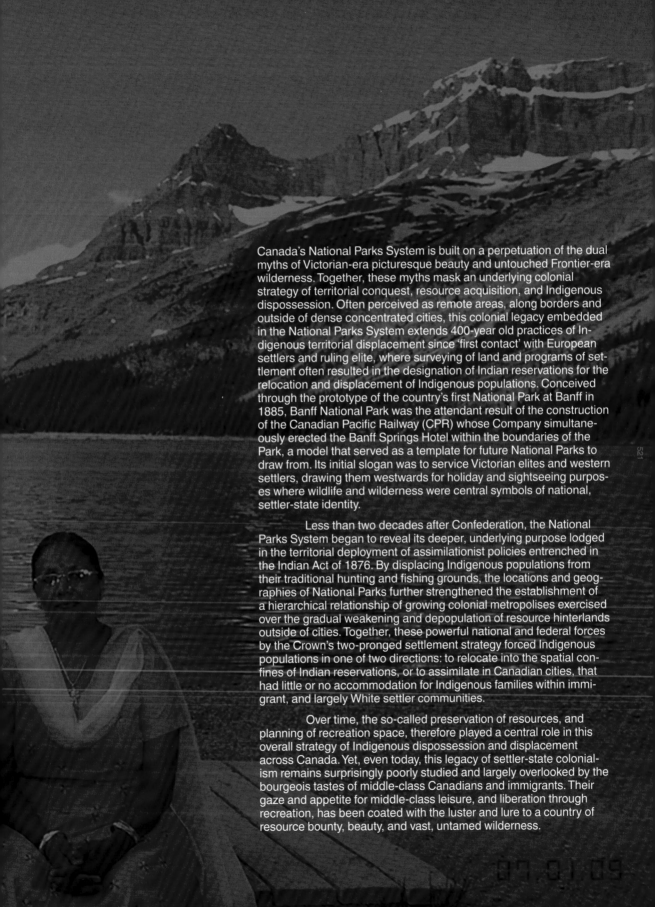

Canada's National Parks System is built on a perpetuation of the dual myths of Victorian-era picturesque beauty and untouched Frontier-era wilderness. Together, these myths mask an underlying colonial strategy of territorial conquest, resource acquisition, and Indigenous dispossession. Often perceived as remote areas, along borders and outside of dense concentrated cities, this colonial legacy embedded in the National Parks System extends 400-year old practices of Indigenous territorial displacement since 'first contact' with European settlers and ruling elite, where surveying of land and programs of settlement often resulted in the designation of Indian reservations for the relocation and displacement of Indigenous populations. Conceived through the prototype of the country's first National Park at Banff in 1885, Banff National Park was the attendant result of the construction of the Canadian Pacific Railway (CPR) whose Company simultaneously erected the Banff Springs Hotel within the boundaries of the Park, a model that served as a template for future National Parks to draw from. Its initial slogan was to service Victorian elites and western settlers, drawing them westwards for holiday and sightseeing purposes where wildlife and wilderness were central symbols of national, settler-state identity.

Less than two decades after Confederation, the National Parks System began to reveal its deeper, underlying purpose lodged in the territorial deployment of assimilationist policies entrenched in the Indian Act of 1876. By displacing Indigenous populations from their traditional hunting and fishing grounds, the locations and geographies of National Parks further strengthened the establishment of a hierarchical relationship of growing colonial metropolises exercised over the gradual weakening and depopulation of resource hinterlands outside of cities. Together, these powerful national and federal forces by the Crown's two-pronged settlement strategy forced Indigenous populations in one of two directions: to relocate into the spatial confines of Indian reservations, or to assimilate in Canadian cities, that had little or no accommodation for Indigenous families within immigrant, and largely White settler communities.

Over time, the so-called preservation of resources, and planning of recreation space, therefore played a central role in this overall strategy of Indigenous dispossession and displacement across Canada. Yet, even today, this legacy of settler-state colonialism remains surprisingly poorly studied and largely overlooked by the bourgeois tastes of middle-class Canadians and immigrants. Their gaze and appetite for middle-class leisure, and liberation through recreation, has been coated with the luster and lure to a country of resource bounty, beauty, and vast, untamed wilderness.

V469/1558, George Noble, Whyte Museum of the Canadian Rockies, whyte.org

Women feeding bear cub ca. 1930

Central to the study of Canada's National Park System is therefore a core colonial strategy and racial realignment of power across a multitude of scales that lie far beyond that of metropolitan environments or provincial jurisdictions, yet reveal inherent corollaries and shared objectives. Colonial effects and spatial scales of influence within Canada's National Park System not only reveal the underbelly of the picturesque beauty and bourgeois recreation as colonizing propaganda that persists even today, but research suggests that critical attention and greater historical understanding of these spatial complexities is warranted to expose how landscape has been shaped, adapted, and manipulated as a tool of colonial representation and Indigenous dispossession. In fact, what is revealed here is that the very nature of the picturesque was not only a tool of colonial ambition, but was in fact, the invention of imperial domination.

The Myth of Wilderness

In true colonial fashion, Parks Canada appropriately tweeted in 2015, that,

> "National parks are among Canada's—and the world's—natural jewels."[2]

Extending the rhetoric of the colonial settler-state, the homepage of Parks Canada further affirmed how the myth of wilderness plays an important part in the image and imagination of the National Park System:

> "Established to protect and present outstanding representative examples of natural landscapes … these wild places, located in every province and territory, range from mountains and plains, to boreal forests and tundra, to lakes and glaciers, and much more."[3]

Opened to the public in 1885, Banff Park was the first National Park in Canada, and third in the world, following the creation of Yellowstone National Park in the United States in 1872 and the Royal National Park of Australia in 1879. Development of Canada's National Park System was largely planned, but its inception and its internal program were also the result of accidental findings.

Over the course of barely three decades, starting in 1871, the Dominion Land Survey mapped over 800,000 square kilometers of land in Western Canada, expanding west from Manitoba to the Rocky Mountains and establishing a method for settling the so-called 'Canadian frontier':

> "The work of early surveyors, whose decisions about how urban and rural land was laid out and used, provided the framework of Canadian settlement. The railroad was part of an economic plan to bring staples to central Canada. Accompanying it was a settlement pattern of towns plopped on the landscape at regular intervals, with identical plans, named in alphabetical order along the railway."[4]

In addition to the colonial layout of gridlines by surveyors demarcating townships and properties for settlement, the work of geographers and territorial mapping was particularly important to the project of nation building:

> "In post-war Canada, many geographers willfully imagined themselves as adjuncts to government nation-building with the special task of understanding and overcoming the country's unique physical challenges … to calculate resources, plan the best transportation routes, and assess the factors limiting and enabling human settlement."[5]

Stampeding Calgary
Mammal migrations transgress the capital economies and infrastructures of the colonial metropolis of Calgary, disrupting the suburban sprawl of settlers, and reclaiming territories as living, synergistic systems in motion.

Furthermore,

> "from the beginning of western settlement, the hinterland residents struggled to improve their status within the system of capitalist expansion and exploitation. Metropolis refers to centres of economic and political control, usually in the larger cities. Hinterland means any comparatively underdeveloped area which exports for the most part raw or semi-processed materials." [6]

Dispossession by Design

Beginning in 1883, Canada's National Parks System took on an unplanned role when workers, building the transcontinental railway west of Calgary, discovered a series of hot springs flowing from a mountainside in the Rockies, near the railway station. Rather than opening the land for private development, the Federal Government—under Prime Minister Sir John A. MacDonald—reserved the land under federal jurisdiction surrounding the hot springs "from sale or settlement or squatting."[7]

As J. B. Harkin, the first Dominion Parks Commissioner noted, "fundamentally…Canadian National Parks were about profit."[8] Harkin further added that,

> "National Parks 'attract an enormous tourist traffic, and tourist traffic is one of the largest and most satisfactory means of revenue a nation can have.'"[9]

However, soon after its opening, deeper underlying intentions were revealed. Indigenous populations were not, and could not, be part of the nation-building project of National Parks. In 1887 for example, when civil engineer and landscape architect George Stewart became the park's first Superintendent, his first Annual Report remarked:

> "It is of great importance that, if possible, the Indians should be excluded from the Park. Their destruction of game and depredations among the ornamental trees make their too frequent visits to the Park a matter of great concern."[10]

The prohibition of hunting in Banff Park was therefore used as a strategy of territorial displacement and Indigenous dispossession, a form of starvation by enforced racial segregation and territorial regulation. Observations from the settler-based hunting community and its settler code of ethics further revealed exclusionary underpinnings:

> "Sport hunters [argued] that no one, not even aboriginal people, had the right to hunt for subsistence. The sportsmen's Code of Ethics stated that 'the value of wild game as human food should no longer be regarded as an important factor in its pursuit.'"[11]

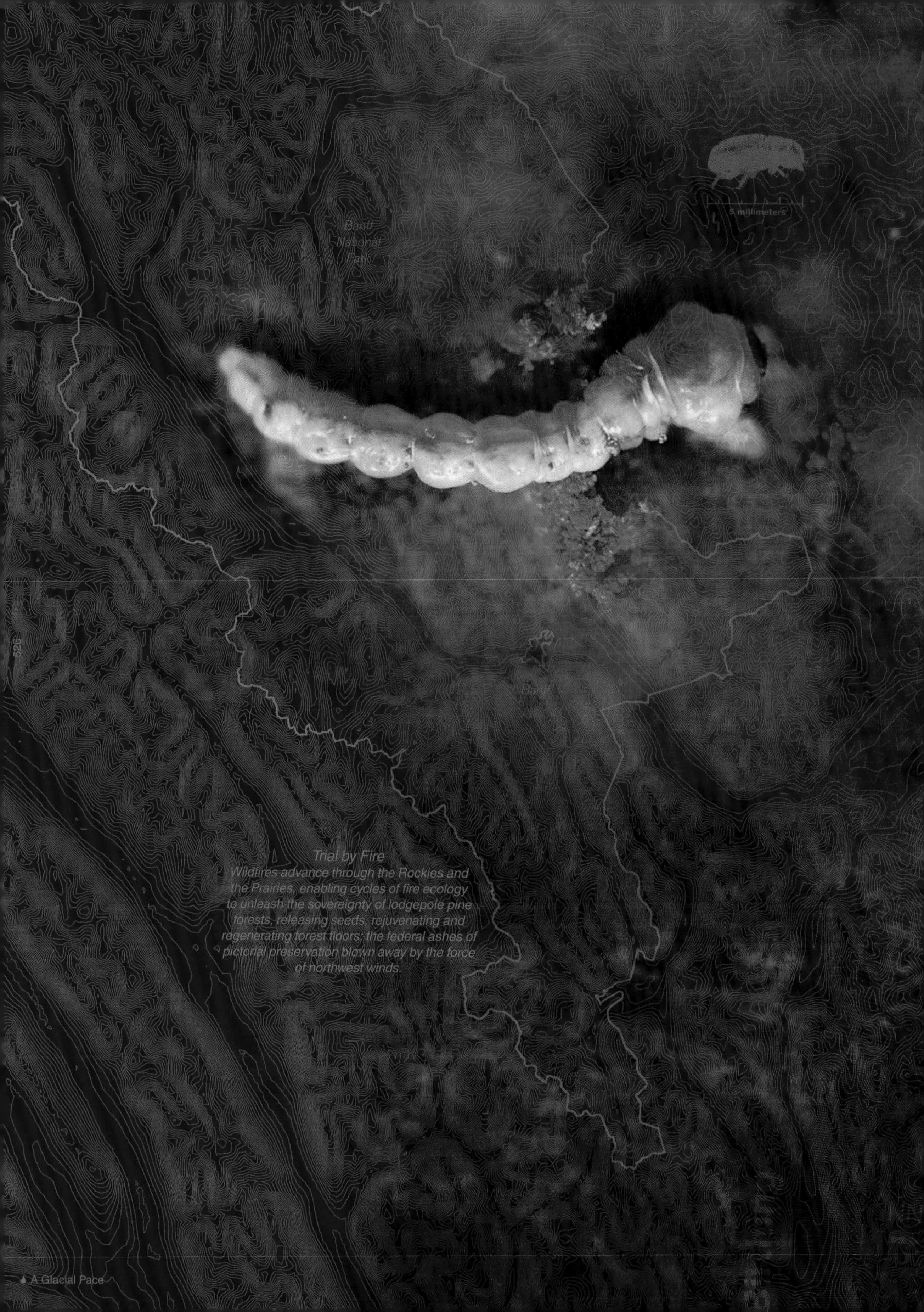

Trial by Fire

Wildfires advance through the Rockies and the Prairies, enabling cycles of fire ecology to unleash the sovereignty of lodgepole pine forests, releasing seeds, rejuvenating and regenerating forest floors; the federal ashes of pictorial preservation blown away by the force of northwest winds.

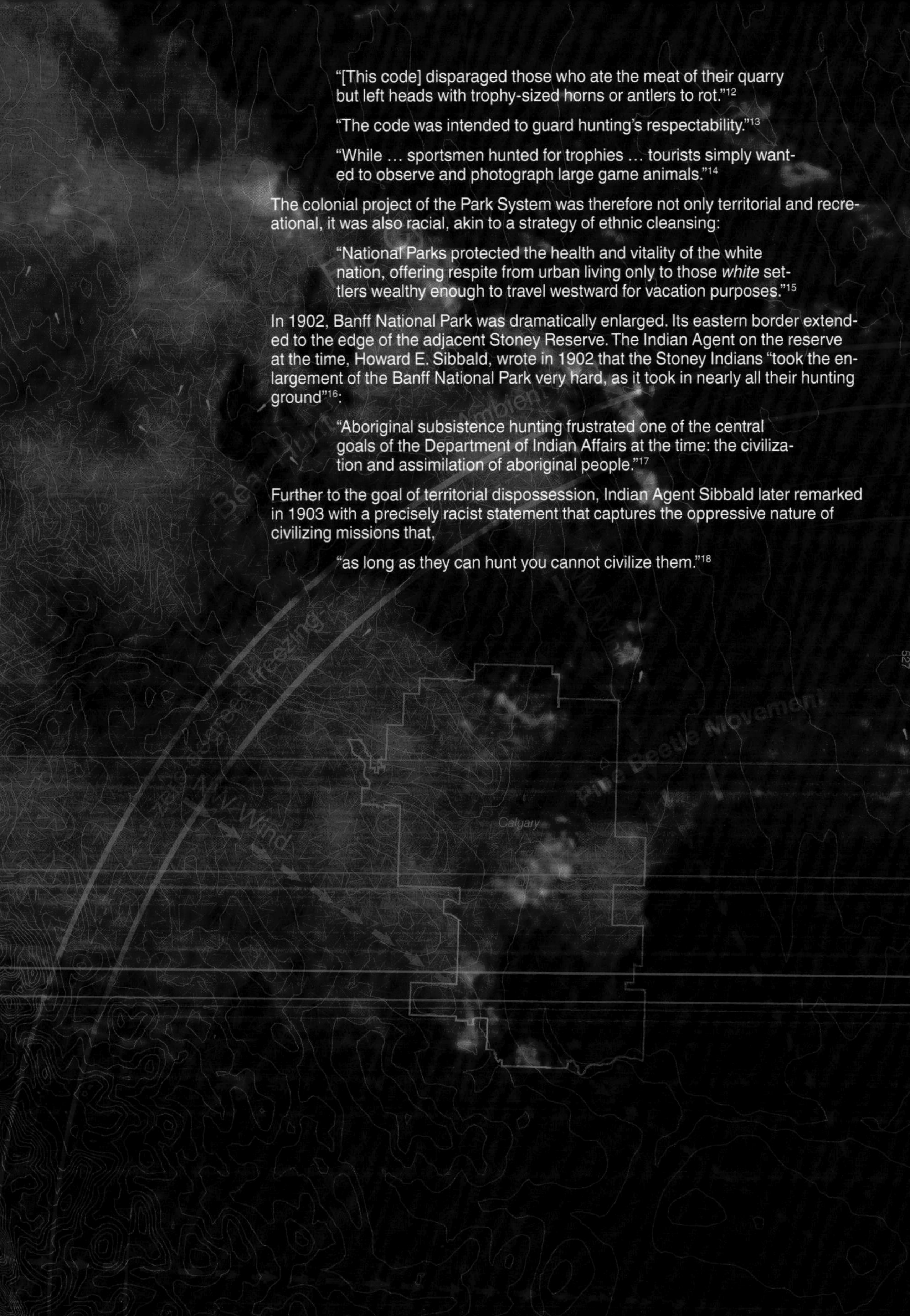

> "[This code] disparaged those who ate the meat of their quarry but left heads with trophy-sized horns or antlers to rot."[12]

> "The code was intended to guard hunting's respectability."[13]

> "While … sportsmen hunted for trophies … tourists simply wanted to observe and photograph large game animals."[14]

The colonial project of the Park System was therefore not only territorial and recreational, it was also racial, akin to a strategy of ethnic cleansing:

> "National Parks protected the health and vitality of the white nation, offering respite from urban living only to those *white* settlers wealthy enough to travel westward for vacation purposes."[15]

In 1902, Banff National Park was dramatically enlarged. Its eastern border extended to the edge of the adjacent Stoney Reserve. The Indian Agent on the reserve at the time, Howard E. Sibbald, wrote in 1902 that the Stoney Indians "took the enlargement of the Banff National Park very hard, as it took in nearly all their hunting ground"[16]:

> "Aboriginal subsistence hunting frustrated one of the central goals of the Department of Indian Affairs at the time: the civilization and assimilation of aboriginal people."[17]

Further to the goal of territorial dispossession, Indian Agent Sibbald later remarked in 1903 with a precisely racist statement that captures the oppressive nature of civilizing missions that,

> "as long as they can hunt you cannot civilize them."[18]

Preservation
As the Bow Glacier melts and flows through rills and river courses, the sovereignty of fresh glacial waters challenge the violence of the Dams and Hydroelectric Power Plants along the Bow River System that have dispossessed the people of the Stoney First Nation. By liberating the contained waters of Lake Minnewanka, the Town of Banff is inundated by the free flow of glacial meltwater. A mountainous valley—that once was, and still is sovereign—reclaimed.
+ 1400m
TransCanada Highway 1
+ 1391m
CN Railway (submerged)
+ 1390m
+ 1380m
+ 1391m
+ 1389m
Town
of Banff
(submerged)
+ 1380m
+ 1378m
+ 1378m
+ 1391m
1380m
+ 1378m
+ 1360m

Naturally, the colonial project of displacement went hand-in-hand with the Victorian practices of trophy hunting and picturesque gazing. Big game, an inheritance of the Victorian Project, exemplified white supremacy and dominion over the animal world. Gordon Hewitt, Dominion entomologist and consulting zoologist, observed in the 1920s that Banff National Park, "will serve as an unrivaled breeding-ground for the big-game animals of the Rocky Mountains region."[19]

A decade later, with the National Parks Act of 1930, the annexation of Banff National Park, along with Jasper National Park to the North and Yoho and Kootenay National Parks to the South, make the Canadian Rocky Mountain Parks today the country's largest, so-called wilderness area and most sought-after destination in the world. Its uncontested success is the dark, shimmering glow of dispossession.

The Persistent Oppression of the Picturesque

Today, this legacy continues, entrenched in the very nature of the Banff Springs Hotel and the spa that draws tens of thousands of patrons annually. The facility sits atop of the original hot water source 'found' by the CPR Company that led to the inception of the municipality bearing its name today:

> "The pulsating waterfalls and the mineral pool in the heart of the Banff spa rejuvenate tired muscles and soothe troubled spirits. The intensity of oxygen inspires clarity of thought and renews strength of purpose. Willow Stream Spa brings to life the unique healing powers of the alpine air and the sacred waters … Drawing energy from the mountains, the quiet sanctuary helps balance the rhythms of your body to reawaken your senses."[20]

It is therefore not surprising that, across the country today, National Parks have not only facilitated the dispossession of Indigenous territories, but also reinforce the relocation of Indigenous peoples to reservations. The system has capitalized, and is based on, the perpetuation of a recurrent illusion: the frontier as wilderness… a virgin place, uninhabited, and untouched by man.

As Ted Binnema and Melanie Niemi remark in their canonical 2006 essay, "Let the Line be Drawn Now,"

> "our ideas of wilderness and nature are socially constructed, and so we should reflect on how humans have attempted to modify the physical reality to conform to our notions of what wilderness and nature ought to be. In the case of the nature that was modified in the national parks, aboriginal people were removed to create landscapes abundant in wild game, as well as for a broader goal of civilizing and assimilating aboriginal societies."[21]

Following the creation of Banff National Park, in the late 1900s, people of the Stoney Nakoda First Nation were displaced from their traditional lands and relocated into a much smaller area—an Indian Reservation located on the eastern boundary of the Park, effectively 'emptying' the land in service of the illusion of uninhabited wilderness:

> "In 1911, the Canadian government passed the Dominion Forest Reserves and Parks Act, which established the Dominion Parks Branch—the world's first national park service—and helped institutionalize the Warden Service of the national parks. It also altered the boundaries of national parks so that areas that were not important tourist destinations were removed."[22]

Foraging and Berry Picking
Mating
River
Downstream Disruption
Cycles of freeze-up and break-up generate periodic water fluctuations, downstream, from the Bow Glacier to Bow River. Rivers form new right-of-ways overtaking highways, railroads, and pipelines; disrupting the colonized corridors of settler industries. Cyclical inundation floods the engineered infrastructures of the City, leaving abandoned, demolished, and submerged the stolen lands of Calgary.

Counter-Colonial Curriculum
The Fairmount Banff Springs Hotel becomes an Indigenous Center for Territorial Learning. Pedagogical programs of food preparation, clothing, and subsistence, buttress a scholarship of transgenerational story-telling. Disfigured and defaced, the lobotomized architecture of the former Grand Railway Hotel Building stands as the prototype for a reclaimed network of territorial centers across the railway system once built to dispossess and disenfranchise First Peoples.

Banff was only a prototype. “By 1930, Canada’s National Park System had expanded to 17 areas.”[23] With increased leisure time in the postwar Fordist economy, national park building went into overdrive:

> “After 1945, post-war trends towards higher pay, increased leisure time, and increased mobility resulted in a tourist boom in the National Parks. … With the boom of downhill skiing in the 1950s and 60s, Banff emerged as a year-round resort.”[24]

Later, in 1967, when Canada celebrated 100 years of Confederation, it was reported by Natural Resources Canada that over two million people visited the park annually. In 2016, nearly a century and a half later, the yearly number of visitors has doubled to nearly four million visitors.

Today, Stoney Reserve 142,143,144, as it is officially known, is cut in two by Canada Federal Route #1, the TransCanada Highway. The truncated tract of land is also divided by the Bow River, a river that is part of a system of glacial tributaries from the Rockies that are extensively dammed for the production of hydroelectricity and the protection of populations downstream.

Ironically, the Bow Glacier, that lies at an elevation of over 2,700 meters within Banff National Park atop of the Wapta Icefield, serves as a barometer of changing climates and warming temperatures. Rising and increasing meltwater from the glacier is contributing to more frequent flooding and overtopping of dams. The glacially-fed Bow River is quickly changing water flows to cities downstream like Banff and Calgary, that lie within a larger watershed of riverside cities like Regina and Saskatoon, whose waters eventually flow to the Hudson Bay.

The Current Climate of Colonization

If, “landscape is a way of framing the world according to aesthetic principles,”[25] as said by John Zarobell, then the question of beauty—that is the premise of the picturesque underlying all landscape practice, is necessarily formed and framed by the absence of political Indigenous societies. The planning of the picturesque is therefore not only aesthetic but is political, as Stephen Daniels and Denis Cosgrove argue,

> “a cultural image, a pictorial way of representing, structuring or symbolizing surroundings.”[26]

As colonial instrument, landscape then is deliberately ordered according to a range of politically-charged aesthetic principles, whereby the underlying assumption is that landscape is always organized, structured, and crafted. While often perceived as ‘natural’ or devoid of presence, the state crafting of such spaces is highly manipulative,

> “in the colonial context, landscape serves an instrumental function, both political and social. Since colonialism is about control of space—literally seizing territory—its complement is the imposition of the colonizer’s understanding of space onto the colonized.”[27]

National Parks are built on and maintained by colonial imaginations. The myth of wilderness serves to operationalize and advance the colonial process of territorial acquisition by strengthening the extractive hierarchical relationship between the city and the frontier:

> “The promotional discourses of frontier tourism invoked repeatedly throughout Banff at the turn of the century played to white fantasies of New World wildness, a kind of wildness that ostensibly became purer and more authentic the further one traveled westward away from the established colonial settlements … Imperial progress across the space of empire is figured as a journey backward in time to an anachronistic moment of prehistory. … For white settlers from the industrialized east, the allure of such a fantasy of time travel hinged upon the possibility of encountering spectacles of wilderness.”[28]

Trout Nation

Through the resurgence of trout in the Bow River, a declaration of soft-bodied sovereignty is cast on the territory from the Bow Glacier to Hudson Bay where rainbow and brown trout roam free…thus feeding and replenishing bodies of brown bears and eagles, in co-existence with minks, muskrats, and the almighty beavers.

Bow Glacier

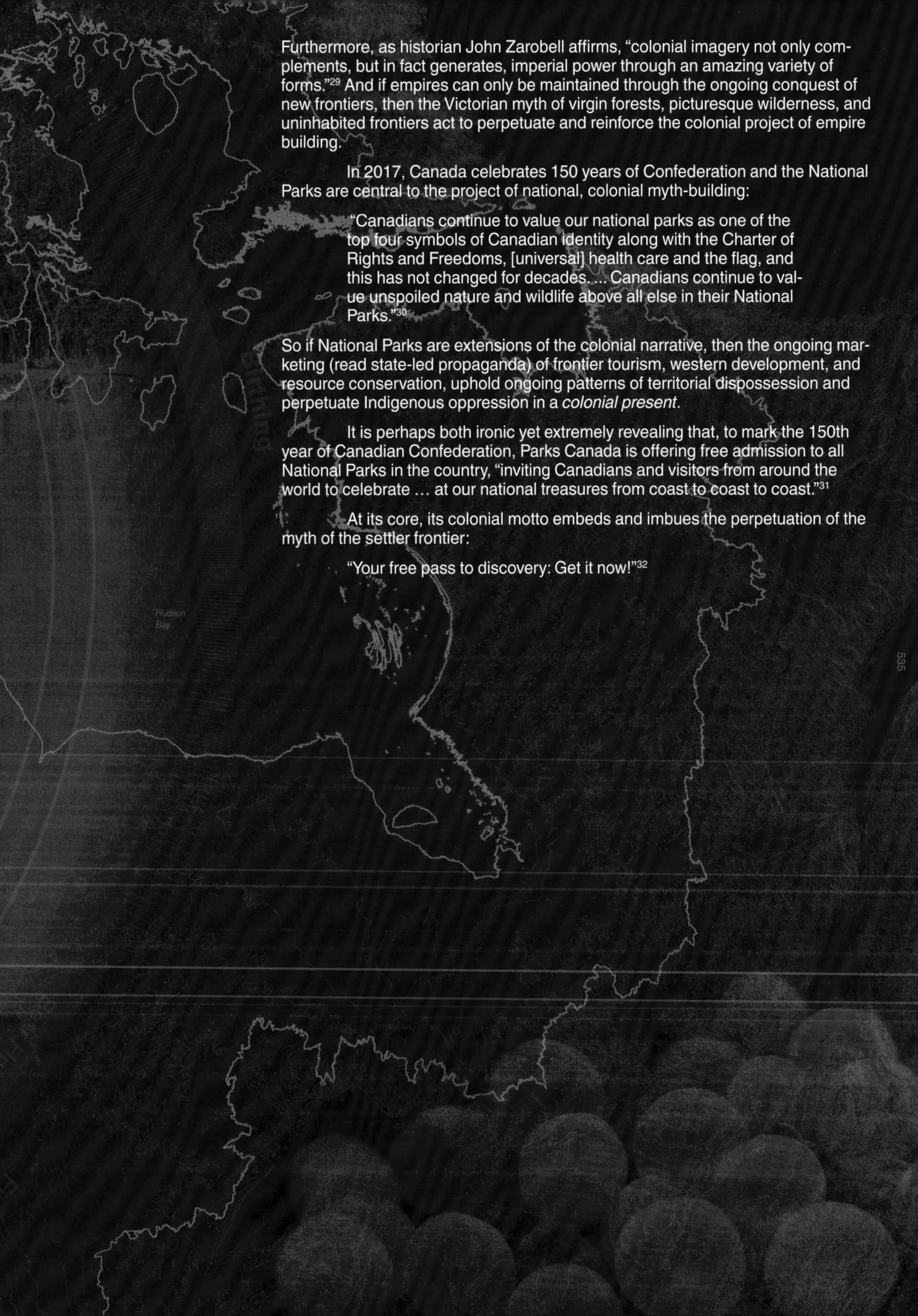

Furthermore, as historian John Zarobell affirms, "colonial imagery not only complements, but in fact generates, imperial power through an amazing variety of forms."[29] And if empires can only be maintained through the ongoing conquest of new frontiers, then the Victorian myth of virgin forests, picturesque wilderness, and uninhabited frontiers act to perpetuate and reinforce the colonial project of empire building.

In 2017, Canada celebrates 150 years of Confederation and the National Parks are central to the project of national, colonial myth-building:

> "Canadians continue to value our national parks as one of the top four symbols of Canadian identity along with the Charter of Rights and Freedoms, [universal] health care and the flag, and this has not changed for decades. … Canadians continue to value unspoiled nature and wildlife above all else in their National Parks."[30]

So if National Parks are extensions of the colonial narrative, then the ongoing marketing (read state-led propaganda) of frontier tourism, western development, and resource conservation, uphold ongoing patterns of territorial dispossession and perpetuate Indigenous oppression in a *colonial present*.

It is perhaps both ironic yet extremely revealing that, to mark the 150th year of Canadian Confederation, Parks Canada is offering free admission to all National Parks in the country, "inviting Canadians and visitors from around the world to celebrate … at our national treasures from coast to coast to coast."[31]

At its core, its colonial motto embeds and imbues the perpetuation of the myth of the settler frontier:

> "Your free pass to discovery: Get it now!"[32]

Capitalism as Kindling

From hinterland to metropolis, a prescribed burning regime is fueled by cycles of fire ecology, where key colonial buildings serve as kindling for capital devolution. Headquartered in the City of Calgary, the sites of Suncor Energy, Canadian Natural Resources Ltd, Encana, Husky Energy, Enbridge, Cenovus, Nexen, Syncrude, Athabasca Oil, and Spectra serve as perfect monuments of a past extractive state.

Preservation as Extraction

Underlying these compounded tenets of the colonial settler-state, the geopolitical history of the National Park System exposes a troubling premise and contested enterprise, both confirming forms of extraction. The federal system of National Parks transmit, convey, and extend these settler-based policies and colonial strategies through the overt, bureaucratic policy of 'natural resource conservation.'

The vast, closed, protected, preserved, and pristine environment of the National Park may invoke here an opposite image of extraction as the continuously-operating open pit resource mine, but it serves exactly the same duplicitous premise. The system goes hand-in-hand, and is dependent on, the perception of territories of extraction as remote hinterlands and uninhabited territories, for the acquisition and exploitation of a dominant, non-Indigenous majority. The image of National Parks not only perpetuates an imagined space of a mythical frontier, it also extends the space of the capitalist economy beyond, and outside of provincial capitals and central cities, that are free and open for leisurely, recreational occupation.

Within the metropolis-hinterland dialectic, these relations of power are embedded in the space and place of resource conservation strategies that National Parks entail. These skewed and contested interrelations confirm what landscape architect Rod Barnett considers as the presence of "power and knowledge differentials [that] are negotiated (but never resolved) through the act of place-making."[33]

As an ecology of power lying in between pristine wilderness and landscape picturesque, the National Parks System proposes then that the strategy of extraction not only derives capital from the raw, unprocessed export of materials from the hinterland to cities for consumption (i.e. the production and conveying of building materials from water to food, minerals to manufactures), it also derives capital from the emptying, evacuation, and liquidation of territory from all forms of political presence that may be communicated, and therefore confirmed, through visible, territorial corpus of Indigenous bodies.

If, then, the existence of economic centers of power—namely, through cities and towns—requires the economic sacrifice of impoverished peripheries, then the perceived bounty of resource hinterlands, and that of large national parks, can only be maintained by the predatory practices of urban centers within this colonial-imperial power structure that persistently overlook and suppress political identities of spatial *others*.

Consequently, the understanding of the foundations of resource extraction as extended extractivism can no longer be limited to what is defined by the mining of resources and their related industries of distribution. This extension of the interpretation of extraction confirms a worldview that defines the contemporary global economy, an economy rooted in the single yet powerful principle of perpetual, unceasing, and infinite growth.

While this growth, or at least the funding of its infrastructure, is concentrated in dense urban centers of economic power, the accumulation of power is always and continuously relational. It comes at the expense of the periphery and other sovereignties. The marginalization of territorial hinterlands and of rural, territorial cultures that depend on the interrelations between land and water; are often expressed and best understood at scales of a watershed, a river valley, an open prairie, a vast forest, a mountain range, a climate zone. In spite of decades, and in some cases centuries of fragmentation, these lands are indivisible from the political presence of Indigenous peoples that once inhabited them, and soon, will reclaim them.

In the case of Canada's National Parks System, growth of the metropolitan center has necessarily entailed the widening marginalization of the periphery, the hinterland, and the 'outside.' Preservation and conservation of vast territorial tracts has been, and continues to be central to the myth of nation building… a myth that encodes racism, Indigenous oppression, violence, and environmental destruction.

The underlying project of extraction is therefore an ongoing relational structure, far beyond a single operation of resource mining and single policy. Here, extraction corresponds to colonial-imperial ideologies that persistently allow for the sustained dispossession of hinterland communities, by the reification of regional capitals and metropolitan cores as centers of state economic power.

Moving forward, the reclamation of this historic legacy of degenerative dispossession, coupled with growing resistance to the skewed representation of the picturesque of the National Parks System, remain imperative as human actions—in spite of a seemingly glacial pace of political change—to understand, internalize, spatialize, and transform, as part of the oppressive regimes and systems of the colonial settler-state in the present.

Notes

1 Rod Barnett, "Designing Indian Country," *Places Journal* (October 2016), under subheading "Landscape as Narrative," par. 2, http://dx.doi.org/10.22269/161011.

2 @Canada, January 22, 2015 (10:01 pm), Twitter, https://twitter.com/Canada/status/558429303059517440

3 "Introduction," Parks Canada, last modified May 23, 2017, https://www.pc.gc.ca/en/pn-np/introduction.

4 Kent Gerecke, "The History of Canadian City Planning," in *The Second City Book: Studies of Urban and Suburban Canada*, ed. James Lorimer and Evelyn Ross (Toronto, ON: Charlottetown Group Publishing, 1977), 150.

5 Matt Dyce, "Canada between the photograph and the map: Aerial photography, geographical vision and the state," *Journal of Historical Geography* 39 (2013): 82.

6 Arthur K. Davis, "Metropolis/Overclass, Hinterland/Underclass: A New Sociology," *Canadian Dimension* 8, no. 6 (1972), 36–38.

7 Parks Canada, introduction to *National Park System Plan* (2016), 3–4, https://www.pc.gc.ca/en/pn-np/plan.

8 Leslie Bella, *Parks for Profit* (Montreal, QC: Harvest House, 1987), 24–27, as quoted in Ted Binnema and Melanie Niemi. "'Let the Line Be Drawn Now': Wilderness, Conservation, and the Exclusion of Aboriginal People from Banff National Park in Canada," *Environmental History* 11, no. 4 (October 2006): 738.

9 J.B. Harkin, "Report of the Commissioner of Dominion Parks," *Annual Report of the Department of the Interior* [ARDI] (1913), 5, as quoted in Binnema and Niemi, "Let the Line Be Drawn Now," 738.

10 George Stewart, *ARDI* (1887), parts 6 and 10, as quoted in Binnema and Niemi, "Let the Line Be Drawn Now," 729.

11 C. Gordon Hewitt, *The Conservation of The Wild Life of Canada* (New York, NY: Charles Scribner's Sons, 1921), 299, as quoted in Binnema and Niemi, "Let the Line Be Drawn Now," 731.

12 Tina Loo, "Of Moose and Men: Hunting for Masculinities in British Columbia, 1880–1939," *Western Historical Quarterly* 32 (2001): 308, as quoted in Binnema and Niemi, "Let the Line Be Drawn Now," 730.

13 Greg Gillespie, "'I Was Well Pleased with Our Sport among the Buffalo': Big-Game Hunters, Travel Writing, and Cultural Imperialism in the British North American West, 1847–72," *Canadian Historical Review* 83 (2002): 555–84, as quoted in Binnema and Niemi, "Let the Line Be Drawn Now," 730.

14 George Colpitts, *Game in the Garden: A Human History of Wildlife in Western Canada to 1940* (Vancouver, BC: UBC Press, 2002), 160–61, as quoted in Binnema and Niemi, "Let the Line Be Drawn Now," 739.

15 Pauline Wakeham, "Reading the Banff Park Museum: Time, Affect, and the Production of Frontier Nostalgia," chap. 1 in *Taxidermic Signs: Reconstructing Aboriginality* (Minneapolis, MN: University of Minnesota Press, 2008), 49.

16 Howard E. Sibbald, Indian agent's annual report in *Annual Report of the Department of Indian Affairs* [ARDIA] (1902), 173; and Ibid. (1903), 192, as quoted in Binnema and Niemi, "Let the Line Be Drawn Now," 735.

17 Binnema and Niemi, "Let the Line Be Drawn Now," 738.

18 Howard E. Sibbald to Frank Pedley, December 23, 1903, Library and Archives Canada RG 10, vol. 6732, f. 420-2, as quoted in Binnema and Niemi, "Let the Line Be Drawn Now," 738.

19 See W.N. Millar, *Game Preservation in the Rocky Mountains Forest Reserve* (Ottawa, ON: Government Printing Bureau, 1915); and Hewitt, *The Conservation of The Wild Life of Canada*, 238, as quoted in Binnema and Niemi, "Let the Line Be Drawn Now," 734.

20 "Willow Stream Spa at Fairmont Banff Springs: Overview," Fairmont Banff Springs, 2017, http://www.fairmont.com/banff-springs/willow-stream/

21 Binnema and Niemi, "Let the Line Be Drawn Now," 740.

22 Ibid., 737.

23 Sheila Robinson, *A Report on the Natural and Human History of Banff National Park* (The Book of Banff), rev. ed. (Prepared for Parks Canada, [1978] 1980), 34.

24 Ibid., 30.

25 John Zarobell, introduction to *Empire of Landscape: Space and Ideology in French Colonial Algeria* (University Park, PA: The Pennsylvania State University Press, 2010), 4.

26 Stephen Daniels and Denis Cosgrove, "Iconography and Landscape," introduction to *The Iconography of Landscape: Essays on the symbolic representation, design and use of past environments*, ed. Stephen Daniels and Denis Cosgrove (Cambridge, UK: Cambridge University Press, 1988), 1, as quoted in Zarobell, introduction to Empire of Landscape, 4.

27 Zarobell, introduction to *Empire of Landscape*, 5.

28 Wakeham, "Reading the Banff Park Museum," 45, 46.

29 Zarobell, introduction to *Empire of Landscape*, 1.

30 Canadian Parks and Wilderness Society (CPAWS), *Protecting Canada's National Park: A Call for Renewed Commitment to Nature Conservation*, (2016), 7, http://cpaws.org/uploads/CPAWS-Parks-Report-2016.pdf.

31 "2017 Free Admission," Parks Canada, last modified July 20, 2017, https://www.pc.gc.ca/en/voyage-travel/admission.

32 Ibid.

33 Barnett, "Designing Indian Country," under subheading "Landscape as Curation," par. 1.

Promises, Promises

A Conversation with Justice Thomas R. Berger

Justice Thomas Berger is a Canadian lawyer, judge, and politician. He was counsel for the plaintiffs in the historic Aboriginal rights case *Calder et al v. Attorney General for British Columbia* and served as commissioner of the Mackenzie Valley Pipeline Inquiry (1974–77). He served as New Democratic Party (NDP) Member of Parliament for Vancouver-Burrard (1962–63), NDP Member of Legislative Assembly (1968–69), and leader of the NDP in British Columbia (1969).

X: You've had a long career fighting for Indigenous rights based on Aboriginal Land Title. Can you explain how this experience has grown and evolved?

JB: I went to Law School in the 1950s and no one, if ever, raised issues about Indigenous rights in the curriculum. We all lived in a province with a very large Indigenous population, sometimes seeing them every day. Looking back, it did not occur to us that there were needs that had to be addressed.

X: Was it ever explicitly or implicitly considered during your legal education that Indigenous peoples—from whom the country was taken as you've mentioned before—had legal and constitutional rights?[1]

JB: When I went to Law School in the mid-1950s, at the University of British Columbia, we never discussed Aboriginal Rights or Titles. We proceeded on the assumption that there were no legal rights enjoyed by them; that we had to worry about, or be concerned with. That was the prevailing attitude of the professors, and probably of the students too. I landed into this area by accident and we began to pursue Aboriginal Title over a period of 50 years, which are now embedded in the law. I was annoyed that we never raised the questions: "Do they have their rights to the Land? So the question on Aboriginal Title was very important to me. It addresses the roots of racism.

From 1963 to 1965, I was asked to take on a case of two men, Clifford White and David Bob, in *R. v. White & Bob*[2] at the provincial level. The men were charged with hunting out of season according to the BC Provincial Game Act, and then jailed for 40 days because they could not pay the $100 fine. They asked me to take the case to the Nanaimo Court of Appeals.

Context here is important. In the course of establishing the *Snuneymuxw Treaty* with the Hudson's Bay Company (HBC) in 1854—a treaty that was signed in *blank* and while James Douglas was Chief Factor for the Hudson's Bay Company. My argument was that if there was no treaty with Governor Douglas as representative of the Crown (1870–1874), this meant that *we* had Aboriginal Title. And, we won the *White & Bob* case. Following Justice Thomas' ruling, the case received a lot of coverage and recognition.[3]

X: Was this just the beginning?

JB: In another case less than a decade later, in *Calder v. the Queen* at the Supreme Court of Canada, I found a case of 'dormant' treaties stuffed in a drawer. One of the judges[4] had explored the question of Aboriginal Land Title in *Fred Calder from the Nisga'a Nation v. the Province of British Columbia*, which began in late 1971 and ended in 1973. The Provincial Archivist Willard Ireland said, "I will talk to Elders," and they said, "we'll find them." He was based in Nanaimo, and he did, in fact, find them.

1 In a keynote address made in 2013, Justice Thomas Berger further explained that "judges in those days had never been trained in the field of Aboriginal rights, so it was difficult to convince them that the Aboriginal peoples possessed rights based on the indisputable fact that they used and occupied vast areas, if not the whole of this continent, before the Europeans came. They had their own institutions, their own laws, but of this fact many lawyers and judges remained unaware. They could not accept that people without an extensive written language would have an elaborate legal system. And as for their Aboriginal Title, how could the court acknowledge it? It was ill-defined, it was recorded in the system of title deeds and land registration; most importantly, it was not usually a form of private property, but often was communal." See Thomas R. Berger, "Modern Treaties: Providing the Foundation for Canada's Political, Legal, Economic & Social Landscapes," *Northern Public Affairs* 2, no. 2 (December 2013): 54–55, (based on the keynote address at the *Land Claims Agreement Coalition Conference* "Keeping the Promise: The Path Ahead to Full Modern Treaty Implementation," Gatineau, QC, February 26–March 1, 2013).

2 According to the *University of British Columbia Legal Citation Guide*, designations of "The Crown" are differentiated by types of legislation. In civil cases, the guide states to "use the common geographical name and omit references to the Crown, Her Majesty the Queen, etc." where an example: "Use Canada rather than Her Majesty the Queen in Right of Canada," or "Use Alberta rather than The Crown in right of Alberta." In criminal cases, "use R, which stands for Rex (The King) or Regina (The Queen), rather than the full name provided in the judgment." For example: "use R v. Seifi rather than Between Her Majesty the Queen and Ex-Private S. Seifi, Accused," or "use R v. McLean rather than William McLean Appellant and His Majesty the King Respondent." This legal differentiation in terminology demonstrates slippage in use of the "Crown" in reference to federal or provincial jurisdictions, as well as the legal, judicial, and territorial complexity in the division of powers of the Crown. See "Law - Legal Citation Guide," University of British Columbia, last modified June 21, 2017, http://guides.library.ubc.ca/legalcitation/cases.

3 Anne Lindsay, "Archives and Justice: Willard Ireland's Contribution to the Changing Legal Framework of Aboriginal Rights in Canada, 1963–1973," *Archivaria* 71 (Spring 2011): 35–62.

4 Ronald Martland, Wilfred Judson, Roland Almon Ritchie, Emmett Matthew Hall, Wishart Flett Spence, Louis-Philippe Pigeon, Bora Laskin.

In an earlier case, I had called Willard Ireland and Wilson Duff as witnesses to prove certain things that required archival and anthropological evidence. As Provincial anthropologist, Wilson Duff was the first anthropologist to testify in these Aboriginal cases. His evidence is registered in the Calder case, with Justice Hall's judgment including existence quotes from Wilson Duff's report, which spoke highly of how Aboriginal societies were organized, what they regarded as the boundary of their own territories before the Europeans came to North America.[5]

The Nisga'a Nation held that they had Aboriginal Title for the Nass Valley and that it would be *cognizable*[6] under Canadian Law. The Crown recognized that the Nisga'a occupied the river valley for thousands of years—since time immemorial, in other words. The decision launched a series of major claims and cases afterwards, after which I became justice in the Supreme Court of British Columbia from 1972 until 1983.

X: There is a "colonial construction of space" at work here, as geographer Cole Harris refers to it.[7] Can you explain the historical background of the Nisga'a Nation's struggle for title recognition?

JB: Yes, dating back to the turn of the last century, representatives of the Nisga'a Nation even went to London (UK) in 1913 to file a petition with King George V. I doubt they ever saw the King but they did get to the palace. The Nisga'a Nation was persistent thanks to a strong culture of leadership. Across different generations, the success of the Nisga'as rested in its leadership structure and its constant assertion of rights in the Nass Valley in relation to the dominant Canadian society. Their struggle dates as far back as 1888. That's why the 1973 Calder Case involving the Nisga'a Nation is a landmark because it proved that Aboriginal Title to their lands were never extinguished.

X: Can you describe the background to the Douglas Treaties?

JB: British Columbia was the "last part" of the colonization of North America. Settlement started in 1840. The gold rush came a little later, during the 1850s in the Fraser River region. Governor James Douglas, who earlier negotiated land agreements on Vancouver Island, was appointed by the Crown to the Colony of British Columbia. Douglas was formerly a major fur trader for the HBC. He saw the change of the incorporation of territory, "Rupert's Land," into a Province of the Crown in 1858, then into Confederation in 1867. A Scot and West Indian, Douglas was of mixed blood himself, growing up in the West Indies, and married himself to a Métis woman, Amelia Connolly, née Nipiy of Cree and French descent. Douglas believed that no lands were to be settled before claims were settled and treaties signed; using the same forms for treaty making as across the Commonwealth.

However, as Douglas was making agreements for land settlements, he began to run out of "forms" in the late 1850s, so he started making blank forms, then signed by Aboriginal chiefs. They were deposed to the Provincial Legislature when it was formed in 1856 during a prolonged period of peace, but problems and disagreements with Indigenous Nations persisted. Attitudes of settlers (farmers, loggers, sawmillers) were changing. They saw less and less 'need' for Indigenous peoples, nor treaties, and after he retired in 1864, the Provincial Legislature never honored them.

X: Does this relate to preemption rights?

JB: Again, this is why the 1973 Calder Case involving the Nisga'a Nation is so important. We claimed that Aboriginal Title was not extinguished, but that it was cognizable by Canadian Law, through a

5 Wilson Duff gave evidence that the lands where the Indians had been hunting were within the traditional tribal territories of the Nanaimo Indians. Willard Ireland produced an Indian Treaty of 1854 dealing with the Nanaimo area. The document was very curious for it had no text, simply a set of signatures. The text, it was concluded, must be identical with the other southern Vancouver Island treaties, it being one of the series of 14. A second curious feature of the treaty was that it was signed by James Douglas as chief factor of the HBC and not as governor of the colony of Vancouver Island. The position of the Crown was that the document was not a treaty because of its form and because it was not signed on behalf of the Crown. The convictions were reversed. Judge Swencisky held that the document of 1854 was a treaty and its protection of hunting rights was effective by reason of a provision of the Indian Act. He went on to hold: 'I also hold that the aboriginal right of the Nanaimo Indian tribes to hunt on unoccupied land, which was confirmed to them by the Proclamation of 1763, has never been abrogated or extinguished and is still in full force and effect.'" See Douglas Sanders, "The Nishga Case," *BC Studies* 19 (Autumn 1973): 10.

6 "Cognizable" is a legal term for "recognition" under Canadian Law. If rights can be proved, they can be asserted. In the Choqoton Case of 2015, they proved they owned a good of the Chilquotum Lands even though these were not treaty lands, and the Supreme Course of Canada supported that assertion but it was also subject to Provincial Law.

7 See Cole Harris, *Making Native Space: Colonialism, Resistance, and Reserves in British Columbia* (Vancouver, BC: UBC Press, 2002): xxi.

technical detail. Those rights existed before the Royal Proclamation of 1763, where the Crown oversaw all lands across what is now known as Canada.

It's important to recognize that, under the administration of Pierre Trudeau in the late 60s and 70s (an opponent of the recognition of Aboriginal Title and of the Nisga'a claim), Gérard La Forest, who later became Justice of the Supreme Court of Canada, advised the Federal Government and the Prime Minister that, with the help of three other judges who represented the sound view of the law, could settle all land claims and settlements while recognizing three Northern Territories in Newfoundland, Quebec, and Labrador. Since treaties were reached between the eighteenth and nineteenth centuries, at least half of the country was subject to Aboriginal Titles by then. So, out of the Calder Case, at least half of Canada's land mass had to be party to these agreements where Aboriginal Title has never been surrendered.

Ironically, today, Justin Trudeau—the son of then Prime Minister Pierre Trudeau—is the one to negotiate those recognized claims.

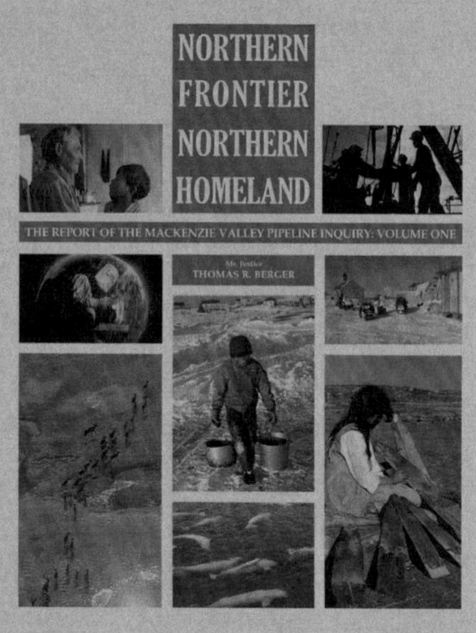

Berger Inquiry, Report Cover, 1974 (Justice Thomas R. Berger, *Northern Frontier Northern Homeland: The Report of the Mackenzie Valley Pipeline Inquiry*)

X: Are colonial assumptions about the 500-years old *Doctrine of Discovery* or notion of *Terra Nullius* relevant here—grounded in fundamental assumptions of racial superiority and European entitlement?

JB: When the British defeated the French at the Battle of the Plains of Abraham in 1759, they virtually acquired all of North America, which included the thirteen colonies under British Rule. By then, the colony of Quebec had already been established by the French. Under the advice of his Privy Council, King George III issued the Royal Proclamation of 1763, which stated that the Indians who lived beyond the colonial boundaries—in the land which later became known as the countries of Canada and United States—are not to be disturbed and cannot be acquired except by the Crown, through a public meeting by consent, as a principle of British Policy. That Proclamation was much resented by the thirteen colonies, and recited in the Declaration of Independence as one of the injustices of King George III that Thomas Jefferson complained about.

That policy was first observed, but then fell into disuse. I took on the Calder Case in 1963 and by the time I argued the case in December 1971, the idea of Aboriginal Title was in the dim past. In Canada, it had to be resurrected.

Going back to the Royal Proclamation of 1763, we were relying on the understanding of Chief Justice John Marshall who, in the 1830s, had written several important judgments. Marshall pointed out that Indians still had a right and title to land. These judgments were the basis of treaty making in the United States.[8] But in Canada, namely in the Maritime Provinces—the Yukon, the Northwest Territories, and British Columbia—there were no treaties to speak of. The Supreme Court determined that Indian Title [Aboriginal Title] was itself embedded in Canadian Law and could be asserted.

X: What are some of the implications of this lack of recognition in the past few decades?

JB: In light of this decision, Pierre Trudeau's administration had to negotiate treaties with all those Indians, Métis, and Inuit in Canada who had never signed treaties because Indian rights were tied to land rights. That process has been going on since 1973, when the federal policy offered its "White Paper" Policy. Up until then, Trudeau did not recognize those rights and titles (or as he called them, the "historical might-have-beens") nor did the Supreme Court had changed its name. So, the idea of Aboriginal Title has become the most prominent part of the advance of Aboriginal people, 45 years ago.

8 "Both Canada and the United States would have been more astute to have paid attention to the pronouncement of U.S. Supreme Court Justice John Marshall, whose 1830s decisions continue to cast a shadow over Indian relations in the two countries. It was a contemporary argument that Indians could hardly be expected to understand what they were signing, and humanitarians and critics of the treaty system alike lamented this failing. This view continues to have adherents today among some Indian groups and among historians seeking explanations. But the governments, which in the twentieth century have been confronted by the obligations they undertook in nineteenth-century treaties, cannot claim ignorance as an excuse. 'The words "treaty" and "nation" are words of our own language,' John Marshall wrote in 1833, 'selected in our diplomatic and legislative proceedings, by ourselves, having each a definite and well understood meaning. We have applied them to Indians, as we have applied them to other nations of the earth. They are all applied in the same sense.'" In Jill St. Germain, *Indian Treaty-Making Policy in the United States and Canada, 1867-1877* (Lincoln, NB: University of Nebraska Press, 2001): 157.

When we adopted a new Constitution and charter of rights under PM Pierre Trudeau in 1982, it specifically provided for recognition of Aboriginal and treaty rights; whereas in the US, the rights of Indians are only derived from the commerce clause in the Constitution.

In 2013, I won a case before the Supreme Court for the Métis, who are descendants of mixed people in the Red River Settlement, the current Winnipeg. The Métis resisted and refused to accept Canada's acquisition of the Plains, what was called Rupert's Land, launching a resistance to Canadian rule. The Prime Minister negotiated 1.4 million acres to lay down their arms, but the Federal Government did not honor the promise. One hundred and forty years had already passed and this rule of Canadian polity had to be redressed, where the Prime Minister Justin Trudeau argued that it had to be honored. Those are the book ends of my career in the field, in these new cases.

There is a new case, probably my last, located in Yukon (argued on March 22, 2017). In 1993, the Federal Government reached a modern treaty as a result of the new policy in 1973 with the Indians of the Yukon (25% of the population) that all land use planning would be done on a cooperative basis. The most recent dispute involved the Peel River watershed, a cause mounted by First Nations and environmental groups, because this is the largest protected area of North America—with 80% of wilderness protection, the size of Scotland. This is a measure of the protection of wilderness which arises out of these modern treaties which provides a grounding for protecting *wilderness parks*, well beyond what was done in the nineteenth century in terms of treaty money, reserves, medicine chest. There is one little wrinkle; when PM Justin Trudeau became the Prime Minister, he gave a mandate to the Attorney General that they were not rigidly opposed to the claims advanced by the Aboriginal claims in courts, the opposite of what the bureaucratic norm has been in the past. So, the Attorney General has intervened on our side. We await the decision shortly.

The Yukon case that we're arguing, where I believe the Supreme Court will agree with us as well, to allow the land use provisions to proceed as it was agreed in 1993, to be interpreted narrowly—which would allow governments to ignore recommendations of independent commissions—would thwart reconciliation, whereas a broad liberal interpretation would advance reconciliation. Reconciliation is always a predicate to these discussions in the long haul.

X: Did the *Calder Decision* overturn the logic of discovery and racial discrimination as set out in the Marshall Court in the US and *St. Catherine's Milling v. The Queen*, and in *Guerin v. the Queen*? Or did it merely sidestep it? I mean to suggest; did it really bring about a radical change or just provided a "crack" in the veneer of Canada's approach to Indian Land?

JB: The *St. Catherine's Milling v. The Queen* Case acknowledged Indian Title, but that was in 1888, and of course, in the Calder Case of 1973, we fleshed that out further.

X: Can you explain the significance of the case of *Guerin v. the Queen* in 1984?

JB: The Guerin Case was about the fiduciary duty of Indian Reserves, which were set aside to protect Indian interests. The Supreme Court had a fiduciary duty to protect Indian interests, and the disposition of any part of their reserve, which was important to address in lands already owned by Indians.

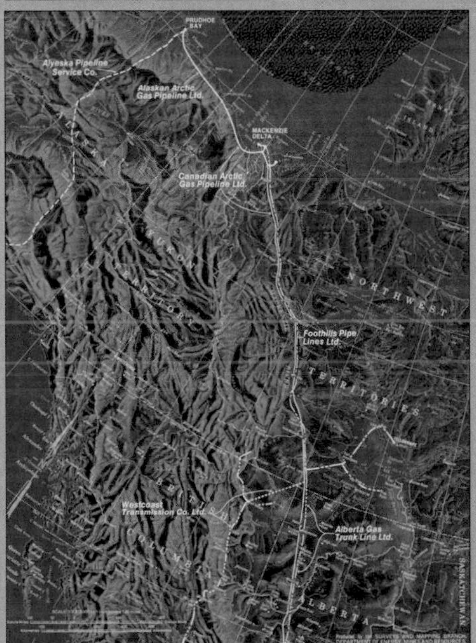

Mackenzie Valley Pipeline Maps and Indigenous Community Hearing with Chief Francois Paulette of Smith's Landing First Nation, ca. 1973 (Justice Thomas R. Berger, *Northern Frontier Northern Homeland: The Report of the Mackenzie Valley Pipeline Inquiry*)

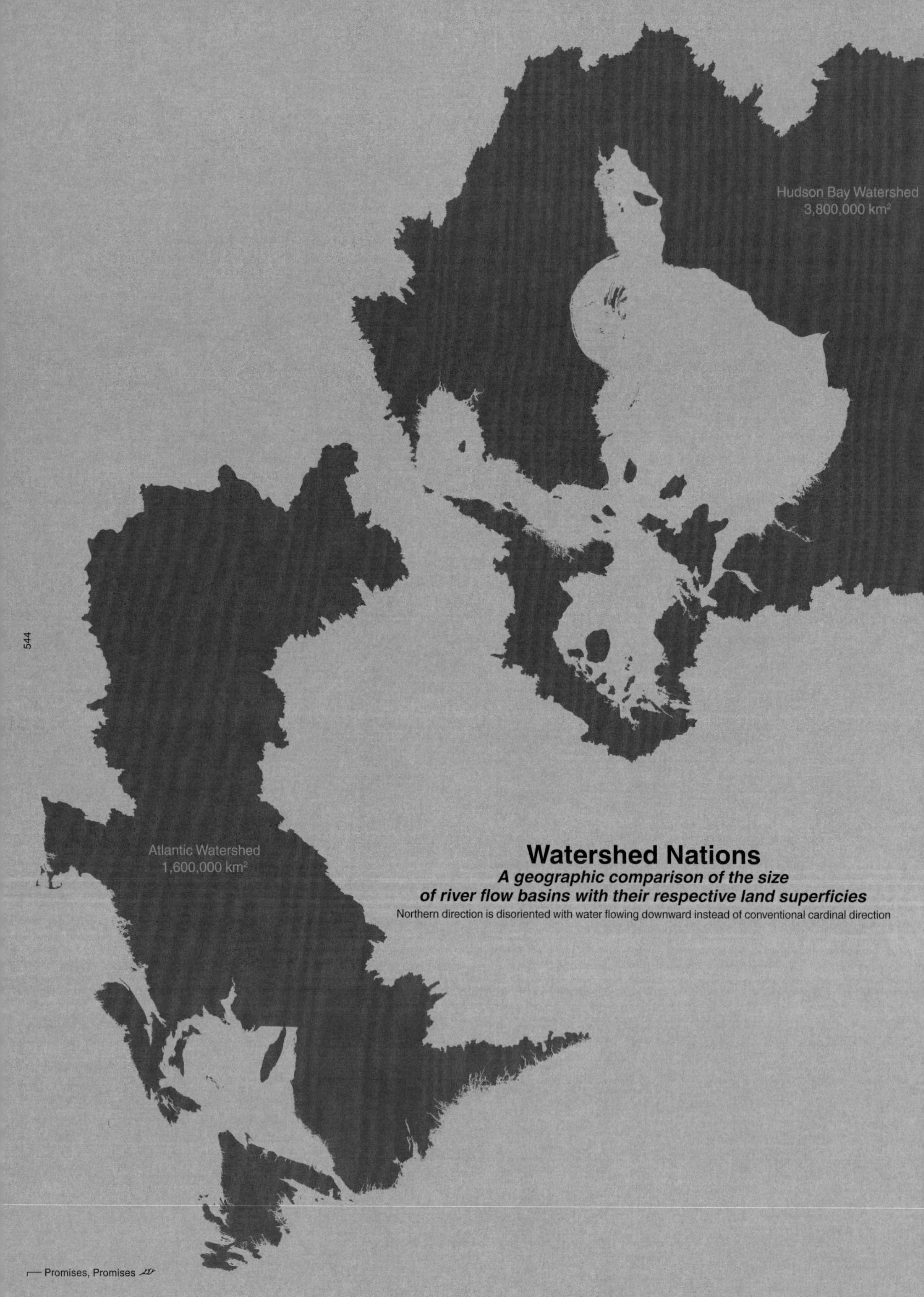
Hudson Bay Watershed
3,800,000 km²
Atlantic Watershed
1,600,000 km²
Watershed Nations
A geographic comparison of the size
of river flow basins with their respective land superficies
Northern direction is disoriented with water flowing downward instead of conventional cardinal direction

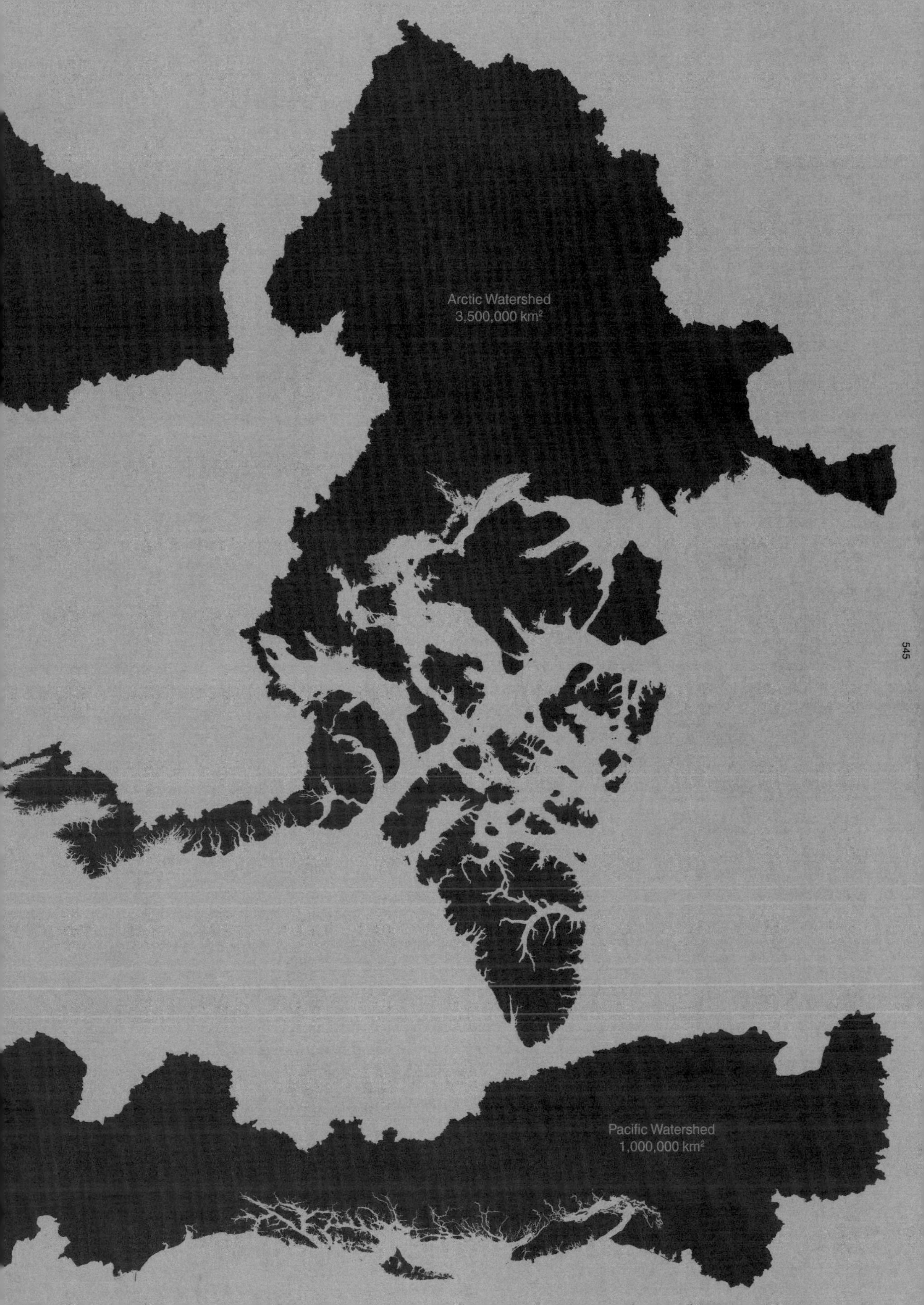
Arctic Watershed
3,500,000 km²
Pacific Watershed
1,000,000 km²

X: Were the treaties always honored?

JB: John A. Macdonald promoted a settlement in 1870. The Métis opposed and resisted the establishment of settlements on their land in the Red River Region from 1869 to 1875. But, they managed to extract a promise to the Government of Canada and lay down their arms whereby 1.4 million acres of land were set aside in the Manitoba Act. However, in *Manitoba Métis Federation v. Canada* (2013), it was shown that Canada has failed and issued a declaration and obligation to do it.

X: Is this where the difference between "rights of occupancy" defined by the Crown under conquest, and "land titles" as defined by Aboriginal culture can be distinguished?

JB: In a case called *The Crown v. Sparrow*, 1980,[9] the Supreme Court of Canada said that when Britain asserted sovereignty over what is now Canada, it meant that the underlying title was held by the Crown, but (following Calder) the burden of the Indian Title which has to be extinguished so that the rights of occupancy (there's an expression of Crown grants or leases which applies better to settlers) that are inherent to Aboriginal Title, are derived from a history of the use and occupancy of that very land. That's the justification. But keep in mind, our whole legal system is based on the idea that the Crown owns the underlying title burdened by the Indian Title, whereas the Indian Title is derived from Indian use and occupancy before the Europeans. Any non-Aboriginal Title must be derived from the Crown. So ordinarily, settlers had to derive land from the Crown, based on the Proclamation of 1763.

9 According to the First Nations Studies Program at the University of British Columbia, "*R v. Sparrow* was a precedent-setting decision made by the Supreme Court of Canada that set out criteria to determine whether governmental infringement on Aboriginal rights was justifiable, providing that these rights were in existence at the time of the Constitution Act, 1982. This criterion is known as 'the Sparrow Test.'" See "Sparrow Case," *Indigenous Foundations*, 2009, http://indigenousfoundations.adm.arts.ubc.ca/sparrow_case/.

Canada's destiny—its *modus vivendi*, is clearly not between English and French, but it is worked as an agreement within and about cultures that acknowledge land rights, languages, histories. It's not a transaction, it is a destiny.

Given that, I believe in equality against discrimination based on race.

Judge Berger listens to Phoebe Nahanni as Jimmy Klondike (left) and Johnny Klondike Sr (right) look on (courtesy of Prince of Wales Northern Heritage Centre)

X: How then do we reconcile special rights for Aboriginals?

JB: It's not a question of blood, it is a political question. They were here first. They had their own governance, they had their own sovereignty. They enjoyed political histories before the Europeans arrived. The majority of the problem for Canadians is the failure to recognize Indigenous People as political societies, and to acknowledge that those structures still exist. That's why a government-to-government relationship is essential.

X: Can you also describe the relationship between the Mackenzie Valley Pipeline Inquiry in the Northwest Territories in 1974 and the Peel River Watershed Case in the Yukon four decades later, in 2014? How did your thinking about long-term land use planning take shape in relation to land claim agreements?

JB: In 1977, I made recommendations that are still unfolding after the *Calder Decision* in 1973, whereas in Alaska by 1985, they settled for 1 billion USD for Indians, Eskimo, and Aleut, which was not successful.

Community Hearing, Berger Inquiry, in Nahanni, NWT (Justice Thomas R. Berger, *Northern Frontier Northern Homeland: The Report of the Mackenzie Valley Pipeline Inquiry*, 1974)

When I oversaw the Inquiry, among other things, I recommended that *wilderness parks* be established to protect the Caribou and Porcupine, to establish an Arctic National Refuge across the Caribou range from the US to Canada (for which I testified before the US House and Congress Senate in 1977–78). Over three decades, there have been two settlements, in 1984 and in 1995, which led to the establishment of a major *wilderness park*. Using his executive powers and presidential authority in 2015–16, President Obama established wildlife status for Arctic National Wildlife Refuge for permanent protection.

For the territorial land claims in the Yukon in 1993, a collaboration lands agreement partnership was developed in Yukon and they are doing it now. In the Peel River watershed, there is a simple formula that 80% is wild and 20% is industrial development. The agreement is based on the protection of the Peel River which is based on a collaborative land use agreement, where the partnership between industry and First Nations is held together with a provision for informed consent as part of the UN Declaration on the Rights of Indigenous People, made in 2013.

The Mackenzie River as it had been in 1977, now has the question to deal with megaprojects.

Francois Paulette, Chief, Smith's Landing First Nation, ca. 1973 (above) and Dene Mapping Project Community Consultation ca. 1974 (below) (©NWT Archives/Rene Fumoleau/N-1995-002: 4198)

X: Is mandatory constitutional consultation with First Nations a pathway towards the adjudication of these rights? With Canada's adoption of the United Nations Declaration on the Rights of Indigenous Peoples (UN-DRIP) by the Government of Stephen Harper in 2007, do you perceive conflicts for, or overlaps with Justin Trudeau's government policies on land claims?

JB: Another example, is the Site C Dam, a megaproject in British Columbia. Under the auspices of the Provincial Crown Corporation, the BC Hydro and Power Authority is building dams, producing electricity and distributing energy to those who are building this dam. In a case by the Supreme Court of Canada, *Haida Nation v. BC*, all projects had to consult with the First Nations commensurate, where the UNDRIP requires consultation. The local First Nation sued and lost their case against BC Hydro. They have to have legal purchase for these cases and no one seems to have it. This is a struggle being waged now with more and more tools, as the Supreme Court has recognized the right of consent, consultation, and accommodation of First Nations. First Nations do not have a veto over these projects (the federal or provincial governments have not acknowledged thus far).

X: Do you think that the partition of powers in the government and its legal system is, or was, strategically organized and purposely administered in order to ensure racial domination and cultural superiority with respect to Aboriginal displacement and territorial dispossession? Or was this merely a by-product, an unintended consequence of internalized cultural assumptions of colonialism? When discussing the notion and implications of land titles, did *Calder v. British Columbia* relate or reform underlying racial and discriminatory principles entrenched in the 500-year old *Doctrine of Discovery*?

Berger Inquiry showing local Dene children peering in, behind Thomas Berger, during the proceedings in the Nahanni Valley (courtesy of Prince of Wales Northern Heritage Centre)

JB: Look, when settlement began in the US and in Canada, it was of course under British Rule (and of course, we still use the idea of the Crown as expression of sovereignty in Canada, the Crown is Sovereign). The Crown i.e. the Federal Government has always taken responsibility under the Constitution for relations with the Aboriginal People of Canada. It did in the US, and to some extent, in the Spanish colonies. Aboriginal peoples have always preferred the Crown, in its federal expression. It should be seen as the interlocutor, in its government to government relationship which was affirmed by the current Prime Minister Justin Trudeau. Settlers, in the United States and Canada, were not inclined to acknowledge rights of Aboriginal people. The Federal Government is still the preferred authority under the Constitution as opposed to the Provincial Governments, which are much closer to the settler mentality and not tasked with assertion of Indigenous rights.

Just over a year ago, in a case of the Federal Court of Appeal in the case *Daniels v. Canada*, the Métis people were under the Federal jurisdiction, which they sought, they did not want to be under Provincial jurisdiction.[10]

10 On April 17, 2014, the Federal Court of Appeal released an appeal decision of a case called Daniels v. Canada, [2014] FCA 101. It ruled that the Métis are within federal jurisdiction because they come within the definition of "Indians" in s. 91(24) of the Constitution Act, 1867. The main plaintiff and organization behind the case is the Congress of Aboriginal Peoples.

X: Attorney Thomas Isaac recently said in a conference on the Daniels case that “it is very difficult to rid the Federal Government of colonial mentality when they misperceive what the law actually is.”[11] Are the divisions of the Crown power—between provincial and federal governments, across the Courts, the Justice System, and Government Departments—part of the conflict or complexity to circumvent? How so?

11 Remarks by Attorney Tom Isaac in “Daniels: In and Beyond the Law” Conference (Edmonton, AB: University of Alberta, Ruperts land Centre for Métis Research in the Faculty of Native Studies, January 26, 2017).

JB: It’s important to note as well that, in terms of the era of modern treaties since 1973, there have been 27 of them. In the Northern territories, Yukon, Nunavut, and the Northwest Territories, the making of modern treaties was simplified by the fact that the Crown Federal is the complete Sovereign; there is no division of power or Crown Sovereignty. Whereas in BC and Alberta for example, there has to be a collaboration with the Crown Provincial (Provincial Government) because if public land (which is in the jurisdiction of the Provinces) is to be transferred to Aboriginal people, there has to be negotiations such as the complicated and important tripartite Nisga’a Treaty of 2000 in British Columbia between the Federal Government, Provincial Government, and Nisga’a First Nation.

When these matters were first argued and litigated in Australia, which occurred concurrently with the Calder Case, and Australian laws progressed in tandem with ours; *Terra Nullius* was first proposed there. *Terra Nullius* argued that Aboriginals had no rights, but the courts in Australia rejected that idea. Whatever the case may have been in different parts of the world, the policy of the British in dealing with Aboriginal peoples was always that treaties had to be made and they had title. That’s why there are treaties all over the US, and in Canada, except in Quebec, where the French did not recognize the title of the Aboriginal. I don’t know if it was called *Terra Nullius*, but the French felt no obligation to the Aboriginals to consult or obtain consent or signed treaties. But no doubt these treaties were signed with people that did not understand what they were signing of course, or what they were agreeing to. But there was imperial policy to obtain consent in the US and Canada before taking over land, and that was for the most part, the theory. Now, there is a big difference between theory and what was practiced. It was honored more in the breach than it was in the observance.

Extraction vs. Harvesting, Berger Inquiry Report Images (Justice Thomas R. Berger, *Northern Frontier Northern Homeland: The Report of the Mackenzie Valley Pipeline Inquiry*, 1974 (above, below, and right)

X: When discussing the notion and implications of land titles, did *Calder v. British Columbia* relate or reform underlying racial and discriminatory principles entrenched in 500-year old views associated with the *Doctrine of Discovery* or *Terra Nullius*?

JB: I don’t think *Terra Nullius* was any official ideology as such, there was an attitude amongst the settlers; and in British Columbia, near the Calder Case, it was part of the last area to be placed in British Columbia, which didn’t really start until the 1850s. The settlers regarded the Aboriginals as having no legal rights under the law that they observed.

X: Do these limitations on sovereignty require a rethinking of colonial structures of occupancy, governance, and property? And how might this play out? What might be the implications, for example?

JB: Regarding reconciliation, the Supreme Court of Canada, and Justice McLaughlin in the judgment of *Manitoba Métis v. Canada* discussed how important reconciliation is and why the failure to observe a promise made 150 years ago, has to be honored today.

X: Are Canadians unable to see the importance of laws made 140–150 years ago?

JB: Yes, of course. In Canada and in Manitoba. That’s why for 140

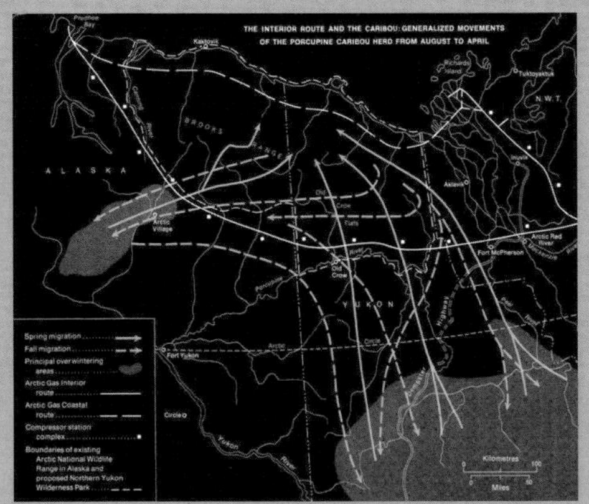

years, the Métis in Manitoba have always said that we never kept our promise, and we kept on responding with "well this is all in the past, c'mon this is Canada, you hackle down and work hard, and you'll make it." These promises were not regarded important as a part of our heritage. The case brought forth by the Métis have changed those attitudes in Manitoba, and the judgment of the Supreme Court Justice was a landmark.

The report made in the Truth and Reconciliation Commission of Canada—and the impact of the Chief Commissioner especially—is remarkable, now at the front row and in the center of Canada. Mind you, their mandate was to look into the Residential Schools, by which the Government was to assimilate Aboriginal peoples. Residential Schools represented a predominant view that Indigenous beliefs, views, languages, economies, laws, and their rights were not even worth discussing, and therefore, the sooner we could get on to the task of converting them into white people, the better off they would be. Well, we now know how that turned out.

X: So it is a problem of this generation, not of the past?

JB: Yes, very much so.

X: Are there international cases that inform your experience on Aboriginal Land Title?

JB: In 1991, I was tasked by the World Bank in India to oversee an inquiry regarding land claims with Bradford Morse.[12] It was the first independent review ever conducted with its own independent budget on the largest water project ever undertaken in India beginning in 1987, in Sardar Sarovar. The task was to assess and evaluate treatment of environmental conditions and Aboriginal rights. At a time when India had become one of the world's most prolific dam builders since gaining independence in 1950, the project involved a large water project including the construction of a high dam on the Narmada River; the creation of a reservoir submerging land in the states of Gujarat, Maharashtra, arid Madhya Pradesh; and an extensive canal and irrigation system in Gujarat. By the early 1990s, India had adopted even more stringent and stronger regulation than in the US and Canada, as I had been dealing with megaprojects in the Mackenzie River Valley Pipeline Inquiry in the 1970s. Whereas India had actually strong environmental regulations, their rules were not being enforced.

12 See Thomas R. Berger, "The World Bank's Independent Review of India's Sardar Sarovar Projects," *American University International Law Review* 9, no. 1 (1993): 33–48.

X: Can you briefly describe how the history of Aboriginal rights is linked to human rights?

JB: In Canada, the experience of human rights really grew out of the atrocities of World War II and was being relatively successful. There was an argument made from the time of the Calder Agreement in the 1970s, as to how can they have special rights, which were not previously acknowledged until after World War II. And we began to find a space for agreement with Aboriginal Peoples.

X: Is this where the 1982 Constitution comes to bear? How so? And, what might be the future implications for this?

JB: As a Provision of the Constitution, Section 35 of the Aboriginal Treaty Rights from 1982 is important. The 1982 Constitution was an attempt to enable First Nations to be themselves, while enjoying rights as Canadians. The working out of these details of land claims, ownership, and recognition of ownership is ok if you work out land claims agreements. It was a large job and experiment.[13] We're still working out these

13 "Under Section 35 of the Constitution adopted in 1982, the rights that Aboriginal people have established and that are written into these land claims agreements, these modern treaties, are protected by the Constitution. That means all of these land claims agreements are constitutional instruments. They are, in a sense, part of the Constitution of Canada. That is an important and far-reaching development that you should keep in mind because there are opportunities from time-to-time before the courts to take advantage of the constitutional status of those promises written in to these land claims agreements." See Berger, "Modern Treaties," 58.

agreements. And the majority of the work is in the details of those land claims in order to establish ownership and enforce their recognition.

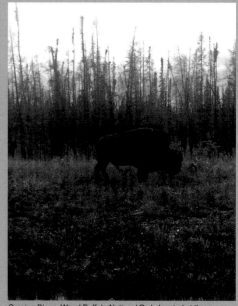

Grazing Bison, Wood Buffalo National Park, located at the convergence of the Peace, Athabasca, and Slave Rivers, 2017 (Pierre Bélanger)

X: Since oversight at the territorial level is largely the responsibility of First Nations, which means they must ensure they avoid being exploited across jurisdictions of the provinces, are you concerned by the level of self-regulation in the resource extraction industry to bestow these rights?

JB: In Section 35 of the 1982 constitutional discussions, Pierre Trudeau left out, at the last minute, the provision of free, prior, and informed consent on extraction on Aboriginal Land. The Federal Government's position regarding this UNDRIP Provision, will be the subject of exploration, negotiation, and discussion for the next 50 years in Canada.

X: What strategies do you think might help non-Indigenous Canadians to dispel the myth of colonial discovery and territorial conquest—embedded cultural myths that ultimately extinguished Aboriginal rights to land—a myth that must be overturned/dispelled if we are to fully acknowledge the political sovereignty of Indigenous First Nations and, as you claim, are needed to end colonial-based, territorial violence?

JB: We now have a new generation of Indigenous students, graduates, teachers of law from across the First Nations, which are contributing a great deal to the success of arguments and cases towards the resurgence and assertion of Indigenous rights. For now, and in the future, we have two roles: never evade the recognition of Indigenous rights and always prevent the exploitation of our laws. ┌──

This conversation took place between Justice Thomas R. Berger, Nina-Marie Lister, and Pierre Bélanger in 2016 and 2017.

Aklavik people ask Berger to save their way of life

By Peter Gorrie

ven after community etings and house to house ts, many people in Akalvik 't understand the propsed kenzie Valley gas pipeline.

they have seen the damage elopment has already caused ne Mackenzie Delta. And they t no more development that destroy the land and their of living

nat was the message given to Justice Thomas Berger ing three days of often tional hearings in Aklavik week--the first of his kenzie Valley Pipeline uiry's 20 informal community rings.

s the first community ring, it was something of an eriment and, although the ssage it conveyed to Judge ger was clear, in the end its ults were inconclusive.

nere were problems with the sical set-up and structure of hearing that frightened some ple from taking part. More ortant, Judge Berger was on the final day of the ring that the entire range of nion in Aklavik had not been sented to the Inquiry. The t of view that was not expssed was the one that dealt st directly with what the uiry is all about.

udge Berger heard from 30 to eople, including some from side the Hamlet of 760 people. s expected, they told him ut their traditional life as ters and trappers; the houses y lived in, the food they ate, way they travelled, the dships they faced, and of their t encounters with the white n. Most simply read tements they had written orehand.

hey described the importance he land and the animals in the development work that has taken place in the Delta, Judge Berger was told. Explosions set off by oil company seismic testing crews have killed muskrat and fish and frightened the caribou.

"When they blast, it does hurt the muskrats," said 64 year old Amos Paul. "Why is it that people are allowed to interfere with another person's life and property? The muskrat fur is valuable and the meat is a livelihood for the families." amilies."

Danny C. Gordon, who came to Aklavik in 1947, described the behavior of caribou in his hunting area: "The last four years there has been a lot of activity in that country where we hunt. Four years ago there weren't as many planes and helicopters and the caribou came down through the flats. Now you see the caribou up in the hills. They want to come down but the planes going by always chase them back. If this little thing can hold the caribou from coming down into the flat, I just wonder what the effect will be when they start building the pipeline."

Fred Joe, who has worked on the Dempster highway, between Inuvik and Dawson, said the moose and caribou refused to cross completed sections of the roadway.

Judge Berger was told that hunting and trapping is more than a sour ce of food and income. It is a way of life the people enjoy and want to pass on to "future generation ," and that they believe will be destroyed by the pipeline and other development.

Not only did development threaten the land and animals, Aklavik residents said, it had already changed their lives for the worse and would bring more serious social problems, while offering few benefits.

"The oil companies want to build a pipeline down the N.W.T., They want to take our gas and our oil We will not even use the gas. It will go past our homes to heat southern Canada and the USA," argued Charlie Furlong, regional director in the delta area for the Metis association.

But the pipeline would bring rape, prostitution and drugs, others said.

About 14 of the speakers, most of them older people, said the pipeline should not be built at all.

A few simply told of their lives and expressed no direct opinion about the pipeline.

Fanny Maring, a young social worker, asked for more time before the pipeline is built. "I would like to see more time and years and the people to understand whats really going on. Everyghing is going too fast and we need time to think and learn about it more and for the younger generation to get an education so they can be prepared for it when it ctmes to get jobs,'" she said.

Almost all of the stements, however, boiled down to the position argued by the three native people's associations, the Brotherhood, the Metis association, and the Committee on Original Peoples' Entitlement (COPE), that the pipeline and all other development should be stopped until a land settlement is reached with the federal government.

"No development before a land settlement" was often tacked onto the end of people's statments like an advertising slogan.

It reflected the influence of the three associations which, funded by part of a $400,000 grant from the Inauiry Commission, prepared the people for the meeting through meetings and house to house visits.

Each eskimo home in the hamlet was visited twice by

Nothing passes for favour here,
  all talk is razor-toothed.

Take nothing from the hand that offers friendship.

In this place
  all promises are bruises
  in good suits.

Treaty

Thomas King

1991

Treaty as Territory

A Conversation with Chief Allan Adam & Eriel Tchekwie Deranger

Athabasca Chipewyan First Nation
Treaty 8 Signatory
August–October 2015
(Part I)

Allan Adam is the Chief of the Athabasca Chipewyan First Nation (ACFN) since 2007. Chief Adam currently holds the Environment and Treaty Lands and Resources portfolios and shares the Justice and Economic Development portfolios for the ACFN.

Eriel Tchekwie Deranger is a Dënesųłiné climate and Indigenous rights advocate, serving as Communications Coordinator for the ACFN. Eriel founded Indigenous Climate Action, pushing for a united Indigenous climate action strategy in Canada and working with the Federation of Saskatchewan Indian Nations, Indigenous Environmental Network, Sierra Club, Rainforest Action Network, and the UN Indigenous Peoples Forum on Climate Change.

"Imperialism, colonialism, torture, enslavement, conquest, brutality, lying, cheating, secret police, greed, rape, terrorism—they are only words until we are touched by them. Then they are no longer words, but becomes a vicious reality that overwhelms, consumes and changes our lives forever."

—Jack D. Forbes, Columbus and Other Cannibals: The Wétiko Disease of Exploitation, Imperialism, and Terrorism, [1978] 2011

X: Can you share background and history of the Athabasca Chipewyan First Nation in relation to Treaty 8?

ED: The 1969 White Paper attempted to assimilate all First Nations into society and remove all of their unique rights. The National Indian Brotherhood, which was the early formation of the AFN (Assembly of First Nations), reacted to this. Some of the founders of the National Brotherhood said, "wait a second here. We have Treaties and Agreements that are the foundations of this country, and you haven't even paid us for the amount of resources you've stripped from our lands. All these [health and education] clauses haven't even been upheld and you're talking about assimilating us into modern society? Not until you're finished paying us, and we live in equality in this country, can you abolish our rights as First Peoples of this country."

After the White Paper failed, they increased funding for First Nations temporarily, but they have been employing tactics of attrition to cut funding significantly to all programs. Now the FNFTA (First Nations Financial Transparency Act) is trying to make First Nations put all their finances in the public domain. This is for two purposes: first, to shame First Nations for atrociously mismanaging their government funding, and second, to prove where it is not being mismanaged, First Nations don't need government money anymore. I think it's just a veil to cover the fact that they've fucked us over for the last 500 years. They've tried to trick us into all these different agreements, and as the Chief said, talking in circles at these meetings.

Map of Treaty No.8 Territory, 1900 (Department of Indian Affairs)

But to a certain extent, Chrétien and Trudeau with the White Paper, as with Harper with the Omnibus Bills during his Administration, did Indigenous people a favor. In those moments of crisis and extreme policy conflict, movements arose. The National Indian Brotherhood was previously in existence, but it came out front and center out of the White Paper. And Idle No More came out of the Omnibus Bills. It sparked an Indigenous movement across the country. There were people even before that fighting fracking, uranium, potash, and hydro, but with the Omnibus Bill and the blatant gutting of legislation, Harper sparked a movement.

Dene families in birchbark canoes, southeast shore, Fort Resolution, NWT, ca. 1900 (Library & Archives Canada)

X: You explicitly and deliberately use the term "tar sands," as opposed to "oil sands." Is your language purposely political, or intended to ground the subject?

AA: We call it "tar sands" for a clear reason. Long before Treaty 8 came into existence throughout the Peace-Athabasca River region, European explorers were exposed for the first time to what was seeping out of the shores of the Athabasca River: tar sands. Natives from the region explained, “we use the tar for vital functions, it is widely available and seeps all over the place.” We used it for patching and coating birch bark canoes. The Natives knew it was a resource, and it meant something to them. When the Europeans came, it meant so much to the Natives that they had to go and show them. They knew it had value, they just didn't know it had value in the way it does today.

Syncrude Operations, Alberta Oil Sands, Northern Alberta (©2013 Garth Lenz)

ED: Everyone called it tar sands, until they decided to try and sell it to the economic centers. When they tried to sell it to the markets, they realized that nobody wants to buy tar, they want to buy oil. In my opinion, it's a political statement on their end to call it oil sands.

X: Do you think that the singular focus on the tar sands region takes away from the complexity of all the resource issues that you're addressing? The issue of tar sands has become so big that it's difficult to talk about uranium, it's difficult to talk about hydro…

AA: The one thing people don't talk about is uranium, but there is an abundance of it in our traditional territories. We're trying to correct the regulatory system so that when, and if, the extraction of the uranium happens, it's not about…

ED: …a matter of dirty or clean, or any of that stuff. It's more about a rights-based approach, and about territory…

AA: It's just a matter of time before they are going to extract it, because there's lots there. It's one of the purest, richest finds in North America. And it's just 16 kilometers from our reserve. They have all these plans, so we're trying to say, “if you don't correct your problem with the oil sands or the tar sands or whatever the frick you want to call it, you ain't gonna get all over there. Because if you're creating a mess over here, you ain't goddarn going to go after that uranium.”

ED: The Caribou Bison Stewardship Plan clearly sets out what we're looking for. We're looking to protect and preserve our traditional territory, on the grounds of treaties and agreements, to have autonomy over our lands and resources. We want to manage hunting, trapping, fishing, and any development in our defined protection zone, to ensure cultural survival. It's not about us trying to gain ownership over the minerals. To come back to what the Chief said, it's to go back to the way of our ancestors.

Treaty 8 Territory
Geographic comparison, to scale, with France and UK

X: If the core of the fight goes to the rights related to Treaty 8, which goes to the core of the constitutional monarchy…

AA: If I was to get to the core of the monarchy, what are they going to do to me? They're going to go all out to protect the monarchy, with full force, right? When all else fails, and negotiations fail, what do most countries do? We are a sovereign nation. We are not scared. We will go to the extreme if we have to.

X: So when you talk about going to the extreme, was Oka in 1990 an extreme?

AA: Oka was an extreme, to an extent. But you have to remember it was only over a golf course. What's it going to be for us?

X: So Oka was nothing?

AA: Nothing compared to the attention that the National Energy Board is giving Fort McMurray these days. Oil is a national commodity, a strategic resource, a national interest—so when you go to the extreme, what do you think they're going to do?

X: I don't think the country wants to see that.

AA: Well, our people are ready and willing to do it. And I would regret to be the chief when that happens, because I know for a fact that I'm going to have to be making some calls. Nothing to be too happy about.

Chief Allan Adam, Athabasca Chipewyan First Nation, 2014 (Toronto Sun)

X: You're fighting a battle that's not only about resources, but also about the environment, health, family, and education. The Athabasca Chipewyan First Nation is at the center of this resource region. It's turning the tables on whoever wants to do business here, by making sure they go through you. This has a cultural implication: you're making yourselves more visible, and empowering yourselves as a cultural force. It sends a cultural message when a territory the size of France and the UK is being protected by a group of people, 1,200 strong.

ED: I definitely think that Allan doesn't give himself a lot of credit. Previously to his leadership, it's not that there weren't people speaking out against this, but they were speaking out in very small numbers, and they were speaking out as individuals, not as a nation.

As a chief, Allan rallied the nation as a whole to act. He has really catalyzed our community, not just to be angry—because a lot of people were angry before—but also to find creative pathways that not only address atrocities but also empower our community to get past those things. It's not just about stopping the extraction of tar sands—if we play our cards right, they're not going to be able to propose uranium mining. If we change the regulatory system, the uranium mine can't happen. Allan is trying to create actual systems change, and to have a Chief willing to take on something that big, is amazing.

As someone who worked in treaty land claims for six and a half years, I realize that you have to create systems change, otherwise history will repeat itself over and over again. If you look at the history of Canada's economic structures, they're based on the reaping of resources directly from the hands of Indigenous populations, and giving them little in return.

Fur Trader Colin Fraser, sorting fox, beaver, and mink furs, at Fort Chipewyan, a Northwest Trading Post ca. 1890 (Library and Archives Canada / C-001229)

The Hudson's Bay Company stepped in with the fur trade, making laws so that the Indigenous peoples could only sell furs to them. The agricultural industry in the south did the exact same thing. The industrial revolution in Canada, and resource extraction, is the exact same thing. It's history repeating itself. If you don't make systems change in what is rooted in the very foundations of the agreements with the colonizers of this country, then nothing is going to change. It's nice to be part of a nation that is willing to do that.

When I was starting this work, I was like, "tar sands are killing us, holy shit, we need to do something!" What do you do? Protest a multinational corporation? You can't do that in the epicenter of energy in Canada. We would have Oka. We would have the military in a day, they would arrest every last one of us, and that would be the end of it.

So then we had to sit down and create an actual strategy. What would be necessary to shift things? Because it wasn't just tar sands. There's uranium, there's Site C Dam, there's quarries, there's the lack of protection and preservation of caribou and bison—we have been trying to get the bison in our territory to be recognized as a species at risk. We are interested in the preservation and longevity of the Athabasca River system. Tar sands is an example of one of the things that violate our rights, but we are actually fighting for intersectionality between all of these things. We are fighting for the recognition of our sovereign, inherent, treaty rights.

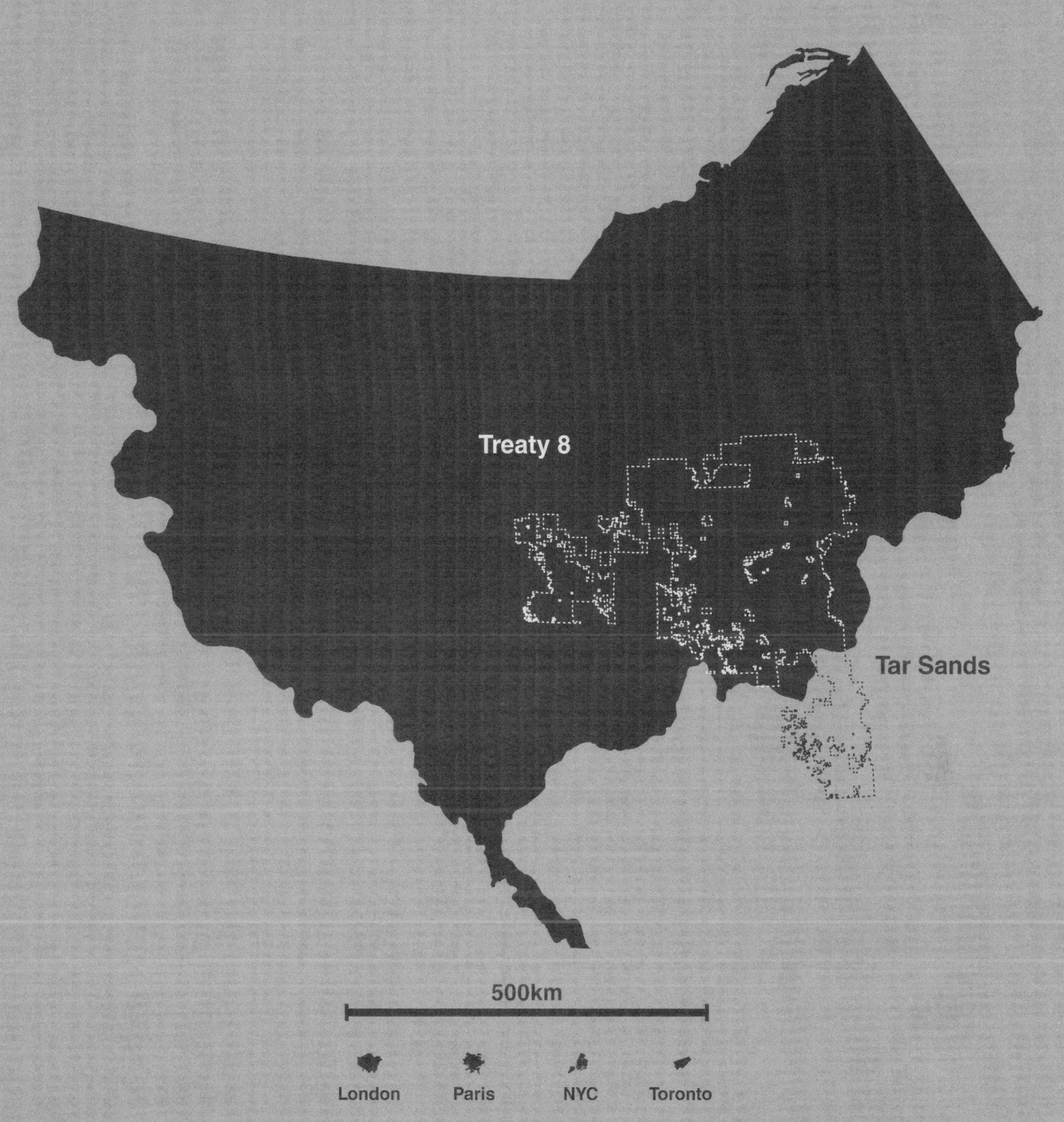
Treaty 8
Tar Sands
500km
London
Paris
NYC
Toronto

X: You have 1,200 people, but there are 35 million Canadians who are happy to live with the separation of surface rights and subsurface rights, who have an illusion of happiness without living off the land. From your perspective, they are the ones who are colonized?

ED: Absolutely. I think that there's a disconnect even within the environmental movement. Part of what I'm trying to do is reflect the ideology that the Chief, Council, and Nation is fighting for—to get that recognized and integrated into the larger narrative around the environmental movement. Most people within the environmental movement live in urban settings. Their solution to climate change is better public transit infrastructure, energy efficient homes, and putting caps and a price on carbon; but none of these solutions get us to the point. So there is a disconnect in the whole narrative.

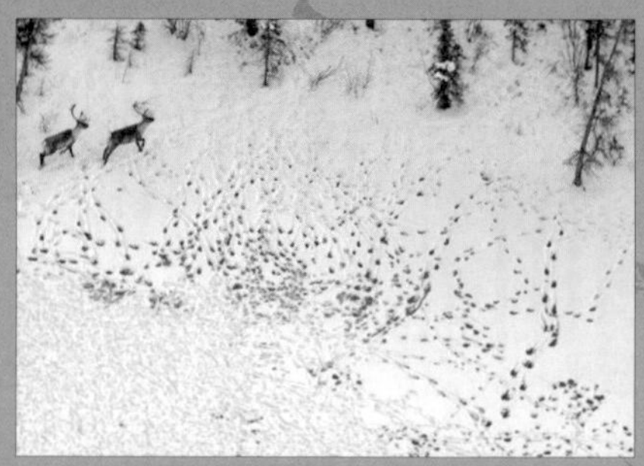

Caribou Migration, Northwest Territories, 2015 (Michelle Valberg)

My mom says that White People are some of the most colonized people because they have lost that connection to their places. As we set all these different standards for environmental protection, it's being done in isolation from the communities that still live in partnership with their ecosystems. Those communities are dependent on their ecosystems not just for trees to build a house, but also clean drinking water, and food to eat directly from the land. They also have a psychological and spiritual connection to those places that drives them to preserve the lands. Having those connections to the environment is different from the mentality of "I own this," because ownership doesn't make you want to protect it, it makes you want to profit off of it.

For us, it's not about owning the land, it's about being a part of the land. I like to say that if you destroy the Athabasca Delta, you destroy the Athabasca Chipewyan First Nation. Who we are is intrinsically linked to those places. For some reason people have this ideology that, following traditions like the Magna Carta and Doctrines of Discovery, we are disconnected and we can go to different places and conquer. So a lot of people say, "if you don't like it, just move." Well, if we move, we will be just like you, and we would be dependent on a system that is detached from the environment. If you moved us to the mountains, we wouldn't know how to live in the mountains because we wouldn't have the millennia of Indigenous knowledge.

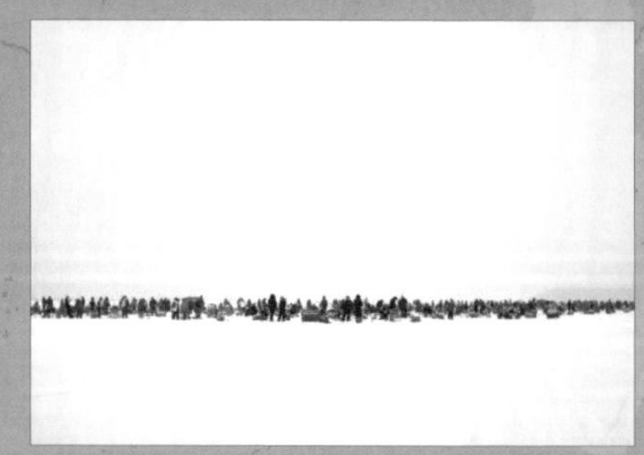

Caribou Harvest, Northwest Territories, 2016 (Life Force Magazine / Aaron Vincent Elkaim)

That's what's missing. People have moved from the places they come from, or they've been forcibly removed through tactics of colonization, and they've lost their connections to the land. When you lose those connections, you go straight into survival mode: I need food, I need water, I need shelter. You don't think, "I need the land to provide me with water, food or shelter, and how do I cultivate a relationship with this land in order to achieve that? You don't have anyone to teach you and pass on crucial knowledge systems, so why would you even know how to do it?

I think what AFCN did with the Surface Water Quantity Framework is groundbreaking. The government comes in, with some ecologists and scientists, and they say "the ecological base flow for the Athabasca River System is 'x' cubic meters of water per second. So can you sign on to this new policy because it's for the betterment of everyone?" But we refused, because our relationship with the land isn't based on qualitative data, it's based on an actual relationship with the land. We took their numbers, and we checked the levels of river systems. But then we did one thing the government didn't do. We got the animals we fish and trap—fish, caribou, bison, muskrats, beaver—and put them in the boat, to navigate back to our community. Guess what? The flow rate of the river system needs to be much higher in order to navigate a boat with a moose in it. Then our surface water quantity baseline became double the 'ecological, scientific' number, because it was based on our intimate relationship with the land.

That's what's missing. This relationship with the land is missing from the way that we define 'healthy' ecosystems. With the climate, in cities, they say that we'll be okay if the global temperature doesn't rise more than two degrees. That's when the climate will be in chaos. Well, actually, the climate will go crazy before that, but it's not going to affect those in cities as much. It's going to affect the ACFN significantly. It's going to affect the people in Fiji, or Tuvalu, or Micronesia. Land-based people across the globe are going to be displaced if we allow for two degrees of warming on the planet. The rights of Indigenous peoples have not been respected, and that's why I feel that we've gotten to where we are.

This conversation took place between Chief Allan Adam, Eriel Deranger, Christopher Alton, and Pierre Bélanger, in Edmonton, Alberta, 2015.

Treaty 9 Adhesion
1905
Treaty 5 Adhesion
1908
Treaty 7
1877
Williams
Treaty
1923
Nunavut
1999
Treaty 3
1873
Treaty 6
Adhesion
1889
Treaty
187
Honor the Treaties
by learning the territories

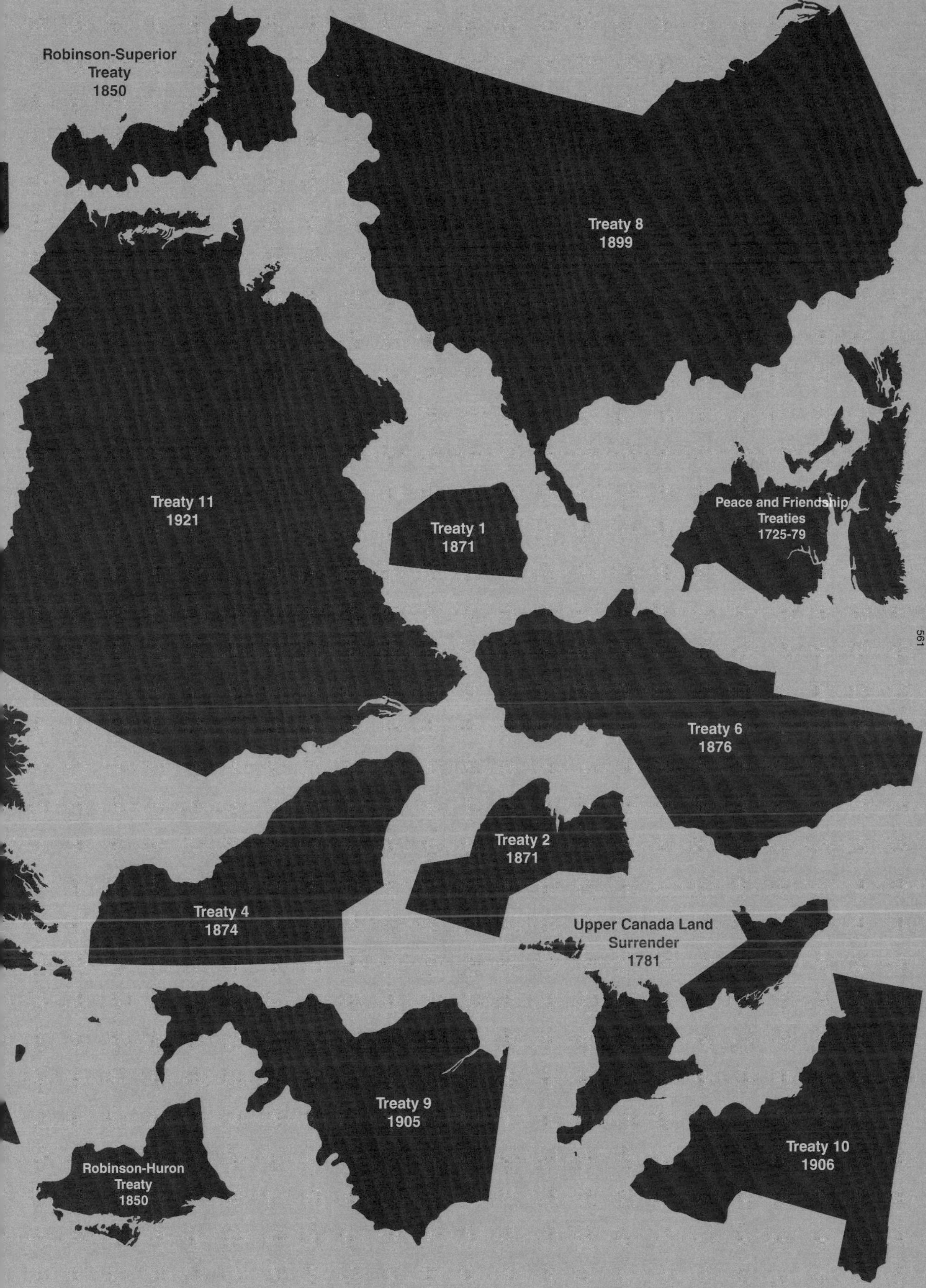

Robinson-Superior Treaty 1850

Treaty 8 1899

Treaty 11 1921

Treaty 1 1871

Peace and Friendship Treaties 1725-79

Treaty 6 1876

Treaty 2 1871

Treaty 4 1874

Upper Canada Land Surrender 1781

Treaty 9 1905

Treaty 10 1906

Robinson-Huron Treaty 1850

The 7th Generation

A Conversation with Eriel Tchekwie Deranger & Kelsey Chapman

June 2016
(Part II)

X: When you're trained in a colonial system—like architects, surveyors, engineers, and cartographers—there is sometimes little difference between what's on paper, and what's on the ground, as if they are the same. They erase the Indigenous politics and identities of space and land. When people talk about the ideas of 'wilderness,' and the 'nationalization of nature,' do these state-based concepts also wipe out any forms of indigeneity in relation to land?

ED: In Jack Forbes' *Columbus and Other Cannibals*, he talks about the Eurocentric values of taming the West, as if nature needed to be conquered, versus Indigenous attitudes of living *in* nature, *with* nature—we weren't ever trying to fight the forces of nature. So that's one of the fundamental differences between Indigenous peoples and non-Indigenous peoples. There are few Indigenous peoples that engage in massive agriculture or husbandry, but even when they do, it is in a more holistic way. The Sámi people are reindeer herders, but they herd the reindeer through their migratory routes.

"The true owners of the land are not yet born."
—Harold Cardinal, Citizens Plus, 1970

On the other hand, Eurocentric ideologies around agriculture and husbandry focus on taming species, controlling them, and making them "better." The cattle industry is so broken, because of this desire to find the best and fastest way to grow meat for consumption, rather than working *with* species. I think that's a huge fundamental difference between Indigenous people versus non-Indigenous people: Indigenous people work within the elements, and are connected with them. In a more—I've heard people say—'backwards' way.

But as humanity is getting more intelligent, we start to realize that we are in fact intrinsically linked to the natural world. Indigenous peoples' connections to the natural world are *real*, as opposed to purely mythical. Indigenous origin stories were devalued during colonization. But it turns out that when groups identify as the Wolf Clan, the Turtle Clan, or identify as connected to certain roots, plants, and flowers, that's because those species are intrinsically a part of those peoples' DNA. In their evolutionary processes in that part of the world, that species, whatever it was, was essential at some point to the survival of those people. So we actually absorb the very essence of the species, as you ingest it and break it into microbiology. Just like the trees in British Columbia have isotopic tracers of the salmon. We pick up those elements of the food that we eat.

Those origin stories weren't just fanciful tales, they're actually based in science. But Western society devalues them because they seem so implausible—because Indigenous people don't have the quantitative data analysis of our knowledge.

When our knowledge systems are devalued, then it's easy for someone coming from a Eurocentric place, to devalue and minimize the significance of identifying as Indigenous or non-Indigenous.

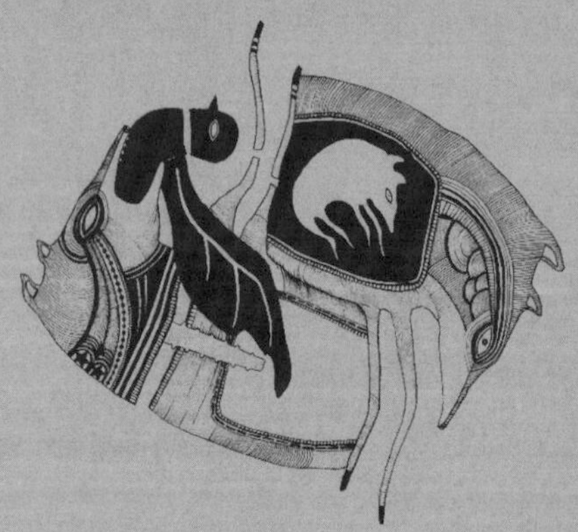

Wétiko
(Design by Veronica Liu, in Jack D. Forbes, *Columbus and other Cannibals: The Wétiko Disease of Exploitation, Imperialism, and Terrorism*, Seven Stories Press, [1978] 2008)

X: So the question, "aren't we all Indigenous" that was recently posed to me at a symposium discussion for the 2016 Venice Biennale by Italian architect Paola Viganò, is a confrontation to an entire way of life, as if it never existed or, it's not legitimate.

ED: Yes. Those people are obviously okay with designations of *Italian*, or *German*, or *Americans*, or *Canadian*—and that's troubling, because it's okay if you're from an accepted identification-place. But if you're not from a place that works within the confines of the colonial conquer-and-conquest construct, then you're just a part of whoever conquested that area.

To say, aren't we all Indigenous because aren't we all human, minimizes Indigenous people as *separate from humanity.* We're not saying we're separate from humanity, we're actually *all* human. But just as women are women, and children are children, there are designations we make, in humanity, to signify different places of power, privilege and access.

Papal Bull, the Romanus Pontifex issued by Pope Nicolas V, 1455 (Arquivo Nacional da Torre do Tombo, Lisbon, Portugal)

X: So, what do you make of the refusal to acknowledge these Indigenous atrocities happening here?

ED: I don't know if it's a conscious refusal. I think that the institutional racism that exists, not only in Canada, but also in every single colonial country, has worked because of its ability to dehumanize Indigenous populations, through segregation policies and such. So I understand what people say when they ask, "aren't we all Indigenous?" I understand that what they're saying is that this classification seems like a way to dehumanize Indigenous people.

But it's not, because what it's doing is trying to give the designation, the recognition, like that of a woman. Feminists constantly have to defend why the designation of a woman—or women's rights—is so important. And it's now accepted, and feminism has come a really long way. Yes, it would be great if we could abolish distinctions, just like it would be great if women received the same pay as men, or if children didn't have to work in factories if they didn't want to. The concept of minority rights was created to recognize the distinct lack of access that certain groups have to power, or the oppressive regimes imposed on them. But for some reason, when you try to create the same recognitions for Indigenous people, people get all up in arms about it.

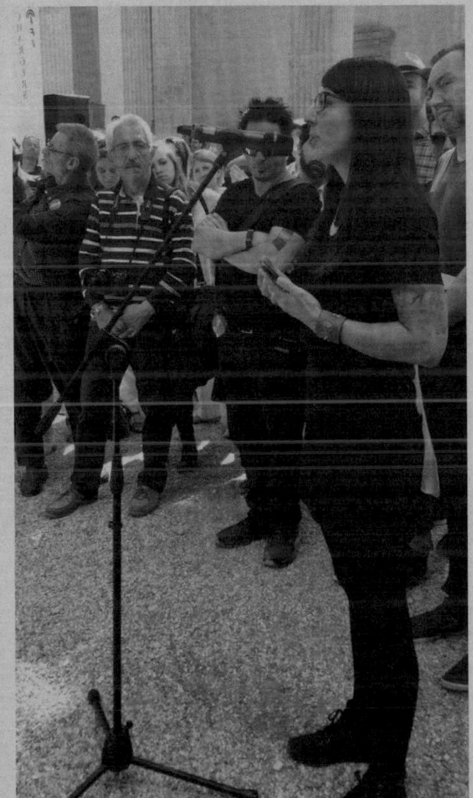

Eriel Deranger, Spoken Word on the 140th Anniversary of the Indian Act, Venice Biennale, Italy, 2016 (Nina-Marie Lister)

But speaking of feminism, even within the feminist movement, there's this devaluation of Indigenous peoples. When First Nations were negotiating treaties with the Europeans, the reason that men were sent wasn't because they were the people in power. It was because the Europeans seemed to be focused on men, so the women told the men to go talk to them, and they also wanted to protect themselves because rape was common during first-wave colonization. So the men and women worked together.

In Indigenous societies, gender and gender roles were much more fluid. Men and women were much more equal. If the full goal of feminism is equality, and Indigenous communities lived in equality pre-contact, then shouldn't feminism look at how Indigenous cultures ran pre-contact?

X: You're making the active choice not to be involved in industry. Dave Tuccaro is an interesting contrast to Indigenous opposition to, or adjustment of, industry, when he's built significant wealth by capitalizing on what's happening in the region.

ED: I say that purposely, because there are those kinds of people. But I'm not condoning their actions, or even saying that that's potentially a way forward.

Development is inevitable, it will always exist in humanity. Even as Indigenous peoples, we developed our lands and resources in a way that supported the advancement and survival of our people. The difference is that Indigenous peoples do it differently from the Eurocentric way of *controlling* those elements. So the ideology behind development of resources differs very greatly—Indigenous peoples are not anti-development. My nation often says, "sure, we'll be a part of tar sands. Why not?" And then it gives a list of all the pre-conditions. For us, we won't be a part of the project in the way that it is developed currently, in the way that it destroys nature, destabilizes the ecosystem, and endangers the life, culture, and identities of Indigenous peoples.

The example of Chief Jim Boucher at Fort McKay, shows that there have been violations of the foundations of the UN Declaration on the Rights of Indigenous Peoples, which is free, prior, informed consent. What happened at Fort McKay is a demonstration of forced assimilation that happened through extreme coercion and manipulation. They tried to fight off industry for years and years, but they couldn't.

They had no potable water, they had some of the highest rates of unemployment, and poverty—they still have a boiled water advisory. They truck in all the water, they don't have a water treatment plant in their community, because they don't think the plant could actually clean the water that they would have to treat.

KC: People get chemical burns from showering.

ED: People with sensitive skin get what looks like chemical burns from bathing or showering. Children have been moved out of the community. There are people that say they have moved out because their daughters had such horrific skin rashes, and the only connection that they could figure out was the bathing water.

KC: But everyone's got a nice truck and a big house. Boucher's done a bit of a disservice, because he's now set a standard for industry and the government, which is that it doesn't matter whether or not your community members have chemical burns and high rates of respiratory diseases, autoimmune diseases, cancers—because Fort McKay gets paid big bucks by the industry, to stay complacent.

So you have an entire community that's complacent. Even people that are not happy and want to speak up end up getting silenced by the leadership, either through a withholding of resources from those people, or by being forced out of the community. Then they're discredited publicly over and over again.

It's like full-on corruption that's sponsored by the oil and gas industry, and sanctioned by the government.

Athabasca River, Townships, Oil Sands Leases, and Industrial Facilities, north of Fort McMurray, Alberta, Canada, 2014 (Terracon Geotechnique)

X: Is the Crown placing too much control in the hands of industry?

ED: I would say that most resource-extractive industries in Canada have become self-regulated. Whether it's Weyerhaeuser or Syncrude, or any other multinational corporation, Canada is like the Wild West—the government invites multinational corporations to come in and exploit the resources of our country. It's no different than "come to the West to get rich and start a new life, and have everything you could possibly desire," without consultation, or recognition of Indigenous peoples' rights. Resource extraction is the new frontier.

And, as much as resource extraction is the new frontier, my call is that within the next ten to twenty years, we're going to be less concerned about oil, gas, and trees, and instead we'll have frontiers around access to clean potable water. Canada knew as early as the late 1890s that water would become an economic base. During the late 1890s into the 1910s, Canada did a huge study on the entire water basin of Canada, and what it would take to export our water to the United States, which didn't have as rich of a water basin.

CANADA'S WORST NEIGHBOURHOOD

"Canada's Worst Neighbourhood," magazine article on Regina, Saskatchewan, 2007 (Maclean's, left and right)

X: You have also mentioned the importance of being able to understand treaty lands, and the fact that most Canadians are taught through political units of the country such as the Provinces. In your opinion, is there a role for mapping, and how can we prevent the inequalities and injustices it often conjures?

ED: In conversations around identifying power, power is largely associated with land. It gets really tricky, because in *Columbus and Other Cannibals*, Jack Forbes says that colonization's desire to consume the natural world is a form of cannibalism. But he also says that even Indigenous people are not immune to this disease of the mind. Even within UN discussions, some of the people that I respect very much get hung up on the territoriality and the boundaries of the maps that have been created by Eurocentrism, and then they want the control, in the same way that the Europeans did.

But borders and territories are important insofar as that the people that hold power follow these Western-Eurocentric ideologies, and that is what we're fighting. So within that context, maps have become important.

In understanding the treaty lands, it's important to know that the people are supposed to have the most power. Within Treaty Lands, First Nations are recognized and so is the Crown, not the corporations. But there's been a perversion of that, through things like the Natural Resources Transfer Agreement from the late 1930s, which gave jurisdiction to the provinces without any consultation with the First Nations. Which is ridiculous. Because they say that "we only gave you rights to the depth of the plow"—what? And then they still do deforestation and forestry work, which is still above ground. I think that forestry fights are more successful because of the fact that it's above ground, but even then, it's tricky. They're always trying to find a loophole and impose their interpretation of the law on us.

North Central Land Use & Zoning Plan, 2017 (City of Regina Planning Department)

X: Going back to the idea of legitimizing and asserting Indigenous forms of knowledge, did you ever consider becoming a lawyer?

ED: I went to university; I never finished my Bachelor's Degree. I came from a very politicized background, we didn't buy pears or grapes from Chile for political reasons—so I grew up with this understanding that certain things were not fair. In university, I took this globalization class, centered on economics, and I realized that it commodified *everything*. Food becomes numbers in a system, even *humans* become numbers in a system.

The valuation of an individual, an individual's salary, depends on whether or not we attain a diploma, a piece of paper that is only accessible to a certain group of people. I was like, "why am I a part of this system? I can't do it—I can't be commodified by a Bachelor's or Master's degree!"

So I continued to take classes, but not for credit. My knowledge wasn't granted to me by institutions that locked me out of its books. That's been a challenge, as I've been constantly asked, why don't you have a Bachelor's, why don't you have a Master's, why aren't you a lawyer. And my response is, "because I don't want to participate in a system that commodifies humans, and that inherently oppresses the weak." Higher education is an institution that allows for the continuation of marginalization, of people that have limited access to those systems. You not only have to have money, but also physical access.

NATIONAL

'It's not the worst neighbourhood'

Regina's honchos gather to defend their city's wounded pride

SHE POINTS OUT COUNTLESS 'DRUG HOUSES' AND YOUNG GIRLS SHE SAYS ARE PROSTITUTES ON A STREET SHE CALLS KIDDIE ALLEY

DANGER AND TAXES

Ottawa tries to claw back a salary boost for soldiers at Level Two

FOR THE RECORD

WE HAVE OURSELVES ALL WRONG

I went to a High School in a neighborhood that Maclean's Magazine called the worst neighborhood of all of Canada. It's called North Central, in Regina, Saskatchewan.

KC: The neighborhood can be pretty scary, especially at night, if you don't know the right streets to go down.

ED: And if you're not Native. But I went to high school in that neighborhood, and I actually think it was the best high school in the city, because teachers built real relationships with the students. The student population was so small, because most people in the neighborhood didn't get

Grade 8 education. So the school was a mixture of people who really wanted to graduate to get out of the neighborhood, and people who were 'mandated' to go to school.

KC: The school has a police department *in* the school.

ED: *In* the school. In a class of fifteen students, in any given class, there were probably between three to six people with ankle bracelets, there were girls that were prostitutes, there was a daycare. So there was a police station and a daycare in the same school.

We were given these fluorescent stickers that were hard to take off cars, because the neighborhood was notorious for child prostitution, and on your walk to school, you get get stopped and people would ask, "how much?"

This is in Regina, Saskatchewan. You don't even have to get to reserves to talk about microcosms of third-world conditions. That reality is the subtext to my pull from a system that didn't serve any of those people.

When you look at voluntourism, and the desire of the Western North American going out there to photograph and save the world, you realize that those exact problems existed right in their *backyards*! Literally across the railroad tracks in the city of Regina.

Mountain of 180,000 buffalo skulls during the last great hunt of buffalo in the 1870s. Collected and stockpiled for bone charcoal at Michigan Carbon Works in Detroit, the skulls originated from Western Canada and Western United States, namely from places such as 'Pile O' Bones' in Saskatchewan, later renamed in 1882 as Regina. (Burton Historical Collection, Detroit Public Library)

X: What's your philosophy on resistance?

ED: Have you ever heard of the seventh-generation prophecies? It varies from region to region, but it's pretty much the same thing. During the first wave of colonization, the reason the people that signed the treaty didn't fight too hard was that if they did, they would all be annihilated. So the prophecy was that they needed to stop fighting, and that for seven generations, they would experience the worst hardships, and people would try to take everything away from them. But Indigenous peoples would hold on to the stories of who they were and the connection to the land, and they would continue to teach that to their children. Then the seventh generation would be able to walk in two worlds, and they would be able to return the people back to the land. The last seven generations has been this perversion of humanity.

X: So you're implying that we're in the seventh generation, that it's happening now?

ED: There are arguments that my children are the seventh generation, there are arguments about who exactly is the seventh generation. But basically—this is our fate, and in order to win, we're going to have to play the long game, and the long game is seven generations.

Eriel Deranger, Highway 63, Fort MacMurray (*Elemental*, 2012)

No one said it was going to be easy. But the other thing I've learned, which is different from some of the other environmentalists that I work with, is that it's not about measures of personal success. Whereas I don't really believe that I'll see the change that I'm working towards in my lifetime. But that's not important to me. What's important to me is that—the success comes in the long game. A measure of success that I can see is my daughter, when she openly and proudly tells people that she's Dene. Without any hesitation or shame. When I was a kid, saying I was Native came with an underscoring of shame. If this is the way this works, then I won't really see it. And that's okay—it's not about trying to achieve that.

Allan Adam's Athabasca Chipewyan First Nation refusal to sell out to the industry is an expression of the long game. It's an indication of "yeah, we could have a lot of money in the next twenty years, but then what?"

So, what's the next generation going to have? They're going to have no money, no land, no culture, no identity. We're robbing future generations so they can have a bigger house and a bigger car, and you want to be like Fort McKay?

That is an indicator of the ideology that is so different from the patriarchal, colonialist, capitalistic ideology.

This conversation took place between Eriel Deranger, Kelsey Chapman, and Pierre Bélanger in Venice, Italy, 2016.

Declaration of the People of Rupert's Land and the Northwest
Louis Riel, 1869

Whereas it is admitted by all men, as a fundamental principle that the public authority commands the obedience and respect of its subjects. It is also admitted that a people when it has no Government is free to adopt one form of Government in preference to another to give or to refuse allegiance to that which is proposed. In accordance with the first principle the people of this Country had obeyed and respected that authority to which the circumstances surrounding its infancy compelled it to be subject.

A company of adventures known as the "Hudson's Bay Company" and invested with certain powers granted by His Majesty (Charles II) established itself in Rupert's Land. AND IN THE NORTH-WEST TERRITORY for trading purposes only. This Company, consisting of many persons, required a certain constitution. But as there was a question of commerce only, their constitution was framed in reference thereto. Yet since there was at that time no government to see to the interests of a people already existing in the country, it became necessary for judicial affairs to have recourse to the officer of the Hudson's Bay Company. This inaugurated that species of government, which slightly modified by subsequent circumstances, rule this country up to a recent date. Whereas this government thus accepted was far from answering the wants of the people, and became more and more so as the people increased in numbers, and as the country was developed, and commerce extended, until the present day, when it commands a place amongst the Colonies. Ever actuated by the above mentioned principles, this people generously supported the aforesaid government, and have to it a faithful allegiance: when, contrary to the law of nations, in March 1869, that said Government surrendered and transferred to Canada all the rights which it had or pretended to have in this territory, by transactions with which the people were considered unworthy to be made acquainted.

And whereas it is also generally admitted that a people is at liberty to establish any form of government it may consider suitable to its wants, as soon as the power to which it was subject abandons it, or attempts to subjugate it without consent, to a foreign power; and maintain that no right can be transferred to the foreign power.

1st. We, the Representatives of the people in Council assembled at Upper Fort Garry on the 24th day of November 1869, after having invoked the God of nations, relying on these fundamental moral principles, solemnly declare in the name of our constituents and in our own names before God and man, that from the day on which the Government we had always respected abandoned us by transferring a strange power the sacred authority confided to it, the people of Rupert's Land and the North West became free and exempt from all allegiance to the said Government.

2nd. That we refuse to recognize the authority of Canada, which pretends to have a right to coerce us and impose upon us a despotic form of government still more contrary to our rights and interests as British subjects than was that Government to which we had subjected ourselves through necessity up to recent date.

3rd. That by sending an expedition on the 1st of November charged to drive back Mr. William McDougall and his companions coming in the name of Canada to rule us with the rod of despotism without a previous notification to that effect, we have but acted conformably to that sacred right which commands every citizen to offer energetic opposition to prevent his country being enslaved.

4th. That we continue and shall continue to oppose with all our strength the establishing of the Canadian authority in our country under the announced form. And in case of persistence on the part of the Canadian Government to enforce its obnoxious policy upon us, by force of arms, we protest beforehand against such an unjust and unlawful course, and we declare the said Canadian Government responsible before God and men for the innumerable evils which may be caused by so unwarrantable course.

Be it known, therefore, to the world in general and the Canadian Government, that before seeing our country coerced into slavery, we shall employ every means of defence that Divine Providence has placed at our disposal. And that is not to see our country which we have so often defended at the price of our best blood against hordes of barbarians (who have since become our friends and allies) invaded by the stranger.

That meanwhile we hold ourselves in readiness to enter into negotiations with the Canadian Government, which may be favorable to its aggrandizement and our good government and prosperity.

In support of this declaration relying on the protection of Divine Providence on oath we mutually pledge ourselves, our lives, our fortunes, and our sacred honor to each other.

courtesy of The Manitoba Historical Society

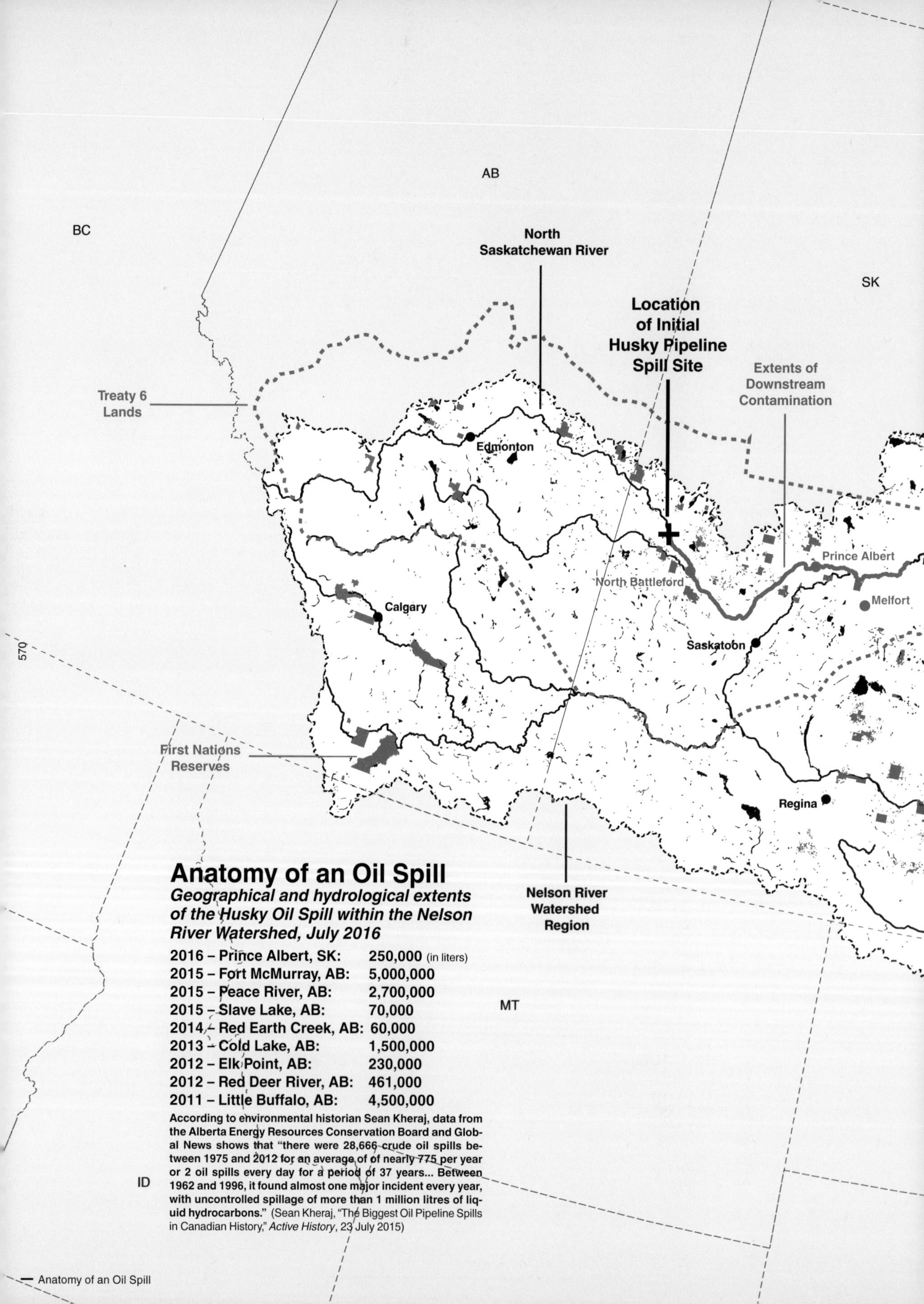

Anatomy of an Oil Spill

Geographical and hydrological extents of the Husky Oil Spill within the Nelson River Watershed, July 2016

2016 – Prince Albert, SK: 250,000 (in liters)
2015 – Fort McMurray, AB: 5,000,000
2015 – Peace River, AB: 2,700,000
2015 – Slave Lake, AB: 70,000
2014 – Red Earth Creek, AB: 60,000
2013 – Cold Lake, AB: 1,500,000
2012 – Elk Point, AB: 230,000
2012 – Red Deer River, AB: 461,000
2011 – Little Buffalo, AB: 4,500,000

According to environmental historian Sean Kheraj, data from the Alberta Energy Resources Conservation Board and Global News shows that "there were 28,666 crude oil spills between 1975 and 2012 for an average of of nearly 775 per year or 2 oil spills every day for a period of 37 years... Between 1962 and 1996, it found almost one major incident every year, with uncontrolled spillage of more than 1 million litres of liquid hydrocarbons." (Sean Kheraj, "The Biggest Oil Pipeline Spills in Canadian History," *Active History*, 23 July 2015)

Hudson's Bay
MB
Lake Winnipeg
ON
Winnipeg
Fargo
Lake Superior
MN
500km

ten little, nine little,
eight little beavers

seven little, six little,
five little beavers

four little, three little,
two little beavers

one little
dead
beaver.

image courtesy of Kent Monkman

Bacchant Beaver, Kent Monkman, 2016

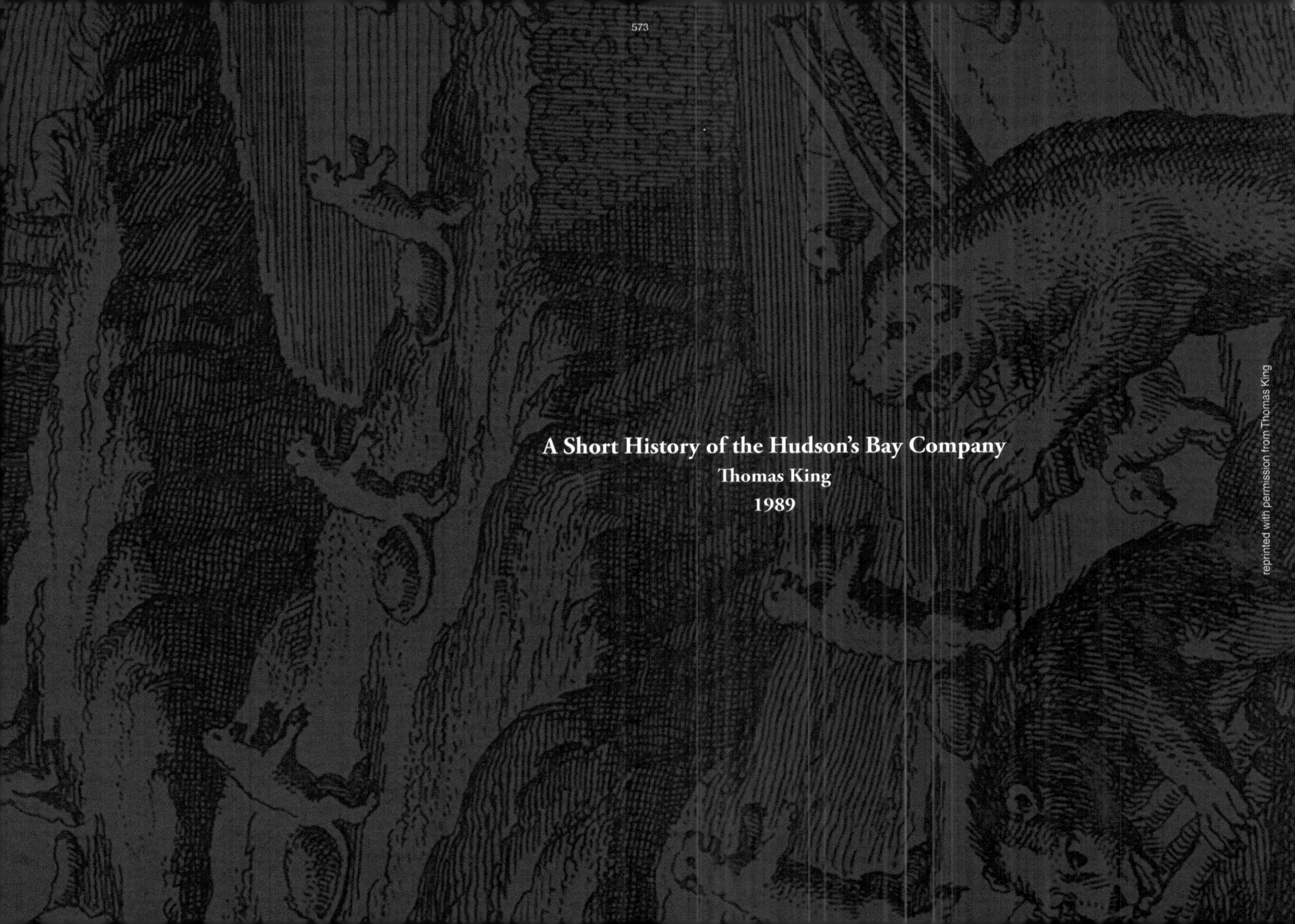

A Short History of the Hudson's Bay Company

Thomas King

1989

Paul Karvanis
Legal Counsel
Tel: 416-256-3347
Email: paul.karvanis@hbc.com

January 6, 2016

VIA EMAIL at info@extraction.ca

EXTRACTION

Attention: Project Commissioner and Project Curator

<u>Re: Infringement and Passing off of Hudson's Bay Company ("HBC") Trade-marks</u>

It has come to our attention that Extraction is using the phrase PRO PELLE CUTEM as its handle on Facebook, Twitter and Instagram. Such use infringes HBC's intellectual property rights.

HBC trademarks are widely known throughout Canada, Europe and the U.S. as well as internationally, recognized through use spanning nearly three and a half centuries. HBC has rights in many jurisdictions over the trademark PRO PELLE CUTEM through both formal trademark registrations as well as through its use.

HBC owns, and Extraction is infringing, the following registered trademarks in Canada:

Reg. #	Description
9553	COAT OF ARMS, PRO PELLE CUTEM (shown at right)
371470	COAT OF ARMS DESIGN
443417	COAT OF ARMS DESIGN

HUDSON'S BAY COMPANY,

We have been using our COAT OF ARMS and our PRO PELLE CUTEM trademarks in Canada since 1680. HBC is a fundamental and primary part of the history of Canada and it is HBC's use of the PRO PELLE CUTEM trademark throughout history that has given the trademark its significance to Canada. Extraction's infringement of our PRO PELLE CUTEM trademark is made worse by Extraction's position as Canada's official representative at the Venice Architecture Biennale in 2016.

A consumer googling PRO PELLE CUTEM would find results that are overwhelmingly connected to HBC. This suggests an affiliation between Extraction and HBC where there is no connection.

We please require that you cease using our PRO PELLE CUTEM trademark as your handle on social media and that you transfer ownership of those profiles to us once you have had a chance to save/delete/transfer your information (which we request that you do promptly but in any event within two (2) weeks).

Our PRO PELLE CUTEM trademark is an integral part of our COAT OF ARMS, one of our defining trademarks. It is our expectation that you will comply with the steps set out above. Please respond within one (1) week confirming that you are proceeding in accordance with the steps set out above.

Should you have any questions, please contact me.

Regards,

HUDSON'S BAY COMPANY

Paul Karvanis
Legal Counsel, Hudson's Bay Company

Settler Colonialism and the Elimination of the Native

Patrick Wolfe

"Independence is not a word which can be used as an exorcism, but an indispensable condition for the existence of men and women who are truly liberated, in other words who are truly masters of all the material means which make possible the radical transformation of society."[1]

—Frantz Fanon, *The Wretched of the Earth*, [1961] 1968

The question of genocide is never far from discussions of settler colonialism. Land is life—or, at least, land is necessary for life. Thus contests for land can be—indeed, often are—contests for life. Yet this is not to say that settler colonialism is simply a form of genocide. In some settler-colonial sites (one thinks, for instance, of Fiji), Native society was able to accommodate—though hardly unscathed—the invaders and the transformative socioeconomic system that they introduced. Even in sites of wholesale expropriation such as Australia or North America, settler colonialism's genocidal outcomes have not manifested evenly across time or space. Native Title in Australia or Indian sovereignty in the US may have deleterious features, but these are hardly equivalent to the impact of frontier homicide. Moreover, there can be genocide in the absence of settler colonialism. The best known of all genocides was internal to Europe, while genocides that have been perpetrated in, for example, Armenia, Cambodia, Rwanda or (one fears) Darfur do not seem to be assignable to settler colonialism. In this article, I shall begin to explore, in comparative fashion, the relationship between genocide and the settler-colonial tendency that I term the logic of elimination.[2] I contend that, though the two have converged—which is to say, the settler-colonial logic of elimination has manifested as genocidal—they should be distinguished. Settler colonialism is inherently eliminatory but not invariably genocidal.

As practiced by Europeans, both genocide and settler colonialism have typically employed the organizing grammar of race. European xenophobic traditions such as anti-Semitism, Islamophobia, or Negrophobia are considerably older than race, which, as many have shown, became discursively consolidated fairly late in the eighteenth century.[3] But the mere fact that race is a social construct does not of itself tell us very much. As I have argued, different racial regimes encode and reproduce the unequal relationships into which Europeans coerced the populations concerned. For instance, Indians and Black people in the US have been racialized in opposing ways that reflect their antithetical roles in the formation of US society. Black people's enslavement produced an inclusive taxonomy that automatically enslaved the offspring of a slave and any other parent. In the wake of slavery, this taxonomy became fully racialized in the "one-drop rule," whereby any amount of African ancestry, no matter how remote, and regardless of phenotypical appearance, makes a person Black. For Indians, in stark contrast, non-Indian ancestry compromised their Indigeneity, producing "half-breeds," a regime that persists in the form of blood quantum regulations. As opposed to enslaved people, whose reproduction augmented their owners' wealth, Indigenous people obstructed settlers' access to land, so their increase was counterproductive. In this way, the restrictive racial classification of Indians straightforwardly furthered the logic of elimination. Thus we cannot simply say that settler colonialism or genocide have been targeted at particular races, since a race cannot be taken as given. It is made in the targeting.[4] Black people were racialized as slaves; slavery constituted their blackness. Correspondingly, Indigenous North Americans were not killed, driven away, romanticized, assimilated, fenced in, bred White, and otherwise eliminated as the original owners of the land but *as Indians*. Roger Smith has missed this point in seeking to distinguish between victims murdered for where they are and victims murdered for who they are.[5] So far as Indigenous people are concerned, where they are is who they are, and not only by their own reckoning. As Deborah Bird Rose has pointed out, to get in the way of settler colonization, all the Native has to do is stay at home.[6] Whatever settlers may say—and they generally have a lot to say—the primary motive for elimination is not race (or religion, ethnicity, grade of civilization, etc.) but access to territory. Territoriality is settler colonialism's specific, irreducible element.

The logic of elimination not only refers to the summary liquidation of Indigenous people, though it includes that. In common with genocide as Raphaël Lemkin characterized it,[7] settler colonialism has both negative and positive dimensions. Negatively, it strives for the dissolution of Native societies. Positively, it erects a new colonial society on the expropriated land base—as I put it, settler colonizers come to stay: invasion is a structure not an event.[8] In its positive aspect, elimination is an organizing principal of settler-colonial society rather than a one-off (and superseded) occurrence. The positive outcomes of the logic of elimination can include officially encouraged miscegenation, the breaking-down of Native title into alienable individual freeholds, Native citizenship, child abduction, religious conversion, resocialization in total institutions such as missions or boarding schools, and a whole range of cognate biocultural assimilations. All these strategies, including frontier homicide, are characteristic of settler colonialism. Some of them are more controversial in genocide studies than others.

Settler colonialism destroys to replace. As Theodor Herzl, founding father of Zionism, observed in his allegorical manifesto/novel, "If I wish to substitute a new building for an old one, I must demolish before I construct."[9] In a kind of realization that took place half a century later, one-time deputy-mayor of West Jerusalem Meron Benvenisti recalled, "As a member of a pioneering youth movement, I myself 'made the desert bloom' by uprooting the ancient olive trees of al-Bassa to clear the ground for a banana grove, as required by the 'planned farming' principles of my kibbutz, Rosh Haniqra."[10] Renaming is central to the cadastral effacement/replacement of the Palestinian Arab presence that Benvenisti poignantly recounts.[11] Comparably, though with reference to Australia, Tony Birch has charted the contradictory process whereby White residents sought to frustrate the (re-)renaming of Gariwerd back from the derivative "Grampians" that these hills had become in the wake of their original owners' forcible dispossession in the nineteenth century.[12] Ideologically, however, there is a major difference between the Australian

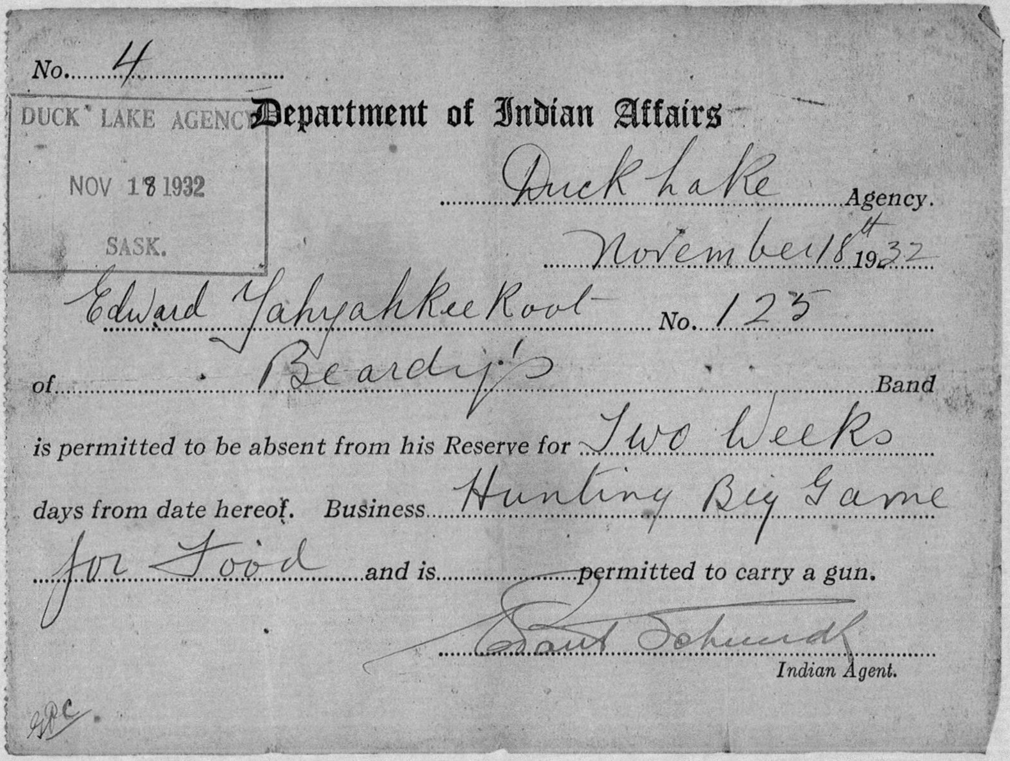

No. *4*

DUCK LAKE AGENCY
NOV 18 1932
SASK.

Department of Indian Affairs

Duck Lake Agency.

November 18th 19*32*

Edward Yahyahkeekoot No. *125*

of *Beardy's* Band

is permitted to be absent from his Reserve for *Two Weeks*

days from date hereof. Business *Hunting Big Game for Food* and is permitted to carry a gun.

[illegible]
Indian Agent.

[illegible]

Pass issued to Edward Yahyahkeekoot
Department of Indian Affairs, 1932

and Israeli cases. The prospect of Israeli authorities changing the Hebrew place-names whose invention Benvenisti has described back to their Arabic counterparts is almost unimaginable. In Australia, by contrast (as in many other settler societies), the erasure of Indigeneity conflicts with the assertion of settler nationalism. On the one hand, settler society required the practical elimination of the Natives in order to establish itself on their territory. On the symbolic level, however, settler society subsequently sought to recuperate Indigeneity in order to express its difference—and, accordingly, its independence—from the mother country. Hence it is not surprising that a progressive Australian state government should wish to attach an Indigenous aura to a geographical feature that bore the second-hand name of a British mountain range. Australian public buildings and official symbolism, along with the national airlines, film industry, sports teams, and the like, are distinguished by the ostentatious borrowing of Aboriginal motifs. For nationalist purposes, it is hard to see an alternative to this contradictory reappropriation of a foundationally disavowed Aboriginality. The ideological justification for the dispossession of Aborigines was that "we" could use the land better than they could, not that we had been on the land primordially and were merely returning home. One cannot imagine the Al-Quds/Jerusalem suburb of Kfar Sha'ul being renamed Deir Yasin. Despite this major ideological difference, however, Zionism still betrays a need to distance itself from its European origins that recalls the settler anxieties that characterize Australian national discourse. Yiddish, for instance, was decisively rejected in favor of Hebrew—a Hebrew inflected, what is more, with the accents of the otherwise derided Yemeni *mizrachim*. Analogously, as Mark LeVine has noted, though the Zionist modernization of the Arab city of Jaffa was intended to have a certain site specificity, "in fact Jaffa has had to be emptied of its Arab past and Arab inhabitants in order for architects to be able to reenvision the region as a 'typical Middle Eastern city.'"[13]

In its positive aspect, therefore, settler colonialism does not simply replace Native society *tout court*. Rather, the process of replacement maintains the refractory imprint of the Native counter-claim. This phenomenon is not confined to the realm of symbolism. In the Zionist case, for instance, as Gershon Shafir has cogently shown, the core doctrine of the conquest of labor, which produced the *kibbutzim* and Histadrut, central institutions of the Israeli state, emerged out of the local confrontation with Arab Palestinians in a form fundamentally different from the pristine doctrine of productivization that had originally been coined in Europe. The concept of productivization was developed in response to the self-loathing that discriminatory exclusions from productive industry encouraged in Eastern European Jewry (in this sense, as Shafir acutely observes, Zionism mirrored the persecutors' anti-Semitism).[14] In its European enunciation, productivization was not designed to disempower anyone else. It was rather designed, autarkically as it were, to inculcate productive self-sufficiency in a Jewish population that had been relegated to urban (principally financial) occupations that were stigmatized as parasitic by the surrounding gentile population—a prejudice that those who sought to build the "new Jew" endorsed insofar as they resisted its internalization. On its importation into Palestine, however, the doctrine evolved into a tool of ethnic conflict, as Jewish industries were actively discouraged from employing non-Jewish labor, even though Arabs worked for lower wages and, in many cases, more efficiently:

> "'Hebrew labor,' or 'conquest of labor' … was born of Palestinian circumstances, and advocated a struggle against Palestinian Arab workers. This fundamental difference demonstrates the confusion created by referring 'Hebrew labor' back to the productivization movement and anachronistically describing it as evolving in a direct line from Eastern European origins."[15]

As it developed on the colonial ground, the conquest of labor subordinated economic efficiency to the demands of building a self-sufficient proto-national Yishuv (Jewish community in Palestine) at the expense of the surrounding Arab population. This situated struggle produced the new Jew as subject of the labor that it conquered. In the words of Zionist architect Julius Posner, reprising a folk song: "We have come to the homeland to build and be rebuilt in it … the creation of the new Jew … [is also] the creator of that Jew."[16] As such, the conquest of labor was central both to the institutional imagining of a goyim-rein (gentile-free) zone and to the continued stigmatization of Jews who remained unredeemed in the galut (diaspora). The positive force that animated the Jewish nation and its individual new-Jewish subjects issued from the negative process of excluding Palestine's Indigenous owners.

In short, elimination refers to more than the summary liquidation of Indigenous people, though it includes that. In its positive aspect, the logic of elimination marks a return whereby the Native repressed continues to structure settler-colonial society. It is both as complex social formation and as continuity through time that I term settler colonization a structure rather than an event, and it is on this basis that I shall consider its relationship to genocide.

⟵

To start at the top, with the European sovereigns who laid claim to the territories of non-Christian (or, in later secularized versions, uncivilized) inhabitants of the rest of the world: justifications for this claim were derived from a disputatious arena of scholarly controversy that had been prompted by European conquests in the Americas and is misleadingly referred to, in the singular, as the doctrine of discovery.[17] Though a thoroughgoing diminution of Native entitlement was axiomatic to discovery, the discourse was primarily addressed to relations between European sovereigns rather than to relations between Europeans and Natives.[18] Competing theoretical formulas were designed to restrain the endless rounds of war-making over claims to colonial territory that European sovereigns were prone to indulge in. The rights accorded to Natives tended to reflect the balance between European powers in any given theatre of colonial settlement. In Australia, for instance, where British dominion

Awareness Series / Barry Pottie

Eskimo Identification Tag
Department of Indian & Northern Affairs, 1940s

was effectively unchallenged by other European powers, Aborigines were accorded no rights to their territory, informal variants on the theme of *terra nullius* being taken for granted in settler culture. In North America, by contrast, treaties between Indian and European nations were premised on a sovereignty that reflected Indians' capacity to permute local alliance networks from among the rival Spanish, British, French, Dutch, Swedish, and Russian presences.[19] Even where Native sovereignty was recognized, however, ultimate dominion over the territory in question was held to inhere in the European sovereign in whose name it had been "discovered." Through all the diversity among the theorists of discovery, a constant theme is the clear distinction between dominion, which inhered in European sovereigns alone, and Natives' right of occupancy, also expressed in terms of possession or usufruct, which entitled Natives to pragmatic use (understood as hunting and gathering rather than agriculture)[20] of a territory that Europeans had discovered. The distinction between dominion and occupancy illuminates the settler-colonial project's reliance on the elimination of Native societies.

Through being the first European to visit and properly claim a given territory, a discoverer acquired the right, on behalf of his sovereign and vis-à-vis other Europeans who came after him, to buy land from the Natives. This right, known as preemption, gave the discovering power (or, in the US case, its successors) a monopoly over land transactions with the Natives, who were prevented from disposing of their land to any other European power. On the face of it, this would seem to pose little threat to people who did not wish to dispose of their land to anyone. Indeed, this semblance of Native voluntarism has provided scope for some limited judicial magnanimity in regard to Indian sovereignty.[21] In practice, however, the corollary did not apply. Preemption sanctioned European priority but not Indigenous freedom of choice. As Harvey Rosenthal observed of the concept's extension into the US constitutional environment, "The American right to buy always superseded the Indian right not to sell."[22] The mechanisms of this priority are crucial. Why should ostensibly sovereign nations, residing in territory solemnly guaranteed to them by treaties, decide that they are willing, after all, to surrender their ancestral homelands? More often than not (and nearly always up to the wars with the Plains Indians, which did not take place until after the civil war), the agency which reduced Indian peoples to this abjection was not some state instrumentality but irregular, greed-crazed invaders who had no intention of allowing the formalities of federal law to impede their access to the riches available in, under, and on Indian soil.[23] If the government notionally held itself aloof from such disreputable proceedings, however, it was never far away. Consider, for instance, the complicity between bayonet-wielding troops and the "lawless rabble" in this account of events immediately preceding the eastern Cherokee's catastrophic "Trail of Tears," one of many comparable 1830s removals whereby Indians from the South East were displaced west of the Mississippi to make way for the development of the slave-plantation economy in the Deep South:

> "Families at dinner were startled by the sudden gleam of bayonets in the doorway and rose up to be driven with blows and oaths along the weary miles of trail that led to the stockade [where they were held prior to the removal itself]. Men were seized in their fields or going along the road, women were taken from their wheels and children from their play. In many cases, on turning for one last look as they crossed the ridge, they saw their homes in flames, fired by the lawless rabble that followed on the heels of the soldiers to loot and pillage. So keen were these outlaws on the scent that in some instances they were driving off the cattle and other stock of the Indians almost before the soldiers had fairly started their owners in the other direction. Systematic hunts were made by the same men for Indian graves, to rob them of the silver pendants and other valuables deposited with the dead. A Georgia volunteer, afterward a colonel in the Confederate service, said: 'I fought through the civil war and have seen men shot to pieces and slaughtered by thousands, but the Cherokee removal was the cruelest work I ever knew.'"[24]

On the basis of this passage alone, the structural complexity of settler colonialism could sustain libraries of elaboration. A global dimension to the frenzy for Native land is reflected in the fact that, as economic immigrants, the rabble were generally drawn from the ranks of Europe's landless. The cattle and other stock were not only being driven off Cherokee land; they were being driven into private ownership. Once evacuated, the Red man's land would be mixed with Black labor to produce cotton, the white gold of the Deep South. To this end, the international slave trade and the highest echelons of the formal state apparatus converged across three continents with the disorderly pillaging of a nomadic horde who may or may not have been "lawless" but who were categorically White. Moreover, in their indiscriminate lust for any value that could be extracted from the Cherokee's homeland, these racialized grave-robbers are unlikely to have stopped at the pendants. The burgeoning science of craniology, which provided a distinctively post-eighteenth-century validation for their claim to a racial superiority that entitled them to other people's lands, made Cherokee skulls too marketable a commodity to be overlooked.[25] In its endless multidimensionality, there was nothing singular about this one sorry removal, which all of modernity attended.

Rather than something separate from or running counter to the colonial state, the murderous activities of the frontier rabble constitute its principal means of expansion. These have occurred "behind the screen of the frontier, in the wake of which, once the dust has settled, the irregular acts that took place have been regularized and the boundaries of White settlement extended. Characteristically, officials express regret at the lawlessness of this process while resigning themselves to its inevitability."[26] In this light, we are in a position to understand the pragmatics of the doctrine of discovery more clearly. Understood as an assertion of Indigenous entitlement, the distinction between dominion and occupancy dissolves into incoherence. Understood processually, however, as a stage in the formation of the settler-colonial state (specifically, the stage linking the theory and the realization of territorial acquisition), the distinction is only too consistent. As observed, preemption provided that Natives could transfer their right of occupancy to the discovering sovereign and to no one else. They could not transfer dominion because it was not theirs to transfer; that inhered in the European sovereign and had done so from the moment of discovery. Dominion without conquest constitutes the theoretical (or "inchoate") stage of territorial sovereignty.[27] In US Chief Justice John Marshall's words, it remained to be "consummated by possession."[28] This delicately phrased "consummation" is precisely what the rabble were achieving at Cherokee New Echota in 1838. In other words, the right of occupancy was not an assertion of Native rights. Rather, it was a pragmatic acknowledgment of the lethal interlude that would intervene between the conceit of discovery, when navigators proclaimed European dominion over whole continents to trees or deserted beaches, and the practical realization of that conceit in the final securing of European settlement, formally consummated in the extinguishment of Native title. Thus it is not surprising that Native Title had hardly been asserted in Australian law than Mr Justice Olney was echoing Marshall's formula, Olney's twenty-first-century version of consummation being the "tide of history" that provided the pretext for his notorious judgment in the *Yorta Yorta* case.[29] As observed, the logic of elimination continues into the present.

The tide of history canonizes the fait accompli, harnessing the diplomatic niceties of the law of nations to the maverick rapine of the squatters' posse within a cohesive project that implicates individual and nation-state, official and unofficial alike. Over the Green Line today, *Ammana*, the settler advance-guard of the fundamentalist *Gush Emunim* movement, hastens apace with the construction of its facts on the ground. In this regard, the settlers are maintaining a tried and tested Zionist strategy—Israel's 1949 campaign to seize the Negev before the impending armistice was codenamed *Uvda*, Hebrew for "fact."[30] As Bernard Avishai lamented of the country he had volunteered to defend, "settlements were made in the territories beyond the Green Line so effortlessly after 1967 because the Zionist institutions that built them and the laws that drove them … had all been going full throttle within the Green Line before 1967. To focus merely on West Bank settlers was always to beg the question."[31] In sum, then, settler colonialism is an inclusive, land-centered project that coordinates a comprehensive range of agencies, from the metropolitan center to the frontier encampment, with a view to eliminating Indigenous societies. Its operations are not dependent on the presence or absence of formal state institutions or functionaries. Accordingly—to begin to move toward the issue of genocide—the occasions on or the extent to which settler colonialism conduces to genocide are not a matter of the presence or absence of the formal apparatus of the state.

While it is clearly the case, as Isabel Hull argues, that the pace, scale, and intensity of certain forms of modern genocide require the centralized technological, logistical, and administrative capacities of the modern state,[32] this does not mean that settler-colonial discourse should be regarded as pre- (or less than) modern. Rather, as a range of thinkers—including, in this connection, W.E.B. Dubois, Hannah Arendt, and Aimé Césaire—have argued, some of the core features of modernity were pioneered in the colonies.[33] It is a commonplace that the Holocaust gathered together the instrumental, technological, and bureaucratic constituents of Western modernity. Accordingly, despite the historiographical energy that has already been devoted to the Holocaust, the genealogical field available to its historian remains apparently inexhaustible. Thus we have recently been informed that its historical ingredients included the guillotine and, for the industry-scale processing of human bodies, the techniques of Chicago cattleyards.[34] Yet the image of the dispassionate genocidal technocrat that the Holocaust spawned is by no means the whole story. Rather, as Dieter Pohl, Jürgen Zimmerer, and others have pointed out, a substantial number of the Nazis' victims, including Jewish and Gypsy (Sinti and Rom) ones, were not murdered in camps but in deranged shooting sprees that were more reminiscent of sixteenth-century Spanish behavior in the Americas than of Fordism, while millions of Slav civilians and Soviet soldiers were simply starved to death in circumstances that could well have struck a chord with late-eighteenth-century Bengalis or mid-nineteenth-century Irish people.[35] This is not to suggest a partition of the Holocaust into, say, modern and atavistic elements. It is to stress the modernity of colonialism.

Settler colonialism was foundational to modernity. Frontier individuals' endless appeals for state protection not only presupposed a commonality between the private and official realms. In most cases (Queensland was a partial exception), it also presupposed a global chain of command linking remote colonial frontiers to the metropolis.[36] Behind it all lay the driving engine of international market forces, which linked Australian wool to Yorkshire mills and, complementarily, to cotton produced under different colonial conditions in India, Egypt, and the slave states of the Deep South. As Cole Harris observed in relation to the dispossession of Indians in British Columbia, "Combine capital's interest in uncluttered access to land and settlers' interest in land as livelihood, and the principal momentum of settler colonialism comes into focus."[37] The Industrial Revolution, misleadingly figuring in popular consciousness as an autochthonous metropolitan phenomenon, required colonial land and labor to produce its raw materials just as centrally as it required metropolitan factories and an industrial proletariat to process them, whereupon the colonies were again required as a market. The expropriated Aboriginal, enslaved African American, or indentured Asian is as thoroughly modern as the factory worker, bureaucrat, or *flâneur* of the metropolitan center. The fact that the slave may be in chains does not make him or her medieval. By the same token, the fact that the genocidal Hutus of Rwanda often employed agricultural implements to murder their Tutsi neighbors en masse does not license the racist assumption that, because neither Europeans nor the latest technology were involved, this was a primordial (read "savage") bloodletting. Rwanda and Burundi are colonial creations—not only so far as the obvious factor of their geographical borders is concerned, but, more intimately, in the very racial

The Quebec Gazette Jan. 29, 1778

ADVERTISEMENTS.

RANAWAY from the Printing-Office in Quebec, on Sunday night the twenty-fifth inſtant, a Negro Lad named JOE, born in Africa, about twenty years of age, about five feet and an half high, full round fac'd, a little marked with the ſmall-pox, ſpeaks Engliſh and French tolerably; he had on when he went away a new green fur-cap, a blue ſuit of cloaths, a pair of grey worſted ſtockings and Canadian macaſſins. All perſons are hereby forewarned from harbouring or aiding him to eſcape, as they may depend on being proſecuted to the utmoſt rigour of the Law, and whoever will give information where he is harboured, or bring him back, ſhall have EIGHT DOLLARS Reward from

THE PRINTER.

Ad for Escaped Slave Joe
The Quebec Gazette, 1778

boundaries that marked and reproduced the Hutu/Tutsi division. As Robert Melson has observed in his sharp secondary synopsis of it, "The Rwandan genocide was the product of a postcolonial state, a racialist ideology, a revolution claiming democratic legitimation, and war—all manifestations of the modern world."[38] The mutual Hutu/Tutsi racialization on which this "post"-colonial ideology was based was itself an artifice of colonialism. In classic Foucauldian style, the German and, above all, Belgian overlords who succeeded each other in modern Rwanda had imposed a racial grid on the complex Native social order, co-opting the pastoral Tutsi aristocracy as a comprador elite who facilitated their exploitation of the agriculturalist Hutu and lower-order Tutsis. This racial difference was elaborated "by Belgian administrators and anthropologists who argued—in what came to be known as the 'Hamitic Hypothesis'—that the Tutsi were conquerors who had originated in Ethiopia (closer to Europe!) and that the Hutu were a conquered inferior tribe of local provenance."[39] Shades of the Franks and the Gauls. In their inculcation with racial discourse, Rwandans were integrally modern. Even the notorious hoes with which some Hutus murdered their Tutsi compatriots symbolized the agriculture that not only encapsulated their difference from their victims. As such, these hoes were also the instruments of the Hutus' involvement in the global market.

Of itself, however, modernity cannot explain the insatiable dynamic whereby settler colonialism always needs more land. The answer that springs most readily to mind is agriculture, though it is not necessarily the only one. The whole range of primary sectors can motivate the project. In addition to agriculture, therefore, we should think in terms of forestry, fishing, pastoralism and mining (the last straw for the Cherokee was the discovery of gold on their land). With the exception of agriculture, however (and, for some peoples, pastoralism), none of these is sufficient in itself. You cannot eat lumber or gold; fishing for the world market requires canneries. Moreover, sooner or later, miners move on, while forests and fish become exhausted or need to be farmed. Agriculture not only supports the other sectors. It is inherently sedentary and, therefore, permanent. In contrast to extractive industries, which rely on what just happens to be there, agriculture is a rational means/end calculus that is geared to vouchsafing its own reproduction, generating capital that projects into a future where it repeats itself (hence the farmer's dread of being reduced to eating seed stock). Moreover, as John Locke never tired of pointing out, agriculture supports a larger population than non-sedentary modes of production.[40] In settler-colonial terms, this enables a population to be expanded by continuing immigration at the expense of Native lands and livelihoods. The inequities, contradictions, and pogroms of metropolitan society ensure a recurrent supply of fresh immigrants—especially, as noted, from among the landless. In this way, individual motivations dovetail with the global market's imperative for expansion. Through its ceaseless expansion, agriculture (including, for this purpose, commercial pastoralism) progressively eats into Indigenous territory, a primitive accumulation that turns native flora and fauna into a dwindling resource and curtails the reproduction of Indigenous modes of production. In the event, Indigenous people are either rendered dependent on the introduced economy or reduced to the stock-raids that provide the classic pretext for colonial death-squads.

None of this means that Indigenous people are by definition non-agricultural. Whether or not they actually do practice agriculture, however (as in the case of the Indians who taught Whites to grow corn and tobacco), Natives are typically represented as unsettled, nomadic, rootless, etc., in settler-colonial discourse. In addition to its objective economic centrality to the project, agriculture, with its life-sustaining connectedness to land, is a potent symbol of settler-colonial identity. Accordingly, settler-colonial discourse is resolutely impervious to glaring inconsistencies such as sedentary Natives or the fact that the settlers themselves have come from somewhere else. Thus it is significant that the feminized, finance-oriented (or, for that matter, wandering) Jew of European anti-Semitism should assert an aggressively masculine agricultural self-identification in Palestine.[41] The new Jew's formative Other was the nomadic Bedouin rather than the *fellaheen* farmer. The reproach of nomadism renders the Native removable. Moreover, if the Natives are not already nomadic, then the reproach can be turned into a self-fulfilling prophecy through the burning of corn or the uprooting of fruit trees.

But if the Natives are already agriculturalists, then why not simply incorporate their productivity into the colonial economy? At this point, we begin to get closer to the question of just who it is (or, more to the point, who they are) that settler colonialism strives to eliminate—and, accordingly, closer to an understanding of the relationship between settler colonialism and genocide. To stay with the Cherokee removal: when it came to it, the factor that most antagonized the Georgia state government (with the at-least-tacit support of Andrew Jackson's federal administration) was not actually the recalcitrant savagery of which Indians were routinely accused, but the Cherokee's unmistakable aptitude for civilization. Indeed, they and their Creek, Choctaw, Chickasaw, and Seminole neighbors, who were also targeted for removal, figured revealingly as the "Five Civilized Tribes" in Euroamerican parlance. In the Cherokee's case, two dimensions of their civility were particularly salient. They had become successful agriculturalists on the White model, with a number of them owning substantial holdings of Black slaves, and they had introduced a written national constitution that bore more than a passing resemblance to the US one.[42] Why should genteel Georgians wish to rid themselves of such cultivated neighbors? The reason why the Cherokee's constitution and their agricultural prowess stood out as such singular provocations to the officials and legislators of the state of Georgia—and this is attested over and over again in their public statements and correspondence—is that the Cherokee's farms, plantations, slaves, and written constitution all signified *permanence*.[43] The first thing the rabble did, let us remember, was burn their houses.

Brutal and murderous though the removals of the Five Civilized Tribes generally were, they did not affect each member equally. This was not simply a matter of wealth

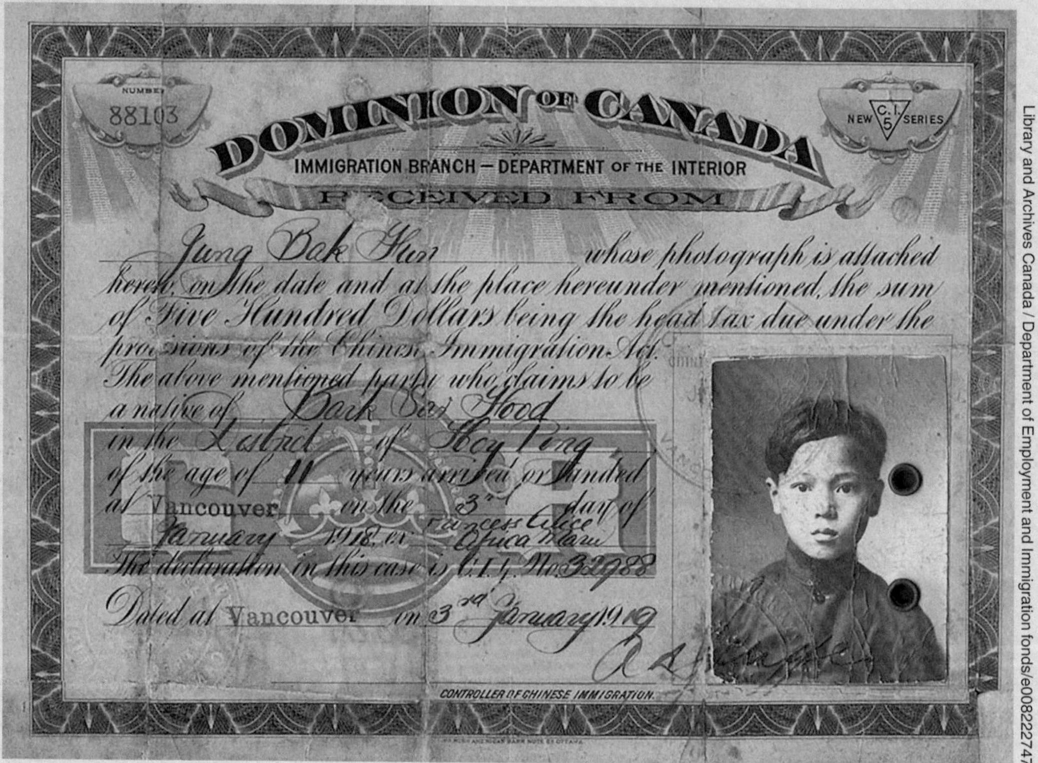

NUMBER 88103

NEW C.I. 5 SERIES

DOMINION OF CANADA

IMMIGRATION BRANCH — DEPARTMENT OF THE INTERIOR

RECEIVED FROM

Jung Bak Hun whose photograph is attached hereto, on the date and at the place hereunder mentioned, the sum of Five Hundred Dollars being the head tax due under the provisions of the Chinese Immigration Act.
The above mentioned party who claims to be a native of Bark Sai Hood in the District of Hoy Ping of the age of 11 years arrived or landed at Vancouver on the 3rd day of January 1918, ex Princess Alice Africa Maru
The declaration in this case is C.I. 4. No 32988

Dated at Vancouver on 3rd January 1919

CONTROLLER OF CHINESE IMMIGRATION.

Library and Archives Canada / Department of Employment and Immigration fonds/e008222747

Chinese Head Tax Certificate for Jung Bak Hun
Department of the Interior, 1919

or status. Principal Cherokee chief John Ross, for example, lost not only his plantation after setting off on the Trail of Tears. On that trail, one deathly cold Little Rock, Arkansas day in February 1839, he also lost his wife, Qatie, who died after giving her blanket to a freezing child.[44] Ross's fortunes differed sharply from those of the principal Choctaw chief Greenwood LeFlore, who, unlike Ross, signed a removal treaty on behalf of his people, only to stay behind himself, accept US citizenship, and go on to a distinguished career in Mississippi politics.[45] But it was not just his chiefly rank that enabled LeFlore to stay behind. Indeed, he was by no means the only one to do so. As Ronald Satz has commented, Andrew Jackson was taken by surprise when "thousands of Choctaws decided to take advantage of the allotment provisions [in the treaty LeFlore had signed] and become homesteaders and American citizens in Mississippi."[46] In addition to being principal chiefs, Ross and LeFlore both had White fathers and light skin. Both were wealthy, educated, and well connected in Euroamerican society. Many of the thousands of compatriots who stayed behind with LeFlore lacked any of these qualifications. There was nothing special about the Choctaw to make them particularly congenial to White society—most of them got removed like Ross and the Cherokee. The reason that the remaining Choctaw were acceptable had nothing to do with their being Choctaw. On the contrary, it had to do with their *not* (or, at least, no longer) being Choctaw. They had become "homesteaders and American citizens." In a word, they had become individuals.

What distinguished Ross and the removing Choctaw from those who stayed behind was collectivity.[47] Tribal land was tribally owned—tribes and private property did not mix. Indians were the original communist menace. As homesteaders, by contrast, the Choctaw who stayed became individual proprietors, each to his own, of separately allotted fragments of what had previously been the tribal estate, theirs to sell to White people if they chose to. Without the tribe, though, for all practical purposes they were no longer Indians (this is the citizenship part). Here, in essence, is assimilation's Faustian bargain—have our settler world, but lose your Indigenous soul. Beyond any doubt, this is a kind of death. Assimilationists recognized this very clearly. On the face of it, one might not expect there to be much in common between Captain Richard Pratt, founder of the Carlisle boarding school for Indian youth and leading light of the philanthropic "Friends of the Indian" group, and General Phil Sheridan, scourge of the Plains and author of the deathless maxim, "The only good Indian is a dead Indian." Given the training in individualism that Pratt provided at his school, however, the tribe could disappear while its members stayed behind, a metaphysical variant on the Choctaw scenario. This would offer a solution to reformers' disquiet over the national discredit attaching to the Vanishing Indian. In a paper for the 1892 Charities and Correction Conference held in Denver, Pratt explicitly endorsed Sheridan's maxim, "but only in this: that all the Indian there is in the race should be dead. Kill the Indian in him and save the man."[48]

But just what kind of death is it that is involved in assimilation? The term "homicide," for instance, combines the senses of killing and of humanity. So far as I know, when it comes to killing a human individual, there is no alternative to terminating their somatic career. Yet, when Orestes was arraigned before the Furies for the murder of his mother Clytemnestra, whom he had killed to avenge her murder of his father Agamemnon, he was acquitted on the ground that, in a patrilineal society, he belonged to his father rather than to his mother, so the charge of matricide could not stand. Now, without taking this legend too seriously, it nonetheless illustrates (as legends are presumably meant to) an important point. Orestes' beating the charge did not mean that he had not actually killed Clytemnestra. It meant that he had been brought before the wrong court (the Furies dealt with intra-family matters that could not be resolved by the mechanism of feud). Thus Orestes may not have been guilty of matricide, but that did not mean he was innocent. It meant that he might be guilty of some other form of illegal killing—one that could be dealt with by the blood-feud or other appropriate sanction (where his plea of obligatory revenge may or may not have succeeded). As in those languages where a verb is inflected by its object, the nature of a justiciable killing depends on its victim. There are seemingly absolute differences between, say, suicide, insecticide, and infanticide. The etymology of "genocide" combines the senses of killing and of grouphood. "Group" is more than a purely numerical designation. Genos refers to a denominate group with a membership that persists through time (Raphaël Lemkin translated it as "tribe"). It is not simply a random collectivity (such as, say, the passengers on a bus). Accordingly, with respect to Robert Gellately and Ben Kiernan (concerning both the subtitle of their excellent collection and their reference, in this context, to 9/11), the strike on the World Trade Center is an example of mass murder but not, in my view, of genocide. Certainly, the bulk of the victims were US citizens. On the scale of the whole, however, not only was it an infinitesimal part of the group "Americans" (which, strictly, is not a consideration), but it was a one-off event.[49] This does not mean that the perpetrators of 9/11 are not guilty. It means that a genocide tribunal is the wrong court to bring them before. Mass murders are not the same thing as genocide, though the one action can be both. Thus genocide has been achieved by means of summary mass murder (to cite examples already used) in the frontier massacring of Indigenous peoples, in the Holocaust, and in Rwanda. But there can be summary mass murder without genocide, as in the case of 9/11, and there can be genocide without summary mass murder, as in the case of the continuing post-frontier destruction, in whole and in part, of Indigenous *genoi*. Lemkin knew what he was doing when he used the word "tribe."[50] Richard Pratt and Phillip Sheridan were both practitioners of genocide. The question of degree is not the definitional issue.

Vital though it is, definitional discussion can seem insensitively abstract. In the preceding paragraph, part of what I have had in mind has, obviously, been the term (which Lemkin favored) "cultural genocide." My reason for not favoring the term is that it confuses definition with degree. Moreover, though this objection holds in its own right (or so I think), the practical hazards that can ensue once an abstract concept like "cultural genocide" falls into the wrong hands are legion. In particular, in an elementary category error, "either/

or" can be substituted for "both/and," from which genocide emerges as either biological (read "the real thing") or cultural—and thus, it follows, not real. In practice, it should go without saying that the imposition on a people of the procedures and techniques that are generally glossed as "cultural genocide" is certainly going to have a direct impact on that people's capacity to stay alive (even apart from their qualitative immiseration while they do so). At the height of the Dawes-era assimilation program, for instance, in the decade after Richard Pratt penned his Denver paper, Indian numbers hit the lowest level they would ever register.[51] Even in contemporary, post-Native Title Australia, Aboriginal life expectancy clings to a level some 25% below that enjoyed by mainstream society, with infant mortality rates that are even worse.[52] What species of sophistry does it take to separate a quarter "part" of the life of a group from the history of their elimination?

Clearly, we are not talking about an isolated event here. Thus we can shift from settler colonialism's structural complexity to its positivity as a structuring principle of settler-colonial society across time.

The Cherokee Trail of Tears, which took place over the winter of 1838–1839, presupposed the Louisiana Purchase of 1803, when Thomas Jefferson had bought approximately one-third of the present-day continental United States at a knockdown price from Napoleon.[53] The greatest real estate deal in history provided the territory west of the Mississippi that successive US governments would exchange for the homelands of the eastern tribes whom they were bent on removing. For various reasons, these removals, which turned eastern tribes into proxy invaders of Indian territory across the Mississippi, were a crude and unsatisfactory form of elimination. In particular, they were temporary, it being only a matter of time before the frontier rabble caught up with them.[54] When that happened, as Annie Abel resignedly observed in concluding her classic account of the removals, "Titles given in the West proved less substantial than those in the East, for they had no foundation in antiquity."[55] Repeat removals, excisions from reservations, grants of the same land to different tribes, all conducted against a background of endless pressure for new or revised treaties, were the symptoms of removal's temporariness, which kept time with the westward march of the nation. In the end, though, the western frontier met the one moving back in from the Pacific, and there was simply no space left for removal. The frontier had become coterminal with reservation boundaries. At this point, when the crude technique of removal declined in favor of a range of strategies for assimilating Indian people now that they had been contained within Euroamerican society, we can more clearly see the logic of elimination's positivity as a continuing feature of Euroamerican settler society.

With the demise of the frontier, elimination turned inwards, seeking to penetrate through the tribal surface to the individual Indian below, who was to be co-opted out of the tribe, which would be depleted accordingly, and into White society. The Greenwood

LeFlore situation was to be generalized to all Indians. The first major expression of this shift was the discontinuation of treaty-making, which came about in 1871.[56] Over the following three decades, an avalanche of assimilationist legislation, accompanied by draconian Supreme Court judgments which notionally dismantled tribal sovereignty and provided for the abrogation of existing treaties,[57] relentlessly sought the breakdown of the tribe and the absorption into White society of individual Indians and their tribal land, only separately. John Wunder has termed this policy framework "the New Colonialism," a discursive formation based on reservations and boarding schools that "attacked every aspect of Native American life—religion, speech, political freedoms, economic liberty, and cultural diversity."[58] The centerpiece of this campaign was the allotment program, first generalized as Indian policy in the Dawes Severalty Act of 1887 and subsequently intensified and extended, whereby tribal land was to be broken down into individual allotments whose proprietors could eventually sell them to White people.[59] Ostensibly, this program provided for a cultural transformation whereby the magic of private property ownership would propel Indians from the collective inertia of tribal membership into the progressive individualism of the American dream. In practice, not only did Indian numbers rapidly hit the lowest level they would ever record, but this cultural procedure turned out to yield a faster method of land transference than the US Cavalry had previously provided. In the half-century from 1881, the total acreage held by Indians in the United States fell by two thirds, from just over 155 million acres to just over 52 million.[60] Needless to say, the coincidence between the demographic statistics and the land-ownership ones was no coincidence. Throughout this process, reformers' justifications for it (saving the Indian from the tribe, giving him the same opportunities as the White man, etc.) repeatedly included the express intention to destroy the tribe in whole.[61] With their land base thus attenuated, US citizenship was extended to all Indians in 1924. In 1934, under the New Deal *Indian Reform Act*, allotment was abandoned in favor of a policy of admitting the tribe itself into the US polity, only on condition that its constitution be rewritten into structural harmony with its US civic environment. A distinctive feature of the model constitutions that the Secretary of the Interior approved for tribes that registered under the 1934 Act was blood quantum requirements, originally introduced by Dawes Act commissioners to determine which tribal members would be eligible for what kind of allotments.[62] Under the blood quantum regime, one's Indianness progressively declines in accordance with a "biological" calculus that is a construct of Euroamerican culture.[63] Juaneño/Jaqi scholar Annette Jaimes has termed this procedure "statistical extermination."[64] In sum, the containment of Indian groups within Euroamerican society that culminated in the end of the frontier produced a range of ongoing complementary strategies whose common intention was the destruction of heterodox forms of Indian grouphood. In the post-World War II climate of civil rights, these strategies were reinforced by the policies of termination and relocation, held out as liberating individual Indians from the thralldom of the tribe, whose compound effects rivalled the disasters of allotment.[65] A major difference between this and the generality of non-colonial genocides is its sustained duration.

Courtesy of Joanne Woo / Multicultural History Society of Ontario

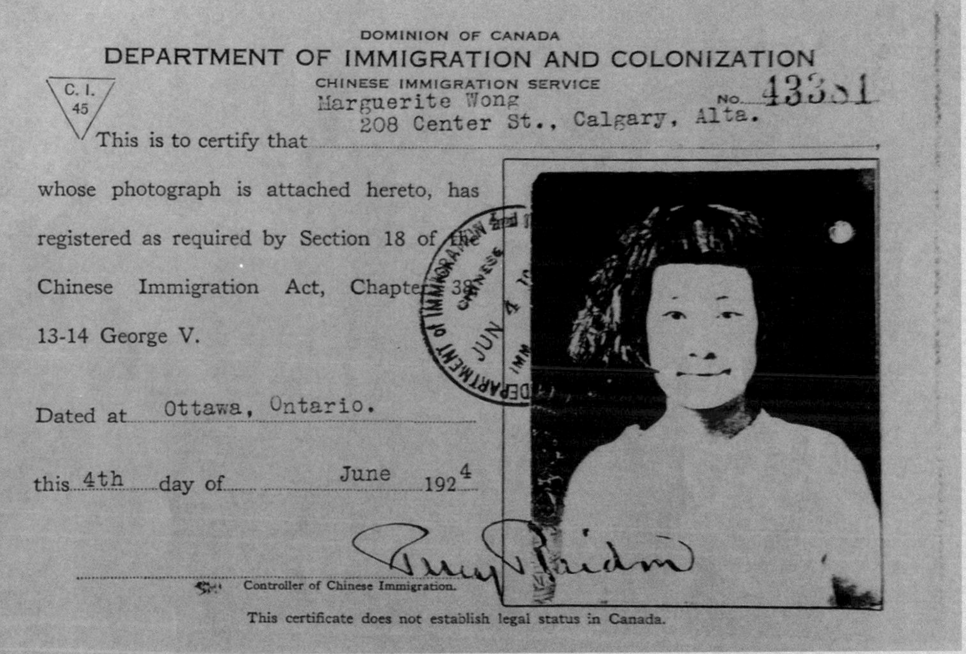

DOMINION OF CANADA

DEPARTMENT OF IMMIGRATION AND COLONIZATION

CHINESE IMMIGRATION SERVICE

C. I. 45

No. 433[illegible]1

This is to certify that Marguerite Wong 208 Center St., Calgary, Alta. whose photograph is attached hereto, has registered as required by Section 18 of the Chinese Immigration Act, Chapter 38 13-14 George V.

Dated at Ottawa, Ontario.

this 4th day of June 1924

Controller of Chinese Immigration.

This certificate does not establish legal status in Canada.

Certificate of Registration for Marguerite Wong
Department of Immigration & Colonization, June 4, 1924

For comparative purposes, it is significant that the full radicalization of assimilation policies in both the US and Australia coincided with the closure of the frontier, which forestalled spatial stop-gaps such as removal. In infra-continental societies like those of mainland Europe, the frontier designates a national boundary as opposed to a mobile index of expansion. Israel's borders partake of both qualities. Despite Zionism's chronic addiction to territorial expansion, Israel's borders do not preclude the option of removal (in this connection, it is hardly surprising that a nation that has driven so many of its original inhabitants into the sand should express an abiding fear of itself being driven into the sea). As the logic of elimination has taken on a variety of forms in other settler-colonial situations, so, in Israel, the continuing tendency to Palestinian expulsion has not been limited to the unelaborated exercise of force. As Baruch Kimmerling and Joel Migdal have observed, for instance, Israeli officials have only permitted family unions "in one direction—out of Israel."[66] The Law of Return commits the Jewish state to numerically unlimited but ethnically exclusive immigration, a factor that, formalities of citizenship notwithstanding, militates against the assimilation of gentile Natives. Thus assimilation should not be seen as an invariable concomitant of settler colonialism. Rather, assimilation is one of a range of strategies of elimination that become favored in particular historical circumstances. Moreover, assimilation itself can take on a variety of forms. In the Australian context, for instance, various scholars have recognized that "the genetic and cultural codes recapitulated each other."[67] Though "softer" than the recourse to simple violence, however, these strategies are not necessarily less eliminatory. To take an example from genocide's definitional core, Article II (d) of the UN Convention on Genocide, which seems to have been relatively overlooked in Australian discussions, includes among the acts that constitute genocide (assuming they are committed with intent to destroy a target group in whole or in part) the imposition of "measures intended to prevent births within the group." Given that the Australian practice of abducting Aboriginal children, assuming its "success," would bring about a situation in which second-generation offspring were born into a group that was different from the one from which the child/parent had originally been abducted, there is abundant evidence of genocide being practiced in post-war Australia on the basis of Article II (d) alone. It is impossible to draw simple either/or lines between culture and biology in cases such as these. Though a child was physically abducted, the eventual outcome is as much a matter of a social classification as it is of a body count. Nonetheless, the intentional contribution to the demographic destruction of the "relinquishing" group is unequivocal.

Why, then, logic of elimination rather than genocide? As stated at the outset, settler colonialism is a specific social formation and it is desirable to retain that specificity. So far as I can tell, an understanding of settler colonialism would not be particularly helpful for understanding the mass killings of, say, witches in medieval Europe, Tutsis in Rwanda, enemies of the people in Cambodia, or Jews in the Nazi fatherland (the *Lebensraum* is, of course, another matter). By the same token, with the possible exception of the witches (whose murders appear to have been built into a great social transition), these mass killings would seem to have little to tell us about the long-run structural consistency of settler colonizers' attempts to eliminate Native societies. In contrast to the Holocaust, which was endemic to Nazism rather than to Germany (which was by no means the only—or even, historically, the most—anti-Semitic society in Europe), settler colonialism is relatively impervious to regime change. The genocide of American Indians or of Aboriginal people in Australia has not been subject to election results. So why not a special kind of genocide?—Raymond Evans' and Bill Thorpe's etymologically deft "indigenocide," for instance,[68] or one of the hyphenated genocides ("cultural genocide," "ethnocide," "politicide," etc.)[69] that have variously been proposed? The apparently insurmountable problem with the qualified genocides is that, in their very defensiveness, they threaten to undo themselves. They are never quite the real thing, just as patronizingly hyphenated ethnics are not fully Australian or fully American. Apart from this categorical problem, there is a historical basis to the relative diminution of the qualified genocides. This basis is, of course, the Holocaust, the non-paradigmatic paradigm that, being the indispensable example, can never merely exemplify. Keeping one eye on the Holocaust, which is always the unqualified referent of the qualified genocides, can only disadvantage Indigenous people because it discursively reinforces the figure of lack at the heart of the non-Western. Moreover, whereas the Holocaust exonerates anti-Semitic Western nations who were on the side opposing the Nazis, those same nations have nothing to gain from their liability for colonial genocides. On historical as well as categorical grounds, therefore, the hyphenated genocides devalue Indigenous attrition. No such problem bedevils analysis of the logic of elimination, which, in its specificity to settler colonialism, is premised on the securing—the obtaining and the maintaining—of territory.[70] This logic certainly requires the elimination of the owners of that territory, but not in any particular way. To this extent, it is a larger category than genocide. For instance, the style of romantic stereotyping that I have termed "repressive authenticity," which is a feature of settler-colonial discourse in many countries, is not genocidal in itself, though it eliminates large numbers of empirical Natives from official reckonings and, as such, is often concomitant with genocidal practice.[71] Indeed, depending on the historical conjuncture, assimilation can be a more effective mode of elimination than conventional forms of killing, since it does not involve such a disruptive affront to the rule of law that is ideologically central to the cohesion of settler society. When invasion is recognized as a structure rather than an event, its history does not stop—or, more to the point, become relatively trivial—when it moves on from the era of frontier homicide. Rather, narrating that history involves charting the continuities, discontinuities, adjustments, and departures whereby a logic that initially informed frontier killing transmutes into different modalities, discourses, and institutional formations as it undergirds the historical development and complexification of settler society. This is not a hierarchical procedure.

How, then, when elimination manifests as genocide, are we to retain the specificity of settler colonialism without downplaying its impact by resorting to a qualified genocide? I

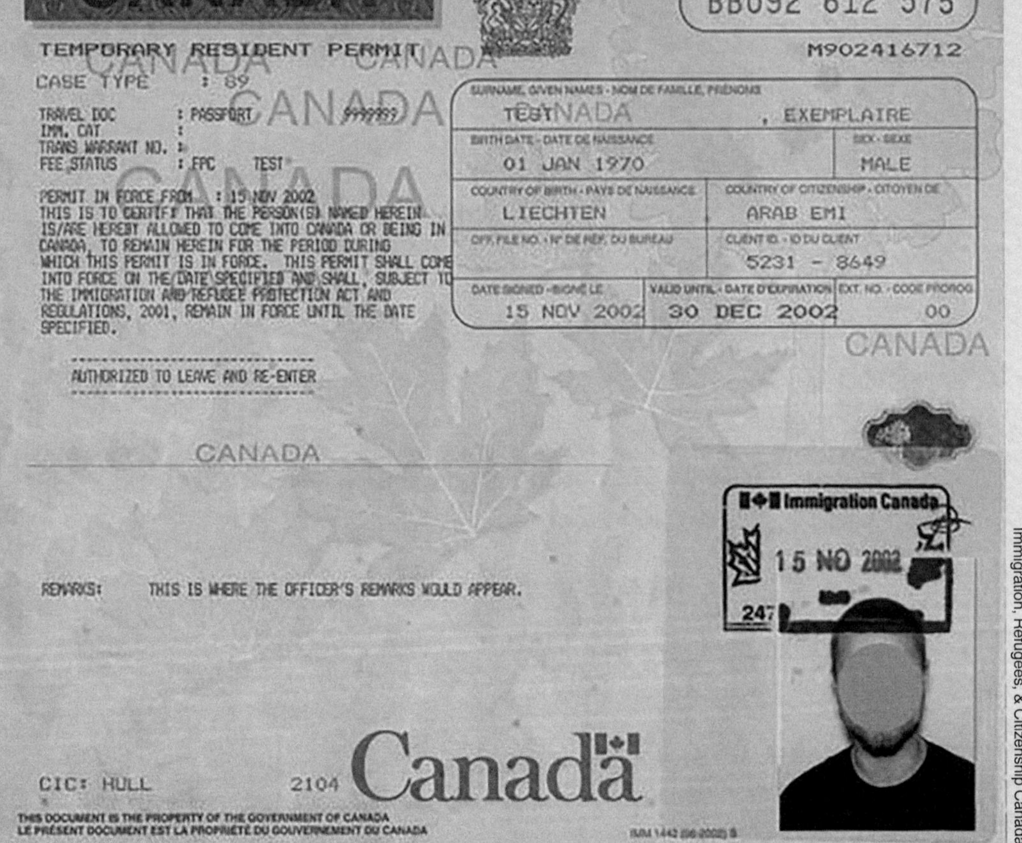

Citizenship and Immigration Canada | Citoyenneté et Immigration Canada

PROTECTED WHEN COMPLETED PROTÉGÉ UNE FOIS REMPLI - B

CANADA

IMMIGRATION

BB092 612 575

M902416712

TEMPORARY RESIDENT PERMIT

CASE TYPE : 89

TRAVEL DOC : PASSPORT 9999999
IMM. CAT :
TRANS WARRANT NO. :
FEE STATUS : EPC TEST

PERMIT IN FORCE FROM : 15 NOV 2002
THIS IS TO CERTIFY THAT THE PERSON(S) NAMED HEREIN IS/ARE HEREBY ALLOWED TO COME INTO CANADA OR BEING IN CANADA, TO REMAIN HEREIN FOR THE PERIOD DURING WHICH THIS PERMIT IS IN FORCE. THIS PERMIT SHALL COME INTO FORCE ON THE DATE SPECIFIED AND SHALL, SUBJECT TO THE IMMIGRATION AND REFUGEE PROTECTION ACT AND REGULATIONS, 2001, REMAIN IN FORCE UNTIL THE DATE SPECIFIED.

SURNAME, GIVEN NAMES - NOM DE FAMILLE, PRÉNOMS		
TEST , EXEMPLAIRE		
BIRTH DATE - DATE DE NAISSANCE		SEX - SEXE
01 JAN 1970		MALE
COUNTRY OF BIRTH - PAYS DE NAISSANCE	COUNTRY OF CITIZENSHIP - CITOYEN DE	
LIECHTEN	ARAB EMI	
OFF. FILE NO. - N° DE RÉF. DU BUREAU	CLIENT ID. - ID DU CLIENT	
	5231 - 8649	
DATE SIGNED - SIGNÉ LE	VALID UNTIL - DATE D'EXPIRATION	EXT. NO. - CODE PROROG.
15 NOV 2002	30 DEC 2002	00

AUTHORIZED TO LEAVE AND RE-ENTER

REMARKS: THIS IS WHERE THE OFFICER'S REMARKS WOULD APPEAR.

Immigration Canada
15 NO 2002
247

CIC: HULL 2104

Canada

THIS DOCUMENT IS THE PROPERTY OF THE GOVERNMENT OF CANADA
LE PRÉSENT DOCUMENT EST LA PROPRIÉTÉ DU GOUVERNEMENT DU CANADA

IMM 1442 (06-2002) B

Temporary Resident Permit
Citizenship & Immigration Canada, 2002

suggest that the term "structural genocide" avoids the questions of degree—and, therefore, of hierarchy among victims—that are entailed in qualified genocides, while retaining settler colonialism's structural induration (it also lets in the witches—whose destruction, as Charles Zika has shown, was closely linked to the coeval transatlantic destruction of Native Americans).[72] Given a historical perspective on structural genocide, we can recognize its being in abeyance (as, mercifully, it seems to be in contemporary Australia) rather than being a thing of the past—which is to say, we should guard against the recurrence of what Dirk Moses terms "genocidal moments" (social workers continue to take Aboriginal children in disproportionate numbers, for example).[73] Focusing on structural genocide also enables us to appreciate some of the concrete empirical relationships between spatial removal, mass killings, and biocultural assimilation. For instance, where there is no space left for removal (as occurred on the closure of the frontier in the US and Australia, or on the Soviet victory on Nazi Germany's eastern front), mass killings or assimilation become the only eliminatory options available. Under these circumstances, the resort to mass killings can reflect the proclaimed inassimilability of the victim group, as in the case of Jews in relation to the "Aryan" blood stock.[74] Correspondingly, assimilation programs can reflect the ideological requirements of settler-colonial societies, which characteristically cite Native advancement to establish their egalitarian credentials to potentially fractious groups of immigrants.[75]

How, then, might any of this help to predict and prevent genocide?

In the first place, it shows us that settler colonialism is an indicator. Unpalatable though it is (to speak as a member of a settler society), this conclusion has a positive aspect, which is a corollary to settler colonialism's temporal dimension. Since settler colonialism persists over extended periods of time, structural genocide should be easier to interrupt than short-term genocides. For instance, it seems reasonable to credit the belated UN/Australian intervention in East Timor with warding off the likelihood of a continued or renewed genocidal program. *Realpolitik* is a factor, however. Thus the Timorese miracle would not seem to hold out a great deal of hope for, say, Tibet.

Since settler colonialism is an indicator, it follows that we should monitor situations in which settler colonialism intensifies or in which societies that are not yet, or not fully, settler-colonial take on more of its characteristics. Israel's progressive dispensing with its reliance on Palestinian labor would seem to present an ominous case in point.[76] Colin Tatz has argued, conclusively in my view, that, while Turkish behavior in Armenia, Nazi behavior in Europe, and Australian behavior towards Aborigines (among other examples) constitute genocide, the apartheid regime in South Africa does not. His basic reason is that African labor was indispensable to apartheid South Africa, so it would have been counterproductive to destroy it. The same can be said of African American slavery. In both cases, the genocide tribunal is the wrong court.

The US parallel is significant because, unlike the South African case, the formal apparatus of oppression (slavery) was overcome but Whites remained in power.[77] On emancipation, Blacks became surplus to some requirements and, to that extent, more like Indians. Thus it is highly significant that the barbarities of lynching and the Jim Crow reign of terror should be a post-emancipation phenomenon.[78] As valuable commodities, slaves had only been destroyed *in extremis*. Even after slavery, Black people continued to have value as a source of super-cheap labor (providing an incitement to poor Whites), so their dispensability was tempered.[79] Today in the US, the blatant racial zoning of large cities and the penal system suggests that, once colonized people outlive their utility, settler societies can fall back on the repertoire of strategies (in this case, spatial sequestration) whereby they have also dealt with the Native surplus. There could hardly be a more concrete expression of spatial sequestration than the West Bank barrier. There again, apartheid also relied on sequestration. Perhaps Colin Tatz, who insists that Israel is not genocidal,[80] finds it politic to allow an association between the Zionist and apartheid regimes as the price of preempting the charge of genocide. It is hard to imagine that a scholar of his perspicacity can have failed to recognize the Palestinian resonances of his statement, made in relation to Biko youth, that: "They threw rocks and died for their efforts."[81] Nonetheless, as Palestinians become more and more dispensable, Gaza and the West Bank become less and less like Bantustans and more and more like reservations (or, for that matter, like the Warsaw Ghetto). Porous borders do not offer a way out.

Indian & Northern Affairs Canada

Indian and Northern Affairs Canada
Affaires indiennes et du Nord Canada

CERTIFICATE OF INDIAN STATUS
CERTIFICAT DE STATUT INDIEN

Registration no./Numéro d'inscription
9997001801

Family Name/Nom de famille
JONES

Given Names/Prénoms
MARY JANE

Alias/Nom d'emprunt
ANNE

Date of Birth/Date de naissance
1980/01/13

Sex/Sexe
F

Date of Issue/Date de délivrance
2010/02/03

Renew Before/Renouveler avant
2015/01/13

Registry Group no. and Name/Nº du groupe de registre et nom
999 - BAND NAME

Certificate of Indian Status
Indian & Northern Affairs Canada, 2010

Notes

This essay was originally published as "Settler Colonialism and the Elimination of the Native," *Journal of Genocide Research* 8, no. 4 (December 2006), reproduced with permission from Taylor & Francis.

1 Frantz Fanon, *The Wretched of the Earth*, trans. Constance Farrington (New York, NY: Grove Press, [1961] 1968), 310.

2 Patrick Wolfe, "Nation and MiscegeNation: Discursive Continuity in the Post-Mabo Era," *Social Analysis* 36 (1994): 93–152; and Patrick Wolfe, *Settler Colonialism and the Transformation of Anthropology* (London, UK: Cassell, 1999).

3 See for example Collette Guillaumin, "The Idea of Race and its Elevation to Autonomous Scientific and Legal Status," in *Racism, Sexism, Power and Ideology* (London, UK: Routledge, 1995), 61–98; Ivan Hannaford, *Race: The History of an Idea in the West* (Baltimore, MD: Johns Hopkins University Press, 1996); and Kenan Malik, *The Meaning of Race: Race, History and Culture in Western Society* (London, UK: Macmillan, 1996). For discussion, see Patrick Wolfe, "Race and Racialization: Some Thoughts," *Postcolonial Studies* 5, no. 1 (2002): 51–62.

4 Robert Manne has missed this point. Responding to a question posed in 1937 by Western Australian Aboriginal affairs functionary A.O. Neville ("Are we going to have a population of 1,000,000 blacks in the [Australian] Commonwealth, or are we going to merge them into our white community and eventually forget that there ever were any aborigines in Australia?"), Manne suggests that, in order to "grasp the genocidal implications" of the question, "we need only replace the words 'blacks' and 'Aborigine' [sic] with the word 'Jew'" and locate the posing of the question in Berlin rather than Canberra. Robert Manne, "Aboriginal Child Removal and the Question of Genocide," in *Genocide and Settler Society*, ed. A. Dirk Moses (New York, NY: Berghahn, 2005), 219–220. Apart from its contrivance, this analogy fails because the Nazi racialization of Jews did not conduce to their assimilation. Rather, the reverse was the case. As Robert Gellately has observed, "Although we can point to some similarities in Nazi plans and actions for Jews and Slavs, there was, and remains one crucial difference: in principle Jews could never be saved, never convert, nor be assimilated." Robert Gellately, "The Third Reich, the Holocaust, and Visions of Serial Genocide," in *The Specter of Genocide: Mass Murder in Historical Perspective*, ed. Robert Gellately and Ben Kiernan (Cambridge, MA: Cambridge University Press, 2003), 262.

5 Roger W. Smith, "Human Destructiveness and Politics: The Twentieth Century as an Age of Genocide," in *Genocide and the Modern Age: Etiology and Case Studies of Mass Death*, ed. Isidor Wallimann and Michael N. Dobkowski (New York, NY: Greenwood Press. 1987), 31.

6 Deborah Bird Rose, *Hidden Histories: Black Stories from Victoria River Downs, Humbert River and Wave Hill Stations* (Canberra: Aboriginal Studies Press, 1991), 46.

7 "One, destruction of the national pattern of the oppressed group; the other, the imposition of the national pattern of the oppressor. This imposition, in turn, may be made upon the oppressed population which is allowed to remain, or upon the territory alone, after removal of the population and colonization of the area by the oppressor's own nationals." Raphaël Lemkin, *Axis Rule in Occupied Europe: Laws of Occupation, Analysis of Government, Proposals for Redress* (New York, NY: Carnegie Endowment for International Peace 1944), 79.

8 Wolfe, *Settler Colonialism and the Transformation of Anthropology*; and Wolfe, "Nation and MiscegeNation," 96.

9 Theodor Herzl, *Old–New Land* [*Altneuland*], trans. Lotta Levensohn (New York, NY: M. Wiener [1902] 1941), 38.

10 Meron Benvenisti, *Sacred Landscape: The Buried History of the Holy Land since 1948* (Berkeley, CA: California University Press, 2000), 2.

11 Walid Khalidi and his team memorialized the obsessively erased Arab past in their undespairing *All That Remains: The Palestinian Villages Occupied and Depopulated by Israel in 1948* (Washington, DC: Institute for Palestine Studies, 1992).

12 Tony Birch, "'Nothing has Changed': The 'Making and Unmaking' of Koori Culture," in *Blacklines: Contemporary Critical Writing by Indigenous Australians*, ed. Michèle Grossman (Melbourne: Melbourne University Press, 2003), 145–58.

13 Mark LeVine, *Overthrowing Geography: Jaffa, Tel Aviv, and the Struggle for Palestine, 1880–1949* (Berkeley, CA: California University Press, 2005), 227.

14 Gershon Shafir, *Land, Labor, and the Origins of the Israeli–Palestinian Conflict, 1882–1914* (Cambridge, MA: Cambridge University Press, 1989), 81.

15 Ibid., 81–82.

16 Quoted in LeVine, *Overthrowing Geography*, 167.

17 For varying analyses and discussions of the principal formulations of the doctrine of discovery, see for example Anthony Anghie, "Francisco de Vitoria and the Colonial Origins of International Law," in *Laws of the Postcolonial*, ed. Eve Darian-Smith and Peter Fitzpatrick (Ann Arbor, MI: Michigan University Press, 1999), 89–107; Andrew Fitzmaurice, *Humanism and America: An Intellectual History of English Colonisation, 1500–1625* (Cambridge, UK: Cambridge University Press, 2003); David Kennedy, "Primitive Legal Scholarship," *Harvard International Law Journal* 27 (1986): 1–98; Mark F. Lindley, *The Acquisition and Government of Backward Territory in International Law* (London, UK: Longmans, Green, 1926); and Robert A. Williams Jr., *The American Indian in Western Legal Thought: The Discourses of Conquest* (Oxford, UK: Oxford University Press, 1990), especially 233–286.

18 This observation unites almost all commentators, whatever their political inclination. Compare for example Anthony Anghie, "Finding the Peripheries: Sovereignty and Colonialism in Nineteenth-Century International Law," *Harvard International Law Journal* 40 (1999): 69; L.C. Green, "Claims to Territory in Colonial America," *The Law of Nations and the New World*, by L.C. Green and Olive P. Dickason (Edmonton, AB: Alberta University Press, 1989), 125.

19 See for example Wilcomb E. Washburn, ed., *History of Indian–White Relations*, vol. 4 of *Handbook of North American Indians*, ed. William C. Sturtevant (Washington D.C., Smithsonian Institution, [1978] 1988), 5–39.

20 As Mr. Justice Johnson put it in his concurrence with Chief Justice Marshall's judgment in *Cherokee v. Georgia*, "The hunter state bore within itself the promise of vacating the territory, because when game ceased, the hunter would go elsewhere to seek it. But a more fixed state of society would amount to a permanent destruction of the hope, and, of consequence, of the beneficial character of the pre-emptive right." *Cherokee v. Georgia*, 30 US (5 Peters) 1, 1831, p. 23.

21 The judgments most often cited in this connection are *Worcester v. Georgia*, 31 US (6 Peters) 515, 1832, *Crow Dog*, 109 US 556, 1883, and *Williams v. Lee*, 358 US 217, 1959. See my article on the limitations of these judgments and of the limitations of US-style Indian sovereignty as a whole, titled "Against the Intentional Fallacy: Legocentrism and Continuity in the Rhetoric of Indian Dispossession," *American Indian Culture and Research Journal* 36, no. 1 (2012): 3–45.

22 Harvey D. Rosenthal, "Indian Claims and the American Conscience: A Brief History of the Indian Claims Commission," in *Irredeemable America: The Indians' Estate and Land Claims*, ed. Imre Sutton (Albuquerque, NM: New Mexico University Press, 1985), 36.

23 The classic accounts from a well-established literature include: Annie H. Abel, "The History of Events Resulting in Indian Consolidation West of the Mississippi River," in *American Historical Association Annual Report for 1906*, 2 vols. (Washington, DC: American Historical Association, 1906), 2:233– 450; Angie Debo, *A History of the Indians of the United States* (Norman, OK: Pimlico, 1970); Grant Foreman, *Indian Removal* (Norman, OK: Oklahoma University Press, 1932).

24 James M. Mooney, *Historical Sketch of the Cherokee* (Chicago, IL: Aldine Transaction, [1900] 1975), 124.

25 The most lively source on the ghoulish enterprise of craniology/craniometry remains Stephen J. Gould, *The Mismeasure of Man* (Harmondsworth, UK: Norton, 1981). For a superbly written account with an Australian focus, see Helen MacDonald, *Human Remains: Episodes in Human Dissection* (Melbourne: Melbourne University Press, 2005).

26 Patrick Wolfe, "The Limits of Native Title," Meanjin 59, no. 3 (2000): 144.

27 Williams, *American Indian in Western Legal Thought*, 269.

28 *Johnson v. McIntosh*, 21 US (8 Wheaton), 543, 1823, p. 573.

29 For discussion of Olney's "tide of history" concept, see Jackie Delpero, "'The Tide of History': Australian Native Title Discourse in Global Context," (MA thesis, Victoria University, Australia, 2003); and David Rittter, "The Judgement of the World: The Yorta Yorta Case and the 'Tide of History,'" *Australian Historical Studies* 123 (2004): 106–21.

30 Ilan Pappé, *The Making of the Arab–Israeli Conflict, 1947–1951* (London, UK: I.B. Tauris, 2001), 187. "In order to justify the inclusion of the Negev in the future Jewish state, eleven new kibbutzim were simultaneously installed in that desert region on October 6th, 1946, in addition to the ten settlements already established there during the War for the same purpose": Nathan Weinstock, Zionism, *False Messiah*, trans. Alan Adler (London, UK: Ink Links, 1979), 249.

31 Bernard Avishai, "Saving Israel from Itself: A Secular Future for the Jewish State," *Harper's Magazine*, January, 2005, 37.

32 Isabel V. Hull, "Military Culture and the Production of 'Final Solutions' in the Colonies: The Example of Wilhelminian Germany," in Gellately and Kiernan, *The Specter of Genocide*, 141–62.

33 In 1902, the renowned English liberal J.A. Hobson was expressing the fear "that the arts and crafts of tyranny, acquired and exercised in our unfree Empire, should be turned against our liberties at home." John A. Hobson, *Imperialism: A Study* (London, UK: Allen & Unwin, 1902), 160.

34 See Enzo Traverso, *The Origins of Nazi Violence*, trans. Janet Lloyd (New York, NY: New Press, 2003); and Charles Patterson, *Eternal Treblinka: Our Treatment of Animals and the Holocaust* (New York, NY: Lantern Books, 2002).

35 "The [Central Government region Jewish] ghetto clearings amounted to wild, day-long shooting sprees in particular sections of cities, at the end of which bodies were lying in the main streets leading to train stations." Dieter Pohl, "The Murder of Jews in the General Government," in *National Socialist Extermination Policies: Contemporary German Perspectives and Controversies*, ed. Ulrich Herbert (New York, NY: Berghahn Books 2000), 99. See also Jürgen Zimmerer, "Colonialism and the Holocaust: Towards an Archaeology of Genocide," trans. Andrew H. Beattie, in Moses, *Genocide and Settler Society*, 48–76. On colonial starvations and the "New Imperialism," see Mike Davis, *Late Victorian Holocausts: El Niño Famines and the Making of the Third World* (London, UK: Verso 2001).

36 Israel is also, of course, a partial exception here, though not so substantial an exception as is asserted by those who claim that Israel cannot be a colonial formation because it lacks a single commissioning metropolis. From the outset, the Yishuv co-opted Ottoman, British and US imperialism to its own advantage, a reciprocated opportunism involving what Maxime Rodinson neatly glossed as "the collective mother country." Maxime Rodinson, *Israel. A Settler-Colonial State?* trans. David Thorstad (New York, NY: Monad, 1973), 76.

37 Cole Harris, "How did Colonialism Dispossess? Comments from an Edge of Empire," *Annals of the Association of American Geographers* 94 (2004): 179.

38 Robert Melson, "Modern Genocide in Rwanda: Ideology, Revolution, War, and Mass Murder in an African State," in Gellately and Kiernan, *The Specter of Genocide*, 326.

39 Ibid., 327–28.

40 "For the provisions serving to the support of humane life, produced by one acre of inclosed and cultivated land, are (to speak much within compasse) ten times more, than those, which are yeilded [sic] by an acre of Land, of an equal richnesse, lyeing wast in common." John Locke, *Two Treatises of Government* (Cambridge, UK: Cambridge University Press [1698] 1963), 312.

41 The new Jew is an enduring Zionist theme. In introducing his terrorist memoir, future Israeli prime minister Menachem Begin announced that, in addition to his Jewish readers, he had also written the book for gentiles: "lest they be unwilling to realise, or all too ready to overlook, the fact that out of blood and fire and tears and ashes a new specimen of human being was born, a specimen completely unknown to the world for over eighteen hundred years, 'the FIGHTING JEW'." Menachem Begin, *The Revolt*, trans. Samuel Katz (London, UK: W.H. Allen, 1979), xxv, capitals in original. For a more recent diasporan example, see, for instance, the Adi Nes photograph used as publicity for the Jewish Museum of New York's 1998–1999 "After Rabin: New Art from Israel" show, at www.thejewishmuseum.org/site/pages/content/exhibitions/special/rabin/rabin_zoom/rabinL1.html.

42 "[John] Ross—the successful self-made Cherokee entrepreneur—was really what white Georgians feared. Their biggest obstacle to acquiring the Cherokee lands was the cultivator's plow and overseer's whip—not the war club, bow, and scalping knife." Sean M. O'Brien, *In Bitterness and in Tears: Andrew Jackson's Destruction of the Creeks and Seminoles* (Westport, CT: Praeger, 2003), 229. For the Constitution of the Cherokee Nation, see the *Cherokee Phoenix*, February 28, 1828.

43 The capacity to achieve permanence was typically put down to European ancestry, as in Andrew Jackson's exasperated disparagement of the "designing half-breeds and renegade white men" who had encouraged Chickasaw reluctance to cede land. Theda Perdue, *"Mixed Blood" Indians: Racial Construction in the Early South* (Athens, GA: Georgia University Press, 2003), 70, 95–96.

44 Foreman, *Indian Removal*, 310.

45 Perdue, *"Mixed Blood" Indians*, 68.

46 Ronald N. Satz, *American Indian Policy in the Jacksonian Era* (Lincoln, NE: University of Nebraska Press, 1975), 83.

47 European anti-Semitism could produce similar results: "What centuries of deprivation and persecution had failed

to do, the dazzling light of [Jewish] emancipation [in France] achieved. Yet the choice was limited. The words of Clermont-Tonnère, a liberal deputy in the French national assembly: *'Aux Juifs comme nation nous ne donnons rien; aux Juifs comme individuels nous donnons tout'* [to Jews as national collectivity we give nothing; to Jews as individuals we give everything] … reveal how restricted was the application of liberty." Isaiah Friedman, *The Question of Palestine, 1914–1918: British-Jewish-Arab Relations* (London, UK: Routledge & Kegan Paul, 1973), 26.

48 From Richard H. Pratt, "The Advantages of Mingling Indians with Whites" (1892), selection in, *Americanizing the American Indians: Writings by the "Friends of the Indian," 1880–1900*, ed. Francis P. Prucha (Cambridge, MA: Harvard University Press, 1973), 261. Ward Churchill's *Kill the Indian, Save the Man: The Genocidal Impact of American Indian Residential Schools* (San Francisco, CA: City Lights, 2004) illuminates the genocidal consequences of Friends of the Indian-style total institutions with dreadful and systematic clarity.

49 So far, at least. If Al-Qaeda were to repeat the procedure a sufficient number of times, then 9/11 could emerge as the onset of a genocide. Definitionally, in other words, as in the case of other patterned or cumulative phenomena, genocide can obtain retrospectively.

50 He had alternatives. Liddell and Scott give "race, stock, family" as primary meanings of genos, with secondary meanings including offspring, nation, caste, breed, gender(!) and "class, sort, kind." "Tribe" is listed as a subdivision of ethnos ("a number of people living together, a company, body of men … a race, family, tribe"). Henry G. Liddell and Robert Scott, *Greek-English Lexicon* (Oxford, UK: Clarendon, 1869), 314, 426. Compare Lemkin, *Axis Rule in Occupied Europe*, 79.

51 Russell Thornton, *American Indian Holocaust and Survival: A Population History Since 1492* (Norman, OK: Oklahoma University Press, 1987), 133.

52 "In 1998–2000, life expectancy for Aboriginal and Torres Strait Islander peoples was shorter by 21 years for males and 20 years for females, compared with the total population … In 1998–2000, the death rate for Indigenous infants was around four times the rate in the total population." Australian Bureau of Statistics, *Australian Social Trends: Health—Mortality and Morbidity: Mortality of Aboriginal and Torres Strait Islander Peoples* (Canberra: Australian Bureau of Statistics, 2002), 1. See also: House of Representatives Standing Committee on Family and Community Affairs, *Health is Life: Report on the Inquiry into Indigenous Health* (Canberra: House of Representatives, 2000); and Neil Thomson, "Trends in Aboriginal Infant Mortality," in *A Matter of Life and Death: Contemporary Aboriginal Infant Mortality*, ed. Alan Gray (Canberra: Aboriginal Studies Press 1990), 1–8.

53 What Jefferson bought was French dominion. The rawly unsettled nature of the Purchase territory (at least, outside New Orleans and its environs and outpost settlements such as Detroit and St Louis) was illustrated by the rapid commissioning of Lewis and Clark's 1803 expedition to chart it.

54 This was the reality behind the mushrooming frontier demographies. "In the decade before 1820, the population of the new state of Alabama increased by a startling 1,000 per cent." O'Brien, *In Bitterness and in Tears*, 221. For an illuminating catalogue of Creek responses to this invasion, see Richard S. Lackey, comp., *Frontier Claims in the Lower South. Records of Claims Filed by Citizens of the Alabama and Tombigbee River Settlements in the Mississippi Territory for Depradations by the Creek Indians During the War of 1812* (New Orleans, LA: Polyanthos, 1977).

55 Abel, "Indian Consolidation West of the Mississippi River," 412.

56 "No Indian nation or tribe within the territory of the United States shall be acknowledged or recognized as an independent nation, tribe, or power with whom the United States may contract by treaty." 16 Stat., 566 (Act of March 3, 1871), c 120, s 1. For discussion, see Vine Deloria Jr. and David E. Wilkins, *Tribes, Treaties, & Constitutional Tribulations* (Austin, TX: Texas University Press, 1999), 60–61; and Francis P. Prucha, *The Great Father: The United States Government and the American Indians*, abridged ed. (Lincoln, NE: University of Nebraska Press, 1986), 165.

57 In particular, *US v. Kagama*, 118 US 1886, p. 375; and *Lone Wolf v. Hitchcock*, 187 US 1903, p. 553.

58 John R. Wunder, *"Retained by the People": A History of American Indians and the Bill of Rights* (New York, NY: Oxford University Press, 1994), 17, 39.

59 The best source on this campaign remains the authoritative report that found its way into the House hearings preceding the *Indian Reorganization Act* of 1934: D.S. Otis, *The Dawes Act and the Allotment of Indian Lands*, ed. Francis P. Prucha (Norman, OK: Oklahoma University Press, [1934] 1973).

60 *Statistical Abstract of the United States* (Washington, DC: US Bureau of the Census, Department of Commerce, 1955), 180.

61 See for example Frederick E. Hoxie, *A Final Promise. The Campaign to Assimilate the Indians, 1880–1920* (Lincoln, NE: University of Nebraska Press, 1989); and Prucha, *Americanizing the American Indians*, passim.

62 Thomas J. Morgan, "What is an Indian?" in *Sixty-Fifth Annual Report of the Commissioner for Indian Affairs* (Washington, DC: Government Printing Office, 1892), 31–37.

63 "Thus the key factor in colonial and 'post'-colonial race relations is not, as some have argued, simple demographic numbers, since populations have to be differentiated before they can be counted. Difference, it cannot be stressed enough, is not simply given. It is the outcome of differentiation, which is an intensely conflictual process." Patrick Wolfe, "Land, Labor, and Difference: Elementary Structures of Race," *American Historical Review* 106 (2001): 894.

64 M. Annette Jaimes, ed. "Federal Indian Identification Policy: A Usurpation of Indigenous Sovereignty in North America," in *The State of Native America: Genocide, Colonization, and Resistance* (Boston, MA: South End Press, 1992), 137. Patricia Limerick is almost as succinct: "Set the blood quantum at one quarter, hold to it as a rigid definition of Indians, let intermarriage proceed as it has for centuries, and eventually Indians will be defined out of existence. When that happens, the federal government will finally be freed from its persistent 'Indian problem'": Limerick, *The Legacy of Conquest: The Unbroken Past of the American West* (New York, NY: Norton 1987), 338.

65 Donald L. Fixico, *Termination and Relocation. Federal Indian Policy, 1945–1960* (Albuquerque, NM: New Mexico University Press, 1986); and Charles F. Wilkinson and Eric R. Biggs, "The Evolution of the Termination Policy," *American Indian Law Review* 5 (1977): 139–84.

66 Baruch Kimmerling and Joel S. Migdal, *The Palestinian People: A History*, rev. ed. (Cambridge, MA: Harvard University Press, 2003), 172.

67 Wolfe, "Nation and MiscegeNation," 111; and Wolfe, *Settler Colonialism and the Transformation of Anthropology*, 180. Scholars who have made this point after me are too numerous to mention. Among those who made it before I did, see for example Jeremy Beckett, ed., "The Past in the Present, the Present in the Past: Constructing a National Aboriginality," in *Past and Present: The Construction of Aboriginality* (Canberra: Aboriginal Studies Press 1988), 191–217; Gillian

Cowlishaw, "Colour, Culture and the Aboriginalists," *Man* 22 (1988): 221–37; and Andrew Lattas, "Aborigines and Contemporary Australian Nationalism: Primordiality and the Cultural Politics of Otherness," in "Writing Australian Culture," ed. Julie Marcus, special issue, *Social Analysis* 27 (1990): 50–69.

68 Raymond Evans and Bill Thorpe, "The Massacre of Aboriginal History," *Overland* 163 (2001): 36.

69 For examples (some of which are actually hyphenated), see Katherine Bischoping and Natalie Fingerhut, "Border Lines: Indigenous Peoples in Genocide Studies," *Canadian Review of Social Anthropology* 33 (1996): 484–85; Robert K. Hitchcock and Tara M. Twedt, "Physical and Cultural Genocide of Various Indigenous Peoples," in *Genocide in the Twentieth Century*, ed. Samuel Totten, William S. Parsons, and Israel W. Charny (New York, NY: Garland Press, 1995), 498–501. For "politicide" ("a process that covers a wide range of social, political, and military activities whose goal is to destroy the political and national viability of a whole community of people"), see Baruch Kimmerling, *Politicide: Ariel Sharon's War Against the Palestinians*, rev. ed. (London, UK: Verso, 2006).

70 Ever alert to the damaging implications in this connection of Israel's invasion of Palestinian territory, Colin Tatz belittles the significance of "a contest for land and what the land held" as merely "explain[ing] away" colonial ethnocide. Colin Tatz, *With Intent to Destroy: Reflecting on Genocide* (London, UK: Verso, 2003), 180. Lower down the page, however, he observes that "we need to remember that Aboriginal Australians were deemed expendable not just because they were considered 'vermin,' or because they sometimes speared cattle or settlers, but because they failed the Lockean test of being a people capable of a polity and a civility, to wit, they couldn't or wouldn't exploit the land they held, at least not in the European sense."

71 Wolfe, "Nation and MiscegeNation," 110–18; and Wolfe *Settler Colonialism and the Transformation of Anthropology*, 168–90. For US examples, see Robert F. Berkhofer Jr., *The White Man's Indian: Images of the American Indian from Columbus to the Present* (New York, NY: Vintage Books, 1979); and Hugh Honour, *The New Golden Land: European Images of America from the Discoveries to the Present Time* (New York, NY: Pantheon, 1975). For responses to the phenomenon, see for example Fergus M. Bordewich, *Killing the White Man's Indian: Reinventing Native Americans at the End of the Twentieth Century* (New York, NY: Anchor Books, 1996); and Ward Churchill, *Indians Are Us? Culture and Genocide in Native North America* (Monroe, ME: Common Courage Press, 1994).

72 See Charles Zika, "Fashioning New Worlds from Old Fathers: Reflections on Saturn, Amerindians and Witches in a Sixteenth-century Print," in *Dangerous Liaisons: Essays in Honour of Greg Dening*, ed. Donna Merwick (Melbourne: University of Melbourne History Department, 1994), 249–81; and Zika, "Cannibalism and Witchcraft in Early-Modern Europe: Reading the Visual Images," *History Workshop Journal* 44 (1997): 77–105.

73 "At June 2002, 22% (4,200) of children in out-of-home care were Aboriginal or Torres Straight [sic] Islander children. This represented a much higher rate of children in out-of-home care among Indigenous children than non-Indigenous children (20.1 per 1,000 compared with 3.2 per 1,000)." Australian Bureau of Statistics, "Children in out-of-Home Care," in *Australia Now* (Canberra: Australian Bureau of Statistics, 2004), s. 2, "Australian social trends, 2003: family and community-services: child protection." An indication of the progress that Indigenous people in Australia have achieved since the darkest days of the assimilation policy is contained in the sentence that follows this excerpt: "In all jurisdictions, the Aboriginal Child Placement Principle outlines a preference for Indigenous children to be placed with other Aboriginal or Torres Straight [sic] Islander peoples, preferably within the child's extended family or community."

74 Given the matrilineal transmission of—and relative difficulty of conversion to—Judaism, this factor indicates vigilance in relation to Palestine.

75 "Assimilated natives would be proof positive that America was an open society, where obedience and accommodation to the wishes of the majority would be rewarded with social equality." Hoxie, *Final Promise*, 34. See also George P. Castile, "Indian Sign: Hegemony and Symbolism in Federal Indian Policy," in *State and Reservation: New Perspectives on Federal Indian Policy*, ed. George P. Castile and Robert L. Bee (Tucson, AZ: Arizona University Press, 1992), 176–83.

76 A drive to replace Palestinian labor with cheap immigrant labor was begun in the early 1990s in response to the first Intifada. Though this policy was officially abandoned as it generated its own problems, around 8% of Israel's population continues to be made up of illegal immigrants (who are, by definition, non-Jewish). See Shmuel Amir, "Overseas Foreign Workers in Israel: Policy Aims and Labor Market Outcomes," *International Migration Review* 36 (2002): 41–58; Eric Beachemin, "Illegal in Israel," Radio Netherlands broadcast, September 15, 2004, at https://www.amren.com/news/2004/09/illegal_in_isra/; Leila Farsakh, "An Occupation that Creates Children Willing to Die. Israel: An Apartheid State?" *Le Monde Diplomatique*, English language edition, November 4, 2003, http://mondediplo.com/2003/11/04apartheid.

77 Though formal legislative power was, for a time, exercised by Blacks in Black-majority Southern states during Reconstruction. See Thomas C. Holt, *Black Over White: Negro Political Leadership in South Carolina during Reconstruction* (Urbana, IL: Illinois University Press, 1977).

78 See W. Fitzhugh Brundage, *Lynching in the New South: Georgia and Virginia, 1880–1930* (Urbana, IL: Illinois University Press, 1993); Leon F. Litwack, *Trouble in Mind: Black Southerners in the Age of Jim Crow* (New York, NY: Knopf, 1998); and Joel Williamson, *The Crucible of Race: Black-White Relations in the American South Since Emancipation* (Oxford, UK: Oxford University Press, 1984), 180–223.

79 "Slave labor could be analyzed in economic, social, and political terms [in traditional histories,] but free labor was often defined as simply the ending of coercion, not as a structure of labor control that needed to be analyzed in its own way." Thomas C. Holt, Rebecca J. Scott, and Frederick Cooper, *Beyond Slavery: Explorations of Race, Labor, and Citizenship in Postemancipation Societies* (Chapel Hill, NC: North Carolina University Press, 2000), 2–3.

80 Though he is too scrupulous a scholar not to acknowledge that "Israeli actions may become near-genocidal." Tatz, *With Intent to Destroy*, 181.

81 "Capital punishment now being an unquestioned, routine penalty for chucking stones at Israelis." Robert Fisk, *The Great War for Civilisation: The Conquest of the Middle East* (London, UK: Fourth Estate, 2005), 546, as quoted in Tatz, *With Intent to Destroy*, 117. I have chosen not to patronize Professor Tatz by quoting approvingly from his otherwise very useful book, from which I have learned a lot, on account of our fundamental divergence over the issue of contemporary Zionism, which I wholeheartedly oppose, and, in particular, of my disdain for his attempts to confuse contemporary anti-Zionism with anti-Semitism (see for example pp. 19, 27, 127). Apart from anything else, these attempts do grave injustice to the real victims of anti-Semitism.

Shame is the reaction requested when they look you in the mouth and say,

"lost her language,"

but I know language well enough to pinpoint each time it's *lost* instead of *stole*,
and that shame alone cannot build homes or sustain bodies.
So I speak the Queen's English, every day

and you must admit it's fun to watch her squirm

as I roll her words on my wild tongue like they're chokecherries
the way my fingers expand those sentences into shapes she doesn't recognize/can't read

Break into her locked cupboards to devour greedily the literatures, philosophies;

get drunk and daring on the poetries
all of those nice, proper words that linger on my lips a bit
too
long. as if they liked it there (imagine the audacity)
Send me to bed early with no supper

I'll keep playing with colonizer's languages
bringing pleasure back to written letters weaponized to rip through flesh like mine

see those syllables m e l t at the touch of my nehiyawiskwew softness
(imagine the audacity) brown softness

in a world of borders
and sharp corners

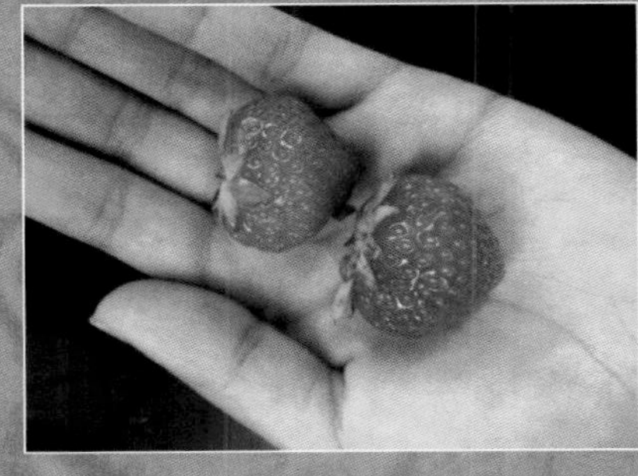

holding strawberries in Italy

The Queen's English

Erica Violet Lee

April 2, 2016

Deed of Surrender

Rupert's Land and North-Western Territory – Enactment No. 3, Part 4, Schedule C to the Constitution Act, 1982. Order of Her Majesty in Council admitting Rupert's Land and the North-Western Territory into the Union

23rd day of June, 1870

WHEREAS the said Governor and Company were established and incorporated by their said name of "The Governor and Company of Adventurers of England, trading into Hudson's Bay," by Letters Patent granted by His late Majesty King Charles the Second in the twenty-second year of his reign, whereby His said Majesty granted unto the said company and their successors the sole trade and commerce of all those seas, straits, bays, rivers, lakes, creeks and sounds in whatsoever latitude they should be, that lay within the entrance of the straits commonly called Hudson's Straits, together with all the lands and territories upon the countries, coasts, and confines of the seas, bays, lakes, rivers, creeks, and sounds aforesaid, that were not already actually possessed by, or granted to, any of His Majesty's subjects, or possessed by the subjects of any other Christian Prince or State, and that the said land should be from thenceforth reckoned and reputed as one of His Majesty's Plantations or Colonies in America, called Rupert's Land; and whereby His said Majesty made and constituted the said Governor and Company and their successors the absolute lords and proprietors of the same territory, limits and places aforesaid, and of all other the premises saving the faith, allegiance and sovereign dominion due to His said Majesty, his heirs and successors for the same; and granted to the said Governor and Company and their successors, such rights of Government and other rights, privileges and liberties, franchises, powers and authorities in Rupert's Land as therein expressed. And whereas ever since the date of the said Letters Patent, the said Governor and Company have exercised and enjoyed the sole right thereby granted of such trade and commerce as therein mentioned, and have exercised and enjoyed other rights, privileges, liberties, franchises, powers, and authorities thereby granted; and the said Governor and Company may have exercised or assumed rights of Government in other parts of British North America not forming part of Rupert's Land, or of Canada, or of British Columbia. And whereas by "The British North America Act, 1867," it is (amongst other things) enacted that it shall be lawful for Her present Majesty Queen Victoria, by and with the advice and consent of Her Majesty's most Honorable Privy Council, on address from the Houses of Parliament of Canada, to admit Rupert's Land and the North Western Territory or either of them into the Union of the Dominion of Canada on such terms and conditions as are in the Address expressed, and as Her Majesty thinks fit to approve, subject to the provisions of the said Act. And whereas, by the "Rupert's Land Act, 1868," it is enacted (amongst other things) that for the purposes of that Act the term "Rupert's Land" shall include the whole of the lands and territories held or claimed to be held by the said Governor and Company, and that it shall be competent for the said Governor and Company to surrender to Her Majesty, and for Her Majesty, by any instrument under Her Sign Manual and Signet to accept a surrender of all or any of the lands, territories, rights, privileges, liberties, franchises, powers and authorities whatsoever, granted or purported to be granted by the said Letters Patent to the said Governor and Company within Rupert's Land, upon such terms and conditions as shall be agreed upon by and between Her Majesty and the said Governor and Company; provided, however, that such surrender shall not be accepted by Her Majesty until the terms and conditions upon which Rupert's Land shall be admitted into the said Dominion of Canada shall have been approved of by Her Majesty, and embodied in an Address to Her Majesty from the Houses of the Parliament of Canada, in pursuance of the 146th Section of "The British North America Act, 1867," all rights of Government and proprietary rights, and all other privileges, liberties, franchises, powers and authorities whatsoever, granted or purported to be granted by the said Letters Patent to the said Governor and Company within Rupert's Land, and which

shall have been so surrendered, shall be absolutely extinguished, provided that nothing in the said Act contained shall prevent the said Governor and Company from continuing to carry on in Rupert's Land or elsewhere trade and commerce. And whereas Her said Majesty Queen Victoria and the said Governor and Company have agreed to terms and conditions upon which the said Governor and Company shall surrender to Her said Majesty, pursuant to the provisions in that behalf in the "Rupert's Land Act, 1868" contained, all the rights of Government and other rights, privileges, liberties, franchises, powers and authorities, and all the lands and territories (except and subject as in the said terms and conditions expressed or mentioned) granted or purported to be granted by the said Letters Patent, and also all similar rights which have been exercised or assumed by the said Governor and Company in any parts of British North America not forming part of Rupert's Land, or of Canada, or of British Columbia, in order and to the intent that, after such surrender has been effected and accepted under the provisions of the last-mentioned Act, the said Rupert's Land may be admitted into the Union of the Dominion of Canada, pursuant to the hereinbefore mentioned Acts or one of them. And whereas the said terms and conditions on which it has been agreed that the said surrender is to be made by the said Governor and Company (who are in the following articles designated as the Company) to Her said Majesty are as follows (that is to say):

1. The Canadian Government shall pay to the Company the sum of 300,000 £ sterling when Rupert's Land is transferred to the Dominion of Canada.

2. The Company to retain all the posts or stations now actually possessed and occupied by them or their officers or agents whether in Rupert's Land or any other part of British North America, and may within twelve months after the acceptance of the said surrender select a block of land adjoining each of their posts or stations, within any part of British North America, not comprised in Canada and British Columbia in conformity, except as regards the Red River Territory, with a list made out by the Company and communicated to the Canadian Ministers, being the list in the annexed schedule. The actual survey is to be proceeded with, with all convenient speed.

3. The size of each block is not to exceed in the Red River Territory an amount to be agreed upon between the Company and the Governor of Canada in Council.

4. So far as the configuration of the country admits, the blocks shall front the river or road by which means of access are provided, and shall be approximately in the shape of parallelograms, and of which the frontage shall not be more than half the depth.

5. The company may, at any time within fifty years after such acceptance of the said surrender, claim in any township or district within the fertile belt in which land is set out for settlements, grants of land not exceeding one-twentieth part of the land so set out; the blocks so granted to be determined by lot, and the Company to pay a rateable share of the survey expenses, not exceeding 8 cents Canadian an acre. The Company may defer the exercise of their right of claiming their proportion of each township or district for not more than ten years after it is set out, but their claim must be limited to an allotment from the lands remaining unsold at the time they declare their intention to make it.

6. For the purpose of the last article the fertile belt is to be bounded as follows: On the south by the United States' boundary; on the west by the Rocky Mountains; on the north by the Northern Branch of the Saskatchewan River; on the east by Lake Winnipeg, the Lake of the Woods and the waters connecting them.

7. If any township shall be formed abutting on the north bank of the northern branch of the Saskatchewan River, the Company may take their one-twentieth of any such township, which,

for the purpose of this article, shall not extend more than five miles inland from the river, giving to the Canadian Dominion an equal quantity of the portion of land coming to them of townships established on the southern bank of the said river.

8. In laying out any public roads, canals or other public works, through any block of land reserved to the Company, the Canadian Government may take without compensation such land as is necessary for the purpose, not exceeding one-twenty-fifth of the total acreage of the block; but if the Canadian Government require any land which is actually under cultivation, which has been built upon, or which is necessary for giving the Company's servants access to any river or lake, or as a frontage to any river or lake, the said Government shall pay to the Company the fair value of the same, and shall make compensation for any injury done to the Company or their servants.

9. It is understood that the whole of the land to be appropriated within the meaning of the last preceding clause, shall be appropriated for public purposes.

10. All titles to land up to the eighth day of March, one thousand eight hundred and sixty-nine, conferred by the Company, are to be confirmed.

11. The Company is to be at liberty to carry on its trade without hindrance in its corporate capacity; and no exceptional tax is to be placed on the Company's land, trade or servants, nor any import duty on goods introduced by the said Company previously to such acceptance of the said surrender.

12. Canada is to take over the materials of the electric telegraph at cost price; such price including transport, but not including interest for money, and subject to a deduction for ascertained deterioration.

13. The Company's claim to land under an agreement of Messrs. Vankoughnet and Hopkins is to be withdrawn.

14. Any claims of Indians to compensation for lands required for purposes of settlement shall be disposed of by the Canadian Government in communication with the Imperial Government; and the Company shall be relieved of all responsibility in respect of them, AND WHEREAS the surrender hereinafter contained is intended to be made in pursuance of the agreement, and upon the terms and conditions hereinbefore stated.

NOW know ye, and these presents witness, that, in pursuance of the powers and provisions of the "Rupert's Land Act, 1868," and on the terms and conditions aforesaid, and also on condition of this surrender being accepted pursuant to the provisions of that Act, the said Governor and Company do hereby surrender to the Queen's Most Gracious Majesty, all the rights of Government, and other rights, privileges, liberties, franchises, powers and authorities, granted or purported to be granted to the said Governor and Company by the said recited Letters Patent of His late Majesty King Charles the Second; and also all similar rights which may have been exercised or assumed by the said Governor and Company in any parts of British North America, not forming part of Rupert's Land or of Canada, or of British Columbia, and all the lands and territories within Rupert's Land (except and subject as in the said terms and conditions mentioned) granted or purported to be granted to the said Governor and Company by the said Letters Patent. In witness whereof, the Governor and Company of Adventurers of England trading into Hudson's Bay, have hereunto caused their Common Seal to be affixed, the nineteenth day of November, one thousand eight hundred and sixty-nine.

Enacted three years after the Confederation of Canada (1867), legislation of the *Deed of Surrender* came into force in 1870. "Schedule C" of the deed presents to the Order in Council of the Imperial (British) Government dated June 23, 1870, "admitting Rupert's Land and the North-Western Territory into the Dominion of Canada." The Deed (National Archive of the UK, ref. CO42/694) contains 14 sections that incorporated and took control of trade agreements and land claims of Indigenous peoples with the Hudson's Bay Company. Queen Victoria oversaw the largest expansion of the British Empire between 1837 to 1901.

Courtesy of Joan Murray, Former Corporate Historian, HBC Heritage Services, Hudson's Bay Company

"The discovery of gold and silver in America, the extirpation, enslavement, and entombment in mines of the Aboriginal population, the beginning of the conquest and looting of the East Indies, the turning of Africa into a preserve for the commercial hunting of black skins… are the chief moments of primitive accumulation."
—Karl Marx, *Capital: A Critique of Political Economy*, 1867

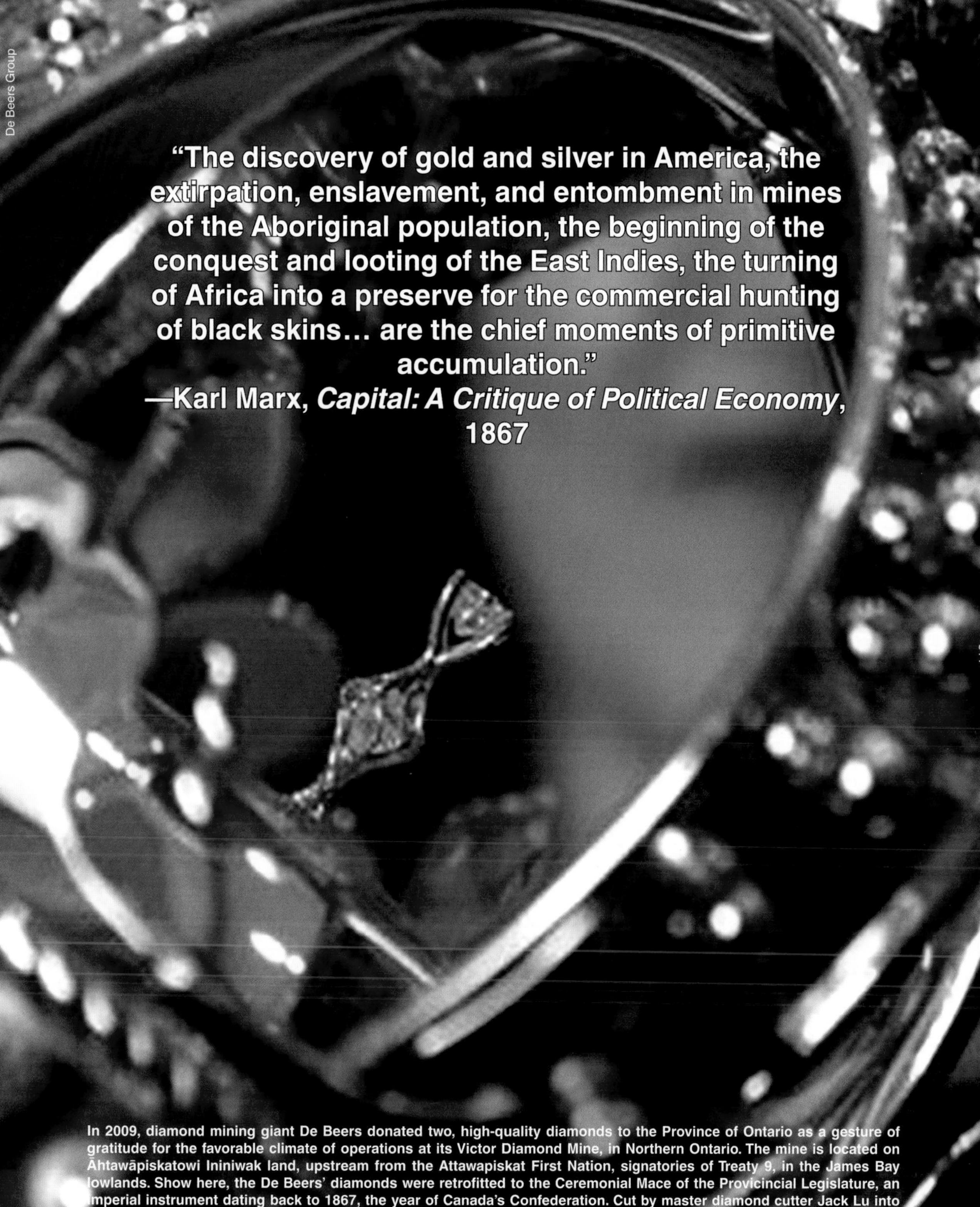

De Beers Group

In 2009, diamond mining giant De Beers donated two, high-quality diamonds to the Province of Ontario as a gesture of gratitude for the favorable climate of operations at its Victor Diamond Mine, in Northern Ontario. The mine is located on Āhtawāpiskatowi Ininiwak land, upstream from the Attawapiskat First Nation, signatories of Treaty 9, in the James Bay lowlands. Show here, the De Beers' diamonds were retrofitted to the Ceremonial Mace of the Provicincial Legislature, an imperial instrument dating back to 1867, the year of Canada's Confederation. Cut by master diamond cutter Jack Lu into the classic 'princess of hearts' shape, the stones were set by Indian-Canadian designer Reena Ahluwalia, with platinum supplied by Vale Inco, and finally then polished by Corona Jewellery Canada. Use of the ceremonial mace in England and France—reportedly a medieval weapon—dates back to the 13th century, as a supposed symbol of the Crown's supremacy

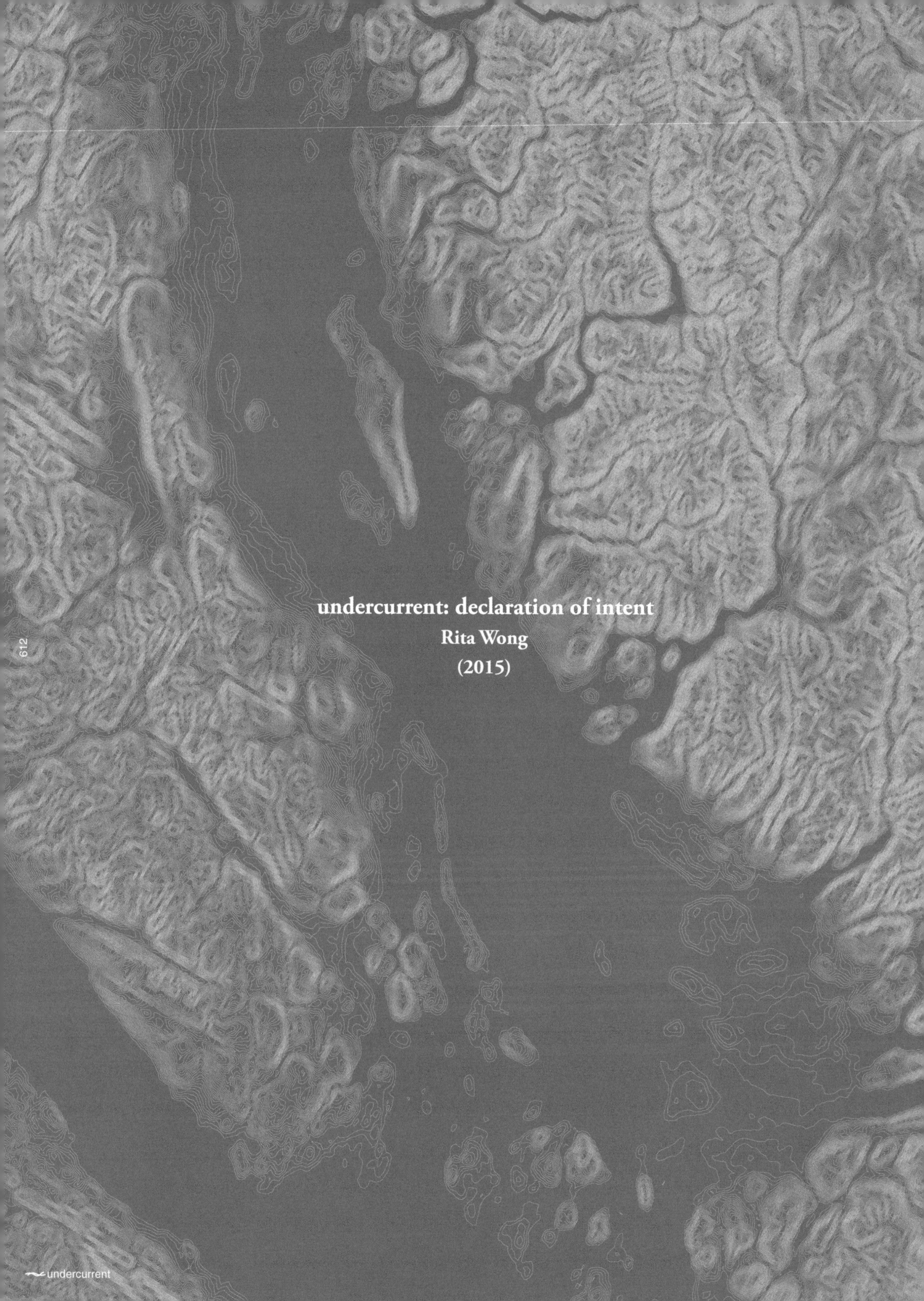

undercurrent: declaration of intent

Rita Wong

(2015)

"water is unstoppable"
— Wes Nahanee, from the Squamish Nation

let the colonial borders be seen for the pretensions that they are
i hereby honour what the flow of water teaches us
the beauty of enough, the path of peace to be savoured
before the extremes of drought and flood overwhelm the careless
water is a sacred bond, embedded in our plump, moist cells
in our breaths that transpire to return to the clouds that gave us life through rain
in the rivers & aquifers that we & our neighbours drink
in the oceans that our foremothers came from
a watershed teaches not only humbleness but climate fluency
the languages we need to interpret the sea's rising voice
water connects us to salmon & cedar, whales & workers
its currents bearing the plastic from our fridges & closets
a gyre of karma recirculates. burgeoning body burden
i hereby invoke fluid wisdom to guide us through the toxic muck
i will apprentice myself to creeks & tributaries, groundwater & glaciers
listen for the salty pulse within, the blood that recognizes marine ancestry
in its chemical composition & intuitive pull
i will learn through immersion, flotation & transformation
as water expands & contracts, i will fit myself to its ever-changing dimensions
molecular & spectacular, water will return what we give it, be that
arrogance & poison, reverence & light, ambivalence & respect
let our societies be revived as watersheds
because i am part of the problem i can also become part of the solution
although i am part of the problem i can also become part of the solution
where i am part of the problem i need to be part of the solution
while i am part of the problem i can also be part of the solution
one part silt one part clear running water one part blood love sweat
not *tar* but *tears*, *e* inserts a listening, witnessing, quickening eye
broken but rebinding, token but reminding, vocal buck unwinding
the machine's gears rust in rain, moss & lichen slowly creep life back
the rate of reclamation is humble while the rate of destruction blasts fast
because we are part of the problem we can also become part of the solution

reprinted with permission from Rita Wong

Prison of Grass

Howard Adams

"Creating reserves made it much easier to manage First Nations and, ultimately, to disrupt kinship and governance systems, as well as removing First Nations form their traditional economies and sources of food. The reserve system was essential to the settlement of Canada and as particularly vital to opening up the Plains for that purpose. Disease and starvation, deliberately weaponized against Indigenous peoples on the Plains, helped to force them onto reserves."[1]

—Chelsea Vowel, *Indigenous Writes: A Guide to First Nations, Métis & Inuit Issues in Canada*, 2016

In my halfbreed ghetto, finding a job was always difficult because the only employers were whites. It mattered little that I did not look truly Indian: all local employers knew whether I was half-breed or white. Seeking employment as a native was more than looking for a job, it was asking to be insulted. The boss did not have to insult me with his words; his actions and attitudes were enough to tell me his racist thoughts. As long as other jobs were available, a native would not apply for jobs he knew were for whites only. Even today Indians and Métis rarely apply for work as postmen, bus drivers, or for any position in which they would meet the public. Those jobs are taboo for natives because we live in a white racist society.

In my youth I therefore applied only for jobs that I knew had possibilities for half-breeds, such as picking roots and rocks, haying, and unskilled laboring jobs in construction. The working conditions of these jobs were even more demoralizing than the ordeal of asking white bosses for such menial work. Furthermore, I knew that I would be paid lower wages than white workers, regardless of how hard I worked. When I settled with the boss for my wages at the end of the job, I was usually cheated out of several dollars. If I protested, he would threaten to call the police and have me thrown out of his place. Since I knew how police regard halfbreeds and Indians and how they support white bosses, I would always leave immediately.

I was extremely fearful of white bosses—they terrified me. I always found them arrogant and cruel. Psychologically, their superiority practically crippled me. As soon as the boss gave me orders, I would become obedient and subservient, thankful for every small favor, no matter how insignificant. These work situations angered me deeply. When I discussed them with my mother—a deeply religious Catholic, who was illiterate but very wise in her own way—she argued that halfbreeds were hard workers only part of the time. She claimed that some had a habit of not being on time, and occasionally they went on a drunk and remained off work. Even though I didn't understand at the time why colonized native people behave the way they do, I felt instinctively that my mother's stereotype argument was false and I had little patience with it.

The hostility I nurtured during my ugly work experiences dominated my thoughts. I needed jobs to get money. The jobs degraded me and destroyed my sense of esteem and humanity, but I had to have money; therefore, there was no way of avoiding these nasty experiences with white employers. It was clear to me that white bosses were using these stereotypes of drunkenness and laziness as excuses to exploit halfbreeds. By classifying us as inferior workers, they could get their work done more cheaply. Native workers are invariably treated in this discriminatory way: branded with the same racial stereotypes—late for work, absent after payday, unreliable on the job—they are then forced to accept poorer wages. In my youth I never solved the puzzle of racism in employment, but today it is clear to me that racism is the product of economics.

The racism that native people encounter today had its origins in the rise of western imperialism during the 1600s:

> "But modern society—Western civilization—began to take on its characteristic attributes when Columbus turned the eyes and interests of the world away from the Mediterranean toward the Atlantic. … The socio-economic matrix of racial antagonism involved the commercialization of human labour in the West Indies, the East Indies, and in America … Racial antagonism attained full maturity during the latter half of the nineteenth century, when the sun no longer set on British soil and the great nationalistic powers of Europe began to justify their economic designs upon weaker European peoples with subtle theories of racial superiority and masterhood."[2]

Businessmen of Europe realized that they would need a large supply of labor to obtain resources from the new continents. Natives furnished this large supply of cheap labor. Since labor was an important item of cost in the production of goods, European businessmen wanted to get the greatest amount of labor for the least possible pay, and the purpose of racism was to reduce native people to a subhuman level where they could be freely exploited. Racism therefore arose from economic factors inherent in capitalism.

In Canada, indentured or semi-slave labor had to be secured and made available for businessmen of the fur-trade industry. Racial stereotypes and prejudices then developed from the realization that Indians provided potentially cheap labor for trapping furs and for whatever other jobs had to be done. Not only that, but they were found to be the most efficient trappers and fur gatherers. So European scholars and clergymen began creating racial theories which showed that the native people of North America and other colonies were primitives, innately inferior and subhuman:

> "Sepulveda [eminent Spanish theologian and university professor of the sixteenth century], then, may be thought of as among the first great racists; his argument was, in effect, that the Indians were inferior to the Spaniards, therefore they should be exploited. …

> Among the Spanish writers of the time (about 1535 onward) who were in rather complete accord with the drastic methods of human exploitation in the New World was Gonzolo Fernandez de Oviedo [sic]. … It was Oviedo's opinion, even after visiting America on a royal commission, that the Indians were not far removed from the state of wild animals, and that coercive measures were necessary if they were to be Christianized and taught the uses of systematic labor."[3]

These scholars stated that, because of their barbaric life style, natives were naturally inferior to white men. Churchmen provided great service in the development of racist ideologies: "A few of the Puritan clergy later asserted that the Indians were children of the devil who might profitably be wiped out and their lands appropriated."[4] Clergymen, particularly Anglican and Catholic clergymen, worked closely with the imperialist companies of Europe in the conquest and exploitation of native people; they were as important as the military in conquering the Indigenous populations of the colonies:

> "With scriptural quotations to support his assertions, Gray [London preacher of the sixteenth century, scholar of St. John's College, Cambridge] proves that Englishmen have a solemn duty to seek out fresh lands to relieve the congestion at home. If these foreign countries are inhabited by savages, then, as the Israelites cast out the Canaanites, so Englishmen must take the land from the idolatrous heathen. But Gray suggests that it would be better if they could first convert the heathen and the peaceably move into the country. The sword is the last resort. … But the land of the heathen must be claimed *at any cost* for the children of God …"[5] [emphasis added] … "During the spring of 1609 several preachers not directly connected with the Virginia Company lent the weight of their influence to the schemes for colonization. Most prominent of these was Richard Crakanthorpe, chaplain to Dr. Thomas Ravis, Bishop of London."[6]

These arguments created a new role for the missionaries; the conversion of heathens to Christianity not only saved their souls but also made them servile and obedient to their economic masters. In addition, university professors developed "scientific" theories that natives were stupid and bestial. These creators of racism argued that natives were mentally and morally inferior to Europeans and incapable of looking after themselves. Europeans refused to accept the fact that Indians had lived in Canada for thousands of years and did not need white masters to look after them. However, at the same time, it was stressed that natives were physically strong and therefore must have been intended for hard labor.

Establishing that native people were little more than stupid beasts of burden allowed easy and uninterrupted exploitation of them as workers, and also denied them all legal or human rights:

"When a philosophy for the dehumanizing of the exploited people has been developed with sufficient cogency, the ruling class is ready to make its grand statement, sometimes implicitly, and to act in accordance with it: The colored people have no rights which the master race is bound to respect. The exploiting class has an economic investment in this conviction and it will defend it with the same vigor as it would an attack upon private property in land and capital."[7]

Thus it became impossible to make appeals for humane and just treatment: the Indian stereotypes created by the exploiters had reduced native people in the eyes of the public to animal-like creatures. Racist schemes of inferiorization claimed that all native people were capable only of the lowest type of unskilled laboring jobs. Scholars did not develop theories that natives were capable of other work than laboring jobs. For example, racial theories did not say that Indians were inherently skilled as priests or merchants. Instead, the stereotypes claimed that natives were good only for work as miners, cotton-pickers, fur-trappers, and so on:

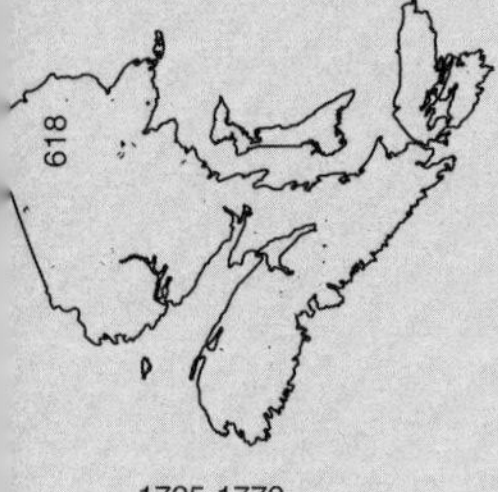

1725-1779
Maritime Peace
and Friendship Treaties

1781-1862
Upper Canada Land Surrenders

1850
Robinson-Superior Treaty

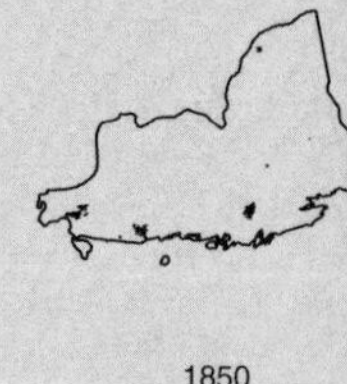

1850
Robinson-Huron Trea

"… the Indians were represented as lazy, filthy pagans of bestial morals, no better than dogs, and fit only for slavery … The capitalist exploitation of the colored workers, it should be observed, consigns them to employments and treatment that are humanly degrading. In order to justify this treatment the exploiters must argue that the workers are innately degraded and degenerate, consequently they naturally merit their condition."[8]

In Canada, racism originated in the imperialist fur-trading industry, and over the centuries it has become deeply entrenched in Canadian society. As a result, assimilation of natives into mainstream society is today not a possibility, at least not in a capitalist society:

"Assimilation diminishes the exploitative possibilities. This social situation is not especially a derivative of human idiosyncrasy or wickedness, but rather it is a function of a peculiar type of economic order which, to re-

> peat, has been developed in the west among Europeans. The exploitation of native peoples is not a sin, not essentially a problem of morals or of vice; it is a problem of production and of competition for markets. Here, then, are race relations ... They are labor-capital-profits relationships; therefore, race relations are proletarian-bourgeois relations and hence political-class relations."[9]

Once natives became integrated into Canadian society it would have been impossible to separate them from other people as a class of special workers. However, as long as Indians were isolated as a special group, they were easily exploited as trappers; isolation or segregation of native people was therefore essential for the fur industry.

The racism created during the centuries of the fur trade cannot be eradicated today. Although cheap Indian labor is unnecessary to the present Canadian economy, the early principles of racism remain as a dominant feature of the Canadian economic system. Canada has a long history of deeply entrenched racism because the fur trade, operating on the

basis of racism, lasted well over 200 years. The Hudson's Bay Company, founded in 1670, also molded certain Canadian social institutions within this racist framework. White supremacy, which had been propagated since the beginning of European imperialism, became woven into Canadian institutions such as the church, the schools, and the courts, and it has remained the working ideology of these institutions. In addition, native people cannot avoid seeing the cultural images and symbols of white supremacy, because they are everywhere in society, especially in movies, television, comic books, and textbooks. Since Indians and halfbreeds cannot live completely outside mainstream society, they are continually subjected to racial stereotypes through their encounters with police, welfare officials, and school authorities.

As soon as native children enter school they are surrounded with white-supremacist ideas and stories—every image glorifies white success. Because they are unable to resist it, they become conditioned to accept inferiority as a natural way of life. They soon recognize that all positions of authority—such as teacher, priest, judge, Indian agent—are held by whites. These people make all the rules and decisions that determine the fate of Métis and Indian people. An aggressive and sophisticated white-supremacist society intimidates colonized people, it makes them self-conscious and withdrawn. As native children grow up, these white-supremacist images become more alive, but natives are powerless to do anything about them. Consequently, the children internalize inferior images as a part of their true selves, often with strong feelings of shame. This partly explains why many native people attempt to hide from their Indianness, while others try to pretend that they are white, French, or Italian. White supremacy dictates that whiteness is beautiful, that mainstream life-styles are the most desirable, and that mainstream life is the only successful way of life. At the same time, white supremacy disfigures not only the native people, but the whole Canadian nation. Because of white supremacy, some natives attempt to abandon their culture and people. Many young persons abandon their parents and relatives and attempt to lose themselves in the mainstream.

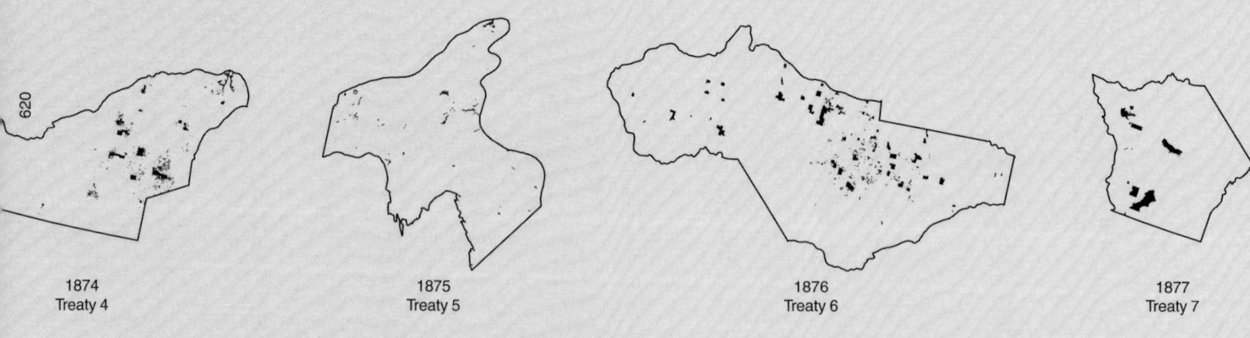

When I left my ghetto as a young man, I made a complete break with my parents and home. To me, everything about them and the community seemed so definitely halfbreed, and therefore ugly and shameful. As a result, I attempted to dissociate myself from everything and everyone that appeared halfbreed. I wanted to be a successful white man in mainstream society. If I maintained a close identification and relationship with my parents, home, and community, they would anchor me to halfbreed society and prevent my success in the white world. I was fully aware of how whites mocked and condemned halfbreeds and their way of life. I wanted to escape from all that ugliness and mockery. Since my parents were precious to me, it was an agonizing experience, yet there was no choice if I wanted to succeed.

In a white-supremacist society, more opportunities and privileges exist for Indians and halfbreeds who "look white"; those who "look Indian" are doomed to stay at the bottom of society. They are forced into the extreme of racism, and they suffer most as a result. It was no accident that I managed to get a good education and a good job, since my appearance is predominantly white. Throughout my school years I was favored, because I closely resembled white students. More privileges were extended to me than to other Métis children who looked more Indian. The white community responded to me in a less racist manner than it did to other halfbreeds. Through this kind of partiality I became aware of the possibility of acceptance in white society and the possibility of success in mainstream life. However, the question arises: What happens to the masses of Indians and halfbreeds who are forced into the deep crevices of the "caste" order because of their Indian appearance and lifestyle? There is no escape from such discrimination. It is understandable why intense racial feelings develop among these Indians and Métis, and it is not surprising that they are the strongest advocates of militancy and red nationalism.

Today, white supremacy is being exposed for what it is—a myth. The "scientific" theories that supported racism throughout the early centuries of imperialism have been

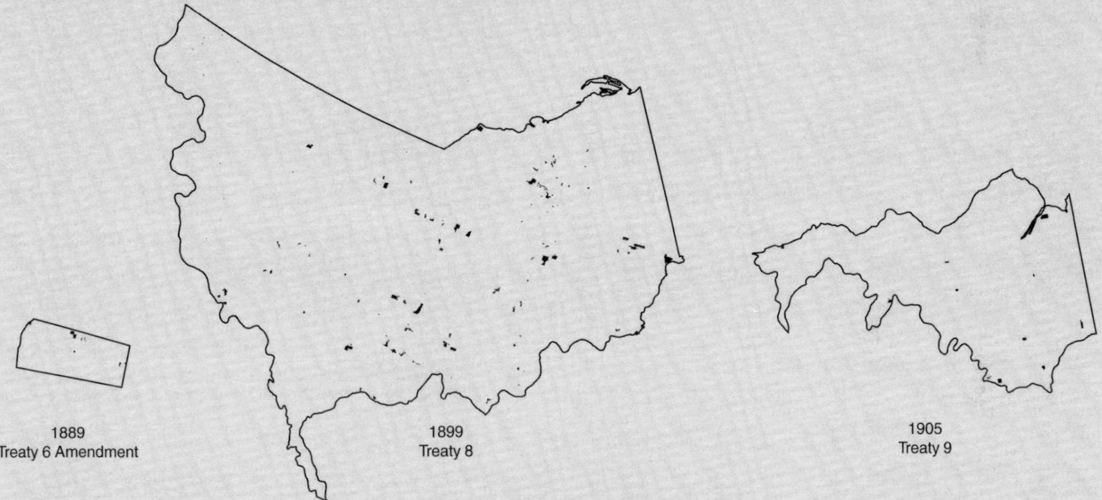

shown to be usually erroneous. In school today, intelligence quotient tests and achievement tests that keep native students at the bottom of the class are being criticized as cultural and racial tests. The tragedy for Indians and halfbreeds is that for so long they have accepted these tests and theories as scientific fact. Even today, many native people still believe that they are inferior to whites and that they are not intelligent enough to become professional workers or to administer their own affairs. Many still regard themselves as awkward and incompetent, in comparison with whites. As a youngster at home in my ghetto, I saw this image very clearly. I tried to gain assurance from my mother that I was not stupid and not inferior to others. If I did not get these assurances, I would react bitterly against my Indian heritage. When I walked through the snow I would look back at my tracks to see if I was pigeon-toed. Whenever I spoke to whites, I was extremely self-conscious about my halfbreed looks, manners, and speech. I was very sensitive about my inferiority because I knew that whites were looking at me through their racial stereotypes and I too began to see myself as a stupid, dirty breed, drunken and irresponsible. It made me feel stripped of all humanity and decency, and left me with nothing but my Indianness, which at the time I did not value. I hated talking to whites because it was such an agonizing experience—their attitudes and the tone of their conversation left no doubt about white supremacy. I would often cut the conversation short so that I could escape from these painful encounters. Not only did my sense of inferiority become inflamed, but I came to hate myself for the image I could see in their eyes. Everywhere white supremacy surrounded me. Even in solitary silence I felt the word "savage" deep in my soul.

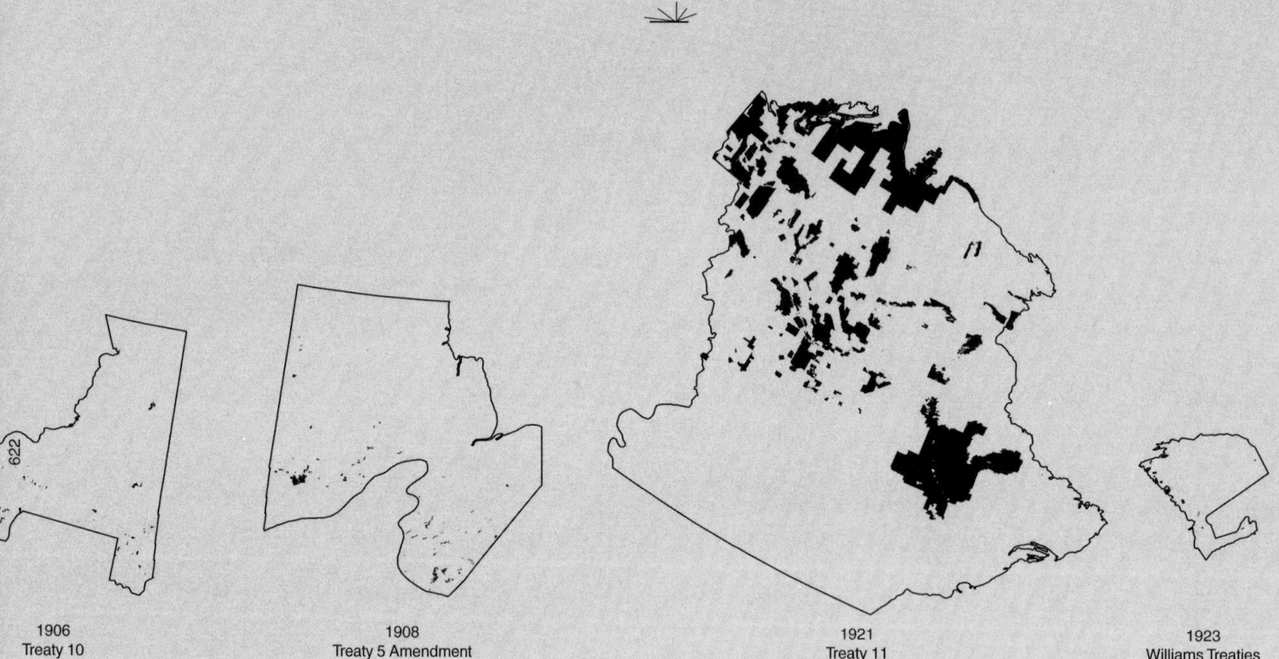

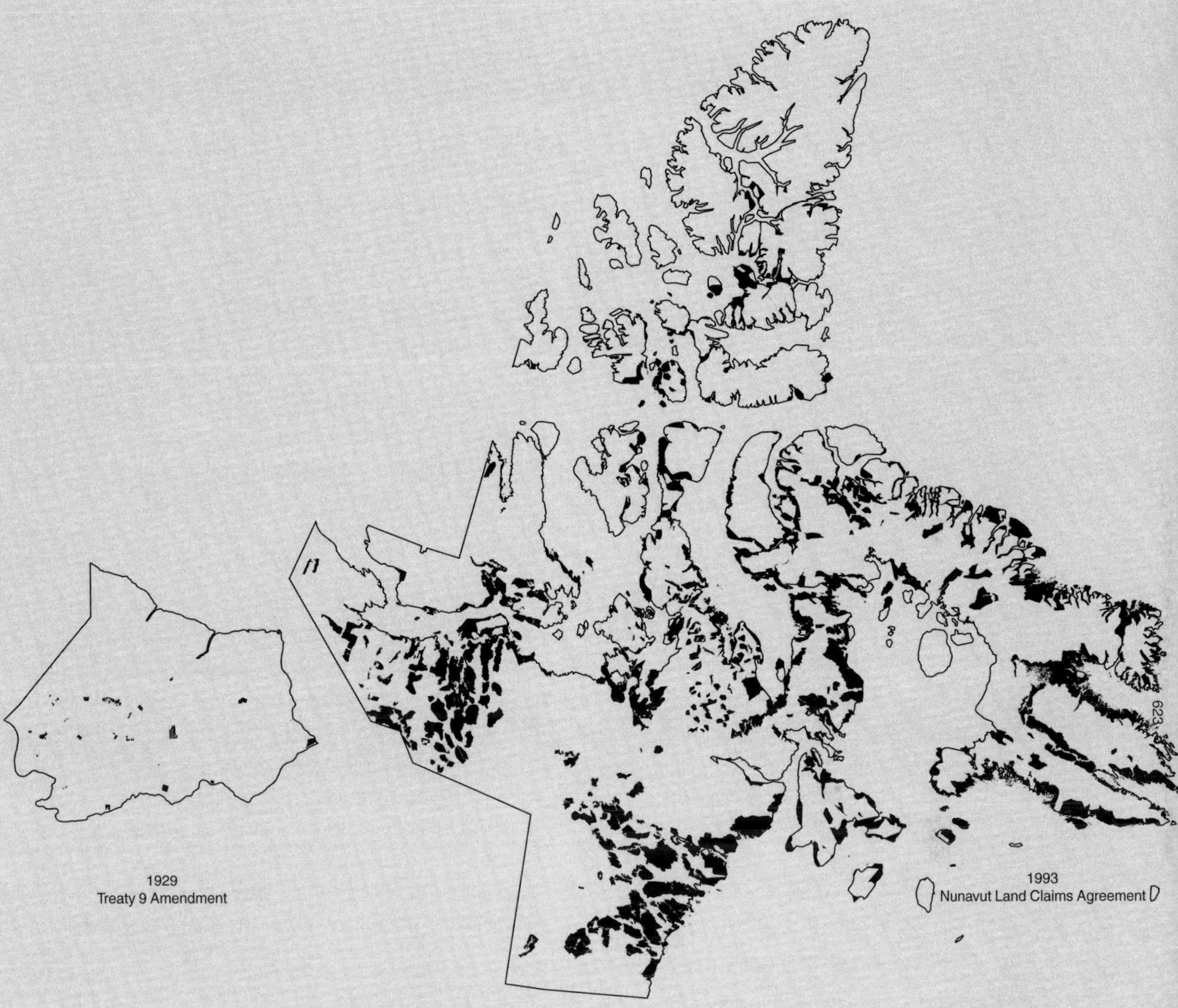

Notes

This essay was originally published as "The Basis of Racism," chap. 1 in *Prison of Grass: Canada from the Native Point of View* (Toronto, ON: New Press, 1975), reproduced with permission from New Press.

1 Parliament of Canada, "Bill C-33," April 10, 2014, http://www.parl.ca/DocumentViewer/en/41-2/bill/C-33/first-reading, as referenced in Chelsea Vowel, *Indigenous Writes: A Guide to First Nations, Métis, & Inuit Issues in Canada* (Winnipeg, MB: Highwater Press, 2016), 264.

2 Oliver C. Cox, "Race Relations—Its Meaning, Beginning, and Progress," chap. 16 in *Caste, Class and Race: A Study in Social Dynamics* (New York, NY: Monthly Review Press, 1959), 330.

3 Ibid., 335, 335n22.

4 Louis B. Wright, *Religion and Empire: The alliance between piety and commerce in English expansion*, 1558-1625 (Chapel Hill, NC: North Carolina Press, 1943), 86.

5 Ibid., 93.

6 Ibid., 96.

7 Cox, *Caste, Class and Race*, 335.

8 Ibid., 334.

9 Ibid., 336.

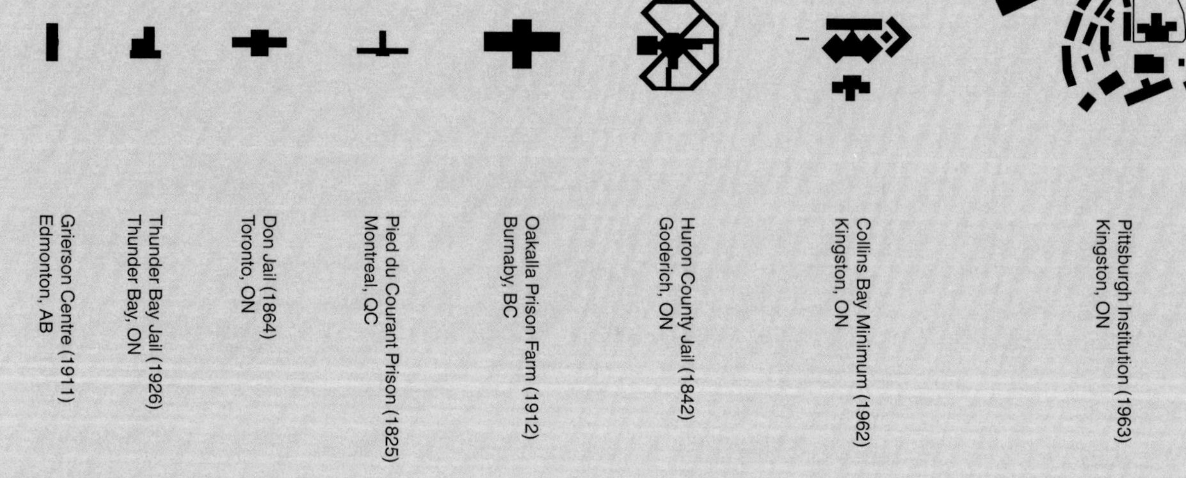

Territories of Incarceration

A Genealogy of Canada's Federal Prisons from 1835–2017

Although prison types have changed over the past 200 years, upon closer examination they are the explicit expression of imperially-based Victorian strategies of spatial incarceration and territorial containment. In spite of modern forms of technological architecture, these Victorian ideologies are inscribed in all facets of prison shapes and sizes, including their form, site, labor, program, surveillance, location, and jurisdiction. More importantly, the construction of Canada's first maximum security prison dates back to 1835, built in the loyalist stronghold of the city of Kingston, Ontario. Preempting the "Act to Encourage the Gradual Civilization of Indian Tribes in this Province, and to Amend the Laws Relating to Indians" (also known as the "Gradual Civilization Act of 1857" and the "1839 Act for the Protection of the Indians in Upper Canada"), prisons and so-called 'correctional facilities' have since served as

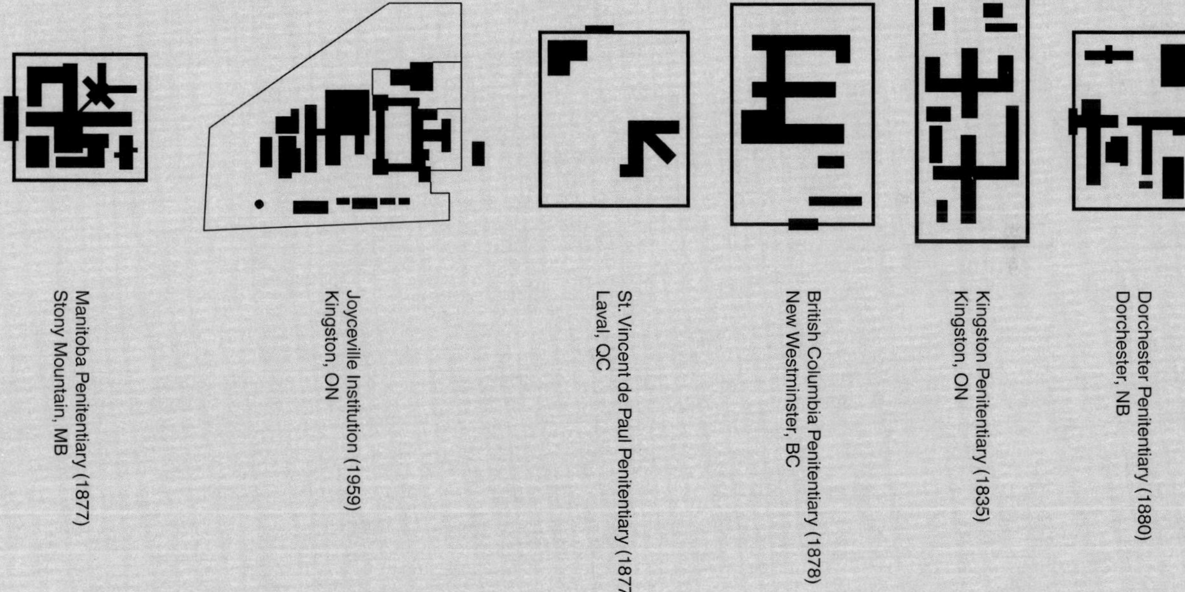

overarching instruments of the State and the Crown in implicit policies of Indigenous assimilation, incarceration, and extermination, much like the method of bodily and territorial containment of Canada's *Indian Reserve System*. Ever since the official enactment of the *Indian Act* in 1876, the form and architecture of prisons or penitentiaries has followed the flow of federal Indigenous policies that have ranged from shame and punishment, to isolation and extermination in a range of mostly remote or peripheral areas of the country. While Indigenous peoples represent approximately 4.3% of Canada's total population, current prison population—according to the Office of the Correctional Investigator at *Correctional Service Canada*—show a staggering level of incarceration of Indigenous men and women, comprising nearly 25% of all inmates across Canada.

Dorchester Penitentiary (1962)
Dorchester, NB

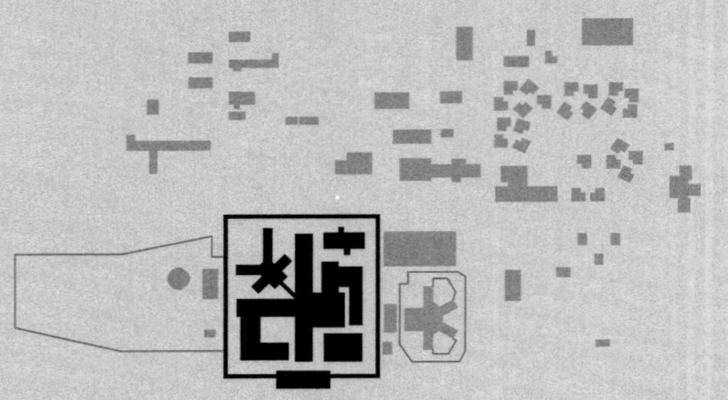

Stony Mountain Institution (2014)
Stony Mountain, MB

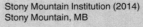

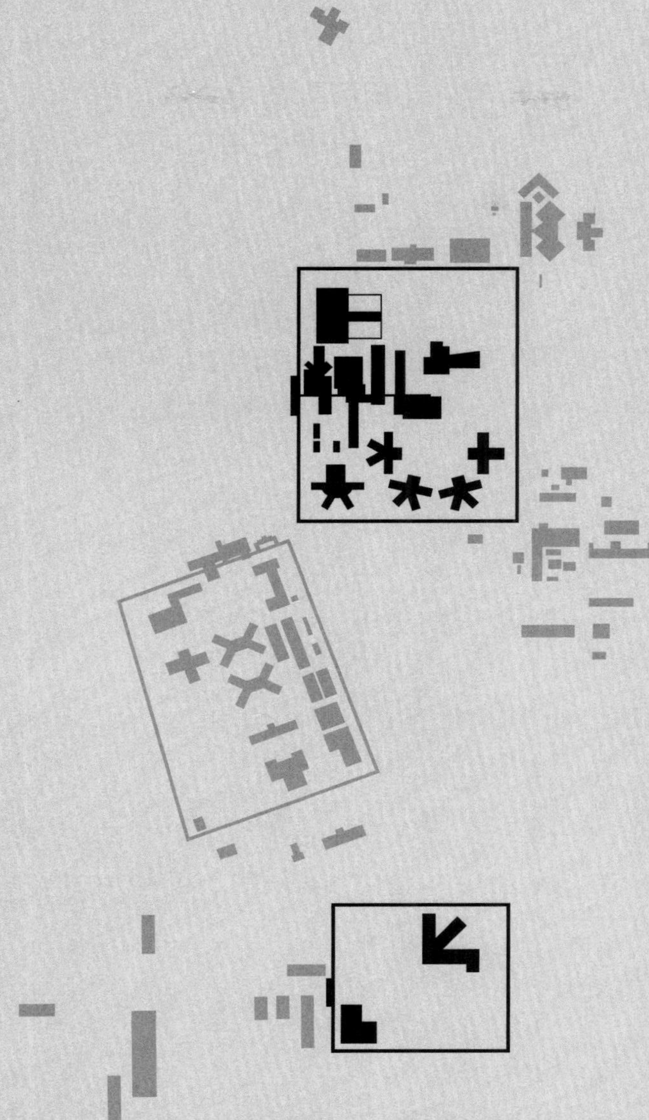

Collins Bay Institution (2014)
Kingston, ON

Federal Training Centre (1963)
Laval, QC

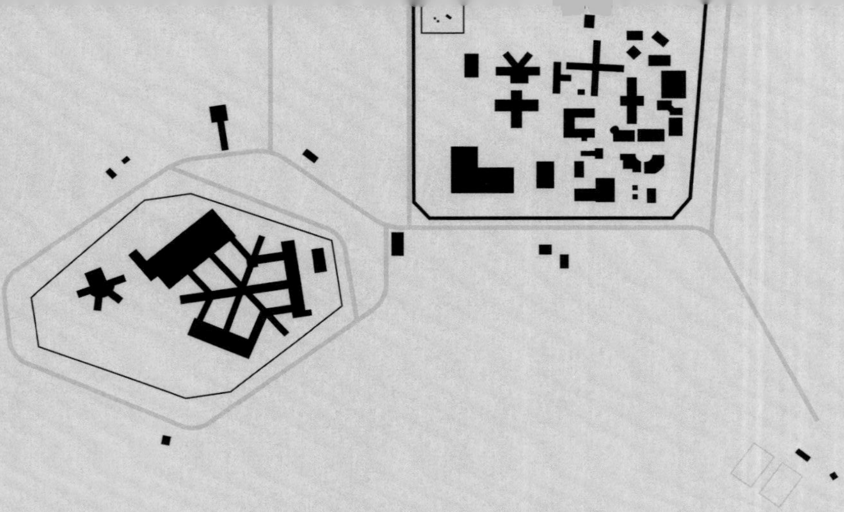

Millhaven and Bath Institutions (1972)
Kingston, ON

Saskatchewan Penitentiary (1962)
Prince Albert, SK

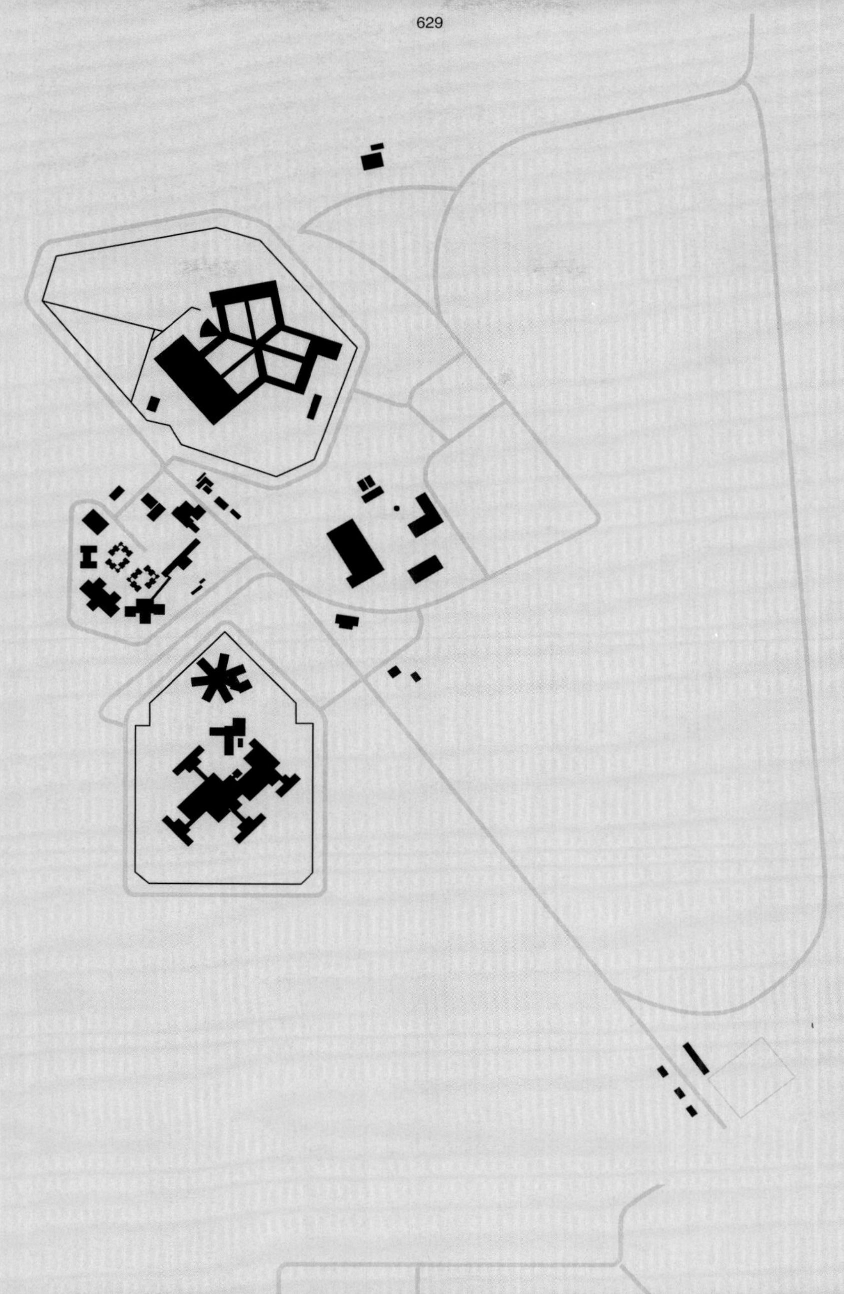

Archambault Institution (1969)
Sainte-Anne-des-Plaines, QC

Victoria's Secret

How to Make a Population of Prey

Mary Eberts

"There is not one moment that persuaded the heart of this nation to care about the incredulous issue of violence against Indigenous women and girls."[1]

—Angela Sterritt, "A Movement Rises," 2015

Being Indian Is a High-Risk Lifestyle

"Indigenous women and girls are far more likely than other Canadian women and girls to experience violence and to die as a result."[2] Young Indigenous women are five times more likely than other Canadian women of the same age to die of violence, and between 1997 and 2000, the rate of homicide for Indigenous women was almost seven times higher than the rate for non-Indigenous women.[3] Indigenous women are more likely than non-Indigenous women to be killed by a stranger, and nearly half the murders are unsolved.[4]

The staggering violence against Indigenous women is a legacy of colonization.[5] [6] In this text, I examine how Canada's *Indian Act*, an instrument of colonization, makes Indigenous women legal nullities, places them outside of the rule of law and transforms them into prey for those who would harm and abuse them.

The Act's historic aim was the assimilation of "Indians." It segregated them from settler society and indoctrinated them until a satisfactory degree of "civilization" had been reached. Such civilization involved stripping Indigenous nations of name, language, culture, and social organization,[7] and re-creating them as small government-dependent politics of "Indians" meant to look like Victorian villages populated by Victorian families. The Scheme discriminates on the basis of both race and gender. It rests on the view that the humanity of the Indian does not measure up to the humanity of the European. Its adoption of the Victorian patriarchal model of the family reflects the belief that the only proper place for women is under the dominion and control of men.

Women subject to the *Indian Act* are doubly diminished. If they fit within the Victorian confines, they have few or no rights. If they do not become (or seem to be) docile Victorian wives, as dictated by the Act, they are branded as deviants (often prostitutes) and considered fair game for mistreatment. The B.C. Human Rights Council found this branding to be discrimination on the combined grounds of race and gender. Valerie Frank of the Comox Indian Bank had never experienced maltreatment when staying at a hotel with her family but was twice roughly evicted from her hotel room, and denied other services, when she visited there by herself.[8] In ruling against the hotel, the Council commented, "What is particularly offensive … is the assumption that she is a prostitute because she is a single Native woman in a hotel by herself."[9]

1000km

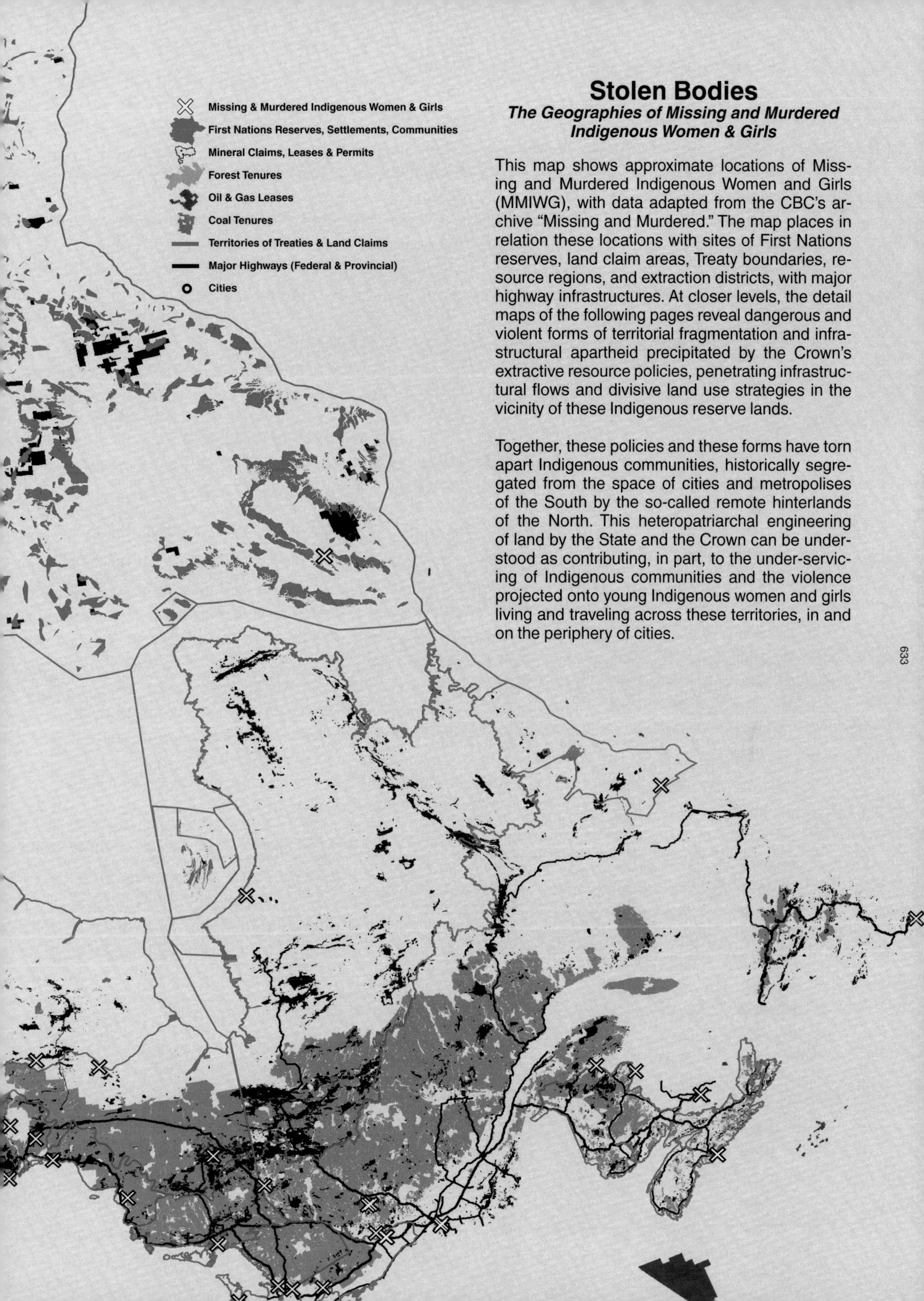

Stolen Bodies

The Geographies of Missing and Murdered Indigenous Women & Girls

This map shows approximate locations of Missing and Murdered Indigenous Women and Girls (MMIWG), with data adapted from the CBC's archive "Missing and Murdered." The map places in relation these locations with sites of First Nations reserves, land claim areas, Treaty boundaries, resource regions, and extraction districts, with major highway infrastructures. At closer levels, the detail maps of the following pages reveal dangerous and violent forms of territorial fragmentation and infrastructural apartheid precipitated by the Crown's extractive resource policies, penetrating infrastructural flows and divisive land use strategies in the vicinity of these Indigenous reserve lands.

Together, these policies and these forms have torn apart Indigenous communities, historically segregated from the space of cities and metropolises of the South by the so-called remote hinterlands of the North. This heteropatriarchal engineering of land by the State and the Crown can be understood as contributing, in part, to the under-servicing of Indigenous communities and the violence projected onto young Indigenous women and girls living and traveling across these territories, in and on the periphery of cities.

The Act's definition of "Indian" is conditioned by the reduction of Indigenous women's identity to primarily, if not exclusively, that of ungovernable sexual beings, appropriately treated as "sub-humans."[10] The *Indian Act* imports into the structure of Indian governance in Canada the widespread stereotypical portrayal of the Indigenous woman as "squaw," described by Métis scholar Emma LaRoque as a being with no human face who is lustful, immoral, unfeeling, and dirty.[11] This stereotype is applied to all Indigenous women, whether they are subject to the Indian Act or not.

Embedding that stereotype into legislation gives it the legitimacy of government approval, and in turn gives Canada an interest in its continued survival. The flourishing of this stereotype in the wider society is difficult, if not impossible, to curtail as long as it remains a centerpiece of official policy. The legislative scheme constructed on that stereotype drives women into exile, separates them from their families and impoverishes them and their children if they do not conform to the model of demure Victorian wife imposed upon Indigenous women.

Violence Against Indigenous Women in Canada

The patterns of violence against Indigenous women in Canada are endemic, pandemic and horrifying. Between 1997 and 2000, the rate of homicide for Indigenous women was 5.4 per 100,000 compared with 0.8 per 100,000 for non-Indigenous women.[12] The Sisters in Spirit research program of the Native Women's Association of Canada (NWAC) found that between the 1960s and 2010, 582 Indigenous women and girls went missing or were murdered in Canada.[13] Two-thirds of the cases are murders, one-fifth are disappearances, and the remainder are suspicious deaths or unknown.[14] The majority of the victims were under the age of thirty-one and many were mothers.[15] An updated report as of March 2013 found a total of 668 missing or murdered Indigenous women and girls.[16]

Further scrutiny has revealed even higher rates of murder and disappearance. Using the same type of public sources available to Sisters in Spirit, researcher Maryanne Pearce found 824 missing or murdered Indigenous women in the years 1990 to 2013.[17] In May 2014, the RCMP stunned observers by revealing, for the first time, statistics it had compiled across all federal, provincial, and municipal police forces in Canada, showing that nearly 1,200 Indigenous women have been murdered or gone missing over the past thirty years.[18] About one thousand of these women are murder victims. This is the first time that official police statistics have been disclosed, and these numbers double those found by Sisters in Spirit. RCMP Commissioner Paulsen stated that although 4 percent of the women in Canada are Indigenous, 16 percent of Canada's murdered women and 12 percent of Canada's missing women are Indigenous, "clearly an overrepresentation." The same day these statistics were disclosed, the Government of Canada once again refused to hold a national inquiry into the murders and disappearances of Indigenous women.[19]

In 2004, Amnesty International Canada kicked off a period of intense international scrutiny of the violence against Indigenous women in Canada with the publication of its *Stolen Sisters* report,[20] characterizing violence against Indigenous women as a violation of their domestic and international human rights. Before that, attention had been drawn to the murder and disappearance of Indigenous women in various ways, including Ryga's 1967 play, *The Ecstasy of Rita Joe*,[21] Amber O'Hare's database of missing and murdered women,[22] the report of the Aboriginal Justice Inquiry of Manitoba into the murder of Helen Betty Osborne in The Pas in 1971,[23] and individual cases like the conviction of two Regina university students for murdering Pamela George in 1995[24] and the conviction of John Martin Crawford for the murder of three young Indigenous women in Saskatchewan in the 1990s.[25]

With increased attention to incidents of violence against Indigenous women, long-buried historical cases came into public view, like the sexual assault of young women at the Cariboo Indian Residential School by Father (later Bishop) Herbert O'Connor in 1961, for which he was charged in 1991,[26] and the sexual assault of two teenage girls by Reform Party MP (and Justice critic) Jack Ramsay in 1969 when he was an RCMP constable in northern Saskatchewan.[27]

The behavior of the justice system in contemporary cases also came under scrutiny, including the decision-making of the Vancouver Police Department with respect to the numerous murders of women in the Downtown East Side of Vancouver (for many but not all of which Robert Pickton was later prosecuted and convicted),[28] and questioning of what was known locally, but whom and when about the abuse of young Indigenous women by sitting judge David Ramsay in Prince George, British Columbia.[29] The perception grew that abuse of Indigenous women is more acceptable to the courts than abuse of non-Indigenous women.[30]

Despite the attention now focused on violence against Indigenous women in Canada, efforts to gather information have been frustrated by government actions. The Sisters in Spirit research program of NWAC was funded from 2005 until 2010, but the funding was not renewed.[31] When a Commission of Inquiry was set up in British Columbia to study police decision-making with respect to the rash of murders in Vancouver's Downtown East Side, the government of British Columbia refused to fund Indigenous groups (including NWAC and the communities of the deceased women) to participate in the Inquiry. The Commission itself declared that it was not in the public interest to deny funding to these groups.[32] The Conservative majority on a Parliamentary Committee blocked it from recommending a national inquiry into the murders and disappearance in a report issued in 2014 after examining the phenomenon for several months.[33]

Indigenous victims of violence are not seen as women with families and communities, or rounded human lives; they are all too often reduced to the stereotypes of women who frequent bars or dangerous urban areas, engage in prostitution or have "high-risk" life-

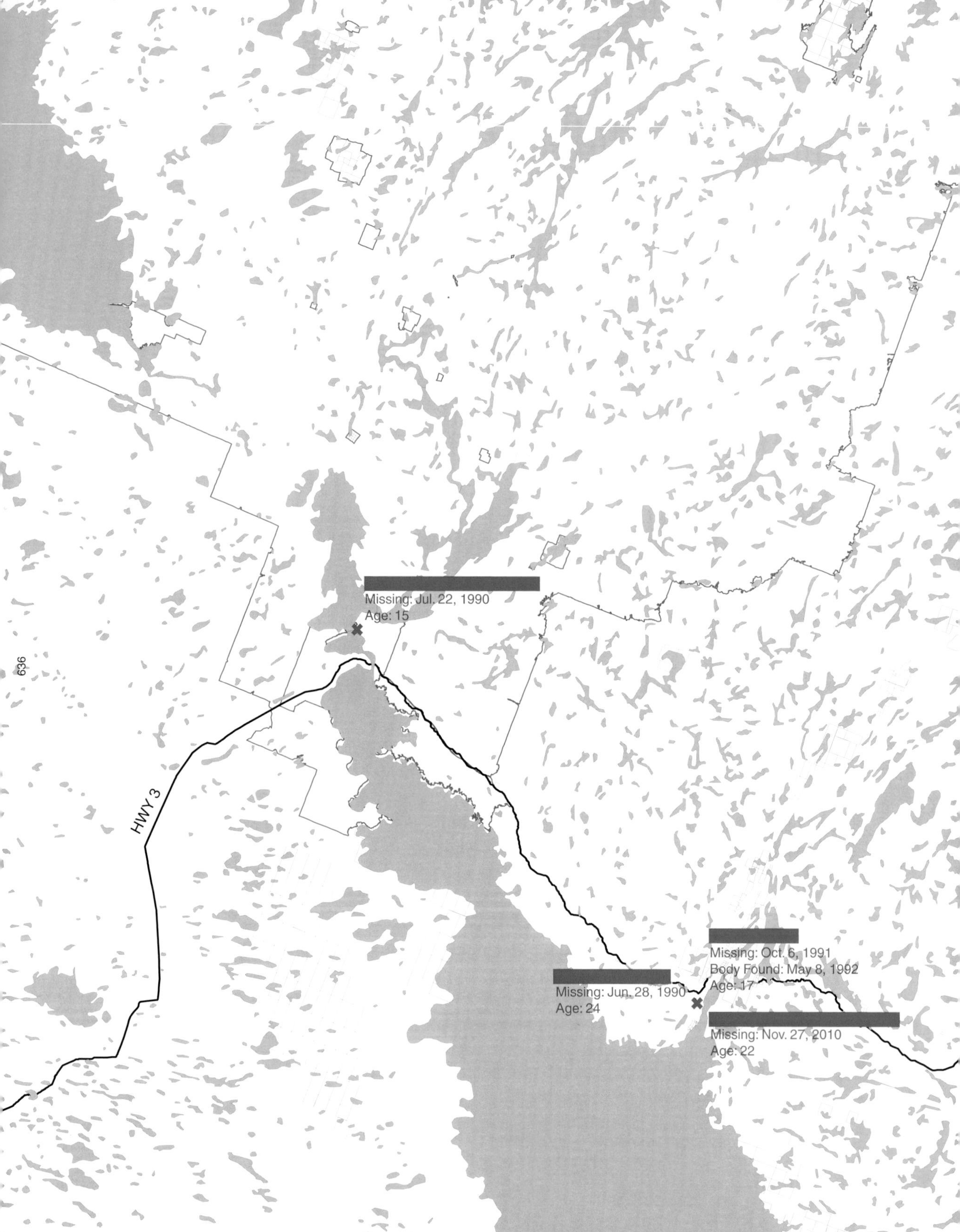
Missing: Jul. 22, 1990
Age: 15
HWY 3
Missing: Oct. 6, 1991
Body Found: May 8, 1992
Age: 17
Missing: Jun. 28, 1990
Age: 24
Missing: Nov. 27, 2010
Age: 22

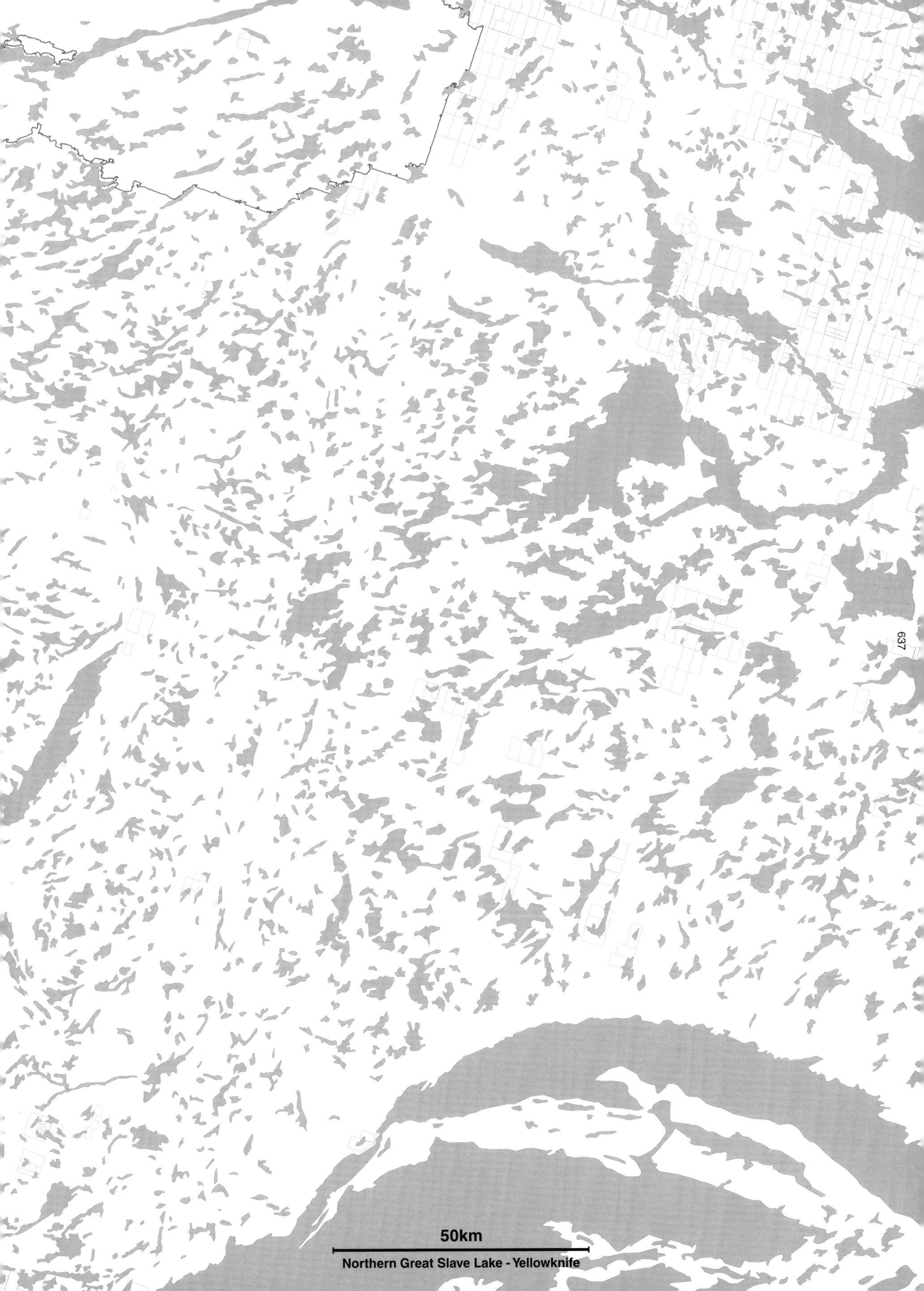
50km
Northern Great Slave Lake - Yellowknife

styles. These characterizations are made by police[34 35] and also by media and public commentary.[36 37] In one case, the court attributed an adult level of sexual agency to a twelve-year-old Indigenous girl sexually assaulted by three white men in their twenties.[38] By contrast, those accused of harming Indigenous women are sometimes portrayed as "normal," and a positive factor in their sentencing is the support they get from family and community.[39 40 41 42] In some cases, the police, or the family or community of those eventually accused, may have known of the crime for some time before it came to light,[43] or families may have helped to conceal it.[44] Police may have been aware of practices endangering Indigenous women but done nothing about them.[45 46 47]

Women disappear from or are found murdered in isolated areas, along highways or on vacant land outside cities, and also in urban areas where the most poor and marginal are forced to live. They may be missing for some time before their families are able to interest police in the disappearance. The Oppal Commission found that barriers in the reporting process contributed to delays in investigation. In some cases, families experienced degrading and insensitive treatment, and in a few, "the barriers were so pronounced as to amount to a denial of the right to make a report."[48] Women and their families do not, and perhaps cannot, expect help from the police, because of indifference, incompetence, or even the involvement of police in misconduct of their own.[49 50 51] It has been suggested by a number of observers that if the missing or murdered woman were white instead of Indigenous, public and police interest would have been more forthcoming.[52 53 54]

There have been many calls for a national commission of inquiry into the missing and murdered Indigenous women. Domestically, these calls come from groups as diverse as NWAC, Amnesty International and the Feminist Alliance for International Action, on the one hand,[55] and the premiers of Canada's provinces and territories on the other.[56] Internationally, pressure has come from the U.N. Human Rights Council (in 2009 and 2013), the Committee to Eliminate All Forms of Discrimination Against Women (2008), and most recently by the United Nations Special Rapporteur on Indigenous Issues, Dr. James Anaya, in the fall of 2013.[57] The Government of Canada repeatedly refuses to undertake an inquiry.[58]

Foundations of the *Indian Act*

Canada is an example of "settler colonialism."[59] Settler colonies practice "internal colonization," where "the dominant society coexists on and exercises exclusive jurisdiction over the territories and jurisdictions that Indigenous peoples refuse to surrender."[60] Tully asserts that the colonizer aims to resolve his contradiction in the long term "by the complete … disappearance of the Indigenous peoples as free people with the rights to their territories and governments."[61] One strategy for accomplishing this objective is that the Indigenous peoples would become "extinct in fact," through dying out, intermarriage, urbanization, or extinguishing their will to resist assimilation.[62]

The *Indian Act* has been preoccupied with extinction in fact. Deputy Superintendent of Indian Affairs Duncan Campbell Scott infamously proclaimed to Parliament in 1920, "Our objective is to continue until there is not a single Indian in Canada that has not been absorbed into the body politic, and there is no Indian question, and no Indian Department, and that is the whole object of this Bill."[63]

S.91 (24) of the *Constitution Act, 1867* gives the Federal Government jurisdiction over "Indians and lands reserved for Indians." Canada has sought to define "Indians" as narrowly as possible, so as to restrict the numbers for whom it is responsible. However, in the Eskimo Reference, the Supreme Court of Canada held that Inuit are included, and in 2014 the Federal Court of Appeal included Métis.[64][65]

The major legislative enactment under s.91(24) is the *Indian Act*. It establishes a comprehensive regime of governance directed at "lands reserved for Indians." The *Indian Act* provides for the existence of bands, which are composed of "Indians," and for allocation to a band of a reserve. Only members of that band may occupy or use the reserve. If the band ceases to exist, then the land is taken by the Crown. A band exists only so long as there are members of the band who are recognized by Canada as entitled to share in its land.

A narrow definition of "Indian" furthers Canada's own land ambitions in several ways: the fewer Indians it recognizes, the less land must be allocated as reserves in the first place; and the more people who are excluded from bands, the more quickly the Indian population will shrink. The faster the bands shrink and ultimately disappear, the more quickly the land may be taken by Canada. Beneficiary programs for Indians, such as education and health care, will also cost the Federal Government less to the extent that the number of Indians is reduced.

Although the Indian Act has never used the term "status," those recognized by Canada for purposes of the Act are commonly referred to as "status Indians."[66][67] Indian status is seen by Canada as inferior to the full citizenship of the "normal" person. However, Indian status is also a form of privilege: as Palmeter[68] observes, "with regard to accessing programs and services, land, natural resources, and seats at self-government negotiating tables, the real question is not whether one is a citizen of a Mi'kmaq, Cree, or Mohawk, but whether one is an Indian and a band member." Sharon McIvor has argued that status confers cultural identity and belonging.[69]

The Victorian view of women and the family embedded in the *Indian Act* features the male as the patriarch and the female his dependent and obedient wife.[70] The relationship between husband and wife paralleled that between master and servant.[71] By depriving married women of property, "the law deprived them of legal existence, of the rights and responsibilities of other citizens, and thus of self-respect."[72] While adopting these elements of the

Missing: Oct. 2007
Age: 24
Body Found: Jan. 27, 2001
Age: 24
Body Found: Jun. 22, 2011
Age: 36
Murdered: Apr. 3, 2005
Age: 13
Missing: May 12, 2004
Body Found: Apr. 19, 2015
Age: 32
Missing: Jun. 1993
Age: 46
Missing: Dec. 9, 2004
Age: 21
Missing: Sep. 2000
Age: 29
Missing: Jul. 6, 19
Age: 21
Missing: Jul. 2013
Age: 25
Missing: Sep. 1, 1976
Age: 14
QE II
Body Found: Mar. 4, 2011
Age: 15
Body Found: Jun. 4, 2014
Age: 35
Victoria's Secret

640

HWY 63
Missing: Oct. 13, 2004
Age: 15
641
HWY 16
50km
North Saskatchewan River - Edmonton

Victorian family, the *Indian Act* made one crucial exception. In Victorian settler society, the mother had responsibility in fact for rearing children, even though she had no legal authority to make decisions about them and no right to their custody. The *Indian Act*, by contrast, provided that Indian children would not be raised by their parents at all.

Nicholas Davin,[73] architect of the residential school system, observed: "The Indian himself is a noble type of man, in a very early stage of development ... the race is in its childhood." Declaring that "one of the earliest things an attempt to civilize them does, is to take away their simple Indian mythology," he continued, "to disturb this faith, without supplying a better, would be a curious process to enlist the sanction of civilized races whose whole civilization, like all the civilizations with which we are acquainted, is based on religion."[74]

Davin[75] urged the government to utilize missionary schools and specified the role that religious women would play: "the influence of civilized women must be constantly present in the early years ... the plan is now to take young children, give them the care of a mother, and have them constantly in hand. Such care must go *pari passu* with religious training."[76] In this plan, Indigenous mothers would be replaced by *civilized white* mothers who would indoctrinate children in the practices and the faiths of the settler regime.

The residential schools run by Canada in conjunction with the major religious denominations had the twin goals of civilizing and Christianizing.[77] The government considered that to achieve these goals, children had to be separated from their families.[78] This policy tore the heart from Indigenous women's role in the family[79][80] as it tore the children from their families and communities.

Among the many reasons why the Indigenous mother was considered an inappropriate influence on her own children was the alleged hypersexuality of Indigenous women, a characterization that was essential to the justification of colonization. Indigenous women were "constructed as lascivious, shameless, unmaternal, prostitutes, ugly, and incapable of high sentiment or manners—the dark, mirror-image to the idealized nineteenth-century visions of white women."[81] Aboriginal women represented "the wild";[82] they are almost wholly sexualized by settler society.[83] They were rarely permitted any other form of identity.[84] Barman attributes much of the white preoccupation with controlling Indigenous women's sexuality to a fear of their exercise of autonomy, the ultimate threat to the Victorian patriarchal family.[85][86] Protecting the Victorian patriarchal family of the settler was, in effect, protecting the white race that was reproduced through that family.

White[87] says French observers "tended to select material that made the women seem merely a disorderly and lewd set of Europeans, not people following an entirely different social logic." One striking example of this is the refusal of colonial society, and the Canadian government, to acknowledge the Indigenous acceptance of divorce and remarriage, an omission that imparted an illicit characterization to any Indigenous conjugal union but the first.[88][89][90] This colonial characterization of Indigenous women was self-serving: it was used to deflect criticism from misbehaving government agents and from police who abused Indigenous women or did not protect them,[91] or to allow settler men to abuse women with impunity.[92][93] This portrayal of Indigenous women as dangerously sexual helped justify the placing of Indigenous peoples on remote reserves and introducing a pass system to confine residents there.[94][95][96]

Differentiating Indigenous women from white settler women was meant to reinforce the divide between Indigenous and white races. This practice emerged when the itinerant fur trade started giving way to established trading posts and with the upswing in settlement, both developments that brought more European women to Canada.[97][98][99] Unions between European men and Indigenous women had been a feature of the fur trade, though they were comparatively rare in the settled eastern part of Canada,[100] and they had underwritten the success of the fur trade by opening up valuable networks and skill sets to the European traders.[101][102][103]

However, such unions were seen to pose grave threats to the establishment of a land-based white settler society. If included in the white population the mixed-race children of such unions could have claims to land set aside for settlers; they could also challenge their white kin for possession of their father's property and estate, compromising whiteness as an entitling factor.[104][105][106] Including too many mixed-race children amongst the acknowledged Indian population could have required additional lands to be added to reserves.[107]

Inter-racial unions threatened not just whites' preferential access to land, but also the racial hierarchy itself. Perry observes that a white man who married an Indigenous woman was seen as "dangerously flirting" with relinquishing his place in the civilized race and becoming deracinated.[108] Rather than upholding the superiority of the white race, the man became a "squaw man," corrupting and degrading the white race.[109]

The answer to all of these challenges was a two-level enforcement by the *Indian Act* of the Victorian family ideal. Indians on reserve were to have conventional Victorian unions, with the husband as the controlling partner and the wife under his dominance. If an Indian woman married a white man, she would follow her husband off the reserve (and the Indian register) like a good Victorian wife.[110] In the words of the Federal Government, this would involve the woman's transition "from dependence upon the Indian community and its special position under our law to dependence upon her husband in the ordinary circumstances of the larger community."[111]

This move from the reserve to the broader world was seen as the achievement of "civilization."[112] For the Federal Government, "off-reserve residence has tended to carry an assumption that the integration process was proceeding satisfactorily."[113] However, in the

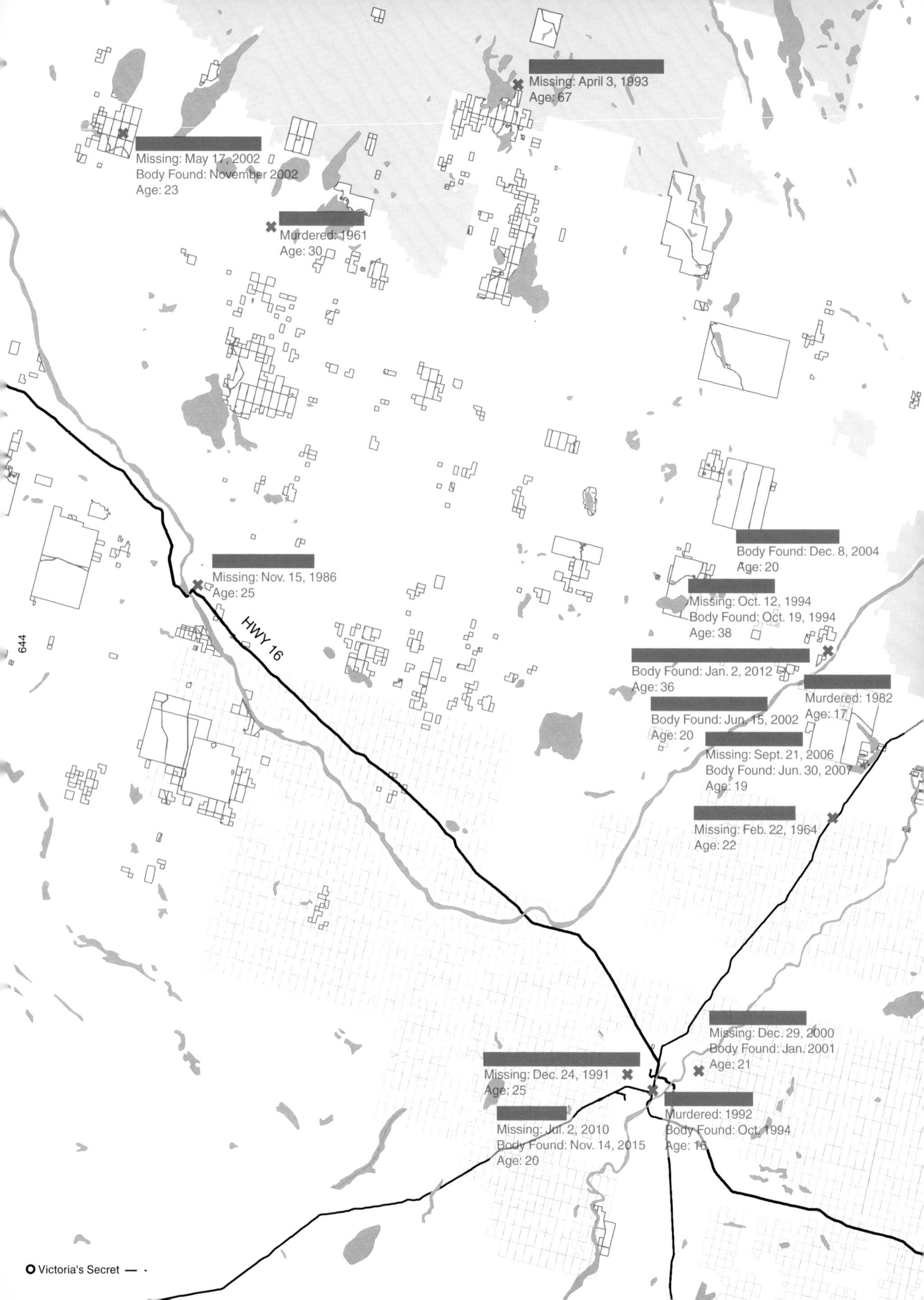

Missing: April 3, 1993
Age: 67
Missing: May 17, 2002
Body Found: November 2002
Age: 23
Murdered: 1961
Age: 30
Body Found: Dec. 8, 2004
Age: 20
Missing: Nov. 15, 1986
Age: 25
Missing: Oct. 12, 1994
Body Found: Oct. 19, 1994
Age: 38
HWY 16
Body Found: Jan. 2, 2012
Age: 36
Murdered: 1982
Age: 17
Body Found: Jun. 15, 2002
Age: 20
Missing: Sept. 21, 2006
Body Found: Jun. 30, 2007
Age: 19
Missing: Feb. 22, 1964
Age: 22
Missing: Dec. 29, 2000
Body Found: Jan. 2001
Age: 21
Missing: Dec. 24, 1991
Age: 25
Missing: Jul. 2, 2010
Body Found: Nov. 14, 2015
Age: 20
Murdered: 1992
Body Found: Oct. 1994
Age: 16

Body Found: Oct. 29, 2013
Age: 40
Missing: Dec. 6, 1989
Age: 26
50km
South Saskatchewan River - Saskatoon

case of women who were forced off reserve for marrying a non-status man, the government was totally indifferent to whether they, and their children, were actually faring well in the new environment.

"To Be an Indian Is to Be a Man"

This opening sentence of Canada's White Paper sums up over one hundred years of Indian policy.[114] Under the rules for determining Indian status that endured in almost identical form until 1985,[115] it only took one Indian parent to make a person eligible for Indian status: the father. A status Indian would confer status on his children, and also on his wife if she were not already a status Indian. By contrast, an Indian woman who married a non-Indian (that is, "married out") would cease to be an Indian, and her children could not be registered as Indians. For purposes of this rule, Canada accepted the validity of marriage between Indigenous women and non-Indigenous men performed according to Indigenous custom: doing so meant that more women would be removed from Indian status.[116]

Although it had been rare for a white woman to marry an Indian man,[117] by the 1960s the practice had become common enough to be cause for concern to Indian women.[118] [119] The 1951 *Indian Act* introduced loss of Indian status at twenty-one for anyone whose mother and grandmother had both acquired status by marrying a status male ('the double-mother rule"). Starting in July 1980 the Minister of Indian Affairs offered bands the option of being exempted from the operation of this double-mother rule, and also from the rule that women who married a non-Indian would lose status.[120] By July 1984, 54 percent of bands had opted for exemption from the double-mother rule, while only 18 percent had asked to be excused from the operation of the marrying out rule.[121] [122] These were choices of a male leadership cadre in Indian bands that was becoming comfortable with the role of Victorian patriarch.

A woman who lost status upon marriage could not live on or visit the reserve, inherit reserve property or participate in the political and cultural life of the reserve. Mary Two-Axe Earley lamented that her marriage to a non-status man meant she could not be buried on her reserve, although it had on its land a cemetery where outsiders could inter their pets.[123] The women's children were similarly excluded. Even after widowhood or divorce, the woman could not regain her Indian status except by marrying a status Indian man.

We do not know how many women and children were excluded this way, for Canada kept no statistics. However, it has been estimated that between 1985 and 1999, changes introduced by Bill C-31 increased the number of Status Indian by about 174,000 individuals.[124] It is also estimated that about 40,000 people will move onto the Indian register as a result of the 2010 amendment to the Act contained in Bill C-3.[125] We can infer from the relatively large numbers of those still alive and seeking to regain status after 1985 that over the previous century the marrying out rule was responsible for a massive exile of women and children. Justice Laskin of the Supreme Court called the rule a "statutory excommunication" and a "statutory banishment" of Indian women who marry non-Indians.[126]

This exile heightened the vulnerability of women and children. Should an exiled woman be deserted, widowed or divorced, she could not return to her family and community. Should she lose the financial support of her husband, Canada would not resume its obligations to her. The marrying out rules, like the residential schools, fractured Indigenous families. Pearce has found that "the most striking" risk factor for violence against an indigenous woman is being separated from her family.[127]

Women who did not lose status on marriage were also discriminated against by the *Indian Act's* embrace of the Victorian family model. Indian women on reserve were legal nullities, having no right to vote or stand for election in band elections until past the middle of the twentieth century. Permission to occupy reserve land was granted preferentially to Indian men. A woman leaving an abusive marriage usually could not get her own reserve residence. Unless they could move in with another on-reserve family member, she and her children would have to leave the reserve, another instance of exile and family fragmentation being caused by the *Indian Act*.

The situation became even worse after the Supreme Court of Canada held in Derrickson[128] and Paul[129] that provincial family law dealing with occupation of the matrimonial home and division of matrimonial real property did not apply on reserve. These rulings left a legal gap, because there was no federal law dealing with these questions.[130] [131] [132] Not only were women left without substantive legal protection, they also suffered from "an equally significant … gap in access to the court system, access to legal aid … policing, enforcement of law … and many other areas."[133] Despite many calls for action,[134] [135] the gap persisted until the *Family Homes on Reserves Act* of 2013.

Although the *Indian Act* did not contain specific provisions allowing department officials to interfere with Indian women's sexual behavior within their families, the Indian agent possessed broad discretion to do so. Sangster found that the Indian Affairs filing system had a whole category dealing with immorality on reserves.[136] In her study based on Agency records for Manitoulin Island and Parry Sound, Brownlie provides details of the Department's campaign between World Wars I and II to confine women to conventional patriarchal marriages. Indian agents used financial threats, like denying women treaty and interest payments on the ground of sexual transgression, and refusing or delaying provision of relief.[137] Agents might also send to residential school the children of a woman regarded as adulterous.[138] Ironically, in the same period, there are many stories of Indian agents and police coercing sex from Indian women in return for rations or other favors.

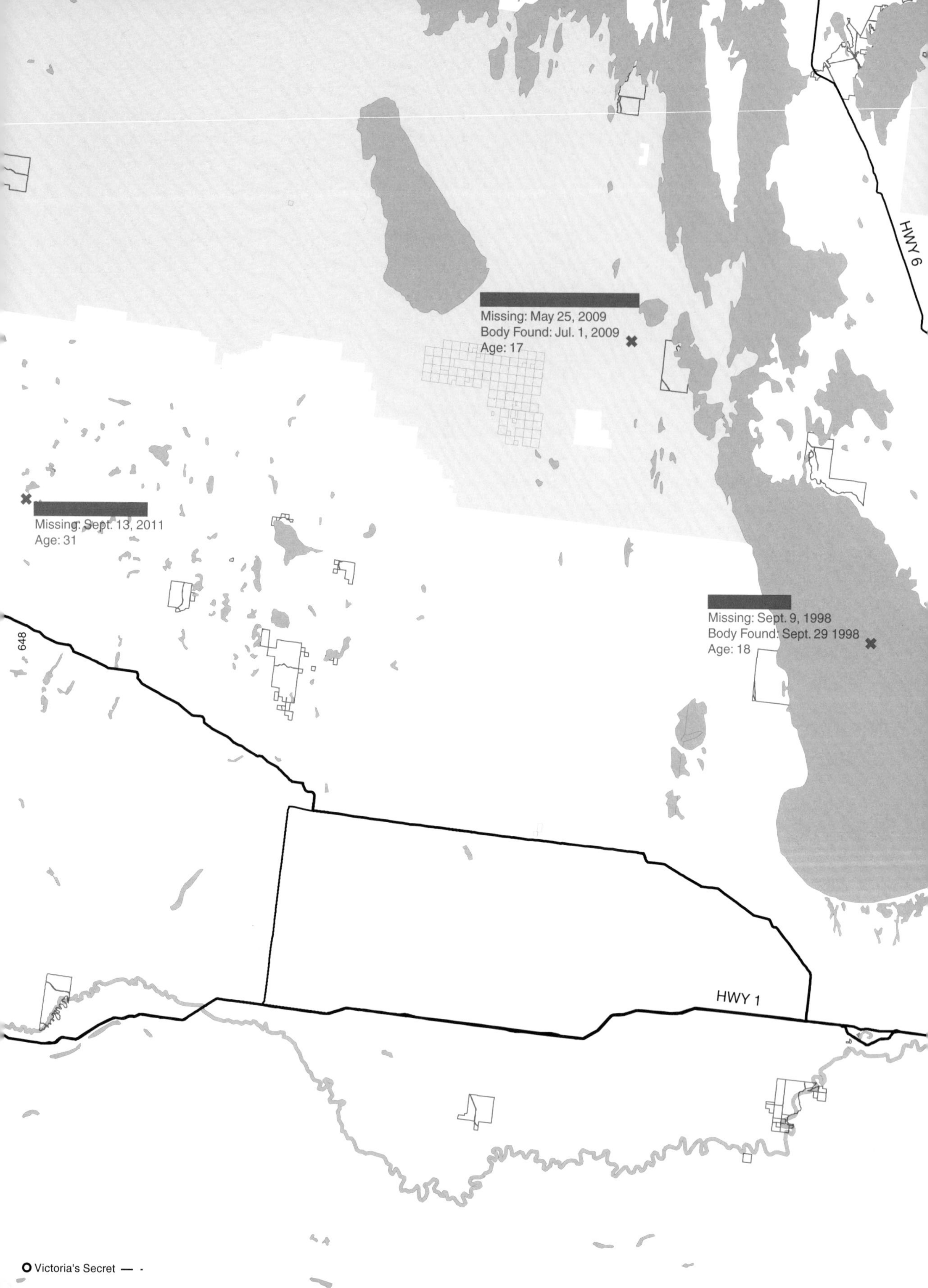

HWY 6
Missing: May 25, 2009
Body Found: Jul. 1, 2009
Age: 17
Missing: Sept. 13, 2011
Age: 31
Missing: Sept. 9, 1998
Body Found: Sept. 29 1998
Age: 18
648
HWY 1
Victoria's Secret

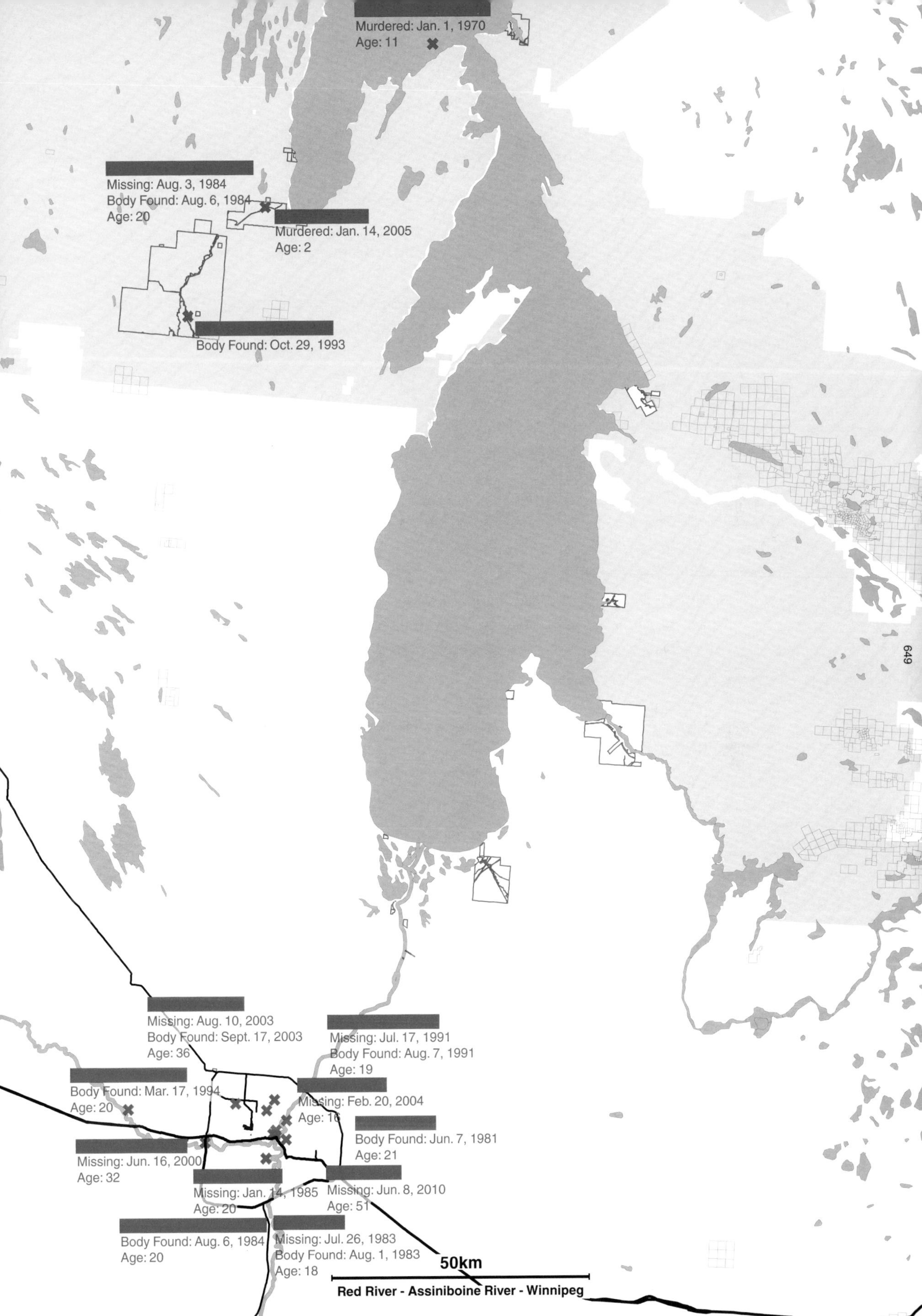

Murdered: Jan. 1, 1970
Age: 11
Missing: Aug. 3, 1984
Body Found: Aug. 6, 1984
Age: 20
Murdered: Jan. 14, 2005
Age: 2
Body Found: Oct. 29, 1993
Missing: Aug. 10, 2003
Body Found: Sept. 17, 2003
Age: 36
Missing: Jul. 17, 1991
Body Found: Aug. 7, 1991
Age: 19
Body Found: Mar. 17, 1994
Age: 20
Missing: Feb. 20, 2004
Age: 16
Body Found: Jun. 7, 1981
Age: 21
Missing: Jun. 16, 2000
Age: 32
Missing: Jan. 14, 1985
Age: 20
Missing: Jun. 8, 2010
Age: 51
Body Found: Aug. 6, 1984
Age: 20
Missing: Jul. 26, 1983
Body Found: Aug. 1, 1983
Age: 18
50km
Red River - Assiniboine River - Winnipeg

649

Indigenous women's activism in the 1960s and 1970s attacked both the *Indian Act's* discrimination against status women on reserve[139] and also the marrying out rules. Indian Rights for Indian Women organizations in Alberta and Quebec secured from the Royal Commission on the Status of Women a recommendation that women should not lose status for marrying a non-status man and should be able to pass status to their children.[140] Jeanette Lavell from the Wikwemikong Unceded Territory on Manitoulin Island and Yvonne Bédard of Six Nations challenged their loss of status upon marriage under the *Canadian Bill of Rights* equality before the law guarantee.

The majority of the Supreme Court ruled against them. Mr. Justice Ritchie observed that rules about the status of Indian women who marry non-Indians were imposed as "a necessary part of the structure created by Parliament for the internal administration of the life of Indians on reserves and their entitlement to the use and benefit of Crown lands."[141] Pointing out that these rules had been in effect for more than one hundred years, Justice Ritchie stated that any change to them had to be accomplished by specific, highly targeted legislation, and not by "broad general language directed at the statutory proclamation of the fundamental rights and freedoms of all Canadians."[142]

Much the same reasoning had been used in the late 1920s when the "Five Persons"—white settler women from the Canadian establishment—argued in the Supreme Court that because "person" includes both male and female under modern-day *Interpretation Acts*, the term "Person" in the *Constitution Act, 1867* should be read to include women, thus permitting their appointment to the Senate. Chief Justice Anglin stated that it would be "dangerous to assume that by the use of the ambiguous term 'persons' the Imperial Parliament meant in 1867 to bring about so vast a constitutional change affecting Canadian women as would be involved in making them Privy Councillors."[143] This holding was almost immediately overturned by the Privy Council in London, which rejected the Supreme Court's protection of the discriminatory status quo.[144] However, this discredited reasoning was still used by the Supreme Court of Canada almost fifty years later to justify the legislated inequality of Indigenous women.

While Yvonne Bédard and Jeannette Corbière Lavell received support from women's organizations and from the Native Council of Canada (representing non-status Indians), major Indian groups like the National Indian Brotherhood and the Alberta chiefs opposed them. Indian leaders feared that a decision making the *Indian Act* subject to the *Canadian Bill of Rights* "would wipe out the *Indian Act* and remove whatever legal basis we had for our treaties."[145] To prevent this, the leadership decided to intervene in the Supreme Court against the women.[146] Despite the Act being upheld by the Supreme Court, Cardinal recounts that the case brought renewed attention to the political urgency of *Indian Act* revisions.[147]

No revisions to the *Indian Act* resulted from the Lavell and Bédard case. What did result was exemption of the *Indian Act* from the *Canadian Human Rights Act* passed in 1977. The exemption was necessary, said the Justice Minister, because the government promised not to revise the *Indian Act* without consulting the National Indian Brotherhood and others.[148 149]

No amendments to the *Indian Act* resulted from this protected process.[150] However, the exemption stayed in the *Human Rights Act* for just over thirty years, in spite of widespread calls for its repeal and its inconsistency with Canada's international human rights obligations.[151 152 153] It was repealed with respect to Federal Government action in 2008 and with respect to band actions in 2011.[154]

This gap in the rule of law is profoundly unconstitutional, as was the gap in family law legislation on reserve. The *Universal Declaration of Human Rights*[155] states in its preamble that it is essential that human rights should be protected by the rule of law. The Supreme Court has held that "constitutionalism and the rule of law" is one of Canada's four foundational constitutional principles. Another is respect for minority rights, which the Supreme Court has interpreted as including the rights of Indigenous peoples.[156] At its most basic, the rule of law "vouchsafes to the citizens and residents of the country a stable, predictable and ordered society in which to conduct their affairs. It provides a shield for individuals from arbitrary state action."[157] The rule of law requires the creation and maintenance of an actual order of positive laws.[158] These two prolonged gaps in the rule of law affecting Indigenous women are clearly contrary to Canada's constitutional imperatives and to its international human rights obligations.

Following the Supreme Court decision in Lavell and Bédard, activist Maliseet women took the marrying out rule to the U.N. Human Rights Committee under the Optional Protocol to the *United Nations Convention on Civil and Political Rights*, with Sandra Lovelace as the case's named complainant.[159] Article 27 of the Convention provides that members of ethnic, religious or linguistic minorities shall not be denied the right, in common with other members of their group, to enjoy their own culture, to profess and practice their own religion or to use their own language. Ms. Lovelace, a fluent speaker of Maliseet, wanted to live on the Tobique reserve, with her son, after her marriage to a non-Indian broke up. She was prevented from doing so because of her loss of status. While holding that Article 27 does not guarantee her the right to live on reserve, the Committee found that her rights under Article 27 were interfered with because there is no place outside the Tobique reserve where she can access a like community. It ruled that to deny Sandra Lovelace the right to reside on the reserve does not seem "reasonable, or necessary to preserve the identity of the tribe."[160]

O Victoria's Secret — ·

Murdered: Dec. 6, 2002
Age: 13
Murdered: Nov. 1986
Age: 51
50km
Frobisher Bay - Iqaluit

In 1985, after the Lovelace decision and coming into force of the Charter's equality guarantees, Canada passed legislation (Bill C-31) to change the registration provisions of the *Indian Act*. A new section, 6(1)(a), affirmed eligibility for status of all of those who had qualified for it under the old legislation. Women who had lost status when they married non-Indians were made eligible to return to status under s. 6(1)(c).

However, Bill C-31 worsened women's situation by replacing the one-parent rule for determining status. Instead of keeping that rule, and allowing either the mother or the father to confer full status on a child, Bill C-31 enacted a new two-parent rule. From 1985 on, under s. 6(1)(f), a person was registrable as an Indian only if both of his or her parents were eligible for status under s. 6(1). The only exception was the new s. 6(2) which allowed registration of those who had one parent registered or registrable under s. 6(1) of the Act. The children of women restored to status under s. 6(1)(c) acquired their status under s. 6(2), as they had only one Indian parent. However, this status was short-term; the section was nicknamed the "second-generation cut-off," because status derived under s. 6(2) did not count for purposes of the two-parent rule. Those restored to status under s. 6(2) would have to parent with a registered Indian in order to be able to pass along status.

Brownlie[161] tells us that by World War II, the Department of Indian Affairs had been forced to abandon its campaign to enforce European-style patriarchal marriage. However, the two-parent rule is actually the crowning achievement of this campaign: it compels an *Indian Act* family modeled after the monogamous patriarchal Victorian family, with two status Indian parents and no crossing of race lines in the production of offspring. It is deeply disturbing that this is the result of "reform" efforts driven by contemporary human rights instruments in Canada and at the international level.

The invidious effect of the two-parent rule is made worse by the way Canada applies the rule in cases where paternity is unstated or unacknowledged. Although nothing in Bill C-31 provided that this be done, Canada changed its policy to require the signature of the father on the birth form and other forms proving paternity. Without his signature, the child's registration would be determined solely on the basis of the mother's entitlement.[162] A mother who is herself registered under s. 6(2) is not able to register her child at all. A mother who is registered under s. 6(1) is able to register her child under s. 6(2), but that child is unable to confer status on their offspring.

Many difficulties have been identified with this practice. If pregnancy was the result of abuse, incest or rape, the mother could be unwilling or unable to identify the father. A father may not want to acknowledge paternity because of concerns about being held financially responsible for the child. Where the relationship has been unstable or abusive, the mother may worry about the father asserting a right to custody or access. Privacy concerns may influence whether the father is willing to disclose paternity, or the mother is willing to ask that he do so, especially if the father is in a relationship with someone else.[163] Difficulties also arise because of lack of knowledge about the requirements or other practical barriers to compliance, which mean that even where the father is prepared to acknowledge paternity, the paperwork may not reach the Registrar in time to permit the appropriate registration.[164]

Most fundamentally, the policy on unstated or unacknowledged paternity is a retreat from previous versions of the Act. Until 1951, community acceptance of a child born out of wedlock would entitle the child to registration. After 1951, the mother's status was sufficient, unless there was actual proof that the father was not an Indian; the burden of proof was on the party seeking to disentitle the child. Now, the government presumes that the father is not a registered Indian if his identity is unknown. Moreover, under the present system, a mother registered under s. 6(2) has no right to transmit status. Under the old Act, any mother with status could pass it on to her child.

Once again, this time under the pretense of compliance with human rights guarantees, Canada has placed the Victorian yoke on the Indian woman's neck. It has done so despite the fact that by 2013, the disabilities imposed upon an "illegitimate" child (and the status of illegitimacy itself) has been removed from the common law and legislation in virtually all jurisdictions in Canada. By contrast, the *Indian Act* is making more severe the consequences for a child of having unmarried parents or an unknown father.

Between April 17, 1985, and December 31, 1999, 37,000 children with unstated paternity were born to women registered under s. 6(1), about 19 percent of the children born to s. 6(1) registered women during that period. About 30 percent of the children with unstated paternity were born to mothers under twenty years of age. It has been estimated that as many as 13,000 children born in this period to women registered under s. 6(2) may be ineligible for registration.[165] By extrapolating this figure forward to 2012, Lynn Gehl[166] calculates that since 1985 as many as 25,000 such children have been unregistered. She describes the unstated paternity policy as genocide.[167]

Sharon McIvor of the Lower Nicola Indian Band, and her son Jacob Grismer, challenged under the *Canadian Charter of Rights and Freedoms* the system of conferring status introduced in 1985. Their litigation is described in detail by Brodsky.[168] Importantly, both the British Columbia Court of Appeal and the Government of Canada responded to the narrowest possible construction of the McIvor/Grismer claim and of the Act's discriminatory history.[169]

The desired change is straightforward and simple. Sharon McIvor, Jacob Grismer, and Lynn Gehl all advocate affirmation of the status of all persons descended from either male or the female line.[170] [171] Instead of accepting this straightforward repair of past discrim-

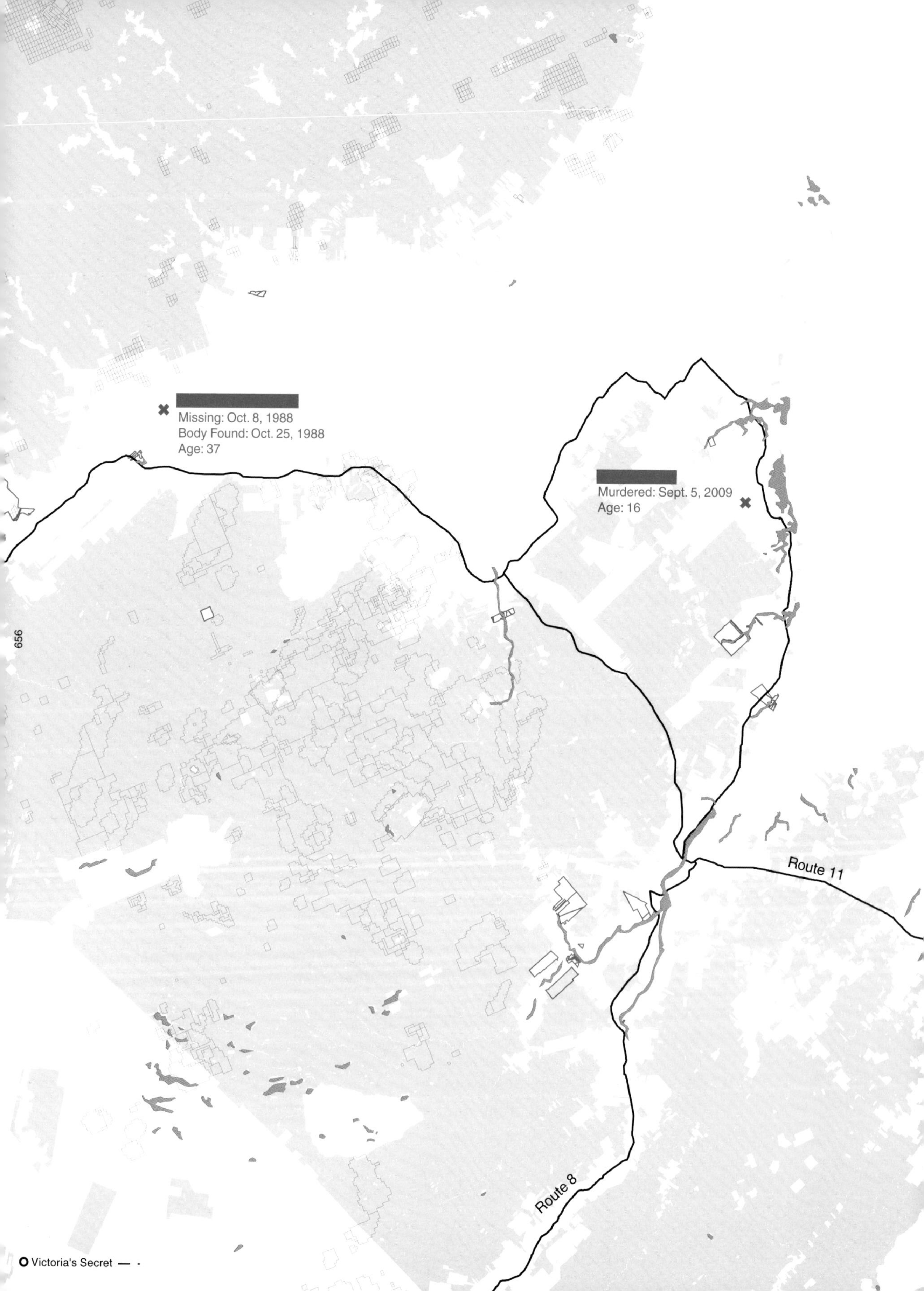

Victoria's Secret — -

For purposes of confidentiality and in respect for families of the victims, the names of Missing and Murdered Indigenous Women and Girls have been concealed.

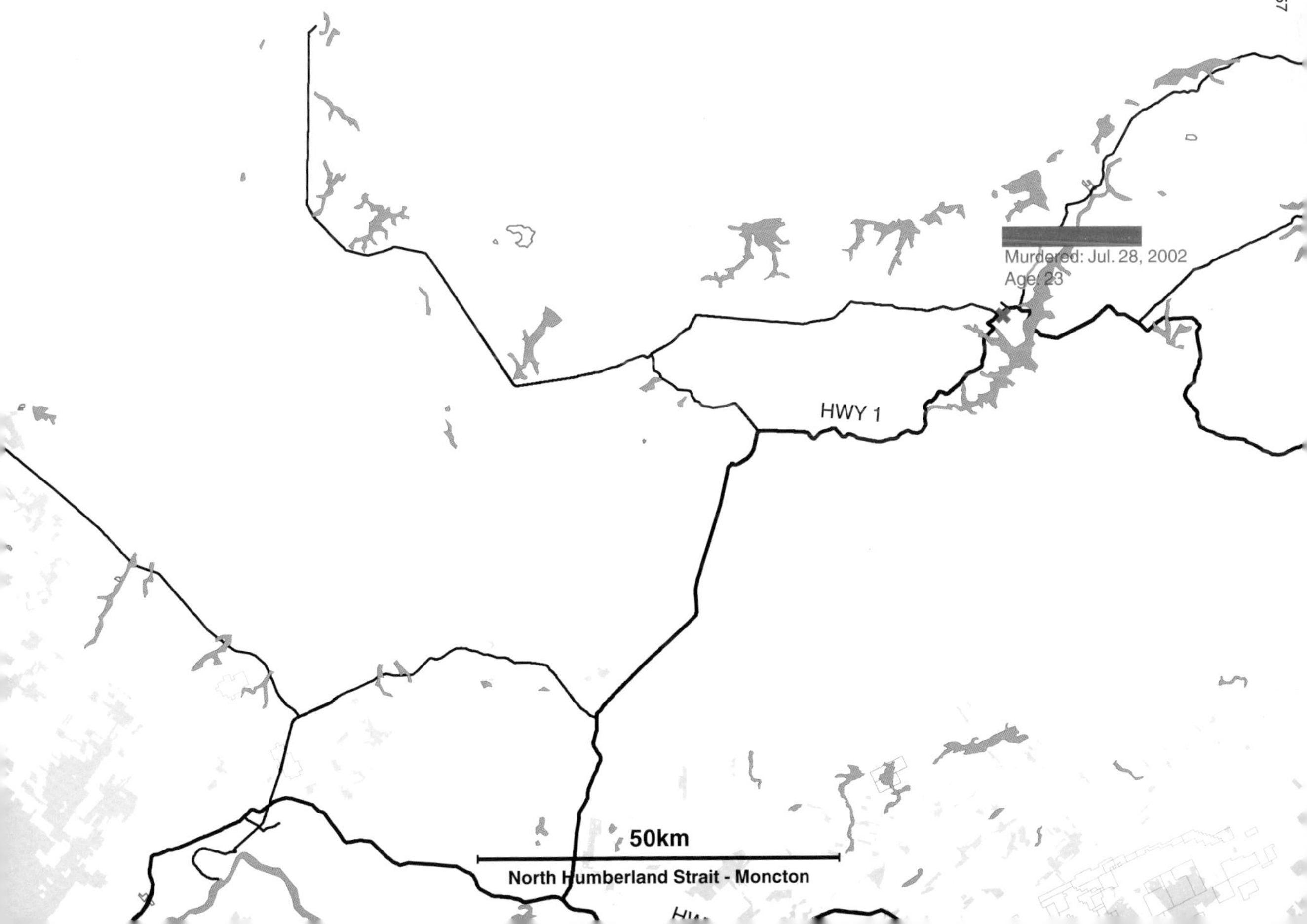

Murdered: Jul. 28, 2002
Age: 23

HWY 1

50km

North Humberland Strait - Moncton

ination, Canada enacted Bill C-3, the *Gender Equity in Registration Act* in 2010, after the decision in the McIvor case. That law merely made one more generation eligible for status, while leaving in place the historical effects of preferring the male line of descent for over a hundred years and the exclusionary effect of the two-parent rule.

Creating a Population of Prey

The profile of violence against Indigenous women corresponds closely to the profile of the treatment accorded women under the *Indian Act*. Under the Act, women have experienced repeated exile, whether to residential school, to remote reserves, or from their reserves when they marry non-Indian men or experience marital violence or breakdown. Children are exiled when their mothers deviate from narrowly prescribed norms. This statutory dispossession both causes and mirrors the dispossession of those who find themselves in urban enclaves of poverty. The theme of exile and dispossession under the Act echoes, as well, the remote and marginal sites where women are murdered, kidnapped, or found; they wind up there in death, just as in life they had to pass through there on the way to, or from, their isolated reserves. The indifference toward missing and murdered women in today's Canada mirrors the official indifference to the fate of Indigenous women who were expelled from their families and communities upon marriage to non-Indians.

That missing and murdered Indigenous women are not seen as members of families by the press and public recalls to mind the strenuous efforts made by Canada to separate women from their extended and their immediate families by means of the *Indian Act's* marrying out rules and the residential schools. The stereotype of "easy squaw,"[172] the "objects with no human value beyond sexual gratification,"[173] attached to Indigenous women who have been assaulted or murdered is exactly the same as the stereotype of rampant sexuality upon which the *Indian Act's* treatment of women is based, as part of colonization. While not the cause of the stereotype, the *Indian Act's* affirmation of it as a basis for government policy has made it virtually immune from social influences that might change or erode it. Moreover, the constricting Victorian family model in the *Indian Act*, justified the image of Indigenous women as needing control, has fractured families and impoverished and oppressed women for over a century, creating conditions of acute vulnerability.

Finally, given the history of the *Indian Act*, it should come as no surprise that Indigenous women are preyed upon in disproportionate numbers: they were expelled from the "rule of law" (however debased) that was the reserve system if they married a non-Indian; they were denied access to legal rights on the reserve right up until 2013, because Victorian wives do not have legal rights, and, for a substantial period, Canada withheld from them access to human rights law as a recourse against gender-based discrimination. Canada has demonstrated for over a century that it does not consider Indigenous women appropriate beneficiaries of the rule of law. Law is to be used only to control and confine them.

The treatment of women under the *Indian Act* is the archetype and model for the treatment of Indigenous women generally and has created of such women a population of prey. Women whose peoples have never been subject to the *Indian Act*, like the Métis and Inuit, suffer the same kinds of violation and oppression as do the women who have been ruled by the *Indian Act* for over a century. The Act captured the racist and misogynist attitudes which Victorian-era settlers and colonial administrators had toward all Indigenous peoples, gave them a safe home and carried them forward through time. Decade after decade, application of the *Indian Act* validated those invidious ideas, which should have been subjected to the change processes of a maturing constitutional democracy. From their secure hideaway in the *Indian Act*, where they were refreshed and renewed year after year, these poisonous notions continued to infect our treatment of all Indigenous women, down to the present day, even though they should long ago have been thrown on the scrap heap. Until Canada signals a change in its official view of Indigenous women by changing the *Indian Act*, Indigenous women will continue to suffer violence to a disproportionate degree.

ooo

Notes

This essay was originally published as "Victoria's Secret: How to Make a Population of Prey," in *Indivisible: Indigenous Human Rights*, ed. Joyce Green (Black Point, NS: Fernwood Publishing, 2014), reproduced with permission from Fernwood Publishing.

1 Angela Sterrit, "A Movement Rises," *OpenCanada*, November 20, 2015, https://www.opencanada.org/features/movement-rises/.

2 Human Rights Watch (HRW), *Those Who Take Us Away: Abusive Policing and Failures in Protection of Indigenous Women and Girls in Northern British Columbia, Canada* (February 2013), 25, http://www.refworld.org/docid/5209e6e94.html.

3 Native Women's Association of Canada (NWAC) and Feminist Alliance for International Action (FAFIA), *Murders and Disappearances of Aboriginal Women and Girls in Canada: Information Update for the United Nations Committee on the Elimination of Discrimination Against Women* (Ottawa, ON, 2013), 6.

4 Native Women's Association of Canada (NWAC), *What Their Stories Tell Us: Research Findings from the Sisters in Spirit Initiative* (Ottawa, ON, 2010), ii.

5 Rauna Kuokkanen, "Confronting Violence: Indigenous Women, Self-Determination and International Human Rights," in *Indivisible: Indigenous Human Rights*, ed. Joyce Green (Black Point, NS: Fernwood Publishing, 2014).

6 HRW, *Those Who Take Us Away*.

7 Martin J. Cannon and Lina Sunseri, *Colonialism and Indigeneity in Canada: A Reader* (Toronto, ON: Oxford University Press Canada, 2011), xvi, xviii.

8 Frank v. AJR Enterprises [carrying on business as Nelson Place Hotel], [1993], 23 Canadian Human Rights Reporter D/228 (British Columbia Council of Human Rights), paras. 6-12.

9 Ibid., par. 31.

10 The Honourable Wally T. Oppal, *Forsaken: The Report of the Missing Women Commission of Inquiry* (Victoria and Vancouver, BC, 2012), IIA:2.

11 Aboriginal Justice Inquiry of Manitoba (AJI), *Report: Vol.1: The Justice System and Aboriginal People* (Province of Manitoba, 1991), 479.

12 HRW, *Those Who Take Us Away*, 25, note 11.

13 NWAC, *What Their Stories Tell Us*, 20–21.

14 Ibid., 18.

15 Ibid., ii.

16 NWAC and FAFIA, *Murders and Disappearances*, 7.

17 Maryanne Pearce, "An Awkward Silence: Missing and Murdered Vulnerable Women and the Canadian Justice System" (Doctorate in Law thesis, Common Law Section, University of Ottawa, 2013), 18, 28.

18 Tonda MacCharles, "1,200 native women murdered or missing," *Toronto Star*, May 2, 2014, A3.

19 Ibid.

20 Amnesty International Canada (AIC), *Stolen Sisters: A Human Rights Response to Discrimination and Violence Against Indigenous Women in Canada* (Ottawa, ON, 2004).

21 George Ryga, *The Ecstasy of Rita Joe and Other Plays* (Toronto, ON: New Press, 1971).

22 Jim Bronskill and Sue Bailey, "Aboriginal Women Fair Game for Predators Amid Public Indifference," *canoe.com*, September 18, 2005, http://cnews.canoe.ca/CNEWS/Canada/2005/09/18/pf-1222622.html.

23 AJI, The Justice System and Aboriginal People.

24 Sherene H. Razack, "Gendered Racial Violence and Spatialized Justice: The Murder of Pamela George," in *Race, Space and the Law: Unmapping a White Settler Society*, ed. Sherene H. Razack (Toronto, ON: Between the Lines, 2002), 123–25.

25 Warren Goulding, *Just Another Indian: A Serial Killer and Canada's Indifference* (Calgary, AB: Fifth House Ltd., 2001).

26 R. v. O'Connor, [1996] CanLII 8458 (British Columbia Supreme Court), sentencing by Oppal J.

27 CBC News, "Ramsay sentenced to probation, community service," October 16, 2001, http://cbc.ca/news/canada/ramsay-sentenced-to-probation-community-service-1.269551; and http://archive.today.M9iB.

28 Oppal, *Forsaken*, vol. IIA, pt. 2; and vol. IIB.

29 HRW, *Those Who Take Us Away*, 31–33.

30 AJI, *The Justice System and Aboriginal People*, I:482.

31 HRW, *Those Who Take Us Away*, 26.

32 Oppal, *Forsaken*, IV:9–10.

33 Special Committee on Violence Against Indigenous Women, *Invisible Women: A Call to Action. A Report on Missing and Murdered Indigenous Women in Canada* (March 4, 2014).

34 Project KARE, *KARE Bulletin: Unsolved Homicides/High Risk Missing Persons* (April 12, 2012), http://www.kare.ca/images/stories/Human%20remains%20identified%20as%20Annette%20Holywhiteman.pdf.

35 Oppal, *Forsaken*, IIA:2.

36 Janice Acoose, *Iskwewak-Kah'Ki Yaw Ni Wahkomakanak: Neither Indian Princesses Nor Easy Squaws* (Toronto, ON: Women's Press, 1995), 96–97.

37 Goulding, *Just Another Indian*, 209–19.

38 Beyond Borders Inc., *Newsletter* 10 (Spring 2007): 7.

39 R. v. Edmonston, [2005] Saskatchewan Court of Appeal 7.

40 ,[2005] Saskatchewan Court of Appeal 7,..

41 Acoose, *Iskwewak-Kah'Ki Yaw Ni Wahkomakanak*, 97.

42 Razack, "Gendered Racial Violence,"123–25.

43 AJI, *Report: Vol. 2: The Deaths of Helen Betty Osborne and John Joseph Harper* (Province of Manitoba, 1991), 1–2.

44 Razack, "Gendered Racial Violence," 139–40.

45 AJI, *The Deaths of Helen Betty*, II:8.

46 AIC, *Stolen Sisters*, 50.

47 Oppal, *Forsaken*, IIB:287.

48 Oppal, *Forsaken*, IIB:284–85.

49 AJI, *The System and Aboriginal People*, I:482.

50 HRW, *Those Who Take Us Away*, 2013, 29, 31–34.

51 Oppal, *Forsaken*, IIB:284–85.

52 HRW, *Those Who Take Us Away*, 37.

53 Acoose, *Iskwewak-Kah' Ki Yaw Ni Wahkomakanak*, 87.

54 Goulding, *Just Another Indian*, 211.

55 Indian Country Today Media Network (ICTMN) Staff, "Open Letter Blasts Canada's Refusal to Convene National Missing-Women Inquiry," September 22, 2013, http://archive.today.M9iB indiancountrytodaymedianetwork.com/2013/09/22/open-letters-blasts-canadas-refusal.

56 Robert Benzie and Richard J. Brennan, "Premiers Support Demands for Probe Into Missing Aboriginal Women," *Toronto Star*, July, 25, 2013, A6.

57 Canadian Press, "James Anaya, U.N. Official, Backs Inquiry into Missing, Murdered Aboriginal Women," 2013, http://archive.today.M9iB huffingtonpost.ca/2013/10/15/ james-anaya-united-nations-inquiry_n_41081.

58 Mike Blanchfield, "Canada Rejects U.N. Call for Review of Violence Against Aboriginal Women," *Globe and Mail*, September 13, 2013, http://www.theglobeandmail.com/ news/national/canada-to-reject-un-panels-call-for-revie.

59 Royal Commission on Aboriginal Peoples (RCAP), *Report, Vol.1: Looking Forward, Looking Back* (Ottawa, ON: Printed from For Seven Generations, published by Libraxus Inc., 1991), part 1, 105.

60 James Tully, "The Struggles of Indigenous Peoples for and of Freedom," in *Political Theory and the Rights of Indigenous People*, eds. Duncan Ivison, Paul Patton and Will Saunders (Cambridge, UK: Cambridge University Press, 2000), 39; see also Patrick Wolfe, *Settler Colonialism and the Transformation of Anthropology: The Politics and Poetics of an Ethnographic Event* (London: Cassel, 1999), 1.

61 Tully, "The Struggles of Indigenous Peoples," 40.

62 Ibid.

63 Brian Titley, *A Narrow Vision: Duncan Campbell Scott and the Administration of Indian Affairs in Canada* (Vancouver, BC: University of British Columbia Press, 1986), 50n55.

64 Canada v. Daniels, [2014] Federal Court of Appeal 101; partially overturning Daniels v. Canada, [2013] Federal Court 6.

65 Until the final submission of this text for publication in October 2014, the Canadian government was appealing this decision; see Eberts, "Victoria's Secret," 148.

66 "Aboriginal Affairs and Northern Development Canada, *Secure Certificate of Indian Status* (SCIS), 2012, http://aadnc-aandc.gc.ca/eng/1100100032380.

67 Aboriginal Affairs and Northern Development Canada, *Terminology*, 2012, http://aadnc-aandc.gc.ca/ eng/1100100014642.

68 Pamela D. Palmeter, *Beyond Blood: Rethinking Indigenous Identity* (Saskatoon, SK: Purich Publishing Limited, 2011), 39.

69 Sharon McIvor and Jacob Grismer v. Canada, [November 23, 2010] Communication Submitted for Consideration under the First Optional Protocol to the *International Covenant on Civil and Political Rights*, before the UN Human Rights Committee; Submission of the Government of Canada on the Admissibility and Merits of the Communication to Human Rights Committee of Sharon McIvor and Jacob Grismer, [August 21, 2010] Communication No. 2020/2010; Petitioner Comments in Response to State Party's Submission on the Admissibility and Merits of the Applicants' Petition to the Human Rights Committee, [December 5, 2011] Communication No. 2020/2010, par. 22.

70 Lee Holcombe, *Wives and Property: Reform of the Married Women's Property Law in Nineteenth-Century England* (Toronto, ON: University of Toronto Press, 1983), 33.

71 Lori Chambers, *Married Women and Property Law in Victorian Ontario* (Toronto, ON: Published for the Osgoode Society for Canadian Legal History by University of Toronto Press, 1997), 10–11.

72 Holcombe, *Wives and Property*, 35.

73 Nicholas Flood Davin, *Report on Industrial Schools for Indians and Half-Breeds* (Ottawa, ON, 1879), 10.

74 Ibid., 14.

75 Ibid., 13.

76 Ibid.,12.

77 Truth and Reconciliation Commission of Canada (TRC), *They Came for the Children: Canada, Aboriginal Peoples and Residential Schools* (Winnipeg, MB, 2012), 10.

78 Ibid., title page.

79 Sarah Carter, "Creating 'Semi-Widows' and 'Supernumerary Wives': Prohibiting Polygamy in Prairie Canada's Aboriginal Communities to 1900," in *Contact Zones: Aboriginal and Settler Women in Canada's Colonial Past*, eds. Katie Pickles and Myra Rutherdale (Vancouver, BC: UBC Press, 2005), 139.

80 Jean Barman, "Taming Aboriginal Sexuality: Gender, Power and Race and British Columbia, 1850-1900," in *In the Days of Our Grandmother: A Reader in Aboriginal Women's History in Canada*, eds. Mary-Ellen Kelm and Lorna Townsend (Toronto, ON: University of Toronto Press, 2006), 286.

81 Adele Perry, *On the Edge of Empire: Gender, Race, and the Making of British Columbia*, 1949-1871 (Toronto, ON: University of Toronto Press, 2001), 49; also see Sarah Carter, *Capturing Women: The Manipulation of Cultural Imagery in Canada's Prairie West* (Montreal, QC: McGill-Queens University Press, 1997), 161; Sarah Carter, "Categories and Terrains of Exclusion: Constructing the 'Indian Woman' in the Early Settlement Era in Western Canada," *Great Plains Quarterly* 148 (1993): 154; Richard White, *The Middle Ground: Indians, Empires and Republics in the Great Lakes Region*, 1650-1815 (Cambridge, UK: Cambridge University Press, 1991), 61.

82 Barman, "Taming Aboriginal Sexuality," 279.

83 Ibid., 277.

84 Ibid., 289.

85 Ibid., 277–86.

86 Jean Barman, "Aboriginal Women on the Streets of Victoria: Rethinking Transgressive Sexuality during the Colonial Encounter," in *Contact Zones: Aboriginal and Settler Women in Canada's Colonial Past*, eds. Katie Pickles and Myra Rutherdale (Vancouver: UBC Press, 2005), 208.

87 White, *The Middle Ground*, 61.

88 Carter, "Creating 'Semi-Widows,'" 140.

89 Sarah Carter, *The Importance of Being Monogamous: Marriage and Nation-Building in Western Canada to 1915* (Edmonton, AB: University of Alberta Press, 2008), 162.

90 Joan Sangster, "Native Women, Sexuality, and the Law," in Kelm and Townsend, *Days of Our Grandmother*, 312.

91 Carter, "Categories and Terrains of Exclusion," 150.

92 Carter, *Capturing Women*, 165.

93 Barman, "Aboriginal Women on the Streets of Victoria," 205.

94 Carter, *Capturing Women*, 187.

95 Carter, *The Importance of Being Monogamous*, 152.

96 Renisa Mawani, "In Between and Out of Place: Mixed-Race Identity, Liquor and the Law in British Columbia, 1850-1913," in *Race, Space and the Law: Unmapping a White Settler Society*, ed. Sherene H. Razack (Toronto, ON: Between the Lines, 2002), 138.

97 Sylvia VanKirk, *Many Tender Ties: Women in Fur Trade Society*, 1670-1870 (Winnipeg, MB: Watson, Dwyer Publishing Limited, [1980] 1999), 174.

98 Jennifer S.H. Brown, *Strangers in Blood: Fur Trade Families in Indian Country* (Vancouver, BC: UBC Press, 1980), 148–49.

99 Mawani, "In Between and Out of Place," 87, 159.

100 Sylvia VanKirk, "From 'Marrying-In' to 'Marrying-Out': Changing Patterns of Aboriginal/Non-Aboriginal Marriage in Colonial Canada," *Frontiers* 23 (2002): 3, 5.

101 VanKirk, *Many Tender Ties.*

102 Susan Sleeper-Smith, *Indian Women and French Men: Rethinking Cultural Encounter in the Western Great Lakes* (Amherst, MA: U. of Massachusetts Press, 2001), 19.

103 Patricia A. McCormack, "Lost Women: Native Wives in Orkney and Lewis," in *Recollecting: Lives of Aboriginal Women of the Canadian Northwest and Borderlands*, eds. Sarah Carter and Patricia McCormack (Edmonton, AB: AU Press, 2001), 61.

104 Carter, *The Importance of Being Monogamous*, 152, 188.

105 Carter, "Categories and Terrains of Exclusion," 158.

106 Mawani, "In Between and Out of Place," 50, 53.

107 Ibid., 65.

108 Perry, *On the Edge of Empire*, 2001, 70.

109 Carter, "Categories and Terrains of Exclusion," 1993, 154.

110 Joyce Green, "Canaries in the Mines of Citizenship: Indian Women in Canada," *Canadian Journal of Political Science/Revue de science politique* 34, no. 4 (2001): 723–27.

111 Attorney General of Canada v. Jeannette Vivian Corbière Lavell, [1972] Supreme Court of Canada, *Factum* Attorney General of Canada, par. 12.

112 Renisa Mawani, *Colonial Proximities: Crossracial Encounters and Judicial Truths in British Columbia, 1871–1921* (Vancouver, BC: UBC Press, 2009), 138.

113 H.B. Hawthorn, *A Survey of the Contemporary Indians of Canada: A Report on Economic, Political, Educational Needs and Policies in Two Volumes* (Indian Affairs Branch: 1966), IAND Publication No. QS-0603-020-EE-A-18, I:250.

114 Royal Commission on the Status of Women in Canada (RCSW), *Report* (Ottawa, ON: Information Canada, 1970), par. 58.

115 See Gwen Brodsky, "McIvor v. Canada: Legislated Patriarchy Meets Aboriginal Women's Equality Rights," and Kuokkanen, "Confronting Violence: Indigenous Women," in Green, *Indivisible: Indigenous Human Rights*, for a more complete description.

116 Carter, *The Importance of Being Monogamous*, 169.

117 VanKirk, "From 'Marrying-In' to 'Marrying-Out,'" 2.

118 Standing Committee on Indian Affairs and Northern Development, March 1, 1985; *Minutes of Proceedings and Evidence … Respecting Bill C-31, An act to amend the Indian Act.* House of Commons, Issue 24, March 26, 1985, 24–33.

119 Janet Silman, *Enough Is Enough: Aboriginal Women Speak Out* (Toronto, ON: Women's Press, 1987), 11, 97, 186.

120 Standing Committee on Indian Affairs and Northern Development, *Minutes of Proceedings and Evidence* (Includes Committee's Sixth Report to the House of Commons, i.e. First Report of the Sub-Committee on Indian Women and the Indian Act, Chair, Keith Penner), Issue 58, September 20, 1982, 10–11.

121 McIvor v. The Registrar, Indian and Northern Affairs Canada, [2007] British Columbia Supreme Court 827, par. 61.

122 McIvor v. Canada (Registrar of Indian and Northern Affairs), [2009] British Columbia Court of Appeal 153, par. 30.

123 Sub-committee on Indian Women and the Indian Act of the Standing Committee on Indian Affairs and Northern Development, *Minutes of Proceedings and Evidence*, Issue 4, September 17, 1982, 461.

124 Stewart Clatworthy, *Re-Assessing the Population Impacts of Bill C-31* (Four Directions Project Consultants, 2001), viii-ix.

125 Indian and Northern Affairs Canada, *Estimates of Demographic Implications from Indian Registration Amendment McIvor v. Canada* (2010), http://www.ainc-inac.ca/bra/is/eod.eng.asp.

126 Attorney General of Canada v. Lavell; Isaac et al. v. Bédard, [1974] Supreme Court Reports 1349, 1386.

127 Maryanne Pearce, "An Awkward Silence," 256.

128 Derrickson v. Derrickson, [1986] 1 Supreme Court Reports 285.

129 Paul v. Paul, [1986] 1 Supreme Court Reports 306.

130 Mary Eberts and Beverly K. Jacobs, "Matrimonial Property on Reserve," in *On Building Solutions for Women's Equality: Matrimonial Property on Reserve, Community Development and Advisory Councils*, eds. Marylea MacDonald and Michelle K. Owen (Ottawa, ON: CRIAW/ICREF [Voix feministes-Feminist voices, No. 15], 2004).

131 Wendy Cornet and Alison Lendor, *Discussion Paper: Matrimonial Real Property on Reserve* (Canada, Department of Indian Affairs and Northern Development, 2002).

132 Wendy Cornet and Lendor Alison, "Matrimonial Real Property Issues On-Reserve," in *Aboriginal Policy Research: Setting the Agenda for Change*, eds. Jerry P. White, Paul Maxim, and Dan Beavon (Toronto, ON: Thompson Educational Publishing Inc., 2004), vol. II.

133 Wendy Grant-John, *Report of the Ministerial Representative, Matrimonial Real Property Issues on Reserve* (Ottawa, ON: Minister of Public Works and Government Services Canada, March 2007), Item 70, 5.

134 Canada House of Commons, *Report: Walking Arm-in-Arm to Resolve the Issue of On-Reserve Matrimonial Real Property* (Standing Committee on Aboriginal Affairs and Northern Development, 2005).

135 Canada Senate, *Interim Report: A Hard Bed to Lie In: Matrimonial Real Property on Reserve* (Standing Senate Committee on Human Rights, 2003).

136 Sangster, "Native Women, Sexuality, and the Law," 313.

137 Robin Jarvis Brownlie, "Intimate Surveillance: Indian Affairs, Colonization, and the Regulation of Aboriginal Women's Sexuality," in *Contact Zones: Aboriginal and Settler Women in Canada's Colonial Past*, eds. Katie Pickles and Myra Rutherdale (Vancouver, BC: UBC Press, 2005), 163, 167.

138 Ibid., 163, 169.

139 Silman, *Enough Is Enough*, 94, 99, 103–104, 175.

140 RCSW, Report, 237–38, 410.

141 Lavell and Bédard, 1359, 1369.

142 Ibid., 1359–60.

143 Reference as to the meaning of the word "Persons" in Section 24 of the British North American Act, 1867, [1928] Supreme Court Reports 276, 287.

144 On appeal to the Judicial Committee of the Privy Council, Edwards v. Attorney General of Canada, [1930] Appeal Cases 124.

145 Harold Cardinal, *The Rebirth of Canada's Indians* (Edmonton, AB: Hurting Publishers, 1977), 110–11.

146 Sally M. Weaver, *Making Canadian Indian Policy: The Hidden Agenda 1968-1970* (Toronto, ON: University of Toronto Press, 1981), 199.

147 Cardinal, *The Rebirth of Canada's Indians*, 115.

148 Standing Committee on Justice and Legal Affairs, *Minutes of Proceedings and Evidence*, Issue 6A, 1977, 23.

149 Wendy Cornet, "First Nations Governance, the Indian Act and Women's Equality Rights," in *First Nations Women, Governance and the Indian Act: A Collection of Policy Research Reports* (Ottawa: Status of Women Canada 2001), 125.

150 Ibid.

151 Canadian Human Rights Act Review Panel, *Promoting Equality: A New Vision* (Ottawa, ON: Canada, Department of Justice, 2000), 39–41, 127–33, and rec. 141.

152 NWAC, *Aboriginal Rights are Human Rights: Research Paper Prepared [by Mary Eberts] for Canadian Human Rights Review* (Ottawa, ON, 2000).

153 Canadian Human Rights Commission, *A Matter of Rights: A Special Report of the Canadian Human Rights Commission on the Repeal of Section 67 of the Canadian Human Rights Act* (Ottawa, ON, 2005), 8–9.

154 An Act to Amend the Canadian Human Rights Act, Statutes of Canada 2008, c.30.

155 Universal Declaration of Human Rights, adopted and proclaimed by General Assembly Resolution 217 A (III), December 10, 1948.

156 Reference re Secession of Quebec, [1998] 2 Supreme Court Reports 217 (*"Secession Reference"*), paras. 49, 82.

157 Ibid., par. 70.

158 Ibid., par. 71.

159 Silman, *Enough Is Enough*, 74, 134–35, 176–77.

160 Sandra Lovelace v. Canada, [1981] Communication No.R.6/24, UN Doc. Supp. No. 40 (A/36/40) at 166, par. 17.

161 Brownlie, "Intimate Surveillance," 167.

162 Michelle M. Mann, *Indian Registration: Unrecognized and Unstated Paternity* (Ottawa, ON: Status of Women Canada, 2005), 1, 5–6.

163 Ibid., 11–12.

164 Clatworthy, *Re-Assessing Population Impacts*, 229–30, 234–43.

165 Mann, *Indian Registration*, 8.

166 Lynn Gehl Gii-Zhigaate-Mnidoo-Kwe, "Unknown and Unstated Paternity and *The Indian Act*: Enough is Enough!" *Journal of the Motherhood Initiative [Motherhood, Activism, Advocacy, Agency]* 3, no. 2 (Fall/Winter 2012): 194.

167 Lynn Gehl Gii-Zhigaate-Mnidoo-Kwe, "Canada's unstated paternity policy amounts to genocide against Indigenous children," *rabble.ca*, January 29, 2013, http://rabble.ca/news/2013/01/canadas-unstated-paternity-policy-amounts-cultural-genocide-against-indigenous-children.

168 Until the final submission of this text for publication in October 2014, this litigation had not been resolved; see Brodsky, "McIvor v. Canada."

169 Standing Committee on Aboriginal Affairs and Northern Development, *Evidence*, AANO, Number 007, April 1, 2010, 3.

170 Standing Committee on Aboriginal Affairs and Northern Development, *Evidence*, AANO, Number 008, April 13, 2010, 3–4.

171 Lynn Gehl Gii-Zhigaate-Mnidoo-Kwe, "The Queen and I," in *Canadian Woman Studies: An Introductory Reader*, eds. Andrea Medovarski and Brenda Cranney, 2nd ed. (Toronto, ON: Inanna Publications and Education Inc., 2006), 186.

172 Acoose, *Iskwewak-Kah' Ki Yaw Ni Wahkomakanak*, 39.

173 AJI, *The Justice system and Aboriginal People*, 52.

"An Awkward Silence" Maryanne Pearce (2013)

In 2000, I was part of the Aboriginal children's health team at Health Canada, in Vancouver for a conference. My colleague and I went for a stroll on the Seawall at low tide. There, we found an eagle feather. And another. And another. In all, I gathered 75 eagle feathers as we walked along the tidal zone. These were not the long feathers used in ceremonies, but short, ragged, and about to be swept out to sea, lost forever. But they were still eagle feathers, still sacred. As it was my colleague's moon time, her tradition did not allow her to handle the feathers; they were meant for me. I packed the feathers in cedar, returned to Ottawa and called an Elder. He advised that the feathers were probably lost during a fight between two eagles. He did not know what it signified for me to have found so many, but said the reason would become clear in time.

Seven years later I sent an email to my mother and sisters about my acceptance to the LL.D. program and my research. Instantly, my sister *Janette* responded: "I always knew you would figure out what those eagle feathers were for." *Janette* has always been able to help me see the forest despite the trees. Like the feathers I found, many of the women I write about experienced violence, were a bit tattered, and may have disappeared without a trace, but they are still sacred.

In September 2009, *Constance Backhouse* suggested I write this section, when she advised that I stop work on the database and focus on writing. I found it very difficult to stop adding to the database; these women's lives were important to me. I know their faces, names, facts about their deaths or disappearances, but also facts about their lives and families. I hear their voices in my nightmares.

Most of the cases are seared into my memory. Reading the court decisions, books and media articles, watching the *Missing Women Inquiry* every day for eight months, writing summaries of serial homicide … it was very difficult to put out of my head. I felt powerless. The women had already disappeared or been murdered; there was nothing I could do. I was haunted by the violence; insomnia and nightmares were common.

During this time, I started rescuing dogs with the *Great Pyrenees Rescue* and *Hopeful Hearts*. It seems a bizarre leap from murdered women to dogs, but it was a concrete way I could make a change in the world. I just needed to save someone, even if that someone had a tail. My husband understood this. For five years, he has endured dozens of huge, hairy beasts that have destroyed carpeting, eaten couches, hats, shoes and books (a rather egregious sin in a house of book lovers), cried all night, were not house trained and were afraid of men. He endured this because he knew my spirit needed it.

I am not certain what to do with the nine crates of files on the individual cases of missing and murdered women in my dining room. I am not sure that I will stop collecting new cases and updates, even though the dissertation is now complete. I do know that I will continue to care about these women, and continue to fight for justice and change. The lyrics of *Warrior*, by Winnipeg band The Wyrd Sisters, speaks to me. I hope they will forgive me for using their song as my battle cry.

"Warrior" (1994)

I was a shy and lonely girl
with the heavens in my eyes
and as I walked along the lane
I heard the echoes of her cries

I cannot fight
I cannot a warrior be
it's not my nature nor my teaching
it is the womanhood in me

I was a lost and angry youth
there were no tears in my eyes
I saw no justice in my world
only the echoes of her cries

I cannot fight
I cannot a warrior be
it's not my nature nor my teaching
it is the womanhood in me

I am an older woman now
and I will heed my own cries
and I will a fierce warrior be
'til not another woman dies

I can and will fight
I can and will a warrior be
it is my nature and my duty
it is the womanhood in me

I can and will fight
I can and will a warrior be
it is my nature and my duty
it is the sisterhood in me

—The Wyrd Sisters

soft ground
and
a shovel.

How to Entrench Native Rights

Thomas King

1989

Making Canadian Indian Policy

The Hidden Agenda (1978)

Sally M. Weaver

"The ultimate failure to include Indians raises the basic question of how the demands of the Indians at the consultation meetings were perceived by the policy-makers inside government. It also requires us to understand how 'the Indian problem' was defined by the policy-makers and the public, for defining the problem that a policy is to solve is the first and the most crucial step in policy-making."

—Sally Weaver, *Making Canadian Indian Policy: The Hidden Agenda 1968–70*, 1981

The Hidden Agenda

Indian Policy and the Trudeau Government

Sally M. Weaver
Dept. of Anthropology
University of Waterloo
Waterloo, Ontario. N2L 3G1

1978

Introduction

In 1969 the federal government came forward with a White Paper on Indian policy which native people rejected and the government subsequently withdrew. Public controversy surrounding the policy was widespread, but much of it centered on the secretive fashion in which the policy was prepared. Despite government commitments to Indians that they would "participate" in the process, the production of the White Paper had many earmarks of political sophistry. To its critics, the policy was at best a perversion of "consultative democracy" and at worst a deception.

To date there has been no recounting of how the White Paper on Indian policy was developed within the federal government. This book attempts to fill this void by providing one microcosm of policy-making during the first Trudeau administration. In a more general vein, it offers some insight into how the government attempted to cope with the application of "participatory democracy" in regard to policy-making for an unorganized and disadvantaged minority. Still broader, it tries to show how the policy-makers attempted to work with the basic political values in our society and apply them to a minority group.

Overview of the Policy

In some respects the recent history of Canadian Indian policy is a familiar feature of Canada's social and political landscape. It is now well known that after a year of consultations with Indians on revising the Indian Act--a process that the Minister called "an attempt at consultative democracy"--the federal government released a White Paper on Indian policy in June of 1969 which proposed a global termination of all special treatment for Indians including the Indian Act. The Act came under heavy criticism by native people themselves, but they also recognized that it enshrined some of their "charter rights" which had historically been granted or promised them since before Confederation. The new policy was an abrupt reversal of the traditional practice of dealing with Indians although termination of special treatment had been the implicit goal of governments since colonial days. In 1947, for example, during the Special Joint Committee hearings on the Indian Act, termination had been explicitly proposed by Diamond Jenness, then Dominion Anthropologist, in his "Plan for Liquidating Canada's Indian Problem in 25 Years" (Jenness 1947:310-311). The federal government's 1969 White Paper on Indian policy had finally brought this implicit policy theme into the open.

The White Paper argued that "equality," or "non-discrimination" as it was often phrased, was the key ingredient in a solution to the problems of the native people, and that special rights for Indians had been the major cause of their problems (DIAND 1969). The goal of equality was to be achieved by terminating the special legislation and bureaucracy that had developed over the past century to deal with Indians, and by transferring federal services for Indians to the provinces.

Henceforth Indians would receive the same services from the same sources as other Canadians after a transitional period in which enriched programs in economic development were to be offered. The large Indian Affairs bureaucracy would be dismantled within five years, and the federal government was to retain trusteeship functions only for Indian lands which would be administered through an Indian Lands Act. By implication, the result of the policy would see Indians with "Indian problems" become provincial citizens with regular citizens' problems. The policy was essentially one of "formal equality," to use Cairns' phrase from the Hawthorn Report, but the question remained as to whether it would foster equality of opportunity for this disadvantaged minority. Cairns had argued three years previously that it would not (Hawthorn 1966:392):

> The equal treatment in law and services of a people who at the present time do not have equal competitive capacities will not suffice for the attainment of substantive socio-economic equality.

Predictably, the policy caused shock and alarm among Indians. Even though an unorganized minority, they responded with a clear cut rejection of the White Paper; "the MacDonald-Chretien doctrine" as Cardinal labelled it after DIAND's deputy minister and Minister (1969:1). Throughout the Indian Act consultation meetings (July 1968 - May 1969) they had expressed a wide range of opinions on how the Act should be revised with some Indian spokesmen calling for its removal. Although there was little consensus on the detailed revisions to the Act that the government had hoped for, there had been an unmistakable consensus within the Indian movement that was evident in the government's verbatim reports of the consultation meetings (DIAND 1968-1969); Indians wanted their special rights honoured and their historical grievances, particularly over lands and treaties, recognized and dealt with in an equitable fashion. Equally important, they wanted direct and meaningful participation in the making of policies that affected their future.

When the policy was released, it was obvious that none of these major Indian priorities, "the Indian option," had been acceptable to the government. The Ministers themselves had actively raised expectations among Indians that their views would be "listened to," but the policy in both its substance and its preparation had indicated otherwise. It had been prepared within government and without Indian participation. The attempt at consultative democracy miscarried and the government's efforts were discredited. Among Indians the policy invited a retrenching of distrust which was, ironically, the very condition the government had hoped to remove.

Indians responded to the policy with a resurgent nationalism unparalleled in Canadian history. Their spokesmen rallied to the moment by preparing their own counter-proposals[1] and by renewing their efforts to build provincial and national organizations through which they could lobby for their own policies. With the press and other sectors of the public supporting the Indian indictment of the policy, the government came under heavy pressure to set the White Paper aside.

The policy was delivered to the public with considerable ambiguity, compounding the problem of its reception. It was unclear whether the policy was simply "a proposal" or whether it was a "firm policy" signifying final government commitment. Statements by Chretien and certain actions by the department only reinforced the confusion. Eventually, on receiving the first counter-proposal, the Red Paper prepared by the Indian Chiefs of Alberta (1970), Trudeau publicly said the government would not press forward with implementing the White Paper (Trudeau 1970). By the spring of 1971 the policy was formally withdrawn by Chretien although some Indian spokesmen today claim, and some civil servants privately concur, that termination remains the "unofficial policy" of the government and is still being implemented.[2]

Social scientists, especially anthropologists, were shocked by the White Paper. Their surprise reflected their disbelief that the government could embark on a policy that had proven so destructive to Indian communities in the United States during the 1950's.[3]

Paradoxically, the termination policy there had subverted its own intended goal of equality (Brophy and Aberle 1966:188-89). It brought such fear and insecurity to tribal communities that it elicited a response of nativism, a process of cultural reaffirmation which often arises when cultural systems are severely threatened. Instead of being disposed to seek "equality," Indian communities reasserted their cultural uniqueness, emphasizing their social distance from the dominant society. "Termination psychosis" continued to dominate Indian-government relations making administration very difficult (Josephy 1969).

More close to home was the government's disregard of the recommendations in the Hawthorn Report (1966,1967), itself a government-commissioned national survey on Canadian Indians. Working between 1963 and 1967, Hawthorn and his co-researchers had produced a lengthy report on the current conditions of Indians in Canada. The team had rejected termination as a policy option (1966:8), arguing instead for a "citizens plus" status for Indians which they did not view as a deterrent to delivering proper provincial services to Indians. They also recommended the role of advocate-ombudsman for the Indian Affairs Branch because many Indian bands lacked the social, economic and political skills of self-defense. As well they refuted the usual constitutional argument that Indians were the exclusive responsibility of the federal government, thereby leaving the way open for the provinces to deliver programs to Indians (see also Lysyk 1967). In their view the thrust of policy should be middle range, with programs emphasizing development on a broad socio-economic scale if Indian poverty and dependency on the government were to be reversed (1966:386-403). They urged the government to recognize the increasing social problems among Indians in cities, but like their other recommendations, this too was disregarded in the new policy.

Although social scientists had no proven solutions to the complex problems of Indian administration and marginality, as the

Hawthorn Report (1966) and the Brophy and Aberle study (1966) among others argued the "answer" lay somewhere in broad-scale social development programs with Indians and changing public attitudes, not in simplistic legal solutions and rapid legislative action.

Both these studies provided guidelines for Canadian policy-makers in assessing the possible impact of various policy alternatives. Furthermore the policy-making approach espoused by the Trudeau government at that time called for rigorous policy research which, theoretically, would have required evaluations of policy experiences in other countries as well as those close at home. When the White Paper was released, however, it became apparent that the Hawthorn Report and the Brophy and Aberle study were either unknown to the policy-makers or disregarded by them.

A combination of all these factors led social scientists to respond to the new policy with considerable incredulity. That the White Paper would backlash on the policy-makers was no surprise.

Making Indian Policy

Indian policy was developed within the upper reaches of the federal government under tight secrecy common to the policy-making process. The policy-makers were Ministers, their advisers and senior civil servants, and although the policy-making group contained almost fifty people at its maximum, less than twenty played a major part in shaping the policy. The process lasted a full year, beginning in the summer of 1968 and ending in June of 1969, just a week before the White Paper was released.

Ideally, Indian policy was to be developed according to the new more rational approach to policy making which Trudeau and his policy advisers in the PMO (Prime Minister's Office) were then instituting in government.* Policies were to be more deliberately and systematically planned, from the initial identification of the problem which the policy was to address, through the selection of policy objectives or goals, to the weighting of the alternative ways of achieving these goals. The implications of each alternative were to be carefully examined in the hope of minimizing undesirable consequences, especially those that would create new problems for the government or the public. Ministers and civil servants were encouraged by the PMO to be far-sighted in their thinking and to consider fundamental changes in policy rather than limiting their horizons to mere incremental changes. If these steps were followed, policies were expected to be better reasoned and more comprehensively planned. In general, Trudeau's approach was intended to improve government's capacity to develop effective policies by undertaking rigorous policy reviews and by engaging in a far-reaching examination of the issues.

Serious efforts were made by the PMO to develop Indian policy along the lines of the new approach, and as Indian policy evolved it did correspond, in a very general way, with the desired procedures. Beginning in the summer of 1968, each succeeding season of policy-making advanced Indian policy a bit further, but the overall process was not characterized by reason and balance. From the outset the development of Indian policy was a tangled and tortured process. Policy-makers became preoccupied with their own internal conflicts, and efforts to straighten out the confusions and hostilities frequently failed. Although the general procedures in rationalizing the policy were relatively similar to those elsewhere in government at that time, Indian policy was considerably more contentious.[5]

Studies of policy-making firmly demonstrate that the process often involves heated exchanges and conflicting values. In Lindblom's experience policies are often fought out rather than thought out (1968:19). This aptly describes the case of Indian policy. What made Indian policy particularly quarrelsome, however, was the fact it tried to cope with the notion of public participation.

The Question of Participation

During the late 1960's Indian spokesmen placed much of the responsibility for ineffective policies in Indian Affairs on the senior officials in the department (eg. Cardinal 1968). In the opinion of Indian critics the bureaucracy had played an excessively influential role in policy matters, but the future lay in opening up the process to Indians themselves. This demand reflected more widespread concern in many sectors of society which saw "big government" and powerful bureaucracies as threatening the responsiveness and accountability of government to the people.

Making the policy process more open to the public had great appeal during the federal election of 1968 when Trudeau promised "to make Government more accessible to people, to give our citizens a sense of full participation in the affairs of Government..." (Trudeau 1968a:6). "Participatory democracy" became the catch phrase to describe greater public involvement, although the meaning of participation was never publicly explained by Trudeau (Doerr 1973:98). Paradoxically in terms of its ultimate fate, the White Paper on Indian policy can be seen as an extension of this concept in several ways.

During the 1960's localized citizen participation movements sprang up across the country with the intention of eradicating the problems of urban blight and poverty (Lotz 1977). "Grass roots activism" was a common feature of Canadian society especially in urban areas and on university campuses. With the public demanding a more powerful role in shaping government policies, the policy-making process itself came under greater public scrutiny (Aucoin 1971).

But grass roots activism in Canada was not confined to the public. During the mid-1960's the federal government adopted the idea of the citizen participation with the most striking examples being The Company of Young Canadians, the Challenge for Change Program and New Start Program (Draper 1971). Less dramatic were the ARDA (Agricultural Rehabilitation Development Act) programs designed to reduce rural poverty and regional economic disparity. Within Indian Affairs the community development program was such an example, albeit a less successful one. An even more direct illustration relating to Indians was the film "Encounter with Saul Alinsky - Rama Reserve" made jointly by the National Film Board and The Company of Young Canadians in 1967. As a professional organizer of underprivileged groups, Alinsky spent a good portion of the film suggesting to Indian discussants how they might focus Indian concerns and more effectively exert pressure on government to support their goals. Finding the native discussants reluctant to consider his activist approach, Alinsky ends the film by stating that Indians are "the God damnest best ally the government has" if they seek to opt-out of society and yet expect politicians to take an interest in their well being. The clear message of the film was that Indians must organize if they expect changes and mobilize public support if their efforts are to succeed.

Federal involvement in the citizen participation movement usually occurred in cases where the government dealt with unorganized sectors of the public which were unable to articulate their needs and bring them to the government in a focused fashion (Kernaghan 1976). Ministers sometimes promoted policies that were designed to enhance public participation and civil servants were often called upon to act as facilitators in focusing public opinion so government could more effectively respond. Under these conditions bureaucratic involvement in the political process increased and some civil servants, whether encouraged by the administration or their own personal philosophies, began to act as agents of social change, a role traditionally assigned to Ministers (Ibid.: 439). But the activism of civil servants was not always appreciated by other officials or Ministers, especially when their efforts to organize the public succeeded in raising demands for change which were directed back to the government in a very forceful fashion (Bregha 1971). In many instances the sponsoring agencies in government, faced with a rejection of their traditional policies by an agitated public, reacted by closing down the programs rather than changing their policies (Dimock 1971). The community development program in Indian Affairs was a classic example of this problem with government sponsored intervention.

The question of public participation and Indian policy-making did not rest with bureaucratic involvement alone. Individual Ministers who embarked on programs to encourage public participation often had to wrestle with the uncomfortable results of raising public expectations when they presented their policy recommendations to the cabinet. A Minister could personally seek greater public involvement in policy-making, but in the end he had to convince the cabinet of the wisdom of this approach. Decision making rested with the collective cabinet, not individual Ministers, and a Minister could run the risk of cabinet rejecting his ideas, as was apparently the case with Hellyer's Task Force on housing (Axworthy 1971). In the case of Indian policy ministerial activism became a contentious issue because there was no agreement between the two Ministers assigned to Indian Affairs on the nature and extent of Indian participation in the policy-making process. In the end Indian policy would test the Trudeau government's commitment to public participation particularly when it came to the question of how much the government would acquiesce to Indian demands.

An early connection between citizen participation and Indian policy occurred in 1967 when the government decided to hold systematic consultation meetings with native people to discuss revisions to the Indian Act. Although started by the Pearson government, the Trudeau administration supported the idea, and Indians were encouraged to make comments on how the Act should be revised. Statements of Ministers to Indians during the consultation meetings (July 1968 to May 1969) raised certain expectations about the nature of the new policy, including the amount of influence Indians should and could have in the policy process itself. The consultation procedure itself led Indians to believe that the government's policy in 1968 was a "positional policy" to use Aucoin's term (1971:25); a policy designed to give Indians a position in the general structure of decision-making. As Aucoin comments on the nature of public concern with policy-making (1971:25):

> A good deal of policy activity by individuals and groups is related not so much to securing (at least in the short run) an allocation of desired values but rather the attainment of desired positions vis-a-vis other individuals or groups. What is sought is a share of the coercive abilities of the government.

As Indian policy took shape, the question of Indian participation in the process became one of the most contentious issues among policy-makers. As they struggled with the notion its meaning took various forms as did the implications of these meanings for Indians and the government.

The theme of participation wove its way through many aspects of Indian policy, including the decision to release the policy as a White Paper rather than draft legislation for parliamentary debate. The policy-making process ostensibly began with consultation meetings with Indians, its development within government was focused on debates as to the importance of participation (at what stage and with what effect), and its delivery as a White Paper was intended to further the process of participation. Participation was said to have taken place, but in fact it did not occur. Indians were not party to the deliberations that produced the White Paper.

This raises the basic question of how the demands of the Indians at the consultation meetings were perceived by the policy-makers inside government. It also requires us to understand how "the Indian problem" was defined by the policy-makers and the public, for defining the problem that any policy is to "solve" is the first and the most crucial step in policy-making.

Footnote 57: Canada's Apartheid

The Duplicitous Diffusion of Canadian Strategies of Indigenous Segregation, Assimilation, and Extermination

Pierre Bélanger *with Kate Yoon*

By bringing together three different time periods—the initial colonization of the Americas, Nazi Germany, and South Africa under racial segregation and subsequent Apartheid—this research note serves to connect a long and important history of settler colonialism. More specifically, this note looks at how the "elimination of the Native"[1] places Canadian policies and strategies at the center and on the circumference of the production, exchange, and transfer of colonial knowledge. This note holds important implication for studies in universal history, as its content proposes a refocusing *of*, and on Canada's role in the development, diffusion, and innovation of strategies of Indigenous assimilation in the context of studies in the histories of colonization, racial segregation, and Indigenous extermination.

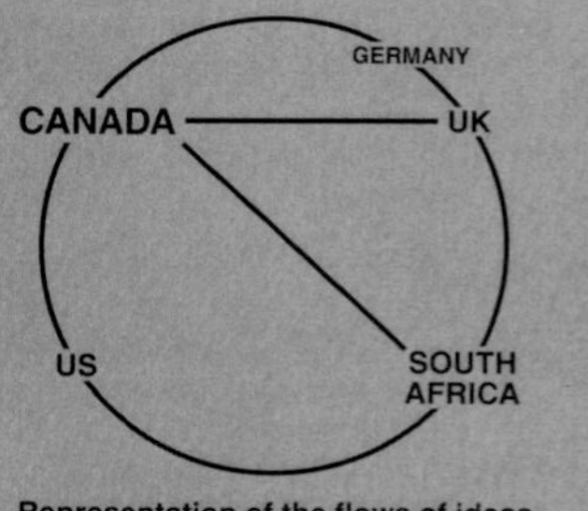

Representation of the flows of ideas and axes of influence regarding the "Native Question" from Canada towards other nation states

I - SOUTHERN CONNECTIONS

Albeit poorly understood, Canada and South Africa have had close diplomatic relations, given shared experiences in the British Empire and shared concerns with Indigenous peoples in the process of establishing colonial settler-states. In fact, there is evidence suggesting that South African government officials received direct information, influence, and inspiration from Canada and its Native reservation system in establishing spatial methods of racial segregation. Historical connections in the transfer of technical information between Canada and South Africa have previously been explored, but the direct connection and exchange of racial policies has never been examined in close detail or great depth.[2] Despite archival evidence on the official exchanges between the two countries, only two dated and somewhat forgotten sources—a popular article written by Ron Bourgeault in 1988[3] and the film *We Have Such Things at Home*, released in 1997[4]—mention Canada's contribution to South Africa's racial policies.

Originating prior to the twentieth century, the first portion of this note brings together a prolonged period of exchanges regarding race relations in Canada and South Africa, from the Parliamentary Select Committee on Aboriginal Peoples in the early nineteenth century, to the Carnegie Corporation's funding of visits by educators and government officials, to direct government correspondences between the two countries.

Previous studies of the Carnegie Corporation's (often racist and segregationist) interests in the study of race relations have often focused on studies within South Africa, rather than the Visitors' Grants.[5] [6] What kinds of information South African visitors sought from the US and Canada, as well as the activities they engaged in and people they met, has not been examined in close detail. This note aims to show that, through interactions and tours of Indian reserves, ideas about both Indigenous segregation and assimilation emerged from racial logics and racist policies that transferred across national borders and originated in Canada.

While the transfer of ideas about "Natives" may seem ad hoc and discontinuous, this note outlines the evidence showing that governments and individuals on both sides of the Atlantic shared models of segregation and oppression regarding "Natives." This note is an attempt to address the "black hole" of knowledge in current scholarship regarding the history of the trans-Atlantic transfer of segregation strategies, and specifically, the important position that Canada occupies in the innovation, past and present, of legal, territorial, and spatial policies.

The trans-Atlantic flow of ideas and transfer of techniques associated with "Natives" during the twentieth century suggest that, contrary to popular knowledge, Canada has made a significant contribution to imperial discourses of the "Indian Problem," or later known as the "Indian Question," as a tried and tested example of segregationist policies. As Deputy Superintendent for the Department of Indian Affairs, Duncan Campbell Scott infamously stated in 1920:

> "I want to get rid of the Indian problem. I do not think as a matter of fact, that the country ought to continuously protect a class of people who are able to stand alone. ... Our objective is to continue until there is not a single Indian in Canada that has not been absorbed into the body politic and there is no Indian question, and no Indian Department."[7]

Canada's Indian reserve system—conceived for extermination through attrition, by means of segregation in remote regions—emerged from a long history of previous (largely unsuccessful) policies of assimilation and enfranchisement.[8] This system, and its accompanying system of policies under the Indian Act, may have become the spatial model and administrative precedent for policies of segregation in other countries, such as South Africa, prior to the advent of apartheid in the late 1940s.

The transfer of ideas, however, turns out to be more complex and more historical. Ideologies related to "Natives" and "Aborigines" flowed between colonies of the British Empire and evolved during a period of two to three centuries, if not longer, prior to the creation of the Indian Act. To what extent ideas, ideologies, and models dealing with "The Native Question" flowed from the UK to Canada, or the UK to South Africa, or directly from Canada to South Africa is unclear;[9] yet as this note argues, there is sufficient historic evidence and records of international visits and meetings during intense periods of colonization between these countries to position Canada's treatment of "Indians," under the guise of administration, as the bureaucratic test bed for other countries. There is significant and substantial circumstantial evidence showing a triangulation of ideas by which the Canadian systems of Indian reservations, pass laws, and residential schools, set a distinctive example of colonial segregation and imperial subjugation. Furthermore, they could be upscaled, downscaled, or transferred to other countries and states.

As Métis sociologist Ron Bourgeault has argued, one of the main roots of the problematization of Indians is the changing role of *labor* that they represented.[10] After periods of war against American invasions, namely following the War of 1812, an important turning point took place vis-à-vis changing relations between Indians and Whites, as settlement of land and centralization of power took place and took off in the late nineteenth and early twentieth centuries.

In his 1988 article "Canada Indians: The South African Connection," Ron Bourgeault investigates the historical connection between South Africa and Canada's treatment of their respective Indigenous peoples. He states that "by the turn of the twentieth century Canada was probably the only advancing capitalist state that had an elaborate system of administration and territorial segregation of an internally colonized Indigenous population, a possible exception being the United States."[11] He goes on to suggest that there was a diffusion of strategies across countries—bluntly put, that "South Africa came to Canada ... to study [how] Indian people were controlled."[12] The connection regarding how to deal with the "Native question" or "backwards peoples" may have spanned centuries and many other countries, including multiple countries in former British colonies as well as the United States.

Canada's involvement in the transfer of ideas, specifically in relationship to the use of "Native reserves," can be traced through several centuries of imperial administration and colonization. Three major phases characterize this exchange of ideas. The first is the diffusion of ideas mediated by the British Empire and its colonial offices during the eighteenth and nineteenth centuries. The second is corporate philanthropy from the US that funded numerous visits and reports regarding "inferior races" throughout former British colonies during the early twentieth century. The third consists of cases of direct Canada-South Africa diplomacy and partnerships during early-to-mid twentieth century.

A. "The Native Problem" and the Colonial Diffusion of Native Policy from the British Empire to South Africa and Canada

The roots of British colonial Native policy date back to the nineteenth century, if not earlier. As early as 1837, the British Parliament appointed a *Parliamentary Select Committee on Aboriginal Tribes*. A Christian missionary organization called the "Aborigines Protection Society" was involved in the development of the Parliamentary Committee, and its first report in 1837 "established [the Society] as the protector of those, who have no power to protect themselves."[13] The report lays out detailed accounts of the various "tribes" in British-controlled territories in the interest of finding what "measures ought to be adopted."[14] While the report further acknowledges the atrocities that have been committed against the "Native inhabitants," the alternative strategies to civilizing missions that are proposed are to Christianize those "barbarous regions."[15]

Moreover, in a rather contradictory manner of bureaucratic double-speak, the Report decries the inhumane subjugation of "Aboriginal tribes," but seems to too easily defend it. On the removal of the entire Indigenous population from Van Diemen's Land following the Tasmanian War of 1832 between the British colonists and the Australian Aborigines, the report states that a "no better expedient could be devised than the catching and expatriating of the whole of the native population."[16] The report covered all territories of the expanding British Empire, from North America, to the Pacific islands, to South Africa, cataloging and classifying the behaviors of different groups—some as "civilized," others as still "savage."[17]

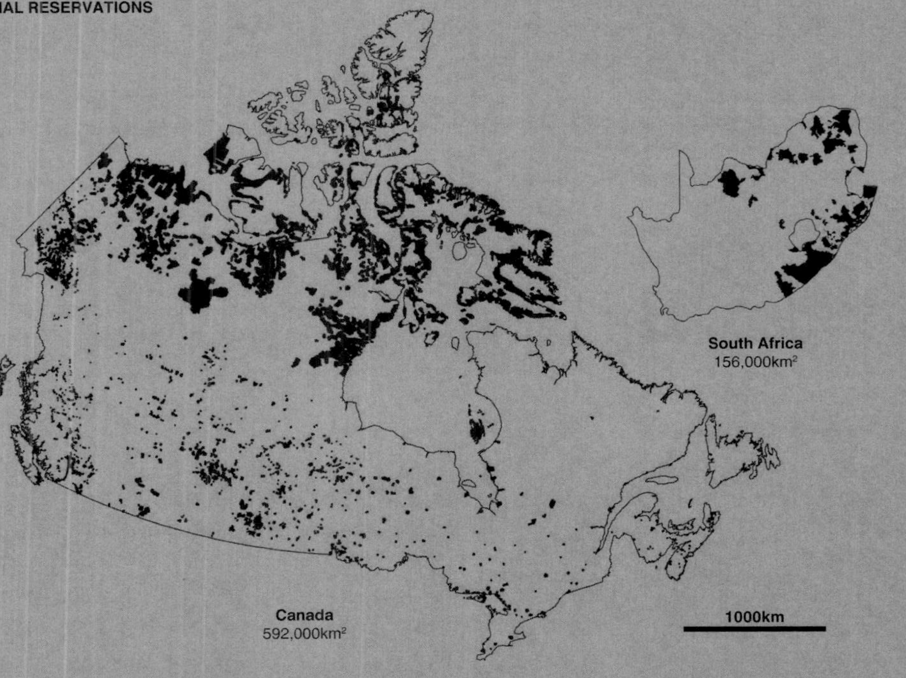

Comparison of Indigenous Reserves with South African Homelands

Records show the British Empire's continued involvement in Native policy, especially linked to the administration of "Natives," in sometimes very specific ways. The notorious *Pass Laws* of South Africa, originating in 1800 and in place in different permutations until 1994, restricting the movement of South African Blacks in different regions, was overseen and approved by the British government. The Colonial Office in the Cape Colony (South Africa) communicated to London in 1884 that a special session of the legislature had been summoned "to consider an amended Native Pass Law."[18] The recipient responds that "Her Majesty's Government learn with satisfaction that President intends to submit amended Pass Law. You might suggest the words Resident Commissioner or other British Officials authorized to issue passes."[19] Although pass laws have existed in South Africa since at least the early nineteenth century, these correspondences confirm that they were distinctly colonial legacies to control the movement of Non-Europeans, either slaves or farm labor. Interestingly enough, parallel research shows how Canada developed its own system of passes, imposed on Indians to limit their movement in and out of reservations or treaty areas, starting within a year of these correspondences with South Africa in 1885. Canada's own illegal pass system and its history have been obscured by the sanctioned destruction of official documents.[20]

B. "The Native Question" and the Economic Diffusion of Indian Policy from Canada to South Africa through the Carnegie Corporation of New York

After the turn of the twentieth century, intellectual leadership and policy innovation related to the international conversation about the "Native Question" was partially transferred to corporate hands under the guise of social philanthropy. Established in 1911 by Andrew Carnegie, the Carnegie Corporation of New York held a separate fund of $10 million allocated to the British dominions (Canada, South Africa, Australia, and New Zealand) and funded travelers' grants for educators from these countries to visit the United States.[21] Most likely related to the growth of infrastructure and expansion potential of these countries that was originally of interest to the growth of Carnegie Steel Company (later to become US Steel Corporation, the world's largest supplier for bridges and railroads), the Carnegie Corporation sponsored many studies on education, poverty, and social policy in South Africa.[22]

In *Waste of a White Skin: The Carnegie Corporation and the Racial Logic of White Vulnerability*, UC Irvine professor Tiffany Willoughby-Herard writes that the Carnegie Corporation produced studies that reinforced "segregationist philanthropy and scientific racism,"[23] and that, "in addition to the long-standing *Dominions and Colonies Fund*, the CCNY supported numerous segregationist philanthropic projects."[24] She writes that,

> "It is little wonder, then, that Pan-Africanists regarding this era understood it as the 'transfer overseas of American patterns of social organization … [and] extensions of the Corporation's domestic grand-making'—dissemination of widespread Jim Crow and racial colonialism."[25]

Among the Carnegie-funded studies, one of the most notorious was the "Poor White Study" from 1928—a study of poverty initiated among whites in South Africa. Francis Wilson, a South African labor economist who conducted the second Inquiry into Poverty (this time including blacks, but much later, in 1989) claimed that this selective attention to the poor white population led to the logic of apartheid:

> "So Carnegie had funded what became a major commission with, I think, five or six commissioners, an economist, a sociologist, a writer, who, in the end, traveled around the country in two Model-T Fords, traveled all over South Africa … and had done an extremely interesting survey of poverty, given the limitation of being for whites only. That had produced five volumes and so on.
>
> But one of the things that had happened with that, is that it had been hijacked by the National party in its rise to power. Hendrik F. Verwoerd, who became prime minister in South Africa in the 1960s—the 1950s, really—had a big conference in Stellenbosch, and said that the Carnegie Inquiry had become a very powerful instrument in the battle against poverty.
>
> Now, this had both a good side and a bad side. The good side was the sort of rise of social welfare, a social welfare department was set up in the government, and real attempts to deal with poverty. At the same time—and this became apparent quite quickly as we analyzed it—the Carnegie study had been part of the intellectual source, if you like, of the movement towards apartheid, because what emerged was that an anti-poverty program could also take the form of excluding other poor."[26]

That the Report supported pro-white stances was not a coincidence, either. Rather, it was written in a way that was specifically "pro-Afrikaner, with the morals of the Dutch Reformed Church and in the fashionably scientific language of eugenics … it is also significant that 'most of the contributors to the Carnegie report … landed up in the nationalistic or ultra-right camp.'"[27]

Besides the research that was conducted in South Africa, the Carnegie Corporation also funded tours of the US and Canada, for white South African officials and educators. In *The Ambiguous Champion: Canada and South Africa in the Trudeau and Mulroney Years*, Linda Freeman suggests the Carnegie-sponsored trips could have led to the design of apartheid, based on Canada's system of Indian Reserves.[28] The evidence is not so conclusive, but does show that Native policy on both sides of the Atlantic was part of a sustained conversation facilitated by corporate philanthropic interests.

Oswin Boys Bull, a South African educator, toured the US and Canada extensively in 1934—with stops at major US and Canadian cities, as well as Tuskegee, Alabama; Brantford, Ontario (to meet the Principals of Industrial College and the Mohawk Institute); and Ottawa, Ontario (to meet the Deputy Ministers for Immigration, Indian Affairs and Labour).[29] His papers are now located in the Carnegie Corporation of New York archives at Columbia University. In the report, he summarized that:

> "The objectives of my study in North America, as stated in my application for a 'Carnegie Visitor's Grant.' were as follows:
>
> To study,
>
> (i) The special problems surrounding Technical Education among backwards peoples …
>
> (ii) The somewhat baffling problem of how trained Native craftsmen are to be enabled to establish and support themselves in a somewhat primitive community …"[30]

Bull observes the "more backward groups of the population, and [tries] to estimate the extent to which [technical education] was really enabling them to win a fair position in an industrial society."[31] As part of his grand civilizing mission, he categorizes the "Retarded Groups" into the "Poor Whites, the Indians, and the Negroes"[32]

In his report, Bull asks, "(a) Is the Black Man really capable of mastering the trades of the modern world? (b) If he can, is he to be allowed to do so?—or (c) Is he to be limited by some sort of Colour Bar"?[33] He concludes that:

> "The hindrance most commonly complained of is the very general lack of any initiative—that weakness so often found amongst retarded peoples whose modes of life for ages have been defined and circumscribed by custom and tradition. It is felt by those responsible for the progress of the Indians that the stimulus of the guidance of the White Man will be needed for a long time to come."[34]

Bull's report foreshadows the many studies relating to how Native populations could be used for the workforce of settler industrial society. Sheila van der Horst, in a Carnegie-commissioned paper in 1955, decries apartheid for the limits it would pose on access to labor. Or, in her words: "This argument has great political appeal, but it ignores the economic integration which has already taken place and the interdependence of white and black in every branch of production."[35]

A few years later in 1949, Peter Cook, of the South African Native Affairs Department, received a $5,000 Carnegie Corporation grant "to study Negro education and administration" in the US for the South African government's Native Education Commission. Correspondences show that a representative of the Carnegie Corporation encouraged sharing knowledge, given that "the Commissions dealing with Native education are very greatly handicapped by the lack of accurate information."[36] Records also show correspondences in 1954 with a Sally Chilver of the Colonial Office and the Institute of Commonwealth Studies, regarding a visit to the US "to discuss CC [Carnegie Corporation] policies in BDC [British Dominions and Colonies] areas."[37]

In 1954, Frederick van Wyk received a grant to "study race relations in the United States and inter-cultural relations in Canada."[38] A newspaper article notes that he will study "the extent of Negro integration in the US, and the Negro employment problems in the stores, services, professions, trade and industry."[39] However, these conversations were not exclusively limited to academic institutions; records for Carnegie-sponsored visits also exist in the Canadian government's archives, as Canadian officials were informed about these visits. Correspondences between the Under-Secretary of State for External Affairs and the Department of Indian Affairs discuss F.J. van Wyk's visit, suggesting that he met the Director of the Canadian Citizenship Branch, as well as the Director of Indian Affairs. A note in the Department of External Affairs lays out the purpose of his visit (which can also be found in the Carnegie Corporation's funding files):

> "He is interested in all aspects of racial relationships and [that] while in Canada, he hopes to observe French-speaking–English-speaking relationships and his plans to spend a portion of his Canadian stay in Quebec are designed for that purpose. He is equally interested, however, in knowing something of the administration of our native populations, especially the steps which are being taken to integrate them into the Canadian economic and social structure…"[40]

Many of these Carnegie-sponsored trips, including those of Oswin Boys Bull and Peter Cook, involved meetings with a central figure in the exchange of knowledge between Canada and South Africa through the US: Professor of Education Charles Templeton Loram at Yale University. Prior to his appointment at Yale, C.T. Loram served in various government positions in education, including Chief Inspector of Native Education in Natal and Superintendent of Education.[41] Loram was, in historian John Whitson Cell's words, a "humane paternalist" who supported Afrikaner nationalist politicians and segregationist programs.[42] Although his main 'area of expertise' was in the assimilationist education of South African blacks, his interests spanned other parts of the world, namely Indigenous peoples in North America. His publications include "The Education of Indigenous Peoples," "The Education of Backwards Peoples, a Suggested Program," and "The Navajo Indian Problem."[43]

Most importantly, in 1939, a Toronto-Yale University joint conference, also sponsored by the Carnegie Corporation and entitled "The North American Indian Today," was held to discuss issues of Native education. C.T. Loram, then Chair of the Department of Race Relations at Yale University, published the outcomes of the conference in a book in 1943. The general sentiment—reflecting many of the Carnegie Corporation-sponsored reports—was that *assimilation* was unavoidable. As a book review of the seminar publication in 1944 summarizes,

> "Valuable as it would be to review a number of the authoritatively presented topics, exigencies of space compel one to search out the basic theme that constitutes the common denominator of them all. What do the vast majority, if not all, of these missionaries, anthropologists, and government officials envisage as the ultimate goal toward which the Indian should travel? 'Save the pure romanticist, no one really believes that Indian culture can ultimately prevail on the American continent. The only question is the rate at which it should be... superseded...' As Dr. Loram wrote, the basic problem is one of 'acculturation,' involving the assimilation of the way of life of the Indian to that of the white man, while preserving his essential rights as a human being. This assimilation is conceived as being not only desirable but quite possible since, as was observed by Dr. T.R.L. MacInnes, of the Canadian Indian Affairs Branch, 'the mental endowment of Indians is not inferior to that of other races.' ... Nevertheless, while recognizing the equal capacity for advancement of Indian and European, anthropologist and government official are agreed as to the difficulty of changing 'the characteristics and habits fixed for generations.'"[44]

This reflects the widespread view in academic circles at the time and shows that Canadian residential schools (of which representatives were also present at this conference) were used as a model of methods of "acculturation." In the introduction to the volume, Loram advocates for the need to exchange knowledge regarding Native peoples between Canada and the U.S., noting differences between the two countries:

> "Perhaps because of our restlessness, our willingness to experiment, our besetting sin in so often mistaking change for progress, our systems of education, all in the United States, officials as well as the general public, have 'views' on the Indian question which we are allowed and even encouraged to make public. In Canada, so it seems to me, the British traditions of reticence, of letting well alone, of hushing up 'scandals,' of trusting officials, are stronger, so that there is apparently not so much interest on the part of the public in the so-called Indian Question."[45]

As to what exactly were the Carnegie Corporation's interests in South Africa, it is plausible that the expansion of the railway fueled the need for cheap black labor. This affirms the comparative difference outlined by Ron Bourgeault at the beginning of this note, with Canada's untethering of Indian labor. Interestingly, 30 years earlier, the Carnegie Steel Company (now U.S. Steel) had tremendous interest in the development of railway infrastructure, namely in the Cape-Cairo railway, touted by imperialist and entrepreneur Cecil Rhodes during British expansionism. The discovery of diamonds and gold in South Africa in the late nineteenth century led to a massive increase of economic interest following major mining-resource extraction projects. A July 1900 edition of *American Manufacturer and Iron World* reports that "The Carnegie Steel Company has booked a 4,000ton steel rail contract for the Cape government railways, the first South African order for railway supplies received in this country since the outbreak of hostilities in that part of the world."[46]

C. "The Indian Question" and the Political Diffusion of Indian Reservation Policy from Canada to South Africa

In the Canadian government's Indian Affairs Central Registry Files (RG10, Volume 8588, 1/1-10-4), a file regarding liaison activities with the Union of South Africa concerning Native Affairs, shows that the Canadian government received reports from South Africa regarding the Department of Native Affairs, including one on the "Resettlement of Natives." In this file, there are also correspondences about visits of South African officials to Canada. One of the most high-profile visits is that by the South African Ambassador to Canada, W. Dirkse van Schalkwyk, in 1962. According to formal government correspondence, "His Excellency has expressed a deep interest in the Indian population and we have already assured him of our willingness to be of service and assistance should he wish to visit reserves..."[47]

In a letter to the Director of the Indian Affairs Branch, Department of Citizenship & Immigration, W.P.B. Pugh, Superintendent of Stony/Sarcee Indian Agency, wrote:

> "I drove [Dirkse van Schalkwyk] out to the Morley Indian Residential School where he toured the school with Mr. R.F. Campbell, principal, seeing the classrooms, dormitory, and general lay-out of the school. We only met one or two Indians for which I was disappointed. There was to have been a barbecue at the school at 3:00p.m., but in good old fashion Indian timing when we left at 4:00p.m., it was still not under way."
>
> "On the return trip to Calgary, I drove him through the Sarcee Indian Reserve from one end to the other, showing him leases, community pasture and new homes, but time did not permit us to visit any of the homes."[48]

The Assistant Regional Supervisor of Manitoba wrote to Ottawa that,

> "... during the trip, His Excellency showed keen interest, and asked innumerable questions concerning the origins, status and customs of the Indian people on these Reserves; their form of band Government, and the social problems prevalent amongst them; and the relationship of our administration to them. I believe his questions were all answered adequately, and it was most interesting to hear his observations on the native people of South Africa and the seeming points of resemblance and dis-similarity between them and our people."[49]

Dirkse van Schalkwyk was reported as being "most anxious to have a quick look at one of the Indian reserves in the Region," visiting Fort Alexander Reserve, Brokenhead Reserve, Duck Lake Agency, Beardy's Reserve, and the Ermineskin Reserve.[50] It is unclear where specifically Dirkse van Schalkwyk's interest in reserves came from, but the year in which these events took place might give us a clue. 1962 was during the height of apartheid—the year in which domestic and international protests escalated, and Nelson Mandela was arrested for conspiracy to overthrow the state. Amidst this turmoil in South Africa was the ever-pressing political discussion of the "Native question," and it seems that Canada suggested potential solutions in its light.

The Canadian National Archives house a surprising abundance of information regarding liaison activities with other countries to exchange knowledge about the "Native question." As John Leslie, historian of the Indian Act, said in a personal phone call on July 6, 2016, Resource Group 10 (Indian Affairs) is the biggest collection in Library and Archives Canada because the influence of Indians, then as is now, pervaded every sector of the economy and the culture of the country: "for a marginalized peoples, they're everywhere."

Finally, a fourth and perhaps more speculative understanding of the evolution of the treatment and administration of "Natives" emerges from the etymology of the word "reserves," "reservations," and "reserved lands" that originated, as noted in the Proclamation of 1763, under the mandate of reserving and setting aside lands exclusively for Natives in Canada by order of the King George III. The spatial process of reservation of land, and later that of "Natives," can be said to have evolved from the original process of reservation applied to forests and other natural resources in earlier centuries. One of the most notable examples, and perhaps one of the more structural cases, is the utilization of forest reserves emerging out of the ground-breaking Forest Charter of 1217, a component of the 1215 Magna Carta, which allocated civil rights and individual freedoms on territories that once belonged exclusive to the King and the Monarchy. The progression and evolution of the idea of *reserving* "forests," *reserving* "resources," *reserving* "land," or *re-*

serving "peoples,"[51] can also be said to be a natural progression of a spatially and legally enforceable technique used in monarchic and military regimes, easily transferred under the ideology of preservation and conservation to specially-designated peoples and Natives of territories invaded and colonized thereafter, over the course of several centuries.

II - EASTERN CONNECTIONS

Much recent scholarship has speculated on the potential connections between American settler colonialism and the Holocaust.[52] American notions of geopolitics and racial entitlement to land crossed the Atlantic and gained devoted supporters in Germany, but the exact effects of this connection are debated. Some, such as Jens-Uwe Guettel, remain skeptical that American influences in Germany lasted until the Third Reich, claiming that the National Socialists, unlike Wilhelmine imperialists, rejected American liberalism and individualism.[53] On the other hand, others such as Carroll Kakel, explicitly characterize Nazi eastward expansion as colonial and inspired by the American precedent. Kakel draws attention to the "mass political violence" of the Holocaust "whose patterns, logics, and pathologies can be located in the Early American project."[54]

The second portion of this note seeks to draw attention to not only specific strategies—the "patterns, logics, and pathologies"[55]—but also the broader colonial imagination and notions of geography that inspired Germans, and specifically German leaders, throughout the early part of the twentieth century to the National Socialist movement. Such images were often inaccurate and internally inconsistent, but the prevalence of references to American tropes nonetheless precedent was a source of influence and potential inspiration for Hitler and other Nazi leaders of his time.

This relationship exposes, in Patrick Wolfe's words, the "logic of elimination" as it relates to the Native—first in America, then later in Germany—and the troubling relationship between settler colonialism and genocide.[56] By putting a broader focus on the deep-seated cultural stereotypes and images that were nonetheless powerful and constantly referenced during Hitler's regime, this note exposes the American influences on Nazi Germany geopolitical ideologies.

Canada may have been a source and precedent of colonial inspiration and Indigenous segregation to not only South Africa, but also to other colonial regimes of subjugation and extermination. Rooted in ideologies of racial segregation are not only relations of labor, but also racial distinctions made possible by who lies and lives outside and inside boundaries of dominant state powers or expanding colonial territories. Inherent to this distinction is the creation and significance of the frontier as Frederick Jackson Turner argued in 1893, "the existence of an area of free land, its continuous recession, and the advance of American settlement westward, explain American [history and] development." More popularly known as the "Wild West," Turner's frontier thesis was contingent on the mythic representation and cultural representation of what was Indigenous lands as free, open, and uninhabited.[57]

While the diffusion of ideas likely occurred in several key directions, southward to South Africa for example (as explained in the first section of this note), one in particular that ward back to Europe, especially towards Germany. While models of colonial administration traveled westward from London and Great Britain to Canada, prolonged periods of peacetime emerged after the War of 1812 made room for the development of a hyper-colonialism of Canada's policies of Indian extermination (later of assimilation and integration). Focused on the management of the so-called "Indian Problem," Indian policy (as represented in the Indian Act of 1876, spearheaded by Sir John A. Macdonald in Canada, and the Indian Removal Act of 1830 under Andrew Jackson in the US) was setting precedent for forms of dispossession and displacement domestically for other countries and other ideologues.[58] One of the least recognized yet possibly one of the most problematic appropriation of Indian policy towards wartime ends of state control and cultural extermination can be found in the models, means, and measures adopted by Adolf Hitler before and during the rise of National Socialism, known as the Nazi Era.

With great interest in stories of 'Cowboys and Indians' from North America's 'Wild West,' there is considerable albeit fragmented evidence that points to Hitler's North American influences and at times 'inspirations,' namely in his writings and speeches. Well before the advent of World War II, there are three distinctive yet intertwined dimensions to the intellectual environment that can be said to have influenced Hitler simultaneously as boy, soldier, ruler—contextually, ideologically, practically—that can be identified. First, his reference to the romanticized Indian and its demise in the novels of Karl May, of whom Hitler was a devoted fan; second, Hitler's frequent use of Indian analogies own version of "Manifest Destiny," and thirdly and lastly, the historical evidence which points to geopolitical thought that transferred from Frederick Jackson Turner to Hitler over the course of several decades.

These references to America in relation to conquest of land and extermination of Indians point to the fact that Hitler (as well as many others in his time) explicitly saw European

Buffalo Bill and his Wild West cast in Venice, Italy, April 1890 (William F. Cody Archive)

settlement on the North American continent as colonial, and moreover as a successful colonial *model* to be applied to Germany's own territorial exploits.

A. Karl May's Wild West Stories and the Exoticization of the Dying "Red Indian"

The ahistorical perception of pioneering advances on the frontier of the American West (which included the westward advance of Canada) was popularized by the tales, fables, and fictions most blatantly perpetuated by William Frederick Cody (1846–1917) of Iowa, the American showman and self-professed scout-cum-hunter-cum-Indian fighter, otherwise known as Buffalo Bill. Thanks to a traveling circus-like show, mythologized images of the 'Wild West' not only traveled across America but also became especially popular in Europe towards the end Bill became a household name as he toured Europe in his Indian War-themed *Wild West* show. In fact, Buffalo Bill's extensive tours across the Atlantic accounted for "almost a third of [his] performative lifespan."[59] The reception of Wild West stories varied across the continent, country by country, but was largely attended on two successive tours. One particular country—Germany—was especially receptive to these stories. In the words of Julia Stetler in her doctoral dissertation, "Buffalo Bill's Wild West in Germany: A Transnational History," German audiences showed an "astoundingly positive response" to Buffalo Bill and similar stories.[60] This, along with the novels of the likes of James Fennimore Cooper and Karl May, served as the backdrop for a large-scale, pan-European misrepresentation of North American history which seemingly justified racially-based ideologies of territorial conquest, racial superiority, and Indigenous subjugation.

Karl May's (1842–1912) bestselling Wild West novels involving the noble adventures of Winnetou, chief of the Apaches, and his companion Old Shatterhand, have left a lasting mark on German popular culture. May is one of the most widely read writers in German history, with distribution estimates at over 100 million copies worldwide, including translations into 28 different languages; and his works have inspired a romanticized image of the American West in German audiences that persists to this day.[61] Interestingly, he had never visited America until after his works about the American West became popular, and his writings reflect an understanding of America that is far from reality. Instead, as Heribert Feilitzsch writes in "Karl May: The 'Wild West' as seen in Germany," he characteristics: A romantic view of nature, 'Fernweh,' the longing for distant places and 'Schulmeisterei,' the tendency to dogmatize out of a perceived feeling of general German superiority … The novels constitute a mixture of stereotypes, chauvinist plots and exotic reality."[62] Feilitzsch characterizes May's popularity as a response to rapid industrialization and Germany's insecurity relative to other strong empires. The fetishization of the "Other"—the dying Indians—in readily consumable romantic stories, expressed a longing for a fabricated, exoticized "past."[63] In fact, Hitler (who exploited these very tendencies in German popular culture himself in his rise to power) recognized that industrialization triggered a romantic impulse in Germany and encouraged the consumption of Indian stories. In his own words, and referencing Karl May, Hitler said that,

> "The industrialization of a country invariably provokes an opposite reaction and gives rise to a recrudescence of a certain measure of romanticism, which not infrequently finds expression in a mania for the collection of *bibelots* and somewhat trashy *objets d'art*.[64] … The only romance which stirs the heart of the North American is that of the Redskin; but it is curious to note that the writer who has produced the most vivid Redskin romances is a German."[65]

Adolf Hitler, who grew up during Karl May's literary heyday, was certainly inspired by these novels from an early age. Probably not unlike many others of his day, Hitler's very notions of world geography were inspired by such highly inaccurate narratives, as Hitler himself was quoted as saying:

> "I've just been reading a very fine article on Karl May. I found it delightful. It would be nice if his work were republished. I owe him my first notions of geography, and the fact that he opened my eyes on the world. I used to read him by candlelight, or by moonlight with the help of a huge magnifying glass. The first thing I read of that kind was *The Last of the Mohicans*. But Fritz Seidl told me at once: 'Fenimore Cooper is nothing; you must read Karl May.' The first book of his I read was *The Ride Through the Desert*. And I went on to devour at once the other books by the same author. The immediate result was a falling off in my school reports."[66]

Cover, *Winnetou*, Karl May's bestselling Wild West novel, first published in 1893 (Karl May Bücher)

It seems that Hitler's appreciation of Karl May's novels inspired in him, and for many Germans, more than just a faulty knowledge of geography, a fantastic image of the West, and a derogatory—if not delusional—representation of Indians. For one thing, Hitler romanticized the violence and conquest portrayed in the novels. Kurt Ludecke, a member of the Nazi party who eventually published a memoir (*I Knew Hitler*), noted that "[Hitler] was delighted to hear that as a boy I had devoured Karl May's stories about the Indians, Old Shatterhand and Winnetou, and said that he could still read them and get a thrill out of them."[67] Similarly, Albert Speer, Hitler's chief architect, remembered that,

> "Hitler would lean on Karl May as proof for everything imaginable, in particular for the idea that it was not necessary to know the desert in order to direct troops in the African theatre of war; that a people could be wholly foreign to you, as foreign as the Bedouins or the American Indians were to Karl May, and yet with some imagination and empathy you could nevertheless know more about them, their soul, their customs and circumstances, than some anthropologists or geographers who had studied them in the field. Karl May attested to Hitler that it wasn't necessary to travel in order to know the world.
>
> Any account of Hitler as a commander of troops should not omit references to Karl May. Hitler was wont to say that he had always been deeply impressed by the tactical finesse and circumspection that Karl May conferred upon his character Winnetou … And he would add that during his reading hours at night, when faced by seemingly hopeless situations, he would still reach for those stories, that they gave him courage like works of philosophy for others or the Bible for elderly people."[68]

Not only do Karl May's books glorify war and conquest, but they are also amenable to a reading that emphasizes the superiority of Germanness and white Christianity. While some have portrayed May's writings in a more sympathetic light, emphasizing his pacifism, Christian values, and overall "positive" view of Indigenous peoples,[69] others, such as Klaus Mann, have suggested that "the Third Reich is Karl May's ultimate triumph, the ghastly realization of his dreams."[70] In fact, *Winnetou*—the first installation of the series that follows the adventures of German émigré Charlie (meaning Karl) and the Apache "noble savage" Winnetou—in its very first sentence deplores the fate of the Indian: "Alas, the red race is dying!" It is "a destiny inexorable."[71] Perhaps more striking is the German nationalism that underlies the plot, as shown in the laziness and weakness of the German protagonist's fellow American plotters. His first ally, one of the rare men who are morally and physically strong, turns out to be a "German immigrant just like [Charlie]."[72] Rather than give a "truthful" account of the American West, May's hyperbolic writings seem to praise the bravery and strength of their thoroughly German protagonist. Moreover, they justify extermination through the pseudo-Darwinian logic that the death of the "red race" is inevitable.

Hitler gained more from Karl May's books than a rough sense of geography and "winner and loser." It seems he even interpreted the books literally and sought direct tactical inspiration: some historians have suggested that Hitler handed out May's books to his generals as recommended reading. Klaus Fischer, writer of *Hitler and America*, reports that "in 1944, despite the shortage of paper, he ordered 300,000 copies of May's books to be printed and distributed among the troops as exemplary military field literature."[73 74] The wildly inaccurate images of the Indian that pervaded German popular culture thus became part of a more sinister narrative, used in Hitler's own version of Manifest Destiny.

B. Conquest and Settlement: Indians as a Metaphor for Hitler's Own "Frontier"

Hitler's almost childlike obsession with Wild West novels aside, he frequently used the imagery and analogy of Indian conquest to his very own "frontiers" of territorial expansion. Speaking of fighting the Russians, he said that "the struggle we are waging there against the Partisans resembles very much the struggle in North America against the Red Indians. *Victory will go to the strong, and strength is on our side.*"[75] This underlines the pseudo-Darwinian logic behind the statement that the death of the "red race" is a "destiny inexorable"—far from benign or sympathetic, such a worldview justifies the frontier battles that Hitler fought himself, in particular in his efforts to expand to eastern Europe. The American precedent confirmed for Hitler that the strong did indeed win out, at whatever cost necessary. Further extending this analogy, Hitler has been quoted as saying that the "Volga must be our Mississippi."[76]

Not only was German victory inevitable (at the cost of others' lives and land), but it was also retroactively justifiable because the masses would soon forget. Elaborating on his plan to settle and populate the "immense spaces of the Eastern Front," and to carefully consider the question of the "Natives,"[77] Hitler again makes an analogy to the American West:

> "In this business I shall go straight ahead, cold-bloodedly. What they may think about me, at this juncture, is to me a

> matter of complete indifference. I don't see why a German who eats a piece of bread should torment himself with the idea that the soil that produces this bread has been won by the sword. When we eat wheat from Canada, we don't think about the despoiled Indians."[78]

Here it becomes explicit that Hitler referred to Canada as a model for "frontier expansion" through extermination, and that he knew that historical judgment favored the side of the victor thanks to collective amnesia. More importantly, on the other side of the "conquest" narrative is that of exploitation (of both people and land) and of gaining access to the bounty of the "free land" that has been made available through it. "There is only one task," he said in October 1941, "to set about the Germanization of the land by bringing in Germans and to regard the Indigenous inhabitants as Indians."[79] In fact, the reference to "wheat from Canada" was far from accidental (coinciding with the world largest land survey in Canada's Western Prairies) as Hitler had the understanding that the land thus cleared would yield material wealth, through agriculture and mining, to Germany. Hitler had a clear, albeit colonial, understanding of settlement, as he expressed in *Mein Kampf*:

> "We ought to remember that during the first period of American colonization numerous Aryans earned their daily livelihood as trappers and hunters, etc., … The moment, however, that they grew more numerous and were able to accumulate larger resources, they cleared the land and drove out the aborigines, at the same time establishing settlements which rapidly increased all over the country."[80]

There was prevalent knowledge of the material riches in North America, and more specifically Canada, at the time. At the infamous Auschwitz-Birkenau concentration camp, prisoners were forced to give up their belongings in an area called "Canada," so named for the famed riches of the country.[81] The prevalent conceptions of North America, and specifically Canada, as a place with abundant resources and riches (distinctly acquired through dispossession, displacement, and extermination) were thus reflected in German culture in some of the darkest ways possible.

Hitler repeated the myth of the frontier through his frequent references to America's vast territory, in which he expressed insecurity vis-à-vis other great empires due to Germany's relative lack of land. He believed—not unlike many scholars in his day—that the strength and spirit of a nation came from its territory. In *Mein Kampf*, Hitler commented that "the gigantic American State Colossus, with its enormous wealth of virgin soil, is more difficult to attack than the wedged-in German Reich."[82] In Chapter 14 of *Mein Kampf*, "Eastern Orientation or Eastern Policy"—also translated as "Germany's Policy in Eastern Europe"—Hitler continues to assert that Germany cannot lay claim to "world power" status due to its meager territory:

> "The German nation entered this battle [World War I] as an alleged world power. I say alleged here, because in reality it was not. Had the German nation in the year 1914 had a different relation between territorial area and population, Germany would really have been a world power, and regardless of all other factors the War could have been happily terminated.
>
> … Germany is no world power today. … In an epoch when the earth is gradually being divided among States, some of which encompass almost whole continents, one cannot speak of a structure as world power the political mother country of which is limited to the ridiculous area of barely five hundred thousand square kilometers."[83]

He goes on to compare Germany unfavorably to other countries, namely the United States of America, which by virtue of their vast territories qualify as *real* world powers:

> "Looked at purely territorially, the area of the German Reich compared with that of the so-called world powers altogether vanishes. … We must also consider as giant States first of all the American Union, then Russia and China."[84]

Hitler claims that German colonial policy was inferior to that of the French or British because it did not "[enlarge] the area of settlement of the German race … [the other powerful states] not only far outstrip the strength of our German population, but … have the greatest support of their position of political power above all in their area."[85] Not only did he think that the vast spaces of this territory were a source of imperial strength, but in the case of America, he also thought it was a source of national spirit; those spaces being (in the public mind) empty and unoccupied. In a speech he gave in June 1943, he said that,

> "One thing the Americans have, and which we lack, is the sense of the vast open spaces. Hence the particular characteristics of our own form of nostalgia. There comes a time when this desire for expansion can no longer be contained and must burst into action. It is an irrefutable fact that the Dutch, for example, who occupied the most densely populated proportions of the German lands, were driven, centuries ago, by this irresistible desire for expansion to seek ever wider conquest abroad.
>
> What, I wonder, would happen to us, if we had not at least the illusion of vast spaces at our disposal? For me, one of the charms of the Spessart is that one can drive there for hours on end and never meet a soul. Our autobahnen give me the same feeling; even in the more thickly populated areas they reproduce the atmosphere of the open spaces."[86]

This "desire for expansion"—both physical and supposedly spiritual—for more space, then justified in Hitler's own words, "wider conquest abroad."[87] Here is Hitler's own version of "Manifest Destiny"; Hitler acknowledged that, whatever legal and political justification can be given to it, in the end colonization was a project and product of asserting white superiority. As early as 1932, he said that:

> "Just in the same way Cortez or Pizarro annexed Central America and the northern states of South America, not on the basis of any right, but from the absolute inborn feeling of the superiority [*Herrengefühl*] of the white race. The settlement of the North American continent is just as little the consequence of any superior right in any democratic or international sense; it was the consequence of a consciousness of right which was rooted solely in the conviction of the superiority and therefore of the right of the white race."[88]

Although Hitler's racist notions of who was "entitled" to a certain land culminated in an extreme project of territorial expansion and racial extermination, his concepts of race and land actually date back to much earlier—and then seen as "scientific"—theories of land and national strength, namely Frederick Jackson Turner's Frontier Hypothesis, where state and frontier were intermingled.

C. The Historical Origins of Frontier-Thinking: German Expansion and "Manifest Destiny" from Frederick Jackson Turner to Adolf Hitler

Hitler's admiration for America's open spaces, combined with an understanding of how they were won (Karl May's factual inaccuracies notwithstanding) culminated in a doctrine of entitlement to land based on racial superiority. In other words, Hitler's images of Indian conquest and America's vast open spaces went hand in hand to produce his own version of "Manifest Destiny." Given that Hitler used American expansion as an analogy and model for his own project, it is unsurprising that even his geopolitical ideologies were influenced by American notions of the frontier. Hitler's expansionism was motivated by the then-popular concept of "Lebensraum"—direct translation "living space"—the pseudo-Darwinian idea that stronger nations were like growing biological species and needed to expand their territory.[89] This concept embodied the rather circular logic that white people were meant to occupy a given space because they were succeeding in doing so. German geographer Friedrich Ratzel was first to suggest this term, particularly in his essay "Der Lebensraum."[90] Ratzel was influenced by his knowledge of America, American history, and most importantly the writings of Frederick Jackson Turner. As Jens-Uwe Guettel writes in "From the Frontier to German South-West Africa: German Colonialism, Indians, and American Westward Expansion":

> "German colonialists were impressed by American expansionism and envied the United States for the easily available ('empty') lands in 'the West.' In fact, Friedrich Ratzel was so impressed that he changed his career after visiting the United States as a journalist in 1874–1875. … In his first book, a travelogue entitled *Sketches of Urban and Cultural Life in North America* (1876), his fascination with both the question of race control and the immense space available for the westward expansion of the United States was already clearly apparent, and America featured prominently in most of his subsequent publications."[91]

Ratzel also openly expressed praise for Frederick Jackson Turner's "frontier thesis," agreeing that the "frontier" was what produced "Americanness."[92] In his work *Politische Geographie*, where Ratzel expresses a detailed interest in and knowledge of America, including the political, cultural, and economic union achieved by the westward expanding railway.[93] He also directly references the concept of the frontier, saying that Frederick Jackson Turner "contrasted the lively border of the American western expansion … with the European border which lies between densely populated countries."[94] In a book exclusively about the United States, Ratzel argued that:

> "In the United States … the greatness of space has counteracted decay … the will and the power to colonize that enlivens so many individuals have time and again enlarged the national economic sphere."[95]

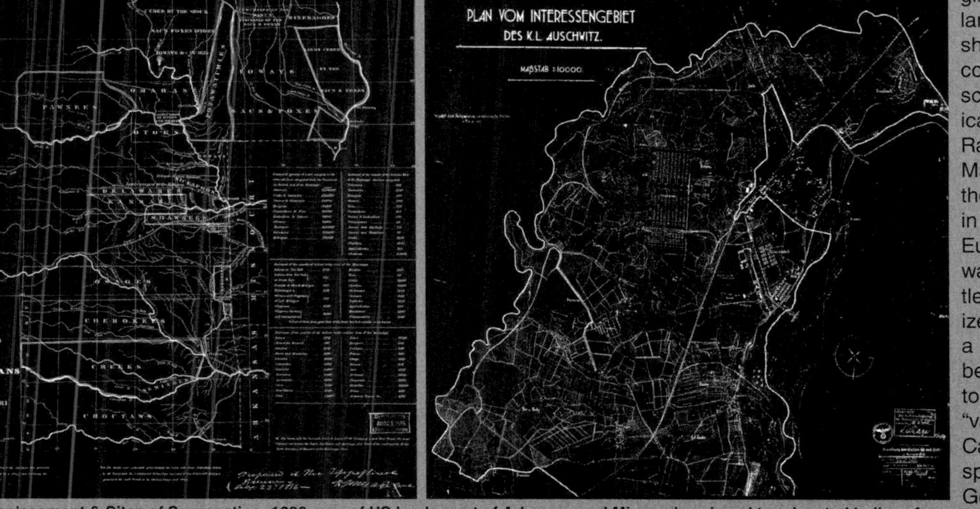

Territorial Displacement & Sites of Segregation: 1836 map of US lands west of Arkansas and Missouri assigned to relocated Indians from Southeastern US under the Indian Removal Act (left; Library of Congress Map Division) and 1941 map of Auschwitz-Birkenau Concentration Camp in the context of Krakow, Poland during World War II under Adolf Hitler's Nazi Regime (right; courtesy of Auschwitz Museum)

David Thomas Murphy, writer of *The Heroic Earth: Geopolitical Thought in Weimar Germany, 1918-1933*, attributes these influences to a broader transnational conversation that engaged the likes of Americans Frederick Jackson Turner, Alfred Mahan, British geographer Halford Mackinder, and Friedrich Ratzel in Germany.[96] Norman Rich, professor emeritus at Brown University, has gone so far as to claim that the "United States policy of westward expansion, in the course of which the white men ruthlessly thrust aside the 'inferior' Indigenous population, served as the model for Hitler's entire concept of *Lebensraum*."[97]

These ideas were later transferred to Hitler via Haushofer (who was a disciple of Ratzel) who taught Hitler's private secretary Rudolf Hess, and later Hitler himself.[98] The relationship between Haushofer and Hitler is chronicled in detail in the book *The Demon of Geopolitics: How Karl Haushofer "Educated" Hitler and Hess*, by Holger H. Herwig. Hitler himself uses the term "Lebensraum" in many of his speeches, and in *Mein Kampf* alone three times, as a justification for Germany's territorial expansion. During his speech at the Dusseldorf Industry Club on January 27, 1932 he said that:

> "We have a number of nations which through their inborn outstanding worth have fashioned for themselves a mode of life that stands in no relation to the life-space (*Lebensraum*) they inhabit in their densely populated settlements. We have the so-called white race which, since the collapse of ancient civilization, in the course of some thousand years has created for itself a privileged position in the world."[99]

In fact, American influences in German geopolitical thought date back much earlier than Hitler's notions of Lebensraum. In "From the Frontier to German South-West Africa: German Colonialism, Indians, and American Westward Expansion," Jens-Uwe Guettel argues that prior to looking to the "East" (i.e. Eastern Europe), Germany applied ideas from America's expansion to the colonization of German South-West Africa. He writes that "colonial administrators actively researched the history of the American frontier and American Indian policies in order to learn how best to 'handle' the colony's peoples." He continues, "the United States as a 'model empire' was especially attractive for Germans with liberal and progressive conviction. The westward advancement of the American frontier went hand in hand with a variety of policies towards Native Americans, including measures of expulsion and extinction."[100]

Alan E. Steinweis, in "Eastern Europe and the Notion of the 'Frontier' in Germany to 1945," says that the American "frontier hypothesis" origins of German expansionism are largely ignored in traditional scholarship regarding Nazi Germany.[101] But contrary to this lack of contemporary scholarly interest, many geopolitical theorists—including Friedrich Ratzel, Rudolf Kjellen, Halford John Mackinder, and Karl Haushofer—thought up schemes, first to expand in Africa, then to expand to Eastern Europe. "Implicit in these schemes was the understanding that the settlement of the frontier would revitalize German society, and serve as a safety valve for a society that has become too densely populated and too urbanized."[102] Thus the imagined "vast spaces" of America (and thus Canada by association) had been inspiring a longing for "living space" in Germany for even a long time before Hitler.

While this connection is not new, this proposes an ongoing easterly flow of ideas from North American Canada specifically. Robert L. Nelson's conclusion comments, in his portrayal of a short official visit of German agrarian economist Max Sering to the Prairies in 1870, offers an important observation on the subjugation of Indians as subtext to the inner colonization of Canada, set against the universal history of genocide:

> "I am not arguing that Canadians were Nazis. I am however more than comfortable in … habits of thought and practice that provided many of the

precursors to one of the most murderous colonizations of the twentieth century, from 1939 to 1944."[103]

Lastly, these complex connections in overt or subdued colonial techniques, from the dispossession of land to methods of displacement (including the Long Walk of the Navajo and the Death Marches during the Holocaust) to finally segregation in designated areas under inhumane conditions intended for ultimate extermination, suggest a more complex yet lesser understood historical link and spatial relationship between the scientific management and bureaucratic administration of extermination policies that were embedded in the planned forms and exclusionary geographies of Indian reservations and the engineered architectures and designed structures of Nazi concentration camps. In his biography of Hitler, historian John Toland wrote that "Hitler's concept of concentration camps as well as the practicality of genocide owned much, so he claimed, to his studies of English and United States history."[104] While the direct relationship between the planning and design of concentration camps in relationship to Indian reservations is lesser known and has been lesser studied, it is possible that the architects of the political regimes of Nazi Germany may have drawn direct spatial and territorial "inspiration" in their designs.[105]

Moreover, as Thomas Kühne explains, it is in "differentiation, not generalization [that is] required … when it comes to inquiring into the roots and reasons of Nazi violence."[106] What differentiates the Canadian example of race-bound segregation of Indians is a form of domestic inner-colonialism, as opposed to outer-colonialism. Both akin to goals of empire-building through accumulation of power with resources (spatial, environmental, human labor) demonstrate "variants of colonialisms and imperialisms. Useful here would be the distinction between 'colony' as an individual place and 'empire' as the assemblage of many of them, or calling the activities (and sufferings) at the peripheries 'colonialism' and those at the metropoles 'imperialism.'"[107] These inner-colonialisms are clearly evident in the substandard conditions of First Nations across today.

Examining the complex origins of Hitler's expansionism shows that, while a singular cause cannot be attributed to Nazi Germany's ideologies of racial superiority, entitlement to land and desire for natural resources and wealth, the case of the colonization of North America, at least, served as an example and a reference point in a shift of universal history that places history of genocide in a unique and unprecedented historical context of Germany and the Holocaust. From the largely fictionalized accounts of the American West by Karl May, which evoked pity for the dying Indian (insofar as he was civilized and Christianized) to the expressed envy of America's vast territory and material riches, such perceptions of America pervaded popular culture and influenced Adolf Hitler, both directly and indirectly. Such influences may since have become largely forgotten, and potentially barely relevant, yet the legacy of "frontier" ideologies remains. By depoliticizing Indigenous land and the Indigenous subject, these views underlie state-sanctioned corporate expansionist policies, race-based sciences, and extraction projects worldwide, the legacy of which rests in our hands today.

This brief record of Hitler's own thought regarding America and Canada, as well as scholarship regarding diffusion of ideas of racial superiority and white supremacy in territorial expansion, spatial strategies of segregation, and ideologies of extermination, serve to uncover the broad impact of the North American legacy and reality of settler colonialism. Finally, this note will hopefully serve as a starting point for future research in studies of decolonization of territorial policies and strategies of segregation that reposition Canada's role as both precedent recipient and progenitor of genocide culture that further reorders the universal history of state-based violence towards Indigenous peoples. This process is only possible if the extractive state of Canada acknowledges, understands, and internalizes the violent atrocities that form its foundation and its perpetuation.

⊗⊗⊗

Notes

Thank you to Ron Bourgeault, John Leslie, and Dirk Moses for their helpful comments during the research process and preparation of this research note.

1 See Patrick Wolfe, "Settler Colonialism and the Elimination of the Native," *Journal of Genocide Research* 8, no. 4 (December 2006): 387–409.

2 See Linda Freeman, *The Ambiguous Champion: Canada and South Africa in the Trudeau and Mulroney years* (Toronto, ON: University of Toronto Press, 1997); and Joan G. Fairweather, "Is this Apartheid? Aboriginal reserves and self-government in Canada 1960-1982" (MA diss., University of Ottawa, 1993).

3 Ron Bourgeault, "Canada Indians: The South African connection," *Canadian Dimension* 21, no. 8 (January 1988): 6–10.

4 *We Have Such Things at Home*, directed by James Cullingham (1997; Toronto, ON: Tamarack Productions, 2014), http://www.tamarackproductions.com/we-have-such-things-at-home/.

5 Tiffany Willoughby-Herard, *Waste of a White Skin: The Carnegie Corporation and the Racial Logic of White Vulnerability* (Oakland, CA: University of California Press, 2015).

6 Edward-John Bottomley, "The Poor Volk," in *Poor White* (Cape Town: Tafelberg, 2012), under subheading "Redeeming the Poor."

7 Public Archives of Canada (PAC), RG10, vol. 6810, file 470–2–3, vol. 7: Evidence of Duncan Campbell Scott to the Special Committee of the House of Commons examining the Indian Act amendments of 1920, pp. 55 (L–3) and 63 (N–3) respectively; see Memorandum for the Special Committee of the House of Commons re. Bill 14 (An Act to amend the Indian Act) from the Secretary, Six Nations Council, March 30, 1920, as quoted in Kahn-Tineta Miller, George Lerchs, and Robert G. Moore, "The Impact of Immigration and WWI: 1906–1927," chap. 7 in *The Historical Development of the Indian Act*, 2nd ed. John Leslie and Ron Maguire (Ottawa, ON: Treaties and Historical Research Centre, P.R.E. Group, Indian and Northern Affairs, 1978), 114, 176n57, http://www.kitselas.com/images/uploads/docs/The_Historical_Development_of_the_Indian_Act_Aug_1978.pdf.

8 For the history of Indian reserves, see Richard Bartlett, [esp. chap. 1 in] *Indian Reserves and Aboriginal Lands in Canada: A Homeland* (Saskatoon, SK: University of Saskatchewan, Native Law Centre, 1990). For a more general discussion of land rights and Indigenous peoples see Wolfe, "Elimination of the Native." See also Miller, Lerchs, and Moore, *Historical Development of the Indian Act*.

9 For South African uses of the pejorative term the "Native question," see for instance Howard Pim, *The Native Question in South Africa: A Paper Read before the Society on the 19th March, 1903* (Johannesburg: Argus Printing, 1903); and Alice Werner, "The Native Question in South Africa," *Journal of the African Society* 4, no. 16 (1905): 441–54.

10 Bourgeault, "Canada Indians," 7.

11 Ibid.

12 Ibid.

13 Aborigines Protection Society, *Report of the Parliamentary Select Committee on Aboriginal Tribes (British Settlements)* (London, UK: William Ball, Aldine Chambers, Paternoster Row and Hatchard & Son, 1837), x.

14 Ibid., 1.

15 Ibid., 3–4.

16 Ibid., 14.

17 Ibid., 4–59.

18 Parliamentary Papers, Accounts and Papers: 1884-1885, Forty-six volumes, contents of the twelfth volume, Vol. 56, "South Africa further correspondence respecting the Cape Colony and Adjacent Territories (in continuation of [C.–3855] February 1884) presented to both Houses of Parliament in Command of Her Majesty, December 1884," [C.–4263] (London, UK: Eyre and Spottiswood, 1884), 1.

19 Ibid.

20 *The Pass System: Life under Segregation in Canada*, directed by Alex Williams (Toronto, ON: Canada Council for the Arts, 2015), 50 min.

21 Maxine K. Rochester, "The Carnegie Corporation and South Africa: Non-European Library Services," *Libraries & Culture* 34, no. 1 (Winter 1999): 28.

22 See Rochester, "Carnegie Corporation South Africa"; and Willoughby-Herard, chap. 1 in *Waste of a White Skin*.

23 Willoughby-Herard, *Waste of a White Skin*, 2.

24 See William Watkins, *The White Architects of Black Education: Ideology and Power in America*, 1865–1954 (New York, NY: Teachers College Press, 2001); Paul Rich, *White Power and the Liberal Conscience: Racial Segregation and South African Liberalism* (Johannesburg: Ravan Press, 1984); and Edward Berman, *The Ideology of Philanthropy: The Influence of the Carnegie, Ford, and Rockefeller Foundations on American Foreign Policy* (Albany, NY: State University of New York Press, 1983), as referenced in Willoughby-Herard, *Waste of a White Skin*, 24.

25 Ellen Condliffe Lagemann, *The Politics of Knowledge: The Carnegie Corporation, Philanthropy, and Public Policy* (Chicago, IL: University of Chicago Press, 1992), 11, as quoted in Willoughby-Herard, *Waste of a White Skin*, 3.

26 Francis Wilson, interview by Mary Marshall Clark, *Carnegie Corporation Oral History Project: Interview Catalogue* (New York, NY: Columbia University

Libraries, 2006) transcript, Session #2, Cape Town, August 3, 1999, 118–19.

27 Bottomley, "The Poor Volk," under subheading "Redeeming the Poor," par. 6.

28 Freeman, *The Ambiguous Champion*, 16.

29 Oswin Boys Bull, *Training Africans for Trades: A Report on a Visit to North America under the Auspices of the Carnegie Corporation Visitors' Grants Committee* (Pretoria: The Carnegie Visitors' Grants Committee, 1935), 9.

30 Ibid., 5.

31 Ibid., 19.

32 Ibid.

33 Ibid., 10.

34 Ibid., 26.

35 Carnegie Collections: Columbia University Library, Rare Book and Manuscript Library, Carnegie Corporation of New York Records, 1872-2000, Series III: Grants [hereafter cited as Carnegie Collections].

36 Ibid.

37 Ibid.

38 Ibid.

39 Ibid.

40 Library and Archives Canada, Vol. 8588, File 1/1-10-4 (1949-1962), MS, RG10 (Indian Affairs Central Registry Files) [hereafter cited as Indian Affairs Record Group 10], Ottawa, ON: December 13, 1956.

41 "Guide to the Charles Templeman Loram Papers MS10," Yale University Library Manuscripts and Archives, New Haven, CT: September 1969, revised May 1998, http://drs.library.yale.edu/HLTransformer/HLTransServlet?stylename=yul.ead2002.xhtml.xsl&pid=mssa:ms.0010&clear-stylesheet-cache=yes, [hereafter cited as Charles Templeman Loram Papers].

42 John Whitson Cell, *The Highest Stage of White Supremacy: The Origins of Segregation in South Africa and the American South* (Cambridge, UK: Cambridge University Press, 1982), 221.

43 Charles Templeman Loram Papers.

44 A.G. Bailey, review of *The North American Indian Today*, by C.T. Loram and T.F. McIlwraith, *The Canadian Journal of Economics and Political Science* 10, no. 1 (February 1944): 110.

45 C.T. Loram and T. F. McIlwraith, ed., *The North American Indian Today: University of Toronto - Yale University Seminar Conference; Toronto, September 4–16, 1939* (Toronto, ON: University of Toronto Press, 1943), 4–5.

46 "Pittsburgh and Vicinity," *Steel and Iron* (National Iron and Steel Publishing Company) 67, July 5, 1900, 31.

47 Indian Affairs Record Group 10.

48 Ibid.

49 Ibid.

50 Ibid.

51 For a discussion of the historical development of Indian reserves, see Bartlett, *Indian Reserves and Aboriginal Lands in Canada*.

52 See David Blackbourn, *The Conquest of Nature: Water, Landscape, and the Making of Modern Germany* (New York, NY: Norton, 2006); Mark Mazower, *Hitler's Empire: How the Nazis Ruled Europe* (New York, NY: Penguin, 2008); Jens-Uwe Guettel, *German Expansionism, Imperial Liberalism and the United States: 1776–1945* (Cambridge, UK: Cambridge University Press, 2012); and Carroll P. Kakel III, *The American West and the Nazi East: A Comparative and Interpretive Perspective* (Basingstoke, UK: Palgrave Macmillan, 2011).

53 Guettel, *German Expansionism, Imperial Liberalism*.

54 Kakel, *American West Nazi East*, 3.

55 Ibid. See also Kakel, *American West Nazi East*, 219n12, where the author explains: "For the purposes of this book, the term 'patterns' refers to recurring empirical patterns of events in both the 'American West' and the 'Nazi East'; the term 'logics' refers to ways of thinking, beliefs, and attitudes (conscious and unconscious) held by political elites and 'ordinary' citizens, in both historical contexts, that influenced behaviour and decisions; and the term 'pathologies' refers to values and patterns of behaviour displayed by political elites and 'ordinary' citizens in both historical contexts."

56 For the centrality of culture in the conception of genocide, see A. Dirk Moses, "Raphael Lemkin, Culture, and the Concept of Genocide," chap. 1 in *The Oxford Handbook on Genocide Studies*, ed. Donald Bloxham and A. Dirk Moses (Oxford, UK: Oxford University Press, 2010), 19–41. See also Wolfe, "Settler Colonialism and the Elimination of the Native," 387.

57 Frederick Jackson Turner, "The Significance of the Frontier in American History," (paper read at the meeting of the American Historical Association during the World Columbian Exposition, Chicago, IL, July 12, 1893).

58 See Jill St. Germain, *Indian Treaty-Making Policy in the United States and Canada, 1867–1877*, (Lincoln, NE: University of Nebraska Press, 2001). See also James Daschuk, *Clearing the Plains: Disease, Politics of Starvation and the Loss of Aboriginal Life* (Regina, SK: University of Regina Press, 2013); and Wayne Dougherty and Dennis Madill, *Indian Government under Indian Act Legislation, 1868–1951* (Ottawa, ON: Department of Indian Affairs and Northern Development, Treaties and Historical Research Centre, 1980).

59 Julia Simone Stetler, "Buffalo Bill's Wild West in Germany: A Transnational History" (PhD Diss., U of Nevada, Las Vegas, 2012), 2.

60 Ibid., iii.

61 Heribert Frhr v. Feilitzsch, "Karl May: The 'Wild West' as seen in Germany," *Journal of Popular Culture* 27, no. 3 (Winter 1993): 173.

62 Ibid.

63 Ibid., 180.

64 Adolf Hitler, *Hitler's Table Talk 1941–1944: His private conversations*, ed. Hugh R. Trevor-Roper (New York, NY: Enigma Books, [1953] 2000–2008), 535.

65 Ibid., 536.

66 Ibid., 240.

67 Kurt Georg Wilhelm Ludecke, *I Knew Hitler: The Story of a Nazi Who Escaped the Blood Purge* (New York, NY: C. Scribner's Sons, 1937), 524.

68 Albert Speer, *Spandau: The Secret Diaries* (New Haven, CT: Phoenix Press, 2000), 347.

69 See Jennifer Michaels, "Fantasies of Native Americans: Karl May's continuing impact on the German imagination," *European Journal of American Culture* 31, no. 3 (2012): 205–18.

70 Klaus Mann, "Karl May: Hitler's Literary Mentor," *The Kenyon Review* 2, no. 4 (Autumn 1940): 400.

71 Karl May and David Koblick, *Winnetou* (Pullman, WA: Washington State University Press, 1999), 1.

72 Ibid., 23.

73 Klaus P. Fischer, *Hitler & America* (Philadelphia, PA: University of Pennsylvania Press, 2011), 21.

74 See also Hans Severus Ziegler, *Adolf Hitler aus dem Erleben dargestellt* (Göttingen: K. W. Schütz, 1964), 72; Robert G.L. Waite, *The Psychopathic God Adolf Hitler* (New York, NY: Basic Books, 1977), 11–12; and Timothy W. Ryback, *Hitler's Private Library* (New York, NY: Alfred A. Knopf, 2008), 179–80.

75 Hitler, *Hitler's Table Talk*, 469 [emphasis added].

76 Blackbourn, *The Conquest of Nature*, 293.

77 Hitler, *Hitler's Table Talk*, 54.

78 Ibid., 55.

79 Adolf Hitler, monologue on October 17, 1941 in *Monologe im Führerhauptquartier 1941–1944: Die Aufzeichnungen Heinrich Heims*, ed. Werner Jochmann (Hamburg: Albrecht Knaus, 1980), 91, as quoted in Blackbourn, *The Conquest of Nature*, 286.

80 Adolf Hitler, *Mein Kampf*, trans. Ralph Manheim (Boston, MA: Houghton Mifflin, 1943), 419.

81 Laurence Rees, "Auschwitz 1940–1945: Corruption" in *Auschwitz: Inside the Nazi State*, series televised by PBS, 2004–2005, http://www.pbs.org/auschwitz/40-45/corruption/.

82 Hitler, *Mein Kampf*, 929.

83 Ibid., 936.

84 Ibid., 937.

85 Ibid., 938.

86 Hitler, *Hitler's Table Talk*, 536.

87 Ibid.

88 Adolf Hitler, "Adolf Hitler addressed the Industry Club in Düsseldorf, January 27, 1932," in Roderick Stackelberg and Sally A. Winkle (eds.), *The Nazi Germany Sourcebook: An Anthology of Texts* (Abingdon, UK: 2002), 106–7.

89 See Jeremy Noakes, "Hitler and 'Lebensraum' in the East," BBC, last modified March 30, 2011, http://www.bbc.co.uk/history/worldwars/wwtwo/hitler_lebensraum_01.shtml; and Woodruff D. Smith, *The Ideological Origins of Nazi Imperialism* (New York, NY: Oxford University Press, 1986).

90 See Friederich Ratzel, *Der Lebensraum: eine biogeographische Studie* (Tübingen: H. Laupp, 1901).

91 Jens-Uwe Guettel, "From the Frontier to German South-West Africa: German Colonialism, Indians, and American Westward Expansion," *Modern Intellectual History* 7, no. 3 (2010): 535.

92 Alan Steinweis, "Eastern Europe and the Notion of the 'Frontier' in Germany to 1945," *Yearbook of European Studies* 13 (1999): 66, as quoted in Kakel, *American West Nazi East*, 21.

93 Friedrich Ratzel, *Politische Geographie* (München: R. Oldenbourg, 1897), 417.

94 Ratzel, *Politische Geographie*, 612.

95 Ibid., 95, as quoted in Guettel, "German South-West Africa," 536.

96 David Thomas Murphy, *The Heroic Earth: Geopolitical Thought in Weimar Germany, 1918-1933* (Kent, OH: Kent State University Press, 1997), 2.

97 Norman Rich, "Hitler's Foreign Policy," in *The Origins of the Second World War Reconsidered: The A.J.P. Taylor Debate After Twenty-Five Years*, ed. Gordon Martel (Boston, MA: Allen & Unwin, 1986), 136, as quoted in Kakel, *American West Nazi East*, 1.

98 Kakel, *American West Nazi East*, 32.

99 Hitler, "Industry Club in Düsseldorf," 106.

100 Guettel, "German South-West Africa," 523.

101 Steinweis, "Notion of the 'Frontier' in Germany," 57.

102 Ibid., 64–65.

103 Robert L. Nelson, "German on the Prairies: Max Sering and settler colonialism in Canada," *Settler Colonial Studies* 5, no. 1 (2015), 15. See also Robert L. Nelson, "From Manitoba to the Memel: Max Sering, inner colonization and the German East," *Social History* 35, no. 4 (2010): 439–57.

104 John Toland, *Adolf Hitler* (Garden City, NY: Doubleday, 1976), 702.

105 This intercontinental comparison parallels Carroll P. Kakel's argument in *The American West and the Nazi East*.

106 Thomas Kühne, "Colonialism and the Holocaust: continuities, causations, and complexities," *Journal of Genocide Research* 15, no. 3 (2013): 346.

107 Ibid., 346–47.

Cyclops

Margaret Atwood

1965–1975

You, going along the path,
mosquito-doped, with no moon, the flashlight
a single orange eye

unable to see what is beyond
the capsule of your dim
sight, what shape

contracts to a heart
with terror, bumps
among the leaves, what makes
a bristling noise like a fur throat
Is it true you do not wish to hurt them?

Is it true you have no fear?
Take off your shoes then,
let your eyes go bare,
swim in their darkness as in a river

do not disguise
yourself in armour.

They watch you from hiding:
you are a chemical
smell, a cold fire, you are
giant and indefinable

In their monstrous night
thick with possible claws
where danger is not knowing,
you are the hugest monster.

The Institutionalization of Indoctrination & Dispossession

An 800-year Timeline of Doctrines and Counter-Movements, from the Magna Carta to IdleNoMore

A comprehensive outline of bureaucratic doublespeak related to the creation of the Indian Act and Treaties cloaking and camouflaging colonial strategies of displacement and dispossession towards imperial ambitions of assimilation, integration, and extermination.

"A barbarous Country must be first broken by a War, before it will be capable of good Government; and when it is fully subdued and conquered, if it be not well planted [settled] and governed after the Conquest, it will oftentimes return to the former Barbarism."

—Sir John Davis, *Historical Relations*, 1733

"Sir Thomas Dale haveinge allmoste finished the foarte and settled a plantacyon in thatt parte, dyvers of his men beinge Idell and nott willeinge to take paynes did Runne away unto the Indyans many of them beinge taken ageine S[i]r Thomas in a moste severe mannor cawsed to be executed. Some he apointed to be hanged some burned some to be broken upon wheles others to be Staked and some to be shott to deathe all theis extreme and crewell tortures he used and inflicted upon them To terrefy the reste for attempteinge the Lyke, And some w[hi]ch Robbed the store he cawsed them to be bownd faste unto Trees and so sterved them to deathe."

—George Percy, *A Trewe Relacyon*, 1625

1749 Dummer's Treaty renewed, then ratified again

1732 **Reservations**
Huron Indian Reserve, Huron Village of Lorette along St. Charles River

1728 Dummer's Treaty ratified less than 3 years later

1725 **Treaty or Trick?**
Dummer's Treaty of 1725, first of the "Peace and Friendship Treaties"; signed between British Crown and Mi'kmaq, Maliseet, and Passamaquoddy First Nations

1670 **Incorporation**
Hudson's Bay Company established claimed ownership of Rupert's Land

1637 **Segregation**
First Indian Reservation established by the French at Sillery Reserve

1605 **Penetration**
Champlain establishes Port Royal on Mi'kmaq Territory

1598 **Royal Assent**
James VI of Scotland declares the Divine Right of Kings, akin to God only

1504 **Fishing Trade**
European fishermen begin trading with Mi'kmaq and Maliseet First Nations

1497 **Giovanni Caboto**
With letters patent in hand, John Cabot arrives in Newfoundland and initiates exploration and conquest of so-called *terra incognita,* lands unknown to the European Christians yet occupied by Indigenous Nations for ten millenia

1493 **Inter caetera**
Pope Alexander XI grants overseas territories beyond the line of the Treaty of Tordesillas (Papal Line of Demarcation of non-Christian territories beyond Europe) in papal bull and supreme order to the Catholic Monarchs of Spain and Portugal on 4 May 1493. The era of Crusades justifies European, territorial conquest, colonization and global expansionism

1492 **Invasion**
Retroactively sanctioned by papal bull, Cristoforo Colombo practically eradicates the Taino peoples on Hispaniola starting in Cap Haïtien, spearheading the era of genocide of the *indios* (Spanish for 'Indians') and of the translatic slave trade

1492 **Romanus Pontifex**
Pope Nicholas V issues Papal Bull to King Afonso V of Portugal to confirm to the Crown of Portugal Dominion over all lands south of Cape Bojador in Africa. The Romanus Pontifex follows *Dum Diversas,* the authorization to consign exclusive sphere of control over non-Christians into 'perpetual servitude'

1217 **The Provisions of Oxford**
A Pillar of written Parliamentary Law, along with the Provisions of Westminster (1259) and the Ancient Statute of Merton (1235)

1217 **Magna Carta II:** ***Charter of the Forest***

1215 **Magna Carta**
Establishes supreme law of the land for civil liberties for individual, English (Anglo-Saxon) freedoms within the boundary of England's Kingdom

1847 Land Claims (New Zealand)
Locations Commission started to settle land and administer reserve system

1847 Upper Canada Statute of 1839
Classified Indian Lands as "Crown Lands" for protection against tresspass and damage

The Durham Report
Recommended unification of Upper and Lower Canada, encouraged immigration from Britain to Canada

1839 British Indian Department
Becomes independent from the Military Department

1832 End of of Black War (Australia)
Virtually wiped out all Aboriginal peoples of Tasmania

1830 Indian Removal Act (United States)
Signed into Law by President Andrew Jackson

1829 Death of Shanawdithit
Last known Beothuk dies of tuberculosis in St. John's, Newfoundland

1828 Start of Black War (Australia)
Resistance in Tasmania to British Colonialists

1821 End of Pemmican War
NWC merges with HBC; HBC gains monopoly over all lands west of Rupert's Land to the Pacific Coast; HBC receives exclusive license to trade with Indigenous peoples in "unsettled" parts of British North America

1812 Trade War
Start of Pemmican War between Hudson's Bay Company (HBC) and North West Company (NWC) following the colonization of the Red River Region

1809 Enslavement
The Hottentot Code by the British Government proclaimed in South Africa enslaving indigenous KhoiKhoi peoples as labour force for Afrikaner farmers

1797 Exclusion
First internal passes (passports) introduced to exclude Natives from Cape Colony

1775 Disposition
Instructions of 1775 establishes hierarchy of British Administration

1764 Forced Surrender
First Land Cession under the Protocols of the Royal Proclamation; signed between British Crown, the Seven Nations of Canada, and the Western Lakes Confederacy (additional 25 Upper Canada Land Surrenders c.1764-1783)

1763 Royal Proclamation
Established a firm western boundary for the colonies, all lands to the West become "Indian Territories"; no settlement or trade was permitted without permission from the British Indian Department

1760 Treaty of Peace & Friendship
Signed between British Crown and LaHave Tribe of the Halifax Region

1755 Militarization of Indigenous Dispossession
British Indian Department established as a Branch of the British Military

1752 Treaties and Articles of Peace & Friendship c.1725
Renewed, reiterated, and forever confirmed

"Their unsettled habitation in those immense regions cannot be counted a true and legal possession; and the people of Europe, too closely pent up at home, finding land of which the savages stood in no particular need, and of which they made no actual and constant use, were lawfully entitled to take possession of it, and settle it with colonies … We do not, therefore, deviate from the view of nature, in confining the Indians with narrower limits."

—Emmerich de Vattel, *The Law of Nations*, 1758

"We can, if need be, ransack the whole globe, penetrate into the bowels of the earth, descend to the bottom of the deep, travel to the farthest regions of this world, to acquire wealth, to increase our knowledge, or even only to please our eye and fancy."

—William Derham, *A Survey of the Terraqueous Globe*, 1713

"On entering into treaty in the Western provinces, those who followed the Indian mode of life on reserves and received annuity and certain other treaty benefits, were known as 'Indians' whether they were of pure Indian blood or mixed. Those who elected to take scrip in lieu of the treaty benefits and to live off the reserves were known as "Half-breeds," although they may have been of pure Indian blood...

...The distinction between the Indian and the Half-breed, from an official standpoint, is not a matter of blood but of the status they elected to assume at the time of the treaty."

—G. M. Matheson, Registrar for the Department of Indian Affairs, "Memorandum: Half-breeds," 1935

1875 **Treaty 5**
Signed between the Crown and the Saulteaux and Swampy Cree First Nations

1874 **Treaty 4**
Signed between the Crown and the Cree and Saulteaux Frist Nations

1873 **Treaty 3**
Signed between the Crown and the Ojibwe First Nations

1872 **Dominion Lands Act**
Encouraged settlement of the Canadian Prairies

1871 **Treaties 1 & 2**
Signed between the Crown and the Chippewa and Swampy Cree First Nations

1870 **Deed of Surrender**
Canada acquires Rupert's Land and NWT from HBC

1869 **Davin Report**
Led to public funding of residential school system

Gradual Enfranchisement Act
Became basis for residential schools; attempt to assimilate and segregate

The Declaration of the People of Rupert's Land and the North West

1868 **Rupert's Land Act**
Enabled the Crown to accept a land surrender from HBC and admit into the Dominion of Canada

1867 **British North America Act**
Gave legislative authority over Indian affairs to federal Parliament

Homestead Act (United States) 1862

1860 **Indian Lands Act**
Centralization of control over Indian affairs

1857 **Gradual Civilization Act**
Introduced concepts of enfranchisement

1850 **Robinson-Superior Treaty**
Signed between the British Crown and Ojibewa Indians of Lake Superior

Robinson-Huron Treaty
Signed between the British Crown and Ojibewa Indians of Lake Huron

An Act for the Better Protection of the Lands and Property of the Indians in Lower Canada
An Act for the Protection of the Indians in Upper Canada from imposition
Two acts intended to protect Indian Lands against tresspass and damage; permitted the Commissioner of Indian Lands to lease lands and collect rent

"The policy of destroying the tribal or communist system is assailed in every possible way, and every effort made to implant a spirit of individual responsibility instead."

—Hayter Reed, Indian Commissioner, *Sessional Papers: Fourth Session of the Sixth Parliament of the Dominion of Canada*, 1890

"Our object is to continue until there is not a single Indian in Canada that has not been absorbed into the body politic and there is no Indian question, and no Indian Department, that is the whole object of this Bill."

—Duncan Campbell Scott,
"An Act to amend the Indian Act,"
1920

1921 Treaty 11
Signed between the Crown and Slave, Dogrib, Loucheux, and Hare First Nations

1913 Native Land Act; Bantu Land Act (South Africa)
Developed reserve system and barred Africans from puchasing land outside of reserves; precursor to apartheid

1911 Mines and Works Act (South Africa)
Established the "Colour Bar" and prevented black mine workers from practicing skilled trades

1911 Tahltan Declaration
Legal declaration of rights, claiming sovereignty over Tahltan land for the Tahltan people

1910 Establishment of the Republic of South Africa

1909 South Africa Act (South Africa)
Created Union of South Africa from separate British colonies modelled on British North America Act (1867)

1909 The Myth of the Wild West
Including the Savages, Barbarous and Civilized Races

1906 Reserve Policy Proposed (South Africa)

General Pass Regulations Act (South Africa)
Denied blacks the vote and limited them to fixed areas

1906 Treaty 10
signed between the Crown and the Chipewyan and Cree First Nations

1905 Treaty 9
signed between the Crown and the Ojibeway and Cree First Nations

1899 Treaty 8
signed between the Crown and the Cree, Beaver, and Chipewyan First Nations

1896 Introduction of New Pass Laws
Required use of metal badges

1894 Glen Grey Act (South Africa)
Foundation for racially segregated South Africa. Limited land that Africans could hold

1889 Treaty 6 Adhesions
Signed between the Crown and the Plain and Wood Cree First Nations

Potlatch Ban Instated

Introduction of Informal Pass System
Encouraged separation of Indians and settlers

1885 North-West Rebellion

1877 Treaty 7
Signed between the Crown and the Blackfeet First Nations

1876 Indian Act
Consolidation of several acts; including the Gradual Civilization Act and Gradual Enfranchisement Act; combined aspects of assimilation, enfranchisement, Indian status, self government

Treaty 6
Signed between the Crown and the Plain and Wood Cree First Nations

"In the never-ending government quest to integrate Indian people into Canadian society an acrimonious public debate concerning the question of individual rights versus collective Native rights—derived from modem land claim settlements and self-government negotiations—has surfaced and threatens to poison Aboriginal relations with the dominant society. Whether contemporary Canadian society—still imbued with liberal democratic values and principles—can accommodate collectivist Native aspirations is a moot point. Until this fundamental political and social policy issue is resolved, Indian policy will remain a highly contentious and problematic field of Canadian public policy."

—John F. Leslie, "Assimilation, Interrogation or Termination? The Development of Canadian Indian Policy, 1943–1963," 1999

	1985	Indian Act Ammendments
	1982	Constitution Act A series of amendments to the Constitution of Canada; including the Charter of Rights & Freedoms and the Aboriginal Rights Clause
	1973	Together Today for our Children Tomorrow Declaration by the Yukon Indian Brotherhood
	1974	Dene Declaration Assertion of Territorial Rights & Sovereignty by the Dene Nation
	1973	Calder Case Landmark Court Case with Justice Berger regarding Nisga'a Treaty
Black Homeland Citizenship Act (South Africa) blacks became citizens of one of ten autonomous territories	1970	Citizens Plus "The Red Paper" Document presented by the Indian Chiefs of Alberta to Prime Minister Pierre Trudeau and Cabinet in response to the 1969 White Paper
		Hudson's Bay Company becomes Canadian Queen Elizabeth II grants new charter for Hudson's Bay Company; headquarters move from England to Winnipeg
	1969	White Paper on Aboriginal Affairs Policy paper proposing the end of the Indian Act, eliminating Indian Status and Reserve Land, abolishing the Department of Indian Affairs and terminating existing treaties
Sharpeville Massacre (South Africa) South African Police open fire on protesters demonstrating against pass laws	1960	Right to Vote Granted to First Nations
Group Areas Act (South Africa) End to diverse areas and determined where one lived according to race	1951	Revision to the Indian Act Removed ban on cultural practices but continued paternalistic elements
Bantu Authorities Act (South Africa) Established separate government structures for blacks and whites		
Population Registration Act (South Africa) Classified all South Africans by race; became one of the "pillars" of Apartheid	1950	
Kanada (Poland) Symbol of Wealth, the Warehouse of Death built at Auschwitz concentration camp	1943	
Lebensraum (Germany) The Mythological License of the Nazis to Divide, Conquer and Kill	1938	
Native Land and Trust Act (South Africa) Completed 1913 Native Land Act	1936	
Representation of Natives Act (South Africa) Removed black voters from voters' roll	1929	Treaty 9 Adhesions Signed between the Crown and the Ojibeway and Cree First Nations
Native Urban Areas Act of 1923 (South Africa) Deemed urban areas as "white" requiring passes	1923	

"We should not aim at merely segregating the Indians and dragooning them into what we consider is respectable living, and preserving them for ever on reserves. We have to let them gather experience and self-control and gradually fuse with our own people so far as they come to desire such fusion and above all we must not seem to them to be forcing them to observe regulations we do not observe ourselves."

—Charles C. Perry, BC Indian Commissioner, "Proposed Amendments to the Indian Act," 1939

"In situations in which sovereignties are nested and embedded, one proliferates at the other's expense; the United States and Canada can only come into political being because of Indigenous dispossession. Under these conditions there cannot be two perfectly equal, robust sovereignties. Built into "sovereignty" is a jurisdictional dominion over territory, a notion of singular law, and singular authority (the king, the state, the band council, tribal council, and even the notion of the People). But this ongoing and structural project to acquire and maintain land, and to eliminate those on it, did not work completely. There are still Indians, some still know this, and some will defend what they have left. They will persist, robustly."

—Audra Simpson, *Mohawk Interruptus*, 2014

"With the creation of these two colonies, land was framed in a new problematic. Colonies entailed settlers, and settlers required land, which could be got only by dispossessing native people. A relationship based on trade was replaced by one based on land. As their land was taken away, native people had to be put somewhere. A solution with many precedents in other settler colonies was to put them on reserves. Dispossession began in the 1850s and continued through the rest of the century. Physical violence, the imperial state, colonial culture, and self-interest all underlay it."

—Cole Harris, *Making Native Space*, 2002

2016 Stolen Bodies
National Inquiry into the alarming rate of violence towards women and girls (including heterosexual, Two-Spirit, lesbian, gay, bisexual, transgendered, queer, and those with disabilities or special needs) that have been murdered or gone missing—robbed of their sacredness

2012 Idle No More
Ongoing movement established to raise awareness for indigenous issues through rallies, social media, and other methods

2008 Reconciliation 3.0
94 recommendations and calls to action officially established by the *Truth and Reconciliation Commission* to acknowledge Indigenous injustices and inhumanities from the transgenerational legacy of residential schools, colonial experiments, and programs of assimilation

1999 Nunavut Act
Created the Territory of Nunavut covering 1.9 million square kilometers

1995 Reconciliaton 2.0 (Australia)
Following the era of "stolen generations" in Australia, the *National Inquiry into the Separation of Aboriginal and Torres Strait Islander Children from Their Families* is formed to reclaim the era of forced removal of Aboriginal children from their homelands

1994 Reconciliaton 1.0 (South Africa)
Following the end of apartheid in South Africa, the first *Truth & Reconciliation Commission* is formed for reparations associated with gross human rights violations

1994 Official End of Apartheid
(South Africa)

1990 Resistance
78-Day Standoff between Mohawk Protestors, Royal Canadian Mounted Police, and Federal Army at Oka, Québec, over land dispute regarding a golf course expansion on Mohawk territory

White Space

A Conversation with Jean Chrétien

Jean Chrétien is a lawyer and was the 20th Prime Minister of Canada. He negotiated the patriation of the Canadian Constitution as well as the Canadian Charter of Rights and Freedoms. As Minister of Indian Affairs and Northern Development, authored the "White Paper," a policy proposal for eliminating the Indian Act in 1969.

X: Yes, hello, I'm trying to reach Mr. Jean Chrétien please?

JC: This is he.

X: Yes, Mr. Chrétien, this is Pierre Bélanger calling from Boston, Harvard University.

JC: Just a second, I'll turn off the television.

Yes, I saw your email.

X: I really appreciate the time you took to consider this call. Do—

JC: Well! If you would like to ask me questions about my career, go ahead!

X: Thanks a lot, I simply have three questions about your career, which obviously start with your collaboration with Pierre Trudeau. And, these questions are about discussing your view on the, almost now, 50 years of policy on economy and territory in Canada. So—

JC: Even more, I was elected in 1963. Which means that it's been 54 years…

X: 54 years.

JC: …Since I became a Member of Parliament. And it's 50 years since I became a Minister. And it's been 49 years since I became Minister of Indian Affairs and Northern Development. We can start there.

X: Yes, that's a good point of departure because it's interesting that with the 1969 White Paper that followed the Hawthorn Report in 1966… Now, certain policies and strategies have been implemented, a little indirectly, over these last 50 years—

JC: It's constantly evolving!

X: Right.

JC: It's nothing etched in stone. It's always evolving. But when I became Minister there were problems that hadn't been dealt with, there were… I'm the one that installed Band Councils that have the power to administrate within their own local governing procedure. Before that it was Superintendents appointed by the Ministry of Indian Affairs.

Next, it was the same thing in the Northwest Territories. There is a democratic point in governance. I was the first to allow the election of all the members of the Northwest—of the Assembly of the Northwest Territories. Even if 90% of the budget comes from the Federal Government. And obviously it is going to evolve a lot. When I started there were only a few Indians who graduated from university, but now it's thousands every year. It's a problem that is constantly evolving and is always very controversial.

X: Now, can you simply reflect on what influence you had on the policies that you had developed with the—

JC: Well, I was influenced by what was certainly evolving at that time. There were documents that were approved by the Cabinet. They are documents that were approved by my Prime Minister and signed by me.

X: Right.

JC: But in the preliminary stages, I had many meetings with my Prime Minister, who was Pierre Trudeau. But he was a friend as well. We had been elected—well, he was elected after me. He was elected as a member of Parliament in '65. I was elected for the first time, at the age of 29, in '63. So we developed policies together but I was the Minister responsible, in charge.

X: And so, obviously, the Hawthorn Report (a few years before the White Paper) was important—you still consider it pretty significant as a lesson?

JC: Yes, the Hawthorn Report, I remember vaguely enough because when you mention my White Paper, called *The Red Paper—The Indian Policy*. It was based on discussions I had within my Ministry. And it was based on discussions I had at Social Services Committees and on conversations I had with Mr. Trudeau. What I proposed was very radical and it was refused by the Indians. Because in the consultations I had with them, the Indians told me, "Ah! You guys are doing apartheid to us. We have Indian reserves. We have a Ministry of Indian Affairs. We have the Indian Act. We want equality."

So I proposed equality. Which meant that the Minister of Indian Affairs— it's very rare that a Minister proposes the abolition of his own Ministry. The abolition of the Department of Indian Affairs, the abolition of the Indian Act, and so… I said, "Govern yourselves like the rest of the citizens do, by local government, municipal government, and more!"

But they didn't want to do that because they were not, in reality, ready to go that far with it. So I was forced to backpedal. But I backed off and I said what they asked me to do. And well then, when they say the consequences of our proposals that they gave me in consultations they said, "Whoops, this is not going to get us what we want!" Because the problem that we had is that we thought for a moment that there was apartheid. That we had second-class citizens that called themselves "Indians."

X: And so, in that vein, if we go back now, almost 50 years after, with requests to terminate the Indian Act now, today, even if it is on terms—

JC: Yes, they are returning to what I proposed. I think that, obviously, they already have part of it. They have their administration, and their difficulties. It's that they aren't, like other citizens, under the power of provincial governments when they are on reserve. When they live in the cities they are treated like the other citizens necessarily.

[A pause while Chrétien has a conversation with someone else.]

X: Yes, hello, so on the other hand, do you think that the White Paper had been created within a reasonable time allotment proposed to be five years, while it's almost 50 years later now. Do you think that it would take—

JC: Ah, in reality, they didn't accept it, and I withdrew my proposal while saying to them "So fine, if you want to continue in the same vein as in the old days, we will continue." So, I withdrew my—effectively my propositions. But the evolution of the issue is bringing us in that direction, except a lot later.

X: Right.

JC: But I respected the wishes of the time. They were a little bit, as happens often with such claims, a little bit in contradiction with themselves. When they said that there was apartheid we said, "Fine. We'll end it." But when we abandoned our policy together, they still have the Indian Act. And they still have reserves. And well, that makes citizens of a different category.

X: Right.

JC: That's what we call apartheid.

X: And so now, if we discuss very briefly, your experience when you went from Minister of Energy, Mines, and Resources, and Minister of Justice, and then of Finance, and then after that you became Prime Minister in 1993. In what way, or can you otherwise discuss your experience a little more because it was there that you had a global, international point of view—

JC: Well, Canada, especially in the domain of resources, we export a lot. We export our gas, our oil to the United States. We export our wood and all our minerals abroad, so we were necessarily interested. In Ottawa, we depend on relations with countries that buy our resources. And it was normal, so I was very implicated in it as Minister of Energy, and also later as Minister of Finance.

X: Yes.

JC: And that's perfectly natural... Obviously, because I was in government so long, that I was in the cabinet for 30 years, or just about. I was from 1967 until 1984, except for Joe Clark's two months. And then, I came back as Prime Minister in 1993, until 2004.

X: Two thousand... and four, is that right?

JC: True. To the final days of 2003, right?

X: Yes. Now, I'm just going to continue with two small questions: while you were Prime Minister it was obviously during the time while we saw the apartheid regime in South Africa fall in 1994. We saw Indigenous policy in Australia transformed pretty radically with land claims and truth commissions. Does that time period—while you were Prime Minister—how do you perceive the evolution of your thinking on the subject of—

JC: Well, when I was Prime Minister, our handling of Indian Affairs during the 10 years that I was Prime Minister had been a pretty calm period, if I can say that.

Relations were all complicated, but if you look at the amount of news about Indians during my 10 years in office as Prime Minister, this position had been a lot smaller than afterwards, or previously. Maybe the fact that I had been the Minister of Indian Affairs for six years allowed my government to be more flexible, or more comprehensive rather, instead than others. My experience with them, and probably also the fact that they know me well, had necessarily established a certain level of trust that never existed before, or since between the Prime Minister and Aboriginals.

Also, I think at this moment that Mr. Justin Trudeau is working very hard to have very good relations with them and that is good. But it is still, always, an extremely complicated subject, because it is full of contradictions, like I explained earlier.

X: Right. Obviously the questions of land and territory are intertwined and complex given that—

JC: Well, for those that have Treaties, it is easy. But, there are a lot of Indians that have no treaty since... so it's, we recognized the rights of what we call the Métis, who were, during that era, not under federal jurisdiction because they weren't Indians. They were mixed, and were only registered by their paternity, so it was changed to their mixed bloodline, rather than to paternity or maternity.

X: Do you think that, between the Oka Crisis in 1990 and the inception of the Territory of Nunavut, this period was still very—

JC: Ah, but that isn't under the Treaties. The Northwest Territories are my creation, and then, what you call the Territory of Nunavut, was created under my administration.

X: Right.

JC: So, it was separating the Territory of Nunavut, the Northwest Territories.

X: And so, from—

JC: And it was passed when I was Prime Minister.

X: And, so in that way, it is during this time—I read that while you were in the Yukon, you consulted with Indigenous peoples in the region and that you stated very clearly that your intentions as Minister of Indian Affairs were not based on a point of view of assimilation, but rather from a point of view of—

JC: Of integration. That is to say, integration into our Multiculturalism Policy for respect. We didn't want that everyone be considered, exactly the same. There is a way to be equal, and to be different, and to respect these differences. You are francophone, I am francophone, so we have a language, and necessarily a different culture than our brothers in Ontario, or in British Columbia.

X: Right.

JC: And, so that makes up the cultural diversities of a country that can be based on language, that can be based on religion, that can be based on color, or even can be based on a region. An anglophone fisherman from Newfoundland is closer to a francophone fisherman from Gaspé than he is to an agricultural worker in the prairies. Because they have fishing culture and the culture of living on the coast, which is different than those that live on the interior.

X: And, given that you grew up, in reality. in the region of, if I understand correctly, the region of Shawinigan, was it those experiences—regional ones, that also helped to form your views—

JC: No, because there were no Aboriginals in the region where I'm from. There were no Indians.

X: Ok.

JC: It's one of the things that Trudeau asked me. I said, "Why do you want to appoint me as Minister of Indian Affairs? I've never met one in my life." But they said, "At least, with you, you don't have any prejudices…"

X: I see. And so—

JC: …But, if you come from a rural part of Quebec and you're from a minority language, maybe you can better understand them than others could. That is how I became Minister of Indian Affairs.

[Pause.]

I should leave you because people are waiting for me.

X: Okay. I'm sorry to keep you Mr. Chrétien. I really appreciate—

JC: Ok… One last question.

X: Yes, thank you. One last question then: now, regarding what is going on right now. Obviously you still believe that it is important, with your point of view of Indigenous peoples in Canada, that integration involves cities because, as you observe, there are infrastructure problems in the Reserves.

JC: A lot. We are far behind in that area. But often economic development in isolated areas, like Reserves, is not easy. And, maybe not sustainable either, because if the population grows, then the resources where there is nothing that can be established as economy. Otherwise, they will just be dependent on government welfare. It's not easy for anyone. There needs to be a way for them to freely and honorably earn their living without being completely dependent on government grants. That doesn't help progress.

X: Should the Federal Government plan the direction of the North for the future?

JC: Ah, well, there are always many. Listen, the other coast is very important. Obviously I retired 13 years ago so—almost 14 years, so I don't tell them what to do. It's up to them to decide. They are responsible and I don't want to be the cranky mother-in-law.

X: Obviously. I imagine that you are still very busy with your work in the sector, I imagine, of resources and international relations.

JC: Ah, well, yes, I'm busy enough, but a bit less than before. Yes, I go to the office every day. I retired, but retirement wants little of me.

X: And regarding your family, two sons and a daughter, if I can ask you one last question, how did they influence you as politician? Your daughter, France, is married to André Desmarais, and your son, an adopted baby boy from Inuvik in the Northwest Territories during the late 1960s—an Inuit, your son, Michel.

JC: Oh, yes.

X: It's very important also as an example of the nature of your family. Can you simply discuss very briefly in what way your adopted son influenced *you* as Minister of Indian Affairs, later as Prime Minister? There is a certain transformation, I imagine.

JC: It just happened after I had already been Minister for many years so… He was a baby, he wasn't able to influence me very much. We took him, he was one-year old. But, we had taken him because, my wife and I, we come from big families and there were a lot of young Aboriginals who were abandoned. So we decided simply to adopt one. And he is 48 years old now, so… That is life, he's my son. That's what he is.

X: I think that there is a certain influence that—

JC: Ah, but this is entirely a family situation, and personal decision.

X: Yes, understood. So, Mr. Chrétien I don't have any—

JC: You, where are you from Mr. Bélanger?

X: I'm from Montreal and I was born in Pointe Claire, but my parents come from—

JC: There's a bit of anglophone in you, isn't there?

X: Yes, I have a mother from Hamilton with family of the Bruce Peninsula, in Ontario. You can tell. And a father from Trois-Rivières, with extended family in the Eastern Townships. I come from a bastardized background, I never learned perfect French, nor English.

JC: No, no. You speak French well! You have a small accent but it isn't bad at all. *Au contraire*. It's just that I detected it because I talk with a lot of people when I have the opportunity.

So listen, good luck with your project, and that's all I can say. I'm sorry but, they told me this would be a few minutes and I have to leave, my family is waiting for me.

This conversation, originally in French, took place between Jean Chrétien and Pierre Bélanger in 2017.

this page
intentionally
left blank

Indian Act Indian: You no longer exist.

One hundred and forty years and we still feel our ancestors in our blood calling us to be proud

Calling us to be strong

Calling us to break down the walls of policy that have caged us in reservations

Caged us in poverty

Caged us in sickness

Caged us in colonization

We are not broken

We will never be broken.

We will decolonize and organize

And we will become stronger and louder

And we will help each and every one of our sisters and brothers over the walls of oppression

We will stand up and return to our lands and territories

We will sing our songs and speak our languages

And we will return to the sacredness of our peoples.

One hundred and forty years

And we are breaking the systems

That tried to break us.

140 Years of the Canadian Indian Act
A Declaration
Eriel Tchekwie Deranger
April 13, 2016

140 years of being told

How to act

How to use our lands

How to govern

How to learn

How to raise our children

How to build our communities

How to live

How to assimilate.

My assimilation is broken because I still have an eagle feather and a smudge bowl

My assimilation is broken because I still know where I come from

My assimilation is broken because I know who my leaders are:

The land, the water, the air, my ancestors on this earth

My assimilation is broken because my knowledge wasn't granted to me from an institution that left me out of its books

My assimilation is broken because my children have always been proud of who they are

My assimilation is broken because I can still hear the sounds that call me home

I can still see the northern lights guiding me to where I'm supposed to be

My assimilation is broken because I am here standing and holding on to the Indigenous Peoples we are.

transcribed with permission from Eriel Tchekwie Deranger

Georges Erasmus, Tim Hortons, Yellowknife, NWT (Pierre Bélanger, July 3, 2017)

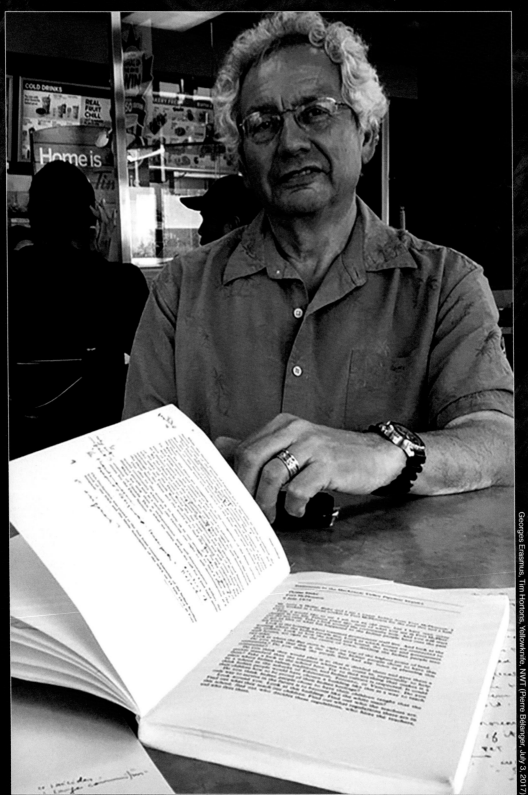

We the Dene

Georges Erasmus

1976

We the Dene of the Northwest Territories insist on the right to be regarded by ourselves and the world as a nation. Our struggle is for the recognition of the Dene Nation by the Government and peoples of Canada and the peoples and governments of the world. As once Europe was the exclusive homeland of the European peoples, Africa the exclusive homeland of the African peoples, the New World, North and South America, was the exclusive homeland of Aboriginal peoples of the New World, the Amerindian and the Inuit. The New World like other parts of the world has suffered the experience of colonialism and imperialism. Other peoples have occupied the land - often with force - and foreign governments have imposed themselves on our people. Ancient civilizations and ways of life have been destroyed. Colonialism and imperialism are now dead or dying. Recent years have witnessed the birth of new nations or rebirth of old nations out of the aches of colonialism. As Europe is the place where you will find European countries with European governments for European peoples, now also you will find in Africa and Asia the existence of African and Asian countries with African and Asian governments for the African and Asian peoples. The African and Asian peoples - the peoples of the Third World - have fought for and won the right to self-determination, the right to recognition as distinct peoples and the recognition of themselves as nations. But in the New World the Native peoples have not fared so well. Even in countries in South America where the Native peoples are the vast majority of the population there in not one country which has an Amerindian government for the Amerindian peoples. Nowhere in the New World have the Native peoples won the right to self-determination and the right to recognition by the world as a distinct people and as Nations. While the Native people of Canada are a minority in their homeland, the Native people of the Northwest Territories, the Dene and the Inuit, are a majority of the population of the Northwest Territories. The Dene find themselves as part of a country. That country is Canada. But the Government of Canada is not the Government of the Dene. The Government of the Northwest Territories is not the Government of the Dene. These governments were not the choice of the Dene, they were imposed upon the Dene. What we the Dene are struggling for is the recognition of the Dene nation by the governments and peoples of the world. And while there are realities we are forced to submit to, such as the existence of a country called Canada, we insist on the right to self-determination as a distinct people and the recognition of the Dene Nation. We the Dene are part of the Fourth World. And as the peoples and Nations of the world have come to recognize the existence and rights of those peoples, who make up the Third World the day must come when the nations of the Fourth World will come to be recognized and respected.

The challenge to the Dene and the world is to find the way for the recognition of the Dene Nation. Our plea to the world is to help us in our struggle to find a place in the world community where we can exercise our right to self-determination as a distinct people and as a nation. What we seek then is independence and self-determination within the country of Canada. This is what we mean when we call for a just land settlement for the Dene Nation.

⊕ ⊕ ⊕

EXECUTIVE OFFICE

Indian Brotherhood of the Northwest Territories

P.O. Box 2338,
Yellowknife, N.W.T.

August 11, 1975

Memorandum to: Dear People of the Dene Nation
From: Georges Erasmus
Re: The Dene Declaration and the Land Settlement

Please find enclosed material for your information as follow up to the recent Joint General Assembly in Fort Simpson. The information is centered around the Dene Declaration with particular reference to our struggle for a just Dene Land claims settlement.

We know the Dene are not white people. We know that the Dene way is different from the white way. We, the Dene, think differently from the way white people do. Although there are many differences between the Dene themselves (for example, the Loucheux and Dogrib speak different languages), we know that these differences are small compared to the difference between the Dene and white people. The Loucheux, Slavey, Hare, Dogrib and Chipewyan have always lived together without interfering with each other's way of life. Trouble for the Dene began with the coming of a very different people, a different "nation"; the white people.

Although the white people recognized the Dene as a nation when they signed the peace treaties with them in 1899 and 1921, they soon forgot. The white people who entered the

North not only did not accept the ways of the Dene, but they took away the rights of the Dene by forcing them to accept the white ways — white laws, language, education, and government. Finally, the white people began to destroy the very life blood of the Dene — our land.

This is why a land settlement is necessary for the Dene. When we Dene talk of "land settlement" we are talking of our right to survive as a nation, living according to our ways on our own land. It is because the government does not understand this, or will not accept it, that they see the land settlement differently. The white people want our land so they can continue to develop in their selfish way. They see land differently. They see land the same way they see ski-doos — something they can buy and sell, something they can "own."

The biggest difference between the land claims issue. Too many people have misunderstood what the land settlement was all about. They think that the difference between the Dene and the government positions is that the Dene are asking for more land and money than the government will give us. This is not the real problem. The truth is that we want to survive as people, the Dene nation, with the right to govern ourselves (make our own laws), to educate ourselves, to develop ourselves on our own lands. This is what the government will not accept. They will not accept the Dene way and its right to survive.

It was to make this point clear to everyone that the delegates at the Fort Simpson Joint General Assembly agreed to the Dene Declaration of Rights. In that declaration they made it clear that the land settlement is truly "our struggle for the recognition of the Dene Nation." The kind of land settlement we are talking about is one which will guarantee the Dene "independence and self-determination within the country of Canada."

The Dene Declaration is therefore a very important statement about the land settlement. It will help other Canadians understand us better. It makes it clear that we are asking only what Canada supports elsewhere in the world. Canada claims to support the rights of nations in Africa, for example, to self-determination. What the Dene Declaration does is to challenge the government of Canada to practice what it preaches at home. The Dene are a nation and as more people realize this they will come to support our land settlement position.

But the Dene Declaration of Rights agreed to in Fort Simpson is just the beginning. As a nation, the Dene must begin to work out together what this means for their future after the land settlement. How will we govern ourselves? How will we keep our nation strong, as a whole, and still recognize the differences between regions, tribes and communities? What kind of education would we give our children if we got back control? What kind of development do we want on our lands? These are questions the Dene Nation must answer in time.

The Dene Declaration will grow from this first statement agreed to at Fort Simpson to include many statements about what we Dene believe in. These statements will help explain to others what the Dene nation is. We will change and add to these statements as we gain more experience. They will become a living history of the development of our nation. For one thing is clear, the Dene nation is not a return to the past, but the building of our own society, firmly rooted in our past, but developing under our control to meet the needs of all our people, both young and old.

Included in this envelope for your consideration is the short Dene Declaration that was approved in principle at the joint general assembly in Fort Simpson – but, also a long version with some specific possible Declaration on terms like Development – ideal and goals of the Dene Nation etc.

Please find the time to go through this material and discuss it with others in your community. Our nation must be preserved and constructively changed or moulded from the communities. Only with direct movement from individuals like yourself will we grow in time with the wishes of local people.

If enough community peoples read and discuss this paper by this October with community meetings etc, perhaps we may bring the long version of the Dene Declaration at the next Indian Brotherhood general assembly - planned for this fall.

If you and others have questions or which more information on the Dene Declaration and how it fits in with our land claims work - or any other question around the concept of the Dene Nation please feel free to contact myself or others at headquarters in Yellowknife - we will be more than pleased to do what we can to explain things to the best of our knowledge. Please call collect or write a letter, what ever.

Take care.

Yours in unity through a strong Dene Nation,

Georges

Georges Erasmus,
Community Development.

Statement On Strategy and Organizing
For Achievement of Our Goals
1976

1. Our goal is maximum independence and self-determination of the Dene Nation within the Country of Canada through a just and equitable land settlement.

2. The struggle for achieving our goals involves organizing and strategy on two fronts: The external front and the internal front.

3. Our struggle is like a war, but a peaceful one. On each front there is an enemy. On each front there are allies.

4. On the external front the enemy is those not a part of the Dene Nation who resist and deny the achievement of our goals such as the government and other people who do not want to see the recognition and self-determination of the Dene Nation.

5. On the internal front the biggest enemy is ourselves, our disunity and lack of organization.

6. There are also Dene who are the enemy. There are Dene who would betray and are betraying their brothers in the struggle for their goal. These are Dene who work for the enemy against their brothers. These are traitors to the cause of the Dene Nation. We must learn to identify such persons.

7. There are also Dene who hold back the cause by forgetting who the real enemy is. These are Dene who prefer to fight amongst themselves rather than against the real enemy.

8. There are Dene who have not yet learned who the real enemy is. It is the duty and responsibility of Dene who have learned to recognize the real enemy to educate their brothers.

9. The struggle involves then the simultaneous battle on two fronts, internally and externally. While we organize and plan to build a strong organization and unity of all our people we must also organize and plan to defeat the enemy without.

10. Organizing and planning on the internal front means building unity and strength. It means breaking down that which divides us and educating our people and organizing so as to defeat the enemy without.

11. In fighting the enemy without we must at all costs keep a united front whatever our differences. We must always keep our differences to ourselves and solve our differences amongst ourselves. We must never fight amongst ourselves before the enemy.

12. We must accept that there now are real differences amongst us and always will be differences. But if we remain committed to our goal our differences will not defeat us. If we constantly remember that defeating the enemy is more important than our differences, we can solve the problems created by our differences.

13. Our differences are real. The most serious differences are between the Treaties versus the Non-Treaty and Métis and between the young and the old.

14. As long as we remember that there are differences between Indian and Métis, but that it is more important to remain united against the enemy than to fight amongst ourselves, we will be in a position to solve our differences ourselves.

15. The old people are our strength and wisdom. They are our roots to our history, tradition and cultures.

16. The young people bring energy and knowledge of the enemy to the struggle. They are the link to the future.

17. But the experience of the young people is much different to that of the old people. Often the old people do not understand and respect the young people.

Often the young people do not understand the old people. What we must always strive for is an understanding and respect of the young for the old and the old for the young. Without that understanding and respect, we will fail, for the past will become separated from the future.

18. In fighting the enemy on the external front we must always remember that the way of the European is different from that of the Dene.

19. We must always remember that the situation is constantly changing. Each one of us must bear the burden of keeping ourselves informed on each change so that we can easily adapt and change our strategies so as to defeat the enemy.

⊕ ⊕ ⊕

Together Today for Our Children Tomorrow

a text collage by Lianne Marie Leda Charlie

YUKON INDIAN PEOPLE

always live off the land
keep the waters clear, the air pure and the land clean.

the Indian people have always been there
if we are to survive
we feel
many deep feelings about our land and about the future of our children

big responsibility

Our way of life was handed down by word of mouth.
The Indian Way

they changed our way of life
sixty thousand Whitemen

the Gold Rush changed the way of life of the Indian people,
, now we see the new mines doing the same thing

We have been brought up to "feel",
you took it from us by putting our children in your schools
separating us from the open land

But we still have not completely accepted your religion
we lived by the sun, moon, and the seasons
Money alone is not enough for every person Indian OR White.

the Whiteman's World

give us back control over our own lives
Being an Indian is something only an Indian can decide
we are the product of our culture –

We feel that you are going to build the Pipeline anyway
. He does not show the same respect as we do.
he pays no attention to the future of the land
he can make a quick dollar from selling it to foreigners

Non-Status

Status

enfranchisement.

One of the most unfair tricks ever used to wipe out a race of people

signing a piece of paper.

northern allowance public school hold title to a piece of land

All of these bribes

discriminating laws

✓ back on the list

over half our families are on welfare,

eighty percent.

"White" jobs

Solutions to Indian problems

must be found within the framework of our culture

CONTROL

RESPONSIBILITY

one-way communication.

Whitemen talk -

~~Indian listen~~

They are all salesmen,

brainwashed

TOMORROW

This is for tomorrow

for our children

our children's children

learn to live in a changing world.

We will not sell our heritage
quick buck / White solutions

what it means to be Indian
living your culture.
like **Our Old People**
We are needed in **Our** village

go back to the bush
where We would be free,
a way of life that We know and understand.

Indian Values
put into practice

This was ~~never~~ possible

Our only defence against "Assimilation" is a strong unified Indian Identity
Our old people
our children
can once again be proud

the police, the welfare worker, the Indian agent,
We have lived without these people breathing down our necks before

one hundred percent drop-out rate,

The whole Yukon is our school
taught on the land
not in Lower Post, B.C.

✓ oil leases
National Parks
✓ Hydro Development
✓ mining
✓ pipeline
✓ White populations
✓ pollution
will stop the way of life of our people

We plan to teach the Whiteman something

respect

who we are

why we are different

Implementation :

- ✓ expert help
- ✓ training courses
- ✓ money

our problems are getting worse, not better

The first five years of implementation will tell

we are strong supporters of development

~~land or~~ money

~~land or~~ money

~~land or~~ money

~~land or~~ money

~~land or~~ money

It will be of only temporary help

who will be eligible to participate?

- ☒ ancestry
- ☒ age
- ☒ blood
- ☒ enrolled
- ✓ proof

Money must be . . . .

harmonious with our traditional way of life

Land must be . . .

held in perpetuity

; or , for

(one year from date of Settlement Agreement)

we will make some of the same mistakes . . .

It is not possible to say at this time . . .

There must be consideration given to . . . .

this list will give some idea . . .

who will administer the land on behalf of the people?
her Majesty the Queen?
this Ordinance?
zoning?
bylaws?
The local people?
the Central Indian Fund?
Senior Governments?
Indian Communities?
the Department of Indian Affairs?
Band Managers and Secretaries?
Urban Corporations?
those who migrate?
Land Management?

Without land Indian People have no Soul — no Life — no Identity — no Purpose.

As each year passes
, more and more land is taken away
given to the Whiteman
(in perpetuity)

Surface rights
sub-surface rights
Mineral rights
Water Rights
Timber rights
Forest Product Rights

(When we use the term
" authority
we mean
"token")

one hundred and
thirty-two million
five hundred and
twenty-eight
thousand
six hundred and
forty acres.

Indian lands
traditionally occupied

we do not feel we should be forced to move

☑ Indian villages
☑ Camps,
☑ Burial Grounds
☑ cabins
☑ Hunting
☑ Fishing
☑ Water
☑ Caribou

☑ Life

we still think of the Yukon as OUR LAND

what would be fair and just
we used to be independent and free
born and raised on land that we always thought of as our land
These are not folk-tales

We will not accept promises
— because we have very little faith any more in Whiteman promises

; respectfully submitted by:

Signature: ______________________________

TOGETHER TODAY FOR OUR CHILDREN TOMORROW

This text collage was crafted from the collection of words and phrases directly from *Together Today for our Children Tomorrow,* an influential document drafted by the Yukon Native Brotherhood on behalf of the "Yukon Indian People" and presented to Prime Minister Pierre Trudeau in 1973. The words and phrases appear in the same order as they originally appeared in the text from pages five to forty. I use spacing, line breaks, symbols, strikethrough, and punctuation (or lack thereof) for added effect and emphasis. Collage, as an artistic practice and as theory, invites us to work with the contradictions and tensions in our political world in new and creative ways. It is deployed here as one way of augmenting our current view of our political context in the Yukon. Collage helps bring seemingly unrelated pieces into new proximities to one another in a way that better reflect our needs, desires, and responsibilities as Indigenous peoples.

Superheroes

A Conversation with A Tribe Called Red
by Yassin 'Narcy' Alsalman

A Tribe Called Red is an Indigenous electronic hip-hop music group from Ottawa, formed by three DJs: Ian 'DJ NDN' Campeau, Tim '2oolman' Hill, and Ehren 'Bear Witness' Thomas. Their albums include *We Are the Halluci Nation* (2016), *Nation II Nation* (2013), and the homonym *A Tribe Called Red* (2012).

YA: It's good to have you guys in town. You guys have a long career as individual artists, but as a crew it's been a crazy year. It's been non-stop, you guys have been touring the world, and your music has taken the industry by storm. What you guys do is bring justice to a lot of things that need to be heard.

So I want to track back to the beginning of history, if you will, the beginning of the history of the Americas, before getting into the history of the group. There's an author called Daniel Quinn who wrote a book called *Beyond Civilization*. And in that book, he says: "Tribal life wasn't something humans sat down and figured out. It was a gift of natural selection, a proven success—not perfection, but hard to improve on. A tribe is nothing but a coalition of people, working together as equals."

Daniel Quinn, *Beyond Civilization: Humanity's Next Great Adventure*, Broadway Books, 2000

When I see you guys doing what you do, it is very much a collaboration, even on stage. It's constantly three brains working together. So what I'd like to know is, why *A Tribe Called Red*, and what's the formation of the group? How did it start? What was the initial *ping* moment?

Bear: [Laughter] I guess it's seven years ago, now going on eight years. We first got together in Ottawa, just with the idea to showcase ourselves, as Indigenous DJs. So Ian, myself, and two other DJs, got together just to start a club night. And that's kind of the beginning of it all.

YA: Electric Pow Wow?

Bear: Electric Pow Wow, yeah.

YA: And was *A Tribe Called Red* something that existed at the time, or did that name come later?

Bear: It actually came quite a bit later. Initially we were called *Red-Handed*. Yeah, that was our name for the first couple of years, and as people came and left, we got *A Tribe Called Red* as the second incarnation.

YA: Amazing. Beyond the music, there's a sense of community behind you guys. When I see the show, I definitely feel that it's more than three people on stage. The narrative you're presenting is way bigger than the music itself.

So I'd like to start with you, Ian—I knew your online personality before I knew your real-life personality. So I knew DJ NDN. You're very proactive with people, and you call out racists straight up, and there's no holding back about what you talk about online. You take the time to address every individual. I've watched it progress from the Change the Name campaign, to the Caucasian t-shirt moment, you know, T-Shirt-gate. [Laughter] Why do you think it's still taboo in public to root for the underdog? The under-represented?

Ian: I don't think it's taboo at all, within media right now. It seems like social media has brought this plateau, this level playing-field to confront racist ideas. It's not so much that society's rooting for the underdog as that it's seeing this confrontation and empathizing a little more.

YA: It used to be fifteen minutes of fame, to whatever it is—a cause, a person—but now it seems like it's "under hundred-and-forty characters" of fame. Things travel much faster, and lose meaning much faster. So how do you counter a larger narrative, by individually addressing people online? Are you really changing people?

Ian: Yes. And there's a lot of different proof of that. Without these hundred-and-forty characters, we wouldn't have the Arab Spring. Without these hundred-and-forty characters, we wouldn't have *Idle No More*. Twitter sparks revolutions. It brings people together in a way, globally and quickly, in a way that we haven't had before. So we're able to organize, we're able to have these protests, and have these demonstrations—all peaceful—and organize, really quickly.

YA: What do you think is more dangerous: that people can organize online more quickly, or that *they're* watching us organize more quickly?

[Laughter]

Ian: That we can organize quickly… because then they're just watching us organize quicker! It's terrifying that we're being watched, but it hasn't stopped these revolutions from happening in the first place, globally.

YA: You know, before we started recording, we were having a conversation about terminology. Being in hip-hop culture, you can't use words like the N-word if you're not black, you can't say the F-word towards people in the LGBT community. It becomes an issue when somebody that's not in the community says these words.

In the media, when I hear stories of Natives in Canada or the United States, they either use "Indian," "Native," "First Nations," or "Indigenous." All of them have a different connotation coming from a white person saying it, in my opinion. But what do you identify as, within those categories? If not one of those, what do you identify as, as an individual from a culture?

Ian: Personally, as Anishinaabe, which is my nation, specifically.

Bear: I talk about being Indigenous, because it really has more of a worldview regarding how we're all connected now. Indigenous people are connected all over the world, because of our shared experience with colonialism. So, really what I'm interested in addressing is that unity.

Tim: For me, it would be either Ganung'sTsne'ha, "People of the Longhouse," since I grew up that way, and Haudenosaunee.

Logo of the Great Tree of Peace, surrounded by the 50 Chiefs of the Haudenosaunee Confederacy, arms linked, protecting their people and their traditional environment (courtesy of the Haudenosaunee Confederacy)

YA: Is there a big difference between the different nations that you belong to, linguistically, culturally, and is there any crossover although you might speak different dialects?

Tim: Bear's Ojibwe, I'm Mohawk, but I speak Cayuga. The language is completely different but within the Six Nations—Mohawk, Cayuga, Onondaga, Seneca, Oneida, and Tuscarora which we're adopted into our nation later—we all lived close to each other, we all have similar dialects within each nation. So Cayuga can have three different dialects. It's about different territories and different places.

Gord Hill

YA: Historically, how did those nations interact with each other? If it wasn't linguistic, there were definitely connections—a spiritual connection to the earth in the stories, the histories.

Bear: A lot of people spoke more than one language. There were people whose job was to go to different communities and trade knowledge, information, techniques. There's knowledge about Pan-American trade routes that ran South, from Chile all the way up North. You can see the remnants of that in the way we grow food. Corn, beans, and squash, being a staple up and down the Americas for food. There's a way of growing them where you put all three beans in one hole and when they grow up, the corn creates shade for the squash, and creates a stalk for the beans to grow up so that they all grow up together harmoniously as well as treating the land properly since it's not a mono-crop and it doesn't deplete the land of minerals. You find that style, up and down the Americas. So we traded that information, sometimes from thousands of miles away and came together to gather and trade.

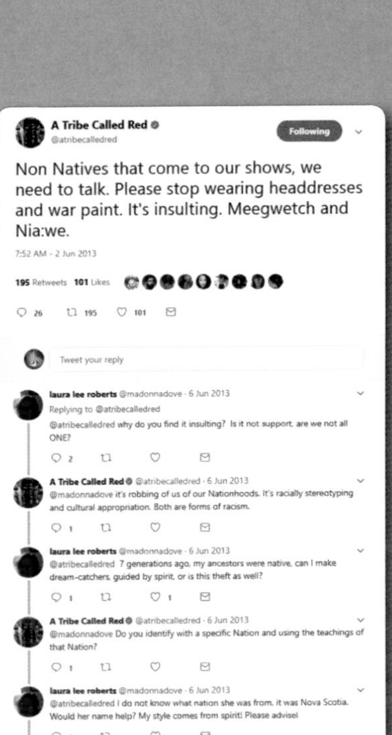

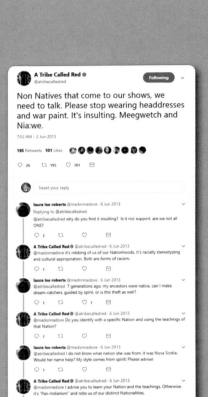

A Tribe Called Red's tweet to end Indigenous appropriation by their non-Native fans, calling out the wearing of Indigenous headdresses and war paint, 2 June 2013 (https://twitter.com/atribecalledred/status/341205589025779712?lang=en)

YA: But if I can go back to culture and appropriation, "cultural appropriation" becomes acceptable to the mass of society after a cultural group has been pillaged, bombed, or basically torn apart, and stripped of its natural rights. It seems like racial iconography and cultural fetishization allows the viewer to disregard their involvement with, or remove their guilt from, that history. So how do you feel *Tribe* counters a hegemonic representation of Indigenous people?

Ian: Again, just confronting it head-on, and addressing it when it happens at our shows. One tweet from 2013 changed a lot of things, it seems. A lot of dialogue started after that tweet "Non Natives that come to our shows, we need to talk. Please stop wearing headdresses and war paint. It's insulting. Meegwetch and Nia:we."

Canadian festivals started banning headdresses full-on, down to one of the biggest concerts in the world.

Bear: Glastonbury, England.

Ian: Glastonbury banned headdresses too, and that sparked from somewhere. I don't want to say that that tweet sparked all of that, but that conversation wasn't happening before. So it seems that when we're confronting all this appropriation, the best thing to do is to just mention it, call it out, until they stop doing it.

YA: While I was hearing you guys blow up on the radio while you were on tour, here in Canada, a lot of the narrative on CBC and the radio was about Native Nations deserving reparations. It was a conversation that was being had, for the first time, publicly. The time that you guys *A Tribe Called Red* came in historically is a very timely moment for the Canadian narrative.

In the media, certain rights take over others, and start "trending." Now it's all about freedom of speech, with everything that happened in France and Canada. But with the openness of the Internet, you start to realize that there are limits to free speech.

Ian: Well, I'm a huge freedom of speech advocate, and people can be assholes as much as they want, but my freedom of speech is to be able to confront those ideas. It works, and it has to continue to work.

YA: What about you, Bear? You work with a lot of already existing visual narrative that's out there. People often focus on the violence, not the root because of the violence in society. How do you change that story and bring attention to these deeper issues at hand through the arts?

Bear: There's always been a place for the artist on the frontline, in any sort of revolution, in any sort of fight for, and by, an oppressed group of people. So being an artist, you have a real responsibility, being an artist in a community that's coming out of five hundred years of genocide. You have a huge responsibility to use your girt to bring attention to the things that aren't being talked about. To give a voice to your community.

For us, specifically, we try to not only give a voice to the community, but also to take control of our image, and how we're portrayed is a huge thing. Before we can even talk about standing up and being recognized in the way that we want to be seen, we have to stand up and to be recognized, we affect the way that we're seen.

YA: It's hard, because—you guys have won a Juno, you won a Native American Music Award, you won four APCMAs (Aboriginal People's Choice Music Awards), you have a lot of accolades publicly. So your industry image doesn't necessarily come from yourselves, it comes from how people perceive you. How have you been able to maintain the image that you want to maintain? Has it been like, "we don't do interviews."

[Laughter]

You know, what has it been?

Bear: I think we've always been really careful about the way that we're portrayed. Really early on, when we got our first newspaper articles, they sent out a photographer, and the photographer wanted to take us out to an alleyway, nobody smiling, mean-mug the camera … it even got to one point where the guy asked us to squat down, and it was like okay are we supposed to start throwing up gang signs? What's going on here?

[Laughter]

And so we started to do our own photo shoots, and hiring a photographer to do our own press photos, because we kept on ending up in these uncomfortable situations. When we did our first press shots, it didn't take long for people to start cracking jokes, and the photographer ended up getting a whole bunch of shots of us laughing. And right away, it was like, okay, those are the images that we put out. That's the image that goes on the cover of the newspaper, that's the image that we want people to see of us.

We could do something to combat the urban image that we had to be stuck in. That really simple turn, of showing ourselves laughing, was huge. And that was a way that we could really easily and quickly affect our own image and the way that people saw us.

A Tribe Called Red (from left to right) Ian 'DJ NDN' Campeau, Tim '2oolman' Hill, Ehren 'Bear Witness' Thomas (Matt Barnes)

YA: Knowing that the voice of public perception is loud, can you speak on how the *Tribe* doesn't use words to vocalize your rage, your anger, your weight? Your music is very positive and powerful, and it's party music, but it carries a certain weight, of rage.

Bear: Some of my biggest influences in music are, going back to my rave'ing and partying days, Congo Natty, or Asian Dub Foundation. These were artists that were bringing politics into the party. They were dropping huge club tracks—they were massive names—but they brought in identity, they brought in culture, and they brought it in an unapologetic way, in your face. But it was still party music, and it was still fun.

I grew up in a very politically active environment, my mother was a member of the American Indian Movement, and very involved in art and politics my entire life. So when I started getting confronted with these issues, I realized that it didn't have to happen in a confrontation, or in a protest … We could use the space of the club, having a good time, but it could sink in a few hours, or days later.

YA: I've seen a couple of your shows. You guys are on stage, sometimes you rock bandanas on your face, I've seen cartoons marching towards you, and you chop it all up. But the crowd doesn't seem to have an emotional reaction towards it, it's a party. So it's very interesting how you blend that consciousness into the seams. When people come to your shows in headdresses, do you think it flies over their heads?

American Indian Movement, 1890-1973, Patch (AIM Texas Chapter)

Ian: Definitely. Whether they're Native or non-Native, there are always experiences and it's good to be a part of that. It's positive, uplifting, educational, how we can build without words, all three of us together, a big picture that everyone can see and feel.

YA: From the outside in, the visuals of your shows are strong, dark moments and happy moments. It's interesting how you blend the consciousness into the scene, but do you think it flies over their head, do they take it in for surface and don't experience your culture?

Bear: We're music producers first and foremost, so if we have people dancing and having a good time, we've done our job. If they get what we're trying to say and what we're trying to do and the politics behind it, that's even better.

YA: Why DJ NDN?

Ian: There are a few reasons. It's an acronym for "Not Dead Native." In reaction to U.S. Army General Sheridan who said, "The only good Indian is a Dead Indian."

YA: What do you 'bear witness' to?

[Laughter]

Bear: The name 'Bear Witness' was given to me. I've been 'Bear' all my life. My parents have called me that since I was a baby. But 'Bear Witness' came when on a beach in a small island off the coast of Tofino, at a solstice party, and a guy named Clooney with same birth date and year as my own, stood up and said "Bear Witness" when he saw me. Everyone looked, and at the time, my DJ name was 'Bear Necessities,' and so as soon as he said that, I said "yeah, that's my name." As I went on in my career doing visual and video work with other arts projects, all of a sudden the name, Bear Witness, took on a new meaning.

YA: 2oolman?

Tim: It's a play off my name 'Tim,' and I've been called that since I've been nine or ten years old, since *Home Improvement*. But when I was making beats, I tried to make my own name, but in the community, even older relatives would call me '2oolman' so I just decided to keep it.

YA: Amazing. So I'm just going to throw a quote out there, and I want to hear your thoughts on this. It's from Bartolomé de las Casas, the 'Protector of the Indians.' He was the 'Protector of the Indians' in the early colonization of the Americas: "Thus husbands and wives are together only once every nine or ten months, and when they met they were so exhausted or depressed that they ceased to procreate. As for the newly born died because their mothers, overworked and famished, had no milk to nurse them, and for this reason, while I was in Cuba, 7,000 children died from lack of milk. Some mothers even drowned their babies from sheer desperation. In this way, husbands died in the mines, wives died at work, and children died from lack of milk, and in a short time this land which was so great, so powerful and fertile, was depopulated. My eyes have seen these acts so foreign to human nature, and now I tremble as I write…"

YA: *Genocide*. *Colonization*. But it is not called so. It is not publicly acknowledged as 'genocide.' You guys even had an incident with the Human Rights Museum in Winnipeg, where that was a reason why you stood by and you didn't perform there. What does it take for society to acknowledge injustices that great? It might take hundreds of years, but what is the breaking point that might make this a reality? And will calling it a genocide make a difference?

Ian: It's a start, I definitely think. As a rapper, you recognize how powerful words are, and how powerful labels are. Within this world construct, I think that *that* word specifically will set off the alarm bells to the atrocities that happened, and atrocities had to happen in order for North America to exist, for it to become Canada and US.

As soon as people confront that, and realize that their patriotism is based on a large pillar of racism, that people will start to confronting them and start empathizing to the realities to what the oppressed people are feeling today.

Bear: I think we might get there on a social level, but as far as a governmental, legislation kind of level, I don't think we can ever see it. The recognized genocides in the world, none of them mess with the colonial construct. And specifically, the genocide of people in the Americas, like Ian was saying, was necessary.

I read a Martin Luther King quote, recently, saying that all of the folklore and the heroes about North America is based on the genocide of Indigenous people, and we'll never be able to admit that. The quote you read Yassin gets to the heart of this. In that moment in time, when colonization was happening, and they were trying to enslave Indigenous people to work in the mines, we made horrible slaves. Indigenous people would just lie down and die, because we were still on our land. We still had the connection to our land, to our community. So you can't break a person in a situation like that. That's why they had to go to Africa, to take Africans away from their home, their land, their culture, their language, and their people, and put them in a foreign place, to break them. If you just look at that moment in time, that's how North America was made, that's the Americas, that's how we get it.

That's why we don't look at it as genocide, and it's such a hard conversation to start.

Ian: Bear hit the nail on the head when he said "It's only genocide if it messes with colonialism."

YA: To bounce off that, what are the colonial remnants that still exists in your community, that you wish need to be broken?

Ian: We all speak 'English.' My reserve is specifically Christianized. There are definitely alcohol problems … the effects of colonization is paramount and it's within everyday actions of what we do. Having the only piece of legislation that's based on race alone—the Indian Act—in North America, which is just for us. We're the only ones affected by race legislation.

YA: If we go into the aesthetics of the crew, you're very responsible for the visual element. In the studio, I see you reading comic books and

Martin Luther King, "I Have A Dream," Patch (AE)

"Our nation was born in genocide when it embraced the doctrine that the original American, the Indian, was an inferior race. Even before there were large numbers of Negroes on our shore, the scar of racial hatred had already disfigured colonial society. From sixteenth century forward, blood flowed in battles over racial supremacy. We are perhaps the only nation which tried as a matter of national policy to wipe out its Indigenous population. Moreover, we elevated that tragic experience into a noble crusade. Indeed, even today we have not permitted ourselves to reject or feel remorse for this shameful episode. Our literature, our films, our drama, our folklore all exalt it. Our children are still taught to respect the violence which reduced a red-skinned people of an earlier culture into a few fragmented groups herded into impoverished reservations."

—Martin Luther King, Jr., "Why We Can't Wait" 1963, as quoted by Malcolm X in his 1987 Autobiography

delving into visual work, and on stage you also challenge the music with visuals. So, there's something very heroic about the work you do. It's anthemic—like superhero shit, when I see you guys on stage. How do you translate the audio experience into visuals? What's the process like—do you take the music and go dig? Or do you have a reserve? What's your process, visually?

Bear: Well, first off, I felt the same way when I saw you perform for the first time, when you came out in the all-white, with the screen in the background. I felt the same way, "this guy's a superhero!" That superhero idea was kind of behind this album, in that sense. We wanted to go out and find all these superheroes to work with, so that we could have our own Brown Justice League.

YA: Like the 2014 collaborative album.

Bear: Exactly. Looking at all of these artists as heroes. But back at your question, I do have a reserve of images, I'm just a huge collector, I'm a massive movie head. I have two separate collections for all my different interests: my general action movies and action toys, then action movies with Indians and toys with Indians in them.

"Billy," *Predator*, played by Sonny Landham, 1987, Film Stills (David Wharton)

YA: Anything I relate to is on the shelves.

Bear: Exactly. So that whole library is what I draw from. Especially in a lot of my work before *Tribe*, I drew from imagery from childhood connections, and the ways that I made heroes out of one-dimensional racist stereotypical images that were being portrayed by Indigenous actors.

YA: Oh wow, so it's like the Arab dude in Iron Man who's speaking the wrong language.

Bear: Yeah. So the one I always go to is a character named Billy, from *Predator*. And Billy is played by an Indigenous actor, named Sonny Landham, he's from down South, and he, his character has all the trappings of the Indian scout. He doesn't speak full sentences, he spends a lot of time holding on to his medicine pouch and staring off into the trees, you know, "there's something in the trees"... [Laughter] but he's a very powerful character in the movie. Whenever they're stuck someplace, and Dutch needs to get them out of a bad situation, he turns to Billy and says, "Billy, get me out of here." And when it comes down to the final scenes, and Dutch is trying to get away, Billy is the only one who stands his ground and goes and faces head-on with the predator. So as a kid, this character was *huge* to me. And he was huge. The movie wasn't about Arnold Schwarzenegger, the movie was about Sonny Landham and Billy. So in my work, Billy has become a really iconic image, and he represents my superhero. I actually did an edit of the movie, where there's nobody else but Billy. And it's just him running around the jungle, re-edited from other movies where he's trying to save a group of young Native boys who are being chased by the army.

YA: Is Billy still alive?

Bear: Yeah, he ran for Governor in the State of Kentucky.

YA: Outside of *Tribe*, Ian, with your process as a DJ, as a producer—did you study this stuff or was this a natural thing you loved doing? What did you study, and how did you get into music?

Ian: I've always been a music fan, always, since I was a kid, and my dad always was an eclectic music collector, and so I grew up listening to a lot of Beatles, and Stones, and that sort of stuff, then I grew into punk rock, eventually, and toured with a band, Montreal's Ripcordz, I was a drummer for the Ripcordz for a year, and then I started bouncing at a club, and then from there I just started DJing, it seemed like.

[Laughter]

YA: One day you find yourself—you're like (scratching noises).

[Laughter]

Ian: Well, I came from bouncing, it was like working in a club, sort of thing…

YA: They're like, "the DJ didn't show up, Ian, we need you!"

[Laughter]

Ian: Well, I knew all kinds of music, right? I was really eclectic, and the club I was working at was a music venue, Barrymore's, in Ottawa, so there would be bands, and every day there would be a different band. So you'd have a Metallica cover band playing one night, and then Swollen Members playing the next night, and then Feist playing the next night. So as a bouncer, I'd go in and DJ between the bands every once in a while, but because I knew all of these different kinds of music, I would be able to play—just because I liked music so much, that they asked me to DJ one night, they had a new 90's night that was opening there, and they asked me to DJ that, and that's where my DJing came from.

"I have decided to scalp you and burn your village to the ground," Addams Family Values, 1993, Film Still (cinespia.org)

YA: Amazing. Tim? How'd you get into production?

Tim: I used to make pause tapes back in the day, so basically, just take breaks from tapes and loop them and make them. I was just fascinated with that, so when I got my first computer, I started messing with the audio recorder and started figuring it out—I've been doing it for a long time, just by myself—it was kind of like a thing I built on my own, and I just kind of did my own thing. I started collaborating with other artists, as far as recording them and learning how to do that, and then that kind of turned into this … but in the last couple of years I just spent most of my time making albums with friends and stuff like that, and then just sort of making full projects, and then actually, the last two years I was starting shopping music labels and stuff like that, where I felt like I was good enough to do it…

All the days I spent in my bedroom, 10 years just doing it. It feels like I've been training for this moment. Ian's knowledge of music, tons of music, teaches me so much. And Bear, is crazy with movie quotes, like how "Burn Your Village to the Ground" came from. The opportunity to work together and collaborate is incredible.

A Tribe Called Red, "Burn Your Village to the Ground," 2014 Single, Cover (courtesy of A Tribe Called Red)

YA: What's more satisfying for you, finding the right snare or playing in front of a huge audience?

Ian: If you would have asked me a year ago, I would have said the snare. But it's the greatest rush ever, it's great to see everybody and have fun up there, on top of it. Being in the studio all that time is awesome, I love it too. It's worth it.

YA: I mean, look. Now that you guys are here at the Phi Centre in Montreal, and you're producing together, this new collaborative record, the first one was kind of an insular project where you guys just came out, and *boom*! Blowed it. Now you're reaching out and the *Tribe* is growing. So *A Tribe Called Red*, beyond the three figureheads, are on stage, and it's really that history that we're talking about, and the future generation. But also the people that you reach out to become a part of *A Tribe Called Red*. And you're all individual artists, and figuring out how to work in a group is difficult, let alone bringing in outsiders. What has the process been like for you guys, in the studio? Has it been smooth, or have you hit new creative roadblocks?

Nation II Nation, Cover Art, A Tribe Called Red, 2013 Album (courtesy of A Tribe Called Red)

Tim: It's actually been like a dream, to be honest. The people that we've been talking about, the people that we've met, it's like what Bear was saying about gathering all the superheroes up. It's kind of like the Brown Justice League. That's pretty much exactly what it is. It's like, we get to pick who we want to work with, and on top of that, the opportunities of working with people that we've looked up to forever, since our music careers started. This is all possible. It's good.

Bear: I really feel like *A Tribe Called Red* isn't the three of us, that's something that we tapped into, we came about at the right moment in time, to be able to articulate something that our community was thinking and feeling, and thought was necessary, and that we're really just handling this energy at the moment, and that it doesn't belong to us, it belongs to our community. So going out there and finding other people to work with is, that's what it's supposed to do. We've brought this here—I don't want to say we've brought this, we got *caught up*. We're caught up in this thing, and its purpose is to create a whirlwind. Its purpose is to bring Indigenous culture into the forefront, in a way that hasn't been able to happen before.

Like I said about representation, being able to take control of our own representation, we want this to represent more than just us. This is about Indigenous art, Indigenous culture, Indigenous politics across the board. Our name, we spent a lot of time trying to figure out a name, that would include all Indigenous people. And when you're talking about *hundreds* of different nations, just in North America alone, never mind Indigenous people all over the world, you're trying to find something that includes *everybody*. That's a tough job, but it's something that we felt was necessary. And I think we conduct this group in a way…

YA: It's getting bigger than you guys.

Bear: So then if it's a tough job, then we're going to do that, because that's necessary. If we have to make a decision that maybe isn't the best from a financial standpoint, or a business standpoint, well we're going to make that decision based on what's best for this entity.

YA: Everybody.

Bear: This *thing*, that's now here.

Yassin "Narcy" Alsalman (courtesy of Cheb Moha)

YA: Ian, you want to chime in on that one?

Ian: Well, just about the bandana thing, and having *A Tribe Called Red* be a bigger thing, I think that us wearing bandanas on stage is a testament to us not being the focal point of the group. We're covering our faces for a reason, it could be anybody up there. And that's kind of the point—that it's not just us three. It's way bigger than that.

Bear: That's really interesting you brought in the superhero thing earlier, because I was going to bring that up with the bandanas before, just because when we started wearing the bandanas, friends of ours had made them and started giving them to us, but it became a group thing really quickly, and not really understanding why we're doing it, it just felt necessary. And then every night, it became essential, that we had to mask up. It became this point in the night when we masked up, and we got pumped up for the show, and got ready. It wasn't just the feeling of, let's put on these bandanas to be *cool*, there was a necessity to it. That, yeah, we were going out there to portray something that made it so that we needed to cover our identity. And if you look at grassroots or Indigenous revolutions across the board, there's that necessity of needing to cover your identity. Because if you're out there fighting for your culture, for your life, with the only thing you have left, which is your body, you still need to protect the people that you care about. And that goes right back to comic books, right? Why does the hero wear the mask? Why does Batman wear the mask? To protect the people that he loves. So, you know, it was all of those things coming back around, and realizing that bandana had become something more than the necessity to protect our identity, or the necessity not to breathe in tear gas or whatever—the bandana has become a symbol, and that symbol is Indigenous, grassroots resistance.

We are the Halluci Nation, Album Seal, A Tribe Called Red, 2016 Album (courtesy of A Tribe Called Red)

This conversation was adapted with permission from "We Are The Medium, Episode 5: A Tribe Called Red," by Yassin 'Narcy' Alsalman, February 2, 2015. Note that "YA" stands for Yassin 'Narcy' Alsalman, "Ian" for DJ NDN, "Bear" for Bear Witness, and "Tim" for 2oolman.

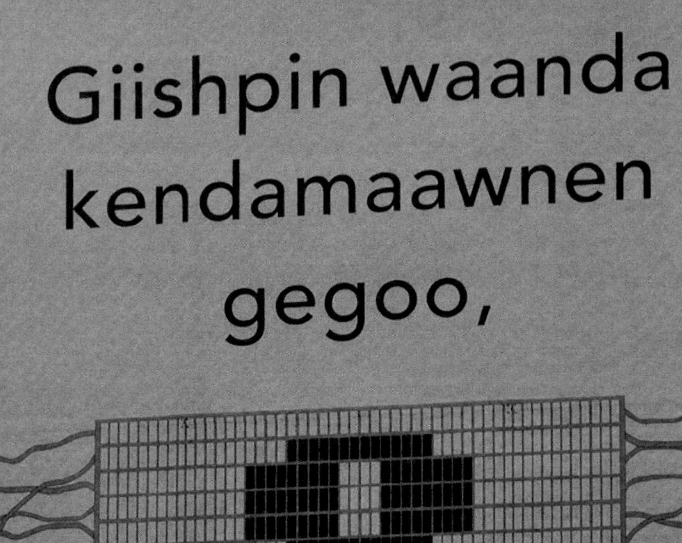

Ogimaa Mikana Project, courtesy of Susan Blight and Hayden King

Colonization & Decolonization

A Manual for Indigenous Liberation in the 21st Century

by Zig Zag

How to Use this Manual: This manual is divided into four sections. The 1st section defines colonialism, its methods, and its history up to today (i.e., the US invasion and occupation of Iraq). The 2nd section details the impact colonialism has had on Indigenous peoples, including sociological and individual impacts. The 3rd section examines the concept of decolonization. The 4th section discusses decolonization within North America. It will be seen that the liberation of Indigenous peoples in N. America is closely connected to a global process of resistance and survival. This manual is intended for self-study as well as for use in training classes. The following are lesson plans that can be used or adapted for the classroom.

"Knowledge makes a person unfit to be a slave." —Frederick Douglas

Three Classes Total:
Sections 1 and 2 can be held as separate 45 min.–1 hour classes; the 3rd and 4th sections on Decolonization can be presented as one X 1 hour class. Total: 3 classes at 1 hour each: 3 hours.

Training Aids:
1st Class: maps of world, flip-chart/ board, graphic poster-boards, video clips of events/news. 2nd Class: flip-chart/board, graphic poster-boards, video clips. 3rd Class: flip-chart/board, graphic poster-boards, video clips, Warrior Unity flag, large map of N. America.

Training Tips:
- Be motivated & enthusiastic: use graphic training aids. History and colonialism can be hard subjects to teach. Instructors must strive to make it interesting, inspiring and relevant to student.
- Know your subject. This material can be difficult to present.

Study & Prepare:
- Emphasize important points and concepts.
- Use intro, body, and conclusion (tell them what you're gonna tell them, tell 'them, tell 'em what you told 'em).
- If practical, hold classes consecutively (3 in a row, with 10–15 min. breaks).

Class Plans

1. Introduction & History
A. Define Colonialism, four Stages of Colonialism.
B. History of Colonialism.
 a. Egyptian
 b. European
 c. Roman
 d. 1492: Invasion of Americas, 1498 North America
 e. Revolts; Settler & Afrikan
 f. Final Phase N. America 1890
 g. Africa, Asia & the Middle-East
 h. War & The Rise of USA
 i. WWII & UN 'Decolonization'
 j. Vietnam/US Domestic Unrest
 k. New World Order—War for Oil
C. Conclusion

2. Impacts of Colonialism
A. European Settler Society
 a. Settler-Nations
 b. Imperialism
 c. Apartheid
 d. White Supremacy
 e. Patriarchy
 f. Neo-colonialism
 g. Pyramid of Power/Social Structure
B. Sociological Impact
 a. Genocide
 b. Loss of Sovereignty & Territory
 c. Assimilation
C. Individual Impact
 a. Post-Traumatic Stress Disorder
 b. Individualism, Identity, Inferiority
 c. Internal Violence
 d. Alcohol, Drugs, Suicide
 e. Health
D. Conclusion

3. Decolonization
 a. Culture
 b. Warrior Culture/Fighting Spirit
 c. Identifying the Common Enemy
 d. Disengaging from Colonial System
 e. Liberation of Mind & Spirit
 f. Active Use of Territory

4. Decolonization in N. America
 a. Decline of Roman/USSR/USA
 b. USA: An Empire Divided
 c. Mexico & US Southwest
 d. Insurgency in Iraq
 e. Crises, Conflict, & Resistance
Conclusion

Introduction

> "Liberation is the task imposed upon us by our conquest and colonization."[1]

Colonialism: The practice of invading other lands and territories, for the purpose of settlement and/or resource exploitation.

When an invading force confronts an Indigenous population already occupying a territory, colonialism becomes a violent conflict between two hostile and opposing ways of life, with one attempting to impose its will on the other. This is a standard definition of war, and colonization itself can be considered a *war for territory* involving all the means used to carry out wars: military, political, economic, psychological, diplomatic, cultural, etc.

Cecil Rhodes, a British colonial official for which Rhodesia (now Zimbabwe) was named, articulated the motives and goals of European colonialism in the nineteenth century:

> "We must find new lands from which we can easily obtain raw materials and at the same time exploit the cheap slave labor that is available from the natives of the colonies. The colonies would also provide a dumping ground for the surplus goods produced in our factories."

Due to its history and culture, European colonialism is characterized by genocidal practices, including wars of extermination, massacres of non-combatants, biological warfare, and scorched earth policies (destroying food and shelter). Other atrocities include the torture of prisoners, rape, and enslavement of Indigenous populations. These acts are fueled by racist and patriarchal ideology (i.e., Christianity and white supremacy), greed, and a psychopathic desire to kill and inflict violence and suffering on others.

> **Psychopath** *n.* A person with an antisocial personality disorder, manifested in aggressive, perverted, criminal, or amoral behavior without empathy or remorse."[2]

Stages of Colonialism

The methods and history of colonization are unique in every case, due to many different variables (geography, population density, resources, etc.). Despite this, there are common patterns that can be easily recognized. In the Americas, Africa, and Asia, colonization generally consisted of four stages: recon, invasion, occupation, and assimilation.

1. Recon: Colonialism begins first with small recon forces that map out new lands or regions and gather intelligence. These are often celebrated today as voyages of "exploration" and "scientific discovery." The 1492 voyage of Columbus, for example, was a recon expedition to find a new route to Asian markets. There were only 3 ships: the Pinta, Niña, and the Santa Maria.

2. Invasion: The second phase is invasion, which begins a period of armed conflict as Indigenous nations resist colonial forces. For example, when Columbus returned to the Caribbean in 1494, he had 17 ships and over 1,000 conquistadors. Invasion can begin immediately after the recon, or may be delayed by a period of trade and settlement that serve as a basis for later invasion (i.e., N. America). In every case, colonial military strategy is genocidal and includes the destruction of food supplies, resources, and shelter, as well as massacres and biological warfare (disease).

3. Occupation: When Indigenous peoples are militarily defeated, the occupation is expanded. A colonial government is set up to control the surviving population of Natives, who are contained in reservations, or enslaved. By the 1700s, many colonial authorities were corporations (i.e., the Hudson's Bay Company, French Senegal Company, etc.). They organized settlement and resource extraction, including the construction of railroads, dams, roads, ports, etc.

4. Assimilation: An important part of imposing control is the indoctrination of surviving Natives into the European system. In order to do this, Indigenous society and culture must be dismantled and erased as far as possible. Colonial violence, including physical destruction and biological warfare, achieve this through depopulation, often during the period of invasion. Once occupation is entrenched, this process becomes institutionalized, with generations of Indigenous youth being removed from their people and forced into government or Church-run schools.

The period of occupation and assimilation are connected, as only through occupation can systems of assimilation be imposed. This phase can be long and drawn out over centuries, as has occurred in the Americas.

1. History of Colonialism

Colonialism is neither new nor limited to any specific historical period (i.e., the 'colonial period' of the fifteenth to nineteenth centuries). Ancient civilizations were the first to begin colonizing other lands and people. When their populations became too large, and as resources became depleted, colonists were sent out to occupy and settle new lands. When these lands were already occupied, military campaigns were carried out to gain control.

When nations and territories were conquered, the survivors were enslaved and forced to provide resources, including human labor, food, metals, wood, spices, etc. The invaders then imposed their own forms of governance, laws, religion, and education. Over time, these populations became assimilated into the culture and society of their oppressors.

Early Egyptian Colonialism

In ancient Egypt (around 1,500 BC, or 3,500 years ago), all the methods of colonialism were already being practiced. An African scholar, Cheikh Anta Diop, described these methods:

> "In some towns, as in Jaffa, the conquered princes were purely and simply replaced by Egyptian generals ... Egyptian garrisons were stationed at strategic points, important towns and ports ... 1,400 years before Rome, Egypt created the first centralized empire in the world.
>
> The children of vassal [conquered] princes were taken as 'hostages' and educated in Egyptian style, at the court of the Egyptian emperor, in order to teach them Egyptian manners and tastes and to assimilate them to Pharaonic culture and civilization...
>
> The Pharaoh [emperor] could at any moment require money, chariots, horses, compulsory war service; the vassal was constantly under the orders of the Egyptian generals ... The vassals enjoyed only internal autonomy; in fact they had lost their international sovereignty; they could not directly deal with foreign lands." [3]

European Colonization

When looking at the world today, we can see that this process still continues, sometimes referred to as imperialism, globalization, or even 'peacekeeping' and 'humanitarian' missions. Whatever term is used, the principles of invasion, occupation, and exploitation remain the same.

Today, the European states and their settler nations dominate the global system. How did this come to be? Why is it that Western Civilization is now the primary economic, political, and military power in the world? The answer to these questions can be found in the history of civilization.

Early civilizations concentrated vast amounts of human and material resources under the control of a central authority. This authority was usually in the form of kings and priests, who based their right to rule on spiritual or religious tradition. They controlled all governance, economic trade, law and order, education, etc. Through religion, mind control was imposed over citizens, which created a culture of obedience, slavery, and war (just as we see today).

The first civilizations were established in northern Africa and Mesopotamia (the Middle-East), comprised of the Egyptians, Sumerians, and Babylonians. Other civilizations also began in India, Asia, and the Americas, but those in Egypt and the Mid-East had a direct influence on Europe.

The Greeks were the main transmitters of civilized culture into Europe, based on both Egyptian and Mid-East models. The Greeks, southernmost in all of Europe, were strategically located to serve just such a role. Prior to this, southern Europe was inhabited by tribal peoples. While Egyptians built massive pyramids and cities, had a written language, advanced science and astronomy, etc., Europeans were still hunting and gathering.

This history tells us that colonization results from a society's culture, not its racial or biological background. This culture, based on expansion, control, and exploitation, arises from civilization. Despite this, it is the European system that now dominates the world, the result of history, geography, and the exchange of culture and technology that occurred throughout the Mediterranean.

Roman Colonization

The first people colonized by Western Civilization were the European tribal peoples, such as the Goths (Germany), the Gauls (France and Spain), etc. They were invaded and occupied by the Roman Empire, beginning around 200 BC (some 2,300 years ago):

> "Conquered territories were divided into provinces ruled by governors appointed in Rome for one-year terms. Governors ruled by army-enforced decree ... Conquered peoples all had to pay extraordinary taxes to Rome." [4]

Early accounts by Romans described these peoples as worshiping Mother Earth, organized in clans and tribes, living as semi-nomadic hunters and gatherers. They were also strong and adept military forces that inflicted numerous defeats against Roman forces, with some regions never being pacified or conquered (i.e., the Scottish Picts).

Despite this resistance, some areas such as present-day Spain, Portugal, and France, as well as parts of Germany and Britain, were occupied by Roman forces for as long as 400 years. Forced to work

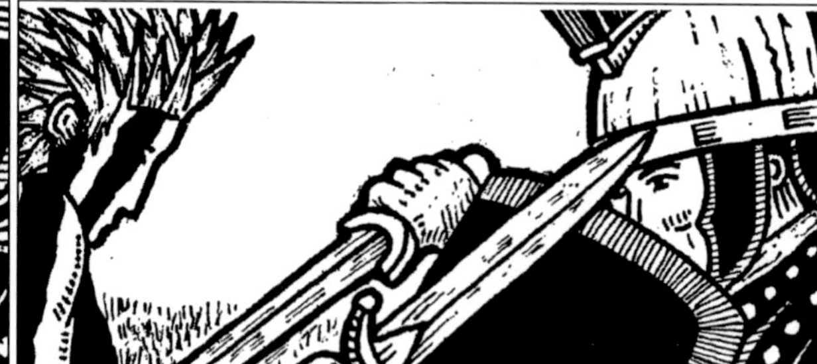

as slaves, to build houses and fortifications, to serve as expendable frontline soldiers, to provide resources and manufactured goods, or as servants (cooks, janitors, barbers, tutors), these conquered peoples were also increasingly assimilated into Imperial Rome.

Tribal chiefs and high-ranking families were targeted for systematic assimilation; often, their children were taken and taught how to speak and read Latin (the language of Rome). Roman clothing and overall culture were imposed. After several generations, these peoples were effectively Romanized or Latinized, with some gaining citizenship and high-ranking positions in the Roman military or political system. These families, along with the Roman governing system and the Christian church, served as the basis for the feudal system which evolved in Western Europe after the collapse of the Roman empire (fifth century BC).

Perhaps more than any other region, Europe stands as a stark example of the effects of colonization and assimilation. Today, very little remains of the European tribal cultures, which were destroyed and assimilated into the Roman imperial system (which explains why European civilization is essentially fascist in nature).

1492: Invasion of the Americas

In 1492, the European colonization of the Americas began with the voyage of Christopher Columbus, in command of the Niña, Pinta and the Santa Maria. This recon expedition arrived in the Caribbean and landed on the island of present day Haiti and the Dominican Republic, which was named Hispaniola. In 1494, Columbus returned with a second, larger force, comprised of 17 ships and 1,200 soldiers, sailors, and colonists.

By 1496, it is estimated that half of the 8 million Indigenous peoples on Hispaniola were dead, killed by a combination of European diseases and massacres. Both priests and conquistadors have left detailed accounts of their atrocities, killing for fun, hunting Indigenous peoples as if they were animals, and devising all kinds of cruel and inhuman methods of torture. Survivors were enslaved and forced to supply gold, silver, and food to the conquistadors. Those who failed to meet their quotas had their hands, ears, or nose cut off. From this strategic location, military campaigns were conducted into nearby islands; by 1510, the Spanish were relocating Indigenous peoples from the Bahamas and Cuba to replace the dying slaves on Hispaniola.

By 1535, Spanish conquistadors had launched military operations into Mexico, Central America, and Peru. Using guns, armor, and metal edged weapons, as well as horses, siege catapults, war dogs, and biological warfare, the Spanish left a trail of destruction, massacres, torture, and rape. Tens of millions of Indigenous people were killed within the first century. The Mexica (or Aztec) alone were reduced from some 25 million people to just 3 million. Everywhere the death rate was between 90–95% of the population.

The European invasion of the Americas was, without question, the most devastating genocide and holocaust in history. Despite this, it is still celebrated today as a 'discovery.' With some exceptions, the history of this holocaust has been minimized or concealed.

The main goal of the Spanish and Portuguese was to take control of the land and enslave the surviving Indigenous people. Settlement was not a main objective. They established huge plantations to grow crops for export to Europe, while vast ranches were set up for cattle raising. Mines were opened to dig for gold and silver. Millions of Indigenous people were enslaved and died working in these mines.

In order to maintain a source of slaves, European traders turned to West Africa. There, Indigenous Africans, engaged in intertribal war, traded prisoners of war with the Europeans, clearly ignorant or indifferent to the long-term effects such actions would have. As many as 15–20 million Africans were shipped onboard slave ships, with an estimated 40 million dying from disease and starvation on the trans-Atlantic crossing.

Despite this high level of violence and destruction, Spanish and Portuguese colonial forces were largely restricted to the coastlines of Central and South America. Many interior regions resisted for 2–3 centuries and were never conquered by the Spaniards. The Maya in the Yucatan Peninsula, for example, withdrew into the forest lowlands, where Spanish forces fell victim to disease and the intense heat. The Maya then launched military attacks and were able to resist total Spanish control.

By 1800, the Spanish laid claim to a vast region encompassing parts of South, Central, and North America. Despite this, it was an empire in decline, faced with ongoing Indigenous resistance, slave rebellions, and even settler revolts. By the mid-1800s, settler independence movements forced the Spanish out of the Americas (with the exception of Cuba and Puerto Rico).

At the time of the invasion of the Americas, Europe was in the Dark Ages, suffering from resource depletion, overpopulation, widespread poverty, and social decline. Colonialism brought new resources and wealth into Europe, while destroying Indigenous nations in both the Americas and Africa. It is from the colonization of the Americas that the European nations were able to further expand and dominate the world.

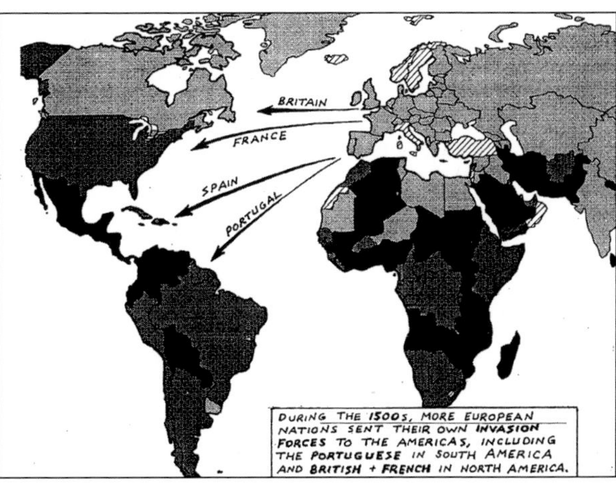

DURING THE 1500s, MORE EUROPEAN NATIONS SENT THEIR OWN INVASION FORCES TO THE AMERICAS, INCLUDING THE PORTUGUESE IN SOUTH AMERICA AND BRITISH + FRENCH IN NORTH AMERICA.

1498: Invasion of North America

In 1498, John Cabot, sailing under command of the English King, claimed the east coast of present-day Newfoundland. He was followed by the French shortly after. Throughout the 1500s, the British and French attempted several colonies on the east coast of the US, but none survived the harsh winters (or, in the south, attacks by the Spanish). Finally, in 1607, a British colony was established at Jamestown, Virginia. It survived due to the help of Indigenous peoples (the tradition of Thanksgiving, adapted from Indigenous peoples, arises from this).

These early British settlers took great care to not engage in any offensive actions, especially as Indigenous peoples were militarily stronger. At first, peace and friendship treaties were made. As colonist's numbers grew, they began to seek greater land and resources, especially the agricultural lands of Indigenous peoples. By the 1620s there was all-out war in the northeast, with colonists carrying out massacres and scorched earth policies. Combined with the effects of biological warfare (smallpox), these attacks gradually broke the ability of Indigenous nations to militarily resist.

Unlike the Spanish and Portuguese in the south, the British and French found little gold or silver with which to finance large-scale invasion. Instead, they relied on trade with Indigenous nations (i.e., the fur trade) as well as the gradual development of agriculture for export to Europe. For this reason, a dual policy of maintaining friendly relations with some, while waging war on others to gain territory, was used. Over time, however, even those that actively collaborated with the settlers were attacked, their lands taken, and their populations enslaved.

A main focus for the French and British was the transfer of large numbers of citizens to the colonies in order to relieve the pressure of overpopulation, as well as to garrison them against other European powers. Settlement was therefore a major factor in the colonization of N. America.

As in South and Central America, Indigenous populations suffered death rates of 90–95% across North America. Although diseases had a major impact, they were most often accompanied by wars of extermination that targeted not only men, but also women and children. Those not killed by disease or massacre suffered starvation, as villages and crops were systematically burned by heavily armed European militias. Extermination of Indigenous people was an official policy of colonialism, limited only by the potential to make money through slavery.

Competition between the French and British led to a series of wars, fought both in Europe and in the American colonies. By 1763, France was defeated and surrendered its colonies to the British (including present-day Quebec). In tum, the British reorganized their colonial system and imposed new taxation on the colonies themselves, to help pay for the costs of war.

Along with this, the British issued the 1763 Royal Proclamation. This law limited the expansion of colonies by imposing a western boundary line (along the Appalachian Mountains). Only British Crown forces could trade, acquire land, and conduct other business in the 'Indian Territories.' This act, which also recognized Indigenous sovereignty to land, served to limit some Indigenous resistance. At the time, the British were faced with an insurgency led by Pontiac, with an alliance of Ottawas, Algonquins, Wyandots, and others. They had captured 9 of 12 British forts and laid siege to Detroit for 6 months.

New taxes and the 1763 Royal Proclamation angered many settlers in the 13 original colonies, especially their exclusion from gaining more land. Real estate had become a huge business, with settlers taking land by violent conquest and selling it or growing cash crops such as tobacco. In response, they organized an armed revolt against the British in order to establish an independent Euro-American empire.

Settler Revolts in the Americas

The Euro-American Revolution of 1775–83 was the first in a series of settler independence revolts in the Americas. Unrestrained by British colonial policy, the new USA began a rapid military expansion westward, killing, enslaving, or relocating Indigenous peoples. At the same time, tens of thousands of European immigrants were brought in. Despite this, it would take over 100 years for Indigenous resistance to be defeated by US forces.

In the early 1800s, inspired by the 'American Revolution,' settler revolutions occurred throughout South and Central America, with new independent nation-states being created (i.e., Bolivia, Chile, Peru, etc.). Although these movements kicked out European colonial powers, they did not liberate the Indigenous peoples. Instead, it was the immigrant European elites and their descendants who assumed power. We do not refer to these as examples of anti-colonial resistance.

By the late 1800s, these settler governments began to take out huge loans from European banks. These loans were used to build roads, railways, dams, ports, etc., in order to better exploit the natural resources. US and European corporations became heavily involved in these countries, where they could make huge profits exploiting cheap labor, land, and resources. This period established the imperial relationship between the 'Third World' and the Western powers, based on debt and repayment of loans.

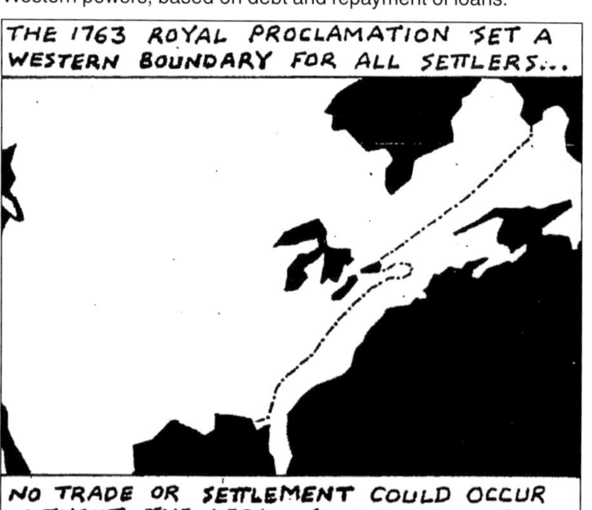

THE 1763 ROYAL PROCLAMATION SET A WESTERN BOUNDARY FOR ALL SETTLERS...
NO TRADE OR SETTLEMENT COULD OCCUR WITHOUT THE LEGAL SURRENDER OF INDIGENOUS LANDS TO THE BRITISH...

Afrikan Slave Revolts

As early as 1526, Afrikan slaves had rebelled against their European 'masters.' In some regions, such as Brazil, escaped Afrikans established liberated zones, defending them against colonial forces. In the Caribbean, Central and South America, escaped Afrikans also found sanctuary among Indigenous peoples.

During the 'American Revolution' in the US, the British offered freedom to Afrikan slaves. As many as 100,000 are believed to have abandoned the slave-plantations and to have fought with the British. Many Afrikan units continued fighting after the British defeat, others went to Canada, and those that didn't were re-enslaved.

While the Euro-American settler elite were planning and executing their continental expansion, Afrikan slaves in Haiti rebelled and defeated French forces in 1791. This had an alarming effect on the US, where some began to realize the dangers of having a large Afrikan slave population. In 1800, a large slave revolt occurred in Virginia. Efforts were made to reduce the numbers of slaves; in 1808, the government banned the import of new slaves.

In 1812, settler vigilantes attacked Seminole communities in Florida in an attempt to re-capture escaped Afrikan slaves, who had gained sanctuary among the Seminole. This began the first phase of the Seminole Wars, which cost over 1,600 dead US soldiers and millions in dollars. Even after the Second Seminole War of 1835, the Seminole and their Afrikan allies remained undefeated.

Meanwhile, slave revolts continued. By the 1820s, many cities had large numbers of Afrikans, often concentrated in certain areas. Although most were slaves, and urban slaves outnumbered those on plantations, an increasing number were also escaped slaves. They were able to find sanctuary in large numbers. As a result, many insurrections and rebellions had their origins in urbanized slaves.

By the 1830s, large numbers of these slaves were being re-located to the plantations. It was also felt that too many Afrikan slaves were being exposed to education and learning "too much" in the cities.

By the 1850s, slavery had become a dividing policy among the southern and northern settler elites. The slavery-based plantation system of the south was now seen as retarding the interests of empire, threatening it with Afrikan revolution while limiting the growth of the northern industrial-capitalist system. What the US needed was a vast army of Euro-settlers to take and hold territory, to work in the factories and farms, to produce and consume.

Between 1830–60, some 5 million European settlers emigrated to the US. By this time, the struggle for power between the north and south erupted into the US Civil War (1861-65). Again, freedom to New Afrikan slaves was promised, this time by the northern forces. Once again, tens of thousands escaped and joined the northern Union army. With this mass withdrawal of slave labor, further strengthening the north, and unable to compete against the economic and industrial capacity of the north to wage war, the south was defeated. New Afrikans in the south immediately organized to defend their freedom. Strikes and armed occupations of land occurred. The new northern government offered limited political, legal, and property rights, while attacking the most militant elements. Union soldiers also disarmed Afrikan army units, or redeployed them to the ongoing 'Indian Wars' on the plains.

But these reforms were too much for southern settlers, thousands of whom joined white supremacist groups such as the Ku Klux Klan to wage a campaign of terror against blacks. Thousands of Afrikans were killed during state elections. The north withdrew its forces and allowed local settler regimes to regain power, who maintained a brutal system of apartheid until the 1950s civil rights struggles.

Final Phase of N. American Indigenous Resistance: 1800s

In 1812, US forces attempted to invade eastern Canada. At this time, Britain was again at war with France, and the US hoped to take advantage of this. They were repelled, however, by a small British force and an alliance of Indigenous warriors. It is generally acknowledged that, had it not been for the involvement of Indigenous peoples, the US would have been successful. This resistance was led by Tecumseh and Blackhawk, who also helped organize insurgencies against European colonial forces throughout this period.

In 1838, US troops forced thousands of Cherokee into prison camps and then, in winter, on the Trail of Tears, a forced relocation during which one in four died. Numerous other nations were also forcibly relocated, including the Choctaws, Creeks, Shawnees, Miamis, Ottawas, Delawares, and others. Many were sent to Oklahoma.

In 1848, the US invaded and took control of northern Mexico, including California, Nevada, New Mexico, Arizona, Texas, Colorado and Utah. That same year, gold was discovered in California, setting off an invasion of settlers that decimated Indigenous nations in that region. In the 1860s, as the US Civil War raged, Indigenous nations on the plains and in the southwest continued to resist their colonization. Apache resistance, led by Cochise and Colorado, began at this time and would not end until the final capture of Geronimo, in 1886.

In 1863, the Shoshone began attacks against invading settlers and military forces in Utah and Idaho. As well, the Dene in New Mexico and Arizona began to carry out attacks against colonists. During this time, US, British, and Russian colonists were also active on the Northwest Coast. In BC, British navy gunboats were used to bomb villages, destroying houses, canoes and food supplies. On the plains, the Lakota, Cheyenne, and Arapaho began to wage guerrilla war against US troops and settlers. From the 1860s to '70s, the US Cavalry suffered numerous defeats (including the defeat of Custer, 1876) and lost several forts. By 1885, however, the last great buffalo herd was slaughtered by settlers, depriving the plains nations of their single most important source of food, shelter, clothing, etc. That same year, the and Cree in southern Manitoba rebelled against British Canadian authorities (led by Louis Riel and Poundmaker). The British were able to use the destruction of the buffalo herds to impose control on the plains nations in Canada, forcing them to sign treaties and live on reserves At the same time as these military campaigns were being carried out, diseases continued to have a devastating impact on Indigenous populations. At times, the intentional use of biological warfare was also used to destroy Indigenous resistance.

By 1890, Indigenous peoples in both Canada and the US were militarily defeated. That year, nearly 300 unarmed men, women, and children were massacred by the US Cavalry at Wounded Knee, South Dakota. At this time, the systematic assimilation of the surviving populations began, with Indigenous peoples relocated to reserves and generations of children forced into residential schools, where they were indoctrinated with European culture and ideology, language, religion, etc. Many suffered physical, mental, and sexual abuse, while tens of thousands would die from diseases such as tuberculosis and influenza.

Africa, Asia, & The Middle-East

Although the Portuguese had begun trading and raiding along the African coasts in the mid-1400s, European colonialism on the continent remained limited for several centuries. The first attempts by Portuguese forces to invade were met with strong resistance and deadly African diseases. The main concerns for Europeans were economic trade and securing slaves for the colonization of the Americas. In North Africa, Europeans were limited by the presence of large and equally powerful Islamic civilizations. Despite the crusades of the tenth–twelfth centuries, Europe was unable to invade and conquer these empires.

Until the mid-1800s, most of the African interior remained unmapped and unknown. It was referred to as the 'Dark Continent,' a land of black 'savages' and deadly diseases. At this time, new antibiotics were developed and more European explorers began to penetrate beyond the coastlines into the heart of Africa. Here they encountered nations already depopulated and weakened after centuries of the slave trade. Europeans were also armed with far more deadlier firearms, cannons, etc. As a result, a new phase of European colonialism began. As in the Americas, millions of Indigenous Afrikans were killed and enslaved, while European nations looted and plundered the natural resources.

By this time, colonization in both North America and Africa were at similar stages. In 1876, the Lakota, led by Crazy Horse and others, destroyed Custer and the 7th Cavalry. In 1879, Zulu warriors completely destroyed British forces at the Battle of Isandhlwana. Likewise, by the 1890s, machine guns were used to massacre men, women, and children, both in N. America and the African Congo.

In the 1920s, as a result of World War I, the Ottoman Empire was weakened. This empire held together the Arab Islamic civilizations of the Middle-East. During the war, the Ottoman rulers had sided with Germany. Following their defeat, the Mid-East was divided up between the Western powers, especially Britain and France. They took control of countries such as Iraq, Iran, etc., as well as their oil and gas resources.

In Asia, Spanish and Portuguese forces had conducted extensive recon missions during the 1400s, establishing trade with the Chinese and other Asian empires. Here again, in the face of an equally powerful civilization, the Europeans were unable to simply invade and occupy (with the exception of islands such as the Philippines, invaded in 1565 by the Spanish). While Europeans were able to take control of international trade in Asia during the sixteenth and seventeenth centuries, it was not until the mid-1800s that many Asian nations came to be controlled by Europeans (primarily the French and British, including India, Burma, Laos, Vietnam, Cambodia, Malaysia, etc.).

World War & the Rise of The USA

By the early 1900s, virtually the entire world was divided up between the (primarily) European and US empires. The final phase of this occurred in Africa, where the Europeans divided up the continent among themselves. After this, there were no new lands to invade and colonize. Having taken possession of the world's peoples and lands, the imperialists turned against one another (as they had for centuries). World War I was the inevitable result of this power struggle for global domination

While Western Europe was devastated from 1914–18, with as many as 20 million killed, the US remained largely untouched. Although there was widespread repression inside the US, including mass arrests and deportations of tens of thousands of European immigrants labeled 'subversives,' the US did not suffer any combat on its own soil. Entering the war only in 1917, the US emerged in a stronger economic and military position than the Europeans.

As a result of the war and forced industrial production (under Martial Law), the US economy expanded. The postwar economic 'boom' of the 'Roaring Twenties' ended in 1929 with the collapse of the stock markets and the beginning of the Great Depression. The party was over. Seen as the result of over-production, this economic crisis forced tens of millions around the world into unemployment, poverty, and starvation.

In response to this crisis, and the threat of communist revolution (i.e., the Russian Revolution of 1917), many Western governments resorted to police repression and, in the case of Italy, Germany, and Spain, fascism. Nazi Germany, established in 1933, was funded and supported by many businessmen and politicians in the US. By 1939, Germany had invaded neighboring countries including Austria and Poland. This aggression was used as a pretext for World War II.

Portrayed as a war to end fascism, WWII was in reality a result of the unresolved power struggle that had initiated the First World War. While Western Europe and Asia were devastated by the war, once again the US emerged unscathed and strengthened. At the same time, the Union of Soviet Socialist Republics (USSR) extended its control over Eastern Europe. As a result, the world was divided into two major blocs: the capitalist West and the communist East.

WWII & UN 'Decolonization'

As a result of the destruction of World War II, former European empires were unable to maintain direct control of their colonies as new anti-colonial movements emerged in Africa and Asia. Many of these struggles were the result of power struggles between the US and the USSR during the 'Cold War.' Decolonization was also promoted by the US as a means of further undermining W. European states and extending US imperialism.

The result was an explosion of anti-colonial insurgencies in Africa and Asia, wars of liberation that succeeded in forcing out European powers. Some of the hardest fought battles were those of Algeria, Vietnam, Mozambique, Kenya, and Rhodesia/Zimbabwe. This period of anti-colonial war extended from the 1950s into the mid-70s.

As a part of this, the United Nations was used to assimilate these new independent nation-states into the global system (based on rhetoric of peace and human rights). The UN, it should be noted, was itself set up by the US in the aftermath of WWII to impose just such a system. The US also provided funding and built the UN headquarters in New York City.

At the same time, the US also established the International Monetary Fund and World Bank, Along with the UN, these groups were used to reconstruct the global system after the war. The main beneficiaries were US corporations. The post-WWII period is often remembered as a US 'Golden Age' of US prosperity and stability.

THIS GLOBAL UPRISING WAS INSPIRED BY THE FIERCE RESISTANCE OF THE VIETNAMESE PEOPLE FIGHTING U.S. INVASION + OCCUPATION (1963-73)...

Since its establishment, the UN has served as a convenient cover for Western imperialism, giving legal and moral sanction to ongoing colonial invasions (including Korea and Vietnam Wars, the Congo, Iraq in 1991, Somalia, Haiti, Afghanistan, etc.). These are termed 'peacekeeping' or 'humanitarian' missions, although their primary purpose is to maintain or re-impose Western control.

After gaining independence, many colonies remained dependent on the western economic system (a legacy of colonialism, including large-scale export of agriculture, petroleum, and minerals). Decolonization, in fact, served to open up these former colonies for penetration by US-based corporations. Others became dependent on the USSR for industrialization and modernization of military forces. Overall, decolonization did not fundamentally alter the imperialist relationship between the Western nations and Africa and Asia.

Vietnam & US Domestic Rebellion

Vietnam was first colonized by the French in the mid-1800s. After WWII, anti-colonial resistance to the French in Vietnam increased. By 1954, Vietnamese guerrillas had defeated the French during the Battle of Dien Bien Phu. The US, which had begun aiding the French in the early '50s, began increasing its involvement. At the same time, the UN partitioned the country in half. The north was controlled by the Vietnamese communists, while the south remained a puppet regime for the US.

As Vietnamese resistance to foreign occupation continued in the south, more US forces became involved. At first, a handful of Special Forces were sent in to train and organize anti-guerrilla forces. By 1968, over 500,000 US troops were in Vietnam.

At this time, resistance movements had emerged around the world, inspired by the anti-colonial-wars of the time. One of the most influential was that of the Vietnamese, which created a climate of insurgency and rebellion. Inside the US itself movements such as the Black Panthers, Puerto Ricans, Chicanos, Indigenous, student, women's, gay and lesbian, and others began. These were also part of a broader, multinational anti-war movement.

During the same period, large-scale urban riots occurred, primarily by blacks, during which National Guard troops were deployed to maintain order. Many civilians were killed, and tens of millions of dollars in damage inflicted. In response to these increasing revolts and organized resistance, the FBI intensified its domestic counter-insurgency campaign (the Counter-Intelligence Program, COINTEL-PRO). Many movement organizers were killed, imprisoned, assaulted, etc.

Meanwhile, US forces in Vietnam became increasingly demoralized. Many citizens and soldiers alike began to question the purpose of the war. Insubordination and drug use became common among US combat forces, with entire units refusing to fight, or avoiding combat. Commanders became the target of 'fraggings' (a term that arose from the practice of using grenades to kill or wound commanders seen as dangerous or reckless).

Many combat veterans returned from the war, traumatized but also angered and disillusioned with their country. Some became involved in resistance movements and added their combat skills and experience to these. By the early '70s, in the face of lethal repression, urban guerrilla groups had formed in the US, including the Black Liberation Movement, Puerto Rican independentistas, and white anti-imperialists. These and many other groups carried out bombings, arsons, and armed attacks against police, throughout the country. In 1973, the 71-day siege at Wounded Knee, S. Dakota, occurred.

Faced with growing internal revolts, and mounting casualties (as many as 50,000 dead), from an increasingly unpopular war, the US had retreated from Vietnam by 1974. This domestic unrest, and the refusal by large segments of the population to support wars of this nature, has been termed the 'Vietnam Syndrome.'

The New World Order

The term 'New World Order' was first used by US President George Bush Sr. in 1990, as the US prepared to invade Iraq. This 'new order' was the result of the collapse of the USSR and, with it, the entire communist East Bloc. With the demise of the USSR, the US emerged as the dominant global power, the strongest economic and military force in the world.

With the threat of Soviet reprisal now removed, the US invaded Iraq in 1991, severely damaging Iraq's military and infrastructure. As many as 200,000 civilians are estimated to have been killed. A UN embargo was then placed on Iraq, limiting imports of food, medical supplies, and equipment necessary to rebuild. The UN also set quotas for Iraqi oil production, continuing the export of oil in exchange for food imports. US/UN forces also established bases around Iraq and carried out systematic bombing campaigns, including cruise missile strikes.

War for Oil & Global Domination

The US/UN siege of Iraq continued until 2003, when the US again invaded. The invasion of Iraq is part of a larger US strategy to take direct control of Mid-East oil, part of its plans for global domination. One official described it as a "stupendous source of strategic power and one of the greatest prizes in world history."

US involvement in the Mid-East increased after WWII, following the retreat of primarily British and French forces during the period of 'decolonization.' Corporations such as Exxon, Gulf Oil, Standard, and Texaco moved in. Israel (established in 1948 through Zionist war and terror) is a vital part of overall US control, serving as a US fortress and a source of instability in the region. Other Arab countries, such as Saudi Arabia, Jordan, and Egypt, are the largest recipients of US military and economic aid in the world.

Planning and preparation for direct US invasion of the Mid-East began in 1973, during the 'Oil Crisis' when Mid-East Arab nations cut oil supplies in protest of US-Israeli military aggression in the region. Following this, US military forces began extensive training and preparation for desert warfare.

In 1979, an Islamic Revolution in Iran overthrew the US-backed dictator (the Shah) and cut off a valuable source of cheap oil to the US. Demonstrators stormed the US embassy and took over 50 Americans hostage. The hostage ordeal was a humiliating and frustrating event for the US, which appeared impotent and helpless. In 1980, an attempted hostage rescue ended in disaster when US special forces crashed in the Iranian desert (the hostages were released in 1981).

In 1980, as Saddam Hussein gained power, the US used Iraq to attack Iran. The war lasted until 1988, with two million Iraqi and Iranian dead. Western nations, such as the US, Britain, and France, supplied arms to both sides, despite widespread atrocities and the use of chemical weapons during the conflict. As the war ended, the US Navy 'accidentally' shot down a civilian Iranian jet, killing nearly 300 passengers.

The Iranian Revolution was a great concern to the US, and it quickly moved to expand its control. In 1980, the US established a Rapid Deployment Force, prepared for short-notice invasion of the Middle-East. From 1980–83, new bases were built in Saudi Arabia and Oman. In 1981, Bright Star annual training exercises began in the Mid-East.

In 1982, nearly 250 US Marines were killed in a truck bombing in Beirut, Lebanon. The marines were part of a UN 'peacekeeping' mission to maintain control of Lebanon. After the withdrawal of Soviet forces from Afghanistan in 1989, the US became an increasing target for Islamic militant groups. It is now common knowledge that these groups had initially been trained, funded, and armed by the CIA during the Soviet war in Afghanistan (including al-Qaeda).

As a result of the September 11, 2001 attacks against the Twin Towers and the Pentagon, the US declared its 'War on Terror,' beginning with the military invasion and (ongoing) occupation of Afghanistan. In 2003, the US invaded Iraq, using the pretext of weapons of mass destruction (none were found). Three years later (2006) the US occupation of Afghanistan and Iraq continues (with Iran in between, part of the 'axis of evil' targeted by Bush, including Syria and N. Korea).

In Iraq, the US faces an organized and expanding insurgency, while in the US itself a growing number of people are becoming increasingly disillusioned with the war altogether. From its origins in ancient Mesopotamia, the Western imperial system has now gone full circle, invading and destroying Iraq, the homeland of Babylon itself.

2. Impacts of Colonialism

A. European Colonial Society

When considering the overall impact of colonialism on Indigenous peoples, one aspect that cannot be neglected is the form of society imposed by Europeans. Although arising from the history of Western Civilization, colonial societies are defined by their oppressive relationship with Indigenous populations and therefore have specific characteristics.

Settler-Nations

Settler-nations are colonies in which large numbers of European immigrants relocated and eventually set up new nation-states. Canada, the US, New Zealand, and Australia are examples of settler-nations.

Settlerism is, by its very nature, parasitical, taking and exploiting not only land and resources, but also Indigenous culture and knowledge. The lifestyle of most settlers is one of extreme material wealth, luxury, and privilege, characterized by an emphasis on entertainment and recreation.

Although the term settlerism is used, Roman citizens lived a similar life of extreme wealth and luxury. This way of life is essentially imperialist, and this can be seen when considering modem settler-nations in relation to the global system. They are imperialist nations and the lifestyle of most their citizens—especially Europeans—reflects this.

Imperialism

Imperialism is a regional or global system arising from colonialism and based on political, economic, and military control. For example, Rome was an empire, and the word imperium (meaning command) is itself Latin in origin. At its height, Rome controlled much of Western Europe, North Africa, and the Middle-East, its influence extending into eastern Asia. Within this vast empire, Rome stood as the capital, with wealth, resources, and slaves flowing to it from all around ("all roads lead to Rome"). Rome controlled all trade, governance, and military forces within the empire.

Today, the US dominates a global imperialist system, which also includes Canada and much of Western Europe (the G7: Canada, France, Germany, Italy, Japan, US, and UK). Combined, these countries enforce political, economic, and military control over the rest of the world. In this, international organizations such as the United Nations, International Monetary Fund, and World Bank play a vital role.

In Africa, Asia, and Central and South America, the majority of the world's population are forced to work in factories, mines, oil fields, and farms, extracting resources and producing goods and food, primarily for export to the imperialist nations.

At the same time, these impoverished countries serve as huge markets for Western corporations selling weapons, pesticides, industrial technology and machinery, etc. Control is reinforced through bank loans with strict conditions for repayment. Through this, the World Bank and IMF, for example, can dictate that a national government cut social services, open up certain industries to foreign investment, etc.

Apartheid

Apartheid means 'apartness' and comes from South Africa, once ruled by a white-minority settler regime. Although racial segregation and white supremacy had already been established for several decades, in 1948 it was made an official state policy. Apartheid imposed strict separation of races, including whites, Africans, Colored (mixed racial groups), and Asians (mainly of East Indian origin). In devising this apartheid system, S. African officials sent delegations to Canada and the US to study N. American models.

Until apartheid was abolished in 1993, Africans could not vote or own land, their movement was controlled by pass laws, as was their residence and place of employment. Although at first living on reserves, ten large Bantustans were established for Africans based on their tribal nations, beginning in the 1960s. They covered some 14% of the national territory (Africans being some 75% of the population). Also known as 'homelands,' these were declared self-governing and even independent, although no one but S. Africa recognized them. In the self-governing Bantustans, Africans lost what limited rights they formerly had as 'S. Africans.'

The Bantustans were usually on poor quality land from which neither food nor industry could be created. Many Africans relocated to urban areas where they lived in townships on the outskirts of the city, serving as a desperate and highly exploited work force. The remainder of the country went to the whites and mineral corporations (worked by poor Africans). Apartheid ended as a result of a long resistance campaign by Africans to overthrow the white regime.

In the southern US, the descendants of African slaves were also subjected to an apartheid system until the 1950s, when the civil rights movement arose and dismantled it (some refer to blacks as an internal colony of the US). Although racism has been officially denounced, and apartheid legal systems repealed, blacks in the US continue to suffer from racism as an oppressed peoples.

In both Canada and the US, formal apartheid still exists for Indigenous people. In Canada, this includes the Indian Act and the Department of Indian Affairs, which comprise a set of separate laws, legal status, and political systems for Indigenous peoples. Native peoples continue to live on reserves, usually on land unable to sustain the population through traditional methods. As a result, many resort to some form of resource exploitation in collaboration with corporations.

One result of apartheid is an overall lack of knowledge by non-Indigenous people as to the social conditions under which colonized peoples live. It is, in fact, the establishment of two separate worlds, or social realities (i.e., colonizer and colonized).

White Supremacy

In European settler-nations, racism is more accurately termed white supremacy, the mistaken and arrogant belief that all things good were made by Europeans, who are inherently smarter, better, more beautiful, and stronger than all others. This message is constantly repeated through a variety of means, from official history to media coverage, from entertainment to the justice system. It is so widespread and pervasive that it is accepted as an unspoken truth.

White supremacy is a foundation upon which European civilization is based, a belief that enabled European colonizers to engage in invasion, genocide, and, slavery. Christianity was an important method by which white supremacy was maintained. During the invasion of the Americas, it was the Christian church which provided both moral and legal authority to colonial forces. It was, in fact, the duty of Christians to conquer and possess the lands of "infidels" and pagans (non-Christians, which at that time really meant non-Europeans).

For centuries, European settlers and colonial authorities proudly proclaimed white supremacy as righteous and as God's will (just as colonialism was). In the 1920s, the Ku Klux Klan had several million members in the US, while tens of thousands joined on the Canadian prairies. Many government officials, mayors, professionals, and police were members of the KKK.

After World War II and the defeat of Nazi Germany, racism as an official government policy became less popular. It was even denounced after the black civil rights struggles of the 1950s. After the rebellions of the 1960s and '70s, governments even began to claim that they were anti-racist! Despite this, white supremacy remains firmly entrenched in Western society, and non-European peoples remain racially oppressed and marginalized (despite some concessions).

This is because white supremacy is deeply rooted in European history, culture, and philosophy. It is not a problem of a few people with bad attitudes, but is instead a systemic problem maintained through social institutions, beliefs, and traditions. White supremacy is similar to patriarchy, a taboo subject that is rarely discussed because it strikes at the very core of Western society.

Patriarchy

Patriarchy means 'male rule.' Most ancient civilizations began as patriarchal systems, in which adult males had all political, economic, and social power. Women had no more rights than slaves. In ancient Rome, the adult male of a household could kill his wife or sell her and their children into slavery.

During the European Middle-Ages, millions of women were killed during the Holy Inquisition. They were accused of being witches and pagans. In Western Europe and North America, white women could not vote until the twentieth century.

Most Indigenous societies in N. America, on the other hand, were matrilineal prior to colonization. Women had far more political, economic, and social power. In many, lines of descent passed through the mother. Abuse was limited by the presence of family and community in daily life. Nor were women considered the property of men.

In the 1960s, a women's liberation movement emerged in N. America, along with other social movements. The women's movement challenged patriarchy both in society and within the movements themselves. By the 1970s, some of this analysis had been absorbed. Eventually, some women were promoted in government and business, and the idea that women were inferior to men became less popular.

Despite this, patriarchy has remained the basis of Western Civilization for over 2,000 years now. Like white supremacy, patriarchy is no longer an official policy and yet it remains firmly entrenched as a way of life. Overall, men continue to enjoy greater economic, political, and social power, even though women now have equal legal rights. As well, women continue to be the target of male violence and abuse. In Canada, there are over 500 dead/missing Aboriginal women; in Mexico, Guatemala, etc. hundreds of women have been found raped and murdered. Prostitution is also a form of male domination and violence.

When Europeans colonized the N. American Indigenous nations, they had to impose patriarchy through laws and policies. In Canada, band councils had to be comprised of 12 male members. Under the Indian Act, Native women who married non-Natives lost their legal rights and status as Natives. They could not get housing or enjoy other benefits provided by the state. Along with assimilation to European ways of life, these measures served to transfer political and economic power to Native males, who today comprise the majority of band chiefs and councilors, as well as businessmen, professionals, etc.

Neo-Colonialism

Neocolonialism means a 'new colonialism.' It involves the use of state-funded Native government, business, and organizations to indirectly control Indigenous people. In Canada, for example, the government spends billions of dollars annually to maintain a system of neocolonialism, funding band councils, Aboriginal political organizations, as well as social programs, arts and culture, etc.

> "Neocolonialism involves the use of Natives to control their own people. In general, it means giving some of the benefits of the dominant society to a small, privileged minority, in return for their help in making sure the majority cause trouble ... the image of successful Aboriginals in government [helps] create the myth that all Natives have a place in the dominant society.
>
> The change from colonialism to neocolonialism is a change only in how the state controls the colonized people. Colonialism is a system in which the colonized people have no control over their lives economically, socially, politically, or culturally. The power to make decisions in these important areas of daily life are almost totally in the hands of others, either the state or corporations and business ... the state is willing to share some of the wealth of a racist system with a few Natives in return for a more effective method of controlling the majority.
>
> The most threatening and effective form of neocolonialism devised by the state has been its efforts to intervene and control popular Native organizations which had been previously independent. They began with core grants to help the associations organize; then the elected leaders of the organizations got larger and larger salaries—making them dependent on the state just as the Native bureaucrats in government were. As the years went by more money was provided to organizations—money for housing, economic development, and service programs, etc.
>
> The most important effect of government funding, or state intervention, is that the state, by manipulating grants, can determine to a large extent what strategy the organizations will use. It is no coincidence that when organizations were independent of government money in the mid-sixties, they followed a militant strategy which confronted government. Now, after twenty years of grants, they are following a strategy that requires subservience to the state." [5]

Pyramid of Power

The structure of European society is, by its very nature, a system of oppression and control. It is organized in a pyramid structure, with a small elite at the top and the masses of people at the bottom. Indigenous peoples comprise the bottom layer of this pyramid, and it can be said that it is literally built on top of them (i.e., in Mexico City, the Presidential Palace is built on top of an Aztec temple).

The pyramid structure is one that reappears throughout civilization, reflecting the oppressive relationships and patterns upon which such societies are based. The patriarchal family unit, the government, the church, the army, the corporation; all share similar organizations of hierarchy, central authority, and control.

In society, one's position in this pyramid is determined by gender, race, and economic class; the global elite are overwhelmingly rich white males. They are the descendants of the European nobility and aristocracy established after the collapse of the Roman Empire. Their rise to global power as a class began with the 1492 invasion of the Americas. This class system is maintained in the interests of the rulers and is protected by national police and military forces (including courts and prisons).

Globally, the pyramid of power exists in the relations between nations; the predominantly Euro-American Group of Seven (the G7: Canada, France, Germany, Italy, Japan, United Kingdom, and United States. With Russia it is the G8) control the international political and economic system. They are the top of the pyramid. Most of the world's countries are poor and impoverished, forming the bottom layers of the pyramid.

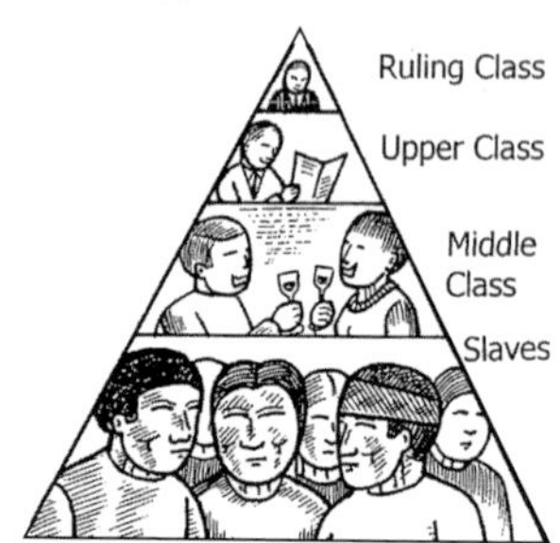

B. Sociological Impact

The sociological impacts of colonialism, those that affect the entire society/nation, include:

Genocide

According to Article 2 of the UN 1948 Convention on Genocide,

> "Genocide means any of the following acts committed with intent to destroy, in whole or in part, a national, ethical, racial, or religious group as such:
> (a) Killing members of the group;
> (b) Causing serious bodily or mental harm to members of the group;
> (c) Deliberately inflicting on the group conditions of life calculated to bring about its physical destruction in whole or in part;
> (d) Imposing measures intended to prevent births within the group;
> (e) Forcibly transferring children of the group to another group."

According to Article 4,

> "Persons committing genocide or any of the other acts enumerated ... shall be punished, whether they are constitutionally responsible rulers, public officials, or private individuals."

Acts of genocide (as defined by the UN) are common to virtually all colonial invasions and occupations of Indigenous or sovereign territories. This can be seen in the patterns of military conflict (massacres, biological warfare, scorched earth), assimilation (residential schools), sterilization of Indigenous women, and fostering out of Native children.

Clearly, genocide has been, and is now being, committed against Indigenous peoples, including those in Canada and the USA. The most visible effects of this are the high rates of suicide, alcohol, and drug addiction, mental disorders, poverty, internalized violence, and imprisonment, among colonized Indigenous populations. These are a direct result of colonial oppression arising from political and economic policies.

That member states of the UN can so blatantly violate its conventions and international law, and yet retain their status and suffer no consequences, is due to the imperialist structure of the global system itself.

Loss of Territory & Sovereignty

Sovereignty is defined as a "supreme authority within a territory," free from external control and dependence. This definition is used to describe the international relations between nations, which are seen as sovereign entities having total independence and control over a certain territory.

A nation is often defined as a group sharing a common ethnicity, language, culture, history, and territory. Today, terms such as country and state are also used to describe nations, but a nation is more correctly defined as a group of people, not a nation-state (which often contains many nations within its borders).

Along with internal governance and independence, a primary aspect of sovereignty is the ability to control who enters the territory. This control is necessary as a means of self-defense and security for the nation itself. In the face of armed aggression, this defense can only be carried out through some form of military force.

Although the term has its origins in European political terms, it is generally acknowledged that Indigenous nations had all the attributes of sovereignty prior to colonization. One of the earliest recognitions of this by colonial powers was in the process of treaty-making (treaties being international agreements between two or more nations).

Today, terms such as First Nations and self-government imply some form of sovereignty. In the US, 'tribal sovereignty' is often used to describe the power of tribal governments on reserves. None of these terms, however, or the policies from which they are derived, have any basis in the actual exercise of sovereign power by Indigenous peoples.

Assimilation

Following the period of military invasion, and once an occupation has been established, surviving Indigenous populations are then subjected to policies of assimilation. This is only possible after their military defeat.

In many colonial situations, a first step in assimilation is to contain the surviving Indigenous populations in a reservation system (i.e., the South African Bantustan, or reserves in North America). This is necessary to 'open up territory for settlement and exploitation, while providing a basis for systematic indoctrination into European society.

In many colonial situations, it is the Church and missionaries who begin the process of indoctrination. A common tactic is the forcible removal of children from their families and communities, and their placement in Church-run schools (i.e., missions, Residential or Industrial Schools, etc.).

A primary target for indoctrination are chiefs or high-ranking families; once converted, they serve as useful collaborators, able to influence their communities and to mobilize resources.

Along with education, all aspects of the colonial society are utilized in a process of assimilation, i.e., political, economic, ideological, cultural, etc. The goal is to eradicate as much of the Indigenous culture and philosophy as possible, and to replace these with those of European civilization.

Assimilation is a final phase in colonization. What distinguishes it from the previous stages of recon, invasion and occupation is its primarily psychological aspects. It is not a military attack against a village, but a psychological attack against the mind and belief system of a people.

As a result of assimilation polices in Canada and the US, generations of Indigenous people have become increasingly integrated into European society. Since the 1970s, more Indigenous people have become professionals (lawyers, doctors, businessmen, etc.), and more have passed through universities or colleges. As a result of this increased training, band councils now self-administer government policies and are more involved in business and resource exploitation that at any time in the past.

While this is promoted as progress (and even 'decolonizaton'), it is actually greater assimilation into the colonial society. Overall, today's generations of Indigenous people show a greater degree of assimilation than previous ones. Some factors that account for this are the effects of residential schools, decline of culture, reduced reliance on traditional ways of life, greater dependence on the colonial system, increased urbanization, and ongoing exposure to Western culture through modem communications (TV, movies, music, printed material, etc.).

C. Individual Impact

Post-Traumatic Stress Disorder

Overall, Indigenous peoples can be said to suffer from Post-Traumatic Stress Disorder (PTSD), both as a group and as individuals.

"Traumatic events involve death or the threat of death; injury or the threat of injury. It is not just the events themselves but the experience of those events that makes them traumatic." [6]

Some examples of traumatic events include war, natural disasters, physical or sexual assault, robbery, kidnapping, etc. PTSD is shared by many survivors of trauma, such as combat veterans, victims of torture, sexual or physical abuse, etc.

Colonization and genocide are examples of collective trauma that impact on a people's culture and identity, On an individual level, Indigenous people continue to suffer traumatic events, including widespread sexual abuse, domestic violence, suicides, police violence, imprisonment, etc.

Symptoms of PTSD include depression, paranoia, panic and anxiety attacks, sleeping disorders, etc. Depression is characterized by strong feelings of worthlessness, hopelessness, fatigue, irritability, irregular sleep, and an inability to feel pleasure. As well, many trauma survivors experience feelings of shame and guilt for being an 'unwilling accomplice' when forced to participate, endure, and/or witness traumatic events. Overall,

"one of the most profound losses trauma survivors experience is the loss of a positive self-image." [7]

Victims of traumatic events may have difficulty forming relationships based on trust, especially if the abuser was a family member. Women assaulted by men may have difficulty trusting any men. Survivors of trauma may have eating disorders and a pre-occupation with body image. Alcohol, drugs, and/or sexual promiscuity are some coping methods commonly used.

PTSD is just .one of several possible reactions to trauma, including somatization (physical illness resulting from anger, pain, etc.), and disassociation (mentally blocking or 'forgetting' the traumatic event). In addition, trauma survivors must deal with disbelief, rejection, and even hostility from family or community members (i.e., blame the victim). If not rejected, survivors of trauma are expected to keep silent about their experiences, especially allegations of abuse involving community/family members.

Many, but not all, of the following individual impacts of colonialism can be traced to some form of Post-Traumatic Stress Disorder.

Individualism, Identity, & Inferiority Complex

With the breakdown of Indigenous society, nations and families also become broken and fragmented. European values of individualism and self-interest (essentially capitalist) increasingly replace traditional Indigenous values of community and collectivity. In fact, the entire fabric of Indigenous culture and society is torn apart:

"Colonial domination, because it is total and tends to over-simplify, very soon manages to disrupt in spectacular fashion the cultural life of a conquered people. This cultural obliteration is made possible by the negation of national reality [loss of sovereignty], by new legal relations introduced by the occupying power [i.e., the Indian Act], by the banishment of the natives and their customs to outlying districts by colonial society [reservations], by expropriation [theft], and by the systematic enslaving of men and women." [8]

Alongside the breakdown of family and community is the loss of culture. When confronted with systematic assimilation into European culture, the result is a loss of identity and feelings of inferiority:

"Every effort is made to bring the colonized person to admit the inferiority of his culture which has been transformed into instinctive patterns of behavior, to recognize the unreality of his 'nation,' and, in the last extreme, 'the confused and imperfect character of his own biological structure." [9]

Internalized Violence

As a result of the physical and psychological effects of colonialism, patterns of internalized violence and crime are established. The colonized tend to attack and victimize their own. These attacks range from violent assaults and murder, to petty theft and vandalism. These patterns are common among colonized peoples (i.e., a leading cause of death among young black males in the US are young black males).

One reason the colonized prey on one another is that of proximity; one's family and community are right there, while the oppressor lives in another world. The physical realities of colonialism, the establishment of reserves and urban ghettos, along with an apartheid system, separates the colonized and the settler communities.

More than the physical proximity of one's own people, however, is the psychological impact of colonization. Not only is the settler community physically distant, it is also foreign and threatening. It is well guarded. The penalties for violating the settler's person or property are more severe than for violating one's own.

Many forms of internalized violence arise from European colonial society itself. Widespread sexual abuse among Indigenous peoples in Canada and the US, for example, was first introduced through the Residential School system. Children who experienced abuse by school staff (priests and nuns) returned to their communities and began abusing their own family members, resulting in intergenerational patterns of abuse that continue to this day.

Alcohol, Drugs, & Suicide

Arising from the oppressive social conditions that colonialism creates (i.e., poverty, loss of identity, feelings of inferiority, etc.), Indigenous peoples suffer from high rates of alcoholism, drug addiction, and suicide, in both rural and urban communities. These are common methods of temporarily escaping the oppressive routines of day-to-day life, of suppressing trauma or tension, or ending feelings of despair and hopelessness (suicide).

High rates of violent death and imprisonment among Indigenous peoples are both attributed to alcohol and drug abuse. In Saskatchewan, a study found that alcohol was involved in 45% of suicides among those 15–34 years of age; 92% of fatal motor vehicle accidents, over 38% of homicides, and over half the deaths by fire and drowning. [10]

Rates of suicide among Indigenous peoples in Canada are estimated at 33 per 100,000 population, compared to the national average of 13 per 100,000. Among Indigenous youth 15–24 years of age the rate is 114 per 100,000, compared to 26 per 100,000 among the general population. [11]

According to the Royal Commission on Aboriginal Peoples, a multi-million-dollar investigation into the conditions of Indigenous peoples in Canada:

> "We have concluded that suicide is one of a group of symptoms ranging from truancy and law breaking to alcohol and drug abuse and family violence, that are in large part interchangeable as expressions of the burden of loss, grief, and anger experienced by Aboriginal people in Canadian society." [12]

Health

A primary argument in favor of colonialism was that it brought the benefits of civilization to Indigenous peoples, greatly raising their standard of living. The genocidal practices of colonialism easily dismiss such claims, and yet they persist, based largely on an incorrect view that Native peoples barely managed to survive, scraping out a meager existence and victim to all sorts of injury, disease and death. In fact,

> "anthropologists have long recognized that undisturbed tribal peoples are often in excellent physical condition." [13]

Colonialism, far from raising the living standard of Indigenous peoples, instead plunges them into economic impoverishment, disease, and rapidly deteriorating health conditions. Drastic changes in diet resulting from limited access (or destruction) of traditional food sources, and dependence on European food items, has caused extensive health problems for Indigenous peoples.

After exposure to white flour, sugar, milk, etc., Indigenous peoples began to suffer rapid tooth decay and mouth diseases. After generations of dependence on European food products, Indigenous and other colonized peoples today suffer from high rates of diabetes, obesity, high blood pressure, and heart problems. Indigenous peoples also suffer the highest rates of diseases such as tuberculosis, pneumonia, cancer, AIDS, hepatitis, etc.

In Canada, the leading cause of death among Indigenous peoples is categorized as resulting from injury, primarily motor vehicle accidents. As the Saskatchewan example shows, most of these deaths result from alcohol abuse. Death by injury accounts for more than one-quarter of all deaths among Indigenous peoples, with a rate of 148 per 100,000, compared to the national average of 46.9 per 100,000.[14]

The next most common cause of death are diseases of the circulatory system (i.e., heart attacks). This accounts for about two thirds of all deaths among Indigenous peoples. Rates of tuberculosis are approximately 47 per 100,000, compared to the national average of 7.2 per 100,0000. Diabetes has been described as one of the most prevalent chronic health problems among Indigenous peoples, causing numerous side effects including heart and circulatory disease, blindness, kidney and nerve damage, and obesity.

Malnutrition, especially from protein deficiency, has become a major problem for Indigenous peoples around the world. In addition, poor health is compounded by conditions of urbanization and poverty, poor sanitation and housing, stress and trauma, as well as high levels of alcohol and drug abuse.

Exposure to industrial pollutants, including chemicals such as mercury, cyanide, and uranium, as well as fertilizers, has also had disastrous impacts on the health of Indigenous peoples. In areas of extensive mining, industrial production (i.e., pulp mills, oil and gas wells), or agriculture, for example, Indigenous peoples suffer high rates of cancers, birth defects, stillborns, etc. This is a result of their proximity to natural resource exploitation, exposure to contamination through water and the food chain (hunting and fishing), and lack of access to proper healthcare.

If You Stand for Nothing...
You'll Fall for Anything

3. Decolonization

Decolonization is the ending of colonialism and the liberation of the colonized. This requires the dismantling of the colonial government and its entire social system upon which control and exploitation are based. Decolonization, then, is a revolutionary struggle aimed at transforming the entire social system and reestablishing the sovereignty of tribal peoples. In political terms, this means a radical de-centralization of national power (i.e., the dismantling of the nation-state) and the establishment of local autonomy (community and region, traditionally the village and tribal nation).

Any discussion of decolonization that does not take into consideration the destruction of the colonial system and the liberation of land and people can only lead to greater assimilation and control. The demand for greater political and economic power by chiefs and councils, although presented as a form of decolonization (i.e., "self-government"), only serves to assimilate Indigenous peoples further into the colonial system.

Just as colonialism enters and passes through various phases, beginning first with recon missions and then the application of military force, so too does decolonization. It would be a mistake to conceive of decolonization as a single event. Instead, it is a process that begins with individuals and small groups. The primary focus in the first phase of decolonization is on disengaging from the colonial system and re-learning one's history, culture, etc. This phase places a heavy emphasis on rejecting European society and embracing all that is Indigenous as good and positive.

Some common steps in this phase include returning to one's community, re-establishing family relations, re-learning culture (inc. art, language, songs, ceremonies, hunting, fishing, etc.). This not only counters the destructive effects of colonialism, but also instills in the Indigenous person a greater respect and appreciation for their own culture and way of life. In many ways it is a struggle for identity and purpose. While this is a crucial first step in any decolonization process, without the infusion of radical and revolutionary analysis, however, the focus on cultural identity in and of itself does not necessarily lead to anti-colonial consciousness. In fact, this focus on 'culture' alone can easily lead to conservative and even pro-colonial sentiments.

Empowered by their renewed cultural identity, one frequently drawn from a mish-mash of tradition, Christianity, and New Age spirituality, the 'decolonized' individual begins to believe that nothing has really changed, that they are in fact living the way of life of their ancestors, albeit in a very different world. They rationalize their relationship to the colonial world as one of supreme adaptation, while at the same time pitting their 'personal power' against the power of the system. In this fantasy world, inspired in no small part by the system itself, the spiritualists believe that through prayer and ceremonies alone everything will work out just fine.

The influence of New Age spirituality and Christianity among Indigenous peoples is not hard to find. Arising from the European slave society, these religious ideologies teach obedience and submission to authority, including the principle of non-violence in the face of violent repression (a tactic of the weak elevated to a moral or spiritual principle). When confronted with the overwhelming oppression and destruction perpetuated by the system, the New Age spiritualists retreat into their fantasy world, where all conflict is resolved and there are no distinctions made between oppressor and oppressed.

From the very outset, then, there is a possibility for co-optation of the decolonization efforts made by individuals. This co-optation is, in part, engineered by the colonial system in the form of funding and publicity for cultural programs, educational materials, etc., and official recognition of high profile collaborators as 'spiritual leaders.'

After the collaborators involved in the Indian Act band council system, the spiritualists are often the most vocal opponents of anti-colonial resistance at the community level. They are influenced and manipulated by the political leaders to fulfill this role, and at the same time act according to their own logic, which is essentially conservative. While they actively oppose organized resistance, they are silent in regards to colonial oppression and advocate maintaining the status quo (conservatism). How is it possible that Indigenous culture, the basis for decolonization, can be so easily co-opted?

Culture & The Struggle for Liberation: Fanon

Frantz Fanon (1925–1961) was an African intellectual and psychologist, involved in Algeria's war for independence in the 1950s. His analysis of colonialism and its effects on colonized peoples have had a profound impact on anti-colonial resistance movements around the world. For Fanon, culture was a vital part of this resistance.

As noted, Indigenous culture is a primary means of decolonization. It is both a link to our ancestral past and to another way of thinking, of seeing the world. It is the essence of our identity as Indigenous peoples and a vital part of challenging colonial ideology. Yet, as Fanon and others have observed, this culture, when not totally erased, is warped and distorted by the colonial society:

> "The colonial situation calls a halt to national culture in almost every field ... By the time a century or two has passed there comes about a veritable emaciation [starvation, or thinning out] of the stock of national culture. It becomes a set of automatic habits, some traditions of dress and a few broken-down institutions. Little

movement can be discerned in such remnants of culture; there is no real creativity and no overflowing life. The poverty of the people, national oppression and the inhibition of culture are one and the same thing. After a century of colonial domination we find a culture, which is rigid in the extreme, or rather, what we find are the dregs [left-overs] of culture, its mineral strata. The withering away of the reality of the nation and the death-pangs of the national culture are linked to each other in mutual dependences." [15]

Here, Fanon describes the effects of colonization on culture. Its natural development, the incorporation of new experiences, etc., are more or less stopped at the point of contact. In many ways, it is the colonial power (or anthropologists, etc.) that comes to define what is traditional and what is not. The colonized, in an effort to retain traditional culture, at the same time also stop its development and impose strict limits on interpretation in an effort to retain an imagined 'purity.' While superficial aspects of culture remain, the essence and vitality of the culture itself are lost or minimized (think pow-wow, or consider the influences of Christianity and New Age 'spiritualism' on Indigenous culture).

An important point Fanon makes is that a people's culture is directly linked to the physical world: the colonial occupation of a nation's territory is total, affecting everything and everyone. According to Fanon, it is the anti-colonial resistance that revitalizes the culture of the colonized:

> "It is the fight for national existence which sets culture moving and opens to it the doors of creation ... We believe that the organized undertaking by a colonized people to re-establish the sovereignty of that nation constitutes the most complete and obvious cultural manifestation that exists. It is not alone the success of the struggle, which afterwards gives validity and vigor to culture; culture is not put into cold storage during the conflict. The struggle itself in its development and in its internal progression sends culture along different paths and traces out entirely new ones for it. The struggle for freedom does not give back to the national culture its former value and shapes; this struggle which aims at a fundamentally different set of relations between [people] cannot leave intact either the form or the content of the people's culture. After the conflict, there is not only the disappearance of colonialism but also the disappearance of the colonized..." [16]

In the process of struggle, the culture of the colonized is transformed into the means of resistance, incorporating new forms of expression and interpreting traditional culture in order to make it relevant to the new colonial reality (and present generations). Fanon uses the examples of literature, oral tradition, crafts, dances, songs and ceremonies, which develop alongside the anti-colonial resistance:

> "While at the beginning the native intellectual used to produce his work to be read exclusively by the oppressor, whether with the intention of charming him or of denouncing him ... now the native writer takes on the habit of addressing his [her] own people ... This may be properly called a literature of combat, in the sense that it calls on the whole people to fight for their existence as a nation...
>
> On another level, the oral tradition—stories, epics and songs of the people—which formerly were filed away as set pieces are now beginning to change. The storytellers who used to relate inert episodes now bring them alive and introduce into them modifications which are increasingly fundamental. There is a tendency to bring conflicts up to date and to modernize the kinds of struggle which the stories evoke, together with the names of heroes and the types of weapons. The method of allusion is more and more widely used. The formula 'This all happened along ago' is substituted by that of 'What we are going to speak of happened somewhere else, but it might well have happened here today, and it might happen tomorrow.' The example of Algeria is significant in this context. From 1952–53 on, the storytellers, who were before that time stereotyped and tedious to listen to, completely overturned their traditional methods of storytelling and the contents of their tales ... Colonialism made no mistake when from 1955 on it proceeded to arrest these storytellers systematically.
>
> The contact of the people with the new movement gives rise to a new rhythm of life ... Well before the political or fighting phase of the national movement an attentive spectator can thus feel and see the manifestation of new vigor and feel the approaching conflict. He will note unusual forms of expression and themes which are fresh and imbued with a power which is no longer that of invocation but rather of the assembling of the people, a summoning together for a precise purpose. Everything works together to awaken the native's sensibility, and to make unreal and unacceptable the contemplative attitude, or the acceptance of defeat ... The conditions necessary for the inevitable conflict are brought together." [17]

If we accept Fanon's analysis as correct, culture is indeed the basis of decolonization. As this resistance grows and expands, it not only re-applies traditional culture but revitalizes it, frequently adapting new forms of expression, as part of the decolonization process.

A primary example of this cultural shift, of new forms of expression and vigor that reveal the "approaching conflict" can be seen in the Mohawk resistance at Kanesatake/Oka in 1990. Similar in many ways to Native blockades and protests which began in the 1960s period, Oka served to renew not only the concept of sovereignty, but in particular a warrior culture charged with the responsibility of defending people and territory.

Extensive media coverage of the Oka Crisis, including images of armed, masked warriors and the Warrior Unity flag, set the tone for Indigenous resistance throughout the 1990s, inspiring many Indigenous people and communities, instilling in them a warrior culture adapted to the realities of modem-day colonialism.

When considering the process of decolonization and the tendency for many Indigenous people to become co-opted even when engaging in traditional cultural practices, a primary element we find lacking is that of the warrior. This occurs for various reasons, including the idea that such a culture is no longer necessary (i.e., the idea that 'modem-day' warriors are now lawyers and businessmen), that such a culture is criminal (state propaganda), or that warriors and the very idea of conflict are somehow anti-spiritual (New Age/Christian pacifism).

By discarding the single most important element of Indigenous culture in regards to self-defense and survival, modern-day spiritualists and reformers reveal their inability to comprehend the full nature and extent of the problem confronting Indigenous peoples and the Earth. The culture they promote is that which Fanon warns us of: superficial, lacking vitality, "rigid in the extreme." Influenced by New Age/Christian ideologies, they also preach submission to oppression and exploitation.

Warrior Culture & Fighting Spirit

Decolonization, then, begins with culture. But in order to meet the objective of decolonization (liberation of land and people), this culture must of necessity include that of the warrior. Without this, all efforts at decolonization will be vulnerable to co-optation and assimilation. This is because the warrior symbolizes and represents resistance and fighting spirit.

When we consider the impacts colonialism has had on Indigenous peoples, including post-traumatic stress disorder, identity crisis, feelings of inferiority, etc., it is no surprise that our communities are afflicted with such high rates of alcoholism, drug addiction, interpersonal violence and overall dysfunction. The most extreme impact is perhaps that of suicide among Indigenous youth.

These are the symptoms of a broken and defeated people, so colonized that they are unable to comprehend the means by which they have, in fact, been colonized. While history and analysis can awaken people to the realities of colonial oppression, it is the suppressed culture of the warrior that can provide the necessary fighting spirit to motivate people into action. This is why warrior societies, organization, and culture were so ruthlessly attacked by early colonial forces, and why today they remain a primary target for state repression and propaganda.

From the very outset, then, the warrior culture must be emphasized as part of any decolonization effort. This includes the traditional responsibilities, organization, and methods of warrior societies, along with songs, crests, regalia etc. Today, this warrior culture also includes camouflage fatigues, masks, and direct action (blockades, occupations, etc.), along with specific groups and a recent history of resistance (i.e., the 1960s Red Power, American Indian Movement, Mohawk Warrior Societies, aka, Gustafsen Lake, Ipperwash, and the Native Youth Movement). All of these should be used to raise the morale and fighting spirit of our people, and especially the youth.

Identifying the Enemy

Along with fighting spirit, identifying the enemy is another important aspect of decolonization. Without this, the root cause of our oppression cannot be clearly understood, nor can the means of liberation be seen. Although we can say that the colonial society is itself the enemy, including all those who participate in and maintain the system, such a view is far too vague and implies that all European settlers are our enemies. It fails to account for internal divisions within the society, particularly those based on economic class, and limits our ability to expand resistance into the lower ranks of the settlers themselves.

If we see Western society as essentially a slave system, then there are rulers and those who are enslaved. Even if many are unaware of their own oppressed condition, the fact that there are rulers and slaves reveals a fundamental division within the society. We should seek to exploit this division by identifying potential alliances, thereby strengthening our forces and weakening those of our enemy.

Our common enemy is clearly the ruling class, which organizes and directs the system of exploitation, oppression, and control. The means by which this is accomplished is through the government and corporations, who work hand in hand to ensure that the entire system continues to function. The government organizes and imposes control over the population, in order for the maximum profits to be made by corporations.

In essence, then, our position is one of class war, and our common enemy is the ruling class based in the government and corporations.

Disengaging from the Colonial System

When we say we are in a war for territory, it must be understood that this is a total war, and that the territory is not only physical. Today, this territory also includes the mental and psychological landscape. In fact, our main enemy at this time is not colonial troops or police, but instead the ideology of the system itself, the primary means by which social control is maintained.

This colonial ideology is transmitted through a variety of means, including schools, TV, corporate media, movies, and pop music (including mainstream rap). Through these, the values and way of life of the oppressor (i.e., individualism, greed, materialism, patriarchy, etc.) are imposed. Even in the most remote reservation communities, the colonial way of life can be seen: a daily routine of watching TV and videos; playing video games, or listening to pop music.

An important step in breaking the mind control that has been imposed is limiting exposure to these forms of communication. Put the TV, the VCR, and the video games away. Turn off the radio. Organize activities that do not center on these forms of 'entertainment' or 'education.' Find positive alternatives such as videos and music that truly educate and inspire.

Another aspect of the daily routine of colonial society is alcohol and drugs. Despite the destructive effects on society overall, alcohol and drugs are both promoted by the system as forms of 'recreation' and the release of stress. We can see this in the 9–5 work routine, in which the tired worker returns home to relax by drinking a six pack (i.e., Homer Simpson), as well as in the five-day work week, in which weekends are seen as 'party time.' Even those without jobs conform to this routine, in which Fridays and Saturdays are seen as 'party nites.'

Breaking these routines, disengaging from the system, is an important first step in decolonizing. It must be emphasized, however, that the objective is to limit exposure, not to cut ourselves off completely. To do so would only isolate us from the outside world and reduce our situational awareness. We need to know what our enemy is saying and doing in order to counter his efforts. At the same time, we must keep 'up to date' with sociological trends and patterns, in order know the spirit of the time and conditions.

Along with this, there is the ongoing need to organize the resistance movement. Isolating ourselves from the colonial society, in an effort to decolonize our minds, bodies and spirits, limits our ability to gather information, communicate, and organize. We will, of necessity, continue to use modem technological tools in our organizing so long as such means continue to function. Our objective is the liberation of land and people, not the life-long pursuit of our individual decolonization. The return of the Indigenous person living in accord with the natural world, living a free, sovereign life, will only occur after the destruction of the colonial system. We should have no illusions about this, and it is in any case a fulfillment of our obligations to the future generations that they live such a life.

A slave cannot live as a free person until he or she has been liberated. The near-constant interaction with the slave-master, the daily routine of slavery, exploitation, and control, these are the conditions that create the need for liberation in the first place. Although we can, and do, struggle for the liberation of mind and spirit, it is a choice few are able to make. The vast majority are so indoctrinated and assimilated that they accept things as they are, both participating in, and perpetuating, the system itself.

Nevertheless, our task is to liberate the slaves and destroy the system. And we can only do this as we *liberate our mind and spirit.* This is why the personal journey of decolonization is not only logical as a starting point, it is also necessary. But it has its limitations. Although the total rejection of all European culture, including reading and writing, TV, etc., is correct in principle, we can see how, in practice, it limits our ability to organize and therefore *fails to meet the primary objective* (liberation of land & people).

Now we turn our attention to this first, crucial step: the liberation of mind and spirit.

Liberation of Mind & Spirit

The pervasiveness of the colonial system, its ability to penetrate virtually all aspects of our daily lives, including relationships, values, beliefs, etc., should not be underestimated. From the moment we are born, the process of socialization begins which has as its goal the production of obedient worker-slaves. Exposure to corporate entertainment, in the form of Hollywood movies, sports, music, etc., for example, begins at an early age. The educational system then provides a systematic means by which indoctrination is imposed.

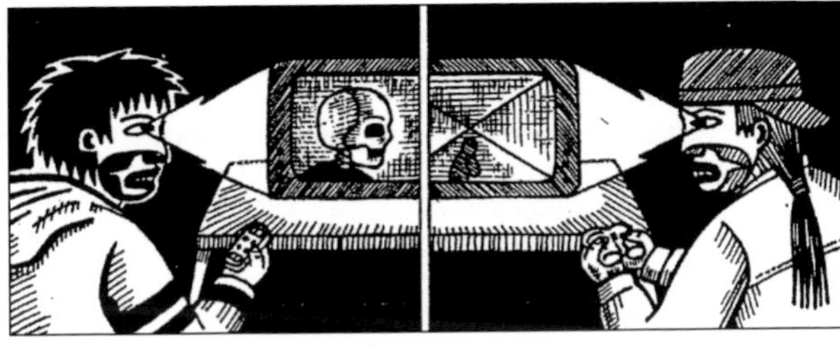

Generally, all throughout one's childhood and youth, one is exposed to the system's ideology and way of life. Only later is the full extent of this indoctrination realized, if at all. Almost everything we are taught or told about society and the world we live is then seen to be a matrix of lies and deception. This is the realization that must be reached in order for us to even consider the concept of decolonization.

In this initial step at decolonization, we seek to disengage from the colonial system and to immerse ourselves in our own culture and way of life. This provides us with a positive alternative to the system, as our struggle is between two opposing and contradictory ways of life. As noted previously, however, there is a danger of co-optation without the influence of our warrior culture and the identification of a common enemy.

Relearning one's culture occurs through a variety of means. Participating in cultural activities, ceremonies, learning songs and dances, language, arts and crafts, traditional skills such as hunting and fishing, and living on the land, are some examples. In the late 1800s, anthropologists made a concerted effort to document our cultures in the belief that we would disappear as distinct tribal peoples. Their research, while serving the interests of the colonial system, also contains vast amounts of ancestral knowledge provided by Indigenous informants. This information, acquired through reading and study, is a source of decolonization that should be exploited.

One of the most important methods of liberating our mind and spirit is participating in ceremonial activities. Many Indigenous people recovering from alcoholism and drug addiction tum to ceremonies for healing, and we should apply the same rationale in regards to decolonization. Common examples in North America include sweatlodges, fasting, potlatches, peyote, yuwipi, cold-water bathing, sundances, etc. Whatever the form, these ceremonies share some common characteristics such as the need to endure periods of discomfort or suffering, isolation, sleep deprivation, etc., all of which serve to alter one's mental state and to open up channels for spiritual communication.

Ceremonies also instill positive values such as sacrifice, discipline, self-control, humility, and the ability to withstand hardship. These values are essential parts of our warrior culture and it can be said that through these a warrior spirit permeated all aspects of traditional Indigenous society. Contrasted with European values of individualism, greed, and materialism, we can see how ceremonies contain within them the potential to radically alter an individual's way of life and perception of the world.

Active Use of Territory

When individuals begin the process of decolonization, disengaging from the system and immersing themselves in their own culture, activities out on the land and away from the urban/suburban environment increase. This is in accord with our strategy of disengagement and immersion in regards to personal decolonization. At the same time, the land is, in and of itself: a powerful method of liberating one's mind and spirit.

One of the most crucial understandings of the colonial system is that it is an alien system, not only in that it comes from an external force (Europe), but also in the way that it is completely alienated and removed from the natural world. Civilization is a man-made system that today permeates all aspects of our lives, alienating us from the natural world.

The land is not only necessary for our survival and sustenance as Indigenous peoples, providing us with food, water, shelter, clothing, tools, etc., it is also the source of our culture. The environment we live in affects our mind and spirit. In the urban culture of Western society, the natural world is something to be feared, conquered, and exploited (this is a basic message of Christianity); only by first alienating its own citizens from the land can such a perspective be accepted as true and logical. In contrast, Indigenous culture is part of the natural world; forces of nature, animals and plants, all these form integral parts of the culture that can only be understood by being out on the land.

Active use of territory is also a method of monitoring, and thereby defending, one's national territory. Regularly patrolling one's territory, by hiking, hunting, camping out, etc., can be seen as a form of asserting sovereignty. Even more so is the active defense of territory through re-occupation camps or the construction of traditional shelters. These types of activities also involve larger numbers of people in decolonization efforts.

4. Decolonization in North America

> "Colonial exploitation, poverty, and endemic famine drive the native more and more to open organized revolt. The necessity for an open and decisive breach is formed progressively and imperceptibly, and comes to be felt by the great majority of the people. Those tensions which hitherto were non-existent come into being. International events, the collapse of whole sections of colonial empires and the contradictions inherent in the colonial system strengthen and uphold the native's combativity while promoting and giving support to national consciousness." [18]

Decolonization, the liberation of an oppressed and colonized people, must ultimately mean the liberation of land and territory. Just as colonialism occurs in ways unique to each situation, so too does decolonization. When we consider this process in regards to our own situation in N. America, it becomes clear that decolonization will be far different than other anti-colonial liberation struggles. It will more closely resemble the collapse of an empire, arising from both external and internal factors. History has many examples of the decline and collapse of imperial systems, including those of the Romans and the Union of Soviet Socialist Republics (USSR).

Decline of the Roman Empire

Ancient Rome, which once ruled from Western Europe to North Africa, rose and collapsed within the span of just over 1,000 years. The period of its greatest expansion into an empire was during 200 AD to 400 BC, a period of just 600 years. In the end, it collapsed due to overextension of its military forces, political corruption and instability, internal social decline and rebellion.

More police and repressive laws failed to stop ongoing revolts by slaves, peasants, and colonies, while more soldiers sent to the frontiers could not stop growing tribal insurgencies (often led by war chiefs formerly trained by the Roman military).

The more imperial-power Rome gained through conquest, the greater the wealth and privilege of its citizens. At the same time, the more it conquered, the more divided its society became as new colonized subjects and territories were assimilated.

In the end, Rome faced both internal and external threats that converged to create a systemic breakdown. Rome itself was invaded, looted, and even occupied for periods of time by the same 'barbarian' tribes it had invaded and colonized.

Collapse of the USSR

In the 1970s, the USSR seemed all-powerful and monolithic. It was the second largest super-power next to the US, dominating eastern Europe and numerous 'client states' in Africa and Asia. In the 1980s, however, it was involved in a losing war in Afghanistan, which cost billions and demoralized much of the domestic population (just as Vietnam had done to the US). Soviet citizens became increasingly disillusioned with the political system, dominated by the corrupt Communist Party. The economy continued to decline as the country experienced a worsening ecological crisis. This resulted from widespread industrial pollution and waste, and included the 1986 Chernobyl nuclear power plant disaster. Entire regions and lakes were turned into deserts or wastelands. By 1991, these factors converged and lead to the collapse of the USSR, when many of the republics seceded and declared independence.

Decline of the USA

Today, the US dominates the global imperialist system, seemingly all-powerful. Nevertheless, it faces both internal and external threats not unlike those found in Rome, but on a far greater scale. Its military forces are spread around the world, in Europe, S. America, Africa, and Asia. Nearly 150,000 combat troops are presently deployed in Iraq and Afghanistan. On its southern border, poverty and exploitation have driven millions of Mexicans to immigrate into the US, while creating insurgent movements within Mexico itself.

Internally, the US is more polarized and divided than at any time since the Vietnam War (a period characterized by widespread rebellion and resistance). Many citizens have become increasingly disillusioned with the political system, ranging from leftists to right-wing 'Patriot' movements. The 2000 US presidential elections, which saw George W. Bush take power, are seen by many as the result of electoral fraud, in which millions of primarily African-American votes were disqualified in Florida.

Following the September 11, 2001, attacks on the World Trade Center and Pentagon, the US federal government immediately passed the USA PATRIOT Act and began forming a new Department of Homeland Security. At the same time, it launched its 'War on Terror,' consisting first of the invasion of Afghanistan, followed by Iraq in 2003.

The Patriot Act and Dept. of Homeland Security have established a domestic police state, a trend set in motion with the 1960s–70s period of domestic counter-insurgency. Today, this police state has established broad, sweeping new powers for police, FBI and intelligence agencies to carry out surveillance, arrests, secret trials, and deportations. Hundreds of primarily Arab nationals remain in jails across the country, while color-coded 'terror alerts' continue to be issued by Homeland Security, at times consisting of border closures and the deployment of heavily-armed police in major cities ('Hercules' teams).

These 'internal security' measures have further polarized US society, with growing numbers of citizens beginning to question the entire premise of a 'War on Terror,' and in particular the US invasion of Iraq, which was based on falsified evidence of weapons of mass destruction and Iraq's ties to al-Qaeda. Despite extensive propaganda campaigns, disillusionment and cynicism continue to grow.

An Empire Divided

Poverty, drugs, crime, and police repression, continue to increase throughout the US, creating greater social tension and conflict, based primarily on racist oppression. In 1970, the number of persons in US federal or state prisons was around 200,000. Over the last 30 years, this number has increased to over 2 million, some 65% of which are black, and 25% Latino.

Blacks in the US have been a major catalyst for domestic resistance since at least the 1950s (i.e., the civil rights struggle, Martin Luther King, Malcolm X). During the 1960s, the black liberation movement (i.e., the Black Panthers) was also a major contributor to social rebellion within the us. This is due to the racist oppression of US society and the size of the black population itself (primarily urbanized).

According to the US Census Bureau, there are approximately 300 million US citizens. African-Americans comprise some 36 million, or 15%. Mexicans, Chicanos, and immigrants from Central America (referred to as Latino/as) comprise some 40 million, or 18% of the total population. They are the fastest growing population in the US (in the 1950s they were an estimated 5 million). There are also an estimated 10 million undocumented 'immigrants' in the US, most from Mexico. Indigenous peoples (not including the Mexica nation) are an estimated 2 million.

Combined, these colonized peoples are some 85 million, over 30% of the entire population. Added to this are millions of Asian and Arab immigrants, many of whom are also impoverished and subjected to racist oppression. The great potential for revolt of this oppressed underclass can be seen in the rebellions of the 1960s–70s (primarily urban blacks), and again in 1992, when major cities saw large-scale rioting after the Rodney King trial in Los Angeles (multinational).

Throughout the 1990s, there was an overall renewal of rebellion in North America, beginning with the 1990 Oka Crisis in Canada; the 1992 LA riots; the 1994 Zapatista uprising in Mexico; and the 1999 anti-WTO riots in Seattle. In 1993, US federal agents massacred over 80 men, women, and children at Waco, Texas. All of these events had major social impacts and revealed the growing potential for revolt throughout society (even within elements of the white settler population).

Despite these sporadic rebellions and confrontations, it is difficult to conceive of mass organized resistance to the system here in N. America under present social conditions. In fact, the situation appears counter-revolutionary and ultimately hopeless. Imagine organizing revolution in Babylon, 'cuz that's what we're trying to do.

It is therefore necessary to broaden our analysis. The most important observation is that the vast majority of the world's population are oppressed and impoverished by the global system, in South and Central America, Africa, and Asia. In these regions we find not only conditions of extreme exploitation, but also a far higher level of resistance and struggle.

These struggles have an enormous impact on the imperialist system, creating economic uncertainty, limiting corporate access to resources, and requiring massive amounts of military aid or intervention (War on Drugs, War on Terror).

Mexico & The US Southwest

In Mexico, the Zapatista rebellion has had a profound effect on Mexican society after 10 years of struggle. It has renewed the fighting spirit of Indigenous peoples and unified many diverse social movements. The Zapatistas have also established autonomous zones in Chiapas, based on principles of political autonomy and self-organization. It has promoted the role and status of women in Mexican society. In several Mexican states, over a dozen new guerrilla movements have emerged. These factors have caused increasing concern for US authorities, which has supplied the Mexican state with funding, training, and new military equipment in order to wage counter-insurgency warfare (most under the pretext of the War on Drugs).

Foremost among US concerns is the threat of economic disruption posed by Mexican insurgents, as well as the danger of contagion and the spreading of Mexico's revolutionary and insurgent culture within the US domestic population. During the Mexican Revolution in 1917, some 35,000 US troops were placed on the border to stop immigration.

Following the '94 Zapatista rebellion, the US intensified efforts to seal off its southern border. The INS greatly expanded its Border Patrol, increasing from 980 agents in 1994 to 2,264 in 1998. The INS annual budget nearly tripled from $1.5 billion to $4.2 billion during the same period.

The result has been the creation of a militarized zone along the US-Mexico border, with joint police-military operations, checkpoints on roads and highways, constant patrolling, use of floodlights to illuminate areas, etc. Along with racist anti-immigrant laws, the result of this official policy has been a substantial increase in human rights violations. INS prisons have also increased their daily capacity from 8,279 in 1996, to 20,000 by 2001 (each year some 200,000 persons are detained by the INS).

Mexico, and by extension the US southwest, is clearly a strategic point to which our enemy devotes considerable resources. The Mexican population, both in Mexico and the US southwest, are seen as hostile and dangerous. Despite this, Mexicanos are a large and necessary part of US society, serving as a highly exploitable source of manual labor. Like New Afrikans, the Mexica/Chicano peoples are a strategic factor in decolonization.

I'M AN INSURGENT. I HAVE DEDICATED ALL MY LIFE AND TIME TO THE CAUSE.*
*MAJOR ANNA MARIA, EZLN

Insurgency in Iraq

Today, one of the most critical regions for US imperialism is the Middle East and, in particular, its oil and gas resources. This region alone contains two-thirds of all known petroleum supplies, and is vital to the Western industrial system as a whole. European nations, as well as those in Asia (i.e., China), are increasingly dependent on these supplies. US plans for global domination requires direct control of Mid-East oil; whoever controls this region exercises control over the global system itself. For this reason, the US invasion of Iraq has not been supported by most European nations, who see it as an attempt by the US to assert control while undermining their own positions.

Faced with a growing insurgency in Iraq, the US is now involved in a war from which it cannot simply withdraw, but which, ultimately, it cannot win. As Vietnam showed, large segments of the US population are unwilling to support wars of this nature and are far less loyal to the system than during World Wars I and II. As economic conditions continue to decline, the US will experience increasing social conflict from within, while resistance in the Middle East and other regions will also expand, requiring ever-greater deployment of police and military forces.

As noted, US society is deeply divided by race, and this is reflected in its military forces. African, Asian, and Indigenous/Latino peoples comprise a disproportionate number of frontline combat troops (and therefore casualties). During Vietnam, racism in the US military contributed to a culture of mutiny, desertion, and even the killing of commanders seen as racist or willing to sacrifice troops ('fragging'). US troops were also demoralized by their inability to stop guerrilla attacks, mounting casualties, widespread drug use, and the blatant injustice of their actions.

Already, US troops in Iraq have become demoralized and disillusioned with their mission. Many citizens have come to question the legitimacy of the war itself. Within months of the invasion, the largest anti-war demonstrations occurred around the world. Not until the late sixties, almost 5 years after escalating US involvement, were there similar mass demonstrations against the Vietnam War. As casualties continue to mount, US public opinion is increasingly turning against the occupation of Iraq, threatening to further polarize-US society.

Crises, Conflict, & Resistance Potential

The global environment is rapidly deteriorating due to industrial pollution and resource depletion. Global warming is already causing extreme weather patterns, including storms, droughts, deadly heat waves, bug infestations, forest fires, melting of polar ice caps and glaciers, etc., all of which have negative effects on the global economic system.

Within the G7, globalization and neo-liberal trade policies have shifted large amounts of industrial production to 'less-developed' countries, where labor and resources are far cheaper. This has resulted in growing unemployment within the most industrialized nations.

Despite decades of economic growth, poverty has expanded around the world and within the G7 nations themselves. Only the rich and certain middle-class sectors have benefited from the process of globalization (another term for imperialism). Today, overall economic and social conditions are worse than in the 1960s, a decade that saw widespread social rebellion.

The convergence of war, economic decline, and ecological crises will lead to greater overall social conflict within the imperialist nations in the years to come. It is this growing conflict that will create changes in the present social conditions, which will create greater opportunities for organized resistance. The rulers are well aware of this, and it is for this reason that state repression is now being established as a primary means of social control (i.e., greatly expanded police military forces, new anti-terror laws, etc).

As these crises deepen, the system becomes more and more vulnerable. When colonial or imperial systems weaken, this has been the time at which colonized nations have advanced. We can see this in ancient Rome, in the post-WWII period, and in the collapse of the USSR.

If this analysis is correct, we are now in a period that can be described as the "calm before the storm," a storm that will eventually shake the very foundations of the imperial system itself: It is this growing potential for social conflict and systemic breakdown that provides the best possibility for decolonization, at both national and international levels.

❦ ❦ ❦

References

1 Chinweizu, *The West and the Rest of Us: White Predators, Black Slavers, & The African Elite* (New York, NY: Random House, 1975), 33.
2 *The American Heritage Dictionary*, s.v. "psychopath," 1415.
3 Cheikh Anta Diop, *Civilization or Barbarism: An Authentic Anthropology*, ed. Harold J. Salemson and Marjolijn de Jager, trans. Yaa-Lengi Meema Ngemi, (New York, NY: Lawrence Hill Books, [1981] 1991), 85–86.
4 Jack C. Estrin, *World History Made Simple*, rev. ed. (New York, NY: Doubleday, 1968), 65.
5 Howard Adams, *Tortured People: The Politics of Colonization* (Penticton, BC: Theytus Books, 1999) 56–57.
6 Aphrodite Matsakis, *Trust After Trauma: A Guide to Relationships for Survivors and Those Who Love Them* (Oakland, CA: New Harbinger Publications, 1998), 339.
7 Matsakis, *Trust After Trauma*, 29.
8 Frantz Fanon, *The Wretched of The Earth*, trans. Constance Farrington (New York, NY: Grove Press, 1968), 236.
9 Ibid.
10 J. Rick Ponting, *First Nations in Canada: Perspectives on Opportunity, Empowerment, and Self-Determination* (Toronto, ON: McGraw-Hill Ryerson, 1997), 86.
11 Ibid., 83–85.
12 Royal Commission on Aboriginal People, 1995, 90, as quoted in Ponting, *First Nations in Canada*, 83.
13 John H. Bodley, *Victims of Progress*, 5th ed. (Lanham, MD: Altamira Press, 2008), 173.
14 Ponting, *First Nations in Canada*, 81.
15 Fanon, *The Wretched of The Earth*, 237–38.
16 Ibid., 244–45.
17 Ibid., 240–41.
18 Ibid., 238.

Bibliography

Adams, Howard. *Tortured People; The Politics of Colonization*. Penticton, BC: Theytus Books, 1999.

Chinweizu. *The West and the Rest of Us; White Predators, Black Slavers & the African Elite*. New York, NY: Random House, 1975.

Fanon, Frantz. *The Wretched of the Earth*. Translated by Constance Farrington. New York, NY: Grove Press, 1968.

Hill, Gord. *500 Years of Indigenous Resistance*. Arm the Spirit/Solidarity, 2002.

Koning, Hans. *The Conquest of America*. New York, NY: Monthly Review Press, 1993.

Matsakis, Aphrodite. *Trust After Trauma; A Guide to Relationships for Survivors and Those Who Love Them*. Oakland, CA: New Harbinger publications, 1998.

Ponting, J. Rick. *First Nations in Canada; Perspectives on Opportunity, Empowerment, and self-Determination*. Toronto, ON: McGraw-Hill Ryerson Ltd., 1997.

Sakai, J. *Settlers; The Mythology of the White Proletariat*. Chicago, IL: Morning Star Press, 1989.

Stannard, David E. *American Holocaust; the Conquest of the New World.* New York, NY: Oxford University Press, 1992.

Note: the purpose of these appendices is not to promote the United Nations, but rather to show the hypocrisy of colonial states and a global system that claims to stand for human rights, peace, and freedom.

Appendix A

UN Declaration on Decolonization

Declaration on the Granting of Independence to Colonial Countries and Peoples Adopted by General Assembly resolution 1514 (XV) of 14 December 1960

The General Assembly,

Mindful of the determination proclaimed by the peoples of the world in the Charter of the United Nations to reaffirm faith in fundamental human rights, in the dignity and worth of the human person, in the equal rights of men and women and of nations large and small and to promote social progress and better standards of life in larger freedom,

Conscious of the need for the creation of conditions of stability and well-being and peaceful and friendly relations based on respect for the principles of equal rights and self-determination of all peoples, and of universal respect for, and observance of, human rights and fundamental freedoms for all without distinction as to race, sex, language or religion,

Recognizing the passionate yearning for freedom in all dependent peoples and the decisive role of such peoples in the attainment of their independence,

Aware of the increasing conflicts resulting from the denial of or impediments in the way of the freedom of such peoples, which constitute a serious threat to world peace,

Considering the important role of the United Nations in assisting the movement for independence in Trust and Non- Self-Governing Territories,

Recognizing that the peoples of the world ardently desire the end of colonialism in all its manifestations,

Convinced that the continued existence of colonialism prevents the development of international economic co-operation, impedes the social, cultural and economic development of dependent peoples and militates against the United Nations ideal of universal peace,

Affirming that peoples may, for their own ends, freely dispose of their natural wealth and resources without prejudice to any obligations arising out of international economic co-operation, based upon the principle of mutual benefit, and international law,

Believing that the process of liberation is irresistible and irreversible and that, in order to avoid serious crises, an end must be put to colonialism and all practices of segregation and discrimination associated therewith,

Welcoming the emergence in recent years of a large number of dependent territories into freedom and independence, and recognizing the increasingly powerful trends towards freedom in such territories which have not yet attained independence;

Convinced that all peoples have an inalienable right to complete freedom, the exercise of their sovereignty and the integrity of their national territory,

Solemnly proclaims the necessity of bringing to a speedy and unconditional end colonialism in all its forms and manifestations; And to this end Declares that:

1. The subjection of peoples to alien subjugation, domination and exploitation constitutes a denial of fundamental human rights, is contrary to the Charter of the United Nations and is an impediment to the promotion of world peace and co-operation.

2. All peoples have the right to self-determination; by virtue of that right they freely determine their political status and freely pursue their economic, social and cultural development.

3. Inadequacy of political, economic, social or educational preparedness should never serve as a pretext for delaying independence.

4. All armed action or repressive measures of all kinds directed against dependent peoples shall cease in order to enable them to exercise peacefully and freely their right to complete independence, and the integrity of their national territory shall be respected.

5. Immediate steps shall be taken, in Trust and Non-Self-Governing Territories or all other territories which have not yet attained independence, to transfer all powers to the peoples of those territories, without any conditions or reservations, in accordance with their freely expressed will and desire, without any distinction as to race, creed or color, in order to enable them to enjoy complete independence and freedom.

6. Any attempt aimed at the partial or total disruption of the national unity and the territorial integrity of a country is incompatible with the purposes and principles of the Charter of the United Nations.

7. All States shall observe faithfully and strictly the provisions of the Charter of the United Nations, the Universal Declaration of Human Rights and the present Declaration on the basis of equality, non-interference in the internal affairs of all States, and respect for the sovereign rights of all peoples and their territorial integrity.

Appendix B

Excerpts from: 1948 United Nations Genocide Convention

The Contracting Parties,

Having considered the declaration made by the General Assembly of the United Nations in its resolution 96(1) dated 11 December 1946 that genocide is a crime under international law, contrary to the spirit and aims of the United Nations and condemned by the civilized world;

Recognizing that at all periods of history genocide has inflicted great losses on humanity; and Being convinced that, in order to liberate mankind from such an odious scourge, international co-operation is required,

Hereby agree as hereinafter provided:

Article 1
The Contracting Parties confirm that genocide, whether committed in time of peace, or in time of war, is a crime under international law which they undertake to prevent and to punish.

Article 2
In the present Convention, genocide means any of the following acts committed with intent to destroy, in whole or in part, a national, ethical, racial or religious group as such:
(a) Killing members of the group;
(b) Causing serious bodily or mental harm to members of the group;
(c) Deliberately inflicting on the group conditions of life calculated to bring about its physical destruction in whole or in part;
(d) Imposing measures intended to prevent births within the group;
(e) Forcibly transferring children of the group to another group.

Article 3
The following acts shall be punishable:
(a) Genocide;
(b) Conspiracy to commit genocide;
(c) Direct and public incitement to commit genocide;
(d) Attempt to commit genocide;
(e) Complicity in genocide

Article 4
Persons committing genocide or any of the other acts enumerated in Article 3 shall be punished, whether they are constitutionally responsible rulers, public officials or private individuals.

The Dene Mapping Project

Phoebe Nahanni

> "The Dene do not have an equivalent of the Dead Sea Scrolls to use as historical reference. They have to rely on their oral tradition. It is therefore critical for the Dene to continue using their languages. Language is the main tool for the transmission of historical and cultural information."
>
> —Phoebe Nahanni, "Dene Women in the Traditional and Modern Northern Economy in Denendeh, Northwest Territories, Canada," 1992

Our way of life is very ancient and enriching. Our economy is based on hunting, trapping, and fishing. Long before any non-Dene ever set foot on our land, our ancestors lived and learned from each other, from the land, and other beings on the land—the animals, birds, and insects. The mysteries of nature reveal themselves more and more through our experience on the land. Our way of life has been happening for a long, long time before any non-Dene ever set foot on this land, and it is still happening today.

From generation to generation our ancestors have passed on information by word of mouth, through legends, and by relating personal experiences. The intricate values of our way of life are most appreciated by those who speak our languages. To the non-Dene such ways of recounting events may be subject to bias, error, misunderstanding, and misinterpretation. We Dene understand these shortcomings to be part of human nature.

Our people have withstood many seasons and many cycles. We consider our way of life to be the most practical for the climate and the environment we live in. Our songs, dances, games, and legends are expressions and reflections of it. Here are the views of some trappers:

Willie MacDonald (65) at Fort McPherson, 13 January 1975:

> "But, wherever we go in the bush, we always see old signs, from long ago. We see old deadfall places and all this. We know that long before us this country been used lots and long time before that. People still using it. I mean the people that were brought up to the life of trapping and hunting. We belong to it, we belong to the land and we look at it like that land is our mother. That's where we're born and that's where we're going to go back. We're going to be part of this land when we die. There is no way we're going to leave it. That's why we think of this land. What else we got? Just land, land is our bank, it's our living, it's everything. Everything we got is this land. What we depend on next year coming is depending on this land. The land is the most precious thing we got."

courtesy of NWT Archives

Dene Geographer Phoebe Nahanni
Nahanni Valley, during the Mackenzie Valley Pipeline Inquiry, ca. 1975–1983

> Alexis Potfighter (75) at Detah, 15 September 1974:
>
> "The wild food gives us strength and faith, as far as it goes, for the whole Indian people. They gave us faith and strength to do anything … Talking about land, we're asking for the land to live on, to do things, like we're doing now, trapping. The land that we depend on must not stop. For we trap, hunt, and fish on the land that is ours. This is why our great ancestors handed to us to protect it from any damage in our land …"
>
> Louis Moosenose (50) at Lac La Martre, 24 September 1974:
>
> "This land was given to us to make our living for food, clothing, and income … The land was given to us to look after it and the land was supposed to be protected. The land, the water, and the animals are here for us to make living on it, and it's not to play with."
>
> Amen Tailbone (60) at Rae Lakes, 30 September 1974:
>
> "The animals is our food, the land is our everything and the water is our drink, so the land is ours to keep as long as the sun is shining. The game wardens should not give hunting licenses to their friends, because they do not kill animals for food, they kill animals for sports. The land, the water, and the animals are not here to play with. It was given to us to look after and protect it from enemies like white people. The white people are up to destroy everything they see, and the land is the only thing, we depend on it for everything, and it is not for sale."

In our culture, emphasis is placed on the importance of human beings, land and environment, and animals and birds. Without the land and environment and the animals and birds, we could not survive. This realization was not derived from nowhere; it was derived from years of experience and adaptation. And the land and animals cannot continue to live and survive without our understanding and respect.

This creation was produced neither from our intelligence nor from our hands. We are part of it. So are all animals, birds, fish, and insects. We all inhabit this environment (and ultimately this planet). Therefore, this realization is encouraged amongst us to ensure in the long run a continuation of resources.

When a human being occupies a space in the environment, his behavior can be destructive or complementary to that space. (The former has been more evident since the white man moved on our land.) We understand that changes occur in human beings, environment, and in animals. These changes show in growth, seasons, and movement, and we are continually learning from this evolution through our experiences from our way of life. Our way of life has been tested over hundreds of years. It makes a lot of sense to us and we like to return to the land again and again, and we want our children to have the option to experience the same things.

Over the years we have observed the changes and the limitations placed on our freedom by the white peoples' government, the white peoples' laws, and the white peoples' teachings. We did not write the laws that govern us and those that affect our way of life, such as the Migratory Birds Convention Act, Indian Act, Territorial Land Use Regulations, Game Ordinances, etc.; our silence in the past does not mean consent. We are participating in the white man's system not because we accept it, but because we have no choice. We regard our participation as being temporary until such time as our right to choose our own institutions is recognized.

There are debates all over Canada between the Indians and white people over many things. When we see that the present white institutions do not even suit many white people, we hope that will demonstrate the inability of these institutions to meet our needs. In many important respects our position in Canada today is a magnification of the experiences of a vast number of Canadians.

While common experiences tie us to certain groups within Canadian society at large, many of our major problems stem directly from white people's ignorance of our differences. We believe a healthy society is not where we pretend such difference do not exist but one which recognizes and appreciates these differences and their contribution to the future development of society in general. A society as wealthy as Canada's which continues to perpetuate, even compound, the miseries of certain of its members is a sick society.

Land-Use & Occupancy Research

For a year and a half the Indian Brotherhood of the Northwest Territories was involved in negotiating a proposal and budget with the Department of Indian and Northern Affairs to document data on traditional land use. At the time it was clear that no agency, government or private, was planning to undertake such a study. We knew that anthropologists and white researchers had previously attempted to integrate a study of our land-based activities into their theses. But our community leaders and community people expressed their dislike of the invasion of their privacy by outsiders who didn't speak their language. We know from our past experiences that government research by white researchers has never improved our lives. Usually white researchers spy on us, the things we do, how we do them, when we do them, and so on. After all these things are written in their jargon, they go away and neither they nor their reports are ever seen again. We have observed this and the Brotherhood resolved to try its best to see that, in future, research involves the Dene from beginning to end.

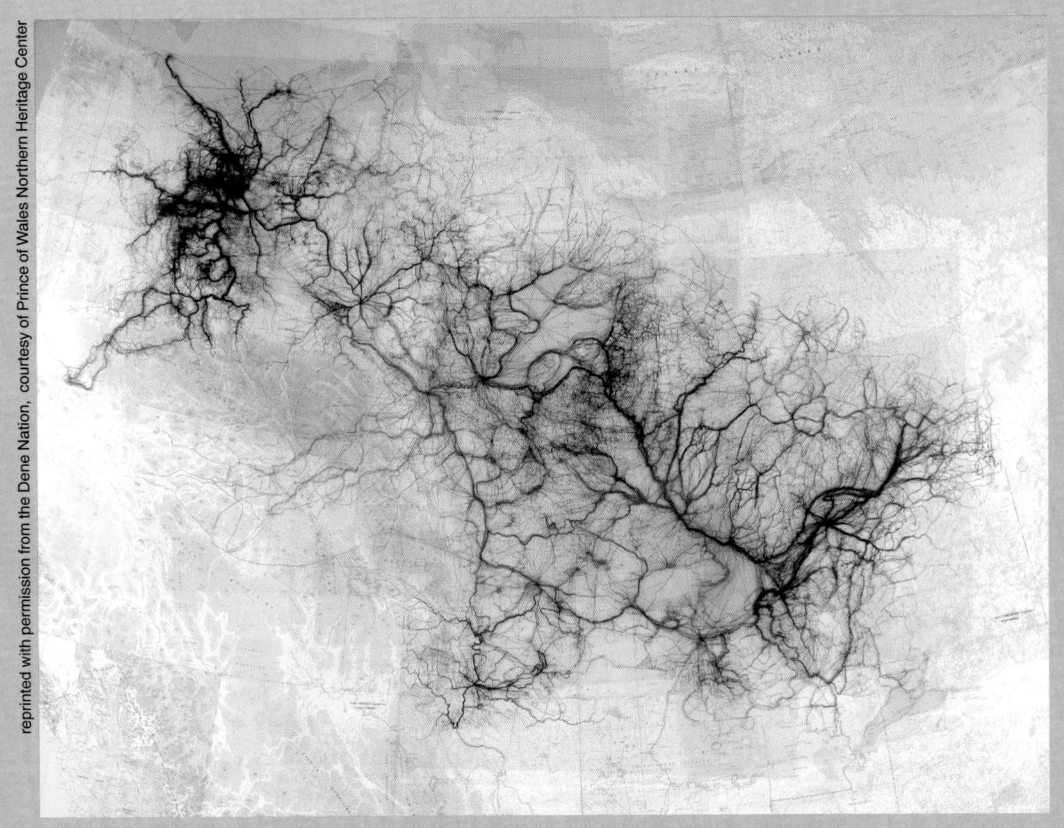

Counter-Colonial Cartography

The compiled 150 mosaic tiles of the Dene Mapping Project showing hunting routes, trap lines, fishing areas, foraging regions, camps, and burial grounds across 200,000 km^2 of Dene Territory, ca. 1983

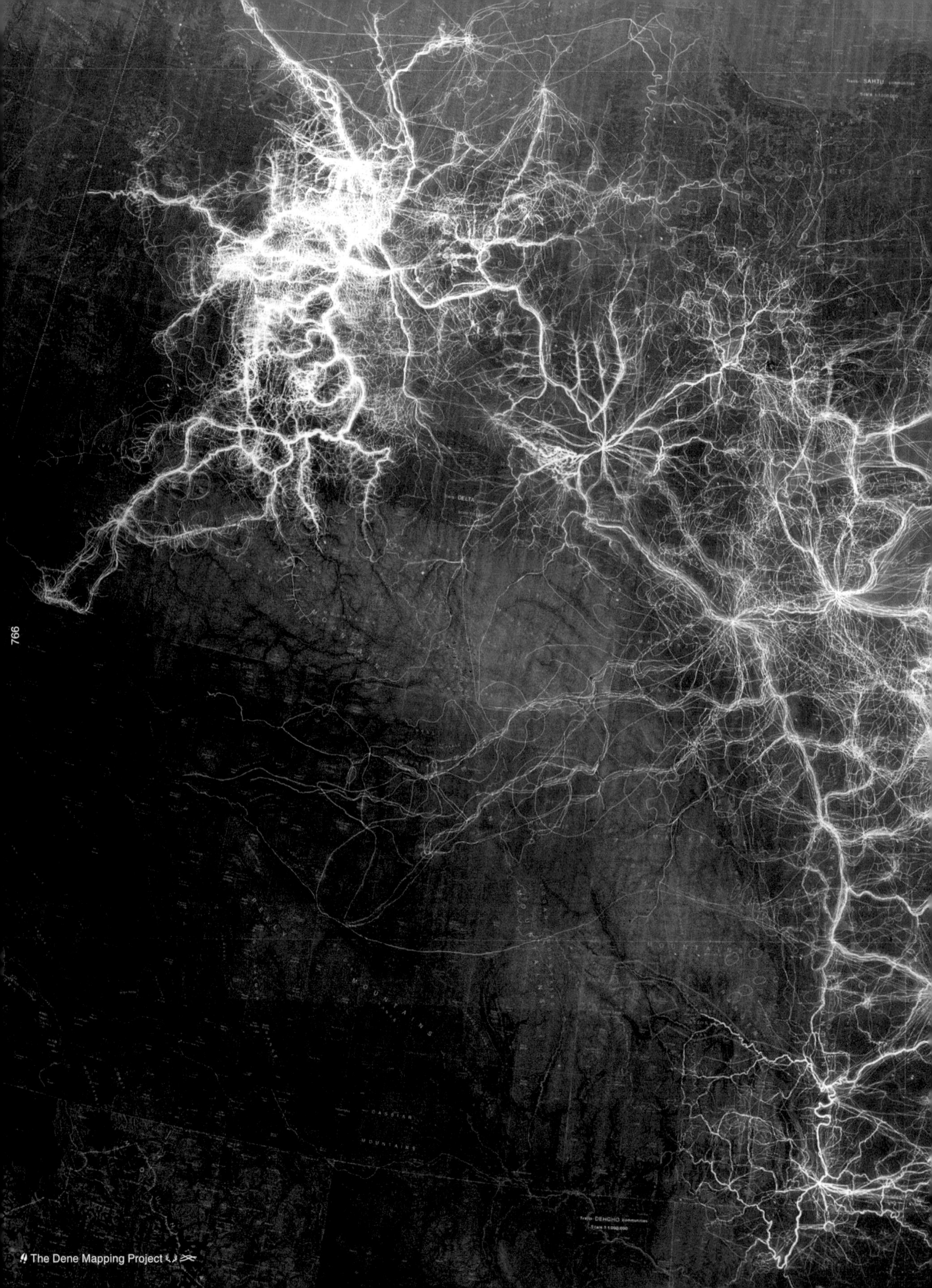
SAHTU
DELTA
DEHCHO
The Dene Mapping Project

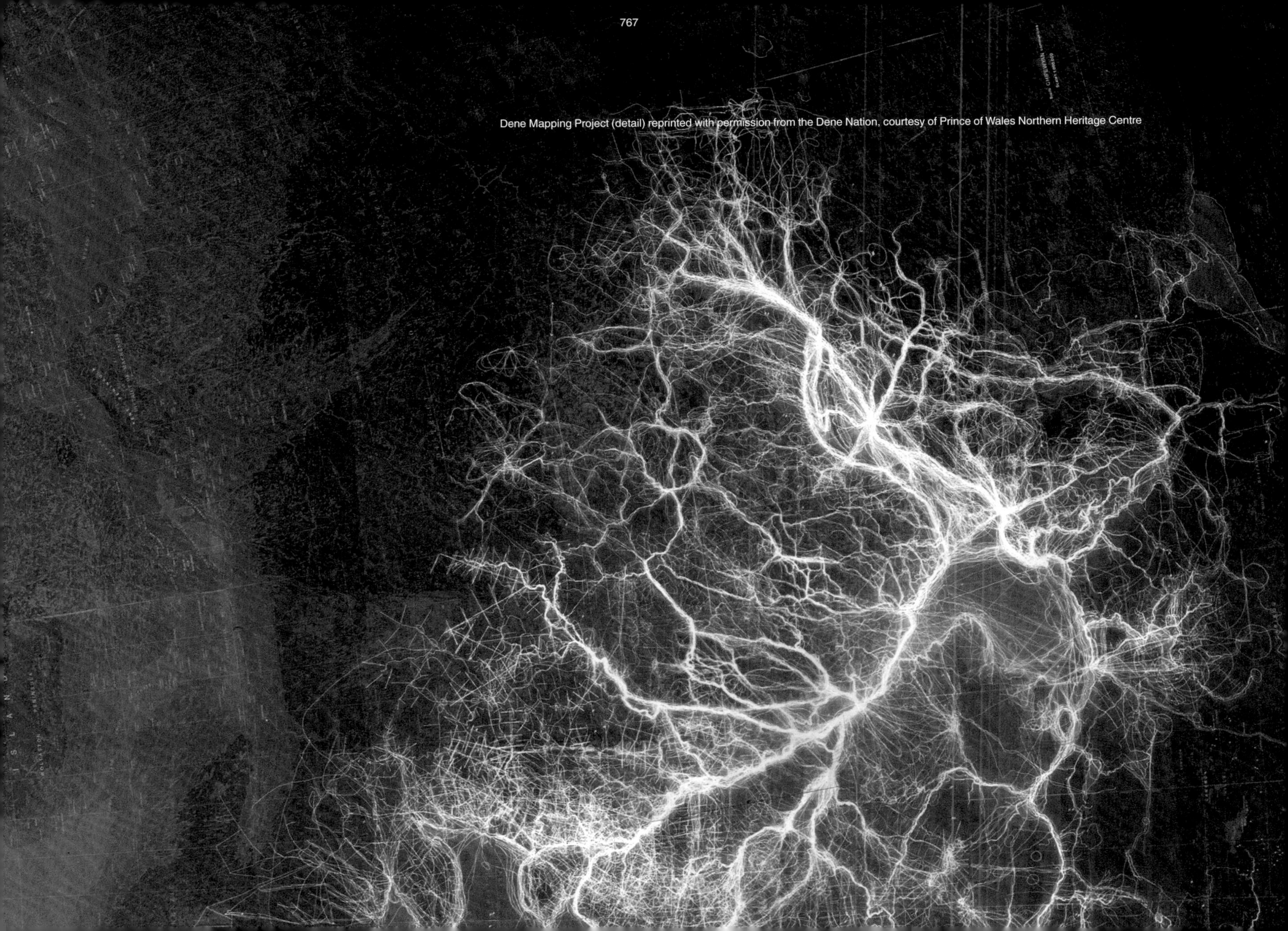

Dene Mapping Project (detail) reprinted with permission from the Dene Nation, courtesy of Prince of Wales Northern Heritage Centre

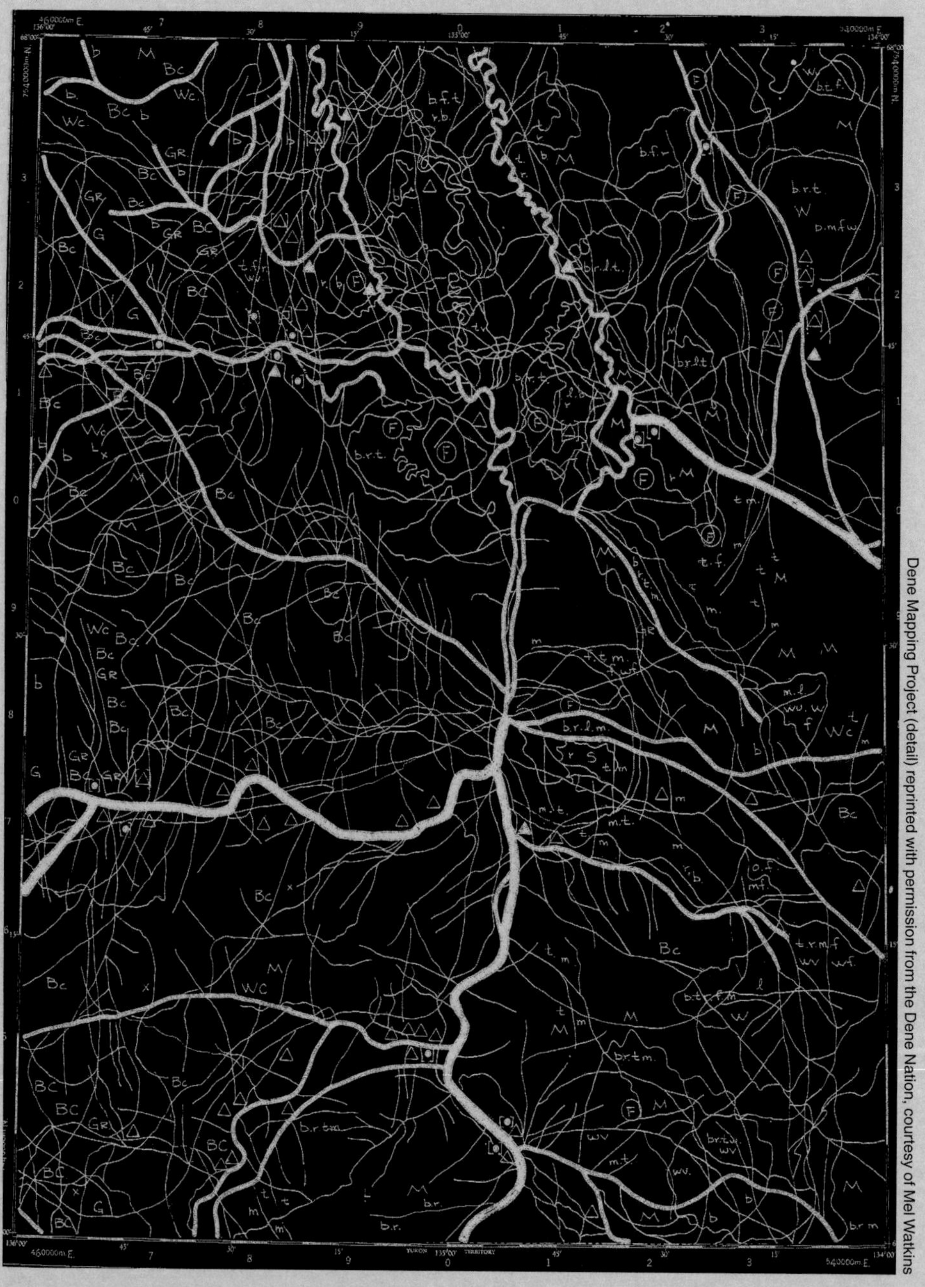

Dene Mapping Project (detail) reprinted with permission from the Dene Nation, courtesy of Mel Watkins

Digitized Dene Mapping Project as Geographic Information System

Sample Map showing trails and traplines around Fort MacPherson. It was compiled by the Mapping Project of the Indian Brotherhood of the NWT. One inch equals about twelve miles.

We got a commitment from DINA to fund the research for 1974–75 and 1975–76. Two years of research on a subject as old and continuous as our land-based renewable-resource economy could not possibly produce an exhaustive study. We therefore decided to work within the time constraints by obtaining a one-third sample, or approximately 30 per cent of the total number of trappers in each of five regions (Delta, North Mackenzie, South Mackenzie, North Slave, South Slave). Those trappers thirty years of age and over, we considered, would be best to provide us, from their substantial experience, with information on our economy. There are some exceptions to this, of course, judging from traplines covered by some hunters and trappers younger than thirty years of age.

The objectives of our land-use and occupancy research project were:

1. To form an information base upon which a just settlement of our grievances and land claims could be built, with an eye to securing our continuing benefit from the use of our land.

2. To provide a medium for re-establishing the bond between our young and their past, a bond essential to our future independence as Canadian citizens.

3. To provide us with material essential to us in deciding any action to be undertaken by us, in the event of a fair settlement, to improve our lot as we see fit.

4. To provide the people of Canada with a record of land use and occupancy in the Northwest Territories by people of Dene descent, from the distant past down to the present.

5. To provide the people of Canada with an understanding of the importance of our land to the integrity of our culture, our identity, and our present way of life, and to our future hopes.

6. To convey to the people of Canada the destruction inflicted on our civilization by the ethnocentric ignorance of the white man.

7. To give a truer appreciation of the costs of the present path of development the white man has chosen in the North, in terms of the Dene as people with a right to survival *as a people*, and with a right to control and direct the changes the present and future force on them.

8. To give us an opportunity to present once again, without the confusion of middlemen, and in accordance with our own aims, the Dene experience and point of view as only we can present it.

Our main information source was to be the hunters and trappers themselves, and we held interviews in the following communities:

Delta Region: Aklavik, Inuvik, Fort McPherson, Arctic Red River

North Mackenzie: Fort Good Hope, Colville Lake, Fort Franklin, Fort Norman

South Mackenzie: Fort Wrigley, Fort Simpson-Jean Marie, Fort Liard, Nahanni Butte, Trout Lake, Fort Providence-Kakisa, Hay River

North Slave: Rae, Edzo, Lac La Marte (Marian Lake, Snare Lake), Rae Lakes, Yellowknife, Detah

South Slave: Resolution, Fort Smith, Snowdrift

We wanted to gather information which could be represented on maps to show all the lands we have occupied at any time as far back as we can remember and also the situation at the present; how we used and use the land; and what the land means to us, as a people, at present. We asked such questions as: When did you first begin to trap alone? Where did you trap? If you moved to other areas, why? What did you hunt and trap? Where did you fish? Where did you set up camps? What kind of camp? Do you go to this camp every year? How much does land mean to you? What are your views on land?

The greatest frustration we encountered in data collection, predictably, was the difficulty of locating the men we wished to interview; much of the time they were in the bush trapping, and were inaccessible to our researchers. This minor frustration was, of course, far outweighed by its positive aspect: the men's absence was an additional empirical proof of the widespread continued use of the land that we had set out to document.

Another limitation we encountered was that it was simply impossible for every trapper interviewed to supply us with complete information on all his land-use activities throughout the years. There were various reasons for this: researchers could not spend as much time with each individual as would have been optimum; or informants grew too tired to continue an interview before they could impart the wealth of information they possessed to the interviewer. Therefore, we have been able to present only an outline of the present land-use patterns of approximately one-third of the men who are currently involved in hunting, fishing, and trapping. When looking at the sample map shown here, you must keep in mind that, even if you multiply in your imagination to three times the number of lines and routes indicated here, it still represents only partially the extent of use and occupancy of Dene lands by the Dene as it occurs today.

Mapping as Political, Participatory Activism
Phoebe Nahanni, Wilson Pelissey, Freddie Greenland, Charlie Snowshoe, Betty Menicoche, Louis Blondin during a Hearing of the Berger Inquiry, 1976

Living with Land
Phoebe Nahanni and her younger sister, at Jean Marie River, 1952

An important conclusion from our research is that there are approximately 1,075 men in five regions actively engaged in hunting and trapping at the present time. We found that we could classify the traditional land uses as follows:

1. Subsistence hunting (all seasons from camps, e.g. moose hunting)

2. Seasonal hunting (e.g. caribou hunting)

3. Part-time hunting (all seasons based from a community)

4. Spring hunting and trapping (end of March through April to May for, e.g., beaver and muskrats)

5. Winter trapping (October–November to February–March for fur-bearing animals)

6. Fishing (all seasons)

7. Fishing (spring and/or summer at fish camps)

Dene people have considerable experience in surveying the environment we live in. Our ancestors traveled throughout the lands and, when the white men traveled on our land, it was with the advice and help of our ancestors. We have our own Dene place names for all our camps, for the lakes, the rivers, the mountains—indicating that we know the topography of our land intimately. Before Mackenzie came and claimed the river to be named after him, for example, we called it Deh-cho.

From the interviews we had with twenty-six young people, it is clear that there is a bond between young and old; that bond is essential to our future independence. The maps clearly show what the Dene have been saying all along before your legal institutions—that we have been here for hundreds and thousands of years; this is our land, and our life. This is the most graphic demonstration of the truth that we Dene own 450,000 square miles of land.

Our field work reinforces the statements made by the Dene at the community hearings of the Berger Inquiry that our attitude towards our land has far more substance than is fully appreciated by the oil and gas companies and government. That the proposed oil and gas pipeline routes and construction sites conflict with our land-based activities is obvious in the cartographic representation of those activities. These routes show no sign of regard for our trails, travel routes, and traplines, and our camps. The implications of such intrusions not only affect the trails, travel routes, and traplines; they also indiscriminately and without discretion affect the animals, fish lakes, and the environment and our way of life.

Producing these maps has been a lot of work. Over two years, about two dozen people have worked on this project and at any one time about a dozen people were working on it. The result is that data now exists on Dene land use that simply were not available before. Through the evidence of our land use and occupancy, we are showing that we have tolerated at great cost to our culture the kind of development thrust upon us, and from here on it is our right to control and direct the changes that affect our survival as a people.

This essay was originally published as "The Mapping Project," in *Dene Nation: The Colony Within*, ed. Mel Watkins (Toronto, ON: University of Toronto Press, 1977), reproduced with permission from Mel Watkins & The Dene Nation Office.

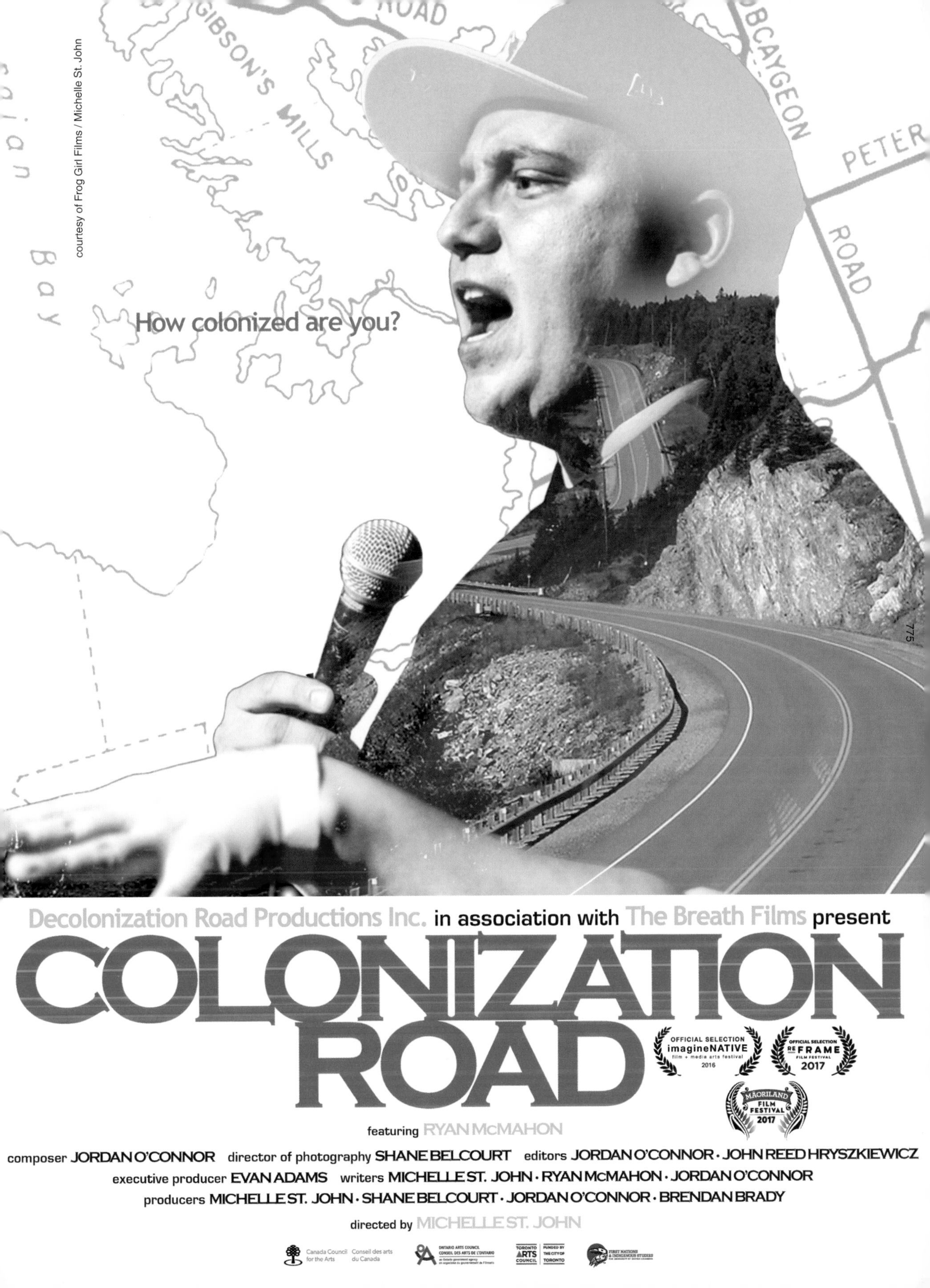

courtesy of Frog Girl Films / Michelle St. John

How colonized are you?

Decolonization Road Productions Inc. in association with The Breath Films present

COLONIZATION ROAD

OFFICIAL SELECTION imagineNATIVE film + media arts festival 2016

OFFICIAL SELECTION REFRAME FILM FESTIVAL 2017

MĀORILAND FILM FESTIVAL 2017

featuring RYAN McMAHON

composer JORDAN O'CONNOR director of photography SHANE BELCOURT editors JORDAN O'CONNOR · JOHN REED HRYSZKIEWICZ

executive producer EVAN ADAMS writers MICHELLE ST. JOHN · RYAN McMAHON · JORDAN O'CONNOR

producers MICHELLE ST. JOHN · SHANE BELCOURT · JORDAN O'CONNOR · BRENDAN BRADY

directed by MICHELLE ST. JOHN

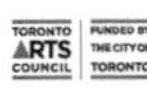

When Colonization Road Ends...

A Conversation between Ryan McMahon, Charmaine Nelson, & Eli Nelson

Ryan McMahon is an Anishinaabe comedian, writer, media maker, and community activator based out of Treaty #1 territory (Winnipeg, MB). He co-wrote and featured the 50-min. documentary *Colonization Road* (2016).

Charmaine Nelson is professor of art history at McGill University, focusing on postcolonial and Black feminist scholarship, critical race theory, Trans-Atlantic slavery studies, and Black diaspora studies. She is the 2017–18 William Lyon Mackenzie King Visiting Professor of Canadian Studies at Harvard University.

Eli Nelson is a PhD candidate in the History of Science department at Harvard University. He studies the history of native science in settler-colonial and postcolonial contexts.

RM: Before we start, let's just acknowledge how ugly the word colonialism actually is. I mean honestly, it has the word *colon* in it. When we think of colonialism, we think of seventeenth-century European men on horseback. We don't often think of white men in business suits driving BMWs. In the film *Colonization Road* that I made with Michelle St. John, I argue strongly that colonialism, its effects and its systems, are alive and well in Canada today. They just wear different clothes. Yes, *Canada*. *Your* Canada. This Canada.

To create Canada, the government had to construct 1,600 kilometers of roads to allow easy passage for settlers to find the land they'd eventually call home. It was an entire system, called *The Colonization Road Act*, and the roads were called "Colonization Roads." It sounds crazy. So, I'm sorry to say, Canada does indeed have a long history of colonialism.

Two Row Wampum, 2014 (Ken Maracle)

It's a real roadblock for me as long as Colonization Road exists, literally or metaphorically. Born and raised in Fort Frances, I spent a lot of time on Couchiching, on my reserve, lands of Treaty 3. I drove on that road, past that road, around that road. I swam off that road into the lake. We don't consider what that street sign might mean. What is the challenge that I could offer to Canadians? To get to understand that a better relationship is possible, that we need you to be a part of that process to ensure that that relationship happens?

EN: Your film is absolutely fantastic and gives me a lot to think about. I was really happy to see people talking about the Two Row Wampum; what that Treaty means, and what that vision of living together looks like. The temporal parameters of the Two Row Wampum are really interesting. Over time, we stopped saying that Treaty is about our canoes (native/settler) next to each other because the truth is, we have a canoe and they have a steamboat. We all have a different way of relating to the water, but the vantage from the canoe and the steamboat are completely different. Interestingly, the Treaty was enacted "as long as the rivers run and the grasses grow." Now, what's really important about this focus on land and how land is shaped, is that *the rivers don't actually flow anymore*. Conditions today have completely changed. Transforming our land will actually affect our treaties… that's incredibly powerful.

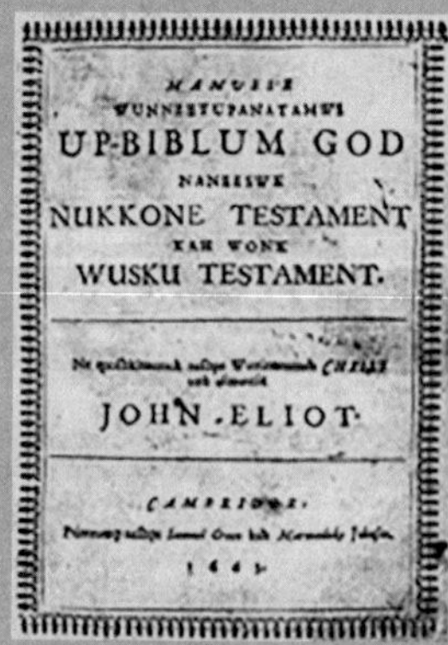

UP-BIBLUM GOD

NUKKONE TESTAMENT

WUSKU TESTAMENT.

JOHN ELIOT.

Eliot's First *Indian Bible*, 1663
(Harvard University Houghton Library)

RM: You're right, I don't believe we can live the same Two Row Wampum anymore in Canada. I don't believe we should change the name of the roads. The underlying project is Indigenous liberation. Full Stop. What Indigenous liberation means, is where nationhood is achieved through self-determining Indigenous Nations that made Treaty with this Nation State. It's fundamentally about *land*. It's funny that Canadians seem to be traumatized every time I show the film. Their backs always have to be rubbed a little on the way out of the theater.

EN: That's very similar to your comedy act where you have to say to the audience, "it's not your fault, it's your ancestors' fault." So, how do you reject *recognition* and enact *refusal*, but still continue to talk to people? This is *the* question that you're asking in the film.

Interestingly, Harvard University's 1650 Charter dedicates the institution to "the education of English and Indian Youth of this country in knowledge and godliness." When the college was struggling financially early on, it accepted funds in exchange for establishing an Indian School to educate Indigenous people to Christianize them. Today, over 300 years later, Harvard doesn't even have an Indigenous Studies department. We only just hired our first full time professor in Indigenous history this past year. Lobbying for recognition, for Harvard to recognize its foundational commitment to "educating Indian youth," automatically forces us to express Indigenous rights and politics on terms that are recognizable to settler institutions. And yet, I think that's very important work. Part of that hard work, like your film, is to take ownership of those institutions—to know them, name them, and reveal them for exactly what they are.

"Indigenous women right now are on the forefront of communicating the complications, violence, and perhaps the love, that we currently are experiencing as part of this colonial project, in what we call the colony of Canada."

—Ryan McMahon, Opening Remarks to Colonization Road Screening, Harvard University, 2017

RM: Bringing people along in this history, is a must. In Canada, we're celebrating the 150th birthday of a very young country this year. It's coined "Canada 150" with hashtags, t-shirts, even hotdog wrappers. Damn, you saw the graphic in the film, over a hundred First Nations communities in Canada right now—about one in five—are currently under boil-water advisories. They don't have clean drinking water, while there was no problem finding half a billion dollars to throw a big birthday party for Canada. So you know, we're still trying to tell people, "um, I'm not going to go to the birthday party, in fact if I do go, what I might do, is spill a little red wine on the white carpet. On purpose, accidentally, I go, "whoops, was that me? Did I ruin your party?" The film opens a doorway for people to walk through, but you can't just like hit 'em with pillows, or bricks, on their way through.

CN: The film floored me. Certainly, it's something that I would love to use to teach my own classes. I love the way you play on the discomfort of the White Canadian audience.

As someone who studies the black diaspora, I study enslavement and the representation of enslaved Africans and people of African descent. When I work on Quebec, I must address the fact that both Indigenous people and people of African descent were simultaneously enslaved together. That history is one that we haven't begun to fully explore, especially in terms of understanding the relationships that were formed between Black and Indigenous people who were enslaved in the same households.

For instance, the founder of my university in Montreal, James McGill, was a slave owner. A lot of people don't know or don't talk about it. McGill enslaved both Indigenous people and people of African descent. McGill became very wealthy from slavery, not just fur trade. When I went into the university archives, I found that he was trading in plantation-produced rum from the British Caribbean, in the period of Trans-Atlantic slavery. He was not just enslaving Indigenous and Africans in Montreal, but was also getting wealthy under the plantation regimes in places like Jamaica, Barbados, Trinidad, etc.

The Harvard Charter, 1650
(Harvard University Pusey Library)

Different practices of colonization and colonialism were inflicted on different groups of people in different ways, depending on the desired outcome. In the context of Quebec, the enslavement of different groups were not necessarily treated the same. Indigenous and Blacks were both considered 'undesirable, but necessary' populations, because white settlers needed to exploit them for labor. When you drill down and go back to things like sale ads for slaves or fugitives, you can see the usefulness and distinctiveness of those populations.

This brings up an issue for me as an educator: how do we teach these topics? At McGill for instance, I'm lucky if I have a single Indigenous student in a class of a hundred. Usually I have ten students of color… ten out of a hundred! It's 90% White. For a class called "The Visual Culture of Slavery," I come at the topic of slavery through art and visual culture. I start by saying, "How many of you know, as Canadians, coming through the door on the first day of class, that slavery happened in Canada?" The answer is: nobody. Since 2003, I have never yet had one Canadian student enter my class knowing that slavery happened in Canada. What they have all been taught, since Grade 2, is that Canadians were the wonderful abolitionists who saved African American enslaved people fleeing North during the Underground Railroad. With the American students, their blind spot is that slavery never happened in the American North, it was just in the South.

So, to this question of attribution or fault, I say to the students: "part of what will get you to engage in this class is your understanding that slavery is *your* history. When you go to your ancestral tree as a White Canadian, you're going to hit someone who was manning the slave ship, a crew of the slave ship, a merchant, someone selling the sugar in a shop… somebody either directly or indirectly implicated. It's not about the black kid sitting beside you—their history, it's everybody's history."

> *"Each time I confronted white colonial society I had to convince them of my validity as a human being. It was the attempt to convince them that made me realize that I was still a slave."*
>
> *—Lee Maracle, I Am Woman: A Native Perspective on Sociology and Feminism, 1996*

RM: In a way, bringing people to that colonial realization, is about looking at the systemic structures put in place to ensure Indigenous people remain impoverished on the Reserve. Our special rights granted through Treaty are not transportable. In Canada, the special benefits or rights of being a "Status Indian" under the Indian Act—the apartheid system of reserves that was adopted in South Africa—are extinguished once you *leave* the Reserve. Today, we are still trying to convey to Canadians how this system exists in our daily lives, and permeates everything that we do. That's the project of the decolonization of education.

Map Showing Line of Possible Routes Between Lake Superior & the Red River Settlement, 1869 (University of Manitoba Archives, Josiah Jones Bell fonds, MC 11 (A.01-54), folder 1, item 2.)

CN: You can also see that in the crafting of scholarship. If you look for slavery studies in Canada, you will find it in the section on Black Canadian Studies. Suddenly, it became *Black history* and nobody else's history. So, I push the students to go further: "go home and talk to some of your elders and see what comes up." One very brave student came back to me once and said, "Professor Nelson, I found out that one of my great-great-great grandfathers was the Governor of Virginia." This student never knew about this, they never talked about it; it was just a secret in the family. And of course, this governor enslaved many people. Other students came back and said, "you know I have an ancestor who was an indentured servant, but people have been hiding this from me because they're ashamed of it." And I respond: "you have to rethink that… if they wouldn't have survived their contract, you would not be here."

So, to this question of who's at fault: for people today of course, it's not your fault but you are inheriting the privilege of that White ancestry. We are still being colonized. We need to rethink how we engage at all these different levels.

RM: Lee Maracle, astutely says in her scholarship, that we actually have to be very careful about what we hand off to Canadians, especially to young Canadians that are engaged in this discourse. Because it's not their fault, but they do benefit from it like their relatives and their relations. But, how much of the blame or the story can they shoulder when they're first learning about it? Maracle has taught me a lot about being delicate in introducing these conversations. Whereas before I probably wasn't so nuanced and just, you know, too angry about everything.

It's an interesting conversation as a father as well. I have two kids. My mother is a Residential School survivor, three of four of my grandparents are survivors. As I carry their stories now through my art, the question for me is how much of this history do I give or share with my daughters? How much do they need to know as young Anishinaabekwe—young Ojibway girls—in terms of who they are.

What they do need to understand is that Indigenous women are disproportionately murdered in our country. In Canada, there's a National

Inquiry into Missing and Murdered Indigenous Women and Girls taking place right now. The estimated number stated publicly is 1,200 missing and murdered Indigenous women since 1980, but behind the scenes and probably more accurately, the number is somewhere around 4,000.

Ryan McMahon & Michelle St. John, 2017 (Kelsey James / The Projector)

Violence on the land is directly tied to violence against Indigenous women and Two-Spirit people. When you're bringing in man-camps to the North or South Dakota for example, and when you're bringing in mining to the territory, there are workers and visitors who don't give a fuck about the people that were there before them. They are at the bars, in the shopping malls. And, it shouldn't be lost on us that Indigenous women and Two-Spirit people have faced the brunt of the colonial project over the last, several hundred years. But moving forward, their voices are at the center and we will start to see substantive change. So teaching young people today about how this colonial system works is dicey, but we have to try to approach these questions with a fair and balanced mind, while doing so in a gentle and loving way.

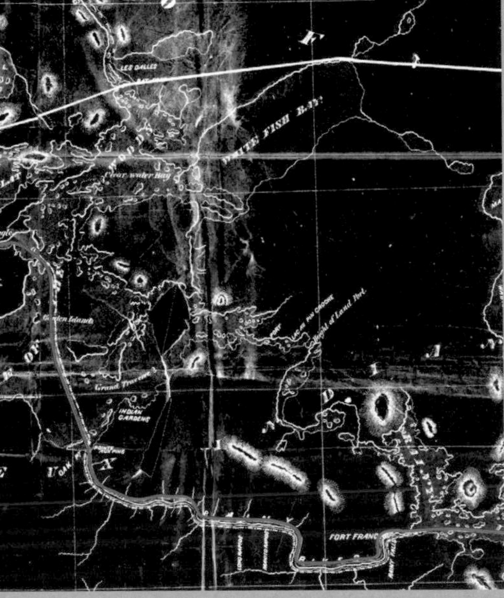

CN: The parallels between Indigenous life and Black life in the Americas are fascinating. When you speak about your daughters and what you share with them at certain stages of their lives, it makes me think of my own experience. I was born in Toronto (Canada), to two Jamaican parents, who immigrated to Canada where a lot of African-Caribbeans immigrated to in the 1960s. Looking back at my childhood, as someone who is now so passionate about histories of slavery, my parents never gave me the 'slavery conversation.' I can't pinpoint when I learned about slavery but it's as if I always knew about slavery. It's not what they directly said to me, but listening to them speak to their friends and commiserate… their remembrance of Jamaican culture as well as their articulation of racism they experienced as Black people entering Canada in the 1960s and 70s. They all said that they moved to Canada to give their children a better life. But then, what did that actually mean to them? What did that mean to us as children, who were confronted in many cases by excruciating forms of racism in White schools? And then on top of that, imagine the types of whitewashed education that we were learning.

I also know that Black, Brown, and Indigenous people have these conversations, especially with their boys. If you're a responsible Black parent, there is a certain moment when you talk to your boys in their 'tween' years about how to deal with the police. If not, you're falling down on the job. That moment is a moment of stolen innocence. Since a kid just wants to play, he doesn't want to know where to put his hands when the police stop him. We know that police often 'age' Black children, so the kid that is actually 11 years old, the police see him as 18. You're forced to have these conversations in order to try to protect your children from the infrastructures of racial violence.

EN: Erasure is also part of the structure of settler-colonialism. To the extent to which someone is raping and murdering these women, these people are not only still benefiting from, but in fact perpetuating, the structures of settler-colonialism. This raises many questions. To what extent that structure is different for different populations? What Indigenous children have to grow up knowing versus what settler children have to grow up learning? And how settler children are allowed to have this innocence going forward. How difficult is that?

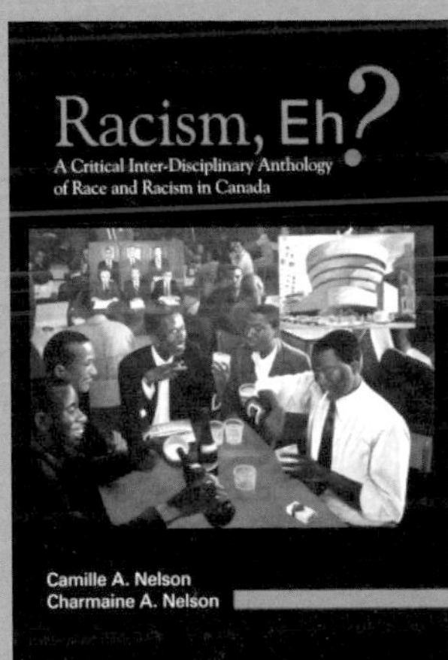

Racism, Eh? edited by Charmaine Nelson & Camille Nelson, 2004 (courtesy of Captus Press Inc. http://www.captus.com)

CN: There are also different practices and forms of colonization. As you said so astutely Eli, settler-colonialism is one of those practices. What does colonialism look like in places like Canada, Australia, and the United States, where the Whites came *en masse* and stayed, where? They became the majority population? In this context, Indigenous people are displaced in their own lands on one hand, and then on the other, the expropriation of Africans who are removed from the continent of Africa and scattered across the Americas.

The Africans, in some places like Canada, have become a minority. But in other places like Jamaica, they are the majority. In the context of

Jamaica, Whites are outnumbered because they simply own the plantation while living in London, England or other places. They were absentee planters. In settler-colonies like Jamaica, you have infrastructures of violence like the work-house and the jail. These were not just places of incarceration, but places where the enslaved fugitive who got caught, was then slated to be punished. You left them to languish in the work-house, where they were collared and forced to build roads. We should think about what colonization *actually looks like* and what that means in terms of the violent infrastructures of colonialism that are imposed upon certain populations.

The film articulates all the instruments and infrastructures of dispossession. Children taken away from their parents and elders. Their language is broken. They're forcibly Christianized. They're changed physically, how they look, by altering their hair or by dressing them a certain way. What does that infrastructure actually look like?

Burial Mound in Koochiching County, MN, ca. 1985–2000, (Cartwheel Company Postcard, courtesy of Koochiching County Historical Society)

The Cuban scholar Antonio Benitez Rojo talks about the maritime infrastructure of imperialism. He calls it *la flota*, 'the fleet system.' That's very similar to what *Colonization Road* talks about. Your film peruses the *inland*, terrestrial infrastructure of colonialism, whereas Rojo talks about the *offshore*, maritime infrastructure: places for ships to dock and to be cleaned, etc. In the age before planes and trains, imperialism was all about the boat. The original infrastructures of Empire and of Slavery were constituted around bodies of water. Cities like Halifax, Quebec City, and Montreal are so important in the scheme of what becomes Canada, because they were port settlements, on the water. It's fascinating (and horrifying) to see then how *inland* colonization evolved with development and technology of roads. These technologies are always discussed as being liberating, good for everybody, but they're actually manipulated strategically to oppress certain populations and to bring benefits to other populations.

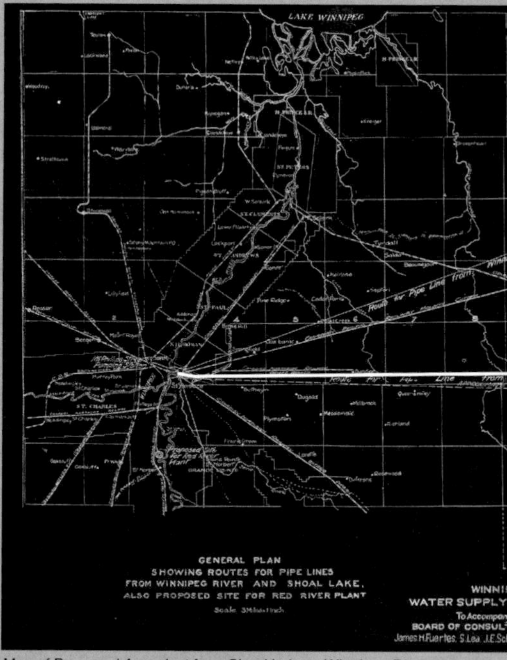

Map of Proposed Aqueduct from Shoal Lake to Winnipeg River, 1907 (University of Manitoba Archives)

EN: That's one of the reasons why the *Colonization Road* film is so important. It shows that this physical manifestation of colonization cannot be erased. Quite literally, it's called "Colonization Road." That's very powerful.

CN: The film also reminds me about scholars who talk about the creation of Canada as a colonial nation state and the different ways that "undesirable populations" are managed, as Scholar Eva Mackey mentions in her 1998 book, *The House of Difference: Cultural Politics and National Identity in Canada*.

RM: In my hometown of Fort Frances, there was a vote very recently. A petition was started where all good discourse happens, on Facebook. They were calling for the renaming of 'Colonization Road.' They got enough signatures from the petition, sent it to the Town Council, but the councilors voted unanimously, 6 to 0, against the name change. The Town Council claimed that the name change would cost too much money to change identification, mailing addresses, and that the Province of Ontario would bungle the process. There were all these so-called big questions. The Town of Fort Frances was up in arms: "Oh, what number will be on the front of my house when the mail comes? ... A hundred and twenty homes would be affected!" But, it turns out, all of these reasons were not true. Changing your ID is free. No cost at all. The provincial government even put out a special letter three years ago stating that the renaming of these roads would be handled using a Special Office. Even after this, the Town Council of Fort Frances said, "Well, still, no!"

I've publicly stated many times that renaming the road is *not* the project for me. Systemic wholesale change is *the* project. There are so many stories like this.

To juxtapose Shoal Lake's need for a road on one hand, with the renaming of a road in Fort Frances on the other, strikes me as totally

"Pure untouched nature can be constructed as the raw material for the civilising work of settlement; it re-affirms the settlers' sense of themselves as those who transform raw nature into developed civilisation. Indeed, in this image, the early nature/Native roots of Canada are then integrated into a narrative of settlement and progress, as the settlers categorise, map, re-shape and civilise nature itself."

—Eva Mackey, The House of Difference: Cultural Politics and National Identity in Canada, 2002

paradoxical. All they want in life at Shoal Lake 40 First Nation *is* a road. They have been physically cut off from the mainland with the damming of their lake by the City of Winnipeg, creating a massive water diversion with a huge 150-kilometer long water aqueduct. The road will reconnect them. I was recently able to travel down the first 12 kilometers of that new road connection. It's beautiful. It's called *Freedom Road*.

CN: For Black communities in Canada, the question of renaming of colonial infrastructures is also very important. We've had a debate recently in Quebec because we have some horribly racist place-names like 'Nigger Rock.' It's actually the name of a well-known cemetery where both free and enslaved black people were interred near the Quebec-Vermont border. Like you Ryan, I am less invested in debating whether or not the name should be changed than the fact that this cemetery is privately owned and it is still being farmed! By living in Canada and by understanding the way colonialism works, I would argue that I don't want the name changed. It would just promote good, old Canadian amnesia that says "Oh well, we never called it that… it's always been called African-Canadian Rock." I'd rather live with the racist name and talk about why that horrible name is there, while dealing with the actual problem of the infrastructure as opposed to the practices of renaming as some kind of band-aid.

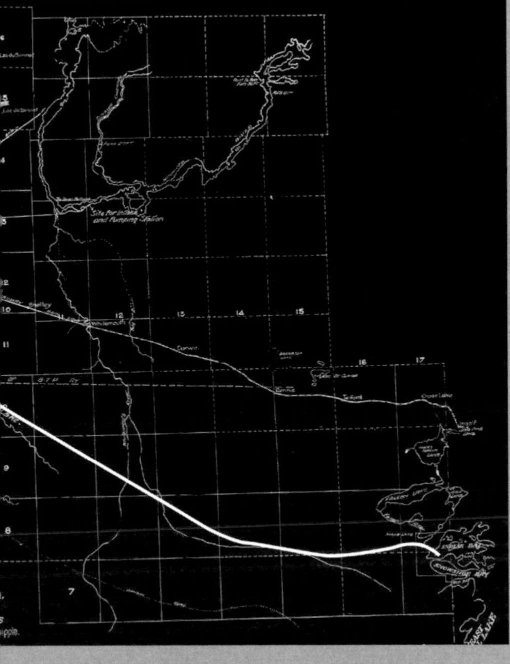

RM: It's similar to the problem and paradox of recognition. What are we going to rename Colonization Road… Anishinaabe Way? The truth of Colonization Road in Fort Frances is that it runs to the foot of our sacred burial mounds, which were bulldozed by the Royal Ontario Museum in the 1950s. I've seen the contents of those burial mounds. Adding insult to injury, the Town of Fort Frances and the Province of Ontario are actually going to court over that land where that burial mound was located because the 100-year lease on that land lapsed in 1997. Indian bands are saying, "Ok thanks, we'll see you later. This is our land again!" And the Town responded, "Oh, no, no, no…this is ours." So now, they're embroiled in a land claim dispute.

If we were working under the guise of 'reconciliation' in Canada, the question of the land claim is one that, in good faith, could close the books. The Town could walk away and say, "Sorry, we're wrong." Instead, they're trying to starve us out, economically. We'll eventually run out of money to pay for our lawyers and we'll lose. The land will then be theirs.

There are a lot of questions and contradictions surrounding the renaming of things. It has started a massive debate across Canada. They're showing this film in towns and cities across Canada to rename things. Take, for instance, Ryerson University in Toronto. Egerton Ryerson was the architect of the Indian Residential School System. The debate is of course happening in the US, in places like Charlottesville. We are finally talking about what these monuments mean, and the film has certainly stirred that debate in Canada.

Civil Engineers at Shoal Lake Aqueduct, Mile 71, 1915 (City of Winnipeg Archives)

There is a now a major discursive shift happening in Canada. In 2008, Canada's Prime Minister Stephen Harper apologized for the Canadian Government and the Church's role in attempts at assimilation of Indigenous people, in spite of denying Canada's history of colonialism a year later. So, once you learn about the lies and the oppression—you can't forget it. It calls into question everything you thought you knew. Now, we're seeing political parties in Canada trying to respond with Indigenous specific platforms, universities and colleges trying to respond to the *Truth and Reconciliation Commission*'s "Calls to Action." Down to the grassroots, we're seeing more and more solidarity being built between communities, faith groups, First Nations, Indigenous and non-Indigenous people, from all walks of life. There is a positive momentum for change.

EN: On the U.S. side, we're seeing a major shift in both Supreme Court cases and Indigenous activism, with the responses to that. A lot of my mentors are saying there is going to be a shift towards violence in a way that we haven't seen in a while, and possibly in a good, productive way. This generation of Indigenous activists, the people you saw at Standing Rock, and Indigenous youth were a key organizing component of the 'Water Protector' Movement. They're putting their bodies on the line in a way we haven't seen in a long time. Recognition and reconciliation might have been the paradigms that emerged on the settler side from the Red Power Movement, where we saw a similar turn toward bodies on the line, decades ago. We may very well see something new on the horizon. That we are, in this moment, in a paradigm shift, as you might say.

The Red Man Laughing Podcast
(Ryan McMahon)

RM: The unique challenge here in the U.S. is the sheer size of the country. I was so excited about Standing Rock and the huge visibility it had. Almost for the first time, the whole world was paying attention. Certainly, we know what happened at Standing Rock and the blockade against the pipeline: the camps were dismantled, water protectors arrested. I really believe though that once people have the information, they can make up their own minds pretty quickly that Indigenous people, North and South of the border, have not been given a fair shake. Change is being brought on by artists on the ground, not the politicians from above. Across media, film, music, and visual art, Indigenous artists are telling and sharing the story. Art is more accessible, it opens a doorway for people to walk through.

That's why my focus is squarely in media, and the podcasting space. Podcasts get millions of weekly downloads, there's great potential for exposure. But for all the talk about the U.S. and the history of architecture, science, or history, Indigenous people are completely erased from the conversation. But, we're so close to breaking through to some major influencers in new, digital media spaces. Once we start giving this conversation to people, there is cause for pause and consideration. I believe in the goodness of people, and they're starting to listen.

EN: Some of the podcasts that you have a hand in, like *Red Man Laughing*, are also some of the few things that make me hopeful too. *Métis in Space* by Molly Swain and Chelsea Vowel is one of my favorite podcast ever. But I can also remember how, just a few years ago, *The Daily Show* brought in a Native American comedy troupe, the 1491s, to talk very calmly with football fans of the Washington Redskins about racist mascots. The audience walked out saying that they felt attacked. There was this sense of extreme anger and rage, because of white, settler entitlement.

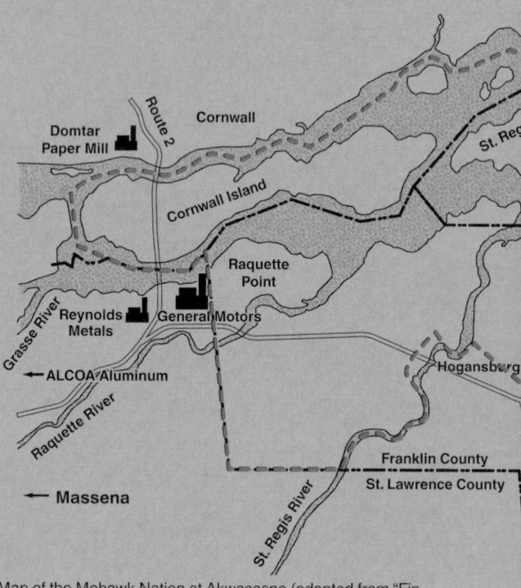

Map of the Mohawk Nation at Akwesasne (adapted from "Fingerprinting PCB Patterns among Mohawk Women," *Journal of Exposure Analysis & Environmental Epidemiology*, 2001)

Beyond podcasting, there have been a lot of interesting partnerships and environmental task forces that have grown out of thinking through changes to the land. A lot of these have been different models of partnership programs—for instance, with the Forest Service and the Environmental Protection Agency—but there are probably more radical routes that we can take.

Alternatively, there are really influential counter-projects like the *Akwesasne Mother's Milk Project* started in 1985 by Mohawk midwife Katsi Cook. That project is led by Indigenous women that carry centuries of weight from harmful, colonial western doctors and industrial practices, but now they're finding ways to heal *with* the land. Some of the ways that we're thinking through and dealing with oppression, pollution, and environmental harm by large infrastructures such as the St. Lawrence Seaway, and the industries it supports, is by focusing on *care*. The consequences to the land from pollution at the General Motors Plant (with ALCOA and Reynolds Metals upstream) in Haudenosaunee country along the Seaway, particularly at Akwesasne for example, have been horrible. When pollution, with neurotoxins like methyl mercury and PCBs were found in the drinking water and fish at Akwesasne for example, the New York Forestry Department stated it was "the worst place in the world to be a duck"; but they didn't even mention that people were still living there.

Métis in Space Podcast
(Molly Swain & Chelsea Vowel)

"A civilization that proves incapable of solving the problems it creates is a decadent civilization. A civilization that chooses to close its eyes to its most crucial problems is a stricken civilization. A civilization that uses its principles for trickery and deceit is a dying civilization."

—Aimé Césaire, Discourse on Colonialism, 1972

A lot of partnerships and projects are now about decolonizing education, which is what I work on. In my case, with the field of Native Science, I work with ways of knowing that actually start us at the root source. It's not only about talking to people educated in the western academy, but taking seriously other ways of doing things and applying those different ideas. Care is the most important thing here.

RM: A lot of the political movements right now with young Indigenous activism and activists are about water, oil, and pipelines. Standing Rock was all about that. We're starting to see more conversations about the environment and how unsustainable capitalism is. Collapse is near… we all feel and fear that. But how that plays out on the ground is still in question. This is why the voices of Indigenous people, with lives bound to land, are so important. These voices are essential to a pathway forward in North America.

CN: Also, there have been many scholars who invest a lot of time in studying the psychology of what we call 'colonialism,' and they always study the people who are colonized. But if you look back at someone like Aimé Césaire, he was really interested in studying what happens to White people. What kind of damage is done to your psyche when you colonize other people over centuries? I think we really need to examine both sides of that, and to really think about the wounding of White people in terms of what happens to you when you brutalize other human beings.

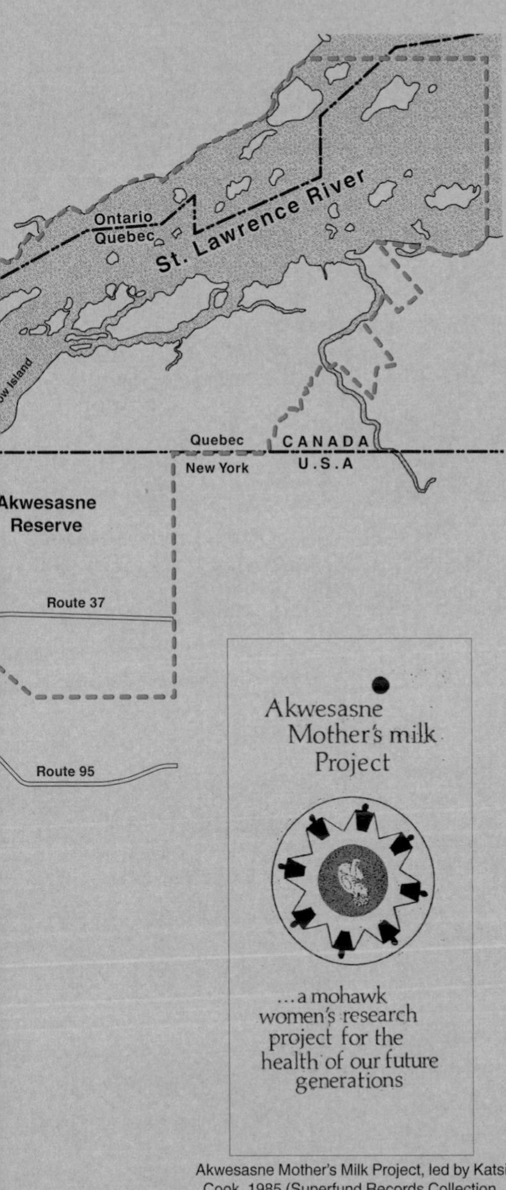

Akwesasne Mother's Milk Project, led by Katsi Cook, 1985 (Superfund Records Collection, Environmental Protection Agency)

RM: What I'm starting to think more deeply about now, is the psyche of those that work in resource extraction towns and the disproportionate violence that you see in man-camps, with the types of lives that they live. Recently, I met someone on a plane working in the Tar Sands at Fort McMurray and they said that they were leaving after "Seven Years of Hell." He moved there for the promise of work. They get paid very well to live up there. There's a remoteness allowance, and all kinds of other special incentives to live up there. But on payday, he and his friends couldn't keep the money in their pockets. Halfway through the month, his entire paycheck was spent. He became an alcoholic and a drug addict. I wonder about that a lot. That the minute you get paid from doing the job, your salary is already spent. Now, the Tar Sands are slowly starting to close and dwindle down, to just a fraction of what they were. The jobs, the money, have dried up.

CN: The other part too is *representation*. When we think of colonial infrastructures and different forms of colonization and the infrastructure of slavery, one of the surest ways to oppress and control Black and Indigenous people, is by denying them access to *self-representation*. Enslaved people were beaten and executed, if they learned to read. If there's a revolution in Haiti, you don't want your enslaved people here in Jamaica to know what's going on over there. If they can't read, they can't represent or portray themselves… because *their* portraits are constantly denigrated by *your* portraits.

It's about the infrastructures of culture that are policed and withheld from people as tools of colonization. To put these mechanisms—media and art making—back into the hands of people who have historically been colonized and disenfranchised, let us represent ourselves. As a Black art historian, I couldn't have been asking the types of questions fifty years ago that I now ask in this discipline. As communities of color and Indigenous people, we need to nurture those artistic seeds in our youth and in our children to do the incredible work that you are doing as well.

RM: To problematize this situation even further, I look White. That really messes with people's first impressions. The other thing about representation that is really weird, is the Indigenous erasure practiced by painters of 'landscape art' from the *Group of Seven*. Canadian art space holds this group up as the preeminent visual artists of our history as a country. They constructed the idea of 'wilderness.' But their paintings actually created *terra nullius*; the idea that wilderness is out there, remote, empty. There are no people in their paintings of the 'wild.'

CN: And they're getting to these 'wild' spaces through the colonial infrastructures that we've been talking about! They had to take the train to get to these places. And once they got there, they didn't represent anybody, not even themselves.

RM: So, with the absence of people from our crown jewels of achievement, when you walk through the National Gallery of Canada, you start to look at this work, and realize pretty quickly what people saw. I think art and representation play a massive role in Indigenous resurgence and the process of decolonization.

CN: What your film does so beautifully too, is talk about the outcomes and horrors of structures of dispossession… for instance, like the ways racism gets internalized. When the female Elder was talking about her father not feeling at peace until a White person told him that it wasn't his fault… as opposed to someone from his own community. But you can understand how that happens, because of the ongoing abuse over decades that the man had internalized.

RM: So now, Canada is reconciliation-drunk. Our Prime Minister, Justin Trudeau, is like the Mayor of a New Town being built called Reconciliation-Ville. Population: You and Me. Now, there are cities, towns, universities, corporations, banks across the country that are all trying their hand at this 'reconciliation.' In the backdrop, there are '94 Calls to Action' from the recent *Truth & Reconciliation Commission* (TRC) that are specifically targeted at certain groups of people in the country to instigate change.

The City of Edmonton has been one of the first cities in Canada to respond in a real tangible way. In Treaty 6 Territory, the City of Edmonton has formed an Elders' Council that works with all departments across the city, including urban planners. When new structures go up in the city, those planners and builders now have to sit with the Elders first. They have to take into consideration what materials will be used, where they're sourced, and the season in which they start building… all of these things are taken under consideration through the Elders' Council at the city level. This is pretty amazing.

There's also a major bike and infrastructure plan being designed in Edmonton. The City is acknowledging important traditional places and place names. Instead of shutting their eyes to that history, they are facing it and saying, "we understand that this is a sacred site, but we're also here. We can't tear down the apartment buildings and remove all of these people that are here." As they're working through their territorial acknowledgement in the Traditional Territory of the Cree, they're also saying: "What does it mean to be here?" In the case of the citywide bike path network, they're making special spaces in traditional places where trade and economy happened. They're planning to open up shops and enable Indigenous artisans to visibly create economies in that space. Indigenous place names are being used along that bike path to acknowledge Indigenous territory, land, and space. These are just a few ways that people are grappling with the meaning of the reconciliation project.

EN: This leads me to the discussion of *where colonization road goes*. How you reach Indigenous land is an experience that continues to entrench settler-colonialism. In my work, I look at how this is enacted during the New Deal Era of the 1930s. As attempts to re-create settler space and reshape the country, the National Parks and the National Forests were ways to recapture the colonizing experience as the defining experience of North Americans that still exist today. This construction of colonial space is in part about resource extraction and where capitalism can thrive, but on the other hand, it is also about perpetuating the settler colonial narrative. These spaces continue to promote a continued relationality of dominance.

"To consider Europeans as intruders is to challenge the ways in which they presented their colonial actions as natural and legitimate; it is to interrogate the very processes of colonization by recuperating the presence of indigenous peoples and questioning the forced removal and transplantation of Africans as 'necessary' supplements for their imperial designs."

—Charmaine A. Nelson, "Interrogating the Colonial Cartographical Imagination," 2017

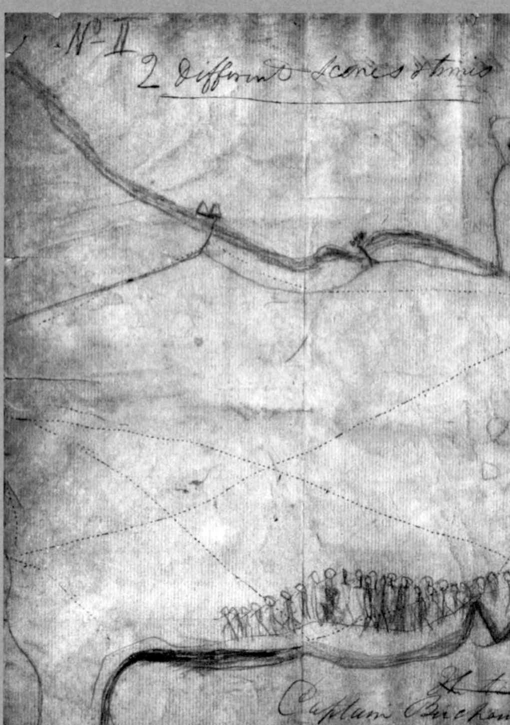

Shanawdithit, Sketch II. Captain Buchan's Visit in 1810–11 at the South Side of the Lake (The Rooms Corporation of Newfoundland and Labrador, Provincial Museum Division)

"Stormy Weather, Georgian Bay" (left), by Frederick Varley, 1921(National Gallery of Canada) in comparison with James Hakewill's "Port Marial" (right), in *A Picturesque Tour of the Island of Jamaica*, 1825

R A N - A W A Y

FROM the Subscriber, on Sunday morning the 24th ult. about four o'Clock, a Negro Lad named *NEMO*, born in Albany, near eighteen years of age, about five feet high, full round fac'd, a little mark'd with the Small pox, speaks English and French tolerably; he had on when he went away a double breasted Jacket of strip'd flannel, old worsted Stockings, and a pair of English Shoes. Also a Negro Wench named *CASH*, twenty-six years old, about 5 feet 8 inches high, speaks English and French very fluently; she carried with her a considerable quantity of Linen and other valuable Effects not her own; and as she has also taken with her a large bundle of wearing apparel belonging to herself, consisting of a black sattin Cloak, Caps, Bonnets, Ruffles, Ribbons, six or seven Petticoats, a pair of old Stays, and many other articles of value which cannot be ascertained, it is likely she may change her dress. All persons are hereby forewarned from harbouring or aiding them to escape, and Masters of vessels from carrying them off, as they may depend on being prosecuted to the utmost rigour of the Law; and whoever will give information where they are harboured, or bring them back to the Subscriber at Quebec, or to Mr. GEORGE ROSS, Merchant at Sorel, shall have TEN DOLLARS Reward for each, and all reasonable charges.

HUGH RITCHIE.

N.B. The Lad was seen at Sorel on Friday morning the 29th ult. and there is reason to believe they are both lurking thereabout.

Quebec, November 2, 1779.

Fugitive Slave Advertisement, 1779
(Montreal Gazette)

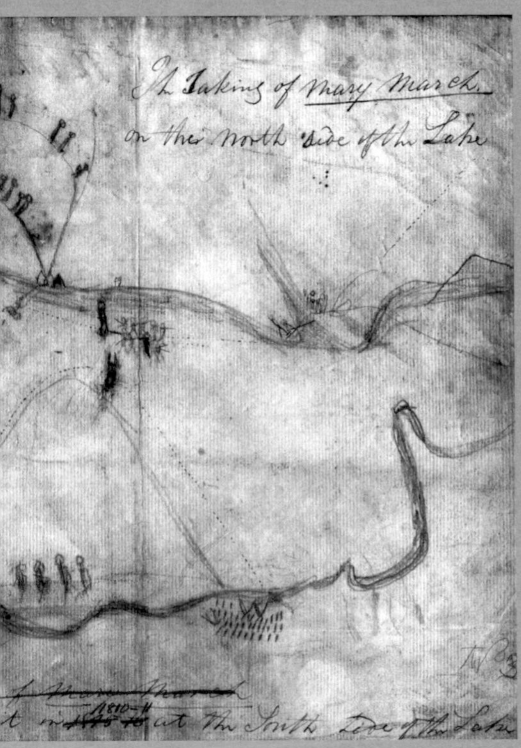

CN: Part of what we have to do is challenge White people to reconfigure their understanding of themselves and their claim to neutrality. They experience their racialization as *privilege*—as opposed to marginalization. In my most recent book, *Slavery, Geography, and Empire in Nineteenth-Century Marine Landscapes of Montreal and Jamaica* (2016), I deliberately positioned White colonialists as *imperialist intruders*—as opposed to people who showed up and were just supposed to be there. Everything they do after, including practices of cartography and geography, is naturalized. But actually, they are instruments of imperialism with two sides. How do I create a map in a way that allows me to dispossess Indigenous people and claim that they weren't here? How do I create a topographical landscape to indigenize Africans?

In places like Jamaica, you erase the Indigenous people, literally through the practice of genocide, and then you expropriate the Africans, drop them in Jamaica, and make Blacks look as if they've always been there. Why? Because you don't have to deal then with the trauma of the Middle Passage.

Enslaved Africans were often used as domestics and treated as luxury items with the perception of their foreignness… regardless of whether they were actually foreign. Then, Africans start to be born in Canada, and you get a population that is actually African-Canadian. But again, in their blackness, there's still this idea of foreignness.

In *The Making of Race and Place in Nineteenth-Century British Honduras*, scholar Melissa Johnson writes about how the reverse scenario played out. Enslaved Black men were considered 'at home,' in the country, because their labor was to identify and harvesting mahogany trees. The knowledge that they built over time was seen as essential to them as Africans, but it was simultaneously a way to police them and keep them out of the urban spaces deemed to be the spaces of White (modern) Hondurans.

As a Black Canadian, this also plays out because I always get asked where I'm from, because I can't possibly be Canadian… even if I speak like a Canadian. Whereas a White Canadian, who maybe literally just got off the plane, is always already assumed to be Canadian because of their whiteness. Also, Black and Brown bodies are seen as belonging in cities like Toronto, Montreal, or Vancouver, not in rural or suburban spaces. Your Black body belongs in a city. We have to understand that those views are artificial constructs. How can we mess with that?

All of that is manifested through the terms of representation in what we call 'western geographic practice.' What happens then when we position Whites as *intruders*, as opposed to people at home in the spaces of the Americas? What does colonization look like in a specific space? What are the needs of the colonizers? How are they going to disenfranchise or marginalize certain populations on the basis of what they need that population to be for them in terms of wealth creation?

None of this is essential. If it was essential, then Black people would always be seen as un-homed in the country. But that's not the case. We need to think about the ways in which we can disrupt patterns and understandings of whiteness. This all has to do with colonial geographies and they need to be mapped out.

This conversation took place between Ryan McMahon, Charmaine Nelson, and Eli Nelson following a screening of the film *Colonization Road* at Harvard University's Weatherhead Center for International Affairs on October 30, 2017. Part of Ryan McMahon's introductory remarks are excerpted from his "12-Step Guide to Decolonizing Canada" for CBC Radio. Note that "RM" stands for Ryan McMahon, "CN" for Charmaine Nelson, and "EN" for Eli Nelson.

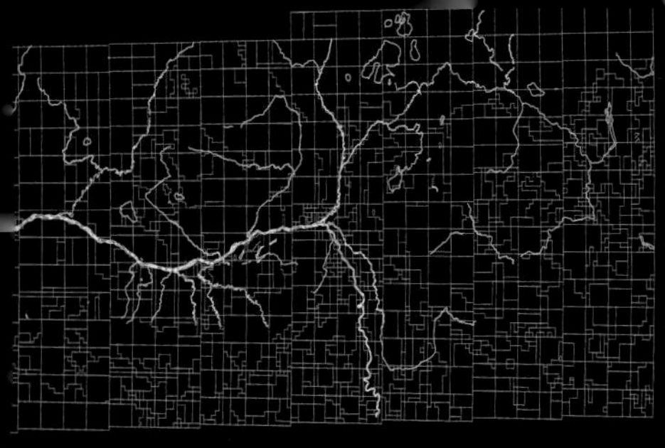

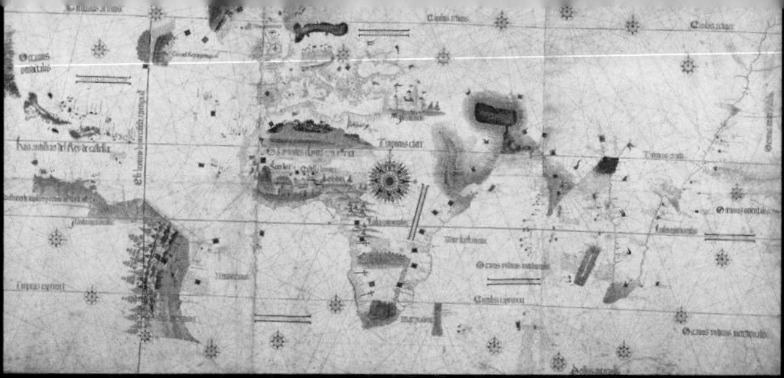

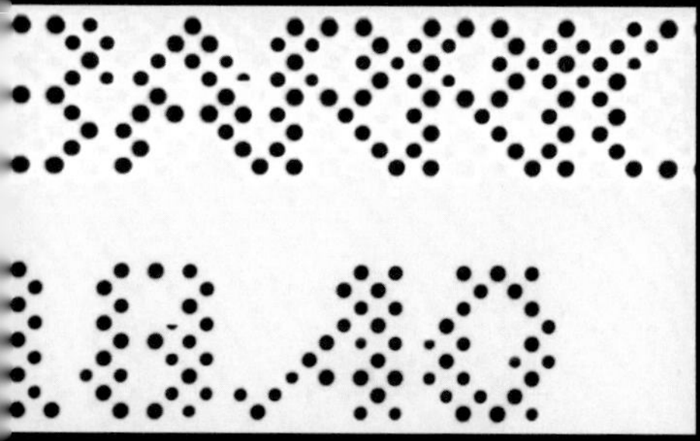

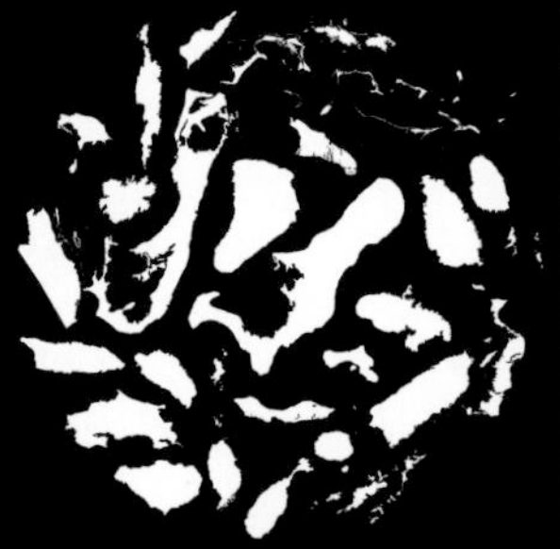

Watermarks & Endpapers. Front / Back: *Athabasca River, Townships, Oil Sands Leases, and Industrial Facilities*, north of Fort McMurray, Alberta, Canada, 2014 (Terracon Geotechnique). Page 28–29: *Doctrine of Discovery & Papal Bulls*, 1455–1502 (Cantino Planisphere, Biblioteca Universitaria Estense, Modena, Italy). Page 32–33: *Upper Canada Survey & Canada Company Dominions*, James Chewett, Assistant Draftsman, Thomas Ridout, Surveyor General, 1825 (via University of Toronto Map & Data Library). Page 34–35: *Mineral Map of the Dominion of Canada*, Department of Interior, 1924. Page 145: *Geologische Karte der Deutsch-Ostafrika*, Bearbeiret van Curt Gagel, 1908 (Universität Greifswald). Page 62–63: *Memorandum of Agreement for the development of the Porgera gold mine between the Independent State of Papua New Guinea and Porgera Landowners*, May 1989 (Columbia Center on Sustainable Investment - Extractive Industries). Page 70–81: *Toronto Stock Exchange (TSX) Ticker*, 2016 (Zannah Matson). Page 90–91: *World Financial Offshore Islands* (Olga Semenovych, 2016–17). Page 92: "In the Rubber Coils," King Leopold as Snake, *The Punch*,1906 (Edward Linley Sambourne). Page 105: *Kingdoms and States situated within the headwaters of the Rivers Atbarah and the Blue Nile*, Africa, ca.1807 (Arrowsmith Map of Africa). Page 162–63: *Aeronautical Map & Air Space*, Lake Ontario, Mohawk Territory, 2016 (skyvector.com). Page 194–95, 204–05: *Mine Tambang to Sumatra*, Indonesia, 1694 (Artokoloro Quint Lox Limited). Page 234–35: Sir Charles Lyell, F.R.S., *Travels in North America, Canada, and Nova Scotia*, 1855. Page 250–51: *Lord Selkirk's Colonists, the romantic settlement of the pioneers of Manitoba*, George Bryce, ca.1909 (Library and Archives Canada / AMICUS No. 5614009). Page 295: *Proposed Plot Layout for Agisanyang*, Botswana, John van Nostrand, 1982 (Old Naledi). Page 252: *Idle No More!* Silkscreen, Andy Everson, K'ómoks First Nation, 2016 (courtesy of Andy Everson). Page 296–97: One of the five, 100-kilogram gold coins valued at over 1 million dollars produced by the Royal Mint of Canada in 2007 (REUTERS / Toru Hanai). Page 552–53: Surrender of the "Mississague tract," 1818 (Treaties, Surrenders and Agreements RG10, Vol. 1842, IT 060 ©Government of Canada, reproduced with permission of Library and Archives Canada, 2017). Page 436–37: *A Sketch showing the Lands occupied by John Small, appropriated for the Government House at York by his Excellency Lieutenant Governor Simcoe*, 1805 (Toronto Reference Library / Baldwin Collection / Ms1889.1.5).

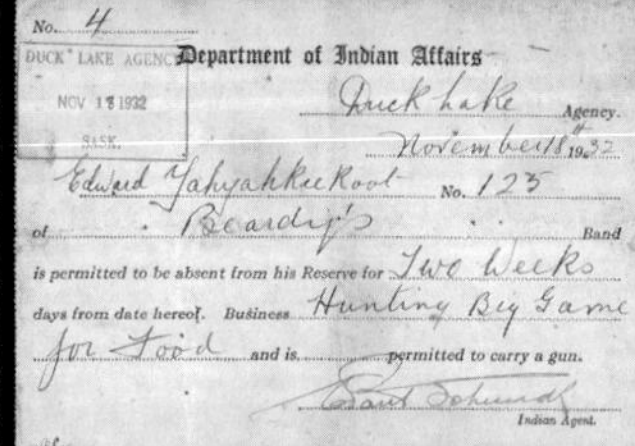

No. 4

DUCK LAKE AGENCY NOV 18 1932 SASK.

Department of Indian Affairs

Duck Lake Agency.

November 18th 1932

Edward Yahyahkeekoot No. 125

of Beardy's Band

is permitted to be absent from his Reserve for Two Weeks days from date hereof. Business Hunting Big Game for Food and is .... permitted to carry a gun.

[signature]

Indian Agent.

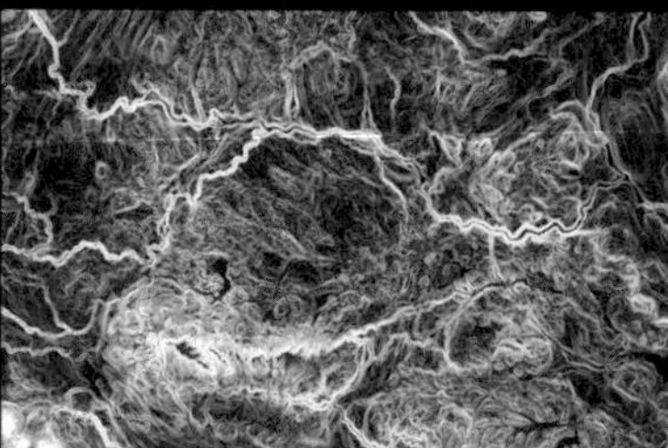

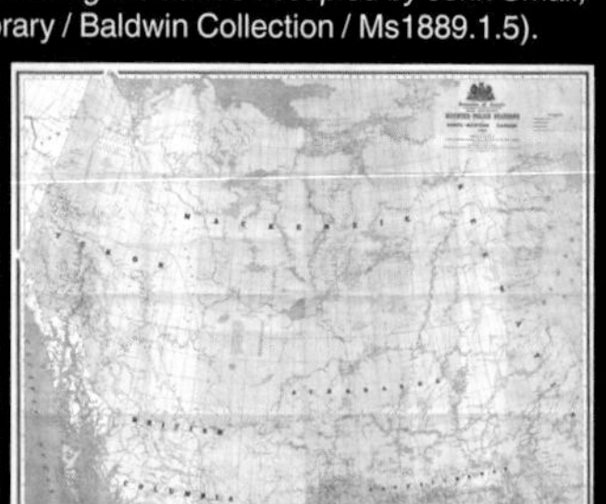

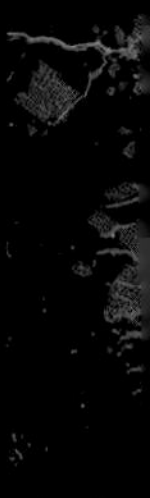

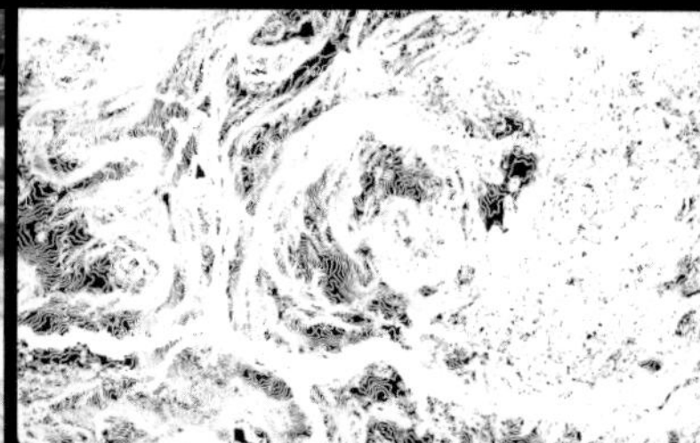

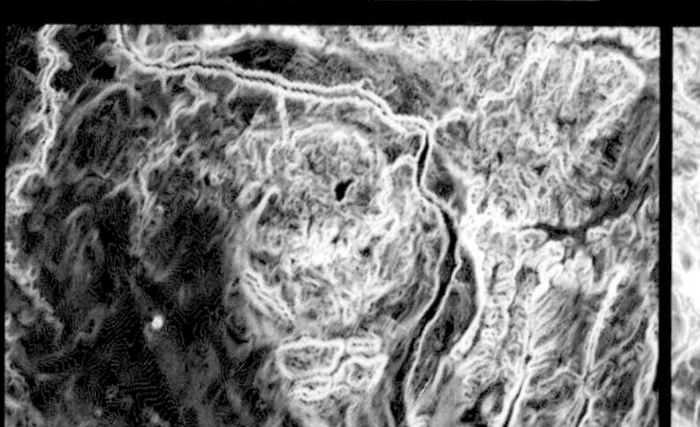

Page 438–39: *Owen Sound Reservation Map*, Charles Rankin, 1855 (Toronto Reference Library / Baldwin Collection / 912.71318 R11). Page 520–521: Rajyaguru Family, Bow Glacier, Banff National Park, 2015 (courtesy of Tushar Rajyaguru). Page 572–73: *A New and Exact Map of the Dominions of the King of Great Britain on ye Continent of North America*, detail, Herman Moll, 1715. Page 576–77: *Request to Leave Reserve Pass for Edward Yahyahkeekoot*, 1932 (Provincial Archives of Saskatchewan, S-E19, File 36). Page 606–7: *Topographic drawing of South Saskatchewan River*, Treaty 6 Lands, 2017 (OPSYS/Tiffany K. Dang). Page 614–15: *Map of Reserves and Royal Canadian Mounted Police Stations of North-Western Canada*, James White, 1904 (University of Toronto Map & Data Library / G3536 .F8 792 1904). Page 630–31: *Map of Highway 16 (Highway of Tears)* showing exploration & extraction regions, reserves & highways between Alberta and British Columbia, 2017 (with data from CBC Archives). Page 612–13: *Topographic drawing of Fraser River Delta*, Coast Salish Territory, 2017 (OPSYS/Tiffany K. Dang). Page 666–67: Taiga groundcover, Slave River, Northwest Territories, 2017 (Pierre Bélanger). Page 668–69: *Making Canadian Indian Policy*, original manuscript, Sally M. Weaver, 1978 (©2017 Dr. Sally Weaver Papers - courtesy of the University of Waterloo Special Collections & Archives). Page 694–95: *Nature and Power in the English Metaphysical Romance of the 19th and 20th Century*, unfinished dissertation, Margaret Atwood, 1973 (Harvard Library). Page 568–69: *Get Riel*, stenciled tag of Louis Riel on utility box facing statue of John A. Macdonald, Kingston, Ontario, 2016 (Pierre Bélanger). Page 708–9: *Topographic drawing of Athabasca River Delta*, Treaty 8 Lands, 2017 (OPSYS/Tiffany K. Dang). Page 710–19: *Topographic drawing of Mackenzie River,* Treaty 11 Lands, 2017 (OPSYS/Tiffany K. Dang). Page 720–25: *Topographic drawing of Yukon River*, Gwich'in Lands, 2017 (OPSYS/Tiffany K. Dang). Page 737: *Ogimaa Mikana Project, Reclaiming / Renaming*, an effort to restore Anishinaabemowin place-names to the streets, avenues, roads, paths, and trails of Gichi Kiiwenging (Toronto) to transform a landscape that often makes invisible the presence of Indigenous peoples, 2016 (Hayden King and Susan Blight). Page 760–61: *Map of Deh-Cho Research Area*, Phoebe Nahanni, 1992 (from "Dene Women in the Traditional and Modern Northern Economy in Denendeh Northwest Territories, Canada," PhD dissertation, McGill University). Page 786–87: *The Triumph of Mischief*, 2007 (Kent Monkman) IIIII

Overwriting Empire

Selected Readings on the Decentering, Decolonization & Deterritorialization of the Settler-State

Adema, Seth. "The Christian Doctrine of Discovery: A North American History." A literary review commissioned by the Doctrine of Discovery Task Force with the support of the Christian Reformed Centre for Public Dialogue, Christian Reformed Church, 2013.

Ali, Saleem H. (Saleem Hassan). *Mining, the Environment, and Indigenous Development Conflicts*. Tucson, AZ: University of Arizona Press, 2003.

Athabasca Chipewyan First Nation (ACFN) and Pat Marcel. *Níh boghodi: We are the stewards of our land.* An ACFN stewardship strategy for thunzea, et'thén and dechen yághe ejere (woodland caribou, barren-ground caribou and wood bison). Fort Chipewyan, AB: ACFN Chief and Council, 2012.

Auld, James and Robert Kershaw, eds. *The Sahtu Atlas: Maps and Stories from the Sahtu Settlement Area in Canada's Northwest Territories*. Yellowknife, NT: Northwest Territories Resources, Wildlife, and Economic Development, 2005.

Baber, Zaheer. *The Science of Empire: Scientific Knowledge, Civilization, and Colonial Rule in India*. Albany, NY: State University of New York Press, 1996.

Beinart, William and Lotte Hughes. *Environment and Empire*. The Oxford History of the British Empire Companion Series. Edited by W. Roger Louis. Oxford, UK: Oxford University Press, 2007.

Berger, Carl. *The Sense of Power: Studies in the Ideas of Canadian Imperialism, 1867-1914*. 2nd ed. Toronto, ON: University of Toronto Press, 2013.

Berger, Thomas R. *Northern Frontier Northern Homeland: The Report of the Mackenzie Valley Pipeline Inquiry*. Ottawa, ON: Ministry of Supply and Services, 1977.

Bhabha, Homi K., ed. *Nation and Narration*. London, UK: Routledge, 1990.

Binnema, Ted and Melanie Niemi. "'Let the Line Be Drawn Now': Wilderness, Conservation, and the Exclusion of Aboriginal People from Banff National Park in Canada." *Environmental History* 11, no. 4 (October 2006): 724–50.

Bliss, Michael. *Northern Enterprise: Five Centuries of Canadian Business*. Toronto, ON: McClelland and Stewart, 1987.

Blomley, Nicholas. "Law, Property, and the Geography of Violence: The Frontier, the Survey, and the Grid." *Annals of the Association of American Geographers* 93, no. 1 (2003): 121–41.

———. *Unsettling the City: Urban Land and the Politics of Property*. New York, NY: Routledge, 2004.

Bordo, Jonathan. "Picture and Witness at the Site of the Wilderness." *Critical Inquiry* 26, no. 2 (2000): 224–47.

Bork, Robert Odell, and Scott Bradford Montgomery. *De Re Metallica: The Uses of Metal in the Middle Ages*. AVISTA Studies in the History of Medieval Technology, Science and Art; v. 4. Burlington, VT: Ashgate, 2005.

Bryceson, Deborah Fahy, Eleanor Fisher, Jesper Bosse Jønsson, and R.A. Mwaipopo. *Mining and Social Transformation in Africa: Mineralizing and Democratizing Trends in Artisanal Production*. Routledge Studies in Development and Society; 37. London, UK: Routledge, 2014.

Buckner, Phillip A. *Canada and the End of Empire*. Vancouver, BC: UBC Press, 2005.

Butler, Paula. *Colonial Extractions: Race and Canadian Mining in Contemporary Africa*. Toronto, ON: Toronto University Press. 2015.

Carroll, William K. *Corporate Power in a Globalizing World: A Study in Elite Social Organization*. Don Mills, ON: Oxford University Press, 2004.

Césaire, Aimé. *Discourse on Colonialism*. Translated by Joan Pinkham. New York, NY: Monthly Review Press, [1972] 2000.

Charland, Maurice. "Technological Nationalism." *Canadian Journal of Political and Social Theory* 10, no. 1–2 (1986): 196–220.

Charlie, Lianne. "Together Today for Our Children Tomorrow: The Next Generation of Yukon Indigenous Politics." *Activehisotry.ca*, January 14, 2016.

Coulthard, Glen Sean. *Red Skin White Masks: Rejecting the Colonial Politics of Recognition*. Minneapolis, MN: University of Minnesota Press, 2014.

Creighton, Donald. *Canada's First Century*. Don Mills, ON: Oxford University Press, 2012.

Daschuk, James W. *Clearing the Plains: Disease, Politics of Starvation, and the Loss of Aboriginal Life*. Canadian Plains Studies; 65. Regina, SK: University of Regina Press, 2013.

Davis, Arthur. *George Grant and the Subversion of Modernity: Art, Philosophy, Politics, Religion, and Education*. Toronto, ON: University of Toronto Press, 1996.

Dawn, Leslie Allan. "How Canada Stole the Idea of Native Art: The Group of Seven and Images of the Indian in the 1920's." PhD diss., Department of Art History, Visual Art and Theory, The University of British Columbia, 2001.

Deneault, Alain. *Offshore: Paradis fiscaux et souveraineté criminelle*. Montreal, QC: Écosocieté, 2010.

——— and William Sacher. *Imperial Canada Inc.: Legal Haven of Choice for the World's Mining Industries*. Translated by Robin Philpot and Fred A. Reed. Vancouver, BC: Talonbooks, 2012.

Dyce, Matt. "Canada between the photograph and the map: Aerial photography, geographical vision and the state." *Journal of Historical Geography* 39 (2013): 69–84.

Edmonds, Penelope. "Unpacking Settler Colonialism's Urban Strategies: Indigenous Peoples in Victoria, British Columbia, and the Transition to a Settler-Colonial City." *Urban History Review* 38, no. 2 (Spring 2010): 4–20.

Evenden, Matthew D. *Fish versus Power: An Environmental History of the Fraser River*. Cambridge, UK: Cambridge University Press, 2004.

Fanon, Frantz. *The Wretched of the Earth* [*Les damnés de la terre*]. Translated by Constance Farrington. New York, NY: Grove Press, [1961] 1968.

———. *Black Skins, White Masks*. Translated by Richard Philcox. New York, NY: Grove Press, [1952] 2008.

Fenn, Catherine J. "Life History of a Collection: The Tahltan Materials Collected by James A. Teit." *Museum Anthropology* 20, no. 3 (1996): 72–91.

Fenton, Anthony P. "Mapping the Ruling Class: Situating Canada's Transnational Corporate Lawyers/Law Firms." MA diss., Department of Political Science, York University, Toronto, ON, 2011.

Forbes, Jack D. *Columbus and Other Cannibals: The Wétiko Disease of Exploitation, Imperialism, and Terrorism*. Rev. ed. New York, NY: Seven Stories Press, 2008.

Fowke, V.C. "The National Policy-Old and New." *The Canadian Journal of Economics and Political Science* 18, no. 3 (1952): 271–86.

Francis, Diane. *Bre-X: The Inside Story*. Toronto, ON: Key Porter, 1997.

Francis, Mark. "The 'Civilizing' of Indigenous People in Nineteenth-Century Canada." *Journal of World History* 9, no. 1 (Spring 1998): 51–87.

Frye, Northrop. *Divisions on a Ground: Essays on Canadian Culture*. Edited by James Polk. Toronto, ON: Anansi, 1982.

Gagnon, François-Marc. "Écrire sous l'image ou sur l'image." *Études françaises* 21, no. 1 (Spring 1985): 83–99.

Gill, Ian. *All That We Say Is Ours: Guujaaw and the Reawakening of the Haida Nation*. 1st U.S. ed. Vancouver, BC: Douglas & McIntyre, 2010.

Glissant, Édouard. *Poetics of Relation*. Translated by Betsy Wing. Ann Arbor, MI: The University of Michigan Press, [1990] 1997.

Golub, Alex. *Leviathans at the Gold Mine: Creating Indigenous and Corporate Actors in Papua New Guinea*. Durham, NC: Duke University Press, 2014.

Gonick, Cy. "Metropolis/Hinterland Themes." *Canadian Dimension* 8, no. 6 (March-April 1972): 24–28.

Gordon, Todd. *Imperialist Canada*. Winnipeg, MB: Arbeiter Ring Pub., 2010.

——— and Jeffery R. Webber. "Imperialism and Resistance: Canadian Mining Companies in Latin America." *Third World Quarterly* 29, no. 1 (2008): 63–87.

Grant, George Parkin. *Technology and Empire; Perspectives on North America*. Toronto, ON: House of Anansi, 1969.

Green, Joyce, ed. *Indivisible: Indigenous Human Rights*. Halifax, NS: Fernwood Publishing, 2014.

Gregory, Derek. *The Colonial Present: Afghanistan, Palestine, Iraq*. Malden, MA: Blackwell, 2004.

Griffiths, Tom and Libby Robin, eds. *Ecology & Empire: Environmental History of Settler Societies*. Seattle, WA: University of Washington Press, 1997.

Guldi, Jo (Joanna). *Roads to Power: Britain Invents the Infrastructure State*. Cambridge, MA: Harvard University Press, 2012.

Haiven, Max. *Cultures of Financialization: Fictitious Capital in Popular Culture and Everyday Life*. Basingstoke, UK: Palgrave Macmillan, 2014.

Hardt, Michael and Antonio Negri. *Empire*. Cambridge, MA: Harvard University Press, 2000.

Harley, John Brian. "Deconstructing the Map." *Cartographica* 26, no. 2 (1989): 1–20.

Harris, Cole. *Making Native Space: Colonialism, Resistance, and Reserves in British Columbia*. Vancouver, BC: UBC Press, 2002.

Harvey, Penny and Hannah Knox. *Roads: An Anthropology of Infrastructure and Expertise*. Ithaca, NY: Cornell University Press, 2015.

Hill, Gord. *The 500 Years of Resistance Comic Book*. Vancouver, BC: Arsenal Pulp Press, 2010.

Houston, Patrick, Zachary Schiller, Sandra D. Atchison, Mark Crawford, James R, Norman, and Jeffrey Ryser. "The Death of Mining: America is Losing One of its Most Basic Industries." Cover Story. *BusinessWeek*, December 17, 1984, 64–70.

Indian Brotherhood of the NWT. *Dene Declaration: Statement of Rights*. Fort Simpson, NT: 2nd Joint Assembly of the Indian Brotherhood of the NWT and the Métis Association of the NWT, 1975.

Indian Chiefs of Alberta. *Citizens Plus* [*The Red Paper*]. Edmonton, AB: Indian Association of Alberta, 1970.

Innis, Harold Adams. *Settlement and the Mining Frontier*. Vol. 9, part 2 of *Canadian Frontiers of Settlement*, edited by W.A. Mackintosh and W.LG. Joerg. Toronto, ON: Macmillan, 1936.

Jacobs, Jane M. *Edge of Empire: Postcolonialism and The City*. London, UK: Routledge, 1996.

King, Anthony D. *Urbanism, Colonialism, and The World Economy: Cultural and Spatial Foundations of the World Urban System*. London, UK: Routledge, [1990] 2015.

King, Thomas and William Kent Monkman. *A Coyote Columbus Story*. Toronto, ON: Groundwood Books, [1992] 2002.

———. *A Short History of Indians in Canada: Stories*. Toronto, ON: HarperCollins, 2005.

Kirsch, Stuart. *Mining Capitalism: The Relationship between Corporations and Their Critics*. Oakland, CA: University of California Press, 2014.

Klein, Naomi. *This Changes Everything: Capitalism vs the Climate*. New York, NY: Simon & Schuster, 2014.

Kühne, Thomas. "Colonialism and the Holocaust: continuities, causations, and complexities." *Journal of Genocide Research* 15, no. 3 (2013): 339–62.

Leslie, John F. "Assimilation, Integration or Termination? The Development of Canadian Indian Policy, 1943–1963." PhD diss. Department of History, Carleton University, 1999.

Levitt, Kari. *Silent Surrender: The Multinational Corporation in Canada*. Montreal, QC: McGill-Queen's University Press, [1970] 2002.

Li, Fabiana. *Unearthing Conflict: Corporate Mining, Activism, and Expertise in Peru*. Durham, NC: Duke University Press, 2015.

Mackey, Eva. *The House of Difference: Cultural Politics and National Identity in Canada*. Toronto, ON: University of Toronto Press, 2002.

Manuel, George and Michael Posluns. *The Fourth World: An Indian Reality*. Don Mills, ON: Collier-Macmillan Canada, 1974.

Maracle, Lee. *I Am Woman: A Native Perspective on Sociology and Feminism*. Vancouver, BC: Press Gang, 1996.

McCullum, Hugh and Karmel. *This Land is not For Sale. Canada's Original People and Their Land: A Saga of Neglect, Exploitation, and Conflict*. Toronto, ON: The Anglican Book Centre, 1975.

McMahon, Ryan. "Dear Canada: You Need A Statement Of Facts if You're Going To Address Indigenous Issues." *Ryan McMahon Comedy* (blog). April 7, 2016. http://www.rmcomedy.com/blog/canada-statement-of-facts.

——— and Brent Brambury. "The 12 Steps of Decolonization in Canada." 6-part mini-series. *Day 6 with Brent Brambury*, 2017. CBC Radio 1.

——— and James Whetung. "The Wild Rice Wars." *Red Man Laughing*, Season 6, Episode 7, May 8, 2017. Podcast, 50:07. https://www.redmanlaughing.com/listen/2017/5/red-man-laughing-the-wild-rice-wars.

Miller, Alexander C. "From the Indian Act to the Far North Act: Environmental Racism in First Nations Communities in Ontario." Independent study, Department of Environmental Studies, Queen's University, Kingston, ON, ca. 2011.

Monkman, Kent. *Shame and Prejudice, A Story of Resilience*. London, UK: Black Dog Publishing, 2017.

Moreton-Robinson, Aileen. *The White Possessive: Property, Power, and Indigenous Sovereignty*. Minneapolis, MN: University of Minnesota Press, 2015.

Nahanni, Phoebe. "The Mapping Project." In *Dene Nation: The Colony Within*, edited by Mel Watkins, 21–27. Toronto, ON: University of Toronto Press, 1977.

———. "Dene Women in the Traditional and Modern Northern Economy in Denendeh, Northwest Territories, Canada." PhD diss., Department of Geography, McGill University, Montreal, QC, 1992.

Nelson, Camille A. and Charmaine A. Nelson. *Racism, Eh?: A Critical Inter-Disciplinary Anthology of Race and Racism in Canada*. Concord, ON: Captus Press, 2004.

Nelson, Charmaine A. *Ebony Roots, Northern Soil: Perspectives on Blackness in Canada*. Cambridge, UK: Cambridge Scholars Publishing, 2010.

———. *Slavery, Geography and Empire in Nineteenth-century Marine Landscapes of Montreal and Jamaica*. London, UK: Routledge, 2016.

———. "Interrogating the Colonial Cartographical Imagination." *American Art* 31, no. 2 (2017): 51-53.

Nelson, Eli. "Repossessing the Wilderness: New Deal Science and American Indian Self-Determination in the Eastern Band of the Cherokee Nation." Paper presented at the Science, Religion and Culture Symposium, Harvard Divinity School, Cambridge, MA, May 2016.

Nelson, Robert L. "From Manitoba to the Memel: Max Sering, inner colonization and the German East." *Social History* 35, no. 4 (2010): 439–57.

Newcomb, Steven T. *Pagans in the Promised Land: Decoding the Doctrine of Christian Discovery*. Golden, CO: Fulcrum, 2008.

Perry, Adele. *Aqueduct: Colonialism, Resources, and the Histories We Remember*. Winnipeg, MB: ARP Books, 2016.

Rickard, Jolene. "Visualizing Sovereignty in the Time of Biometric Sensors." *South Atlantic Quarterly* 110, no. 2 (2011): 465–86.

Said, Edward D. *Culture and Imperialism*. New York, NY: Vintage Books, [1993] 1994.

St. John, Michelle, Ryan McMahon, and Jordan O'Connor. *Colonization Road*. Toronto, ON, Colonization Road Productions Inc., 2016. DVD, 50 min.

Schwartz, Joan M. and James R. Ryan. *Picturing Place: Photography and the Geographical Imagination*. London, UK: I.B. Tauris, 2003.

Simpson, Audra. *Mohawk Interruptus: Political Life Across the Borders of Settler States*. Durham, NC: Duke University Press, 2014.

Simpson, Leanne Betasamosake. *Islands of Decolonial Love: Stories & Songs*. Winnipeg, MB: Arbeiter Ring Publishing, 2013.

———. "Land as pedagogy: Nishnaabeg intelligence and rebellious transformation." *Decolonization: Indigeneity, Education & Society* 3, no. 3 (2014): 1–25.

Smith, Linda Tuhiwai. *Decolonizing Methodologies: Research and Indigenous Peoples*. London, UK: Zed Books, 1999.

Spivak, Gayatri Chakravorty. *A Critique of Postcolonial Reason: Toward a History of the Vanishing Present*. Cambridge, MA: Harvard University Press, 1999.

———. "The 2012 Antipode AAG Lecture: Scattered Speculations on Geography." *Antipode* 46, no. 1 (2014): 1–12.

Sterrit, Angela. "A Movement Rises." *OpenCanada*, November 20, 2015, https://www.opencanada.org/features/movement-rises/.

Thomson, Don W. *Men and Meridians: The History of Surveying and Mapping in Canada*. 3 vols. Ottawa, ON: Queen's Printer, 1966–69.

———. *Skyview Canada: A Story of Aerial Photography in Canada*. Ottawa, ON: Energy, Mines, and Resources Canada, 1975.

Toledo Maya Cultural Council. *Maya Atlas: The Struggle to Preserve Maya Land in Southern Belize*. Berkeley, CA: North Atlantic Books, 1997.

Tough, Frank. "Conservation and the Indian: Clifford Sifton's Commission of Conservation, 1910–1919." *Native Studies Review* 8, no. 1 (1992): 61–73.

Tsing, Anna Lowenhaupt. *Friction: An Ethnography of Global Connection*. Princeton, NJ: Princeton University Press, 2005.

Tuck, Eve and K. Wayne Yang. "Decolonization is not a metaphor." *Decolonization: Indigeneity, Education & Society* 1, no. 1 (2012): 1–40.

Veltmeyer, Henry and James Petras. *The New Extractivism: A Post-neoliberal Development Model or Imperialism of the Twenty-first Century?* London, UK: Zed Books, 2014.

Vowel, Chelsea. *Indigenous Writes: A Guide to First Nations, Métis & Inuit Issues in Canada*. Winnipeg, MB: Highwater Press, 2016.

Wainwright, Joel. *Decolonizing Development: Colonial Power and the Maya*. Antipode Book Series. Malden, MA: Blackwell, 2008.

Watkins, Melville, ed. *Dene Nation: The Colony Within*. Toronto, ON: University of Toronto Press, 1977.

———. *Staples and beyond: Selected Writings of Mel Watkins*. Edited by Hugh Grant and David Wolfe. Carleton Library Series; 210. Montreal, QC: McGill-Queen's University Press, 2006.

Weaver, Sally M. *Making Canadian Indian Policy: The Hidden Agenda 1968–1970*. Toronto, ON: University of Toronto Press, 1981.

Wells, Jennifer. *Bre-X: The inside Story of the World's Biggest Mining Scam*. London, UK: Orion Business Books, 1998.

Wensing, Ed and Libby Porter. "Unsettling planning's paradigms: towards a just accommodation of Indigenous rights and interests in Australian urban planning?" *Australian Planner* (2015): 1–12.

Wildcat, Matthew, Mandee McDonald, Stephanie Irlbacher-Fox, and Glen Coulthard. "Learning from the land: Indigenous land based pedagogy and decolonization." *Decolonization: Indigeneity, Education & Society* 3, no. 3 (2014): i–xv.

Wolfe, Patrick. "Settler Colonialism and the Elimination of the Native." *Journal of Genocide Research* 8, no. 4 (2006): 387–409.

Yukon Indian People. *Together Today for Our Children Tomorrow: A Statement of Grievances and an Approach to Settlement by the Yukon Indian People*. Whitehorse, YT: The Council for Yukon Indians, 1977.

Yuxweluptun, Lawrence Paul. *Unceded Territories*. Edited by Karen Duffek and Tania Willard. Vancouver, BC: Figure 1 Publishing, 2016.

Zarobell, John. *Empire of Landscape: Space and Ideology in French Colonial Algeria*. Philadelphia, PA: Penn State University Press, 2010.

Zeller, Suzanne. "The Colonial World as Geological Metaphor: Strata(gems) of Empire in Victorian Canada." *Osiris* 15 (2000): 85–107.

Contributors

A Tribe Called Red is an Indigenous electronic hip-hop music group from Ottawa, formed by three DJs: Ian 'DJ NDN' Campeau, Tim '2oolman' Hill, and Ehren 'Bear Witness' Thomas. Their albums include *We Are the Halluci Nation* (2016), *Nation II Nation* (2013), and the homonym *A Tribe Called Red* (2012).

Allan Adam is the Chief of the Athabasca Chipewyan First Nation (ACFN) since 2007. Chief Adam currently holds the Environment and Treaty Lands and Resources portfolios and shares the Justice and Economic Development portfolios for the ACFN.

Howard Adams was an Indigenous political leader, educator, and writer dedicated to the struggle against the colonization of Aboriginal peoples in Canada and throughout the world. As the first Canadian Métis to receive a PhD, he authored *Tortured People: The Politics of Colonization* (1995), *Prison of Grass: Canada from the Native Point of View* (1975), and *The Education of Canadians 1800–1867: The Roots of Separatism* (1968).

Yassin 'Narcy' Alsalman is a musician, multi-media artist, and educator. He founded multi-media company "The Medium" in Montreal, to facilitate international collaborations amongst thinkers, artists, and brandeurs worldwide. He also teaches at Concordia University in Montreal.

Christopher Alton is an urban planner and PhD student at the University of Toronto School of Geography and Planning, focusing on resource extraction and planning theory.

Pedro Aparicio is a Colombian architect, activist, educator, exhibition curator, and partner of *Altiplano estudio de arquitectura.* He is Fulbright scholar for regional development, design critic at the Rhode Island School of Design, and research associate at the Harvard Graduate School of Design.

Margaret Atwood has authored more than forty books of fiction, poetry, and critical essays. As one of Canada's most acclaimed writers, her contribution to contemporary literature has received international recognition. Her work includes dystopian trilogy *MaddAddam* (2013), *Oryx and Crake* (2003), and *The Year of the Flood* (2009); as well as *The Handmaid's Tale* (1985).

Aaron Barcant is a Trinidadian researcher at Concordia University and assistant to Kari Levitt. He worked as a researcher with Alain Deneault on the Caribbean offshore.

Réal V. Benoit is a Canadian author, singer, songwriter, and composer from Quebec. Miner by profession, he started writing songs as a hobby while working in a mine in his home town. Some of his albums include *Voilà - Chansons pour mineurs et adultes* (1971), *Sérieusement - La ballade des bills de $20* (2005), and *Trésors retrouvés* (2007).

Justice Thomas Berger is a Canadian lawyer, judge, and politician. He was counsel for the plaintiffs in the historic Aboriginal rights case *Calder et al v. Attorney General for British Columbia* and served as commissioner of the Mackenzie Valley Pipeline Inquiry (1974–77). He served as New Democratic Party (NDP) Member of Parliament for Vancouver-Burrard (1962–63), NDP Member of Legislative Assembly (1968–69), and leader of the NDP in British Columbia (1969).

Hernán Bianchi Benguria is an architect and urban planner from Chile with experience in land and environmental public policy. He is a graduate research fellow for the Canada Program at Harvard University's Weatherhead Center for International Affairs.

Paula Butler is a professor in the Department of Canadian Studies at Trent University, focusing on race, gender, neoliberalism, Canadian internationalism, colonial psychologies, and extraction. She is the author of *Colonial Extractions: Race and Canadian Mining in Contemporary Africa* (2015).

David Chancellor is a documentary photographer based in South Africa, whose trajectory has increasingly focused on the commodification of wildlife. His work includes the photography series *With Butterflies and Warriors* (2014), *Intruders* (2011), and *Hunters* (2010).

Lianne Marie Leda Charlie is a political science instructor at Yukon College in Whitehorse and PhD candidate from the University of Hawai'i at Manoa. As an artist, she creates digital photo collage through Indigenous methodology. Lianne is Tagé Cho Hudän (Big River People) and Northern Tutchone speaking people of the Yukon.

Jean Chrétien is a lawyer and was the 20th Prime Minister of Canada. He negotiated the patriation of the Canadian Constitution as well as the Canadian Charter of Rights and Freedoms. As Minister of Indian Affairs and Northern Development, authored the 1969 "White Paper," a policy proposal for eliminating the Indian Act.

Tiffany Kaewen Dang is a landscape architect and territorial scholar from Edmonton, Alberta.

Alain Deneault is a philosopher and teacher of political science at the University of Montreal. He is author of several books, including *Canada: A New Tax Haven: How the Country that Shaped Caribbean Tax Havens is Becoming One Itself* (2015) and *Offshore: Tax Havens and the Rule of Global Crime* (2011).

Eriel Tchekwie Deranger is a Dënesųłiné climate and Indigenous rights advocate, serving as Communications Coordinator for the Athabasca Chipewyan First Nation. Eriel founded Indigenous Climate Action, pushing for a united Indigenous climate action strategy in Canada and working with the Federation of Saskatchewan Indian Nations, Indigenous Environmental Network, Sierra Club, Rainforest Action Network, and the UN Indigenous Peoples Forum on Climate Change.

The *Diaguitas Huascoaltinos* Community is composed by 250 farming and herding Indigenous families who have ancestrally occupied the fertile and biodiverse Huasco valley in the southern fringes of the hyper-arid Atacama Desert, Chile. This is one of the few Indigenous communities with an ancestral property title over their land that is recognized by the Chilean State.

Mary Eberts is a constitutional lawyer who practices public law litigation and advocacy under the Canadian Charter of Rights and Freedoms. With expertise on constitutional and Charter law, Indigenous law, and human rights, Mary has been counsel for the Native Women's Association of Canada for over two decades and co-founded the Women's Legal Education and Action Fund (LEAF).

Genevieve Ennis Hume is a Canadian designer, intercultural communicator, fair-mining specialist, and co-founder of jewelry studio *Hume Atelier* in Vancouver, BC.

Georges Erasmus is a Dene politician and Indigenous leader. He served as the National Chief of the Assembly of First Nations from 1985–91, guiding First Nations through constitutional talks and the Oka Crisis. He was also Head of the Royal Commission on Aboriginal Peoples in 1991.

Pierre Falcon was an eighteenth-century Métis poet and song-writer from Red River, Manitoba. His songs documented the daily life of settlers, voyageurs, and hunters, as well as major incidents such as the clashes between Métis and settlers during the Red River Rebellion of 1869–70.

Evan Farley is an American designer, architect, and artist.

Alex Golub is a political anthropologist and associate professor at the University of Hawai'i at Mānoa, focusing on the relationship between grassroots people and the mining and hydrocarbon industries in Papua New Guinea. He wrote *Leviathans at the Gold Mine* (2014).

David Hargreaves is Vice President of Surveillance and Intelligence at MacDonald, Dettwiler, and Associates Ltd. (MDA), a British Columbia based company offering information services and systems solutions to customers in maritime defense and security, land surveillance and intelligence, space and remote sensing, aviation, energy, and mining.

Daniel Hemmendinger is an American designer and architect.

Gord Hill (Zig Zag) is a writer, artist, and political activist in Indigenous resistance, anti-colonial, and anti-capitalist movements. Author of graphic novel *The 500 Years of Resistance* (2010), Gord is also editor and publisher of Warrior Publications. He is a member of the Kwakwaka'wakw nation on the Northwest Coast.

James Hopkinson is a complex environment risk management advisor at Assaye Risk with experience in Africa, the Middle East, and South Asia. Following a career in the British military and the Ministry of Defense's Operations Directorate, he joined Blue Hackle, an international risk management provider, later becoming its Chief Operating Officer.

Hume Atelier is a luxury custom jewelry firm based in Vancouver, BC. Founded in 2008 by Kevin Hume and Genevieve Ennis, they have collaborated on design and worked extensively in international mining policy, with a focus on eliminating conflict minerals and supporting artisanal miners.

Michael Ignatieff is a Canadian politician, writer, broadcaster, and educator. He is currently serving as Rector and President of Central European University in Bucharest after teaching at the Harvard Kennedy School (2013–16). As Member of Parliament (2006–11), he served as Leader of the Liberal Party of Canada (2009–11). His books include *Empire Lite: Nation Building in Bosnia, Kosovo, Afghanistan* (2003), *Virtual War: Kosovo and Beyond* (2000), and *Blood and Belonging* (1993).

Thomas King is an Indigenous novelist, short-story writer, essayist, screenwriter, and photographer of Cherokee, German, and Greek descent. He is professor emeritus of English at the University of Guelph. His written work includes *Back of the Turtle* (2014), *The Inconvenient Indian* (2014), *A Short History of Indians in Canada* (2005), *The Truth About Stories* (2003), and the children's book *A Coyote Columbus Story* (1993).

Naomi Klein is a Canadian author and journalist, syndicated columnist, social activist, and filmmaker. She is the author of international bestsellers *This Changes Everything: Capitalism vs. The Climate* (2014), *The Shock Doctrine: The Rise of Disaster Capitalism* (2007), and *No Logo* (2000).

Erica Violet Lee is a Nēhiyaw student at the University of Saskatchewan, Indigenous feminist, and community organizer from inner city Saskatoon, Saskatchewan. She is part of the Canadian Youth Climate Coalition Delegation to the COP 21 climate conference in Paris and COP 22 in Marrakesh. She writes on her blog "Moontime Warrior."

Kari Polanyi Levitt is a development economist and professor emeritus at McGill University, researching the impact of foreign direct investment in host countries and dependent development and industrialization in the Caribbean. Kari also dedicated her career on the literary legacy of her father, economic historian Karl Polanyi.

Nina-Marie Lister is a planner and ecologist, associate professor and director of the Ecological Design Lab at Ryerson

University, and founding principal of PLANDFORM studio. She co-edited *Projective Ecologies* (2014) and *The Ecosystem Approach: Complexity, Uncertainty, and Managing for Sustainability* (2008).

Ryan McMahon is an Anishinaabe comedian, writer, media maker, and community activator based out of Treaty #1 territory (Winnipeg, MB). He has recorded three National comedy specials and two taped Gala sets at the Winnipeg Comedy Festival. In 2012, he became the first Native comedian to ever record a full mainstream comedy special with CBC TV. He co-wrote and featured the documentary *Colonization Road* (2016).

Chris Meyer is an American designer and architect.

Ossie Michelin is a freelance journalist from North West River, Labrador. Coming from a lineage of storytellers, he focuses on northern and Indigenous issues. While working with the Aboriginal Peoples Television Network, covering the Elsipogtog First Nation's battle against fracking, a picture he took of Amanda Polchies before a line of riot police in Rexton, NB in 2013, went viral, inspiring paintings, prints, and appearing in books and magazines worldwide.

Kent Monkman is a Cree and Canadian artist working in mediums such as painting, film, performance, and installation. His award-winning video works have been screened at national and international festivals. Kent's work is exhibited in private and public collections such as the National Gallery of Canada, the Denver Art Museum, Montreal Museum of Fine Arts, Museum London, the Glenbow Museum, the Museum of Contemporary Canadian Art, the Art Gallery of Ontario, the Smithsonian's National Museum of the American Indian, and the Vancouver Art Gallery. He published *Shame and Prejudice, A Story of Resilience* in 2018.

Doug Morrison is President and CEO of Holistic Mining Practices at the Centre for Excellence in Mining Innovation (CEMI) non-profit in Sudbury, Ontario. Prior to joining CEMI in 2011, Doug had a long career in mining and engineering consulting industries.

James Murray is an American designer trained as an architect.

Joan K. Murray has been a corporate historian for Hudson's Bay Company after careers in publishing and records management. She co-authored the 2011 graphic catalog *Hudson's Bay Company* with Mark Reid and Graydon Carter.

Charmaine A. Nelson is professor of art history at McGill University, focusing on postcolonial and Black feminist scholarship, critical race theory, Trans-Atlantic slavery studies, and Black diaspora studies. She is the 2017-2018 William Lyon Mackenzie King Visiting Professor of Canadian Studies at Harvard University. Her edited volumes include *Racism Eh?* (2004, with Camille A. Nelson) and *Ebony Roots, Northern Soil* (2010). Her single-authored books are *The Color of Stone* (2007), *Representing the Black Female Subject in Western Art* (2010), and *Slavery, Geography, and Empire in Nineteenth-Century Marine Landscapes of Montreal and Jamaica* (2016).

Eli Nelson is a PhD candidate in the History of Science department at Harvard University. He studies the history of native science in settler-colonial and postcolonial contexts. His dissertation traces the history of twentieth-century indigenous engagements with western scientific hegemony in the United States. Eli analyzes different modes of engagement with science and the role native and western sciences and epistemes have played in nation building and decolonization projects.

George Osodi is a freelance photographer based in Lagos, Nigeria, and a member of Panos Pictures. He was a photojournalist for the Comet Newspaper in 1999, before working with the Associated Press in Lagos between 2001 and 2008. His work was exhibited at *Documenta* 12 in Germany (2007) and in 2008, he published *Delta Nigeria: The Rape of Paradise.*

Maryanne Pearce serves as Manager at the Royal Canadian Mounted Police. She is known for her PhD research conducted at the University of Ottawa, creating an extensive database of missing and murdered Indigenous women in Canada.

Barry Pottle is an Inuk photographer and artist from Nunatsiavut. He currently works with the Ontario Aboriginal arts community, capturing the essence of Inuit life in Ottawa. Pottle's travelling and rotating exhibitions include the "Awareness Series," "Urban Inuit," "At Home and Away," and "Decolonize Me." His work has been published by the National Gallery of Canada (2010), *Makivik Magazine*, *Inukitut Magazine*, and *Inuit Art Quarterly.*

Moura Quayle is a landscape architect, educator, and institutional leader in British Columbia; currently serving as Chair of the Board for the Canadian International Resources and Development Institute (CIRDI) and Genome Canada. She is Director of the Liu Institute for Global Issues at the University of British Columbia.

Louis Riel was a nineteenth-century Métis leader and politician, founder of Manitoba Province and central figure in the Red River 1869–70 and North-West 1885 resistances. He is recognized for his legacy of building Métis nationalism and political independence.

Design firm RVTR is led by Geoffrey Thün, Kathy Velikov, and Colin Ripley. Founded in 2007, RVTR is a professional architectural practice and an academic design-research platform. The work of RVTR was published in 2015 with the monograph Infra Eco Logi Urbanism, A

Project for the Great Lakes Megaregion, accompanied by a solo exhibition that has traveled throughout Canada and the United States.

Olga Semenovych is an urban planner from Canada and lecturer at the University of Waterloo, School of Planning.

Michelle St. John is a two-time Gemini Award winning actor with more than 35 years of experience in film, television, theatre, voice and music. Michelle was a Co-Founder of Turtle Gals Performance Ensemble, Producer and Host for *Red Tales*, a weekly Native literary show on Aboriginal Voices Radio and President of Dr. E Entertainment, creating media content featuring Dr. Evan Adams. She has also been a producing partner in Frog Girl Films, and The Breath Films. *Colonization Road* is her first documentary.

Maurice Strong was responsible for the creation of the UN Environment Programme, the organization of the 1972 Conference on the Human Environment in Stockholm, and the 1992 Rio de Janeiro Earth Summit. As an international businessman in the oil and mineral industry, he served as President of the Power Corporation of Canada, and CEO of Petro-Canada and Ontario Hydro. He published *Where on Earth Are We Going?* (2000).

Ashley C. Thompson is a designer, researcher, activist, adventurer, and a US Air Force veteran.

Anna Lowenhaupt Tsing is professor of anthropology at the University of California at Santa Cruz, leading an interdisciplinary research project called "Living in the Anthropocene." Her books include *The Mushroom at the End of the World* (2015), *Friction: An Ethnography of Global Connection* (2005), and *In the Realm of the Diamond Queen* (1994).

John Van Nostrand is an architect, planner, and founding principal of SvN in Toronto. His practice focuses on planning and design of affordable housing and community infrastructure, including a number of major mine-related housing projects in Africa, Latin America, and Canada.

Mel Watkins is a political economist and activist. He is professor emeritus of economics and political science at the University of Toronto. His published work includes *Staples and Beyond: Selected Writings of Mel Watkins* (2006) and *Dene Nation: The Colony Within* (1977).

Sally M. Weaver was a Canadian anthropologist and professor at the University of Waterloo, as well as the first Canadian woman with an anthropology PhD. She wrote *Making Canadian Indian Policy: The Hidden Agenda 1968–70* (1980) regarding the 1969 "White Paper," and contributed to *A Canadian Indian Bibliography 1960–1970* (1974).

Patrick Wolfe was an Australian anthropologist and ethnographer whose work used theories of colonialism and indigenous resistance to generate new and different ways of viewing Australia's history, challenging the standard triumphal narrative of civilizing the frontier through pioneering individualism. His book *Settler Colonialism and the Transformation of Anthropology* (1999) launched a major academic reconsideration of the role of settlement in colonization.

Rita Wong is a Canadian poet and associate professor in the Critical and Cultural Studies department at the Emily Carr University of Art and Design on the unceded Coast Salish territories (also known as Vancouver). Her poetry books include *monkeypuzzle* (1998), *forage* (2007), *sybil unrest* (2008, with Larissa Lai), and *undercurrent* (2015).

The Wyrd Sisters is a folk music group originally formed in 1990 by Kim Baryluk, Nancy Reinhold, and Kim Segal in Winnipeg, MB. Their most acclaimed albums include *Inside the Dreaming* (1996), *Raw Voice* (1998), and *Sin & Other Salvations* (2002).

Kate Yoon is an independent writer, research assistant, and college student at Harvard University.

Suzanne Zeller is a Canadian science historian focusing on Victorian culture. She is a professor at the Faculty of Arts at Wilfrid Laurier University, and author of *Land of Promise, Promised Land: The Culture of Victorian Science in Canada* (1996) and *Inventing Canada: Early Victorian Science and the Idea of a Transcontinental Nation* (1987).

Acknowledgments

In Paulo Freire's 1970 *Pedagogy of the Oppressed*, "every relationship of domination, of exploitation, of oppression, is by definition violent, whether or not the violence is expressed by drastic means. In such a relationship, dominator and dominated alike are reduced to things—the former dehumanized by an excess of power, the latter by lack of it... and things cannot love."

This project was built from both love and rage. With the sweat equity of an incredible editorial team and a wonderful circle of contributors, this project started with a simple idea conceived over the course of almost a decade. That evolution was coupled with lifetime experiences of so many powerful voices, creators, activists, and changemakers. It is their cherished engagement, love for shared knowledge, and willful collaboration that kept the fire and energy of this project lighting the way for so long. This project is also built out of angst and anger towards current colonial, territorial conditions in Canada. Outrage serves as a powerful motivator and generator that has sustained so many of the contributors in this book. A brief but meaningful sojourn on Blatchford Lake at Dechinta Bush University in the Northwest Territories during the summer of 2017, formed a lasting experience on lands of the Dene Nation—Treaty 11 Signatories. Immense gratitude and respect is due to the kind invitation from Erin Freeland Ballantyne, political ecologist and co-founder of Dechinta U, and to the initial impetus of Eriel Tcheckwie Deranger from the Athabasca Chipewyan First Nation—Treaty 8 Signatories. The trust that Erin and Eriel placed in the intersectionality of this project has been a continuous source of inspiration and humility. With the powerful conversation between Ryan McMahon, Charmaine Nelson, and Eli Nelson during the screening of Michelle St. John's film *Colonization Road* at Harvard University's Weatherhead Center for International Affairs in Fall 2017, these experiences have shaped emerging outcomes from this book project towards future collaborations.

In spite of the project's southern, metropolitan origins, its territorial commitment towards land and the anti-extractive state remain at its core. Along with meaningful exchanges with instructors Glen Coulthard, Leanne Betasamosake Simpson, Berna Martin, Melaw Nakehk'o, Siku Alloloo, and Gordie Liske at Dechinta, their influence moves in so many ways—the Dene way—as they shared with us. It affirms and asserts what the future holds. Together, the care, trust, and love that they shared can never be fully returned. But, we hope that this project, and its future directions, does justice to the radical activism and deep sense of transgenerational time they aspire and inspire for others to come. We will always be ready, unconditionally, to stand with you.

So many lives, minds, and experiences shape the body of knowledge and action presented in this book. The core editorial teamwork that originally began with lead researchers Christopher Alton and Zannah Matson—accomplished doctoral scholars at the University of Toronto—was later extended by the important efforts and involvement of Tiffany Kaewen Dang, Hernán Bianchi Benguria, Ghazal Jafari, and Sam Gillis at Harvard's Graduate School of Design. They all brought the exhausting production of the book to its final completion with deft and unquestionable commitment. Their trust and belief in the project—a massive undertaking for all of us, from start to finish, cannot be acknowledged enough. From the very beginning of the project, friends and colleagues Nina-Marie Lister and John Van Nostrand provided much intellectual support and creative advice on the content and structure of the book. They did so amidst their own hectic practices and teaching schedules, sustaining the evolution of the content and structure of the book.

The voices of an important group of contributors of text and images form the core component of this book. Many conversations and exchanges with authors and friends, old and new, throughout the past few years have been enlightening as much as they have been sobering. Contributors created a distant family of collaborators that will always be without substitute. Every single author met the challenge of this project by sharing part of their life work and lived experience, in the form of original contributions or permissions to reproduce former writings, with a shared sense of purpose: Mel Watkins, Alain Deneault, Suzanne Zeller, Paula Butler, Anna Lowenhaupt Tsing, Naomi Klein, Eriel Tchekwie Deranger, Kate Yoon, Christopher Alton, Nina-Marie Lister, Tiffany Kaewen Dang, Réal V. Benoit, RVTR (Geoffrey Thün, Kathy Velikov, Colin Ripley), Margaret Atwood, Thomas King, Mary Eberts, Maryanne Pearce, Genevieve Ennis & Kevin Hume, Georges Erasmus, Erica Violet Lee, Lianne Charlie, The Wyrd Sisters, Yassin Alsalman, Rita Wong, Gord Hill, and the Diaguitas Huascoaltinos Community.

Conversations and interviews were central to the shaping of the book's contemporary content: Ryan McMahon, Charmaine Nelson, Eli Nelson, Kari Levitt, Alex Golub, Michael Ignatieff, David Hargreaves, Douglas Morrison, David Chancellor, James Hopkinson, Moura Quayle, Maurice Strong, John Van Nostrand, Chief Allan Adam, Eriel Tchekwie Deranger & Kelsey Chapman, Thomas Berger, Jean Chrétien, A Tribe Called Red (Ian 'DJ NDN' Campeau, Bear Witness, Tim 2oolman). Several of these conversations were led by an active group of graduates from the Harvard Graduate School of Design during an intensive seminar course on territory and power held between 2015 and 2017, looking critically at the consequences of settler-based colonialism through the lens of resource extraction and Indigenous resurgence. Interviewers and researchers include Hernán Bianchi Benguria, Pedro Aparicio, Olga Semenovych, Genevieve Ennis Hume, Ashley C. Thompson, James Murray, Evan Farley, Daniel Hemmendinger, and Chris Meyer.

We're also indebted to the spirit of authors passed, including Howard Adams, Patrick Wolfe, Sally M. Weaver, Maurice Strong, Pierre Falcon, Louis Riel, Jessie Oonark, and Phoebe Nahanni whose work we consider as proposals for future action. Their voices remain as significant as they were during their original time of writing. A few contributors chose to remain anonymous in order to protect their identity for fear of reprisal. We are grateful for the risk they took from the 'inside' and cherish the confidence they shared in sharing their views. These anonymous voices opened an understanding of unspoken spaces and unseen places so seldom discussed in the state of extraction.

For their experience and expertise, a special note of gratitude is owed to Tom Andrews, Alan Berger, Max Haiven, Robert Wright, Patrick Stewart, Kent Monkman, Souie Gorup, Heidi Nast, Joan K. Murray, Megan Red Shirt-Shaw, Cannon Ivers, Michèle Champagne, Daniel Boucher, Joanne Woo, Samantha Eldridge, Chelsea Vowel, Molly Swain, Barry Pottle, Ossie Michelin, George Osodi, Terrance Galvin, Julio Cañas, David Pearce, Nick Richbell, Chris Cornelius, Michelle St. John, Ron Bourgeault, Cy Gonick, John Leslie, Bruce Hartley, and Francois & Lesley Paulette. Their insight and their guidance were an important part of new links made and old connections forged throughout the process of research and evolution of the content.

The creators of the 800 images featured in this book deserve immense respect. We are honored by their generous contributions and creative efforts that compose and generate the rich graphic fabric of the project. Together they resist and challenge tensions that exist between the colonial foundations of maps and the lived experiences of territories.

During the course of this research, several institutions and organizations helped us in direct ways with archival reproductions, image retrievals, and research information: the Dene Nation Office (Yellowknife), Dechinta Centre for Research and Learning, Prince of Wales Northern Heritage Centre, Hudson's Bay Company, Library & Archives Canada, Haudenosaunee Development Institute, Koochiching Historical Society, University of Waterloo Special Collections & Archives, Harvard University Archives, Multicultural History Society of Ontario, University of Toronto Special Collections, and the Northwest Territories Archives.

Sincere gratitude is due to Executive Editor Roger Conover and Production Manager Janet Rossi at MIT Press for their untiring patience and editorial advice over the course of the past three years.

This book is, in part, the outcome of an exhibition in 2016 featured at the Venice Architecture Biennale. The team included: Catherine Crowston (Commissioner, Art Gallery of Alberta), Zannah Matson, Christopher Alton, Tiffany Kaewen Dang, Hernán Bianchi Benguria (OPSYS), Nina-Marie Lister (Ecological Design Lab), Geoffrey Thün, Kathy Velikov, Colin Ripley with research assistants Jen Ng, Andrew Wald, Dan Tish, Ya Suo (RVTR), Geneviève Ennis Hume & Kevin Hume (Hume Atelier), Eriel Tchekwie Deranger (ACFN), Kelsey Blackwell (Blackwell Studio), Steven Beites (Beites & Co), with collaborators, Alessandra Lai, Troels Bruun (M+B Studio), Massimo Benedetti, Dr. Michele Caria (IGEA S.p.A.), Jacob Moginot, Pedro Aparicio, Boram Lee, Irene Chin, Sam Gillis, Michael Awad, Lindsay Fischer, Olga Semenovych, Emanuela Innelli, Juan Santa María, Aya Abdallah, Natalia Woldarsky Meneses, Hamed Bukhamseen, Chella Jade Strong, Takuya Iwamura, Jane Zhang, Carlo Urmy, Mark Jongman-Sereno, and Kate Yoon.

Special acknowledgments are due to friends of the project: Malkit Shoshan, Charles Taylor-Foster, Pierre & Jannelle Lassonde, Peter McGovern, Jasmina Zurovac, Mohsen Mostafavi, Pat Roberts, Phyllis Lambert, Edward Burtynsky, Jennifer Baichwal, Nick de Pencier, Peter Munk, Kelvin Dushnisky, Alanna Heath, Simon Brault, Sylvie Gilbert, Brigitte Desrochers, Sacha Hastings, Geneviève Vallerand, Brigitte Shim & Howard Sutcliffe, Robert Enright, David D'Arcy, Toon Dressen, Alexander Reford, Ted Kesik, Chris Petersen, Peter Buchanan, Lise Anne Couture, Léa-Catherine Szacka, Karen Colby-Stothart, Anne Eschapasse, Cecelia Paine, Doris Chee, Shelley Ambrose, Patrick Schein, Charlie Fischer, Joanne Cuthbertson, Sarah McCallum, Jim Carter, Michèle Lamont, Chelsea Spencer, Elise Hunchuck, Mano & Baldeep Duggal, Micol Soleri, and Manuela Lucadazio. Support for the Biennale project was received from the Canada Council for the Arts, RBC Foundation, Harvard Graduate School of Design, Landscape Architecture Canada Foundation, Ontario Association of Architects, Ryerson University, University of Michigan Taubman College of Architecture, The Walrus Foundation, Border Crossings, IGEA SpA, Gloria Irene Taylor, Barrick Gold, Verona Libri, The Northern Miner, Canadian Architect, Azure, Graham Foundation, Air Canada Cargo, and MIT Press.

Finally, during the process of research for this book, meaningful exchanges with Indigenous & Black scholars, practitioners, activists, elders, and youth forged a sense of alliance in the fight against oppression and project of resistance. Since then, their voices have grown stronger and stronger against the unspoken violence and aggression that Canada's *extractive state* represents and exercises. During this time, their visions filled the air with life and vitality, during late conversations, archival searches and many travels, channeling energy and strength. Through their respective medium, we learned ways that follow from simply learning how to listen. At times, this process has been unsettling and uncomfortable, but it has opened important ways of reading, learning, living, and sharing. Not only has it been transformative, it is transfigurative.

To all our families, who sacrificed time and shared energy to make space in our lives for this project to come to fruition, this book is dedicated to your unending love. Thanks to you, anything is possible.

extraction
empire

Undermining the Systems, States, & Scales of Canada's Global Resource Empire

edited by Pierre Bélanger

The MIT Press
Cambridge, Massachusetts
London, England